Two-Dimensional Imaging

PRENTICE HALL SIGNAL PROCESSING SERIES

Alan V. Oppenheim, Editor

Two-Dimensional Imaging

RONALD N. BRACEWELL

Lewis M. Terman Professor
of Electrical Engineering Emeritus

Stanford University

PRENTICE HALL
ENGLEWOOD CLIFFS, NEW JERSEY 07632

Library of Congress Cataloging-in-Publication Data

Bracewell, Ronald Newbold
    Two-dimensional imaging / Ronald N. Bracewell.
      p.   cm. — (Prentice Hall signal processing series)
    Include bibliographical references and index.
    ISBN 0-13-062621-X
    1. Image processing. I. Title. II. Series.
TA1637.B73   1995
621.36'7—dc20   94-25811
                CIP

Acquisitions editor: Linda Ratts
Production editor: Irwin Zucker
Copy editor: Robert Lentz
Buyer: Bill Scazzero
Cover and jacket design: Doug Delucca and Bruce Kenselaar
Editorial assistant: Naomi Goldman

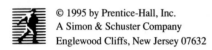

© 1995 by Prentice-Hall, Inc.
A Simon & Schuster Company
Englewood Cliffs, New Jersey 07632

The author and publisher of this book have used their best efforts in preparing this book. These efforts include the development, research, and testing of the theories and programs to determine their effectiveness. The author and publisher make no warranty of any kind, expressed or implied, with regard to these programs or the documentation contained in this book. The author and publisher shall not be liable in any event for incidental or consequential damages in connection with, or arising out of, the furnishing, performance, or use of these programs.

Printed in the United States of America

10 9 8 7 6 5 4 3 2 1

ISBN 0-13-062621-X

Prentice-Hall International (UK) Limited, London
Prentice-Hall of Australia Pty. Limited, Sydney
Prentice-Hall Canada Inc., Toronto
Prentice-Hall Hispanoamericana, S.A., Mexico
Prentice-Hall of India Private Limited, New Delhi
Prentice-Hall of Japan, Inc., Tokyo
Simon & Schuster Asia Pte. Ltd., Singapore
Editora Prentice-Hall do Brasil, Ltda., Rio de Janeiro

Contents

5   TWO-DIMENSIONAL CONVOLUTION   174

**6   THE TWO-DIMENSIONAL CONVOLUTION
      THEOREM   204**

**7   SAMPLING AND INTERPOLATION IN TWO
      DIMENSIONS   222**

8   DIGITAL OPERATIONS   267

Preface

Digital signal processing has a short history and a future worth thinking about. The subject grew out of numerical analysis as once practiced with pencil and paper and later with electric calculators. On the digital side, familiar operations of finite differencing, numerical integration, correlation, convolution, statistical analysis, and spectrum analysis were all well developed in the days of hand calculation, as witnessed by publications of the Census Bureau, the Actuarial Society, and the Nautical Almanac Offices of seafaring nations. On the signaling side, use of electrical signals for information transfer goes back to telegraphy, where digital methods were present from the beginning; development was mainly in the telecommunications industry.

Binary digital signals have been in use since the predecessors of Morse code. They gained ground with the advent of the double-triode flip-flop, an electronic circuit that was stable in either of two states and could therefore be used as a counter. All the numerical operations could then be implemented and digital signals could be processed with the full versatility and speed of the modern computer. Naturally, the teaching of digital signal processing has developed within electrical engineering departments. Equally naturally, the focus has been on the waveforms arising in telecommunications (including television), radar, navigation, and computers.

Meanwhile the disciplines of radio astronomy, remote sensing, and space telemetry have developed electronic digital imaging practice that surpasses optical photography in both resolution and dynamic range. Applications in medical and other biological imaging, resource management, geographic systems, business, typography, art, geophysics, meteorology, and oceanography are expanding rapidly. Since the possessors of the necessary techniques are trained in electrical engineering, digital image processing is a growing sector of the old digital signal processing pie. This growth is bound to continue and will diversify in directions discussed in the Introduction (Chapter 1). Future imaging applications will call for new two-dimensional mathematical background going beyond basic electrical signal processing, which will nevertheless continue to be a prerequisite. Image engineering will become a substantial subfield of electrical engineering and computer science, with a distinctive flavor given by the indispensable infusion of finite (discrete) geometry, psychophysics of vision, and imaging practice from the many fields of application. Accommodating students from earth sciences and life sciences will be the hallmark of a basic course in image engineering as distinct from subsequent specialized courses in medical imaging, seismic imaging, radar imaging, etc., which need to be up-to-date in their specialty but do not need to be fundamental.

So wide are the applications of Fourier analysis to different forms of imaging that a fundamental treatment must begin by building a two-dimensional superstructure on whatever background of spectral analysis one already possesses. For a generation, electrical engineers have found that learning about Fourier's idea was one of their most durable investments, having application well beyond the confines of signal analysis. So also will it be found that two-dimensional Fourier analysis gives entrée to fields unconnected with imaging.

In preparing this book I typeset the text in TEX and used a Hewlett-Packard Integral PC with LaserJet IIIP printer for most of the drawings. The problems were tested on graduate students over the course of several years in a course that developed from origins in radio astronomical imaging. I had the privilege of associating with Joseph Pawsey's group at the Radiophysics Laboratory, Sydney, and with Martin Ryle's group at Cambridge, and I came to know the radio astronomers in France, Germany, Italy, the Soviet Union, and the United States. Pawsey and I coauthored the first book on radio astronomy. Many advances in image construction came from the explosive development of this subject. At Stanford, colleagues who shared my interest in image construction and influenced my understanding include Govind Swarup, A. Richard Thompson, Zvonko Fazarinc, T. Krishnan, L.R. d'Addario, Werner Graf, E.K. Conklin, Jacques Verly, John Villasenor, Domingo Mihovilović, and many students. What I know about painting was learnt from Reginald Earl Campbell. My colleagues Sylvia Plevritis and Ramin Samadani kindly criticized several chapters. I thank my wife Helen for her support and extraordinary patience.

It is a privilege to have worked in a department with a distinguished lineage of textbook writers, especially Frederick E. Terman and Hugh H. Skilling, whose work I have admired. Skilling used to express thanks for the liberal policy of Stanford University in encouraging publication, and I am happy to thank Joseph W. Goodman for maintaining that atmosphere.

R. N. Bracewell
Stanford, California

1

Introduction

Telecommunication by radio shrank the world to a global village, and the satellite and computer have made imagery the language of that village. The creation of images was once mainly in the hands of artists and scribes. Their art works—the famous stone-age cave paintings, the engravings on stone, bone and tooth, and the paintings on bark and skin that have been lost—go back to the distant past. Their inscriptions are also images, though of a different kind. They are also very old, as witnessed by hieroglyphic writing on the walls of tombs and on papyrus in Egypt. In modern times artists and scribes were joined by photographers and printers as creators of images. A variety of artisans, playing a secondary role as artists, created a third kind of image: ornament manifested on pottery, in woven fabrics and carpets, in tile patterns, and in painted friezes. Builders and cartographers created a fourth kind of image, once relatively rare but nevertheless going back to antiquity: construction plans and maps.

Two-dimensional images also occur naturally: a shadow, the dappled light pattern under a tree, the optical image on a retina. Nature provided the motif for much abstract ornamentation, and our written letters and ideograms trace back to representations of nature.

The graven image is a three-dimensional generalization. Examples in art range from primitive ornamental abstract carving to sophisticated modern sculpture. Writing provides three-dimensional examples ranging from cuneiform impressions in clay tablets in Iraq to formal hieroglyphs incised in granite by the Egyptians, jade calligraphy in China, and inscriptions in marble in the elegant monumental alphabet devised by the Romans. Three-dimensional images today have become an important and growing part of the digital world. Examples include volumetric data sets produced for medical diagnosis and other purposes by x-ray tomography, magnetic resonance, positron-electron scanners, ultrasound, confocal microscopes, laser scanners, and supercomputer simulation.

Three-dimensional holographic images, which are of increasing importance, are presented by suitable illumination of holograms. Holograms themselves are usually two-dimensional analogue images (an exception is thin thermoplastic film embossed with corrugations). The construction of analogue holograms from image data sets is a reality, and no doubt composite digital holograms are not far behind. Video processing and multimedia products for mass consumption are also capable of three-dimensional image presentation and will encourage the development of stereoscopic techniques.

The main topic of this book is the two-dimensional image. The book explores the idea that a two-dimensional image is much the same as a mathematical function of two variables, and it explores a variety of phenomena that are susceptible to mathematical reasoning. Some of these quantitative phenomena, such as grey levels, dynamic range, resolution, spatial spectrum, perspective projection, and convolution, have wide significance and are therefore dealt with first. The canvas may be continuous or discrete. Discussion then branches into various specific kinds of image associated with techniques such as diffraction, tomography, interferometry, and range-Doppler mapping.

The creation of images has been moving progressively into the hands of those equipped with new tools: the cathode-ray tube and the laser printer under computer control. Many of us have learned to wield these tools without the accumulated wisdom of artists, typesetters, photographers, advertisers, and movie-makers, all of whom prepare visual material for the attention of others. Much of this wisdom is being reacquired, often painfully, and clearly a new synthesis is in the making. Michelangelo, if he were alive today in his workshop in Florence, might be cutting a titanium monument of extraordinary beauty with a laser-controlled interactive program that he was developing, causing severe competition for the younger generation. We get an inkling of things to come from the fractal images of Mandelbrot and the animated movies of Stephen Spielberg.

As Descartes taught, we should be skeptical. When we question things, our minds are stimulated to creativity. The questioning reader may already have asked, "Is an image really reducible to a function of two variables?" This question leads us to consider the color image. Of course, we know from color printing with ink that a set of four functions of two spatial variables gives the desired description; the four functions are, respectively, the densities of the yellow, cyan, magenta, and black imprints. This is a reassuring minor generalization, bolstering confidence that the mathematical approach is competent. But when you ask, "Are the inks deposited on the paper yellow first and black last and if so why?" you are raising sticky questions whose answers may make the mathematical part of you nervous. Modern practice was not arrived at by logical deduction but by empirical trial, sometimes so lengthy that the early steps have been forgotten. If and when the procedures of this sort of incrementally developed technology come to be understood, not only physics but psychophysics will prove to have been a significant factor. Only occasional allusions to psychophysical and psychological aspects of images can be made in this book, but the topic is introduced here to warn about the incompleteness of the mathematically based material for one wishing to become competent in image handling.

As a reminder of the limitations, take the case of a camera, which has difficulty making a realistic image of a gold ring with a pearl. The camera may capture a function of two variables—photographic density in x and y—very faithfully. It fails to capture surface gloss, metallic sheen, and translucency. The eye is very well aware of these additional variables. How the eye can see something that the camera cannot record will be discussed. Interestingly, some oil painters have been able to convey these nonphotographic surface qualities.

These are very interesting matters that image engineers are going to learn more about. Just as a hint of the phenomena involved, here are some suggestions for action. Get a flower, spray some water droplets on the petals, and put it in the sun; add a fluorescent garment and diamond ring if available. Take a photo and place a print alongside the original arrangement. Look at the solid object and the flat image from different directions; repeat while holding a card with a single pinhole close to your eye. Hold a card with a 5-mm square hole at arm's length, focus on the plane of the card, and use this tool to compare corresponding areas of the arrangement and its photograph. Catch a *camera obscura* image of an outdoor scene inside a cardboard carton and view it through an opening in the bottom. Turn your back on the scene and view it through your legs! Have the shadows changed color? Does the sky seem to have a different color when your head is upside down? Training in physics alone does not prepare you for this, or for what happens when you mix black and yellow paints, or for understanding why Newton reported seven colors in the solar spectrum. Get a prism and a slit and see these color bands for yourself. From experiments one can learn a lot about perception and prepare for the richness of experience that work with images will bring.

SUMMARY OF THE CHAPTERS

Chapter 2, which introduces the picture plane, mentions some of these interesting matters but is mainly concerned with the presentation of graphic images, with elementary operations on images, and with establishing terminology. Most of the operations are applicable to discrete images (those defined on a lattice of points rather than on the plane of continuous x and y) and to quantized images (which have digital function values); consequently, discrete examples are introduced freely, but the mathematics is kept general. The chapter makes relatively light reading. However, this book is written mainly for people equipped to handle computer images using more advanced mathematical methods, especially the Fourier transform in both its continuous and discrete forms. Some previous experience with Fourier analysis of signals will thus be helpful. Two-dimensional Fourier analysis, though, is somewhat different from analysis of signals into temporal vibrations, so the material has been written to be accessible to mathematically inclined students whether they have had previous experience with transform methods or not. The topics of the other chapters will now be summarized.

Several kinds of image involve Fourier transformation either directly, as with optical and radio images, or computationally, as with image processing, or as a way of understanding imaging techniques, such as tomographic reconstruction. But before transforms are taken up in **Chapter 4**, impulsive functions on the plane are explored in **Chapter 3**. One reason for this order is that the delta function is the appropriate mathematical entity for handling discrete elements in an image. The other reason is that delta functions have become indispensable in extending the reach of Fourier analysis to cover entities such as point sources, uniform backgrounds, and the sinusoidal corrugations of amplitude that are set up by diffraction of light through a pair of pinholes. It used to be taught that the Fourier integral did not exist in some circumstances. Indeed, restrictions on the applicability of Fourier analysis gave rise to repeated advances in pure mathematics throughout the nineteenth century. It is understandable, then, that texts spent some time on the nature of the Fourier integral and its convergence. But the point source of light, represented by a delta function, and the pair of pinholes, represented by two delta functions, were then excluded from the domain of Fourier analysis, which was a serious limitation. Likewise, the uniform background (spatial dc) and the sinusoidal corrugation (spatial ac) could not be embraced. One way of comprehending this equally serious exclusion (of functions that are not absolutely integrable) is to note that the seemingly innocuous functions $f(x) = \text{const}$ and $f(x) = \cos kx$ are connected, through diffraction, with one pinhole and two pinholes, respectively, and are thus tainted by association with delta functions.

The delta function itself did not possess a Fourier transform because the notion of integral could not accommodate an entity $\delta(x)$ which was not a function. (The usage, delta "function," simply reflects the modern experience that $\delta(x)$ can, with its own rules, be treated like a function of a continuous variable.) The concept of generalized function supplied these rules; this is one reason for treating the delta function before the Fourier transform. Another reason is that the impulsive concept has to be extended to two dimensions in order to apply to image theory. The two-dimensional point source, implied by terms such as pinhole, can be constructed as $\delta(x)\,\delta(y)$ but is an entity in its own right, as reflected by the symbol $^2\delta(x, y)$, where the preceding superscript 2 means "two dimensions."

The generalization to two dimensions goes beyond the concept of point impulse, for there are also straight-line impulses and curvilinear impulses having little correspondence with functions of one variable, such as waveforms depending on time alone. Much of this development is not self-evident to one trained on waveforms in circuits; if it seems intuitively obvious that $^2\delta(x, y) = \delta(x)\,\delta(y)$, what does your intuition tell you about $\delta(x^2 + y^2)$? More perplexing is $\delta(xy)$. Does it seem reasonable that $\delta(x) + \delta(y)$ should represent crossed slits in a plane? A systematic development following the approach used for generalized functions makes these matters transparent and creates an essential tool for image theory, embracing as it does slits and rasters, pixel arrays, and matrix representations of images. This material has not previously been published. It brings digital, as distinct from continuous, images within the scope of familiar continuous-variable theory. **Chapter 3** is the basis for linear filtering as applied to such digital images.

Chapter 4 builds on previous study of waveforms and spectra, with which most readers are familiar. There is a conceptual block to going from the single time dimension alone to two spatial dimensions. This is handled by presenting examples of one-dimensional transform pairs in conjunction with a two-dimensional generalization. By the time these easily verifiable pairs are absorbed, the conceptual block has evaporated for most readers. This material serves as a useful brush-up for those needing it; others can read it quickly.

Although the mathematical aspects are supposed to come first—applications following in later chapters—it is effective in lectures to illustrate the transform pairs, as they arise, by reference to whatever physics background the students have: optical diffraction from a pair of pinholes, acoustic diffraction from a pair of loudspeakers, propagation of water waves. The power spectrum of a waveform consisting of a pair of impulses, which is easy to work out but not particularly obvious, exemplifies many interesting correspondences between different branches of physics based on common underlying mathematical relations. Thus everyone knows that the diffraction pattern of a pair of pinholes is sinusoidal, and why; but they do not know at the same level of physical understanding that the spectrum of the impulse-pair waveform is sinusoidal. They do know that a cosinusoidal waveform $\cos 2\pi f t$ has a spectrum with just two lines, at f and $-f$, but they may be unaware that the light incident on a grating that imposes spatial amplitude modulation $\cos 2\pi k x$ is diffracted toward just two directions. Many interesting analogies between different fields of application can be used to physically illustrate the interpretation of the basic transform pairs while they are being presented in their distilled form.

Meanwhile, for those still awaiting the flash of comprehension that self-study seems to generate more often than lectures, more time is allowed while the two-dimensional versions of the theorems are surveyed. All these theorems can be presented in physical embodiments. For example, the shift theorem, whose mathematical derivation is rather dry, and so easy that little comment may seem necessary, gains life when interpreted as prism action. Likewise, the phase delay caused by a thickness of glass that increases linearly with distance from the prism edge is associated with a shift in direction of the light beam. Students can be encouraged to think in terms of physical interpretation of mathematical relations and to practice translating these interpretations between fields. There are always two interpretations, depending on whether the left-hand side is taken to refer to the object domain or to the transform domain; thus the second interpretation of the shift theorem alludes to what happens to the beam leaving a prism when the incident beam is shifted laterally by a fixed distance. Some students have a background in statistical signal analysis and probability theory—what does the shift theorem mean to them? It means that when a probability density function is shifted to a new mean, the characteristic function acquires a phase factor while retaining the original absolute values.

By this stage the appearance of lattice objects among the illustrations is noticeable. The elementary pair of pinholes is the exact equivalent of a binary object, albeit a simple one. The concepts acquired thus far will apply readily to more elaborate digital images and to their modification by linear filtering.

Chapters 5 and **6** deal with the important topic of two-dimensional convolution, which arises in a variety of ways with images. Images constructed by the eye, the camera, and the radiofrequency radiometer all represent the outcome of convolution between the response of the instrument to a point source on the one hand and the scene being imaged on the other. Whatever the scene, the point response itself is a convolution between some aperture distribution function and the same function rotated half a turn. Cross-correlation, an operation often applied to images, for example in matched filtering, is expressible as convolution. Digital filtering, image smoothing, image sharpening, and other image operations all involve convolution. For these reasons it is essential to have a good grasp of convolution. The theory is neither mathematically nor conceptually difficult. Some numerical experience, however, especially when gained by hand on small arrays of discrete data, is indispensable and firmly recommended. The insights gained from numerical drill in the various manifestations of convolution are obscured by user-friendly canned programs, especially those whose listings are inaccessible. For repetitive computation professional programs are generally to be recommended (it is usually unwise to rewrite polished algorithms); however, it helps to have numerical understanding of a topic in order to be able to formulate tests on a long or complicated available program.

Familiarity with convolution greatly enlarges the range of Fourier transforms that one can deal with. Numerous exercises have been supplied, at the end of **Chapter 6**, to strengthen the ability to use Fourier transforms as a tool.

Just as a waveform depending on continuous time can be sampled, so also can an image (**Chapter 7**). All images processed numerically involve sampling. The matrix representation of an image can be thought of as a sample set, as can the operator used for digital filtering. And just as the string of equispaced unit impulses, the shah function $\mathrm{III}(x)$, is its own transform (thanks to the concept of transform-in-the-limit that rests on generalized function theory) so, in two dimensions, the infinite bed-of-nails function $^2\mathrm{III}(x, y)$ is its own transform. I gave the proof of this in 1956. The effects of sampling, which can be expressed as a product of an image with a two-dimensional shah function, or infinite bed of nails, can thus conveniently be discussed in the Fourier transform domain with the aid of the convolution theorem. The perils of aliasing when undersampling occurs are made clear this way, as well as the virtue of undersampling by a factor 2 in appropriate circumstances.

Chapter 8 builds on sampling ideas to introduce digital smoothing, a form of filtering that can be applied to an image represented discretely, as with a matrix. (Sharpening, which can also be done digitally, is discussed at greater length in **Chapter 13** under the title of restoration.) Numerical interpolation and differencing are also dealt with in **Chapter 8**. Filtering in two dimensions is broader than in one dimension; nevertheless, image filtering is often performed in real time. For this reason, background in digital signal processing as applied to time series is supplied for those who may need it. Wherever convolution applies, whether discrete or continuous, there is linearity; but convolution fails when space invariance fails, as it often does, even though linearity remains. Transform methods then give way to image-plane operations. The morphology of binary objects, which is fundamental to robot vision, introduces interesting nonlinear operations. Determination of the boundary of a binary object defined on a lattice, lattice-path distances from the interior to the boundary, and lattice distances between separate masses belong to morphology, as do erosion (the peeling of the boundary) and dilation (which resembles crystal growth by adhesion of atoms). These fundamental operations, unlike linear filtering, require new mathematics because they pertain specifically to images defined on a lattice, while linear filtering does not. Ideas from set theory and logical relations are applied to illustrate some of the morphological operations and reduce them to computer instructions. Morphological smoothing is one of the applications.

Chapter 9 introduces special features that appear when circular symmetry exists. Such symmetry most often arises from instrumental design and is therefore important. Bessel functions, jinc functions, Hankel transforms, and Abel transforms are the currency of this subdomain, and because of the simplicity of instruments the discussion can assume more of an analytic nature. Nevertheless, some computational aspects are mentioned. A painful fact of life is the need to fit the circular symmetry of analysis and the machine shop into the straitjacket of the square grid. At this point one could go directly to the projection-slice theorem (**Chapter 14**) if desirable.

Images formed by direct physical convolution with the response of some instrument, as distinct from computational convolution, arise in a variety of ways, as described in **Chapter 10**. In this world the transfer function is the thinking tool. Distinctive types of transfer function characterize different situations, and the different types affect the possibilities for restoration against resolution loss. Laboratory devices such as the microdensitometer possess one such distinctive type of transfer function. If higher resolution is wanted, it can be obtained, at the cost of some acceptable trade-off, by changing the design of the instrument. Many research instruments, however, are fundamentally limited by diffraction. Long-baseline interferometers, which have been extended to intercontinental dimensions, are an example. In these cases it is costly to improve resolution. Even if it *is* improved, the basic situation remains qualitatively the same.

Chapter 11 first establishes how a diffraction pattern set up by a plane aperture distribution is rigorously deducible through a Fourier transform relationship. Then it examines special cases of diffraction-limited sensors. Finally, the occultation technique, which was used in the discovery of quasars, and which fits into the same framework, is treated. Occultation technique is ripe for application to many new fields.

Diffraction-limited sensors include cameras, telescopes, arrays of ultrasonic transducers, microwave slot antennas, geophone and hydrophone arrays, and in addition, in all these diverse fields of application, interferometers. An interferometer is a sensor whose aperture consists of two or more well-separated parts. Thus, if the middle elements of a linear array of sensors are suppressed, leaving two well-separated parts, we have an interferometer. The consequence will be that the point response of the instrument, instead of consisting of a central peak with perhaps some weak sidelobes, becomes oscillatory with many peaks. This viewpoint, which subsumes interferometry within what is already known about single compact apertures, is very helpful. The practice of covering the aperture of a telescope with an opaque screen containing two slits (which does seem like a retrograde step) exactly illustrates this way of thinking about interferometry. More typically, an interferometer is built in discrete parts to avoid the cost of filling in the gaps in the aperture.

Chapter 12 begins by asking: Where in the plane of the input pupil of an instrument does the directional information about the image of a remote source reside? As an example of this unfamiliar topic, consider where on the surface of the moonlit eye the information is to be found that says the moon is half a degree in diameter. By the time an optical image has been formed on the retina by the lens of the eye, it is apparent that the angular diameter, as well as other directional data, have been effectively extracted; but that is after the instrument has operated. Where in the field incident on the eyeball, prior to entry into the eye, is the directional information? The answer lies in the coherence of the field at pairs of points as a function of the vector spacing of each point pair. This way of looking at imaging is not restricted to interferometry; it is a fundamental alternative to the conventional directional description and indispensable as a way of thinking about a wide range of instrumental phenomena. The relationship between the two views is again rigorously expressible mathematically in terms of the Fourier transform. The fundamental relationship is not the same as that between the aperture distribution and the Fraunhofer diffraction pattern, but it is closely related. It supplies a further example of the extraordinary power acquired from a grounding in Fourier transforms.

Transfer functions arising from diffraction-limited sensors of finite spatial extent possess a cut-off which is extremely deep. Nothing can be done, on the basis of observational data alone, in the presence of normal noise, about the spatial frequencies that are cut off. But in the pass band or bands, the response is sufficiently uneven to encourage the attempt to restore, or equalize, the original balance between the Fourier components of the object being imaged. In **Chapter 13** various practical methods of restoration are considered for use in different circumstances. The chord construction, for example, originated as a graphical method for giving a quick view of what restoration might do in some given case; it generalizes to a compact computational procedure in both one and two dimensions. The method of successive substitutions offers a conservative computational approach which accommodates well to the presence of noise and also gives insight into the effects of noise. This method is applicable to a broad range of circumstances. Curiously, when the series of "approximations" does not converge, the results may nevertheless be usable. A convergence criterion, which can be established for successive substitutions, gives a tidy mathematical view of restoration as seen from the Fourier transform domain. The rather elementary theory sheds light on asymptotic series, which are somewhat abstruse as traditionally presented in the function domain; one application is to correct for running means. Wiener filtering, which is widely practised in the world of stochastic time signals specified by probability distributions, has an extension to two dimensions. Deterministic images, which are the more usual, may allow a foothold based on specific characteristics. Högbom's CLEAN and the Gerchberg-Saxton algorithm are mentioned as instances.

Many of the most useful theorems relating to the Fourier transform have simple one-line derivations. It is often easier to follow the derivation than to understand the enunciation, especially when the various ins and outs are taken into consideration. The famous projection-slice theorem (**Chapter 14**), which is derived merely by substituting zero for y in a definition formula, is like that: apparently trivial mathematically, and yet a most powerful thinking tool. One of its best-known applications is seen in the derivation of the modified back-projection algorithm of x-ray tomography, a technique which originated in a radioastronomical data-reduction problem analogous to tomography. When applied to the restricted case of circular symmetry, the projection-slice theorem produces a very useful cyclical relationship between the Fourier, Abel, and Hankel transforms. It provides a synoptic framework for numerous mathematical relations that one usually encounters piecemeal, especially in connection with optical devices, where circular symmetry is common. **Chapter 14** leads naturally into the next chapter, but also leads in other directions as well, and so it has been presented separately. Since it does not depend on the immediately preceding chapters, it could be read with **Chapter 9** if desired.

Chapter 15 combines the strict geometry of projection with the domain of arbitrary images to show how a desired image can be formed when only projections are available. The mathematical formulation in terms of integrals along manifolds was worked out by Johann Radon (1887–1956) and published in 1917 in the reports of the Saxon Academy of Sciences. Meanwhile, corresponding problems arose in many fields and were dealt with in isolation. One was to deduce the radial density distribution of stars in a globular cluster, containing perhaps a million stars, from the area density measured on a photographic plate. Another was to find the radial variation of brightness of solar microwave emission from scans made with a fan-beam antenna having only one-dimensional resolution. The mathematical aspect received much exposure when half of the 1979 Nobel Prize for Physiology or Medicine was awarded to Allan M. Cormack for "discovering the theoretical principles that made CAT scanners possible." Thereupon much prior history emerged, soon summarized by Cormack himself (*Proc. Symposia Appl. Math.*, vol. 27, pp. 35–42, 1982). The idea that images can be reconstructed from projections has proved to be extremely far-reaching. It has made tomography a fast-growing technique, extending well outside the range of astrophysics and medical imaging into various forms of nondestructive testing and into various realms of atmospheric physics, geophysics, and oceanography. The difficult mathematics of Radon's time has been so illuminated by the Fourier approach that the theory is no longer difficult. Indeed, it was insight gained from Fourier analysis that led to the simple modified back-projection algorithm. In consequence, the inversion algorithm now need not actually require numerical Fourier transformation at all.

Synthetic-aperture radar mapping gained public attention when remarkably detailed maps of the surface of Venus were returned by *Pioneer Venus* in 1990, showing novel geological formations in great variety. The impact of this fine imagery was heightened by the knowledge that the atmosphere of Venus is practically opaque and that cameras parachuted to the surface had sent back nothing by comparison. The technique has roots in World War II radar. It was developed in many essential details by radar astronomers working with the moon or Venus and by military specialists looking down at the earth's land and ocean surfaces. Clearly, future applications will lie in remote sensing of the earth environment, with no apologies to photography, even though air transparency and angular resolution tend to favor optical images. The features observable by radar imaging carry information that is different from what is seen under incident light, making radar a valuable complement to photography. Microwave radiometry, although using the same part of the electromagnetic spectrum as radar, provides information that is more akin to the optical, differing in that the signal received is emitted by the object rather than coming from an illuminant. Synthetic-aperture radar, treated in **Chapter 16**, is one more example of an indispensable modern technique that evolved in astronomical research under the stimulus of open international competition. And, as with the even more extraordinary images of radio astronomy, both of these radio techniques found their origins in the burst of technological activity engendered by the rush to develop military radar.

Most images are contaminated by noise. Some, such as radiant images, could be said to be composed of noise. Two-dimensional distributions of noise may, of course, be thought of as images in their own right. A relatively simple example would be the pepper-and-salt scene forming the background of a weak television image, or forming the whole picture when the television set is tuned to an unoccupied channel. The interesting thing here is that noise in two dimensions is much richer than one-dimensional noise. Examples are given to emphasize the variety that one may be required to generate. Spectral and statistical analysis are important. Autocorrelation is a useful tool, especially the economical clipped autocorrelation, a binary object with visual interpretability. Applications to texture analysis, pattern recognition, and restoration (improvement of signal-to-noise ratio) are among those that depend on **Chapter 18**.

Halftone images originated in the printing industry in 1852 as a photomechanical way of emulating the hand-shading techniques of lithography and wood-block printing. A halftone screen made of crossed perpendicular ruled gratings was placed on the printing block to break up the photographic image into dots whose size depended on light intensity. This analogue printing technique has reached a stage of maturity that approaches the quality of photographic prints (though certainly not that of photographic transparencies). The trend, however, is toward dispersal of machines offering adequate halftone printing into nearly every office. The laser printer under digital control offers both opportunities and problems to the engineer. Some computer screens can reputedly display 1024 shades of grey; the problem is to convert the display to a hard-copy black-and-white original that can be published. This is one of the aspects of image design discussed in **Chapter 19**.

The general plan of the chapters has now been sketched; some information follows about notation, selection of topics for a course, the problems at the ends of the chapters, and the relationships between imaging system design, image acquisition, image processing, and image interpretation.

NOTATION

Good notation is helpful as a thinking tool, but it is just as well to be sparing with innovations. The use of a preceding superscript for dimensionality, as with $^2\delta(x, y)$, has already been seen above. Other special notations needing mention are the sinc function and its circular counterpart, the jinc function:

$$\operatorname{sinc} x = \frac{\sin \pi x}{\pi x}$$

$$\operatorname{jinc} x = \frac{J_1(\pi x)}{2x}.$$

In two dimensions,

$$^2\operatorname{sinc}(x, y) = \operatorname{sinc} x \operatorname{sinc} y = \frac{\sin \pi x}{\pi x} \frac{\sin \pi y}{\pi y}.$$

The Fourier transform of the sinc function is the unit rectangle function, which is defined by

$$\text{rect } x = \begin{cases} 1, & |x| < \frac{1}{2} \\ 0, & |x| > \frac{1}{2}. \end{cases}$$

In two dimensions the Fourier transform of the two-dimensional sinc function is the two-dimensional rectangle function $^2\text{rect}(x, y)$,

$$^2\text{rect}(x, y) = \text{rect } x \text{ rect } y.$$

The value of rect(0.5) is usually immaterial, but if a value is desired, 0.5 is recommended. Other conventions differ from each other by only a null function, which integrates to zero and therefore has no physical impact. The virtue of choosing $\text{rect}(0.5) = 0.5$ is that inconsistencies at the rim do not arise when an expression such as $1 - \text{rect } r$ is introduced to describe the complement to rect r (a circular hole in an opaque screen versus an opaque disc on a transparent screen). In rectangular coordinates, the function $[\delta(x + 0.5) + \delta(x - 0.5)] \text{rect } x$ has a Fourier transform that can be calculated by the convolution theorem; it is the same as that of $\frac{1}{2}\delta(x + 0.5) + \frac{1}{2}\delta(x - 0.5)$, which confirms the wisdom of adopting $\text{rect}(0.5) = 0.5$.

A more contentious choice arises with the Heaviside unit step function $\mathbf{H}(x)$ defined by

$$\mathbf{H}(x) = \begin{cases} 1, & x > 0 \\ 0, & x < 0. \end{cases}$$

The value of $\mathbf{H}(0)$ is sometimes taken as 1 and sometimes as 0, but 0.5 is recommended; otherwise the even part $\frac{1}{2}[\mathbf{H}(x) + \mathbf{H}(-x)]$ does not come out well and one cannot write $\text{rect } x = \mathbf{H}(x + 0.5) - \mathbf{H}(x - 0.5)$. With discrete functions $f(x)$ which are defined only at integral values of x, one often needs a step function that sets values for $x \leq 0$ to zero and preserves the values where $x > 0$. In principle, one could write $f(x)\mathbf{H}(x - 0.5)$, and in fact this is a way of getting the right answer using the algebra of continuous variables, but it is convenient to have a discrete unit step function $\mathbf{u}(x)$ defined by

$$\mathbf{u}(x) = \begin{cases} 1, & x > 0 \\ 0, & x \leq 0, \end{cases}$$

understanding that x is an integer. This symbol is useful for expressive notation but needs caution. For example, its odd part $\frac{1}{2}[\mathbf{u}(x) - \mathbf{u}(-x)]$ is $\frac{1}{2} \text{ sgn } x$, which is good, but its even part $\frac{1}{2}[\mathbf{u}(x) + \mathbf{u}(-x)]$ may not be what is wanted.

Sometimes it is convenient to use $\Pi(x)$ instead of rect x; its autocorrelation is the unit triangle function $\Lambda(x)$, which has unit height and unit area. Expresions such as $\text{rect}[f(x)]$, $\mathbf{H}[f(x)]$, $\text{sgn}[f(x)]$, $\Lambda[f(x)]$ present unfamiliar algebraic notation but allow for compact computing and drawing of pulse modulation waveforms, various codes, Walsh functions, etc. In two dimensions expression, such as $\text{rect}[f(x, y)]$, have analogous applications to binary images.

I like to use $\supset$ to mean "has Fourier transform" because it takes less ink. With this relational sign

$$\text{sinc } x \supset \text{rect } s,$$

and in two dimensions

$$\text{sinc}\,x\,\text{sinc}\,y \ ^2\!\supset \text{rect}\,u\,\text{rect}\,v,$$

or $^2\text{sinc}(x, y) \ ^2\!\supset \ ^2\text{rect}(u, v)$. To abbreviate or not to abbreviate is a question that I settle as follows: if the unabbreviated form is going to appear fewer than five times, then spare the reader the definition and do not abbreviate.

If all the transforms in a paper or report are two dimensional, then of course the preceding superscript may be gracefully dropped. But be careful: $\delta(x)$ has a one-dimensional transform, which is just 1, as well as a two-dimensional transform that is different! The superscript notation enables us to write

$$\delta(x) \supset 1,$$

$$\delta(x) \ ^2\!\supset \delta(v).$$

The Fourier transform variables u and v, which are usually spatial frequencies, often appear in the literature as k_x and k_y, k_1 and k_2, or s_x and s_y. Instead of $f(x, y)$ it is common to see $x(n_1, n_2)$ when n_1 and n_2 are restricted to nonnegative integer values; with this convention the Fourier transform is usually written $X(k_1, k_2)$. These and several other variants will be encountered in the literature.

The asterisk for convolution has slowly gained currency since the days of Volterra. It is extremely valuable as a way of focusing on the two functions entering into the operation of convolution. Thus $f * g$ has ultimate economy as an abbreviation for the full convolution expression with its integral sign, its independent variables (the independent variables are actually irrelevant to the concept), and its limits of integration. When the functions f and g are two dimensional, the same notation is suitable for personal use, but there are occasions when both one-dimensional and two-dimensional convolution occur in the same context. For two-dimensional convolution a double asterisk,

$$f ** g,$$

is proposed. Cross-correlation needs to be distinguished from convolution, and there is a tradition of several decades of using a pentagram. With this convention, the autocorrelation of the real function $f(\ ,\ )$ is

$$f \star\star f.$$

Cyclic convolution is also important. In one dimension it may be thought of in the way Fourier thought of heat conduction in a bar. He brought the left and right ends into contact by bending the bar into a ring, so large that within the time interval of interest, the heat flowing to the left and the heat flowing to the right did not arrive in significant amount at the remote contact. This enabled him to simplify any initial function into a trigonometric series with a fundamental period equal to the circumference of the ring. Cyclic convolution on a ring needs to be distinguished from convolution with infinite limits of integration when the combined length of the convolution factors exceeds the perimeter of the ring. The symbol proposed is ⊛. Cyclicity in two dimensions means that the left and right edges of a map are brought into contact by forming a cylinder; then the circular ends are joined to form a toroid. This bizarre distortion (which you may have encountered in the dangerous game of toroidal chess) regularly arises as a consequence of sampling on a grid.

The rectangle function of radius, rect r, is equal to unity over a circle of unit diameter and zero elsewhere. Its two-dimensional Fourier transform is the jinc function:

$$\operatorname{rect} r^{2} \supset \operatorname{jinc} q,$$

where

$$\operatorname{jinc} x = \frac{J_{1}(\pi x)}{2x}.$$

As previously mentioned, the various factors of π that enter into the definitions of sinc and jinc result from the choice of unit height and width for the unit rectangle function. The beneficial outcome of dealing with these transform-generated definitions of sinc and jinc is that innumerable other factors of π are absorbed by the definitions.

TEACHING A COURSE FROM THIS BOOK

As there is more material in the text than can be covered in one course, it is logical to select from the later chapters as suits one's needs. If the essentially mathematical **Chapters 3** to **8** are dealt with thoroughly, and illustrated on the way by whatever applications interest the class, as mentioned above in connection with **Chapter 4**, a good basis will be established for future specialist courses or for self-study. Fourier ideas have proved to be so versatile for approaching many different fields that getting an appreciation of this branch of mathematics has a clear priority for attention. The later chapters then offer a diverse choice of topics through which one can advance more rapidly without spending time learning the same fundamentals over again using different symbols and different terminology. These chapters, or equivalent lectures on topics the instructor wishes to discuss, are introductory to more advanced treatments and give the flavor of what lies in store.

THE PROBLEMS

Many of the problems in the earlier chapters are couched in practical language to convey suggestions as to where the material may lead. There are also plain exercises; in many cases such exercises have been used as a repository of simple results that are not developed in the text but are collected at the end of the chapter for reference. The instructor in a lecture course may assign some of these, perhaps with variations, to bring them to your attention; doing the right problem at the right time is a very important way to learn. Years ago I read about the virtues of problems that follow the plan "Smith says so-and-so, but Jones says something else. What do you think?" Sometimes what Smith says is true, sometimes incomplete or not true at all (but persuasive), and sometimes both statements are wrong. This really makes the student think, and the solutions submitted often bring out nuances that are not raised for discussion by problems of the form "Prove that $X = Y$." It is not difficult for the instructor to construct hearsay problems. To begin with, one does not have to be sure in advance that all the necessary conditions for the presumably true opinion are mentioned. This avoids the embarrassment that results when the instructor realizes that X does not equal Y at all, or only under certain conditions. Opinions expressed in verbal debate between engineers are very often fuzzy, so there is an air of practical reality in what the participants are quoted as saying. There is a place here for real-life anecdotes when the official solution comes out. I can confidently recommend the hearsay question as a means of enlivening dry homework. Problems whose solutions appear in the Appendix are terminated with a pointer ☞

Problems can also be used as a vehicle for introducing information from useful subjects that the student cannot study formally for want of time. Many students of electrical engineering do not know how to calculate the deflection of a beam or membrane, but may be very grateful to encounter a formula, perhaps because it may have a bearing on a moving part of a silicon micromechanism. Up-to-date numerical magnitudes are very good, too, for relieving the tedium of questions posed algebraically. It is better for a student to read, "A buried optical fiber 100 km long has an attenuation of 20 dB at $\lambda = 1.55\ \mu$m," than to see that a transmission line of length L has an attenuation αL at frequency ω.

Some homework consists of exercises whose purpose is to convert something you understand into something you can do; the work may be essentially a repetition of a previously seen example, but with changed numbers. Exercises are not necessarily easy, but if an exercise looks easy in retrospect that is because the purpose was achieved; you can create exercises for yourself and reap the benefits. When the circumstances are changed, as well as the numbers, the purpose is to illustrate that applicability is not limited to the context of the original example. Thus, mastery of the theory of a resonant electric circuit may be followed by a numerical exercise concerning electrical inductance, capacitance, and resistance and then perhaps by a problem on mechanical vibration of an automobile, given some values of mass, spring stiffness, and damping. Even so, it may not occur to the student that the same underlying theory applies to acoustics, mufflers, or water waves; each new application brings in something of its own. Without prior experience it is hard for a student to create illustrative problems of this type.

Most homework stops at exercises and problems—that is, assignments that can be graded—but I would like to say a word in favor of open-ended assignments and puzzles. A reason for not assigning a puzzle for grades is that the time needed by the best students may approach infinity while another student, not noted for high grades, may crack it, which is a bit embarrassing. I cope with this situation by giving 10 out of 10 to everyone, or zero for no attempt. A puzzle may be overtly cross-disciplinary. For example, the project leader in an electronic surveying instrument group asks a recently recruited Ph.D. in electrical engineering to find out the percentage change in electrical resistance when a steel surveyor's tape, calibrated by the National Bureau of Standards, is stretched 1 percent. Other puzzles involve something outside the ostensible field, but you don't know what—an element that occurs in professional life. Puzzles develop creativity in casting about for ideas to try out, and build up reservoirs of first-hand personal experience. Puzzles may involve intense competition, engage more attention than graded homework, and also be entertaining.

Owing to the explosion of subject matter, only small segments of the curricular pie can now be devoted to some very worthy subjects. Contour integration is studied by a smaller fraction of students than formerly, while finite mathematics is an expanding segment, which is appropriate. But geometry also is a shrinking subject, which is not helpful at all for people concerned with imaging. The shading of an image is inherently geometrical and not apt to be handled by adeptness at statistics and grey-level histograms. It is very helpful indeed to cultivate visualization of objects in three-dimensional space, a faculty that used to be instilled in engineers by the rigors of descriptive geometry and the discipline of the drawing board. Skill in communication by drawing complements those other two modes of communication, giving a talk or writing a report, and is still taught, usually in mechanical or civil engineering. Inability to visualize, like color blindness, is a defect of which the victim may be unaware. So, if some of the geometrical notions in this book give you a feeling of high-school-level déjà vu, consider brushing up on geometry.

ASPECTS OF IMAGING

Imaging embraces the *design* of imaging systems, image *acquisition*, image *processing*, image *analysis*, and image *interpretation*. The design of an optical imaging system requires a knowledge of optics, propagation of light through the atmosphere (or other medium), photodetectors, and other special topics. Other systems depend on corresponding knowledge of radiophysics, acoustics, underwater sound, seismic waves, x-rays, and other fields of physics relevant to the application. The reader cannot know all this, and yet some physical background is indispensable, otherwise a student might think that an image is just a mathematical object. This book supplies background where needed and frequently alludes to a physical context as a reminder that images are not disembodied entities but come from some part of science or technology. Knowledge of the origin will distinguish the expert from the technician.

Image acquisition, which may entail observation on the one hand or experiment on the other, refers to the use of observing and measuring instruments and techniques. Angular resolution, noise level, and measurement artifacts are examples of the factors that may influence one's judgment as to what to do with an image. The character of the noise is highly relevant. Gaussian additive noise, which is common in textbooks, is not common in images, and its mention is usually an admission of ignorance pending accumulation of more experience. Numerical values for resolution and noise are introduced in the problems. When you see them specified by only a single parameter, such as "one-degree beamwidth" or "r.m.s. noise level 1 μvolt," you may be sure that these phrases are first-order characterizations of situations where more detailed experience would be brought to bear in practice.

Students usually come with a special interest in some one topic, such as telemetry, microwave radar, radiometry, optics, medical imaging, ultrasound, or geophysical exploration. Studying the aspects of imaging common to several different fields and working out a few problems couched in the jargon of some unfamiliar field gives confidence that expertise in imaging may be transferable to other fields by self-study.

Image processing, in its digital aspect, is essentially mathematical and closely related to the discrete signal processing widely taught in electrical engineering curricula. It is assumed that time series, as encountered in communications, and operations such as power-spectrum extraction and autocorrelation are known to the reader. There is a jump from one-dimensional operations on temporal data to spatial operations on images, where one must not only imagine an independent-variable axis that is perpendicular to the familiar time axis, but also allow for rotation and other additional operations. Consequently, most stress is laid here on the spatial phenomena, and less emphasis is given to numerical operations that reduce to the one-dimensional material dealt with in elementary digital signal-processing courses. Analogue image processing is just as important as digital image processing, and indeed takes precedence. Photographic, ultrasonic, x-ray, and other images are processed in analogue form by lenses, mirrors, gratings, and the devices appropriate to the branch of physics. The product may then be finally displayed in analogue form on paper, film, or a cathode-ray screen. But an analogue image may also be digitized for the purpose of further computer processing. Digitization of images, which forms a link between analogue and digital, is therefore also a part of image processing. While image processing thus forms only part of the whole of imaging, it is a large part.

Another aspect of the analogue/digital dichotomy arises on the mathematical side as a contrast between continuous and discrete independent variables. In the field of probability, complete textbooks used to be written in pairs, one for each case. Fortunately, the delta function has enabled us to do theory in terms of a continuous variable while retaining applicability to the discrete case. This feature passes over into two dimensions, where a discrete sample of value $A_{1,1}$ at the point $(1, 1)$ can be precisely represented by $A_{1,1}{}^2\delta(x - 1, y - 1)$; yet x and y are continuous and the algebraic expression is amenable to the ordinary operations of mathematical analysis. To be sure, a set of impulses is not exactly the same as a set of samples; the point is that the information content of the two entities is identical, and we do not now need a separate formalism for discrete samples.

Other forms of image processing include stretching and compressing, redigitizing to a different grid, distorting so as to conform to a spherical surface, forming mosaics from subimages that do not fit at the edges, and removing instrumental aberrations. Correction for spherical aberration received public attention when the Hubble Space Telescope was found, after launch, to be focusing a point source into a tiny uniform disc instead of into the smaller Airy diffraction pattern that was wanted. The launching and installation of an analogue repair kit reminds us that digital image processing is applicable for some purposes but yields to analogue methods for others.

Image analysis involves processing that goes beyond the simpler mathematical treatments; edge detection and pattern recognition are examples of such processing. In nature, edge detection is vital for survival, because information about a scene resides largely in the edges, or outlines, that yield quick clues to the presence and nature of things in the field of view, some of which, if present, call for immediate action. The vertical and horizontal, needed for locomotion, are apprehended through edges, both upright and level, that are discernible in most natural habitats. Pattern recognition in general, which is more sophisticated than edge recognition, accounts for our remarkable ability to find and sort small objects, and is clearly inherited from a long time ago, judging from the behavior of birds pecking seeds from the soil and fish ignoring inedible debris in the water. Even the simple detection of straight edges is tricky. If you are presented with an algorithm, such as a gradient detector, that purports to detect edges, it is an interesting exercise to construct a simple image with a visible edge that defeats the algorithm, and it is not hard to do so. Straight moiré fringes are conspicuous to the eye, but it requires a tailor-made algorithm to find them. Even that algorithm is defeated by fringes that alternate between light and dark along their length but which are just as conspicuous to the eye. For this reason, pattern recognition, robot vision, feature extraction, image segmentation, and image encoding by reference to features depend intimately on the field of application.

Image interpretation represents an even higher level of endeavor, which this book does not try to deal with. It is worth noting, though, that a toad knows very well when to open its mouth and flick a fly out of the air with its tongue while ignoring a piece of debris on a similar trajectory, so fast image interpretation is possible and can conceivably be approached to some level of success by a computer system. The level of difficulty and expense are daunting; witness the case of collision-avoidance radar in the civil aviation industry. Military defense against missiles is the area where much of the effort at image interpretation will be expended.

COMPUTER CODE

Segments of computer code are provided here and there. Since people accustomed to using one of the popular languages can have trouble translating from one of the others, I use pseudocode, which in my classroom experience is readable by users of C, FORTRAN and PASCAL. But unless a segment of code has been subjected to the discipline of actually running on some machine there is risk of error; for this reason my "pseudocode" is actually Hewlett-Packard Technical Basic. There is only one feature to be watched. The @ sign is a statement separator for use in multistatement lines; users unaccustomed to multistatement lines need to know that a statement following an IF statement on the same line is not executed if the condition is not met. Thus

```
IF i>N THEN total=0 @ image=image+1
```

increments the variable image only if the condition i>N is met. This snappy syntax is easily expanded into your favorite language.

LITERATURE REFERENCES

A moderate number of literature references used to suffice in textbooks, but now rather long lists are seen not only in monographs for the specialist but also in textbooks for students, who will be lucky to be given time to read the whole book let alone the literature cited. The trend to longer lists of references faces an endless proposition, as detailed lists soon become dated. Consequently, this book does not attempt to present a literature survey but contents itself with occasional references as needed. But it would be a pity not to encourage students of technical subjects to read current papers in their chosen field. I have found it effective to issue a homework assignment as follows: "Peruse current issues of such-and-such a journal, find an article that interests you, and provide a one-page abstract that I could use for a five-minute presentation to the class." Unlike ordinary weekly assignments, this one is issued three weeks before the due date; it has several interesting facets and is recommended. Strange as it may seem, many undergraduates have never had the experience of examining current issues of periodicals in their own field. After this imposed experience they can be introduced to the "Science Citation Index," a tool which, combined with regular perusal of two or three periodicals, is indispensable. Here are some of the journals where current research relating to imaging may be found.

> *Computer Vision Graphics and Information Processing: Graphical Models and Image Processing*
>
> *IEE Proceedings: Vision, Image and Signal Processing*
>
> *IEEE Transactions on Image Processing*
>
> *IEEE Transactions on Signal Processing*
>
> *Electronics Letters*
>
> *Digital Signal Processing: A Review Journal*
>
> *Journal of Computational Physics*
>
> *Journal of Electronic Imaging (SPIE)*
>
> *Journal of the Society for Information Display*
>
> *Journal of Visual Communication and Image Representation*

The following texts, which are frequently cited below, have special abbreviations.

A&S (1964) = M. Abramovitz and I.A. Stegun, eds., *Handbook of Mathematical Functions, with Formulas, Graphs, and Tables of Mathematical Functions*, National Bureau of Standards, Washington D.C., 1964.

Erd (1954) = A. Erdélyi ed., *Tables of Integral Transforms*, vol. I, McGraw-Hill, New York, 1954.

FTA (1986) = R. N. Bracewell, *The Fourier Transform and Its Applications*, McGraw-Hill, New York, 1986.

GR (1965) = I. S. Gradshteyn and I. M. Ryzhik, *Tables of Integrals, Series, and Products*, 4th ed., Academic Press, New York, 1965.

NumRec (1986) = W. H. Press, B. P. Flannery, S. A. Teukolsky and W. T. Vetterling, *Numerical Recipes, the Art of Scientific Computing*, Cambridge University Press, 1986.

RECOMMENDATION

The study of imaging now embraces many major areas of modern technology, especially the several disciplines within electrical engineering, and will be both the stimulus for, and recipient of, new advances in information science, computer science, environmental science, device and materials science, and just plain high-speed computing. It can be confidently recommended as a fertile subject area for students entering upon a career in engineering.

2

The Image Plane

Picture plane is a traditional name in descriptive geometry for the location of a two-dimensional image. It corresponds with the plane of the grid used by Albrecht Dürer (1471–1528) as illustrated in Fig. 2-1 and with the film plane in a camera. However, not all images arise from projection: *image plane* is a more general term for the domain of a plane image. Images may be presented in many ways, such as by photography, television, or printing, and there are other modes of presentation that may be equivalent to images. Examples of equivalents include an array of numbers, a voltage waveform (from which a television picture could be generated), or a contour diagram of the kind introduced in 1728 by M. S. Cruquius and most familiar (Fig. 2-2) in the representation of surface relief on maps. One chooses according to end use: a passport photo rendered in grey levels is more appropriate than the equivalent contour diagram or matrix of numerical values, if it is a matter of identifying the bearer. All of these forms, in one way or another, are equivalent to a single-valued scalar function of two variables, and therefore the theory of such functions must be basic to an understanding of images, even though a full understanding of images may go farther.

A colored image may require a three-valued function of two variables for representation. One could think of this as a set of three simple images, such as those made with magenta, yellow, and cyan inks in color printing. In principle, three colors could suffice, but when available ink is used on available paper, a fourth overprinting of black is help-

Figure 2-1   A woodcut by Albrecht Dürer showing the relationship between a scene, a center of projection, and the picture plane.

ful. Color reproduction of paintings that are to be reproduced faithfully requires seven or more ink images. Surface attributes such as gloss, transparency, fluorescence, and tactile texture raise interesting additional topics. Three-dimensional images such as graven images, or sculpture, are not dealt with here except insofar as the representation of relief in two dimensions only is concerned.

After dealing with the representation or display of images, this chapter introduces a variety of elementary operations on images that are basic to image processing.

MODES OF REPRESENTATION

Various traditional ways of representing functions of two variables will now be compared as regards dynamic range and resolution.

Halftone printing, photography and television offer familiar examples of the representation of a two-dimensional function $f(x, y)$. In a photographic transparency, which is viewed by transmitted light, transmittance is the physical quantity (ratio of transmitted to incident light) representing $f(x, y)$; transmittance can vary from 0 to 1 and thus represent a very large dynamic range. In a photographic print or in printing with ink, we are dealing with reflectance, which might range from 0.1 ("black") to 0.8 ("white").[1] On a television screen we deal with brightness.

---

[1] The whitest snow has a reflectance of about 0.9 and white paper about 0.8, but even a reflectance of 0.6 may give the subjective impression of white. Of course, if a second object of higher reflectance is introduced into the field of view the second object will seem white and the first will then look grey. The appearance of black depends not only on reflectance but also on the level of illumination and on the brightness adaptation of the eye at the moment, and so black is a more complicated phenomenon than white. Thus, a white projection screen is capable of looking like a portion of a black automobile. If a reflectance less than 0.1 were the definition of black, then the moon would be black.

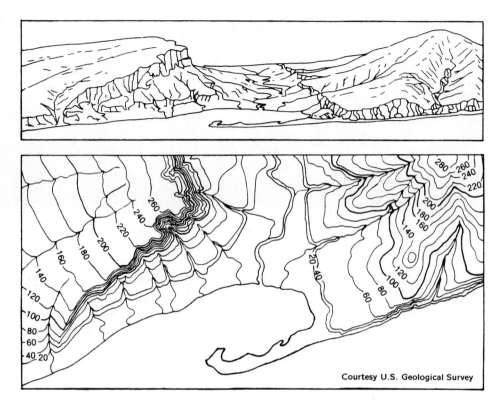

Courtesy U.S. Geological Survey

Figure 2-2   An artist's impression (above) contrasted with a contour map (below).

Dynamic Range

Dynamic range refers to the separation of the extremes of light and dark. For example, the dynamic range is 8 to 1 in the case of printing, where the blackest black shows one-eighth as much light as the whitest white. Another way of expressing the concept is in terms of the number of grey levels. For example, suppose that 10 different grey inks are used whose reflectance values, spaced approximately one decibel, are 0.10, 0.13, 0.16, 0.20, 0.25, 0.32, 0.40, 0.50, 0.63, 0.80. Then we can say that the printer is using 10 grey levels, which is a pactical way of expressing the dynamic range. However, under poor or adverse illumination the human eye might not be able to discriminate these 10 shades. Here we enter the realm of psychophysics, which is of extreme importance in connection with the viewing of images. Suffice it to point out that dynamic range may refer to an object without reference to illumination, or it may include illumination without reference to a viewer, or it may include reference to a viewer's perception. It is quite possible to print 256 grey levels, but the number of distinguishable grey levels may be fewer. About eight levels may be suitable for newsprint and books, while three or four may be adequate for many purposes (as witnessed by photocopiers; refer to Fig. 2-3). In the case of objects

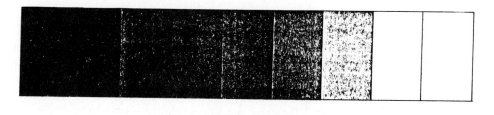

Figure 2-3 Copy of a reflectance chart with nine patches whose reflectances R were originally in geometric progression. After photocopying in a machine where the reflectance R' of a copy is a nonlinear function of the original reflectance R, fewer levels will be distinguishable. The quality achievable with even so limited a dynamic range may be judged from the engraving of Augustin Fresnel (1788-1827), from the frontispiece of his *Collected Works.*

which are themselves light sources, such as computer screens, the number of discernible levels depends on the display brightness control.

The dynamic range of a commercial television picture is comparable with that of printing; thus, as regards dynamic range, both printing and commercial television are poor by comparison with a photographic transparency.

Density and Shade of Grey

In photography absorption is expressed in terms of the photographic density D, which is defined, with respect to a transparency, as the transmission loss measured in bels. Thus if the intensity I_0 of the light transmitted through an unexposed part of the transparency is 10 times the intensity I transmitted through an area of interest, the loss is one bel (or 10 decibels), and the density D of that area is 1. In general

$$D = \log_{10} \frac{I_0}{I}.$$

When light falls on paper, a certain fraction of the light is absorbed. The established usage for transparencies could be applied to the reflectance of opaque paper by taking I to be the intensity reflected back from an inked area of interest and I_0 to be the intensity reflected from an uninked part. In these terms relative reflectance $R_r = I/I_0$. The corresponding "reflective density" $D_{refl} = -\log_{10} R_r$ would range from about 10 dB or more for a very dark surface down to nearly zero for ordinary white paper. For computer-driven printers a linear scale is suitable, because it relates to the fraction of the paper on which ink is deposited. When 50 percent of an area is inked, it is half shaded; when the area is fully covered, it is fully shaded; and when no ink is deposited, it is not shaded at all. The shade thus ranges from 0 to 1. Since it is not possible to put down less than no ink, or to ink more than the whole surface, shades of 0 and 1 are understandably referred to as white and black. Intermediate shades are referred to as grey levels, often as a percentage of coverage by black ink. The term *grey level* is convenient but is not always normalized; one reads of grey levels ranging from 0 to 7, or from 0 to 255 (optimistically). When normalized so as to range from 0 to 1, the term may be ambiguous; a grey level of 0.8 in one major graphics language is called a grey level of 0.2 in another. The term *shade* has an established meaning; a shade of green is a color that a painter would make by adding a little black paint. A green *tone* is made by adding white paint. It therefore makes sense in black-and-white computer graphics to use the term *shade*, or *shade of grey* for the fraction of area covered by black.

Halftone

Dot-matrix and laser printers do not produce grey dots but simulate different shades by varying the dot size so as to control the fraction of paper covered by ink. The resulting discrete set of greys emulates the halftone photographic process well established in the printing industry. Figure 2-4(a), which is presented on a coarse scale so that the structure can be studied, shows how grey levels may be constructed using composite square dots containing 1×1, 2×2, 3×3, ... elements distributed on a diagonal grid. There are nine grey patches shown, the middle one of which is patterned like a chess board. If the illustration is viewed from a distance, so that the structure is invisible, eight distinct grey levels, plus black and white, are clearly distinguishable, provided the light is good (in poor artificial light some of the levels will merge). The four paler shades have squarish dots with relative areas that are nominally 1, 4, 9, and 16, as they are composed of square clusters of the smallest dots available with the laser printer used. The four darker shades have white squarish dots on a black ground. Examination with a lens will show that the smallest dots are not perfectly square but they are small enough that, when more than one hundred are combined to form a single letter, the exact shape of the dot is immaterial. With this plan for generating nine greys plus black and white, an attempt to represent linear variation of brightness from one end to the other results in the bands of unequal width shown. This is because the only levels available under the plan are 2, 8, 18, 32, 50, 68, 82, 92, and 98 percent.

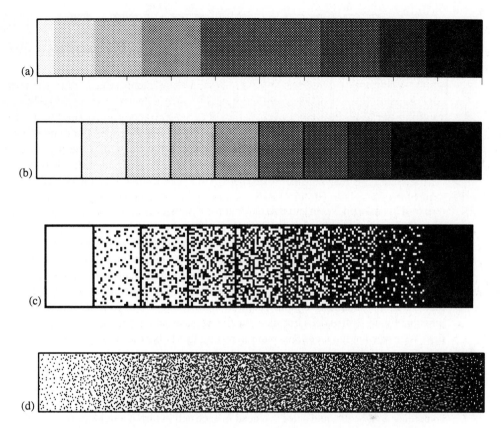

Figure 2-4 (a) Shades of grey constructed from nominally square dots containing 1, 4, 9, 16, 25 pixels, do not advance linearly as regards the fraction of inked area. Consequently the available grey-shade resolution is not uniform. (b) The available grey levels of (a) shown on patches of uniform width. Insertion of separators has had the effect of suppressing the contrast enhancement at the junctions, enabling the grey-shade resolution to be examined more easily. (c) Randomly placed dots, shown magnified, allow finer control over grey shades. (d) Quasicontinuous grey-level variation implemented with randomly placed dots. A useful feature that comes with the relatively coarse texture is tolerance to repeated photocopying.

Figure 2-4(b) is a specimen chart exhibiting the available greys as patches of equal size. Since the density of ink in the second zone is four times greater than in the first zone, while that in the third zone is 9/4 times that in the second, the grey steps are unequal. At the dark end, white dots replace black dots in a symmetrical way. It will be noticed, though, that the elementary white dots have disappeared; an elementary black dot always prints as a speck of some sort, no matter how small, but a white dot is always intruded upon by adjacent inked areas. The reflectance resolution, expressed as a fractional increment dR_r/R_r, is best at the dark end; at the light end the fractional jumps in reflectance are

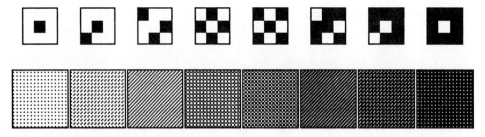

Figure 2-5   Eight grey levels (below) provided by replication of preselected fillings of a 3 × 3 box (above).

coarse. This may not seem satisfactory, but for commercial applications such as pie charts the selection of shades is often sufficient.

Figure 2-4(a) a shows the interesting phenomenon of edge enhancement at a junction between different shades, the lighter side appearing lighter toward the junction and the darker side darker. A consequence is that the plane surface appears fluted or corrugated. This might be undesirable in some applications and could be averted by grading; intermediate levels are obtainable by allowing a little diffusion of dots of different size across the junction. Figure 2-4(b) does not show the fluting, also called venetian-blind effect, because lines were ruled at the junctions—an interesting fact to bear in mind and ponder about.

Figure 2-4(c) shows, on a coarse scale so that the details can be inspected, how a random pattern of dots can be used to reach a much wider range of variation in ink coverage. The technique is to make the probability for dot occurrence at any dot location equal to the desired shade. If 50 shades are wanted, the picture is divided into boxes containing 7×7 cells, and a dot is placed in each cell in turn if a number chosen at random from 0 to 1 is less than the shade wanted. For midgrey, the expected number of dots in a box when the probability is set to 0.5 is 24.5, but the actual number deposited may vary above and below. The appearance is better if the random filling deposits exactly 25 distinct dots, as described in a later chapter.

Figure 2-4(d) is a version of the random-scatter technique on a much finer scale. Boundaries between grey levels are easily made undetectable by this technique. A good dynamic range is achievable, and the illustration is resistant to degradation by photocopying. The appearance is quite elaborate compared with ordinary halftone, in some applications attractively decorative and in others distracting from one's purpose.

If round or squarish dots are not a prerequisite, then eight grey levels plus black and white are obtainable with boxes of 3×3 cells using predetermined filling patterns. Figure 2-5 shows a selected set of 3×3 boxes and the eight shades resulting from replication in the horizontal and vertical directions. Unlike the scheme based on variable-size square dots, this scheme provides sets of equally spaced grey levels. However, this nominal linearity does not suit all tastes. For one thing, the contrast between the first two levels constitutes a big two-to-one jump, while the middle levels are closer together and perhaps even hard to distinguish.

Since it is possible to put k dots into n cells in $^nC_k = n!/(n - k)!k!$ different ways, the samples shown are not unique. But, although four dots can be put into nine cells in 126 different ways, only 14 different patterns are generated by horizontal and vertical replication. If one does not distinguish the changes associated with rotation or reflection, the number of different patterns is two in the case of two or seven dots, four in the case of three or six dots, and five in the case of four or five dots. When circumstances limit the number of grey levels that can be presented or distinguished, the availability of two patterns with the same grey level extends the range of distinguishable greys. In a sense, the use of diagonal hatching in perpendicular directions on engineering drawings is an instance of this.

If Fig. 2-5 is reproduced by an office copier, and the copy is copied repeatedly, some deterioration sets in, which forms the basis for comparing different copiers. The reproducibility of different patterns by the same copier can also be examined. Useful experience can also be gained by making reduced copies. The diagram has about 25 dots per inch; you can ask how many dots per inch your copier can handle at various grey levels.

Resolution

In addition to dynamic range, a second characteristic bearing on the quality of image presentation is resolution. In a photographic emulsion, grain size sets the limit and may approach 1 μm. In printing with ink, resolution is much coarser and is set by the reproducing technique. For example, in letterpress halftone reproduction a screen may be used to break a picture into 32 dots per centimeter (80 dots per inch, dot spacing 300 μm). A good laser printer works at 600 dpi or more (40 μm spacing or better), while a count of 1200 dpi or more is available in the printing industry. Further improvement may be expected; samples of cotton cloth from ancient Egypt already had 320 threads per inch in the days of Rameses III in the second millennium BC, while diffraction gratings could be ruled mechanically with 20,000 lines per inch by 1885.

The angular resolution of the human eye approaches one minute of arc or 1/3000 radian. That is because the part of the lens of the eye effective in forming a sharp image is about 3000 wavelengths in diameter. Thus for material examined close up, say at 30 cm, the resolution permitted by the eye is at best 100 μm ; consequently newspaper halftones with a dot spacing of 300 μm are resolvable. Even if you cannot resolve the individual dots (many people cannot resolve them because the quality of eyes varies a lot), you may still be aware of graininess in the reproduction. For better reproduction the printer can go to more expensive paper, to ink that is more expensive to use, or to a finer screen, say of 150 μm spacing (6 lines/mm), or to other available processes.

The deep importance of the psychophysics of vision in practical matters may be appreciated if you turn a newspaper picture through exactly 45 degrees (Fig. 2-6). The dot structure becomes much more noticeable. In effect, the printer has been able to substantially relax the demand made by resolution, simply by rotating the screen. To predict this effect, you need to know that the retina is connected to the optic nerve so as to emphasize resolution of tilted lines which are near the vertical and horizontal, an advantage bought at the expense of sensitivity to the tilt of lines inclined near 45 degrees. If you hold a newspaper picture at a distance such that you can only just see the dots when the picture is at 45

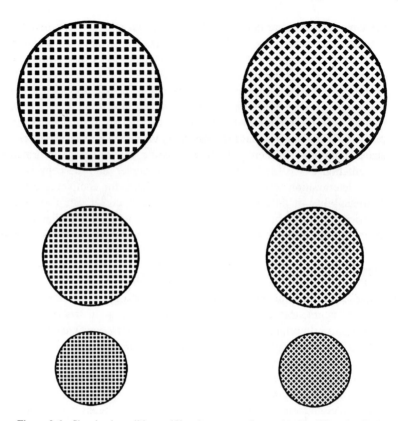

Figure 2-6   Showing how tilting at 45° makes a screen less noticeable. When the tilted screen (right) is held at such a distance that it becomes imperceptible, the eye can still see the untilted screen.

degrees, then you may not be able to notice the dots at all when the picture is upright. No doubt this peculiarity of vision was of evolutionary value, being perhaps connected with the need to know the vertical as your ancestors swung from branch to branch through the forest. Printers discovered how to orient the dot matrix a long time ago. You might find it interesting to take a hand lens to the Sunday comics and find out what printers did about dot orientation in color printing.

A television picture would permit the same relaxation of line spacing, or conversely would show improvement in appearance for a fixed line spacing, if the scan line were at 45 degrees to the horizontal. The resolution of a television picture is rather interesting by comparision with photography, because the horizontal resolution is set by a combination of video bandwidth, properties of the screen phosphor, and many other factors all the way back to the lens of the camera. The vertical resolution is set at 50 μm by the 525 lines/frame raster. Obviously it is a rather subtle thing to compare horizontal and vertical resolution when the mechanisms are so different – at the very least an agreed definition of resolution would be desirable. But even if we adopted a definition from some authority, there would be no guarantee that equal resolutions in both directions (under the definition) would mean that a viewer would report subjective equality. However, a test can easily be imagined that would elicit opinions from a viewer; for example, one might handle some suitable object in front of a television camera and ask the viewer of the screen whether the amount of detail visible, or the legibility in the case of a lettered object, depended on orientation. Ideas for reducing the horizontal striation have included lenses, blurring the beam vertically, and dithering the beam vertically. Experiments show that when a knob is provided that smoothes out the scan lines, viewers adjust the knob to make the lines as sharp as possible. Evidently the brain is better able to extract information from the raw presentation.

Contours

Representation by contours, already referred to, completely relieves the burden imposed by dynamic-range requirements. No matter how much dynamic range you want, a set of suitable contour levels can be selected. We could say, then, that the dynamic range approaches infinity. Resolution, on the other hand, has a character unlike that in either photography or television. Along the direction of the contour very fine detail can be indicated. Perpendicular to the contour, however, resolution is much less. Since contours may run in any direction, there is no one direction of good resolution. Furthermore, the quality of resolution transverse to the contours, depends on the contour spacing at each point. In terms of topography, transverse resolution is good on steep areas and poor on nearly level areas. Thus it is possible to gain an inkling of the fine structure of $f(x, y)$ by studying the wiggliness or featurelessness of the contours but it is also possible for an important feature to fall between adjacent contours and be lost entirely. As producers of contour diagrams know, there is an art in selecting the zero level and the contour interval; appropriate choice can emphasize or suppress a particular feature. There is also a choice between linear and logarithmic contour levels, giving further freedom to choose what to emphasize. Analogous facts are well known to the maker of black-and-white photographic prints.

Contour plotting has been practical in various applications, especially surveying, even though skill must be acquired to read contours freely, and heavy labor may be involved to create the plot. Map contouring, which has been applied to a good part of the whole world, was originally based on land survey carried out on foot. This was largely superseded by examination of stereoscopic aerial photographs, which can now be done automatically. The effort that has been put into the preparation of contour maps and the demand for them speaks for their virtues—namely, readability combined with quantitativeness and unlimited dynamic range. The visual impact of a contour diagram is not the same as that of a halftone representation using a grey scale. Choosing the more suitable mode is a matter of judgment, based on what information is to be conveyed.

Contouring of functions that have nothing to do with topography is now also widely practiced, as for example in Fig. 2-7. The two principal tops are shaded, while shallow bottoms are identified by downward-pointing ticks. Computers facilitate what used to be tedious labor, but it will be found that the computing and plotting still present a noticeable load.

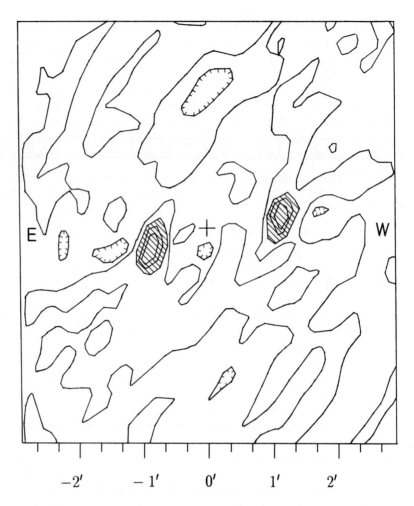

-2'   - 1'   0'   1'   2'

Figure 2-7   The double radio source Cygnus A, the first "radio star" to be discovered (1946), represented by equally spaced contours of brightness at 2.8 cm wavelength (Stull et al., 1975). The angular resolution, indicated by the scale of arcminutes, was achieved by rotation synthesis (a form of aperture synthesis) with the five-element minimum-redundancy array at Stanford.

Labeling of contour levels presents a complication when the contours are intricate, especially on halftone renderings where a separate key to levels may be necessary. Unlabeled contours can be ambiguous, since the level may be either above or below that of the adjacent contour, but a widely practiced convention resolves this by the use of a fringe of ticks pointing downward into depressions. Omission of labels is an advantage in situations where reduction to book size will either make the numbers illegible or obscure the data. An example of an unambiguous unlabeled contour diagram is shown in Figure 2-7.

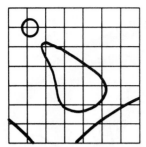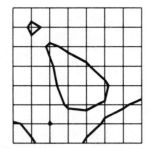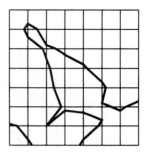

Figure 2-8   A smooth hand-made contour plot of traditional appearance (left), and two different polygonal contours of the kind made by computers (right), all from the same data.

The caption can mention the contour interval, the level of the lowest contour, and helpful features such as use of a heavy zero-level contour.

Contour Plotting by Computer

Since it has been traditional in subjects as far apart as topographic mapping and electro-magnetism to represent two-dimensional functions by means of isolines (contour lines of constant height, electric potential, or electron density, for example), it is natural to ask that the data matrix resulting from a computation should be presented in this familiar manner. A peculiarity of contour generation is that the contour for any one level may have more than one part. There may be one conspicuous closed loop, an arc that enters at the side and leaves by the same or another side, and a tiny loop in one corner. There may also be a loop that has shrunk to a single point. To find all these segments and interpret them reasonably can be difficult. Figure 2-8 shows, on the left, a contour diagram such as may have been made by hand and an example of what a computer program might do with the identical data.

The middle diagram is a fair approximation to what could be achieved by an experi-enced hand; the main defect is the intrusion of sharp polygons where it is known that the contours should be smooth. Of course, such knowledge is drawn from experience, not from the data given to the program. One way to fix this is to construct from the data an array of twice the dimensions, introducing three interpolated data values for each of those given. The same program will still produce polygons, but the sides will be shorter and the visual impression may be more acceptable. This approach imposes some penalty as to comput-ing time and storage space. The improvement may also be illusory; packaged contouring programs can produce relatively gross variations, such as changing the number of peaks, when the data set is reflected or rotated or when a constant is added to the data set.

The right-hand diagram shows an alternative interpretation of this kind. The most noticeable difference is the treatment of the saddle in the southeast. The ambiguity of treatment may be an essential consequence of sampling the data in the first place, and it may be impossible to write a program to make the ambiguity go away. One can test a packaged contouring program for the occurrence of such special areas by repeating the plot

Figure 2-9   A set of computed contour polygons in the presence of noise.

with reflected or inverted data. Another difference is the linking together of the two peaks, denying separate existence to the smaller peak. Of course, this is the same phenomenon. Likewise, an outlying microloop has attracted an unreasonable-looking projection from the main peak. All these defects can be found in practice no matter how finely the data may be sampled. They may be dealt with on a case-by-case basis, provided one looks for them.

A scheme for minimizing defects is as follows. At the stage where the program is about to draw a straight line to the vertex (x_i, y_i), stay the pen and first compute the next vertex (x_{i+1}, y_{i+1}). Then draw the line to the midpoint of the two vertices. This technique has proved to be fast and satisfactory on thousands of published microwave temperature maps of the sun. A contour-plot program CPLOT having provision for this midpoint technique is given at the end of the chapter.

A completely different approach to contouring is to color each pixel in accordance with its value. The extreme example, which is very useful, is to color each pixel black if its value exceeds an even contour and falls below or on an odd contour, otherwise color it white. This requires only one pass through all the data. The contour shapes are immediately put in evidence and while the curves may not be smooth, or even usable, at least the uninterpretable areas are put clearly in evidence. The effects of noise become apparent through characteristically wiggly outlines and give a basis for judgement as to the desirability of representation by smooth contours. Figure 2-9 gives an example of this representation for $f(x, y) = 7(1 - x^2 - 3y^2) + R(x, y)$, where the $R(x, y)$ are independent random numbers between 0 and 3 and represent a heavy perturbation of the contours levels, which are 1, 2, . . . , 8. The nominally elliptical contours are less distorted where the slope of $f(x, y)$ is steep.

Discrete Numerical Presentation

The rectangular matrix is a further familiar mode of presentation of a two-dimensional function. The dynamic range is unlimited. Resolution, which is set by the element spacing, is better horizontally and vertically than it is at 45 degrees. On the surface it seems that the resolution is good and may be made as high as one pleases, but in fact the resolution in real space is poor. Even a contour map, which has poor resolution compared to a newspaper illustration, has better resolution than a matrix printed on the same piece of paper. One to whom the concept of real space is unfamiliar may object that the resolution of a matrix may be increased indefinitely by making the element spacing represent a smaller and smaller distance; of course, a larger sheet of paper or other medium will then be required to write the matrix. But if a larger piece of paper were allowed, then the contour map made on that larger scale would preserve its advantage over the matrix. Even a matrix written on a silicon chip at a microscopic scale does not have the resolution that would be available if one cared to record a contour diagram on the same medium.

In any case, the convenience of the rectangular matrix for computation makes it indispensable. Other matrices, such as hexagonal matrices used in some kinds of color printing and polar-coordinate matrices, suffer from lack of existence of suitable matrix algebra.

Staggered Profiles

A representation that was made famous by Jahnke and Emde (1945), and has gained popularity because it is easy to implement by computer plotting, consists of staggered profiles in a set of parallel equidistant vertical planes. Sometimes a perpendicular set of vertical profiles is shown as well. An immediate appeal is made to one's experience with surface shapes; consequently, people who cannot read a contour diagram receive a clear impression from a set of staggered profiles. The impression, though definite, may be misleading. Experiments with representation of the same surface with different profile spacings and different vertical scales show that different impressions are received. Since the brain is good at interpreting the shapes of rounded surfaces from the shading exhibited, it may be that the amount of ink per unit area, which can change drastically with viewpoint, contour interval, or vertical scale modulus, is subconsciously factored in as the light and shade associated with illumination. For example, the virtual shading in Figure 2-10, on the left, may give the impression of a ridge running down from a peak toward the viewer. The center diagram shows the same surface, but to a reduced vertical scale. Of course, some staggered-profile diagrams completely obscure finite areas of the picture plane, as in the present case. The view from the other side (on the right) shows what is possibly a cliff. It is not really clear whether there are two tops or only one; a side view would

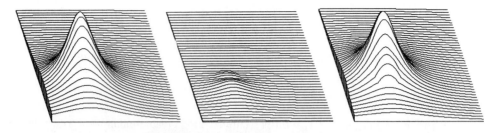

Figure 2-10   Three representations of the same function by staggered profiles. An apparently simple peak (left), the same with vertical scale reduced to one-sixth (center), and the view from the other side.

clarify. For serious quantitative description of this surface a contour plot or brightness map would be better. The staggered profile technique is analogous to terraced perspective. The profiles themselves are faithful but cues conveyed by hidden profiles can be misleading. For example, the middle view of Figure 2-10 suggests a steeper viewing angle whereas all that has been changed is the vertical scale modulus.

Offset contour diagrams have the same immediate appeal as staggered profiles. They are made by raising each contour above the base plane by an amount proportional to the contour level. One's experience with contour diagrams carries over to this sort of diagram so that, for example, tops and saddle points can be located readily, which is not necessarily the case with staggered profiles. However, people who can easily program staggered profiles, complete with suppression of hidden lines, have trouble programming offset contours. As a result, this representation is seen less often. Figure 2-11 was made by Dr. A. R. Thompson. As an artifact of relatively coarse pixellation, Fig. 2-10 faintly reveals a set of offset contours that can be pencilled in with care to give additional detail about the surface shape. For example, the elliptical shapes of the foreground contours do not show any evidence of a near-side ridge. The staggered-profile program can be modified to put dots along the raised contour loci.

Hachuring

Marvelous methods of hand hachuring to show mountainous terrain were brought to a high pitch of perfection around 1900. This type of work can be seen in old tourist maps of the Alps, old military maps and elsewhere (Sloane and Montz, 1943). Methods of hand shading with water color were used in atlases. Some of these old techniques would now be susceptible to computer implementation. One scheme uses short strokes between the contours with greater density of ink (achieved by varying stroke thickness or spacing) where slope is greatest (Skelton 1958). Another makes the ink denser where there would be shadow, assuming a source of light in the north-west. The shadow edge cast by a curved hilltop on the flank of another hill is not particularly easy to compute but the brain is extremely good at interpreting a surface shape from just such a cue. Mining into historical techniques of topography has potential for expanding the repertoire of current computer plots. A computer method distilled from oil painting technique will now be described.

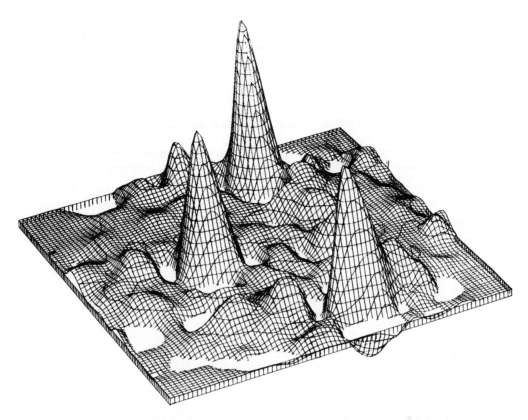

Figure 2-11 A presentation exhibiting two sets of staggered profiles in addition to offset contours.

Surface Relief by Shading

An artifact such as an airplane fuselage which is fully specified by engineering drawings, or a mathematically specified surface, or a fully specified topographic surface, should be easier to apprehend, at least in some ways, if presented as it would appear to the eye under illumination. Architectural draftsmen have a long tradition, launched by Palladio (1508-1580), of rendering shadows in order to convey to the client what is not apparent from the construction plans. (Three-dimensional models for this purpose presumably go back further.) Machines drawn by Leonardo da Vinci (1452-1519), including machine parts such as screws and gears with twisted surfaces, regularly include illumination, as do his paintings. The great oil painters of portraits, landscapes and still lifes carried the use of light for visual representation virtually to perfection in the centuries following the 15th century diffusion of linear perspective. To capture some of this technical achievement mechanically, without pretension to great art, here is the idea. Fit the known curved surface by a network of facets, too small and numerous to be obtrusive. Let the light fall from a point source, and decide on the position of the eye. Assign to each facet a brightness value (zero for black, unity for white) as seen from the direction of the eye and print the corresponding conventional density of ink where that facet appears on the paper. To compute the brightness of an arbitrarily oriented facet illuminated from one direction and viewed from another, one needs a scattering law.

Scattering Law

When light falls on a nonglossy facet at normal incidence, some of the light is absorbed in accordance with the surface color, and the rest is scattered over the hemisphere. Where Lambert's law is obeyed, Fig. 2-12(a), the scattered light intensity varies as the secant of the scattering angle (measured away from the normal), the law being independent of the direction of illumination. Since the projected area per unit solid angle as seen by the viewer goes as the secant, we see that Lambert's law is an assertion that the brightness of an illuminated plane facet is the same in all directions. This is often verifiably true for everyday surfaces, but usually there is a cone of extra brightness centered on the direction of specular reflection. Since this is relatively easy to take into account, once the program is running, it is sufficient to describe the simpler case.

　　When the light falls on the facet at an angle of incidence i, less light is intercepted (by a factor $\cos i$), but what is scattered may still be distributed isotropically, as before. That would be what we see with a polished steel ball in sunlight; it looks equally bright as the observer moves around it, even though the angle of incidence of the sunlight is varying. The surface of milk approximates this behavior, being a suspension of spherical droplets in a transparent medium. A specular component reflected at the milk surface is detectable, but the reflection is weak, as may be verified by looking for the reflection of a neighboring dark object against the sky. The froth on beer has spherical caps showing solar glints, but in numbers dependent on the position of the viewer and the unevenness of the surface; frothy beer is an example of a substance not adequately described by a scattering law. The surface of paper is also more or less equally bright from all directions of view,

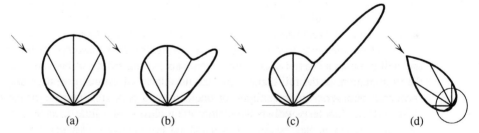

(a)　　　　　　　(b)　　　　　　　(c)　　　　　　　(d)

Figure 2-12　Various scattering laws: (a) Lambert's theoretical cosine law as approached in various degrees by fresh snow, matte surfaces such as paper and milk, (b) partial specular reflection from a slightly glossy surface, (c) enhanced specular reflection from a glossy surface, (d) the moon's surface; the light of the half moon (first and third quarters) goes back toward 45° and 225°, while the light of the full moon goes back toward the sun (135°). The four diagrams are normalized to equal amounts of scattered light, allowing for the moon's low albedo of 0.11.

the specular component depending on surface finish. By placing a piece of newsprint, a sheet of notepaper, and a sheet of photocopy paper level in early morning sunlight, and viewing from different locations, one can learn a lot. The cone angle of specular reflection, generally just a few degrees, can be seen and estimated. By squeezing the paper up into a low hump retained by two thin books, one can examine the brightness decline associated with changing angle of incidence of low sunlight. Somehow, the brightness seems to fall off more sharply toward the shadow edge than a cosine law would suggest. Of course, the shaded side does not have zero brightness because of illumination from the sky. The shaded side of the paper hump must have a blue tinge, but one is unaware of that because the brain tends to cancel information about the illuminant in favor of information about the object. The brain can be deceived into reporting the objective color if one looks through a large white card with a 1-cm hole, held at arm's length, and focuses attention on the plane of the card so as to destroy texture cues to the nature of the shaded surface. The same card can be used to inspect landscape colors, which are generally not the same as the colors of the same objects seen close up. The blue of shadows on snow sometimes jumps into awareness, possibly because the texture of a snow surface is often difficult to appreciate. An extreme case of what happens when texture cues are much reduced is encountered when one tries to ski in moonlight.

Powdery surfaces may exhibit strong backscatter. That is, they look brighter when viewed with the light behind one's head – and the brightness is more or less independent of the angle of incidence of the illumination on the surface. An example is the lunar surface which, apart from the dark maria, appears at full moon about as bright at the center (where sunlight falls at normal incidence) as it does at the limb (where the incidence is grazing). Strong backscatter toward the source implies less light scattered into other directions; thus the half moon, which is viewed at right angles to the direction of illumination, appears only one-ninth as bright as the full moon. The lunar brightness in decibels depends on the angle α between the sun and the earth as seen from the moon in accordance with the expression $-0.104\alpha - 1.6 \times 10^{-8}\alpha^4$. This amounts to about -3 dB at $\alpha = 30°$.

One explanation for strong backscatter is that shadows are hidden from a viewer looking from the source direction. This effect can be demonstrated by spreading out in the sunlight a handful of flour or a paper table napkin of the sort that is embossed with fine dimples. Another explanation depends on properties of the scattering substance.

The requirements on the exhibiting of surface relief by shading are rather modest and will often be achieved if Lambert's law is applied regardless of angle of incidence of the illumination. In the full shadow one can use black, which will give the stark effect of spacecraft photographed in sunlight, or one can use a dark grey suitable for the purpose at hand. Empirical tests will show whether the inclusion of a little specular reflection in the scattering law makes the faceted image better for the user's purposes. There is a small sector of activity in computer imaging aimed at rivalling competent painters at realism. To succeed means taking account of color, space-variable scattering laws according to the nature of the surfaces, and secondary illumination by scattered light. It is doubtful whether it makes sense to factor in all the necessary physical data merely to compute what a camera might record. More likely, future development along these lines will be in the hands of experienced artists who can also compute. Considerations beyond those

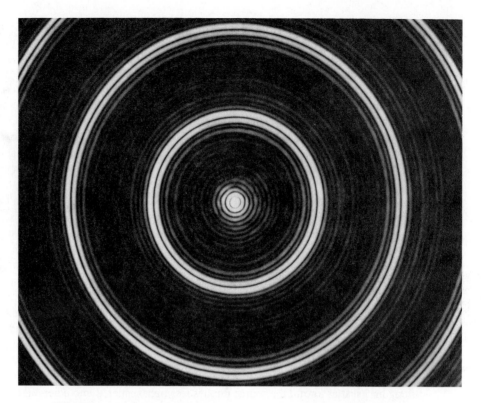

Figure 2-13 The Fraunhofer diffraction pattern of nine concentric equispaced slits, when reproduced photographically, leaves much to be desired. (See Tichenor and Bracewell 1973).

already mentioned include the subjective phenomena of brightness contrast, color contrast, and brightness compression, none of which the camera can handle and which require the experience ordinarily possessed by a painter.

Comparison of Modes

There is a moral to be drawn from the diversity we have discovered by looking at several familiar topics such as photography, television, mapping, and matrices and recognizing that they are all doing the same thing—namely, representing some scalar $f(x, y)$. The moral is that when a new project calls for some $f(x, y)$ to be represented, then alternatives should be considered. Very often that is not done. Optical diffraction patterns are portrayed in textbooks, usually by a photograph. Sometimes the photograph is poor and poorly reproduced (Fig. 2-13). Antenna patterns, on the other hand, which after all are diffraction patterns, are never presented photographically. This is merely telling us that optical people

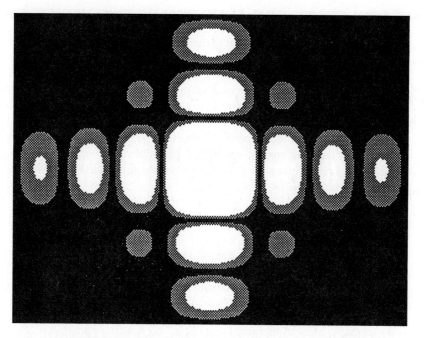

Figure 2-14   An optical diffraction pattern as represented by two contours.

had photo labs and antenna engineers did not. Obviously there is a question yet to be faced: "What is the best way of conveying a diffraction pattern to a student?"

Interestingly, a poor photographic reproduction of a diffraction pattern, which shows only black, white, and one shade of grey is replaceable by a contour diagram with *only two* contours (Fig. 2-14). That is very interesting, because generations of students and authors have accepted these familiar photos of interference fringes, Airy diffraction rings, and so on; therefore it is clear that far fewer contours than one might have thought may be capable of doing some important jobs. As it is easy to calculate and present only two contours, perhaps a lot of information about diffraction could be succinctly conveyed by giving thought to which contours to choose. Although the preceding remarks about poor photographs may seem to be critical, we have to remember that the photos appearing in print represent the ones that the photolab technician, and then the author, chose as crossing over from black to white at an intensity level they thought was about right. This raises the fascinating thought that some contours may contain more information than others. That is indeed the case. In a typical optical diffraction pattern the contour at or near 1 percent of maximum intensity is very rich. It is the contour that lies in the narrow grey zone of the contrasty diffraction pictures. Probably authors are not consciously aware of this 1 percent rule; they just know when an exposure shows what they want it to show.

SOME PROPERTIES OF A FUNCTION OF TWO VARIABLES

To become acquainted with some of the basic features of two-dimensional scalar functions it is convenient to fix on a concrete interpretation of the mathematical theory. Here we have chosen topography, mainly because everyone has an intuitive feeling about land surfaces. Computer-graphics terms such as *gradient detector* draw on this background. The theory of topography has a special interest for electrical people because of a distinguished paper by Maxwell (1831–1879) that appeared in the *Philosophical Magazine* in 1870. Maxwell is famous for his prediction of electromagnetic waves, for quantitative experiments on the three-color theory of vision, and for the kinetic theory of gases. His work on the theory of Saturn's rings empowered him to derive the differential equations of the presumed electromagnetic field (Bracewell, 1992).

Surface Shape

According to Maxwell (1965), a small number of intrinsic features characterize a land surface. Tops and bottoms are respectively maxima and minima and saddles are points such that the contour at the level of the point intersects itself there. Slope lines are orthogonal to the contours. If a slope line is followed downhill, it will in general reach a bottom, and if followed uphill, it will in general reach a top, but in special cases a saddle will be reached. The whole surface can be divided into dales, a dale being a district all of whose slope lines run to the same bottom. A second approach divides the surface into districts referred to as hills, which share common tops. Dales are divided from each other by watersheds, and hills are divided by watercourses. The watercourses are slope lines that mount to a saddle rather than to a top, and the watersheds are slope lines that descend to a saddle rather than to a bottom.

Imaginary islands exhibiting the various features are helpful in understanding the possibilities. Figure 2-15 has a contour line *A* that cuts itself twice and encloses two tops. Figure 2-16 has a contour *B* that forms a figure eight and encloses tops, but it has another contour *C* that cuts itself in such a way as to generate a top and a bottom rather than a pair of tops. Not all contours of special interest are shown because a discrete contour level would coincide with the exact level of a saddle point only by accident. For example, there would be a special contour through the vicinity of *D* that crosses itself at the saddle to the right and encloses three tops and one bottom in all. To study these islands in more detail it is helpful to sketch in the slope lines. To one familiar with contour maps it is evident that the islands are imaginary, because there is something unreal about the appearance of the contours; it would be interesting to discover what patterns are intuitively recognized as real land forms. Another question suggested by the islands is the relationship between watersheds and what a mountaineer would describe as a ridge line. Our main interest in this subject, however, is to lay a foundation for contour diagrams and their terminology.

Smooth contours such as those in Fig. 2-15, obtainable by spline interpolation, have cosmetic value in public relations and advertising and are appropriate representations of regular mathematical functions. In serious science and engineering, an author who is representing data engenders more confidence in his work if the spacing and precision of the

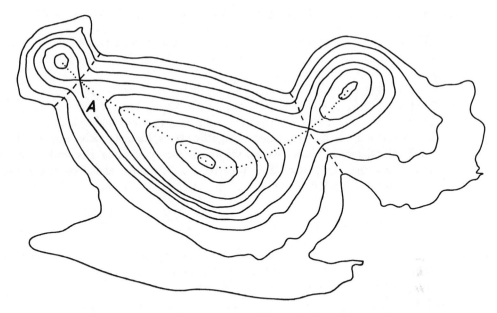

Figure 2-15   An island with three unequal tops and two saddles at the same height. Four watersheds and four watercourses are indicated. There are no bottoms above sea level.

data are allowed to show; smooth contours withhold the openness to falsifiability that is characteristic of reliable work.

It is not immediately apparent that a photograph is like a topographic region. And yet the photographic transmittance $t(x, y)$ is a continuous, single-valued function of x and y ranging between 0 and 1. Therefore, to every photograph there corresponds an imaginary island whose height at any point (x, y) is $t(x, y)$. If it is customary and useful to think of topographic surfaces in terms of contours, the same approach may make sense in thinking about images. In fact any fundamental property of functions of two variables may be useful to know about when dealing with problems of image processing and presentation.

Curvature

Surface curvature offers an example. The curvature of a continuous plane curve $y = f(x)$, defined as the rate of change of direction per unit arc length $(d\theta/ds)$, is calculable as $(d^2y/dx^2)/[1 + (dy/dx)^2]^{3/2}$. If the slope is small, the curvature is approximately equal to the second derivative. At a point P on a topographic (continuous, single-valued) surface $z = f(x, y)$, there are two axis-dependent curvatures $\kappa_x = (\partial^2 z/\partial x^2)/[1 + (\partial z/\partial x)^2]^{3/2}$ and $\kappa_y = (\partial^2 z/\partial y^2)/[1 + (\partial z/\partial y)^2]^{3/2}$. These two curvatures might both be negative, as would be seen by someone on a hilltop looking first east and then north; but they might be both positive or they might be of opposite sign. If the doubly convex regions were colored red and the doubly concave regions blue, the display, though elementary, might be very informative. If the (x, y)-coordinate system were rotated continuously about the z-axis, the

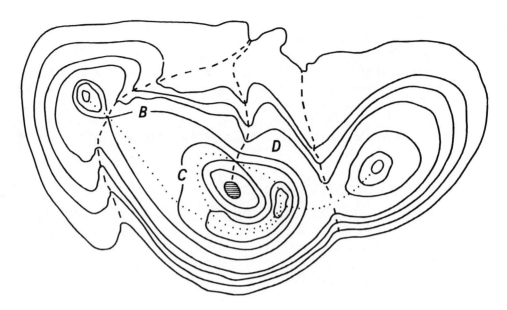

Figure 2-16 An island with three roughly equal tops, three saddles at different elevations and one crater lake (shaded) containing a bottom. Five watersheds and six watercourses are shown. There are three well-defined hill districts and one well-defined dale.

curvatures along the new axes (x', y') would rise and fall, defining characteristic directions and associated extreme curvatures $(\kappa_{x'})_{\max}$ and $(\kappa_{y'})_{\min}$.

In differential geometry one works not with the z-axis but with the normal to the surface at P. Then the algebraic maximum and minimum curvatures are known as the principal curvatures κ_1 and κ_2, and they occur in the two principal directions, which are at right angles. The product $K = \kappa_1 \kappa_2$ is the Gaussian curvature. When K is positive, the surface at P is synclastic (convex or concave), and where K is negative, the surface is anticlastic (saddle-shaped). The locus $K = 0$ is independent of the choice of coordinate system and so is an invariant line on the surface. Another invariant feature is an umbilical point, where $\kappa_1 = \kappa_2$. However, parameters defined with respect to the z-axis will be more significant than invariants in many image applications.

To understand that in image analysis we are often concerned less with intrinsic properties of the object than with properties of oriented surfaces, think about the shadow boundary when a surface is illuminated from a given direction. If we are more interested in whether a surface $z(x, y)$ representing an image is concave upward or convex upward (i.e. in the positive z-direction) rather than whether it is synclastic (there is a difference), there are two eminently computable quantities. A primitive axis-dependent convexity parameter K_{ij} for a digitized image f_{ij} is the product of the two second differences $\Delta_{ii} f_{ij}$ and $\Delta_{jj} f_{ij}$, thus $K_{0,0} = (f_{-1,0} - 2f_{0,0} + f_{1,0})(f_{0,-1} - 2f_{0,0} + f_{0,1})$. Where K_{ij} is positive, the x- and y-sections are both concave upward or both convex upward; where K_{ij} is negative, the second differences are mixed. The sample points, if colored, are likely to be economically diagnostic of surface character. To be sure, the parameter K_{ij} at a sample point is a property of the orientation and spatial phase of the samples.

A second primitive parameter is C_{ij}; at the point $(0, 0)$ it is the amount by which $f_{0,0}$ exceeds $f_{\text{mean}} = \frac{1}{4}(f_{1,1} + f_{-1,1} + f_{-1,-1} + f_{1,-1})$. In geometrical terms, $C_{0,0} = f_{0,0} - f_{\text{mean}}$ is the height of the surface at $(0, 0)$ above the twisted plane (hyperboloid of one sheet) through the four points diagonally adjacent to $(0, 0)$.

PROJECTION OF SOLID OBJECTS

The word projection has recently come to be used in a special sense derived from the practice of tomography, where projection means a set of line integrals, but here we establish some terminology involving previously established usage of the word.

Perspective Projection

The representation of a three-dimensional solid object on a page requires reduction to two dimensions. One way of doing this is to select a picture plane suitably located with respect to the object and choose a coordinate system (u, v) in the plane. From a point $O(x_O, y_O, z_O)$, imagine a straight line connecting to any point (x, y, z) in object space. Then that line penetrates the picture plane at a point (u, v) which is the linear perspective projection of the point (x, y, z) with respect to the center of projection. The straight line in object space that connects two points $P_1(x_1, y_1, z_1)$ and $P_2(x_2, y_2, z_2)$ projects into the straight line joining (u_1, v_1), the projection of P_1, to (u_2, v_2) the projection of P_2. The reason that a straight line in space projects into a straight line on the picture plane is that any two planes that intersect (OP_1P_2 and the picture plane in this case) do so along a straight line (Fig. 2-17).

The idea of establishing a center of projection O by means of a small hole in a fixed card, and interposing a removable card in front of the scene, was introduced by Filippo Brunelleschi (1377-1446), well known as the builder of the great cupola in Florence and many other elegant works of architecture. The implementation by Dürer, where the picture plane is occupied by a wire coordinate grid and the picture itself does not have to be repeatedly swung out of the way but may rest conveniently on a table, was illustrated in Fig. 2-1. Consider a coordinate system (x, y, z) whose origin is at table level at the bottom center of Dürer's grid (shaded in Fig. 2-18). Coordinates in the picture plane defined by the grid, although coinciding with x and z, will continue to be called u and v. In general (see below), the picture plane need not pass through the origin of spatial coordinates, nor even be vertical, but there is a good reason for making the choice shown. The center of convergence O (location of the eye) is at $(0, -d, h)$. The equations relating $Q(u, v)$ to $P(x, y, z)$ are

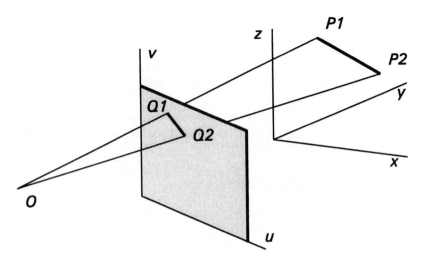

Figure 2-17   How picture plane coordinates (u, v) depend on space coordinates (x, y, z). The rod P_1P_2 in (x, y, z)-space projects to the line Q_1Q_2 on the picture plane.

$$u = \frac{xd}{y + d}$$

$$v = h + \frac{(z - h)d}{y + d}.$$

From these equations, objects specified in three-dimensional space can be rendered on the plane in correct linear perspective for whatever distance d to the eye and height h of the eye are chosen.

Linear perspective involves distortion—for example, the scale is different in different directions—but this is the convention imposed by the camera and television camera and is familiar from everyday illustration. The eye does something similar, in that the image on the retina involves linear projection onto a curved surface, but what is apprehended by the consciousness is profoundly modified by head and eye movement and by processing carried on at the retinal level and on up.

Linear perspective was introduced to medieval painting by Brunelleschi and widely advertised by L. B. Alberti (1404–1472) in a book entitled *Tractatus de Pictura*, which was translated almost immediately into the vernacular and is still in print.

Before Alberti's book came out in 1435 artists used (and still use) other cues to depth, such as obstruction by nearer objects. Aerial perspective indicates distance by various techniques (color, contrast, line quality, texture resolution) that suggest the quantity of blue light scattered by the intervening air molecules. Textural elements such as distant crowded trees in landscapes, are altered by the length of the air path. At shorter distances,

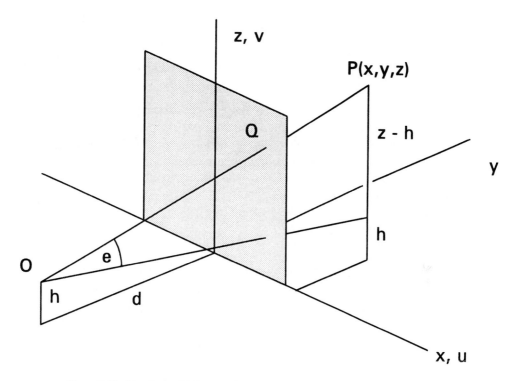

Figure 2-18 The relationship between the space coordinates (x, y, z) of the object point P and the picture-plane coordinates (u, v) of the projected point Q, when the (x, z)-plane is chosen to coincide with the (u, v)-plane.

where the amount of air is negligible, depth can be conveyed by fine texture, which is both reduced in scale and differently resolved as distance increases.

Terraced perspective, which places objects higher on the page in accordance with distance, is familiar from Chinese ornaments. In fourteenth century European paintings, groups of people have their heads at higher levels toward the rear. In the next century a conspicuous difference brought about by Alberti's book places heads at the same level but puts the feet higher for those at the rear.

Many artists, including Leonardo da Vinci (1452–1519) made conspicuous use of the new geometrical theory. On the other extreme Fra Angelico (c.1400–1455), whose brilliant frescoes adorn the cells of San Marco in Florence, understands geometrical perspective but paints as if he did not. Botticelli (1455–1510) exhibits some academic features but deliberately flouts the rules for artistic effect. Piero della Francesca (c.1420–1492) produced a work at San Sepolcro, sometimes claimed to be the world's finest painting, which incorporated two eye positions. Piero wrote the first geometrical treatises on linear perspective.

Change of Viewpoint

A perspective drawing is intended more to appeal to the eye than to convey the quantitative detail contained in the plan and elevation, or equivalent representation, from which a computed perspective rendering starts. Consequently, when the perspective sketch is viewed, ways of improving it can be seen that were not evident at the beginning. You may want to rearrange the component objects if there are more than one, turn the object over, or alter the lighting if shadows are to be shown. Most frequently, you want to adjust the viewpoint. The three coordinates specifying the center of perspective in the object domain can easily be changed. However, that is usually insufficient. For one thing, there will be a change of scale that may be unwanted. Often the picture plane needs to be moved, too. When an object is photographed from a new position the center of perspective moves and the film plane moves with it. To describe the repositioning of a camera takes six coordinates: three to give the new lens location, two to specify the new direction of aim, and a position angle affecting the orientation of the image on the film plane. But with computed perspective we do not necessarily want to keep the picture plane rigidly attached to the center of perspective, because we often want to keep the size of the drawing about the same while adjusting some other property, such as the degree of foreshortening. If you reduce the foreshortening by moving the camera farther away, you get a smaller image. The size adjustment that could be made with a zoom lens is more in keeping with the computing requirement to control one overall property. The image-size requirement is most simply met by putting the origin of the (x, y, z)-system in the middle of the object space and passing the picture plane through that origin. This is quite unlike the historical approach (which puts object space entirely on the far side of the picture plane) and means that changing the distance d does not much affect the image size.

To change the azimuth α of viewing, leaving the screen where it was, one can work out new expressions for u and v when the eye is located at $(d \sin \alpha, -d \cos \alpha, h)$. These are

$$u = \frac{(x + y \tan \alpha)d \cos \alpha}{y + d \cos \alpha}$$

$$v = h + \frac{(z - h)d \cos \alpha}{y + d \cos \alpha}.$$

But when the image plane passes through the middle of the object as recommended, it is often convenient simply to rotate the *object* through an angle α, transforming the space coordinates by

$$x' = x \cos \alpha + y \sin \alpha$$

$$y' = y \cos \alpha - x \sin \alpha$$

$$z' = z.$$

Sometimes one wants to increase the height h of the eye to gain a better view of some horizontal surface. And the question arises whether the picture plane should be tilted forward out of the vertical plane (as would happen automatically to the film plane of a camera which was tilted downward from a raised vantage point). The fact is that such tilting is usually avoided when buildings or groups of people are being photographed from above, because we prefer to see verticals in space come out vertically on the paper. In the event that an angle of depression β is decided on, the desired effect is obtainable by tilting the object; transform the space coordinates by

$$x' = x$$
$$y' = y \cos \beta - z \sin \beta$$
$$z' = z \cos \beta + y \sin \beta$$

Inverse Problems in Perspective

So far we have considered only the elementary problem of deducing image-plane coordinates when the space coordinates are given. An inverse problem is to start with the plane projection and work out what the objects in space were, or, if some properties of an object are known, to find out its three-dimensional position. There is a rich variety of problems of this kind, because in both science and engineering the available data often come from photographs or other plane records. Only one example will be considered here. A photograph records an object known to be a cube; the problem is to set up on the photograph the projection of an (x, y)-grid on a horizontal ground plane, and to construct a vertical scale of z, which can be used for reading off positions of other objects. Figure 2-19 is a tracing from a photograph taken at night and apparently showing a bale of wool, known to

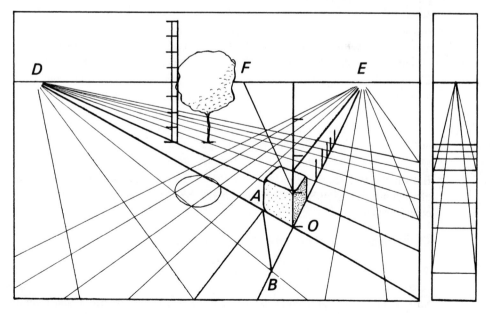

Figure 2-19 Features traced from a photograph and how to determine coordinate grids from them.

be a 1-m cube, and a shrub. The term *horizon* will be used for the straight line on the image plane where the horizontal plane through the camera lens (or in general the center of convergence) cuts the image plane. A *vanishing point* is the point of concurrence (on the image plane) of the projections of a set of lines which, in space, are parallel; the vanishing point of a set of horizontal parallel lines will lie on the horizon.

From the intersections of two sets of three parallel edges of the cube, two vanishing points D and E can be established. In the figure only four of these six lines are drawn, but in practice the three-way intersections would contribute useful information about precision. The join of D and E fixes the horizon line. The parallelism of the vertical edges of the cube says that the camera axis is level. The height of the camera can be determined by marking off equal intervals on the vertical line containing the near edge of the cube and is seen to be four times the length of a vertical side, or 4 m. Ticks 1 m high placed at increasing distances show how the vertical scale depends on distance; it will be seen that the height of the horizon is always four tick lengths above the ground position of a tick. The ground positions of the ticks are determined by the following reasoning. One of the diagonals of the top face of the cube, if prolonged, cuts the horizon at F. This must also be the case for the corresponding (hidden) diagonal on the base of the cube and for all other parallel diagonals, including the one labeled AB just in front of the cube. The point B, which is found at the intersection of FA and EO, allows construction of the grid line DB, and adjacent grid lines through D can be constructed similarly. The grid lines through E will pass through intersections of BF with the lines through D. This construction sets up a square grid in the ground plane. Although this grid is oblique, it may be sufficient for some purposes. For example, the base of the tree is found to be located at (2, 7) with respect to an origin at O, and the tree's height is 6 m, as determined from the previously established tick lengths. Reference grids in any other orientation may be constructed by iterative reference to the vanishing points; a grid with lines parallel to the horizon and spaced 1 m is shown on the right. The scales may be verified by comparison with the small ellipse which is the projection of the circle $\sqrt{2}$ m in diameter which circumscribes one of the oblique 1 m square cells. The principal axes of this ellipse give the left-right and up-down scale moduli.

Much of the literature of linear perspective is based on synthetic geometrical construction as in Fig. 2-19. The resulting algorithms are suitable for manual drawing and, once provided, can be translated for computer implementation without much trouble. But it is not obvious how the constructions involving vanishing points could have been arrived at from a trigonometrical foundation such as was set up in Fig. 2-17. Clearly, a grasp of traditional projective geometry is a useful tool for computer graphics.

Orthographic Projection

When the center of projection P, the vertex of the conical pencil of rays, is an infinite distance away, it is possible to place the picture plane perpendicular to all the rays. We then have an orthogonal projection. If, in addition, the rays are perpendicular to an important plane of the object, the projection is *orthographic*. Sets of three orthographic projections are the worldwide standard for engineering drawings. They constitute an indispensable

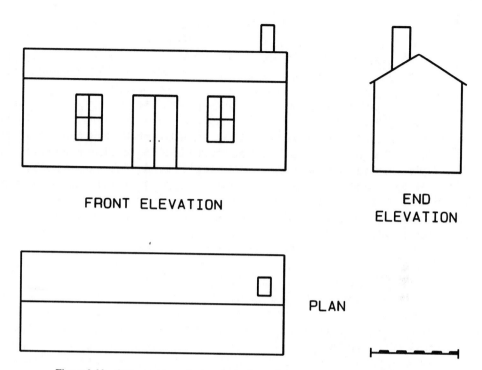

FRONT ELEVATION

END ELEVATION

PLAN

Figure 2-20 Orthographic projections in the form of plan, elevation, and side elevation.

means for conveying and receiving information, fully on a par with those other two essential channels, the spoken and the written word. Figure 2-20 reminds us of conventional names for these three orthographic views and of the strict relative positioning that gives value to standardization. It is a good idea to know the conventions of engineering drawing, especially the rules for hidden outlines, and to be able to adhere to them, because of the respect you gain where you may need it. Correct grammar and spelling convey a message over and above the nominal content, and so does a correct sketch.

Isometric Projection

An *isometric* projection is an orthographic projection made from a special direction, chosen so as to be equally inclined to three mutually perpendicular axes within the object. The useful practical consequence is that equal distances along these axes, or along any edges parallel to these axes, are represented by equal distances on the drawing and may be correctly measured with a ruler. Other measurements, such as diagonal distances, will be somewhat in error. The angle between any axis and the direction of projection is arctan $\sqrt{2} = 54.7°$, and the angle between any axis and the picture plane is $35.3°$.

An isometric projection does not contain as much information as a full set of orthographic projections, but what information it does contain is assimilable by a vastly greater audience, as witnessed by its common use in connection with installation and repair of

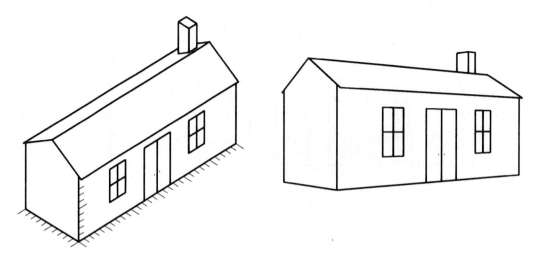

Figure 2-21 An isometric drawing (left), where correct distances can be measured to the same scale, in the three principal directions, and a perspective drawing (right), which looks natural but is not suited to measurement of dimensions.

household appliances. It is also quantitative. Therefore, the following simple program for making isometric drawings of polyhedra is worth memorizing. In Fig. 2-17 when the x-coordinate of P_1 increases, u increases; but when y increases, u decreases. So write down the transformation

$$u = x - y.$$

Now v increases when either x or y increases, but also when z increases. So write

$$v = x + y + z.$$

This is all you have to remember. Figure 2-21 shows the result of applying this technique to a set of coordinates obtained from three orthographic views. It is not necessary to remember where to put the sines and cosines of the various inclination angles, because only the aspect ratio of the final picture will be affected. That ratio will have to be adjusted to fit the different plotters or screens one will encounter through life. A logical approach is to notice, or insert, a cube in the object space and make one final adjustment causing the edges to be equal.

As we have seen, there is more than one way to achieve three-dimensional appearances on the plane. Examine the sixteenth-century example in Fig. 2-22 with a view to finding out what type of perspective was being used.

IMAGE DISTORTION

Images may be distorted by an instrument, by propagation through a medium, by noise, or by processing. A perfectly faithful aerial photograph of the ground may be deemed to

Figure 2-22 Mine machinery drawn by Georgius Agricola (1494–1555). Agricola thought mines were inhabited by elves, who were untidy but did not interfere with mining.

be distorted because a field known to be square does not appear square. Deformation may be deliberately applied in the processing of an image in order to compensate unwanted distortion. Stretching, compressing, shearing, and perspective projection are examples of simple distortion in which straight lines remain straight.

 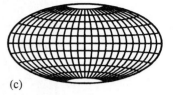

(a)   (b)   (c)

Figure 2-23   (a) Stereographic projection of a hemisphere onto a circle. This preserves
the shape of small elements (the conformal property) but not area. (b) Mollweide's
projection of a sphere onto an ellipse. (c) Aitoff's projection of a sphere onto an ellipse.

Map Projection

Some representations of the earth's surface on a plane make use of linear perspective
projection onto a plane (stereographic projection) or onto cylindrical or conical surfaces
which can then be developed onto a plane. Other map projections do not derive from linear
projection at all but depend on coordinate transformation equations that impose a desired
property, such as that any rhumb line on the sphere (path followed when sailing on a
fixed course) should be represented as straight (Mercator projection). If the transformation
equations are linear, then there is an equivalent linear perspective interpretation; but not
conversely.

Map projections have been the subject of study for many years. The material is
relevant to modern imaging by telemetry and remote sensing, where mosaic scenes of
planetary surfaces are formed from component pictures that cannot be fitted together at
their edges without deliberate distortion. For a useful source of up-to-date applications
consult Pearson (1990). Figure 2-23 illustrates some well-known global projections.

The stereographic grid is a perspective projection of a hemisphere onto the plane of
the great circle bounding that hemisphere relative to a vertex of perspective at the pole
of the great circle. The resulting map is orthomorphic (small shapes on the hemisphere
project into the same shape on the plane). As a special case infinitesimal circles project
into circles, but the area ratio varies (the stereographic is not an equal-area projection).
It falls into the class of zenithal projections defined by the property that equispaced great
circles through some point are represented by equispaced radial lines on the plane. The
scale along the radial lines is the same for all and may be chosen to produce an equal-area
projection. The equations are

$$x = \frac{\sin \phi \cos b}{\cos \phi \cos b + 1}$$

$$y = \frac{\sin b}{\cos \phi \cos b + 1},$$

where ϕ is longitude, ranging $\pm 90°$ from the central longitude, and b is latitude ranging
$\pm 90°$ from the central latitude.

Mollweide's projection presents the whole sphere within an ellipse of width 4 and
height 2. The meridians are equispaced ellipses, while the circles of latitude are straight

and crowded toward the poles, as they would appear to be from a viewpoint at infinity in the equatorial plane. The center of the map is visibly afflicted with anisotropy, or unequal scale factors in orthogonal directions. However, the equations are simple:

$$x = (\phi/90°)\cos b$$
$$y = \sin b,$$

and, because of the straight parallels, the grid is fast to draw. The parallels suffer an illusion of curvature which is the opposite of the gravitational bending shown by Shepard (1990).

The Hammer-Aitoff projection fits the whole sphere into the same 4×2 ellipse with isotropic scaling at the center, and treats the poles more gracefully. Naturally it is difficult to handle the left and right edges, but areas of less interest can sometimes be relegated to the edges. The compression of the meridians is compensated for by the curved latitude lines to produce an equal-area projection. The equations are

$$x = \frac{2\sqrt{1 - \cos b \cos \frac{1}{2}\phi}}{\sin\left[\arctan\left(\sin \frac{1}{2}\phi / \tan b\right)\right]}$$
$$y = \tfrac{1}{2}x \tan \tfrac{1}{2}\phi,$$

$b \neq 0$. This projection can be thought of as derived from a zenithal equal-area projection of the hemisphere by stretching the circular outline into an ellipse and relabeling the meridians to range from $-180°$ to $180°$. The stereographic projection may be stretched in exactly the same way to give an alternative orthomorphic whole-sphere projection with equations

$$x = \frac{2 \sin \frac{1}{2}\phi \cos b}{\cos \frac{1}{2}\phi \cos b + 1}$$
$$y = \frac{\sin b}{\cos \frac{1}{2}\phi \cos b + 1}.$$

Conformal Transformation

Let a point $P(x, y)$ in an image be represented by the complex number $z = x + iy$. Let some transformation move P to a new position $P'(x', y')$ represented by $z' = x' + iy'$. Then the bilinear transformation

$$z' = \frac{a + bz}{c + dz},$$

where a, b, c, and d are constants, is the most general complex transformation in which there is one-to-one correspondence between the two images. In other words only one P' corresponds to any P, and conversely. Since images are single-valued, the one-to-one property is interesting. The bilinear transformation transforms circles into circles. Therefore, as a special case of a circle of infinite radius, any straight line becomes a circle.

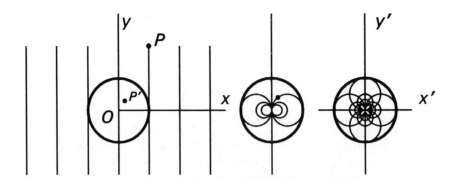

Figure 2-24   The grille in the (x, y)-plane (left) inverts into the circles (and one straight line) in the (x', y')-plane (center). The inversion circle (heavy) has unit radius and inverts, point by point, into itself. The full (x, y)-grid transforms into two sets of circles in the (x', y')-plane (right).

From the theory of complex variables we also know that the bilinear transformation is conformal, angles of intersection remaining unchanged.

The special case

$$z' = \frac{1 - z}{1 + z}$$

is familiar as the basis of the Smith chart used for calculations on transmission lines to convert an impedance z to a reflection coefficient z', while

$$z' = \log(z - 1) - \log(z + 1)$$

generates the field pattern associated with a pair of parallel cylindrical conductors.

Inversion is related to the bilinear transformation $z' = z^{-1}$. Under inversion, a point $P(x, y)$, where $x = r \cos \theta$ and $y = r \sin \theta$, moves to $P'(x', y')$, where $x' = r^{-1} \cos \theta$ and $y' = r^{-1} \sin \theta$. If O is the origin of coordinates, then $OP.OP' = 1$. The grille of seven lines (left side of Fig. 2-24) inverts into six circles and one vertical straight line (center). Each of the two grilles composing a square grid in the (x, y)-plane becomes a family of touching circles, plus one vertical and one horizontal line.

The complete square grid inverts into two families of circles which then constitute a curvilinear coordinate system that can be used to map images from one plane to the other. Figure 2-25 shows a fish transformed by inversion. In accordance with the conformal property the grid intersections remain perpendicular in the curvilinear system.

Complex functions of a complex variable

$$z' = f(z)$$

produce a variety of image transformations in accordance with the choice of function and many are well known. Functions such as z^2, z^k, e^z, $\ln z$, ..., supply a repertoire of elegant transformations that are accessible in textbooks. Since the conformal property implies that local scale change is isotropic, the distortion at any point is fully specified by

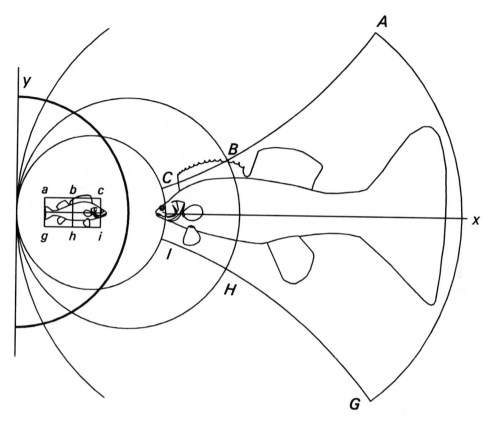

Figure 2-25   A river perch in the (x, y)-plane (left) and a fish (right) related to it by inversion in the heavy circle. The box $abcghi$ inverts into the curvilinear box $ABCGHI$. The perch being entirely within the inversion circle, the inverted fish is consequently enlarged as well as being distorted.

a magnification factor M and a rotation θ. These parameters are rather simply ascertained from

$$Me^{i\theta} = \frac{df(z)}{dz}.$$

In addition to distortion there is translation, which is calculable from $z' - z\,dz'/dz$.

Reciprocal Polar Transform

The conformal transformation $z' = R^2/z$ takes a point $P(x, y)$, whose polar coordinates are (r, θ), into a point P'' whose coordinates are $(R^2/r, -\theta)$. In projective geometry the straight line through P' (Fig. 2-26) that is perpendicular to PP' is known as the *reciprocal polar line* of the point P. An object composed of points thus maps into a set of straight lines, and conversely. An interesting property is that collinear points transform into concurrent straight lines, and conversely.

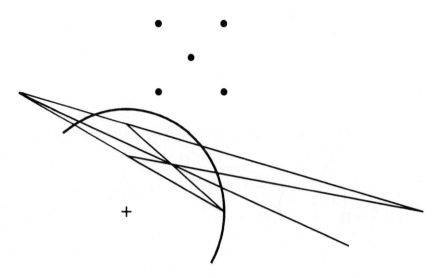

Figure 2-26   The reciprocal polars of the five points arranged as on a die. The two sets of collinear points transform into concurrent line triplets. The line common to both triplets is the reciprocal polar of the point common to both collineations.

Continuous curves can be transformed point by point to yield a family of straight lines defining an envelope. Alternatively, the tangents to a given curve can be transformed to give a set of points delineating the same envelope.

The equation of the polar reciprocal $P(x_1, y_1)$ is conveniently expressed in terms of its intercepts X and Y as

$$\frac{x}{X} + \frac{y}{Y} = 1,$$

where $X = R^2/x_1$, and $y = R^2/y_1$. In Fig. 2-26, which shows the reciprocal diagram for a set of five points arranged as on a die, the polar line of any point is identifiable as the one perpendicular to the join of that point with the origin.

Affine Transformation

A conformal transformation sets up a one-to-one, or biunique, relationship between an image plane (x, y) and a transformed plane (x', y'), but a straight line in the image plane does not necessarily remain straight. An affine transformation is also biunique but takes straight lines into straight lines (as does a perspective projection of one plane onto another); in two dimensions the equations are

$$x' = M_x x + S_x y + X$$
$$y' = S_y x + M_y y + Y,$$

where the six coefficients are constant over the plane. The constants X and Y simply represent displacements of the map and may be spirited away by transposing the equations to obtain

$$x' - X = M_x x + S_x y$$
$$y' - Y = S_y x + M_y y.$$

Since the equations are linear, one can then call on familiar matrix notation to write

$$\begin{bmatrix} x' - X \\ y' - Y \end{bmatrix} = \begin{bmatrix} M_x & S_x \\ S_y & M_y \end{bmatrix} \begin{bmatrix} x \\ y \end{bmatrix}.$$

One can also introduce vector notation where the transformation assumes the form

$$\mathbf{x}' = [M]\mathbf{x} + \mathbf{e},$$

which is familiar in control theory, a subject where the vector **x** represents a time series with a large number of elements. As we have only two elements here, not much is gained by this condensed notation. Various manipulations suggest themselves. For example, the transformation can be inverted to obtain (x, y) from $(x' - X, y' - Y)$. Putting $D = M_x M_y - S_x S_y$ for the determinant of the matrix $[M]$, the inverse relation is

$$\begin{bmatrix} x \\ y \end{bmatrix} = \frac{1}{D} \begin{bmatrix} M_y & -S_x \\ -S_y & M_x \end{bmatrix} \begin{bmatrix} x' - X \\ y' - Y \end{bmatrix}$$

or

$$x = \frac{(M_y x' - S_x y' - M_y X + S_x Y)}{M_x M_y - S_x S_y}$$

$$y = \frac{(-S_y x' + M_x y' + S_y X - M_x Y)}{M_x M_y - S_x S_y}$$

The organization of the matrix and summation operations is displayed in Fig. 2-27.

The adjective affine means a close relationship that is finite—in other words, a one-to-one correspondence where any finite point transforms into a finite point. The finite-to-finite property is not possessed by projection from one plane to another or by the transformations specified by functions of a complex variable. Affine transformations are important in linear algebra, usually in multidimensional form, and in projective geometry. In the case of images, where two dimensions suffice, it is useful to have an intuitive feeling for the coefficients. The displacements, or translations, X and Y have already been interpreted geometrically. The coefficients M_x and M_y are, respectively, magnification factors in the x- and y-directions, while S_x and S_y describe shears. The nature of these effects, taken separately, is shown in Fig. 2-28, together with an illustration of composite shear and of a transformation involving both magnification and shear, all four of the coefficients M_x, M_y, S_x, S_y retaining the previous numerical values.

The application of shear can introduce rotation. Thus when a square cell is sheared into a lozenge (Fig. 2-29), the major diagonal may no longer be at 45°. However, the lozenge is not simply a rotated square; there has been compression along the short diagonal and elongation along the long one. If two area-preserving, anisotropic strains of this kind are applied in perpendicular directions, a net effect of rigid-body rotation can be produced.

Area magnification is one of the outcomes of an affine transformation and, in the

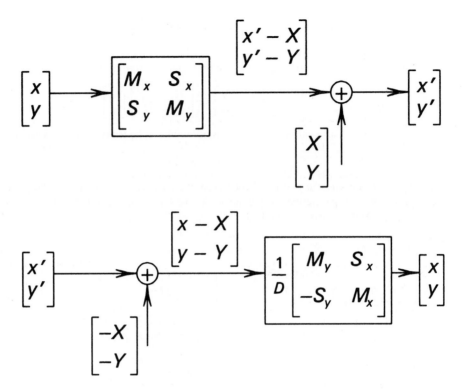

Figure 2-27 A flow diagram illustrating the affine transformation as a matrix operation followed by a summing operation (above) and a corresponding diagram for the inverse transformation using the inverse matrix (below).

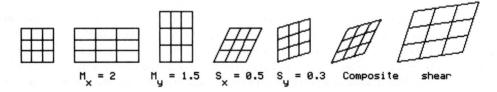

$M_x = 2$ $M_y = 1.5$ $S_x = 0.5$ $S_y = 0.3$ Composite   shear

Figure 2-28 Illustrating the magnification and shear coefficients M_x, M_y, S_x, S_y, composite shear, and the affine transformation depending on the chosen values of the coefficients. Translation is not considered.

absence of shear, $M_{\text{area}} = M_x M_y$. Shear in itself does not alter area; consequently unidirectional magnification followed by shear, or vice versa, also produces an area magnification $M_x M_y$. However, affine transformation does not apply the magnification and shears separately and sequentially; in general

$$M_{\text{area}} = M_x M_y - S_x S_y.$$

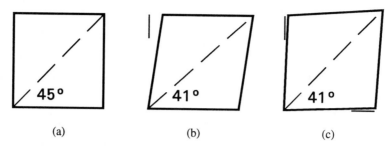

|(a)|(b)|(c)|

Figure 2-29   A diagonal of a square cell (a), initially having an inclination of 45 deg, rotates to 41 deg under the x-shear shown (b). The diagonal can be rotated back to 45 deg (c) but can also be returned to 45 deg by subsequent application of y-shear (not shown), which preserves the area but not the shape of the square.

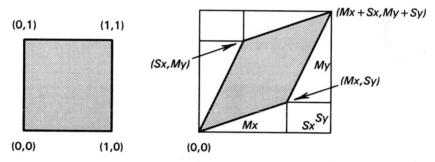

Figure 2-30   A unit square cell before and after affine transformation. If the displacements X and Y were not taken to be zero, the origin of the lozenge would be at (X, Y).

This relation may be verified from Fig. 2-30, which shows how the coordinates of the corners of a unit square are affected by affine transformation. In the illustration, where $M_x = M_y = 1$, $S_x = 0.5$ and $S_y = 0.3$, the area of the cell is reduced by a factor 0.85. If the shear coefficients are small and $S_x S_y << M_x M_y$ then shear does not much affect the area magnification, which remains approximately equal to $M_x M_y$. It is possible, however, for the shear to dominate and even produce negative area magnification.

If the lozenge in Fig. 2-30 is subjected to one-dimensional compression in the direction of the long diagonal the shape can be squeezed back into the outline of a square, but the square will be rotated 41° relative to the original square. This observation suggests that a suitable combination of magnifying and shearing coefficients could produce simple rotation. This is indeed the case when $M_x = M_y = \cos\theta$, $S_x = \sin\theta$ and $S_y = -\sin\theta$. The area magnification $M_x M_y - S_x S_y$ is seen to be unity.

It follows that shear may be dispensed with if desired and that one may think instead in terms of rotation through an angle θ combined with unidirectional elongation or compression expressed by orthogonal magnifications M_ξ and M_η. We obtain the relevant factorization by requiring that

$$\begin{bmatrix} M_x & S_x \\ S_y & M_y \end{bmatrix} = \begin{bmatrix} M_\xi & 0 \\ 0 & M_\eta \end{bmatrix} \begin{bmatrix} \cos\theta & \sin\theta \\ -\sin\theta & \cos\theta \end{bmatrix};$$

whence

$$M_\xi = \sqrt{M_x^2 + S_x^2}$$

$$M_\eta = \sqrt{M_y^2 + S_y^2}$$

$$\cos\theta = \frac{M_x}{\sqrt{M_x^2 + S_x^2}}$$

$$\sin\theta = \frac{S_x}{\sqrt{M_x^2 + S_x^2}}.$$

With these relations an affine transformation expressed in terms of the six standard coefficients can be decomposed into rotation and compressions or elongations. If one wishes to design a transformation starting from rotation and compressions and then convert to the standard linear coefficients, then the relations, obtained by multiplying the two matrices above, are

$$M_x = M_\xi \cos\theta$$

$$M_y = M_\eta \cos\theta$$

$$S_x = M_\xi \sin\theta$$

$$S_y = -M_\eta \sin\theta.$$

Repeated Transformation

Repeated application of a particular affine transformation is a popular technique in computer graphics; examples are shown in Fig. 2-31, where the familiar shape of a fish is used to convey the character of the shape modification. Each stage is produced using exactly the same coefficients. The central fish is *Perga fluviatilis*; the others are plausible fish designs. This sort of diagram has historical interest, dating back to a remarkable demonstration given by d'Arcy Thompson (1966) in 1917 that different species of fish of related genera can have body forms that are derivable from each other by rather simple mathematical transformations. This discovery raises thoughts about how such parameters are encoded in the DNA. Further examples of the effect of repeated affine transformations are shown in Fig. 2-32, starting from a simple square. In (b) each stage is rotated, stretched in the x-direction, and squeezed in the y-direction, keeping the area unchanged. The shape changes from rectangular if the push and pull are applied with respect to an arbitrary orientation. Coefficients for a rotating rectangle of changing aspect ratio can be obtained by sandwiching the push-pull between two opposite and unequal rotations; the coefficients will not be the same from one stage to the next. Iteration, which is a principle for achieving pleasing

Figure 2-31 A sequence of fish exhibiting (from top to bottom) progressive anisotropic magnification and forward rake (or shear).

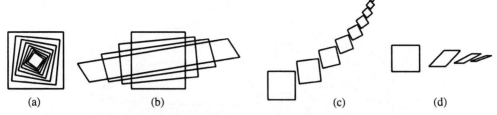

(a)   (b)   (c)   (d)

Figure 2-32   Sequences formed by starting with a square outline and applying the following operations in tandem: (a) Rotate and shrink. (b) Rotate and push-pull. (c) Rotate, translate, and shrink. (d) Shear, translate, and shrink.

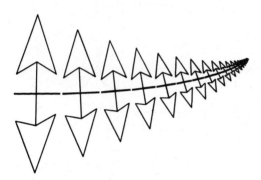

Figure 2-33   Composite structures may be produced by a sequence of identical transformations applied to different original polygons. Although complicated, the patterns are economical to compute.

architecture, endows repeated affine transformation with the capacity to produce harmonious ornaments.

When more complicated polygons are transformed sequentially, rather complicated shapes can be economically generated, especially structures resembling those occurring in nature. In Fig. 2-33, which shows the principle used, the initial polygon on the left has two arrowheads and has been separated just a little from the transformed version to its right, so that the initial seed can be distinguished. Repeating the same transformation many times produces the composite structure, which itself defines a polygon. If this polygon is taken as the initial seed, a more complex bipinnate structure will be generated.

Since it is apparent that an affine transformation of another affine transformation is itself an affine transformation, combinations of various special cases applied in tandem can synthesize wanted effects. Some of the simple operations tabulated below are also expressible as a combination of two others; for example, rotation, as we have seen, can be created by two successive shearing stages.

Figure 2-34 repeats the first part of the previous figure starting with a heptagon instead of a square. The result is rather striking, considering that it consists of nested, disconnected heptagons. Those interested in esthetics can produce a series of related pictures by changing the parameters and will find that viewers will have no hesitation in picking out the most pleasing one. One can then think about how to originate pleasing pictures.

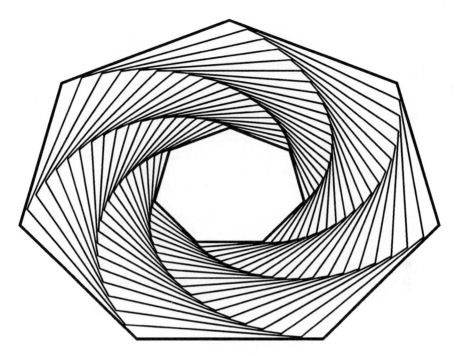

Figure 2-34   An ornament produced by repeated affine transformation of a heptagon.

Identity operator $\begin{bmatrix} 1 & 0 \\ 0 & 1 \end{bmatrix}$ Collapse $\begin{bmatrix} 0 & 0 \\ 0 & 0 \end{bmatrix}$

Isotropic magnification $\begin{bmatrix} \sqrt{M}_{\text{area}} & 0 \\ 0 & \sqrt{M}_{\text{area}} \end{bmatrix}$ Rotation $\begin{bmatrix} \cos\theta & \sin\theta \\ -\sin\theta & \cos\theta \end{bmatrix}$

x-stretch $\begin{bmatrix} M_x & 0 \\ 0 & 1 \end{bmatrix}$ y-stretch $\begin{bmatrix} 1 & 0 \\ 0 & M_y \end{bmatrix}$

x-shear $\begin{bmatrix} 1 & S_x \\ 0 & 1 \end{bmatrix}$ y-shear $\begin{bmatrix} 1 & 0 \\ S_y & 1 \end{bmatrix}$

Area-preserving push-pull squeeze $\begin{bmatrix} M_x & 0 \\ 0 & M_x^{-1} \end{bmatrix}$

General Continuous Transformation

Many kinds of everyday distortion, such as the barrel distortion and pincushion distortion in optical instruments do not fall into the categories described above, nor does simple stretching. Figure 2-35 illustrates various kinds of distortion by showing how a square grid would be affected. Barrel and pincushion distortion arise with lenses. The diagrams show how the area magnification varies over the field and how a square element may undergo

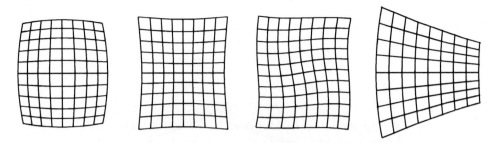

Figure 2-35   Grids that start off cartesian in the middle but acquire various kinds of distortion (pincushion, barrel, spiral, Smith chart) as radius increases.

progressive shear or rotation away from the center of the field. Sometimes it is necessary to correct for these fundamental distortions, and sometimes (as with letters and line drawings of faces) it is interesting to introduce minor distortion. The first three types are illuminated by replotting in polar coordinates, which will show that concentric circles about the origin remain concentric but slowly become more or less tightly packed as radius increases. If the displacement in radius δr increases quadratically as κr^2, then $\delta x \approx \kappa r x$ and $\delta y \approx \kappa r y$. Pincushion and barrel distortion may therefore be introduced or removed by these simple displacements of the points of a square grid, choosing κ as a small negative or positive coefficient. Spiral distortion results with $\delta x \approx \kappa r x$ and $\delta y = -\kappa r y$. The full Smith chart is calculable from the conformal transformation $z' = (1 - z)/(1 + z)$; the part shown is the vicinity of $z = 1 + i0$ plus or minus 20 percent.

A general one-to-one continuous distortion of $f(x, y)$ is likely to be representable locally by affine coefficients that will themselves be spatially variable. It may be necessary to analyze a distorted map on the (x', y')-plane of a known original on the (x, y)-plane and find the coefficients in the relations

$$x' = M_x(x, y)x + S_x(x, y)y + X(x, y)$$
$$y' = S_y(x, y)x + M_y(x, y)y + Y(x, y).$$

This problem can be attacked empirically by determining the local translations, magnifications, and rotations or shears, whichever are most convenient to extract.

Distortion in General

Consider a transformation where

$$x' = f(x, y), \qquad y' = g(x, y)$$

and $f(\ ,\ )$ and $g(\ ,\ )$ are arbitrary functions subject to being single-valued and differentiable. The immediate neighborhood of a point $P(x, y)$ maps into the neighborhood of a point $P'(x', y')$; let the neighborhood of P have a small square boundary centered on P. Then the corresponding boundary around P' will have a different area (described by magnification M), and may be compressed or elongated, rotated, or sheared. Trapezoidal distortion need not be considered in the infinitesimal limit where the square is small enough. There will be a displacement of P' with respect to P, but this will not be classed as distortion. The distortion will in general depend on position; however, in a given small neighborhood the situation will be describable by an affine transformation

$$x' = ax + by + c$$
$$y' = dx + ey + f,$$

where the displacement (c, f) will not be discussed and the coefficients a, b, d, e will be functions of x and y and will fully describe the local distortion.

As has been seen, magnification, anisotropic compression, rotation and shear can all be expressed in terms of the affine coefficients. Therefore we ask how to obtain a, b, d, e from the general transformation functions $f(\ ,\ )$ and $g(\ ,\ )$. By differentiating the affine transformation equations, we find that $a = \partial x'/\partial x = \partial f(x, y)/\partial x$. This and the similar equations for the other three coefficients may be condensed into the matrix equation

$$\begin{bmatrix} a & b \\ d & e \end{bmatrix} = \begin{bmatrix} M_x & S_x \\ S_y & M_y \end{bmatrix} = \begin{bmatrix} \frac{\partial f}{\partial x} & \frac{\partial f}{\partial y} \\ \frac{\partial g}{\partial x} & \frac{\partial g}{\partial y} \end{bmatrix}.$$

The right-hand side is familiar in connection with the often-needed conversion between cartesian and polar coordinates in two and three dimensions. For example, when

$$\xi = r \cos \phi = f(r, \phi)$$
$$\eta = r \sin \phi = g(r, \phi)$$

then

$$\xi = ar + b\phi + c$$
$$\eta = dr + e\phi + f,$$

where

$$\begin{bmatrix} a & b \\ d & e \end{bmatrix} = \begin{bmatrix} \cos \phi & -r \sin \phi \\ \sin \phi & r \cos \phi \end{bmatrix}.$$

From the values of the affine coefficients one verifies that the area magnification equals the absolute value of the Jacobian determinant $|ae - bd| = r \cos^2 \phi + r \sin^2 \phi = r$.

Two-dimensional transformations that are expressible in terms of a function of a complex variable are not fully general, and consequently local distortion is simpler to describe. Thus if

$$z' = f(z),$$

then df/dz, which is itself complex, has a magnitude which is the linear magnification and a phase angle that expresses the rotation of the transformed neighborhood. The magnification is the same in all directions at a point, so there is no anisotropy, and there is no shear. A small square transforms into another small square of different area and different orientation. This is the conformal property, a term which summarizes the special character that falls short of full generality.

Stereograms

Visual depth perception depends on numerous cues, including obscuration by foreground objects and parallax associated with motion, but for distances out to a few meters binocular vision dominates, and it continues to play a role well beyond 100 m. That this was the evolutionary value of having two eyes was not understood until an experimental demonstration in 1833 by Charles Wheatstone (1802–1875) using a stereoscope that with two mirrors combined two photographs into a three-dimensional impression. For a time Wheatstone's conclusion, which now seems obvious, was contested by physicists. Wheatstone also invented the electric telegraph (patented in 1837) and many acoustic and optical devices including the concertina and a polarization-dependent sundial for telling the time when the sun was not out.

Vast numbers of stereophotographs were sold commercially, as well as elaborate stereoscopes, which are still recycled as antiques. No longer in use for drawing room entertainment, stereoscopes are still commonplace for professional use in county land offices for viewing aerial photographs. The economy version consists of a pair of lenses of 100-mm focal length spaced 60 mm apart (with adjustment) and supported on folding legs 100 mm above the table. Two photographic images made with spaced cameras fuse to exhibit the same relief that would have been seen from where the stereophotographs were taken; or the relief may be more pronounced, if the camera lenses were spaced further apart than the spacing of the human eyes.

A simple stereopair, shown in Fig. 2-36, is intended for use with a stereoscope. Some people can learn to dispense with the instrument by directing their gaze at a point beyond the paper. Alternatively, by crossing their eyes to converge on a point halfway to the paper they will see a complementary object in which elevations become depressions. Inquiring from friends who can do this may be helpful in acquiring the knack; however, there are

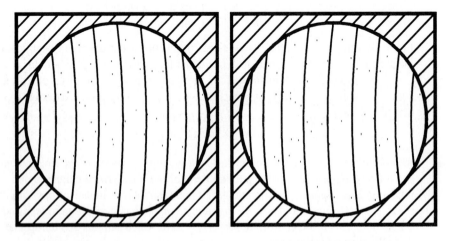

Figure 2-36 A stereogram representing a dome rising above a level diagonally hatched floor.

people whose eyes cannot combine stereo images without a stereoscope. Other kinds of stereograms, which are described later, may be easier to resurrect.

OPERATIONS IN THE IMAGE PLANE

There are a number of elementary operations in the image plane about which little needs to be said. One of these is translation, according to which $f(x, y)$ becomes $f(x - a, y - b)$. Either of the displacements a and b may, of course be zero, in which case the translation would be parallel to the y-axis or x-axis. There is one small thing to be remembered when $f(x, y)$ is a finite image, such as one that could be stored as an array in a computer. In this case a decision has to be made regarding the elements of the image that are shifted out of the span of the array. A second decision has to be made about the array positions on the opposite side from where elements have been shifted out. One decision would be to abandon the elements that fall over the edge and to supply zero values for the vacated positions. Alternatively, the vacated positions could be filled with values computed from a formula that might be available for $f(x, y)$ or from separately stored data. Another possibility is to move the elements that fall over the edge back to the vacant position created on the opposite edge; the rule for this is the same as in the game of toroidal chess, where the top and bottom edges of the board are supposed to be in contact and also the left and right edges.

Magnification and reduction are other elementary operations that may be rather simple but require appropriate decisions when the image is of bounded extent or consists of array elements. In the following discussion continuous variables are implied for the most part, and it will suffice here simply to mention that choices of the type mentioned may have to be made when the time comes for computer implementation.

Reversal and Rotation

In the time domain one can delay or advance a waveform without change of form, and it is clear that in the image plane the corresponding operation is translation without change of form. Of course, there are now two degrees of freedom, because the net translation can have both x- and y-components. But a new possibility exists that is not a generalization of anything in one dimension, and that is rotation without change of form. If an image is rotated half a turn about an axis perpendicular to the image plane so that left and right are interchanged, it is a reminder that there does exist one other form-preserving operation in the time domain—namely, time reversal. One might say that reversal is the one-dimensional aspect of rotation by half a turn. In two dimensions we have the following situations.

Reversal about y-axis: $f(x, y)$ becomes $f(-x, y)$
Reversal about x-axis: $f(x, y)$ becomes $f(x, -y)$
Reversal about both axes: $f(x, y)$ becomes $f(-x, -y)$

Reversal about both axes is the same as rotation by half a turn about an axis perpendicular to the image plane, whereas reversal about the x-axis produces a result that cannot

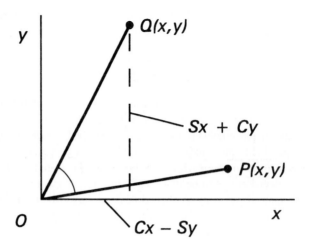

Figure 2-37 In its rotated position Q the coordinates of point P become $(Cx - Sy, Sx + Cy)$, where $C = \cos\theta$ and $S = \sin\theta$ and θ is the angle from OP to OQ.

be reached by rotation of the original $f(x, y)$ in its own plane. However, $f(x, -y)$ may be reached from $f(-x, y)$ by rotating half a turn. In three dimensions, reversal with respect to all three axes yields an object which in general cannot be brought into superposition with itself by rotation about real axes.

The operation described here as reversal could also be called reflection. Reflection in the y-axis would mean that to each point (x, y) we assign the value at $(-x, y)$, that is, at the point having the same value of y but on the opposite side of the y-axis in the position of the mirror image.

Rotation counter-clockwise through an angle θ means that

$$f(x, y) \quad \text{becomes} \quad f(Cx - Sy, Sx + Cy),$$

where $C = \cos\theta$ and $S = \sin\theta$ (Fig. 2-37).

Rotational Symmetry

Considerations of symmetry furnish a powerful tool for thinking about problems. An I-beam has bilateral symmetry; a starfish is organized differently but still has a kind of symmetry. Here is an example of an argument using symmetry principles. Suppose that a structural member, for architectural reasons, has been designed to have a cross section like a starfish with five arms as in Fig. 2-38 (middle of fifth row). We know that an I-beam is much stiffer standing upright than when lying on its side; the question is, what is the stiff orientation for the five-spoked beam? The cross-sectional shape has five-fold rotational symmetry. That is to say, if the shape is rotated through an angle $2\pi/5$ about its center, the original shape repeats. Therefore, the moment of inertia about a horizontal axis through the section will be the same after rotation by one-fifth of a turn. But the moment

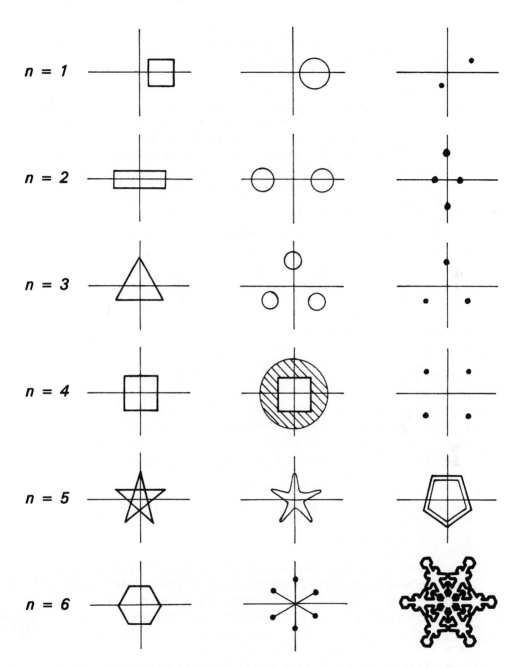

Figure 2-38 Examples of *n*-fold rotational symmetry ranging from $n = 1$ (no symmetry) to $n = 6$, which includes snow crystals.

of inertia has two fold symmetry, because no rotation by half a turn (as happens when a beam is turned upside-down) can change a moment of inertia which depends on the square of the height of each element above the horizontal axis. If $I(\theta)$ is the moment of inertia in orientation θ, you will recall that a polar coordinate plot of $r = I(\theta)$ gives the ellipse of inertia. The only way for a twofold symmetrical shape such as an ellipse to simultaneously possess fivefold symmetry is for there to be circular symmetry. Thus the starfish-shaped beam has equal stiffness in all orientations. The same is true of a hollow triangular beam, for example of the kind made of lattice construction in the booms of cranes and in factory ceilings. If you had to work out the moment of inertia of an equilateral triangle about axes in all directions through the centroid, you would confirm that the result is independent of direction, but you would have done much more work. The correct way to obtain answers that in themselves are simple is to use a direct and simple line of reasoning, in this case a symmetry argument.

We say that $f(x, y)$ has n-fold rotational symmetry if the function obtained by rotating $f(x, y)$ through an angle $2\pi/n$ about the origin is the same as $f(x, y)$, i.e.,

$$f(Cx - Sy, Sx + Cy) = f(x, y),$$

where $C = \cos(2\pi/n)$ and $S = \sin(2\pi/n)$.

Changing Dimensionality

There are many ways of extracting one-dimensional functions from two-dimensional ones. Examples include cross sections, raster scans, and projections. Conversely, from a one-dimensional function one may construct two-dimensional functions. Stacking one-dimensional segments and back-projection provide examples.

Often there is an advantage in changing dimensionality, and very often such a change is required because the dimensionality of the desiderata is not the same as the dimensionality of the data. Likewise in computing, two-dimensional arrays are stored one-dimensionally.

At a higher level, a two-dimensional view is referred to as a projection of a three-dimensional world. Conversely, the brain converts from a two-dimensional image on the retina to a three-dimensional formulation in the brain.

Cross-section

As an example of a cross section, consider the common case of section along a straight line through the origin. Let x' and y' be coordinate axes that are inclined at an angle θ to the x- and y-axes (Fig. 2-39). Then the radial section $s(x')$ is given by:

$$s(x') = f(x'\cos\theta, \ x'\sin\theta)$$

Sections through two-dimensional functions may be expressed rather simply as products with a line impulse of unit strength lying along the line in question. Line impulses are discussed in the next chapter. Such a product is still a function of two variables

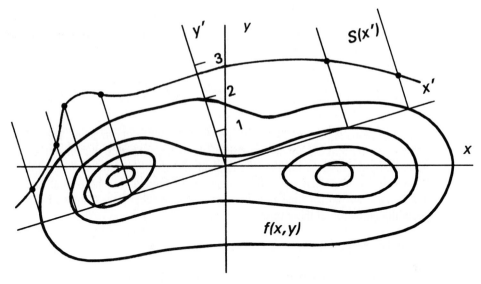

Figure 2-39   A function $f(x, y)$ and its section along the radial line $y' = 0$ or $y \cos \theta - x \sin \theta = 0$. The section $s(x')$ is shown cross-hatched and rotated through 90° into the (x, y)-plane.

and therefore is not the same kind of entity as the one-dimensional section, but the information content is the same. Thus the section along the straight line $x = a$ is $f(a, y)$, a function of the single variable y, but the product $f(x, y)\delta(x - a)$, while it is a function of the two variables x and y, contains the same information as the cross section $f(a, y)$.

Raster Scan

Consider a function $f(x, y)$ that is nonzero over a central square of side W; i.e., $f(x, y)$ is zero where x or y exceeds $W/2$. A raster of N horizontal lines of spacing s crosses the square area. The cross sections on the raster lines, starting from the top, are $f(x, W/2)$, $f(x, W/2 - s)$, $f(x, W/2 - 2s), \ldots$, and the join of these cross sections is the output function that results from raster scanning. As an example, let:

$$
\begin{array}{cccc}
a & b & c & d \\
e & f & g & h \\
i & j & k & l \\
m & n & o & p
\end{array}
$$

represent $f(x, y)$; the top row is traced out as x increases, while y is held constant. Then the raster scan is *abcdefghijklmnop*. Of course, this is the left-to-right top-to-bottom raster scan, which was the standard for writing in European manuscripts and remains so today.

Other scanning conventions are encountered, and it may help to have descriptive terms for them, especially as scanning is a common source of mistakes. Here are some suggestions for terminology:

European	Semitic	Old Chinese	Matrix	Boustrophedon	Cartesian
a b c d	*d c b a*	*m i e a*	*a e i m*	*a b c d*	*m n o p*
e f g h	*h g f e*	*n j f b*	*b f j n*	*h g f e*	*i j k l*
i j k l	*l k j i*	*o k g c*	*c g k o*	*i j k l*	*e f g h*
m n o p	*p o n m*	*p l h d*	*d h l p*	*p o n m*	*a b c d*

The terms *landscape* (4 × 3) and *portrait* (3 × 4) refer to shape. Reverse landscape and reverse portrait are rotated 180° in their plane; with this nomenclature we can say that a Cartesian raster is reverse Semitic, and give names to other rasters, while (inverted) European and Cartesian are mirror images with respect to a horizontal line. European and Semitic are mirror images with respect to a vertical line. Computer-driven plotters generally allow for all these raster variations.

Coordinate Conventions

The cartesian convention arises naturally when the abscissa x is plotted to the right and the ordinate y is plotted upward. It might have been better if Descartes had followed the European convention for writing, and plotted y downward, because printers with graphics capabilities naturally plot y downward. We thus have a clash between cartesian graphics language descended from moving-pen plotters (now used to drive laser printers to make fast line drawings and vector graphics) and picture-control language descended from text printers (now used for slower bitmap plotting, or raster graphics). There is also a clash between cartesian coordinates and another mathematical tradition. In the matrix $[a_{r,c}]$ the column index c advances to the right in each row, while in each column the row number r increases downward; this is the matrix convention, which is followed by FORTRAN. To encode a function of discrete x and y for presentation on the cartesian convention one has to write $[a_{-y,x}]$, which causes the necessary 90° rotation.

Vector Graphics

Raster graphics, as described above, applies to text, tabulated data and digitized photographs. But objects such as line drawings, which form a major part of the illustrations in technical books, can be stored in less space and transferred faster by vector graphics. The technique harks back to XY-plotters, where a pen (or pens) moving on paper is guided along specified polygons, being lowered or raised (writing or not writing) as instructed. Curved lines are treated as polygons with very short sides. The same instructions that are sent to a plotter can be used to guide the electron beam in a cathode ray tube, turning the moving beam on and off according as the beam is to write or not write. Vector graphics language is a set of low-level instructions for moving the pen. Many of the illustrations in this book were made using Hewlett-Packard Graphics Language. The instructions can be generated from a standard high-level computer language, or from applications software. A raster graphics screen consisting of discrete pixels cannot display continuous, inclined, straight lines, but Graphics Language can be sent to a raster-based screen or printer, where

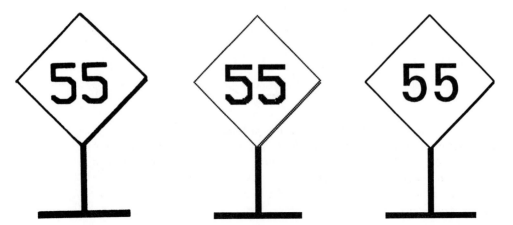

Figure 2-40 A vector plot made by a pen writing on paper (left), a corresponding raster representation on a screen, reproduced by a laser printer (center), and the raster graphics output from a laser printer that understands a vector graphics language (right).

appropriate sequences of pixels are activated to simulate the trace of an imaginary pen with specified width (Fig. 2-40). The speed advantage of vector graphics is then realized. The straight-line algorithm that provides this simulation is dealt with later.

Comparable results are obtainable with pen plotters, screen graphics, and raster graphics in appropriate circumstances. The first two have the advantages, in development work, that the image forms before the eyes, the program can be paused, and the language is simple. Vector graphics as transferred directly to the drum of a laser printer allows easy access to a wide variety of fonts and filling textures and is free from the noticeable staircase effect of a coarse screen raster. However, it would be premature to suppose that the resolution limits of plotting devices have been reached. A digitally controlled milling machine, which can be thought of as a rather robust XY-plotter with an engraving tool in place of a pen, can do better than the 20,000 lines per inch reached by Rowland's ruling engines over a century ago. The atomic force microscope can write at atomic dimensions. This says a lot for mechanical devices (including the scanning mechanism inside a laser printer).

Stacking Data

The converse of raster scanning is to segment a one-dimensional function $f(x)$ into lengths X and to stack them with a vertical spacing Y. The resulting two-dimensional function $g_{\text{stack}}(x, y)$ is given, for stacking in the fourth quadrant, by:

$$g_{\text{stack}}(x, y) = f(x - Xy), \quad 0 \leq x \leq X, \ y = 0, -Y, -2Y, \dots$$

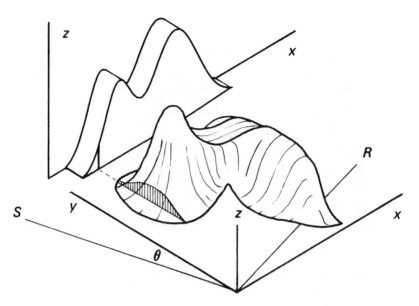

Figure 2-41 The projection operator $\mathcal{P}_\theta$. A two-dimensional function $z = f(x, y)$, represented here by a double-humped mound of clay, generates a one-dimensional function $\mathcal{P}_0 f(x, y) = \int f(x, y) \, dy$, which is a function of x when the clay is projected back at an angle $\theta = 0$ with the y-axis onto the (x, z)-plane. When the projection takes place in a direction making an angle θ with the y-axis, the result is $\mathcal{P}_\theta f(x, y) = \int f(x, y) \, dS$, which is a function of the rotated coordinate R.

Projection of an Image

Imagine a pile of clay as in Fig. 2-41 whose height is described by $f(x, y)$. If all the clay is shoveled straight back against the wall in such a way that the ordinate y is unchanged for any particle, and if a plasterer trowels the clay to a layer of uniform thickness, then the new function $g_0(y)$ describing the height of the layer is referred to as a projection of $f(x, y)$. Such projections may be taken in other directions but this one will be called the zero-angle projection and may be written $\mathcal{P}_0 f(x, y)$, where

$$g_0(y) = \mathcal{P}_0 f(x, y) = \int_{-\infty}^{\infty} f(x, y) \, dy.$$

In general, the projection at angle θ is $g_\theta(R)$, given by

$$g_\theta(R) = \mathcal{P}_\theta f(x, y) = \int_{-\infty}^{\infty} f(x, y) \, dS,$$

where R and S are the rotated coordinates resulting from rotating the (x, y)-system counterclockwise through an angle θ.

From the definition given it is apparent that the numerical value of the projected function is equal to the area of a plane section (such as the one shown shaded) at the appropriate abscissa. The height of the clay layer would, of course, depend on the layer thickness and would merely be proportional to the projected function as strictly defined by the area integral. In a particular problem, $f(x, y)$ might or might not be dimensionless. But if f has dimensions, or if the coordinates have dimensions, then it is worth noting that the dimensions of the projection will not be the same as those of the original function $f(x, y)$. For example, if $f(x, y)$ is measured in meters, as in the case of a pile of clay, and x and y are measured in meters, then the unit of $P_\theta f(x, y)$ is a square meter. Since the amount of clay is conserved, the volume under $f(x, y)$ equals the area under the projection, in the proper units.

If the given function is circularly symmetrical, all projections are the same. Take the case where $f(x, y) = \exp(-\pi r^2)$. Then the projection is

$$\int_{-\infty}^{\infty} [-\pi(x^2 + y^2)]\, dy = \exp(-\pi x^2) \int_{-\infty}^{\infty} \exp(-\pi y^2)\, dy = \exp(-\pi x^2).$$

In this example the projection turns out to be the same as the original cross section. If $f(x, y) = \operatorname{rect} r$, a pillbox function, the projection is the semiellipse $y = \sqrt{1 - 4x^2}\, \operatorname{rect} x$, which we know as the Abel transform of $\operatorname{rect} r$.

A projection, in the sense of a set of line integrals, is called a *scan* in tomography, where the integrals constituting the projection are (or were) acquired as a function of time by scanning a beam, such as a beam of x-rays, in a plane containing a function $f(x, y)$ which is to be found. When there is risk of confusion with established meanings of the word projection, as in map projection or perspective projection, terms such as *line integral* or *scan* are useful.

Projection of a Three-Dimensional Object

A function of three variables $f(x, y, z)$ can be projected onto a plane to give a two-dimensional image. Projections may be made in any direction; the projection onto the (x, y)-plane made in the z-direction is $\int f(x, y, z)\, dz$. The familiar chest x-ray is an example of a plane image derived by projection from the three-dimensional distribution of matter within the chest. In this case the function $f(x, y, z)$ is a function of the density and composition of the tissues. By dividing the three-dimensional object into the two-dimensional distributions on the planes, $y = \text{constant}$, one sees that the projection from three dimensions amounts to a set of the one-dimensional projections from two-dimensional images described previously.

Silhouettes

The silhouette is another kind of image derived from a three-dimensional original, but it has only two function values, say 0 and 1, corresponding to black and white. It is a binary image. At one time it was fashionable to have your profile projected onto a translucent screen to be photographed; so in the everyday sense of the word, silhouettes were formed

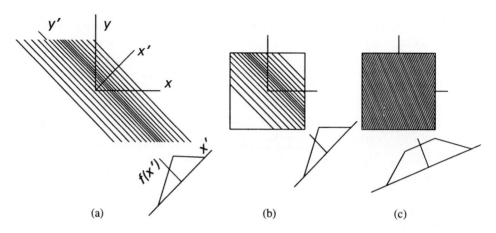

Figure 2-42   (a) Back projection, (b) truncated back projection, (c) conservative back projection.

by *projection* but not in the sense of integration along a line. However, if the shadow-throwing lamp is at infinity and the transmittance of the head is zero, then the two usages of the word *projection* converge.

Back Projection

Sometimes the operation of projection has to be reversed, but it is quite clear that projection is not a reversible operation – information is lost in the process. The situation rather resembles the making of silhouettes by photography. But if projections in many different directions were available, the situation would be different. *Back projection* is a way of creating a function of two variables from a given function of one variable that is thought of as a projection.

Fig. 2-42 illustrates what is meant. From a given projection $f(x')$ we generate a function on the (x', y')-plane that is in fact independent of y' and can be represented by a surface whose contours are all straight and parallel to the y'-axis. If $\mathcal{B}$ is the back-projection operator, then by definition

$$\mathcal{B}f(x') = f(x').$$

This rather deceptively simple equation has to be interpreted with the understanding that on the left-hand side $f(x')$ means a function of one variable, but on the right-hand side it represents a distribution over the (x', y')-plane.

Sometimes back projection is used to fill a finite region, usually circular or square, and no meaning attaches to locations outside that region. It then seems immaterial whether back projection continues on to infinity or not, but in fact it is desirable to distinguish these back-filling operations. Let us use the term *truncated back projection*. Then back projection onto a square area of side W could be described in terms of the full back-projection operator $\mathcal{B}$ by

$$\mathcal{B}_T f(x') = [\mathcal{B} f(x')] \operatorname{rect}(x/W) \operatorname{rect}(y/W).$$

In other words, we first back project to infinity and then gate out the unneeded parts. Of course, we are still working on the infinite plane and have assigned zeros outside the working area, so this procedure is not precisely identical with filling a finite domain.

Some users of back projection for array filling practice conservation. When assigning values within a finite region, they take into account the length of the line segment over which values are to be redistributed and use a multiplying factor inversely proportional to the length of the line segment. In this way the area under a projection to be used for back filling equals the volume under the two-dimensional function resulting from back projecting. If the two-dimensional function were mass area density and the projection represented line density, then mass would be conserved; and so we may refer to this procedure as *conservative back projection*. It is important for the student to be aware of these three different usages of the term back projection, as some writers neglect to specify what back projection means to them.

BINARY IMAGES

A binary image $f(x, y)$ is one which has only two function values, usually 0 and 1, or -1 and 1. A ternary image, which has three values, will result if two binary images are added or subtracted. Naturally, all the operations described for real-valued functions are applicable to binary functions, with the exception of differentiation. What distinguishes binary functions is morphology: information about image shape, which is what remains when the nonzero function values are collapsed into a single constant. Although a lot of information may be sacrificed, a lot resides in shape alone; coastlines, silhouettes, and printing types are examples.

Differencing

Although the familiar column of differences in its present form goes back to classical times with astronomical and trigonometrical data tabulated at discrete intervals, differences can also be calculated as a function of the continuous variables x and y. The span (a, b) is the vector displacement over which the difference is taken. Thus

$$f(x + a, y + b) - f(x, y)$$

is the difference (or first difference) of $f(x, y)$ over the span (a, b). Differencing displaces things by half the span; if displacement is not wanted, the centered first difference $f(x + \frac{1}{2}a, y + \frac{1}{2}b) - f(x - \frac{1}{2}a, y - \frac{1}{2}b)$ is used. As an example, consider the binary function that is equal to unity inside the ellipse $(x/15)^2 + (y/10)^2 = 1$ and zero outside. The centered first difference over a span $(1, 0)$ is shown in Fig. 2-43(b), where black represents 1 and grey represents -1; call it $\Delta_x f(x, y)$. Thus the difference is a ternary image. In the spirit of morphology, which is to deal with shapes rather than with values, we convert back to a binary object by taking absolute values. This is illustrated in (c), which shows $|\Delta_y f(x, y)|$, the absolute value of the first centered difference in the y-direction.

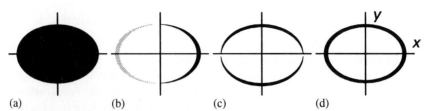

Figure 2-43 (a) A binary object, (b) its first difference in the x-direction, (c) the absolute first difference in the y-direction, (d) the sum of the two absolute differences.

It is clear that the absolute-difference operator is telling us something about the edge of the original binary object. We can form a new binary function $E(x, y)$ which is equal to 1 wherever $|\Delta_x f(x, y)|$, or $|\Delta_y f(x, y)|$, equals 1, as shown in (d). In the notation of binary logic, which is appropriate,

$$E(x, y) = |\Delta_x f(x, y)| \text{ OR } |\Delta_y f(x, y)|.$$

Different results come from different choices of span, which must be based on judgment.

Other functions of $\Delta_x f(x, y)$ and $\Delta_y f(x, y)$ could be contemplated that relate to the boundary—for example $\sqrt{(\Delta_x f)^2 + (\Delta_y f)^2}$, a ternary function that is computationally slower than the edge representation $E(x, y)$ above.

The boundary of the support of a simply connected binary function is its main morphological feature. Some other shape-related parameters, such as centroid, radii of gyration, and orientation of the principal axes of inertia, are determined as for other real-valued functions.

Union and Intersection

Two binary functions $f(x, y)$ and $g(x, y)$ determine a variety of other binary functions corresponding to the Venn diagrams of set theory. Thus if f and g describe islands, then the nation comprising both islands is the union of f with g, written $f \cup g$. Likewise, if f and g are territories where, respectively, French and German are official, then the bilingual area where both languages are spoken (Alsace and elsewhere) is the intersection $f \cap g$. This notation, which is familiar from Boolean algebra, is applicable because the binary functions $f(x, y)$ and $g(x, y)$ share with logical constants the property of being two-valued. It is useful to be aware that things previously learned have relevance; however, the Boolean notation is not recommended here. Suppose that $g(x, y)$ represents a dry lake in the middle of an island $f(x, y)$; there is Boolean notation for the binary object $h(x, y)$ describing the situation when the lake fills with rainwater, but it is simpler to write $h(x, y) = f(x, y) - g(x, y)$, because this line is also valid as computer code. Likewise

$h(x, y) = f(x, y) \text{ OR } g(x, y)$	Union (island nation)
$h(x, y) = f(x, y) \text{ AND } g(x, y)$	Interxection (bilingual area)
$h(x, y) = f(x, y) \text{ EXOR } g(x, y)$	(monolingual area)

are lines that are understood by a computer. For this reason, the symbols $\cup$ and $\cap$ (cup and

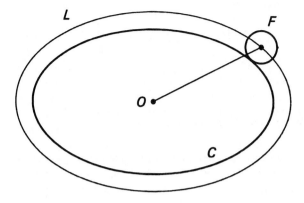

Figure 2-44 A cam follower F rolls on a
cam C to trace a given locus L. The design
of this everyday mechanism is analogous
to taking a binary object bounded by L and
finding an interior binary object C bounded
by the envelope of the set of circles.

cap) for union and intersection will not be used again, now that it is understood that the
logical concepts themselves are fundamental to morphology. As mentioned earlier, subtraction and addition may result in ternary functions; consequently $f(x, y) - g(x, y)$ is
not always the same as the binary expression $f(x, y)$ AND NOT $g(x, y)$, nor is $f(x, y) +$
$g(x, y)$ always the same as $f(x, y)$ OR $g(x, y)$. EXOR is the Boolean exclusive-or operator (.XOR. in FORTRAN).

These ideas carry over into the discrete binary objects that computer graphics deals
with. In preparation for other topics of discrete morphology of binary functions we introduce a simple problem that has important ramifications later.

Border Zones

A binary function is bounded by a curve B, and we wish to find two further boundaries
defined by points that are not more than a given distance from B. Problems of this kind
arose in mechanical engineering with the design of cams, where the center F of a circular
cam follower has to trace out a given locus L. The cam must lie inside L by a distance
equal to the radius of the wheel that rolls on the cam. Occasionally the problem can be
solved analytically, but it is normally done by graphics. The same problem arises with
property lines that are specified by distance from a lake edge or the mean high tide line.
Fishing boundaries set at 200 nautical miles from land constitute a serious example, while
the oil-fired dispute over the boundary between the United States (which extends 12 miles
out to sea) and Alaska (which extends 3 miles) took over a decade of study. In Fig. 2-44 the
curve C is at a given depth d within the locus L, and there is another curve C' (not shown)
that is the same distance d outside L. Now, starting from C, where is the curve that lies
at a distance d outside C? If C is an ellipse then the other curve is not, and conversely.
When problems of this type come to be dealt with by computer graphics, the fact that a
metric quantity d has been injected leads to fundamental questions of distance definition
on a discrete grid with deep and pervasive applications in image processing. The operation
of dilation, which adds a border to a binary mass, and erosion, which peels off a rind, are
discussed further in the chapter on binary operations.

OPERATIONS ON DIGITAL IMAGES

Digital images are distinguished by dependence on two spatial variables, say i and j, that assume only discrete values, normally integers. The dependent variable may be quantized, too, as for example with an 8-bit digital image, but often is not, except as limited by machine representation. So, although many digital images are quantized, others are not; the elements of a digital image may include square roots or other irrationals. Digital images do not exist on media such as paper, film, or the screens of computers but rather are mathematical representations of images. To be brief, a digital image is normally an integer matrix $f(i, j)$. A semiconductor chip carrying a visible array of microscopic lasers, each of which was either on or off, could not be a digital image but would be representable for certain purposes by a matrix of 0s and 1s. We call this a *binary digital image* $B(i, j)$. The following sections introduce some operations on digital images. Integer functions of continuous x and y have a right to be called digital images and are occasionally encountered.

Translation, Rotation, and Reflection

Translation of an image $f(i, j)$ by a displacement (k, l), where k and l are integers, produces the new digital image $f(i - k, j - l)$. This is less simple than the notation suggests: provision has to be made for the locations that are vacated and for the values whose shifted locations are beyond the edges of the original matrix $f(i, j)$. One way is to define $f(i, j)$ with a surround of 0s. A method of recycling values that go over the edge back into the vacant locations is illustrated in Fig. 2-45(c).

There are three rotated versions, $f(-j, i)$, $f(-i, -j)$, and $f(j, -i)$, for counterclockwise rotation through 90°, 180°, and 270°, respectively. Rotation through angles that are not multiples of 90° is often wanted, but requires interpolation, which will be discussed as a separate topic.

There are four versions obtained by reflection: top/bottom, left/right, NE/SW, and

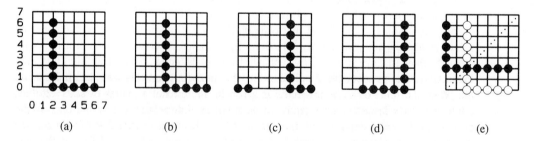

Figure 2-45 (a) The usual 5×7 dot-matrix representation of the letter L (shown on an 8×8 field), a realization of a binary digital image, (b) translated one unit to the right, (c) translated three units to the right; cyclic translation is shown, otherwise at this limited field size, part of the foot would be lost, (d) reflection in the vertical axis $i = 0$ under the cyclic convention, otherwise the object would be lost to the field, (e) reflection NW to SE about the dotted diagonal (the original object is shown in outline).

	0°			90°			180°			270°		
	1	2	3	3	6	9	9	8	7	7	4	1
	4	5	6	2	5	8	6	5	4	8	5	2
	7	8	9	1	4	7	3	2	1	9	6	3

	T/B			L/R			NE/SW			NW/SE		
	7	8	9	3	2	1	1	4	7	9	6	3
	4	5	6	6	5	4	2	5	8	8	5	2
	1	2	3	9	8	7	3	6	9	7	4	1

Figure 2-46   A matrix (top left) and its rotated and reflected versions.

NW/SE. These are, respectively $f(i, -j)$, $f(-i, j)$, $f(j, i)$, and $f(-j, -i)$. (See Fig. 2-46.)

Negative indices have been used here for clarity, but they imply translation as well as rotation or reflection. When negative indices are not allowed, or are not desired, then replace $-i$ and $-j$ by $N - i$ and $N - j$ for $N \times N$ matrices and $i = 0, 1, \ldots, N - 1$. If i runs from 1 to N, then replace $-i$ by $N + 1 - i$. If the matrix is not square, there is a further complication that can be met by extending the original matrix with 0s to make it square.

If rotation and reflection are expressed by operators, then some compact relations can be stated. For example, two successive reflections about perpendicular axes are equivalent to rotation through 180°, and top-to-bottom reflection followed by NW/SE reflection is equivalent to rotation through 90°.

Other elementary operations can be carried out on digital images, the simplest of which is shearing. For example, $f(i, j - i)$ exhibits horizontal shear. Vertical shear and diagonal shear can also be applied.

Translation by a fraction of a unit, reflection with respect to other axes, and rotation through arbitrary angles are operations that do not produce results that fall on the original grid, and so interpolation is required. The technique for interpolation is best explained later in terms of convolution.

Projection of a Digital Image

Projection of an image defined as a function of continuous variables was discussed above, but when the function to be projected is given in the form of a matrix of sampled data, additional considerations arise from the geometrical interrelations between the sample points and the line along which the integral is to be taken. Figure 2-47 illustrates the projection of a digital map along four directions spaced 45°. Although different numbers of elements enter into the various sums, and the spacing of contributing elements changes, the projections all have the same area. At other angles of projection one might simply sum the elements falling within strips of unit width, but one might also think of compensating for the varying number of such elements. A guiding principle would be to make sure that such "scans" have the standard area. From four scans alone, it is possible to gain some

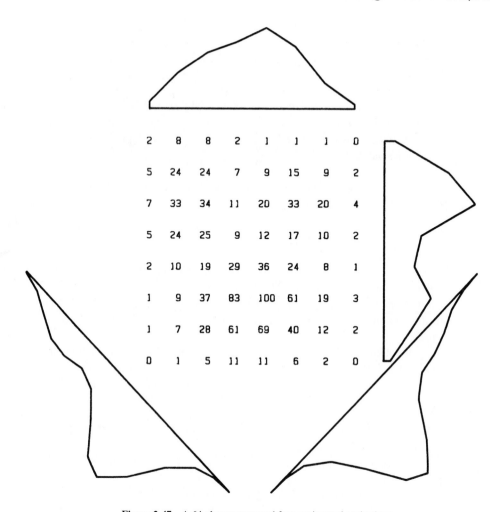

2	8	8	2	1	1	1	0
5	24	24	7	9	15	9	2
7	33	34	11	20	33	20	4
5	24	25	9	12	17	10	2
2	10	19	29	36	24	8	1
1	9	37	83	100	61	19	3
1	7	28	61	69	40	12	2
0	1	5	11	11	6	2	0

Figure 2-47   A 64-element map and four equispaced projections.

impression of the whereabouts of the conspicuous features in the original map. However, with 64 unknowns and only about eight measured data per scan, algebra would suggest that eight scans would be needed if we wished to solve for all the original map elements. Inversion of projections is the subject of a later chapter.

Simple Dilation and Erosion

Consider the border of 0s surrounding the 1s of a binary digital image. Some of these 0s are connected to a 1 by a single step in a cardinal direction (N, S, E or W); if these 0s are changed to 1s the image has been dilated. The procedure can be repeated and will be found, in general, to lead to simplification of the shape of the original boundary. If, on the contrary, those 1s that are connected to a 0 by a single cardinal step are changed to 0s, the image has been eroded. Repeated erosion will ultimately eliminate all the 1s. More elaborate rules for dilation and erosion can be used and will produce different results. The operations are easy to compute and must be regarded as among the fundamental operations on binary images. Compound operations, such as dilation followed by erosion, are also important and bring about modifications that are frequently encountered or desired. These operations extend to binary objects in general and are mentioned again later under cross-correlation and as needed in connection with both digital images and functions of continuous variables.

REFLECTANCE DISTRIBUTION

Reflectance ranges from zero, which means black, to unity, which is white, but it does not follow that, in a given picture, the full range of reflectance will occur. And even if the full range is present in a picture, it does not follow that there will be a general impression of high contrast. Think of a photograph of a twilight street scene mostly composed of murky greys, but with one distant street lamp constituting the only speck of white, and one open window being the only black. Grey levels in the whole of the rest of the picture might be limited to the range 0.3 to 0.5. Such a picture would be susceptible of considerable improvement in the darkroom where the print was made from the negative, either by the photographic technician or by the operator of a machine for converting negatives to video signals for cathode-ray tube display or digital recording. Some improvement could also be made starting from the poor print, given a suitable camera and illuminating equipment, but deterioration of the signal-to-noise ratio would have to be accepted. After the necessary analogue treatment, followed by digitization, computer enhancement could be considered.

As a basis for computing, the frequency distribution $p(R)$ of reflectance R forms a useful beginning. The reflectance distribution $p(R)$ is defined so that, in a given picture, $p(R)dR$ is the fraction of reflectance values lying between R and $R + dR$. As an example, the left side of Fig. 2-48 shows how $p(R)$ might look for the picture described above. As with probability distributions, $\int_0^1 p(R)\,dR = 1$.

Reflectance distribution applies to continuous variation of R but if R may assume only quantized levels R_i, then the concept is still applicable. On the right of Fig. 2-48 is a reflectance distribution referring to ten quantized grey levels placed at $0.05, 0.15, \ldots$, 0.95. Such a diagram would apply after numerical quantization was applied. Of course, a computer could easily handle, let us say, 16-bit representation of grey levels, which for almost all purposes would be essentially continuous variation of reflectance. However, as far as images prepared for viewing by the eye are concerned, a very modest number of grey levels suffices. Where four-bit representation is adequate, the computer storage requirement for a large image can readily be reduced by packing four grey levels into one 16-bit word.

Cumulative reflectance distributions $C(R)$, as shown in Fig. 2-49, are also commonly used, where $C(R)$ is the fraction of reflectance values less than R. Thus $C(0) = 0$ and $C(1) = 1$. As before, quantized values of R are included.

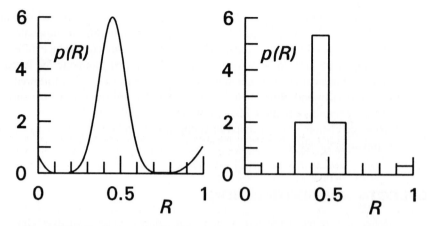

Figure 2-48 A reflectance distribution $p(R)$ for a murky picture with tiny patches of black and white, in both continuous (left) and quantized form with ten grey levels (right).

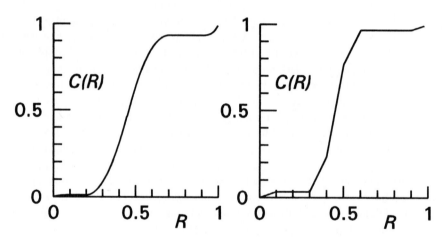

Figure 2-49 Cumulative reflectance distribution $C(R)$, continuous (left) and quantized (right).

Examples of Reflectance Histograms

Consider a black-and-white line drawing, such as a block diagram. When it is digitized to levels from 0, representing nominal black, through 7, representing nominal white, a good proportion of all the values will be 7, because there is a lot of white paper. If the digitizing cell is smaller than the line width used for the drawing, then the remaining values will be mostly zero, representing black, but there will be a few intermediate values that arise from cells straddling the edges separating black from white. Such a histogram is shown in Fig. 2-50(a).

If the digitizing cell is equal in size to the line width, the only full blacks will arise

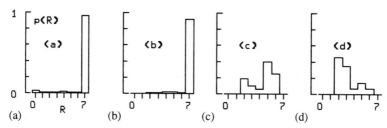

(a)   (b)   (c)   (d)

Figure 2-50   Reflectance histograms for eight reflectance levels as they might appear
for (a) a line drawing with fine resolution elements, (b) a line drawing with resolution
elements the same size as the line width, (c) a landscape painting with lots of sky, (d) a
classical portrait with dark background.

by accident; the small fraction of cells that intercept lines at all will receive reflectance
values spread more or less uniformly over 1 through 6. This case is shown in Fig. 2-
50(b). Reproduction from such coarse sampling would be unacceptable for many purposes,
because the lines would be represented by variable greys and would look like staircases.

A reflectance histogram for a printed page would be of the same nature as for Fig. 2-
50(a) except that the fraction of full blacks $P(0)$ would be greater, perhaps as high as 20
percent.

A landscape painting might have a large area of sky, putting more than half the
reflectance values at 5 or 6. The remaining values might be distributed over 2, 3, and 4.
A portrait painting might have most of the values in the dark range at 2 and 3, with the
information-bearing elements in highlights, say at 5.

Modifying the Reflectance Distribution

If the medium for image display offers only a small number of reflectance levels, ranging
from 0 to $N - 1$, but a particular quantized image has levels ranging from L_{min} to L_{max},
where $L_{min} > 0$ and $L_{max} < N - 1$, then clearly even the limited capacity of the medium
is not being fully utilized. The situation would suggest expanding the reflectance levels
to occupy the full range, a step that could readily be taken in a computer. It would only
require some transformation from the old level L to a new level L' to be found, that would
convert L_{min} to zero and L_{max} to $N - 1$. The transformation

$$L' = \Im\left[\frac{1}{2} + \frac{(N - 1)(L - L_{min})}{L_{max} - L_{min}}\right]$$

would perform the desired operation, where $\Im(\cdot)$ means "integral part of" and $\Im(\frac{1}{2} + \cdot)$
rounds off to the nearest integer. Of course, such a transformation is not unique; the
one shown here is based on linearity (although not really linear in outcome because of
roundoff).

Perhaps the linear relation would not be the best. To examine this suggestion consider Fig. 2-51, where $N = 10$, $L_{min} = 2$ and $L_{max} = 7$. Although the given levels 2
through 7 are equispaced, the output levels L' are not. Even though the range has been expanded to cover 0 through 9, there are still only six levels, and the spacing of these levels is

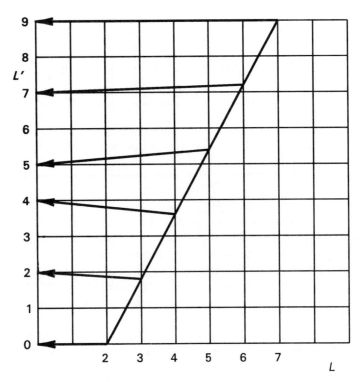

Figure 2-51 A transformation that expands an initially restricted range of levels L $(2 \leq L \leq 7)$ into a full range of modified levels L' $(0 \leq L' \leq 9)$. In this example $N = 10$ reflectance levels.

nonuniform. When the reflectance distribution histogram is now taken into account, it may turn out that severe damage has been done to the image, even though the contrast (range of reflectance values) has been enhanced. Consequently, any computer rule for reflectance distribution modification has to take into account the transformed histogram, and the overall assessment will have to involve subjective judgment. Successive trials and assessment may be required.

Clearly the effects of requantization when small numbers of levels are involved should be avoided if possible. Thus, if a low-contrast photographic print were to be dealt with, it would be desirable to quantize initially to, say, $3N$ levels or more before modifying the distribution.

Suppose that $L_{\min}$ is to be converted to 0 and $L_{\max}$ to $N - 1$ but that $L_{\mid} = \frac{1}{2}(L_{\min} + L_{\max})$ is to be converted to a level that exceeds $\frac{1}{2}(N - 1)$ by ϵ (Fig. 2-52). A bias of this kind may be helpful in expanding a heavily populated part of the reflectance histogram into the middle range of greys while compressing a lightly populated part, so that none or few L' levels are used up on a minor area of the picture. A suitably biased transformation is

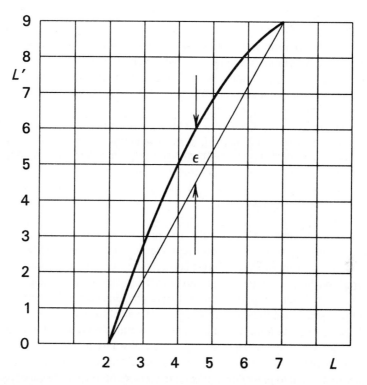

Figure 2-52 A biased reflectance level transformation that both expands the range of L' and raises the original midvalue $\frac{1}{2}(L_{min} + L_{max})$ to a level above the midvalue of L'. More detailed contrast stretching or compression can be based on the reflectance histogram, especially for sets of similar images where space-invariant modification is seen as an improvement.

$$L' = \Im\left[\tfrac{1}{2} + (N-1)\frac{(L - L_{min})}{L_{max} - L_{min}} + 4\epsilon\,\frac{(L - L_{min})(L_{max} - L)}{(L_{max} - L_{min})^2} \right].$$

Analogue Modification

Techniques of contrast modification have been well known in black-and-white photography for a long time and facilitated by the use of variable-gamma printing paper. Some techniques involve sharp or defocused, positive or negative transparencies to be laid on or near the original negative for a final print. Photographic practices can be studied for inspiration as to lines that might be followed in computing.

Human vision also offers food for thought. Earlier a murky street scene was mentioned where there was one black window. After a photograph had been taken, it might be impossible to discern any detail inside the window. However, the interior might very well be interpreted by the eye of a person in the street. The eye can function in both sunlight and moonlight, which are in the ratio of about a million to one in level of illumination, and can readjust to different parts of the same scene. Landscape painters, whose medium offers about the same range of contrast as a photographic print, can acceptably modify the reflectance histogram from place to place so as to convey cloud detail in a bright sky while at the same time representing detail in the shade in the foreground, a feat which would be beyond the reach of a camera. How to convey the sensation of large dynamic range is regularly taught to artists (not in those words), as well as much other accumulated knowledge. To some extent, such knowledge can be gained from reading books and looking at paintings, but the traditional method, as with scientific research, is by apprenticeship. Those equipped with the necessary skills to carry out reflectance modification by computer soon learn that nonmathematical skills are also significantly involved.

Contrast Compression

Since the brightness in a natural scene may vary from direction to direction by a factor of 10^2, or even very much more, reproductions in newspapers, books, and photographic prints necessarily involve contrast compression. Usually the compression will be nonlinear. Recording on photographic film allows a higher dynamic range than ink on paper, but at low light levels proportionality is lost (a phenomenon known as *reciprocity failure*) and at high light levels the medium saturates. A valuable experiment can be conducted with a photocopying machine. We know that full white is reproduced with whatever reflectance R_2 is possessed by the paper supply, which may have a value of, say, 0.8 (Fig. 2-53). Pure black is reproduced less faithfully with a reflectance R_1, which may be around 0.3 and may vary from point to point depending on the texture of the item being copied. The reproduced reflectance R' will depend on the initial reflectance R through the transformation

$$R' = R_1 + (R_2 - R_1)R$$

if the relation is linear. If the copy is then copied, the transformation may be applied twice to give

$$\begin{aligned} R' &= R_1 + (R_2 - R_1)R' \\ &= R_1 + (R_2 - R_1)[R_1 + (R_2 - R_1)R] \\ &= R_1(1 + R_2 - R_1) + (R_2 - R_1)^2 R. \end{aligned}$$

Let $\Delta = R_2 - R_1$. Then the reflectance of the nth copy will be related to the original reflectance R by

$$R^{(n)} = R_1(1 + \Delta + \Delta^2 + \cdots + \Delta^{n-1}) + \Delta^n R.$$

This is still a linear relationship, but now it ranges from a minimum reflectance $R^{(n)} = R_1(1 - \Delta^n)/(1 - \Delta)$ when $R = 0$ to a maximum of $R_2^{(n)} = R_1^{(n)} + \Delta^n$ when $R = 1$. With the sample values mentioned above, $R_1 = 0.3$, $R_2 = 0.8$, $\Delta = 0.5$, the third copy would have reflectance values ranging from 0.52 to 0.65. There would be a very restricted range of grey levels, and the mean level would have risen to slightly below 0.6. However, if you make three successive photocopies of a picture with a good range of greys, the

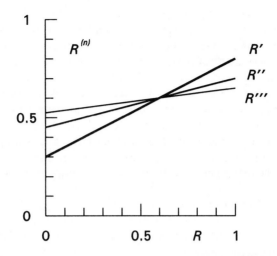

Figure 2-53   A linear reflectance compression characteristic (heavy line). showing the reflectance R' of the first copy. Pure black ($R = 0$) becomes less black, and white ($R = 1$) less white. If the copy is recopied, the medium line results, and if that copy is recopied, we get the thin line.

result is better than the above reasoning would suggest. By copying a sample with grey levels ranging between the best black and white you can obtain, you can find out how the designer of the machine has planned the reflectance transformation nonlinearly to improve the acceptability of repeated copying.

An interesting contrast transformation that comes up in the maximum-entropy method of image restoration is

$$R' = R - R \log R.$$

The integrated quantity $- \int \int R \log R \, dx \, dy$ is the entropy of an image; the transformation results in contrast compression at bright levels ($R > 1/e$), while the contrast at dimmer levels is expanded.

DATA COMPRESSION

Let us agree to measure the amount of data in a digital image by the product of the number of matrix elements with the number of bits representing the matrix values. If the image has been digitized from a newspaper, 3 or 4 bits (8 to 16 grey levels) would suffice to represent the matrix values, but the digitizing procedure might routinely record 256 different levels (the fact that there are only about 10 significantly different grey levels in the object does not mean that intermediate gradations of grey do not occur). Thus the amount of information can be much less than the amount of data. Therefore data compression to suit the amount of information will ease demands on a communications link. The retina receives an image with an immense amount of data but immediately selects out that part of the image falling on the fovea, which amounts to less than 1 percent of the whole retina, for high-resolution use. Spatial chains of retinal cells firing in coincidence allow linear features to be handled by a single neuron, again reducing the data to be passed to the optic nerve. Further reduction relating to illumination levels is achieved by nonlinear thresholding.

Consider a black-and-white facsimile image whose raster lines may be encoded by strings of 0s (representing white) and by 1s (representing black). The first raster line might be

01000001 01000011 01000111 01000111 01001111 01001111 01001111 01000000,

where spaces have been introduced to call attention to the groups of eight. Each 8-bit group corresponds to an ASCII character; in this example the whole line reduces to the string "ACGGOOO@", which is certainly briefer than the string of 0s and 1s, as is the hexadecimal string "414347474F4F4F40". However, no data compression has been achieved relative to the electrical signal that the image displays, or to the image itself, which on the screen looks like

.   ..   .

Very often a digital image has much more white than nonwhite and thus has long strings of 0s and many fewer 1s. One can see how to compress the data in a black-and-white facsimile image, whose raster lines consist of long strings of 0s (representing white) and many fewer 1s (representing black), by replacing a string of 0s by its length. A television frame, which largely repeats the previous frame, can be compressed down to the amount of data sufficing to update the previous frame. Finally, statistical information about an ensemble of images, perhaps relating to the relative occurrence of different spatial frequencies, may allow further economy. In commercially critical applications, such as the introduction of color into a television band 6 MHz wide, a bandwidth that was selected as adequate for black and white when originally allocated, the cost benefit of data compression by only a modest factor may warrant extreme refinement of the technology. The development of high-density television with improved resolution is in this category. Some simple techniques will now be described, and methods based on spectral analysis will be mentioned later.

Run-Length Encoding

An image formed on a grid by uniform black dots on a white field (or vice versa), such as the images on a black-and-white television image or a computer screen, is representable, row by row, by a string of 0s and 1s. *Raster graphics* is the term used for images constructed this way. As an example, the first 64 characters of the first row might be

00000011 00000100 00000100 00000100 00001001 00001001 00000110 00000111,

where spaces have been artificially introduced as before to show a division into bytes. Converting groups of eight into decimal notation shortens the representation into

003 004 004 004 009 009 006 007

Ordinary images very often contain runs of the same number repeated over areas of uniform grey level. Instead of repeating that number, it suffices to precede it by the run length, or number of repeats. The example would then read

$$000\ 003\quad 002\ 004\quad 001\ 009\quad 000\ 006\quad 000\ 007,$$

which is actually a little longer; but clearly this coding becomes much briefer when there is lots of repetition.

TIFF Encoding

As an alternative to run-length encoding there is Tagged Image File Format (TIFF) encoding where the run length is indicated by a preceding tag equal to 256 minus the number of repeats (as far as 127). Then, to save too frequent repetition of 000, where a byte is not repeated, a preceding tag in the range 1 to 127 indicates how many of the following bytes are to be read literally. The example would become

$$001\ 003\quad 254\ 004\quad 255\ 009\quad 002\ 006\ 007$$

Delta Row Encoding

Not only do bytes often repeat along a row, but chunks of a row may be repeated in the following row. After the first row is encoded, it is sufficient to say where bytes are to be replaced and what the fresh bytes are. This is done on a group of eight bytes by indicating the number of bytes to be replaced, their positions, and their values.

While these procedures result in substantial savings, they could not be described as optimal data compression. But the development of commercial printers involves additional facets of optimization that are outside the scope of information theory. If you can think of a better way of encoding graphics, the market allows you to offer your product for sale.

SUMMARY

Ways of presenting images have been described together with elementary image operations that may be desired, or that may be unwanted and need compensation. The remainder of the book is concerned with a variety of ways in which images are acquired or constructed, and then processed for the attainment of some purpose. Spatial spectral analysis is basic to several forms of image processing, both for theoretical understanding and for numerical computation. The next chapter deals with two-dimensional impulses, or delta functions (both dots and lines), which are fundamental to imaging and also provide an essential basis for spectral analysis by the Fourier transform.

APPENDIX: A CONTOUR PLOT PROGRAM

Let a single-valued real function $z(x, y)$ be represented by samples at integer values of x and y in the range $1 \leq x \leq L_x$ and $1 \leq y \leq L_y$. A contour diagram is to be drawn with

N contour levels, not necessarily equispaced, specified as $L(i)$, $i = 1$ to N. The contour lines will be polygons based on linear interpolation as follows. Let $ABCD$ in Fig. 2-54(a) be adjacent points defining a square cell in the (x, y)-system ($AB = BC = 1$), with values $z_A = 25$, $z_B = 15$, and $z_C = 35$ at A, B, and C respectively as shown in square brackets in Fig. 2-54(b). Interpolating linearly between A and B we find that the point P in Fig. 2-54(b), with value 20, lies in between, at $h_x = 0.5$. If the 20 contour *enters* ABC at P it must leave at a point Q on BC, one-quarter of the way up from B, where $h_y = 0.25$. The straight line PQ is one side of the polygonal contour at level 20. By linear interpolation $h_x = ([L(i) - z_A]/(z_B - z_A)$, $h_y = [L(i) - z_B]/(z_C - z_B)$, where $L(i) = 20$ in this example, and the side PQ runs from $(x_A + h_x, y_A)$ to $(x_C, y_C + h_y - 1)$, where (x_A, y_A) and (x_C, y_C) are the locations of A and C respectively.

Higher order interpolation makes the contours appear smoother but the cosmetic effect may be misleading in view of the discrete character of the data. If *independent* data on a finer grid are available, smoother contours may result *but also may not*, a reminder that numerical interpolation does not add information. *Sometimes* interpretability is improved if contours are smoothed, for example, by use of a spline.

If the contour entering at P leaves via a point Q on the diagonal AC, as in Fig. 2-54(c), then the side PQ runs from $(x_A + h_x, y_A)$ to $(x_C + h_y, y_C - h_y)$, where $h_y = [z_C) - L(i)]/(z_C - z_A)$. For example, with $z_C = 19$, $h_y = 1/6$ as illustrated.

It is possible, if the value z_A at A equals the contour value (say 20), for a contour to enter the triangle ABC precisely at A. But if in addition $z_B = 20$ it would be more reasonable to have the contour enter midway between A and B. If further, $z_C = 20$, more thought is required. Some nasty configurations can arise for example, if $z_A = z_B = 20$ and all the nearby values are less, one might wish to show a line segment AB on the grounds that it represented a degenerate contour. The judgment and programming required to handle the accidents of integer function values and the possibility that there is more than one entry and one exit for a triangle are really irrelevant in contexts where measurement error exists. A practical approach is to shift the contour level from 20 to say $20 - 10^{-6}$. The consequence of this will be to exhibit degenerate contours in a way that does not ignore the presence of peaks that rise just above the contour. Conversely, use of $20 + 10^{-6}$ will suppress these features if they are deemed unwarranted. Negative peaks react oppositely. Clearly, if square cells were chosen instead of triangular, ambiguous exits would raise even further complications.

The exit point Q from ABC becomes the entry point to a new triangle which shares a side with the previous one; thus if Q lies on CA, the new triangle is ACD in Fig. 2-54(a).

To facilitate iteration the new triangle is relabeled using three rules; AB remains the entry side, the labeling goes counterclockwise, and all diagonals are parallel. Thus in Fig. 2-54(d), the triangle formerly referred to as ACD becomes $A'B'C'$ (the primes are only included for clarity of the figure). The contour now exits at R on $B'C'$, proceeds to S on $B''C''$, and so on.

We notice that

$$x_{A'} = x_A, \qquad x_{B'} = x_C \qquad x_{C'} = x_B - 1$$
$$y_{A'} = y_A, \qquad y_{B'} = y_C \qquad y_{C'} = y_B + 1.$$

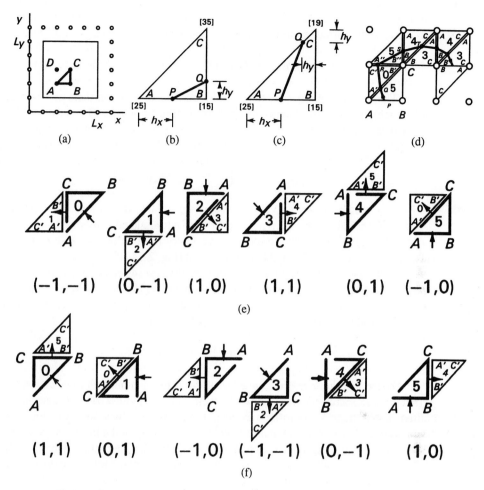

Figure 2-54 (a) A plotting area surrounded by a fringe of dummy values (hollow circles); (b) a contour of level 20 entering a triangle ABC at P and exiting on BC at Q; (c) exit via CA at Q; (d) a chosen contour begins on AB and exits from the first triangle ABC on CA; type numbers, determined by direction of entry to successive triangles, are 5, 0, 5, 4, 3, 4, 3, . . . ; (e) types 0 to 5 with exit on CA; the tabulated vector shifts (V_x, V_y) show the displacement $\mathbf{C}' - \mathbf{C} = (x_{C'} - x_C, y_{C'} - y_C)$ as the current vertex C moves to C'; (f) types 0 to 5 with exit on BC and the associated vector shifts (V_x, V_y) of the third vertex C.

Thus one end of the exit side remains where it is with the same letter (A), while the other end (C) changes its letter (to B') but remains where C was; the new entry side is again AB, as agreed. The choice to keep A and rename C to B' is in accordance with the CCW rule.

In vector notation $\mathbf{A} = (x_A, y_A)$, the three changes are

$$\mathbf{A} \leftarrow \mathbf{A}, \qquad \mathbf{B} \leftarrow \mathbf{C}, \qquad \mathbf{C} \leftarrow \mathbf{C} + (-1, 0).$$

In Fig. 2-54(d) there is a point S where the contour enters at AB and exits on BC (at T). In this case the correction to $\mathbf{C}$ is $(0, 1)$, and in general for exits on CA the changes are

$$\mathbf{A} \leftarrow \mathbf{A}, \qquad \mathbf{B} \leftarrow \mathbf{C}, \qquad \mathbf{C} \leftarrow \mathbf{C} + (V_x, V_y).$$

The first two changes are the same for all exits on CA, but there are six distinguishable configurations determining the shift of the third point C. The six types of case are defined as follows.

Type	0	1	2	3	4	5
Entry from	SE	E	N	NW	W	S

The vector shifts (V_x, V_y), shown in Fig. 2-54(e) are stored. The type definitions have been chosen so that the type number T of a triangle advances by 1 at each stage, or $T \leftarrow (T + 1) \bmod 6$. In the initial example, entry from below, we had $T = 5$. The next triangle, entered from the southeast, had $T = 0$ (Fig. 2-54(d)).

If the exit is on BC instead of on CA, the type number for ABC is the same as in Fig. 2-54(f), but the type number of the next triangle advances in accordance with $T \leftarrow (T - 1) \bmod 6$. For exit on BC a second set of vector shifts (V_x, V_y) needs to be stored for use with the change relations, which are now

$$\mathbf{A} \leftarrow \mathbf{C}, \qquad \mathbf{B} \leftarrow \mathbf{B}, \qquad \mathbf{C} \leftarrow \mathbf{C} + (V_x, V_y).$$

When a contour reaches the edge of the map one could terminate the sequence of triangles and search for new segments at the same contour level, but there is an elegant alternative, which is to surround the data by a fringe (hollow circles in Fig. 2-54(a)) to which are assigned dummy values set just below the level of the lowest contour. As a result all contours on the expanded data will be closed. Stretches of contour lying inside the unit-width border will have no significance and may be suppressed by a clipping command that raises the pen outside the plotting area.

It is convenient to assign a serial number K to each point (x, y) on the map (allowing for the dummy fringe), defined by $K = 1 + x + y(L_x + 1)$. Conversely $x = K \bmod (L_x + 1)$ and $y = K \operatorname{div} L_x$. As the contour is followed, values K_A, K_B, K_C are updated, but the entry value of K_A is preserved as K_1.

A loop of the contour closes when the initial triangle of type 5 is reentered from below. Whenever $T = 5$ occurs the left end A of the entry gate AB is tagged with the number i of the current contour and subsequent checks are made to see whether the current vertex A has a serial number K_A equal to K_1. If so, that loop terminates. There may, however, be another loop at that contour level, so the search for a new entry gate resumes. Such further searches terminate when the right end of the top row of the matrix $z(x, y)$ is reached; then the contour level is raised and the process repeated for all required levels.

The system of serial numbers K for a 5×5 plotting area is shown in the following matrix on the (x, y) plane. The first map value $z(1, 1)$ is at $K = 8$ and the last at $K = 36$.

37	38	39	40	41	42
31	32	33	34	35	36
25	26	27	28	29	30
19	20	21	22	23	24
13	14	15	16	17	18
7	8	9	10	11	12
1	2	3	4	5	6

The commented subprogram that follows, written in H-P Technical BASIC, translates smoothly into FORTRAN or PASCAL. The only things to watch are the multistatement lines (containing @ used as a statement separator). An IF statement followed by @ causes the concatenated statement to be ignored; this conveniently saves parentheses, which will have to be reintroduced in translation to some languages. The only graphics statements are MOVE x,y, which moves the pen to (x, y), and DRAW x,y, which draws a line from the current pen position to (x, y). For program development I find it convenient to place high-level graphics statements in the subprogram for direct display on the screen. For repetitive use the vertex coordinates can be stored for subsequent separate processing by a printer control language.

The calling program contains a line

```
CALL "CPLOT"(LX,LY,z(,),N,L(),ROUGH)
```

whose parameters correspond to L_x, L_y, etc. Putting ROUGH=1 produces the polygonal contours defined by the data $z(x, y)$, while ROUGH=0 produces smoothing that modifies degenerate features such as bristles. The calling program specifies the number of levels N, the array of level values $L(i)$, and positions the plotting window 1 to L_x, 1 to L_y, on the plotting area.

LITERATURE CITED

L. B. ALBERTI (1976), *On Painting*, Greenwood Press, Westport, Ct.

R. N. BRACEWELL (1992), "Planetary influences on electrical engineering,"*Proc. IEEE*, vol. 80, pp. 230–237.

E. JAHNKE AND F. EMDE (1945), *Tables of Functions with Formulas and Curves*, 4th ed., Dover Publications, New York.

J. C. MAXWELL (1965), "On hills and dales," in *The Scientific Papers of James Clerk Maxwell*, vol. 2, pp. 232–240, Dover Publications, New York.

F. PEARSON, II (1990), *Map Projections: Theory and Applications*, CRC Press, Boca Raton, Fl.

ROGER N. SHEPARD (1990), "Mind Sights," W.H. Freeman, New York.

R. A. SKELTON (1958), "Cartography," in C. Singer et al., eds., *A History of Technology*, vol IV, pp. 596–628, Oxford University Press.

R. C. SLOANE AND J. M. MONTZ (1943), *Elements of Topographic Drawing*, 2d ed., McGraw-Hill, New York.

```
SUB "CPLOT" (LX,LY,z(,),N,L(),ROUGH)
INTEGER c,FRST,I,J,KA,KB,KC,K1,LAST,SMOOTH,T,Tnxt,XA,XB,XC,X1,YA,YB,YC,Y1
INTEGER LX1,LY1,NOSTRADDLE,LOOPEND,TAG(1000),VX(12),VY(12)
 REAL DUMMY,HX,HY,LEV,X,X0,Xnew,XCold,Y,Y0,Ynew,YCold,ZA,ZB,ZC,SER(1000)
SMOOTH=NOT ROUGH    Midpoint smoothing is default unless ROUGH specified as 1
LX1=LX+1 @ LY1=LY+1
FRST=LX1+2 @ LAST=LX1*LY1       Serial nos. of first and last elements of z(,)
                                                    Initialize shift vectors
   FOR J=0 TO 2 @ VX(J),VX(5-J),VY(J+1),VY((6-J) MOD 6)=J-1 @ NEXT J
                             Add DUMMY fringe at N, S and W of z(,) and fill SER()
                                 c is index counter for SER() from 1 to LX1*(LY1+1)
DUMMY=L(1)-1
FOR c=1 TO FRST-2 @ SER(c)=DUMMY @ SER(c+LAST)=DUMMY @ NEXT c          S & N
FOR c=FRST-1 TO LAST-LX STEP LX1 @ SER(c)=DUMMY @ NEXT c                  W
FOR Y=1 TO LY @ FOR X=1 TO LX @ SER((LX1)*Y+X+1)=z(X,Y) @ NEXT X @ NEXT Y
                                     Initialize TAG() to zero, LX1*LY1+1 values
FOR c=FRST-1 TO LAST+1 @ TAG(c)=0 @ NEXT c
FOR I=1 TO N                                           Trace all N contours
   LOOPEND=0                                    Set up cue for subprogram EXIT
   K1=FRST-2 @ X1=-1 @ Y1=1 @ LEV=L(I)+1e-006           Finesse integer data
NEXTROW:  ZA=SER(K1)                                           Left z value
NEXTCELL: ZB=SER(K1+1)                                        Right z value
   NOSTRADDLE=(ZA>=LEV)=(ZB>=LEV)                Contour does not cross AB
   IF NOSTRADDLE THEN GOTO NEXTENTRY               No entry by cell floor
   IF TAG(K1)=I THEN GOTO NEXTENTRY                     Loop terminates
         If entering contour is found then follow the contour until it closes
   HX=(LEV-ZA)/(ZB-ZA)        Fractional part of abscissa of point of entry
   X=X1+HX @ Y=Y1 @ T=5             Fix point of entry and set T to 5
   X0=X @ Y0=Y                      Remember point of entry to contour
   MOVE X,Y                             Move to entry point on AB
   XA=X1 @ XB=X1+1 @ YA=Y1 @ YB=Y1 @ XC=XB @ YC=Y1+1     Define 1st triangle
   KA=K1 @ KB=K1+1 @ KC=K1+2+LX           Label triangle C1 C2 C3
   ZC=SER(KC)                Assign serial number to the new vertex C
NEXTSIDE:                    Search for exit on CA or else on BC
   XCold=XC @ YCold=YC             Remember for midpoint interpolation
   IF (ZB-LEV)*(ZC-LEV)>=0 THEN GOTO EXIT_CA              Contour crosses CA
EXIT_BC: HY=(ZC-LEV)/(ZC-ZB) @ XA=XC @ YA=YC @ ZA=ZC @ KA=KC      crosses BC
   Tnxt=(T-1) MOD 6 @ J=(T+3) MOD 6       Is type number of next triangle
   GOTO CONTINUE
EXIT_CA: HY=(LEV-ZC)/(ZC-ZA) @ XB=XC @ YB=YC @ ZB=ZC @ KB=KC   Redo triangle
   Tnxt=(T+1) MOD 6 @ J=T                  Update type T and remember
CONTINUE: Xnew=XCold+SGN(XB-XA)*HY @ Ynew=YCold+SGN(YB-YA)*HY
   XC=XC+VX(J) @ YC=YC+VY(J) @ KC=1+XC+YC*LX1 @ ZC=SER(KC) @ T=Tnxt
   IF T=5 THEN TAG(KA)=I              Tag left end A of AB for type 5
   IF T=2 THEN TAG(KB)=I              Tag left end B of AB for type 2
   X=(Xnew+X)/2 @ Y=(Ynew+Y)/2                 Midpoint interpolation
   IF SMOOTH THEN DRAW X,Y ELSE DRAW Xnew,Ynew    Draw to midpoint if SMOOTH
   X=Xnew @ Y=Ynew                          Remember unsmoothed vertex
   IF KA=K1 AND T=5 THEN GOTO CLOSE ELSE GOTO NEXTSIDE    Check against entry
CLOSE:   DRAW X0,Y0 @ LOOPEND=1               Close loop and take note
NEXTENTRY: ZA=ZB @ X1=X1+1 @ K1=K1+1      Scan R for next gate, advance count
   IF X1<LX THEN GOTO NEXTCELL                          Moving  L to R
   Y1=Y1+1 @ K1=K1+1           Reset as for top of main loop & advance count
   IF Y1<LY+1 THEN X1=0 @ GOTO NEXTROW                  Next cellrow up
   IF LOOPEND=0 THEN GOTO EXIT       Map values do not reach the next level
NEXT I                                                       Next contour
EXIT: SUBEND
```

Figure 2-55   Program for contour plotting.

M. A. STULL, K. M. PRICE, L. D. D'ADDARIO, S. J. WERNECKE, AND C. J. GREBENKEMPER (1975), "Study of the brightness and polarization structure of extragalactic radio sources," *Astronom. J.*, vol. 80, pp. 559–569.

D'ARCY W. THOMPSON (1966), *On Growth and Form*, Cambridge University Press, England.

D. TICHENOR AND R. N. BRACEWELL (1973), "Fraunhofer diffraction of concentric annular slits," *J. Optical Soc. Amer.*, vol. 63, pp. 1620–1622.

FURTHER READING

K. R. CASTLEMAN (1979), *Digital Image Processing*, Prentice-Hall, Englewood Cliffs, N.J.

E. R. DOUGHERTY AND C. R. GIARDINA (1987), *Matrix Structured Image Processing*, Prentice-Hall, Englewood Cliffs, N.J.

E. R. DOUGHERTY AND C. R. GIARDINA (1987), *Image Processing—Continuous to Discrete*, vol. 1, Prentice-Hall, Englewood Cliffs, N.J.

D. E. DUDGEON AND R. M. MERSEREAU (1984), *Multidimensional Digital Signal Processing*, Prentice-Hall, Englewood Cliffs, N.J.

SAMUEL Y. EDGERTON, *The Renaisance Rediscovery of Linear Perspective*, Basic Books, New York, 1975.

PIERO DELLA FRANCESCA (1984), *De Prospectiva Pingendi*, Florence, Italy.

E. L. HALL (1979), *Computer Image Processing and Recognition*, Academic Press, New York.

M. KEMP (1990), *The Science of Art*, Yale University Press, New Haven.

J. S. LIM (1990), *Two-dimensional Signal and Image Processing*, Prentice-Hall, Englewood Cliffs, N.J.

W. K. PRATT (1978), *Digital Image Processing*, John Wiley & Sons, New York.

A. ROSENFELD AND A. C. KAK (1982), *Digital Signal Processing*, Academic Press, New York.

PROBLEMS

2–1. *Shading of sphere.* A white sphere of radius R with a matte surface that scatters light in accordance with Lambert's law sits on an infinite horizontal black plane and is illuminated by a uniformly bright hemisphere of sky. What would be the grey level, as a function of y and z, seen by a distant viewer on the (horizontal) x-axis? ☞

2–2. *Visual experience.*

    (a) Form a sheet of white notepaper into a cylinder by joining the short sides and stand it vertically in early morning sunlight outdoors. Examine the shadow zone. According to Lambert's law, the brightness should fall to one-half at an angle of incidence of sunlight equal to 60°. Is what you see consistent with this?

    (b) Fold a sheet of paper in two along the short direction, open the fold to a right angle, and stand it near the cylinder. When the sunlight falls within the right angle, describe the brightness distribution you see. Are the two sides equally bright and does direction of viewing matter? Why is the brightness not spatially uniform?

    (c) Is the shadow zone on the cylinder much affected by proximity of the angled screen?

    (d) Can you verify the blue component by looking into the full shadow through a hole?

(e) Does a color photograph of the assembly show any blue, or any other discrepancies with visual inspection? ☞

2–3. *Snow.* Look at a fall of fresh snow on an overcast day when the snow can be viewed in outline against cloud cover. The clouds look grey, but the snow looks white; but can even the whitest snow appear brighter than does its diffuse illuminant, the cloudy sky? ☞

2–4. *Grey shade boundaries.* It is claimed that the boundaries between zones of uniform dot density in Fig. 2-4(d) are not detectable. Of course, the word "uniform" refers to the population from which the dots are drawn at random, not to the actual outcome. The boundaries between zones are equally spaced horizontally, but the zones at the left and right ends are not exactly the same as the zones in-between. Try to discredit the undetectability claim by finding where the boundaries are. ☞

2–5. *Distinguishable greys.* A computer science student named Hal, who has yet to take a lab course, says, "My new PC has 256 grey shades and over a million colors." Ivan, who is a student in experimental psychology and eats in the same place, says, "There are only 20,000 different colors." A physics undergraduate volunteers that there are an infinite number of different wavelengths in the visible spectrum alone. It is conceivable that the advertised 256 grey shades are created from distinct dot patterns and that shades 137 and 138 are indeed distinguishable, but that the eye separates them by texture rather than by shade. With a view to persuading Hal to reconsider the grey-level count, construct a row of eight visually distinguishable grey squares, all of which have the same 50 percent shade, as verifiable experimentally by viewing from a distance. If your demonstration image is on paper, find out whether the critical viewing distance depends on illumination (whether you are indoors or out in the sun) or is set by the resolving power of the eye. If your demonstration is on a screen, find out whether ambient illumination makes a difference. ☞

2–6. *Grey-shade spacing.*
(a) An N-level grey scale ranges from $g_0 = 0$ (white) to $g_{N-1} = 1$ (black). Show that the grey shades g_n will be equally spaced if $g_n = n/(N-1)$.
(b) A software designer wishes to have the shades increase from g_1 to 1 by a constant factor k such that $g_{n+1} = kg_n$ so as to create a visual *perception* of equal spacing rather than have equal increments in coverage by black ink. What would the values of g_n be for $N = 8$?
(c) Comment on the value of this idea, with special reference to the fact that stimuli producing sensibly different perceptions often differ by roughly a decibel. If you think it desirable, construct the equal-increment scale for visual inspection.

2–7. *Funny log scale.* Quantities such as brightness are appropriately dealt with in logarithmic units because of the Weber-Fechner law. The use of photographic density D is an example of this. A useful scale of density might run from 0.01 to 1 in two decades. To show the unimportant range from 0.001 to 0.01 would require 50 percent more space, which is disproportionate, and to go to zero density a scale would have to reach to minus infinity. And yet zero density is important to us, and not perceived as infinitely remote – the Weber-Fechner law is no longer justified. It is proposed to introduce the funny log function defined by

$$\text{flog } x = \begin{cases} \log x & x \geq 1 \\ (x-1)/e & x < 1 \end{cases}.$$

For the case above we would use $\text{flog}(100 I_0/I)$, running from $I/I_0 = 0$ to $I/I_0 = 1$.
(a) Make a graph of $y = \text{flog } x$, $(0 < x < 10)$, for reference, with a graduated scale of $\text{flog } x$ $(0 < x < 1)$.

(b) Give any positive or negative comments on the utility of this idea. ☞

2–8. *Visual design.* Design a card, no more than three inches long, to be placed with geological and botanical specimens photographed in their natural habitat, so that dimensions can be read from the photograph. Allow for metric dimensions also. ☞

2–9. *Tops and bottoms.* An island resembling the one in Fig. 2-15 has T tops, B bottoms, and P passes (or saddles). Is it true that $T + B - P = 1$?

2–10. *Image properties.* Set up a 20×20 integer array by adding together ten Gaussian humps of various widths and/or orientations and various amplitudes in the range 1 to 10. Normalize your array so that the greatest element value is 99. Print your array at 10 characters per inch and 3 lines per inch in accordance with the typewriter standard.

   (a) To get a feel for the character of the terrain that you have created (treating your matrix as representing topography rather than photography, tomography, etc.) draw contours *by hand* at contour levels 10, 20, . . . , 90.

   (b) How many peaks are there and how many saddles? Put a △ at each peak and a × at each saddle.

   (c) Color the areas that are convex upward red and those that are concave upward blue (or crosshatch them differently).

   (d) Trace out some ridge lines in red and some watercourses in blue.

   (e) Comment on the transfer of information, paper to brain, of your artwork, relative to the information transfer from matrix printout to brain.

   (f) Experiment with any other image-processing technique that may be available to you such as machine contouring, grey-scale representation, staggered profiles, hachuring, etc. If you try a contour routine package, repeat with the same matrix flipped about an axis and note whether you can get the contours to alter.

   (g) Experiment with additive random digits ± 1; if 1 percent errors are present, how far is a contour likely to be shifted (digitizing interval = 1)? Comment on any wisdom you acquire.

   (h) Suppose that your map has north at the top but, like maps of the Stanford campus, it has to have north $17°$ away from the bottom. Redigitize your image, starting from the digits you have (not from the sum of Gaussians) to get a new 20×20 array that meets Stanford's requirements.

2–11. *Topographic features of matrix.* A large matrix of numerical values $X(I, J)$ representing a land surface is handed to you by your superior in your new job. He says, "I don't believe in newfangled things but the management has decreed that this department should be computerized. See if you can figure out how to extract the coordinates (I_m, J_m) of the hilltops and the (I_s, J_s) of the saddles." Later, you write out the following conditions: "Maxima: $X(I_m - 1, J_m)$ and $X(I_m + 1, J_m)$ and $X(I_m, J_m - 1)$ and $X(I_m, J_m + 1) < X(I_m, J_m)$ Saddles: $X(I_x - 1, J_s)$ and $X(I_s + 1, J_s) > X(I_s, J_s)$ and $X(I_s, J_s - 1)$ and $X(I_s, J_s + 1) < X(I_s, J_s)$, or conversely." Just as you are about to turn in your piece of paper, a fellow worker cautions you, "Be careful, the old boy has seen that before. He'll shred you." Discover situations that defeat your approach and think of something better to say. ☞

2–12. *Saddle point in matrix.* The brightness of an image is represented by a matrix of elements x_{ij}. A computer programmer is asked to find the saddle points and proposes the following. Run along all the rows until a maximum is found, excluding maxima in the first and last columns; if such a row maximum is smaller than its neighbors above and below in the same column, then we have found a saddle; interchange rows and columns and start again to find other saddles. The person responsible for digitizing the image complains, "Your procedure should be independent of the orientation of the camera. If I pointed the camera at the same scene, but rotated it 45 degrees before digitizing again, the rows and columns would be quite different and different results would be obtained for the reputed saddlepoints." Confirm or reject the criticism, giving an example. ☞

2–13. *Pseudorandom topography.* Generate a 10×10 array of 100 digits from the expression $IP(0.5 + 10 * FP(1000 \sin(n/10)))$ MOD 10), where IP stands for "integral part of" and FP stands for "fractional part of" and n ranges from 0 to 99. If $n/10$ is taken to be in degrees, the first row of the array will begin 0, 7, 5, . . . , and the tenth row will end with 5, 2, 9. (You may read these pseudorandom digits from a table of four-figure natural sines by taking the last figure of each four-figure entry in the ten rows from $0°$ to $9°$.) Smooth this random map by replacing each value by the sum of the nine digits centered on that value, assigning zeros to those of the nine that are outside the 10×10 field. The new array will be 12×12 because of the zeros assigned to the bordering fringe.
(a) Draw contours for 10, 20, 30, 40 and 50.
(b) Locate all the tops; how many are there?
(c) Comment on other features you have found. ☞

2–14. *Pseudorandom topography.* Repeat the previous problem using the base-ten logarithm function to generate a 10×10 array beginning 0, 3, 6, . . . and ending 5, 7, 9. Consider the usefulness of other presentations, such as staggered profiles and grey-scale renderings, for conveying a more intuitive feel for the topography represented.

2–15. *Hill climbing.* One way of finding the maxima of a function of two variables is called hill-climbing, by analogy with topography (hill climbing in fog is intended). The technique is often described in terms of continuous functions of two continuous variables. The previous problem, which involves an experiment with a matrix of only modest size, offers an opportunity to experiment with hill climbing on a matrix. Experiment with an algorithm that will find the tops, not worrying about computational efficiency, and report the results. ☞

2–16. *Convex regions.* For the 12×12 array generated as above from the sine function determine the boundary of the convex region and compare with the contour whose level is the mean value of the array elements. Comment on the finding.

2–17. *Convex regions.* Repeat the previous exercise for the 12×12 array based on the logarithm function.

2–18. *Topography.* The accompanying array of data (Fig. 2-56) describes the microwave sun as it was on March 10, 1970. Each number refers to the point midway between the units and tens digits, and the unit is 1000 kelvins.
(a) Superimpose contours at 20, 40, 60, ...
(b) Indicate all the tops with small triangles.
(c) Indicate the concave region by shading.
(d) Knowing that the map represents the sun, can you estimate approximately the element spacing in minutes of arc?

2–19. *Facet orientation.* The picture plane (x, y) is divided into equilateral triangles, one of which has vertices at $(0, 0)$, $(0.866, 0.5)$, $(0.866, -.5)$. Consider the triangular facet constructed by raising the three vertices above the picture plane to heights z_1, z_2, z_3, respectively. The normal to the facet is tilted at an angle χ to the z-axis, and the projection of the normal onto the (x, y)-plane makes an angle α with the x-axis. Show that the tilt χ and azimuth α are given by

0	0	0	0	1	2	4	7	10	12	12	12	14	17	20	18	19	22	21	12	3	-1	0	0	0
1	0	0	0	1	6	17	27	29	24	19	20	24	26	25	21	21	26	26	16	4	-1	0	0	0
2	1	4	1	6	14	34	57	67	54	39	34	37	38	35	29	29	28	24	12	2	-3	-1	-1	0
6	4	4	8	15	25	46	72	87	73	52	41	41	41	38	35	35	32	23	12	4	0	-1	-1	0
10	9	9	16	27	37	47	61	68	52	36	30	42	49	47	43	40	37	29	20	13	6	1	-1	0
11	13	16	29	53	50	47	43	36	25	22	33	53	70	73	65	50	41	33	28	20	12	4	1	-1
10	14	24	40	55	58	50	38	28	25	34	48	66	84	96	91	68	47	35	31	29	26	15	5	-3
9	15	30	47	58	57	50	42	35	39	41	46	50	69	90	94	77	53	36	27	34	46	40	19	-4
7	14	31	46	54	53	49	45	45	41	29	17	19	38	62	73	62	43	29	23	44	75	80	44	3
7	16	33	43	48	51	52	49	42	33	31	39	51	50	53	56	53	27	25	21	44	85	103	71	19
8	18	38	47	49	51	54	53	45	43	80	145	195	162	104	69	63	55	40	37	53	82	96	70	26
13	25	47	57	56	53	54	55	56	56	112	223	316	274	162	86	80	81	68	60	66	75	74	54	22
13	23	56	70	70	62	51	49	60	62	98	190	292	278	168	83	70	79	69	65	73	74	60	36	14
6	24	56	75	76	67	50	43	54	58	60	94	157	168	110	54	41	46	45	50	61	66	53	33	15
-6	9	41	71	74	68	50	40	41	50	45	41	56	66	55	35	24	23	25	34	45	49	41	27	14
-10	-5	21	51	60	54	45	41	35	38	36	33	35	42	45	38	28	25	28	36	35	31	24	16	9
-6	-9	5	29	43	42	38	38	35	32	30	31	34	41	51	50	40	33	34	35	29	20	14	7	4
-1	-5	1	11	24	27	28	31	33	31	30	30	32	37	45	43	35	27	28	27	20	11	6	3	2
0	-1	2	5	11	16	20	25	28	29	31	32	31	30	29	26	22	19	13	14	8	4	2	1	1
0	0	1	2	4	7	10	17	22	26	26	25	24	23	18	13	8	7	7	5	3	1	0	0	0
0	0	1	0	1	2	4	7	11	15	14	13	12	12	9	7	4	4	3	1	0	0	0	0	0

+   +

Figure 2-56   Digital map of the solar brightness at 9.1-cm wavelength on March 10, 1970.

$$\tan \chi = \sqrt{\frac{1}{3}(z_3 + z_2 - 2z_1)^2 + (z_3 - z_2)^2}$$

$$\tan \alpha = \sqrt{\frac{1}{3}\frac{z_3 - z_2}{z_2 + z_3 - 2z_1}}.$$

2–20. *Facet shading.* The radar brightness of a hill standing on an extensive plane, irradiated from above, is a function of surface tilt χ only. Let us say that the surface condition of this particular hill makes the brightness proportional to $\cos \chi$. The height of the terrain is given by $z = 0.4(x - 1)(y + 1)(x + 1)^2(y - 1)^3$ or $z = 0$, whichever is greater.
  (a) Simulate the radar image.
  (b) Comment on the comparison between the visual impression and the known surface relief.

2–21. *Extraterrestrial messenger.* If a messenger probe arrived in the solar system, the first thing we would wish to know is where it came from. The builders of the probe, anticipating our curiosity on this point, could satisfy us and at the same time totally circumvent language difficulties by equipping the probe with a television transmission picturing the constellation they lived in and arranging for their home star to flash on and off.

(a) How would we know that their radio signal was intended to be displayed on a TV screen?

(b) Might not the whole idea fail because their frequency of transmission, number of lines per inch, frames per second, and other parameters would be most unlikely to be compatible with our TV sets?

(c) Bearing in mind that we scan from top to bottom, left to right, and paint a picture 4/3 wider than it is high, how could one avoid getting the picture upside-down, back-to-front, or with the wrong ratio of vertical to horizontal scales?

(d) An array of $M \times M$ picture elements capable of showing stars of six different magnitudes contains $3M^2$ bits of information. However, if there are only N stars and most of the field is empty, it would be more economical just to give the coordinates of the stars (the number of bits being reduced to $2N \ln_2 M$) as a single radio message rather than go to television. These thoughts occurred to a radio amateur who intercepted an unusual radio message that was repeated many times at intervals and which he recorded as follows, using an arbitrary line length of 100 characters.

```
0000000000000000000000000000000000000000000000000000000000000000010000000000000000000000000000000000
0000000000000000000000000000000000000000000000000000000000000000000000000000000000000000000000000000
0000000000000000000000000000000000000000000000000000000000000000000000000000000000000000000000000000
0000000000000000000000000000000000000000000000000000000000000000000000000000000000000000000000000000
0000000000000000000000000000000000000000000000000000000000000000000000000000010000000000000000000000
0000000000010000001000000000000000000000000000000000000000000000010000000000000000000000000000000000
0000000000000000000000000000000000000000000000000000010000000000000000000000000000000000000000000000
0000000000000000000000000000000000000000000000000001000000000000000000000000000000000000000000000000
0000000000000000000000000000000000000000000000000000000000000000000000000000000000000000000000000000
0000000000000000000000000000000000000000000000000000000000000000000000000000000000000000000000000000
0000000000000000000000000000000000000000000000000000000000000000000000000000000000000000000000000000
0000000000000000000000000000000000000000000000000000000000000000000000000000000000000000000000000000
0000000000000000000000000000000000000000000000000000010000000000000000000000000000000000000000000000
000000000000000000000000000000000000000000000000
```

Working on the hypothesis that the digits represent a raster scan, construct the picture that the amateur arrived at. ☞

2–21. *Image invariants.* A land surface has been modeled in plaster by a surveyor in connection with a building development. When I tilt the model, the points on the model that were "tops" are no longer tops.

(a) What operation on an image is equivalent to tilting a topographic surface?

(b) Mention some properties of a surface that are not changed when it is tilted. ☞

2–23. *Centroid of image.* Let the centroid of an image $f(x, y)$ be at $(\bar{x}, \bar{y})$.

(a) Explain the circumstances under which each of the following is appropriate.

(i) $\bar{x} = M^{-1} \int \int x f(x, y) \, dx \, dy, \qquad \bar{y} = M^{-1} \int \int y f(x, y) \, dx \, dy,$
where $M = \int \int f(x, y) \, dx \, dy,$

(ii) $\bar{x} = M^{-1} \sum_1^N x_i m_i, \qquad \bar{y} = M^{-1} \sum_1^N y_i m_i,$ where $M = \sum_1^N m_i,$

(iii) $\bar{x} = N^{-1} \sum_1^N x_i, \qquad \bar{y} = N^{-1} \sum_1^N y_i.$

(b) A horizontal plane lamina balances on a knife edge not parallel to either x- or y-axis. Does this knife edge pass under $(\bar{x}, \bar{y})$? (c) Is the statistical concept of median likely to have useful application to images? ☞

2–24. *Depth perception.* The impression of a quiet landscape in Fig. 2-57 is created with a minimum of means, but how? Addition of only horizontal and vertical lines to a flat page creates

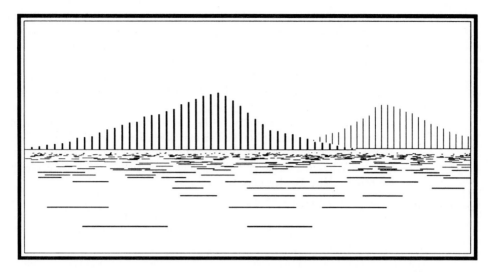

Figure 2-57   A drawing giving an impression of depth.

the perception of a receding horizontal plain and mountains at different depths. Does the frame contribute to the impression of depth by suggesting that you interpret the image as a picture? Is the obstruction of one mountain by another significant? Alternatively, is a distance cue encoded by the shading, and, if so, how does that work? Make photocopies and make alterations that might help answer these questions. Conceivably more than one cue may be relevant. For example, the left mountain is hatched with lines that are thicker and spaced further apart, has a baseline that is lower on the page, partly occults the other mountain, and has a subtly rougher skyline. All these four cues can be reversed; therefore it is possible to rank their influence on depth perception, and, in so doing, you may hit upon other factors. Invent other experiments, such as rotating the image, altering the viewing distance, etc.

2–25. *Perspective paradox.* Imagine ten thousand spheres, all 1 meter in diameter, spaced 5 meters apart between centers along a horizontal straight line. An artist says that if a photographer places his camera 20 meters from the line and photographs the spheres, the print will show the images of the spheres to be not only different in size but also in shape. What is more, she says, the nearer spheres will print narrower, left to right, than the remoter spheres. Intrigued by this, a photographer sets up his camera alongside a straight freeway and takes a photo of the traffic in the near lane, using a wide-angle lens so that he can get lots of cars. The wheels of the distant cars do not seem to him to be noncircular, and he sees no indication at all that the distant wheels are wider. Can both these experts be right? Explain.

2–26. *Circle in perspective.* In a perspective representation of coordinate axes the z-axis is shown vertical and the x- and y-axes are shown equally inclined to the horizontal, separated by an angle θ. A circle lying in the (x, y)-plane is to be shown as an ellipse.

   (a) Explain why the axis ratio b/a cannot be chosen at will but has to have a particular value to be compatible with the prior choice of θ.

   (b) What is the correct value of b/a, as a function of θ?

   (c) If you use a larger value of b/a, what will be the effect produced?

(d) In an isometric drawing θ is $60°$. What should the value of b/a be for a horizontal circle, and what is the angle of elevation above the horizontal of the corresponding line of sight?

(e) Is the answer to (b) dependent on the distance to the eye of the beholder?

(f) A centrally situated circle will appear as an ellipse whose major axis is horizontal on the page; is that true of circles anywhere else in the plane? ☞

2–27. *Correcting for perspective.* An oblique aerial photograph of an area of land is divided into 64×64 squares, and land parties are sent out to drive temperature probes to a depth of 100 mm at the center of each square. From a measured temperature matrix T_{mn}, which refers to a cartesian grid ruled on a perspective projection, how can one form a corrected matrix $T'_{\mu\nu}$ which refers to a square grid on the ground and thus eliminates the convergence and foreshortening characteristic of perspective views? ☞

2–28. *Orthographic projection.* Models of the five regular polyhedra, all fitting into the sphere of unit radius are placed on a horizontal plane, each resting on a face. Draw orthographic projections on the plane, adhering to the convention that invisible outlines are to be shown as broken lines. ☞

2–29. *Orthographic projection of cube.* The largest possible square tunnel has to be bored through a cube of unit side. Draw an orthographic projection of the cube looking along the axis of the tunnel and determine the length of the square side.

2–30. *Map projection.* A map has to be made of the area between latitudes equal to $30°$ and $40°$ N and $118°$ and $123°$ W.

(a) Construct a grid at $1°$ intervals using the transformation

$$x' = \rho \sin \theta$$
$$y' = \cot y_0 - \rho \cos \theta,$$

where x and y are longitude and latitude measured from their midvalues x_0 and y_0, $\rho = \cot y_0 + y_0 - y$, and $\theta = (x/\rho) \cos y$.

(b) Determine empirically whether the $1° \times 1°$ cells have equal area, whether the right angles between meridians and parallels have been preserved, and whether the height and width of the cells has been kept constant throughout. ☞

2–31. *Mosaic of distorted maps.* A fleet dispersed in the Atlantic between longitudes $-5°$ and $5°$ and latitudes $-5°$ and $5°$ is photographed from a synchronous satellite vertically above latitude $b = 0°$ and longitude $l = 0°$.

(a) Show that a ship at (b, l) appears on the photographic plate approximately at $x = k \cos b \sin l$, $y = k \sin b$, where x and y are measured to the east and north, respectively.

(b) Part of the area is obscured by cloud, but a second photograph is obtained subsequently from another satellite also above the equator but at longitude $-10°$. Think about how these two photographs could be pieced together. ☞

2–32. *Coordinate conversion.* A camera points horizontally to the south point of the horizon.

(a) Verify that the sky appears in the film coordinate plane (x, y) so that $x = f \tan az$ and $y = f \tan alt \sec az$, where f is the focal length of the lens, azimuth is measured west from south, and alt stands for elevation angle (known to navigators as altitude). Explain the factor $\sec az$.

(b) The sun, or other celestial object, moves from east to west along a path determined by its declination δ and will be found at a point depending on its hour angle H (the number of hours elapsed since the object crossed the meridian, multiplied by $15°$ per hour). Verify that

$$\text{alt} = \arcsin(\sin b \sin \delta + \cos b \cos \delta \cos H)$$
$$\text{az} = \arccos(\sin \delta \sec b \sec \text{alt} - \tan b \tan \text{alt}),$$

where b = latitude of the observer. Draw lines of constant δ at intervals of 5° from $-30°$ to 30° and lines of constant H at intervals of 5° from $-120°$ to 120°, for your own latitude, on a plot of altitude versus azimuth. This personal nomogram and the corresponding one looking north are worth keeping for occasional reference over the years.

(c) Why is the alt-az coordinate system recommended rather than the (x, y)-system? ☞

2–33. *Reciprocal polars.* The diagonal lines of the 5-dot pattern on a die intersect at right angles. What corresponding property is possessed by the two points constituting the poles of the two diagonals? ☞

2–34. *Perspective and affine.* Does perspective include the affine transformation? In other words, given two planes related by an affine transformation, can they be positioned in space so that the joins of corresponding points are concurrent?

2–35. *Affine nonlinearity.* Pearl says, "An operator is linear if the outcome of operating on the sum of two different inputs is the sum of the outcomes of operating separately on those inputs." She then asks isn't the affine transformation linear, explaining as follows. "A first operand is a circular blob at the origin; a given affine transformation distorts the blob in a certain way and displaces its center to let us say (ϵ_x, ϵ_y). A second operand is a square blob at $(3, 0)$; the same affine transformation rotates it and so on and displaces it by the same amount (ϵ_x, ϵ_y). These are the two separate outcomes; add these outcomes together. Now add the two separate inputs to get a composite image consisting of a circular blob and a square blob off to the right. Apply the affine transformation; surely the transform of the sum is the sum of the two separate transforms. Consequently it seems to me that the affine operator meets the test for linearity." Ruby looks dubious about this and then says, "I don't know much about this, but doesn't the definition of linearity say something about *for all inputs*? Maybe your example only acts linearly because the two inputs were compact. Had they overlapped, nonlinear interaction might have become apparent." Explain clearly who is wrong and where the error lies.

2–36. *Dual of Pascal's theorem.* Six points $ABCDEF$ on an ellipse, labeled in any order, define a hexagon. Label the intersections $(AE, BF), (BD, CE)$ and (AD, CF) as $P, Q,$ and R. According to Pascal's theorem, $P, Q,$ and R are collinear. Enunciate the dual theorem about six straight lines that are tangent to an ellipse.

2–37. *Numerical affine inversion.* An affine transformation is expressed by

$$x' = 1.2x + 0.1y + 1.3$$
$$y' = 1.1x + 0.15y + 2.$$

What are the inverse relations for x and y in terms of x' and y'?

2–38. *Transformed quadrilateral.* The six successive quadrilaterals in Fig. 2-58 were formed by repeated application of the same rule. If this rule was an affine transformation, determine the constants approximately; if not, what was the rule?

2–39. *Bilinear conformal transformation.* A locus $\phi(x, y) = 0$ on the (x, y)-plane is to be transformed into a locus $\Phi(u, v) = 0$ on the (u, v)-plane, where u and v are related to x and y by

Figure 2-58   A transformation to be determined from data.

the complex function

$$w = \frac{\alpha + \beta z}{\gamma + \delta z},$$

where $z = x + iy$ and $w = u + iv$. Show that for each point (x, y) the coresponding point (u, v) can be obtained from

$$u = \frac{(\alpha + \beta x)((\gamma + \delta x) - \beta \delta y^2}{(\gamma + \delta x)^2 - \delta^2 y^2}$$

$$v = \frac{(\beta \gamma - \alpha \delta) y}{(\gamma + \delta x)^2 - \delta^2 y^2}. \quad \text{☞}$$

2–40. *Two half-turns.* An image $g(x, y)$ is given a half-turn about the origin and then a second half-turn about the point (a, b). What is the resulting image $f(x, y)$? ☞

2–41. *Symmetry.* A function $f_1(x, y)$ has two fold rotational symmetry and $f_2(x, y)$ has four fold symmetry. What can be said about $f_1(x, y) + f_2(x, y)$? ☞

2–42. *Symmetry.* Two functions $f_1(x, y)$ and $f_2(x, y)$ both possess four fold rotational symmetry. What can be said about $f_1(x, y) + f_2(x, y)$? ☞

2–43. *Letter symmetry.* Devise an evolutionary explanation for the paucity of pairs of letters that are left-right (or up-down) mirror images of other letters.

2–44. *Letter symmetry.*
  (a) Examine all the letters of the alphabet using upright sans-serif capitals and sort them into groups in accordance with their symmetry.

ABCDEFGHIJKLMNOPQRSTUVWXYZ

   Arrange the groups in order ranging from least to most symmetry.
  (b) Find an English word whose image when you hold it up to a mirror is still an English word.
  (c) Find a word with the same property when it is printed top to bottom, as in some street signs.
  (d) Find a word which is still readable when you look at it in a hand mirror placed on the page parallel to the lines and perpendicular to the paper. ☞

2–45. *Emulate St. Cyril.* Invent an alphabet of 26 characters different from Latin and Greek letters or rotated versions of these or of themselves. To keep the shapes simple, make them capable of representation on a 5×7 matrix. ☞

2–46. *Symmetry of autocorrelation.* If a function $f(x, y)$ has n-fold rotational symmetry, what can

be said about the rotational symmetry of its autocorrelation function $f(x, y) \star\star f(x, y)$?
☞

2–47. *Flip definition.* A student says, "A function has n-fold rotational symmetry if it can be flipped about n different axes in the plane of the paper and still remain the same." Is that true? ☞

2–48. *Dimensionality of data.* While it is true that a two-dimensional 128×256 array of discrete data can be regarded as a one-dimensional string of 32,768 elements, is it possible that a doubly infinite set of function values defined continuously over the square area $0 < x < 1$, $0 < y < 1$ can be regarded as a function of a single continuous variable z that ranges over a finite range such as from 0 to 1? ☞

2–49. *Mystery projection.* What circularly symmetrical function $f(r)$ has a projection $(1 - |x|) \operatorname{rect}(x/2)$? ☞

2–50. *Projection of pyramid.* The function $p(x, y)$ represents the upper surface of a pyramid of unit height on a square base bounded by $x = \pm1$, $y = \pm1$. Find
(a) the zero-degree projection $\mathcal{P}_0 p(x, y)$ and
(b) the 45-degree projection $\mathcal{P}_{\pi/4} p(x, y)$. ☞

2–51. *Projection of a cone.* Find the projection of the right circular cone whose equation is $f(x, y) = (1 - |r|) \operatorname{rect}(r/2)$. ☞

2–52. *Projection of the normal probability function.* Find the projection of the circular probability distribution $(2\pi)^{-1}\sigma)^{-2} \exp(-r^2/2\sigma^2)$. ☞

2–53. *Projection of a hemisphere.* Show that the projection of the hemispherical boss described by $f(x, y) = (a^2 - x^2 - y^2)^{1/2} \operatorname{rect}(r/2a)$ is the parabola $0.5\pi(a^2 - x^2)$. Verify that the area under the parabola equals the volume of the hemisphere. ☞

2–54. *Projection of a paraboloid.* Find the projection of the paraboloid $f(x, y) = (a^2 - r^2) \operatorname{rect}(r/2a)$.

2–55. *Border zone.* A circle of unit radius rolls on the outside of an ellipse of width 3 and height 2.
(a) What is the curve traced out by the center of the rolling circle?
(b) Is this curve the locus of points lying at unit distance from the ellipse, or is the ellipse the locus of points lying at unit distance from the curve? ☞

2–56. *Projection as a function of two dimensions.*
(a) Instead of regarding the projection $g_\theta(R)$ as a function of R that is obtained while θ is kept fixed, suppose that we keep R fixed at a value R_1 and vary θ. Give an expression for $g_\theta(R_1)$ as a function of θ derived from $f(x, y)$.
(b) As $g_\theta(R)$ is a function of two variables R and θ, what is the reason that we do not write $g(R, \theta)$? ☞

2–57. *Projection of Bessel function.* Show that the projection of $J_0(r)$ is $2 \cos x$.

2–58. *Distortion.* An image $f(x, y)$ is distorted by moving the function value at (x, y) unchanged to another plane where the new coordinates are given by $\xi = x + y/10$, $\eta = y + x/10$. If the original image is a uniformly bright disc of unit diameter, describe the distorted image as to brightness and shape.

2–59. *Centroid.* The centroid of a scan $g_\theta(x')$ of a function $f(x, y)$ has an abscissa $\langle x' \rangle$ defined by

$$\langle x' \rangle = \frac{\int_{-\infty}^{\infty} x' g_\theta(x') dx'}{\int_{-\infty}^{\infty} g_\theta(x') dx'}$$

and the centroid of $f(x, y)$ is at the point where, if $f(x, y)$ were the area density of a plane lamina, the lamina could be balanced on a sharp point.

(a) Prove that the abscissa of the centroid of $f(x, y)$ equals the abscissa of the centroid of the scan when $\theta = 0$.

(b) What is the relation between the abscissa of the centroid of $f(x, y)$ and $\langle x' \rangle$ when $\theta \neq 0$?

2–60. *Areal median point.* The median abscissa $x_m(\theta)$ of a scan $g_\theta(x')$ is defined so that

$$\int_{-\infty}^{x_m(\theta)} g_\theta(x')dx' = \int_{x_m(\theta)}^{\infty} g_\theta(x')dx'.$$

What point is it in the original distribution $f(x, y)$ that projects into the median, regardless of direction? ☞

2–61. *3D projection.* An imaginary isotropic snowball has a density $f(x, y, z) = \exp(-\pi r^2)$. What is the projected density per unit area on the (x, y)-plane?

2–62. *3D projection.* A huge nugget thought to be of pure gold about the size of a football but filled with air pockets is x-rayed for inspection. Let the density distribution be $\rho(x, y, z)$. Can the image on the x-ray plate be represented as an integral of $\rho(x, y, z)$? ☞

2–63. *Silhouette.* A sculptor poses her subject on a piano stool and takes N photographs of his silhouette as the stool is rotated π/N degrees at a time. The prints are mounted on stiff cardboard, cut out to the outlines, and assembled into a solid resembling a collapsible, concertinalike, paper ornament. Then the sculptor fills the spaces with clay. How well does the final product resemble the sitter? Of course, the resolution is limited by the finite value of N, but what is the mathematical situation in the limit as $N \to \infty$?

2–64. *Projecting the back projection.* Since projection and back projection are in a sense inverse operations does it follow that

(a) $\mathcal{BP}_\theta f(x, y) = f(x, y)$ and

(b) $\mathcal{P}_\theta \mathcal{B} g(x) = g(x)$? ☞

2–65. *Image status.* Is the following entity a binary image? Is it continuous?

$$
\begin{array}{ccccc}
0 & 0 & 0 & 0 & 0 \\
0 & \frac{1}{2} & \frac{1}{2} & \frac{1}{2} & 0 \\
0 & \frac{1}{2} & 1 & \frac{1}{2} & 0 \\
0 & \frac{1}{2} & \frac{1}{2} & \frac{1}{2} & 0 \\
0 & 0 & 0 & 0 & 0
\end{array}
$$

2–66. *Three-mile limit.* In the days of muzzle-loading cannon, national sovereignty extended three miles out to sea, thus extending to the effective range of shore batteries. Clearly the sea border so defined is not the same shape as the coast, or zero-level contour, but is in some sense smoother. Let the coast be defined by a sequence of N points (x_i, y_i) not necessarily equidistant but chosen so as to reasonably define the coastline.

(a) Propose a simple algorithm that comparably defines the three-mile limit in terms of $2N$ points (u_i, v_i). Doubling the number of points is suggested in order to cope with capes.

(b) Test the simple algorithm and list imperfections that would need attention. ☞

2–67. *Degree of rotational symmetry.* The letters **I** and **O** both have twofold rotational symmetry, while **S** almost does, but not quite. These observations suggest the concept of degree of rotational symmetry of an image, or image segment, $f(x, y)$. Let $f_\theta(x, y)$ be what $f(x, y)$ becomes when rotated through θ. The integral $I = \int \int f(x, y) f_\theta(x, y) dx dy$ is periodic in

θ, with period $2\pi/n$ in the case that $f(x, y)$ possesses n-fold rotational symmetry. Discuss the merits of the parameter $\rho_R = (I_{\max} - I_{\min})/(I_{\max} + I_{\min})$ as a measure of rotational symmetry.

2–68. *Isochronism, dilation, and erosion.* Instead of using a pendulum as the periodic element, an inventor uses a precision, hardened steel ball that rolls on a curved guide. If the bob rolls along a cycloid instead of in a circular arc, the period does not depend on amplitude: it takes the ball the same time to roll down to the bottom of the guide no matter where it is released. The inventor finds the new clock to be better than the corresponding simple-pendulum clock; there is no air friction from the string and no out-of-plane deflection on a moving vehicle. The inventor realizes that rolling on a cycloidal track does not result in the center of the ball following a cycloid: the shape of the track has to be corrected. Although the difference is very small, at the end of the day time errors are magnified by $\sim 10^5$; consequently, a millisecond error counts.

(a) Explain how the problem relates to erosion and dilation.

(b) The parametric equations of the cycloid are $x = \phi + \sin\phi$, $y = 1 - \cos\phi$, $(-\frac{1}{2}\pi < \phi < \frac{1}{2}\pi)$ referred to an origin at bottom dead center. What form should the track take for a ball of radius 0.05? ☞

2–69. *Rotational symmetry.* What rotational symmetry is possessed by

$$f(x, y) = \text{rect}(x - 2)\,\text{rect}(y - 2)? \quad ☞$$

2–70. *Fiber optic TV.* A silica fiber is transparent enough at a wavelength of 1.33 μm to be able to transmit a light beam 100 km without amplification. The diode laser source can be switched on and off at 1 GHz so as to transmit 10^9 bits per second. With such a channel would it be possible to transmit a regular television program? ☞

2–71. *Notation.* "In this journal," said Yan, "there is an equation containing a term TRANSLATE $(k, l)\,f(i, j)$, which seems to mean what you get when you shift the digital image $f(i, j)$ by a vector amount (k, l), where k and l are integers." Tyan said, "That's just a waste of ink and goes against parsimony because it's the same as $f(i - k, j - l)$." Dick said, "It's OK, I see it all the time." What justification is there for such unwieldy notation? ☞

2–72. *Drawing techniques.* Select an illustration from Chapter 2 that you think would be a little difficult for you to draw yourself. Make your own black-and-white original, suitable for submission to a publisher, that is as close as you can get, in exact detail, to the chosen illustration. Write a program, use commercial software, or draw by hand.

2–73. *Tiled triangle.* The traditional tile design composed of white hexagons and black triangles has inspired an architect to create a custom design for a triangular floor (Fig. 2-59). A continuous transformation seems to have been applied to the traditional design. What is this transformation? It may be found helpful to extrapolate the design beyond the bounding triangle.

2–74. *Dot size.* The least dot spacing s that can be addressed on a particular laser printer is one centipoint (a hundredth of a point or 3.5 μm). Suppose that the dot diameter d is also one centipoint. The laser beam is programmed to draw a circle of diameter D. Show that the printed image falls within a circle of diameter $D + d + 2\sqrt{2}s\lfloor D/2\sqrt{2}s + 1\rfloor$, where $\lfloor x \rfloor$ is the integer next less than x. ☞

2–75. *Screen resolution.* The screen of a laptop computer is to have 3074×2048 pixels in the form of light-emitting solid-state devices formed, together with their circuitry, in silicon on a glass substrate 264×176 mm. Calculate some of the parameters of this display and compare with

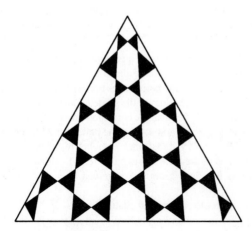

Figure 2-59   An artistically tiled triangle.

a screen that you are familiar with. For example, if the display is to be refreshed every one-thirtieth of a second, what will be the cycle rate in pixels per second? What will be the size of the pixels be? If the angular resolution of the eye is one minute of arc, how close to the screen will you have to be to distinguish the pixels? Compare the resolution in dots per inch with print on paper.

2–76. *Perspective.* An artist glancing at the daily computer-drawn record of a construction site says, "There is something wrong with the program because where you can see the open end of that pipe the oval rim doesn't look perpendicular to the pipe." A mathematics student says, "Each generator of the cylindrical surface is perpendicular to the plane of the rim in 3-space, but right angles do not in general project into right angles in 2-space. That's why the outline of the cylinder is not exactly perpendicular to the major axis of the ellipse." A mechanical engineering student says, "Yes, but there are two sides of the pipe, inclined equally to the axis. Michelangelo means that the rim should look perpendicular to the axis," which elicits the following. "Remember two things: the axis of the cylinder, if it were visible, would not coincide exactly with the line joining the centers of the elliptical ends; also the major axis of the ellipse does *not* in general coincide with the chord joining the points of tangency of the straight outlines to the ellipse." Sort out the true and false statements.   ☞

3

Two-Dimensional Impulse Functions

Just as in one dimension we need entities that are concentrated at a point in time, so in two dimensions we find it useful to introduce corresponding entities that are concentrated at a point in the (x, y)-plane. But in two dimensions there are two generalizations, the second being concentration on a line. One class of examples is furnished by the familiar mechanical notion of pressure; in mechanics, the oldest branch of physics, the concept that we require is well established in connection with the theoretical idea of a point force or point load applied to a surface. We recall that in one dimension the term *impulse* itself derives from mechanics, where it signifies the time integral of a force that is applied for a time that is more or less short. It is therefore useful, in introducing the theory of impulse functions on the plane, to appeal to the subject of mechanics, where the essential concepts are already familiar. First we deal with the two-dimensional point impulse or dot and then with a variety of impulse arrays and with impulsive entities that are not restricted to a point but are distributed along straight or curved lines. The δ-notation of P. A. M. Dirac (1902–1987) generalizes to allow convenient representation of the two-dimensional concepts.

THE TWO-DIMENSIONAL POINT IMPULSE

The point load on a surface, and the point charge of electricity on a surface, are established ideas of the kind to be discussed; point masses and point charges in space are the three-dimensional equivalents.

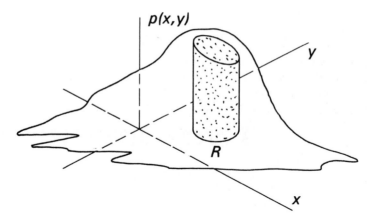

Figure 3-1 A cyclinder of sand suggesting a pressure distribution $p(x, y)$ over the (x, y)-plane applies a downward force $F = \int \int_R p(x, y)\, dx\, dy$ to the region R.

Point Load in Mechanics

Let $p(x, y)$ be the pressure distribution over the (x, y)-plane; then of course the force applied to a region R is given by $\int \int_R p(x, y)\, dx\, dy$ (Fig. 3-1). The pressure might be measured in newtons per square meter and the force in newtons. If a force F is applied at the point (a, b) how would one describe the distribution of pressure? We would have to say that the pressure is zero everywhere except at (a, b). And some might say that the pressure is infinite right at the point of application of the force, sacrificing quantitativeness for picturesqueness. The two-dimensional unit impulse symbol $^2\delta(x, y)$ furnishes convenient notation for this situation, where we write

$$\text{pressure} = F\,^2\delta(x - a, y - b).$$

The symbol $^2\delta(x, y)$ itself stands for the pressure distribution due to a unit force applied at the origin $(0, 0)$ and $F\,^2\delta(x, y)$ represents the pressure distribution set up by a force F at the origin. Since the integral of the pressure distribution must equal the applied force F,

$$\int_{-\infty}^{\infty} \int_{-\infty}^{\infty} F\,^2\delta(x, y)\, dx\, dy = F.$$

From this requirement it follows that

$$\int_{-\infty}^{\infty} \int_{-\infty}^{\infty} {}^2\delta(x, y)\, dx\, dy = 1$$

is a necessary property of the two-dimensional unit impulse as introduced above.

If x and y are measured in meters, then $^2\delta(x, y)$ cannot be dimensionless but must have units m^{-2}. This is an important property to bear in mind when calculations are being checked, and is analogous to the property of $\delta(t)$, whose unit is s^{-1}.

Point Charge

A force applied at a point, or a point load, is commonplace in engineering but is not the only example of a two-dimensional impulse. Electric charge distributes itself on surfaces with a charge density $\sigma(x, y)$ measured in coulombs per square meter, and a charge of Q coulombs concentrated at a point is a special but common example of a theoretical surface charge distribution. If the charge Q is at $(0, 0)$ on the (x, y)-plane we may write

$$\sigma(x, y) = Q\,^2\delta(x, y).$$

As in the previous example,

$$\int_{-\infty}^{\infty} \int_{-\infty}^{\infty} \sigma(x, y)\, dx\, dy = Q \int_{-\infty}^{\infty} \int_{-\infty}^{\infty}\,^2\delta(x, y)\, dx\, dy = Q.$$

The extension to three dimensions is obvious. Thus the volume charge density $\rho(x, y, z)$ measured in coulombs per cubic meter associated with a charge Q at $(0, 0, 0)$ may be written $Q\,^3\delta(x, y, z)$. A mechanical example would be a point mass M at (a, b, c), a situation describable as a mass density distribution $M\,^3\delta(x - a, y - b, z - c)$ kilograms per cubic meter.

These examples remind us that point entities are commonly introduced without fanfare in applied mathematical subjects and are not a cause of mathematical difficulty. On the contrary, the purpose and the result of the approach is to achieve simplification. Other examples such as magnetic poles and point sources of light and sound spring to mind.

Point Load Interpreted as a Sequence

We understand very well, of course, that a "point" load at the center of a simply supported beam (Fig. 3-2), to quote a textbook situation, does not produce infinite pressure. Even if the load is applied by a polished steel sphere in contact with a hardened steel plane, the pressure is not going to go much above 10^9 Nm^{-2}, which is far from infinite. If we wished to study the microscopic deformation of convex elastic surfaces forced together, we could do so. The textbook problem is, however, aimed at a quite different deformation, namely the flexure of the supporting beam, a very much larger effect than local compression at the "point" of contact. The point load approach is implicitly stipulating that the area of contact is small, but only small enough that the precise contact area makes no significant difference to the flexure of the beam.

Suppose that the central load W is composed of three bricks as in Fig. 3-3. If they are arranged end-to-end the curve of deflection is not the same as if a *point load*, also of amount W, is applied at the center; but if the three bricks were *stacked* at the center, the agreement with point loading would be closer; and if the three bricks were stacked *on end* at the center, perhaps the agreement would be within the accuracy with which the deflection is needed. But what could one mean by the term "point load," bearing in mind that a given load applied to a sufficiently small area might raise the pressure to the ultimate strength of the materials and destroy the beam (or the load)? Evidently we appreciate the fact that, long before the area is reduced to where contact pressure builds up to levels

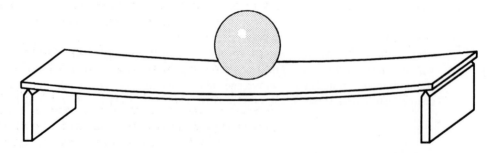

Figure 3-2   A point load W at distance x from the center of a simply supported beam of length L produces a deflection $Wx \times (3L^2 - 4x^2)/48EI$ at a distance x from the left support, where E is the Young's modulus of the material of the beam, I is the moment of inertia of the cross-section, and $0 \le x \le L/2$. The exact deformation at the point of contact is not significant in this problem.

that change the problem, the deflection settles down, for all intents and purposes of the investigation, to a steady curve. Once this steady result is reached, it would not matter if the load were stacked somewhat higher on a more concentrated area. The thing that counts is the integrated load W. The impulse notation enables us to gloss over the irrelevant detail of the pressure distribution and to focus on the total load and location, parameters that are meaningful for the deflection problem.

Exactly the same philosophy applies when we calculate the effects caused by point charges, point masses, and point sources of light. We mean that the degree of concentration is enough that further concentration would make no further significant difference to the effect under consideration. Just how concentrated things need to be depends upon the circumstances of each problem; these details are exactly what we wish to avoid. The impulse symbol allows us to represent point sources, point charges, and their kin, with an economy of notation appropriate to the problem.

In Fig. 3-3 the load distributions are rectangular, as expressible by $\tau^{-1} \operatorname{rect}(x/\tau)$ for a sequence of diminishing values of τ, the width of the base on which the fixed weight rests. In two dimensions we think of a base of area τ^2, a height proportional to τ^{-2}, and a fixed volume as a sequence of distributions is generated as τ diminishes.

Rectangle Function Notation

By analogy with the gate function of signal theory, which selects a finite-duration segment of a waveform, it is useful to be able to select a segment of the picture plane. In one dimension, the unit rectangle function $\Pi(x)$, or $\operatorname{rect} x$, is customary for this purpose and is defined by

$$\operatorname{rect} x = \begin{cases} 1, & |x| < \frac{1}{2} \\ 0, & |x| > \frac{1}{2}. \end{cases}$$

The two-dimensional generalization is

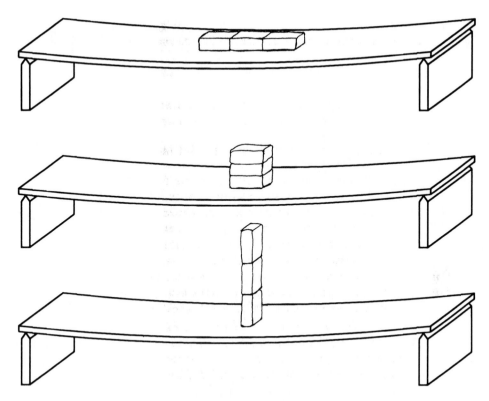

Figure 3-3   A sequence of pressure distributions suitable for defining a point load W. The applied pressure increases in theory *without limit* as we run through the sequence of taller stacks piled on smaller bases but the response of the plank approaches a limit.

$$^2\text{rect}(x, y) = \text{rect } x \text{ rect } y.$$

With this notation we can use

$$^2\text{rect}\left(\frac{x - a}{X}, \frac{y - b}{Y}\right) f(x, y)$$

to select that part of $f(x, y)$ centered at (a, b) within width X and height Y and reduce anything outside to zero.

RULES FOR INTERPRETING DELTA NOTATION

The concept of a sequence of pressure distributions producing a sequence of effects that approach a limit provides a three-step basis for interpreting mathematical expressions containing impulses.

1. Wherever we see $^2\delta(\xi, \eta)$ we first replace it by $\tau^{-2}\,^2\text{rect}(\xi/\tau, \eta/\tau)$, a unit-volume function of height τ^{-2} sitting on a square base of area τ^2.

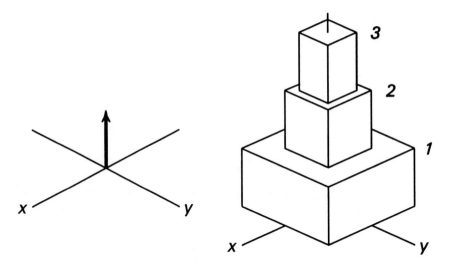

Figure 3-4   An arrow of unit height is the conventional representation of $^2\delta(x, y)$. The concept is defined by a sequence of functions 1, 2, 3, ... Note that the $^2\delta(x, y)$ is not the limit of the defining sequence which, at the interesting point $(0, 0)$, has no limit.

2. Next we evaluate the expression, obtaining in general a result that depends on τ.

3. Finally we ask whether this result approaches a limit as $\tau \to 0$, i.e., as the unit-volume function runs through a sequence of forms, becoming even higher and concentrating on a smaller and smaller base centered on $\xi = 0, \eta = 0$. If such a limit exists, we take it as the meaning of the original expression containing the impulse.

The sequence of load distributions, charge distributions, brightness distributions, may not itself have a limit as the parameter τ approaches zero. Indeed the central value may increase without limit toward infinity (Fig. 3-4). In spite of this, the deflection, or the electric field, or the illumination produced—that is, the sequence of consequences of the sequence of causes—may settle down to a finite value. This fact is what gives utility to the terms point load, point charge, and point source of light, even though the entities described are physically impossible. The point response function, point spread function, impulse response, Green's function, whatever it is called in various fields, is itself a finite and eminently observable thing. Introduced into classical electromagnetism by George Green (1793–1841) in 1828, the concept continues to play a role in quantum electrodynamics, where the term *Green's function* remains current.

Rule 1 calls for a unit rectangle function but could instead use the unit-area Gaussian function $\exp(-\pi x^2)$. That would be theoretically convenient for coping with situations where it might be necessary to differentiate, seeing that $\exp(-\pi x^2)$ has derivatives of any order; but in particular cases it is easier to integrate a product when one of the factors is constant over a finite range and zero outside those limits. In the occasional case where first-order differentiation is required, using a unit-area triangle function instead of a rectangle function will suffice.

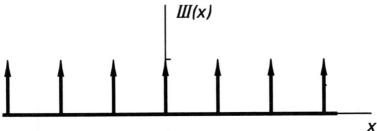

Figure 3-5   The shah function III(x).

GENERALIZED FUNCTIONS

In the forties and fifties, delta function notation, introduced by Dirac in his famous book on quantum mechanics, was frowned on by some mathematicians, but the underpinning has been put in place by respected authors and it is no longer defensible to complain about the δ-notation as such. The basic difficulty is that δ-functions are not functions as introduced in mathematical analysis, and therefore one cannot logically apply concepts of calculus to these entities that do not form part of the structure of calculus. A completely new and independent groundwork must be built. See Temple (1953) for a strict but readable account and Lighthill (1958) for a good book which systematically uses Gaussian sequences. The approach in terms of sequences of rectangle functions (FTA, 1986, chap. 5) has the virtue of being in the continuous tradition of physics and is rigorous mathematically. For all practical purposes, including applications in two or more dimensions, the rectangle-function sequence has disposed of the mathematical dilemma, is teachable, and is productive of results. Occasionally, where derivatives arise, sequences of triangle functions (or higher self-convolutions of the rectangle function) may be helpful; but to go to the Gaussian function on the grounds that it possesses all derivatives, even those higher than the fourth, does seem like overkill. If one really means to evaluate the integrals, integrands containing rectangle functions are much easier to handle, (a) because the limits of integration are always finite, and (b) because the integrand within those limits remains as simple as it already was. The rectangle-function technique also generalizes readily to two dimensions.

THE SHAH FUNCTIONS III AND ^{2}III

In one dimension the shah function performs a dual role, sampling at regular intervals and representing periodic functions. In two dimensions the situation is much richer. To start with, sampling, which arises through *multiplication* with the shah function, can take place not only on a regular two-dimensional lattice of points but also on a regular grille of parallel straight lines. In addition one can have sampling on spokes regularly spaced in angle or on circles regularly spaced in radius. There is also sampling on a lattice of nonuniformly spaced points, such as sampling points in polar coordinates. The representation of periodic functions, which arises from *convolution* with the shah function, also leads to much variety, as seen in later chapters.

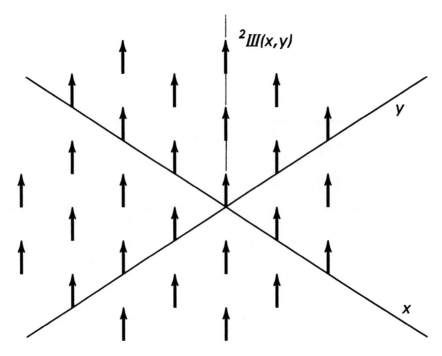

Figure 3-6   The bed-of-nails function or two-dimensional shah function $^2\text{III}(x, y)$.

Let us recall that

$$\text{III}(x) = \sum_{n=-\infty}^{\infty} \delta(x - n).$$

We can introduce a two-dimensional shah function, or bed-of-nails function, defined by

$$^2\text{III}(x, y) = \sum_{m=-\infty}^{\infty} \sum_{n=-\infty}^{\infty} {}^2\delta(x - m, y - n).$$

The product of $f(x, y)$ with $^2\text{III}(x, y)$ is a set of spikes whose strengths are sample values of $f(x, y)$ at unit intervals of x and y. If we wish to have samples spaced X and Y, we write

$$\frac{1}{|XY|} \, ^2\text{III}(\frac{x}{X}, \frac{y}{Y}) f(x, y).$$

To understand this expression it is necessary to remember that the shah function obeys a scaling rule analogous to $\delta(x/a) = |a| \, \delta(x)$. Thus

$$^2\text{III}\left(\frac{x}{X}, \frac{y}{Y}\right) = |XY| \sum_{m=-\infty}^{\infty} \sum_{n=-\infty}^{\infty} \delta(x - mX, y - nY).$$

Consequently, in order to represent a two-dimensional array of *unit* delta functions, it is necessary to write $\mid XY \mid^{-1} \text{III}(x/X, y/Y)$. When we come to take transforms, the factor associated with the scaling will be found to behave exactly as we would wish.

One might like to limit attention to a finite set of point impulses. Take, for example, the 64 unit impulses indexed from 0 to 7 in both dimensions. This set could be expressed as a product of $^2\text{III}(x, y)$ with a suitably sized and positioned rectangle function as follows:

$$^2\text{III}(x, y) \; ^2\text{rect}\left(\frac{x - 3.5}{8}, \frac{y - 3.5}{8}\right).$$

A one-dimensional row of impulses along the x-axis, represented by $\text{III}(x)\delta(y)$, is discussed below.

Logically there is a third generalization of the shah function, namely an impulsive grille, an entity that is zero everywhere except along the lines $y = $ integer. We shall see that $\text{III}(y)$ represents something that is just right for this. But first it is necessary to look into single line impulses and especially into how to express their strength.

LINE IMPULSES

In two dimensions phenomena are richer than in one dimension. Thus, in addition to the point impulse we have, in two dimensions, line impulses, which are needed to describe concentrations along lines, straight or curved. Typical examples would be a line charge and an illuminated slit. With a line charge, what matters is the number of coulombs per meter; we have in mind a narrow band to which the charge is confined, a band so narrow that the effects produced would be essentially the same if the band were made even narrower, only provided that the same number of coulombs remained packed on each meter of length. With a slit it is the quantity of light per meter that counts. The slit will look the same to the distant eye, in appropriate circumstances, if it is actually made narrower, provided that the brightness is correspondingly increased to maintain the amount of light transmitted per meter of slit. A related optical feature is a grey line on a photographic transparency.

Just as a mechanical load applied to a microscopic area may ultimately damage the material, so new electrical phenomena set in as the degree of concentration increases. Electric charges of the same sign repel each other, and it may be very difficult to bring mechanical forces to bear to compress a band of charge within an even narrower band. Furthermore, electric field strength is proportional to surface charge density, which means that one cannot get much above $10^{-5}\,\text{C}\,\text{m}^{-2}$ without tearing apart the adjacent air molecules; and other limits are set by physical properties of the other materials of which the world is composed. These phenomena are mentioned in detail as a reminder that corresponding phenomena will be encountered in every physical problem, though the details will be different. If you were successful in your attempts to make a point mass, the material would collapse into a black hole and disappear (possibly taking you with it).

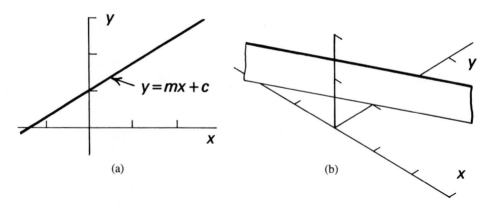

Figure 3-7   (a) A line charge $\delta(y - mx - c)$ on the (x, y)-plane lying on the line of slope m and y-intercept c. (b) A representation showing the line charge as a blade of height equal to the strength [in this case $\cos(\arctan m)$ or $1/\sqrt{1 + m^2}$].

The Straight Line Impulse

It will be sufficient at this stage to deal with the uniform straight line impulse. Suppose we wish to represent a line charge having a certain number of coulombs per meter along the straight line $y = mx + c$ (Fig. 3-7). The surface charge density $\sigma(x, y)$ would be represented in the form

$$\sigma(x, y) = k\delta(y - mx - c).$$

Interestingly, it has not been necessary to use the area impulse ${}^2\delta(\ ,\ )$, because the ordinary one-dimensional impulse suffices. To see this, consider what meaning should be attached to

$$\sigma(x, y) = k_0\delta(x),$$

where k_0 is a coefficient whose meaning will now be ascertained. Evidently we have a charge distribution that is independent of y and confined to where $x = 0$, in other words a uniform line charge along the y-axis. What is its strength in coulombs per meter? Consider the length L lying between $(0, 0)$ and $(0, L)$; we will calculate the charge Q_L there and divide by L. Thus

$$Q_L = \int_0^L \int_{-\infty}^{\infty} k_0\delta(x)\, dx\, dy$$

$$= \int_0^L k_0\, dy = k_0 L$$

and $Q_L/L = k_0$. So k_0 is the linear charge density in $\mathrm{C\, m}^{-1}$. We see that the units check,

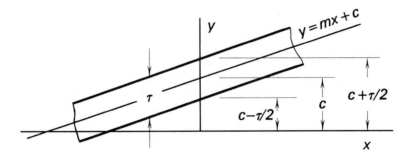

Figure 3-8 The two straight lines $y = mx + c \pm \tau/2$ are separated by a distance τ, measured *vertically* on the page.

because $\delta(x)$ has units m^{-1} and $k_0 \delta(x)$ therefore has units $\text{C}\,\text{m}^{-2}$, which is correct for the surface charge density $\sigma(x, y)$.

We have found that $\delta(x)$, interpreted as a function of the two variables x and y, is a uniform line impulse on the y-axis and that it has unit strength. Unit strength for line impulses means that an area integral $\int\int \ldots dx\,dy$ embracing unit length will be unity. Thinking of an integral of $f(x, y)$ over a specified area as the volume situated on that area, we can say that $\delta(x)$ has unit "volume under it," per unit length.

Correspondingly $\delta(y)$ represents a uniform unit-strength line impulse along the x-axis. By a coordinate transformation we could apply these conclusions to the question of $\delta(y - mx - c)$. We would expect to be dealing with a line impulse confined to the line $y - mx - c = 0$, which we arrive at by equating the argument of $\delta(\cdot)$ to zero. In general we can always say that $\delta(\text{something})$ is located where that something equals zero; ascertaining the strength is done separately.

We may now ask what the strength is and in particular whether it is unity. Let us answer this question by taking the area integral over unit length. A direct way of proceeding is to apply the following rules for interpreting delta symbols (FTA, 1986, p. 70), in this case one-dimensional rules, namely

1. Replace $\delta(\cdot)$ by $\tau^{-1}\,\text{rect}(\cdot/\tau)$
2. Perform the operation indicated
3. Proceed to the limit as $\tau \to 0$.

Making the replacement, we have $\tau^{-1}\,\text{rect}[(y - mx - c)/\tau]$. Now rect u falls from unity to zero where $u = \frac{1}{2}$ and also where $u = -\frac{1}{2}$. Therefore the lines

$$(y - mx - c)/\tau = \tfrac{1}{2} \qquad \text{and} \qquad (y - mx - c)/\tau = -\tfrac{1}{2}$$

are the boundaries of a strip within which the expression $\tau^{-1}\,\text{rect}[(y - mx - c)/\tau]$ represents a value of τ^{-1} and outside which it represents zero.

The width of the strip, in the y-direction, is evidently τ, as shown in Fig. 3-8. The width measured perpendicular to the direction of the strip is less than this and is given by $\tau \cos \alpha$, where $\alpha = \arctan m$. The perpendicular cross section is a rectangle of base $\tau \cos \alpha$

and height τ^{-1} and therefore has area $\cos\alpha$. The volume under a length L is $L\cos\alpha$ and the strength, as measured by volume under unit length, is just $\cos\alpha$, a result that is the same regardless of the value of τ. Taking the limit as $\tau \to 0$ therefore leaves us with the conclusion that $\delta(y - mx - c)$ is indeed a uniform line impulse along the straight line $y = mx + c$ but that the strength is less than unity. This may have been unexpected so we should check this conclusion. Put $c = 0$. As m becomes very small, $\cos\alpha \to 1$, and the strength approaches unity. Of course, the expression is approaching $\delta(y)$, which we know to be a unit-strength line impulse along the x-axis $y = 0$. Going the other way, letting m become large (while keeping $c = 0$), our expression approaches $\delta(-mx)$, which is equal to $|m|^{-1}\delta(x)$, which we know to be a uniform line impulse on the y-axis of strength $|m|^{-1}$. The calculated strength $\cos(\arctan m)$ equals $1/\sqrt{1 + m^2}$, which, for large m, approximates $|m|^{-1}$. So the conclusion is verified that the strength need not be unity. This elementary example reminds us that intuition may be misleading when we are confronted with delta notation; fortunately, the three rules for interpretation are simple and always work.

REGULAR IMPULSE PATTERNS

The two-dimensional shah function $^2\text{III}(x, y)$ is the prime example of a regular pattern of impulses in two dimensions. Two other repetitive structures are the impulse grille, a set of parallel equispaced line impulses, and the picket fence or row of collinear equispaced point impulses. Finally, there are many regular arrangements, such as arise in crystals and tiling patterns, that may be expressed in impulse notation and applied in sampling theory and spectral analysis.

Impulsive Grille

Consider a set of unit-strength line impulses concentrated on the lines $x = 0, \pm 1, \pm 2, \ldots$. The line impulse lying on $x = 0$ is represented by $\delta(x)$, and the full set is $\sum_{n=-\infty}^{\infty}\delta(x - n)$, which is simply $\text{III}(x)$. In this context, $\text{III}(x)$ is understood as a function of two variables that is independent of y. The rules of interpretation, if applied to $\sum_{-\infty}^{\infty}\delta(x - n)$ generate a finite function $f_\tau(x, y) = \sum_{n=-\infty}^{\infty}\tau^{-1}\text{rect}[(x - n)/\tau]$. The sequence as $\tau \to 0$ defines the impulses on the separate lines of the grille.

Row of Point Impulses

A row of unit impulses $\sum {}^2\delta(x - n, y)$ situated on the x-axis at integral values of x as in Fig. 3-9 plays a role where regular samples are taken along a cross section of a function. In such a case the row of impulses enters as a factor in a product. Although two-dimensional convolution is yet to be introduced, it is convenient to record here that, as a factor in a convolution, the row of impulses generates regularly spaced replicas along a straight line.

In addition to thinking of the row of impulses as a sum of terms we may think of it as a product of a shah function of one coordinate and a delta function of the perpendicular coordinate:

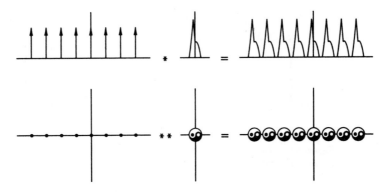

Figure 3-9 The row of impulses and its role in representing regular replication along a straight line, in one dimension (above) and in two dimensions (below).

$$\sum_{n=-\infty}^{\infty} {}^2\delta(x-n, y) = \text{III}(x)\delta(y).$$

The first factor is an impulse grille, the second is a line impulse crossing it at right angles. Confirmation of the relation can also be demonstrated by substituting rectangle functions in the usual way. This symbology is helpful in manipulating expressions in which the row of impulses occurs.

Lattices, Grilles, Spokes and Bull's-eyes

Patterns of point impulses arranged in regular lattices such as hexagonal lattices, patterns of straight line impulses arranged in grilles or on radial spokes, nested circles, and other curves arise frequently and are described in a later chapter.

INTERPRETATION OF RECTANGLE FUNCTION OF $f(x)$

Most applications of the rectangle function are in terms of a simple variable, e.g., in such forms as $\text{rect}\, x$, $\text{rect}(w/W)$, $\text{rect}(t-T)$. One interprets such expressions virtually at sight by making use of general experience with scaling and shifting of the independent variable. In other words, once one has assimilated the definition and the associated graph it is not *necessary* to refer back to the definition when interpreting $h\,\text{rect}[k(t-T)]$; instead one sees immediately a rectangle function of height h, compressed by a factor k horizontally, and shifted by an amount T to the right along the t-axis (Fig. 3-10). This shortcut approach is effective in most occurrences of rectangle function notation.

However, when we first encounter forms such as $\text{rect}(\sin t)$ or $\text{rect}(t^2-1)$, it is necessary to refer back to the definition, although here also quick ways exist of seeing what is intended. Taking these two examples formally, we write

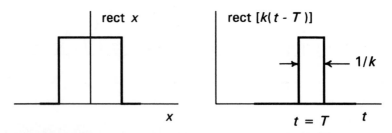

Figure 3-10   The unit rectangle function rect x (left) and a rectangle function of a simple expression (right).

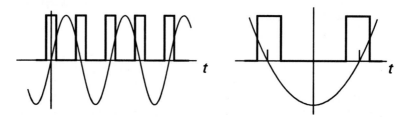

Figure 3-11   Illustrating rectangle functions of a sine wave and of a parabolic waveform.

$$\text{rect}(\sin t) = \begin{cases} 1, & |\sin t| < \frac{1}{2} \\ 0, & |\sin t| > \frac{1}{2} \end{cases}$$

$$\text{rect}(t^2 - 1) = \begin{cases} 1, & |t^2 - 1| < \frac{1}{2} \\ 0, & |t^2 - 1| > \frac{1}{2}. \end{cases}$$

Converting these statements to graphical form (Fig. 3-11) we see that rect$(\sin t)$ represents a regular train of positive unit-height rectangle functions. The frequency of the periodic train of pulses is twice the frequency of $\sin t$. The expression rect$(t^2 - 1)$ represents a pair of rectangle functions which, on careful inspection, will be found *not* to be centered exactly on $t = \pm 1$.

In the above discussion the name of the independent variable was imperceptibly shifted from x to t in order to distract attention from a point that commonly causes error. Clearly the definition of rect$[f(x)]$ must be

$$\text{rect}[f(x)] = \begin{cases} 1, & |f(x)| < \frac{1}{2} \\ 0, & |f(x)| > \frac{1}{2}, \end{cases}$$

where the two inequalities on the right are written in terms of the argument of rect$(\cdot)$, i.e., everything that comes between the parentheses following the rect; in the case of rect$[f(x)]$, $f(x)$ is the argument of rect$(\cdot)$. Thus the conditions $|x| > \frac{1}{2}$ and $|x| < \frac{1}{2}$ that appeared earlier would not be correct here. A procedure for rapidly handling all cases is illustrated in Fig. 3-12. First graph, or imagine a graph of, the argument $f(x)$ and draw in

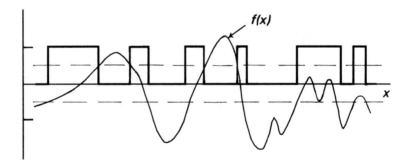

Figure 3-12   Illustrating the rectangle function of an arbitary function.

two horizontal lines at distances 0.5 above and below the horizontal axis. The intersections with $f(x)$ fix the edges of unit-height rectangle functions. Some examples to practice on are given later in the problems.

The illustration shows that the jumps occurring at $f(x) = \pm\frac{1}{2}$ often result in unit rectangle functions that straddle zeros of $f(x)$, but are not necessarily exactly centered on the zeros. The width of the rectangle function will be the narrower, the more steeply the function passes through zero. Thus the width tends to be inversely proportional to $f'(x_n)$, where $f(x_n) = 0$, and so the roots x_n of $f(x) = 0$ are potentially significant features. However, not all zeros result in an extra rectangle function, and conversely rectangle functions can be generated without $f(x)$ going to zero, as may be seen from the illustration.

INTERPRETATION OF RECTANGLE FUNCTION OF $f(x,y)$

In two dimensions the generalization of the fast construction is as follows. Imagine $f(x, y)$ in contour form and fix attention on the contours at $\pm\frac{1}{2}$. Often these contours will parallel a null contour running along between them, but not necessarily. Shade the area between the contours at $\pm\frac{1}{2}$. In that area rect$[f(x, y)]$ is unity and elsewhere zero. The meaning of "between" has to be determined by reference to the definition

$$\text{rect}[f(x, y)] = \begin{cases} 1, & |f(x, y)| < \frac{1}{2} \\ 0, & |f(x, y)| > \frac{1}{2} \end{cases}$$

but as a rule is obvious, especially if a null contour runs between the $\pm\frac{1}{2}$ lines, because null contours are always within the shaded area.

Examples of this simple procedure are given in Fig. 3-13.

GENERAL RULE FOR LINE DELTAS

Consider the expression $\delta(lx + my + c)$, which represents a line impulse situated on the straight line $lx + my + c = 0$ but whose strength (mass per unit length) remains to be investigated. Assume that $l^2 + m^2 = 1$, as would be the case if l and m were respectively

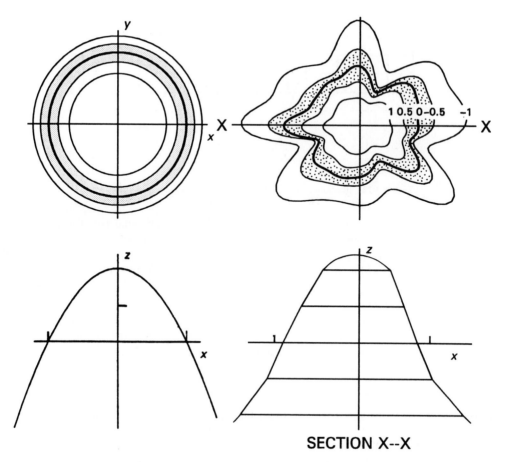

Figure 3-13   Functions $f(x, y)$ (above) with the zone between the contours at ± 0.5 shown shaded. Cross sections (below) show how the width of the shaded zone where it cuts the x-axis is connected with the steepness of the section. In each case $\text{rect}[f(x, y)]$ is unity in the shaded zone and zero elsewhere.

the cosines of the angles between the straight line and the x- and y-axes—in other words, if l and m were direction cosines.

Rules evolved for interpreting operations on delta symbols were as follows.

1. Replace $\delta(\cdot)$ by $\tau^{-1} \text{rect}(\cdot/\tau)$
2. Perform the operation indicated
3. Proceed to the limit as $\tau \to 0$.

These rules may be applied to determine the strength of a line impulse by evaluating the integral of $\delta(lx + my + c)$ along a cross section normal to the line. Call the rotated coordinate in this normal direction x' (Fig. 3-14), where $x' = lx + my$. Thus we wish to evaluate

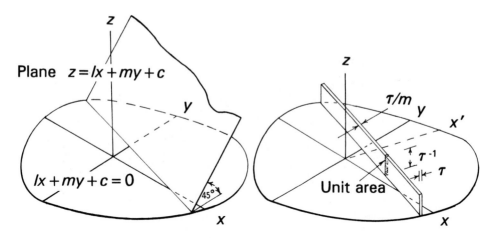

Figure 3-14 The expression $\delta(lx + my + c)$ is impulsive where the plane $z = lx + my + c$ passes through the zero level $z = 0$. The strength of the line impulse, arrived at by substituting narrow unit-area rectangle functions, is the reciprocal of the slope of the plane. If $l^2 + m^2 = 1$, this slope is unity.

$$\int_{-\infty}^{\infty} \delta(lx + my + c)\, dx'.$$

For the rotated coordinate y' we know that

$$\delta(lx + my + c) = \delta(y' + c).$$

Therefore the integral will be unity, because a shift c will not affect the integral; but let us proceed by the formal rules and make the replacement to obtain, first of all,

$$\int_{-\infty}^{\infty} \frac{1}{\tau'} \operatorname{rect}\left(\frac{lx + my + c}{\tau'}\right) dy'.$$

In Fig. 3-14 we give a pictorial interpretation to the equation $z = lx + my + c$ which represents a plane passing through $lx + my + c = 0$. Now we may also illustrate $\tau'^{-1} \operatorname{rect}(z/\tau')$, which appears as a straight wall of height τ'^{-1} and thickness τ'. Since $\operatorname{rect} u$ has its discontinuities at $u = \pm 0.5$, the edges of the wall lie where $(lx + my + c)/\tau' = \pm 0.5$. These edges are parallel to the line impulse, and distant $\tau'/2$ to each side because the plane has unit slope. The function $\tau'^{-1} \operatorname{rect}[(lx + my + c)/\tau']$ is illustrated in Fig. 3-15.

The second step is to perform the integral along y' perpendicular to the direction of the line impulse. As this integral is merely the area of a rectangle of height τ'^{-1} and width τ', the area is unity.

The third step is to proceed to the limit of the integral as $\tau' \to 0$, but in this application of the rules the integral does not vary with τ', and so the limit of a sequence of values, all of which are unity, is itself unity. This confirms our expectation.

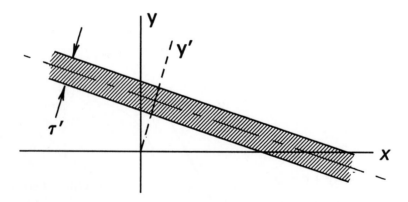

Figure 3-15 The function $\tau'^{-1}\text{rect}[(lx + my + c)/\tau']$ is equal to τ'^{-1} in the shaded area of perpendicular width τ' centered on the line impulse and equal to zero elsewhere.

THE RING IMPULSE

Consider an area density distribution $f(x, y)$ over a plane expressed as

$$f(x, y) = \delta(r - R).$$

From previous experience telling us that an impulse is located where the argument of the function is zero we expect the density to be zero everywhere except on the circle $r - R$ (Fig. 3-16) and we would know everything that is to be known about $\delta(r - R)$ if we knew either the linear density or the total mass. It helps to grasp the nature of the entities involved if we mention the following consistent set of units.

Quantity	Unit
Area Density $f(x, y)$	kg m^{-2}
Linear Density $w(s)$	kg m^{-1}
Mass $\int \int f(x, y)\, dx\, dy$	kg
Mass $\int w(s)\, ds$	kg

In the table s is arc length measured along a curve, in this case along the circle of radius R centered on the origin, and $w(s)$ is the linear density or strength at any point on the curve.

To discuss $\delta(r - R)$ we fall back on the rules for interpreting expressions containing $\delta(\cdot)$, which are

1. Replace $\delta(\cdot)$ by $\tau^{-1}\text{rect}(\cdot/\tau)$
2. Perform the operation indicated
3. Proceed to the limit as $\tau \to 0$.

On making the replacement, we see that

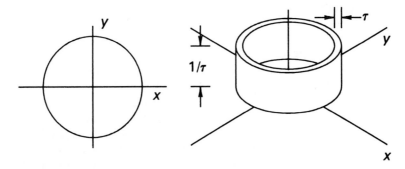

Figure 3-16   The ring impulse on the (x, y)-plane (left) and the circular wall for discussing the ring impulse (right).

$$\delta(r - R) \qquad \text{becomes} \qquad \tau^{-1} \text{rect}[(r - R)/\tau].$$

We can understand the nonimpulsive expression on the right as representing a circular flat-topped wall of thickness τ and height τ^{-1} centered on $r = R$ (Fig. 3-16). In other words, the expression is equal to τ^{-1} between the two concentric circles $(r - R)/\tau = \pm\frac{1}{2}$ and zero outside. The operation to be carried out is, let us say, to determine the mass expressed by

$$\int \int \delta(r - R)\, dx\, dy.$$

In accordance with the second rule we evaluate

$$\int \int \frac{1}{\tau} \text{rect}\left(\frac{r - R}{\tau}\right) dx\, dy.$$

Changing to polar coordinates, we have

$$\int_0^{2\pi} \int_0^\infty \frac{1}{\tau} \text{rect}\left(\frac{r - R}{\tau}\right) r\, dr\, d\theta = 2\pi \int_0^\infty \frac{1}{\tau} \text{rect}\left(\frac{r - R}{\tau}\right) r\, dr$$
$$= 2\pi R.$$

One could, of course, have evaluated this integral geometrically, since it is just the volume of the circular wall, in the form

$$\frac{1}{\tau}[\pi(R + \tfrac{1}{2}\tau)^2 - \pi(R - \tfrac{1}{2}\tau)^2],$$

and have obtained the same answer $2\pi R$. In this example, when the operation of integration has been carried out, the result turns out not to depend on τ any more; therefore, taking the limit as $\tau \to 0$ in accordance with rule 3 makes no further difference. The mass of the ring impulse $\delta(r - R)$ is thus $2\pi R$. From the mass we can get the linear density by dividing the mass by the length of the perimeter. The result is that $\delta(r - R)$ is seen to have unit linear density.

IMPULSE FUNCTION OF $f(x,y)$

The most general curvilinear impulse in the (x, y)-plane is represented as $\delta[f(x, y)]$. From the preceding special cases it is apparent that the line $f(x, y) = 0$ is the locus of the line impulse. Of course, $f(x, y) = 0$ is the general equation of a curve in the (x, y)-plane and may very well possess extreme features such as cusps or points of self-intersection or other more or less drastic singularities. But, having decided where the curvilinear impulse is located, for the moment we are occupied only with the elementary question of finding out the strength of the line impulse at each point along its length. We have seen that the strength at any point on $f(x, y) = 0$ is the reciprocal of the absolute slope of $f(x, y)$ at that point. The direction of maximum slope is normal to the direction of the null contour. If $f(x, y)$ is given analytically, we can calculate the two slope components $\partial f(x, y)/\partial x$ and $\partial f(x, y)/\partial y$ and combine them to obtain $\partial f/\partial n$, the slope in the direction of the normal to the contour:

$$\frac{\partial f}{\partial n} = \sqrt{\left(\frac{\partial f}{\partial x}\right)^2 + \left(\frac{\partial f}{\partial y}\right)^2} = |\text{grad } f|.$$

Then

$$\text{strength} = \left|\frac{\partial f}{\partial n}\right|^{-1}_{f=0} = \frac{1}{|\text{ grad } f |_{f=0}}.$$

Just as the delta functions expressed by $\delta[f(x)]$ have strengths inversely proportional to the absolute slopes at the zero crossings of $y = f(x)$, so the strength of the curvilinear impulse lying along the zero contour $f(x, y) = 0$ is equal to the reciprocal of the absolute gradient at each point.

As an example consider

$$f(x, y) = \sqrt{\frac{x^2}{a^2} + \frac{y^2}{b^2}} - 1.$$

Viewed as a topographical feature, this would be a conical crater with elliptical contours. The zero-height contour $f(x, y) = 0$ would be the central ellipse with semiaxes equal to a and b. It is on this ellipse that we expect to find the curvilinear impulse $\delta(\sqrt{x^2/a^2 + y^2/b^2} - 1)$. Now

$$\frac{\partial f}{\partial x} = \frac{x}{a^2[f(x, y) + 1]}$$

and

$$\frac{\partial f}{\partial y} = \frac{y}{b^2[f(x, y) + 1]}.$$

As we are only interested in slopes at the null contour $f = 0$, we can write

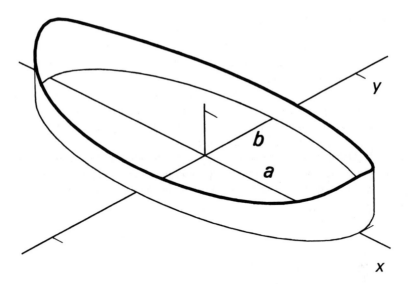

Figure 3-17   The elliptical line impulse $\delta(\sqrt{x^2/a^2 + y^2/b^2} - 1)$ is not of uniform strength.

$$\left(\frac{\partial f}{\partial x}\right)_{f=0} = \frac{x}{a^2}$$

$$\left(\frac{\partial f}{\partial y}\right)_{f=0} = \frac{y}{b^2}.$$

Then

$$\left|\frac{\partial f}{\partial n}\right|^{-1} = \left(\frac{x^2}{a^4} + \frac{y^2}{b^4}\right)^{-1/2}.$$

Fig. 3-17 shows a special case where $a = 2$ and $b = 1$. A little thought makes it clear why the strength of the elliptical line impulse is twice as great at the ends of the major axis. The slope at $x = a$, $y = 0$ is a^{-1}, while the slope at $x = 0$, $y = b$ is b^{-1}. Since the strength of the impulse is the reciprocal of the slope, the strengths at the ends of the major and minor axes are a and b, respectively.

SIFTING PROPERTY

When a two-dimensional unit impulse situated at (a, b) is multiplied by some test function $f(x, y)$ and the product is integrated to infinity over the whole plane, all information about function values at points other than (a, b) is lost because the impulse function is zero at those locations. Only the value $f(a, b)$ can show up in the result; in fact, the product is exactly equal to $f(a, b)$. Thus

$$\int_{-\infty}^{\infty} \int_{-\infty}^{\infty} {}^2\delta(x - a, y - b) f(x, y) \, dx \, dy = f(a, b).$$

As a special case, when the impulse is at the origin,

$$\int_{-\infty}^{\infty} \int_{-\infty}^{\infty} {}^2\delta(x, y) f(x, y) \, dx \, dy = f(0, 0).$$

The operation of multiplying by an impulse function and integrating sifts out the function value at the location of the impulse. To derive the sifting property, we use the rules for interpreting impulse notation, replacing the integral by

$$\int_{-\infty}^{\infty} \int_{-\infty}^{\infty} \tau^{-2} \, {}^2\text{rect}[(x - a)/\tau, (y - b)/\tau] f(x, y) \, dx \, dy$$

$$= \int_{a-\tau/2}^{a+\tau/2} \int_{b-\tau/2}^{b+\tau/2} \tau^{-2} f(x, y) \, dx \, dy = f(a, b) + \text{correction}.$$

The expression represents the volume of a prism of square cross section such as the one shown in Fig. 3-18. If the top of the prism were plane, the volume would be $f(a, b)$, but in general there will be a correction. However, after evaluation of the integral in accordance with the second rule we will go to the limit as $\tau \to 0$, and the correction will tend to zero. Of course, if $f(x, y)$ is not continuous at (a, b) but tends to different limits as (a, b) is approached from different directions, then instead of $f(a, b)$ we will get some combination of those limits.

A different sifting property arises when we multiply a curvilinear impulse $\delta[g(x, y)]$ with a test function $f(x, y)$. If the curvilinear impulse has unit strength all along its length, the double integral of the product will give the cross-section area of $f(x, y)$ along the curve $g(x, y) = 0$. For example, if a level road cutting has to be excavated through a mountain $z = f(x, y)$ along the line $y = 0$, then $\int \int \delta(y) f(x, y) \, dx \, dy = \int f(x, 0) dx$ will be proportional to the amount of rock to be removed. This particular result follows from the sifting property in one dimension. But if the road is to follow a curve C, then $\int \int \delta[g(x, y)] f(x, y) \, dx \, dy$ will equal $\int_C S(s) f(x, y) \, ds$, where ds is the element of arc length and $S(s)$ is the strength of $\delta[g(x, y)]$ at position s.

Example. To illustrate the role of strength of a curvilinear impulse we suppose that a photograph is made with a flash to obtain an image $f(x, y)$. Because the intensity of the flash falls off with distance, we might be interested in the mean brightness of the image in the annulus at radius R from the center of the image. The expression $\int \int \delta(r - R) f(x, y) \, dx \, dy$ would represent the total light from the narrow annulus centered on $r = R$, because $\delta(r - R)$ has unit strength; and we could divide by the circumference $2\pi R$ to get the mean brightness. The expression $\pi^{-1} \int \int \delta(r^2 - R^2) f(x, y) \, dx \, dy$ would give the mean brightness directly, because, as one can verify, $\delta(r^2 - R^2)$ has strength diminishing inversely as R, which is appropriate to our interest.

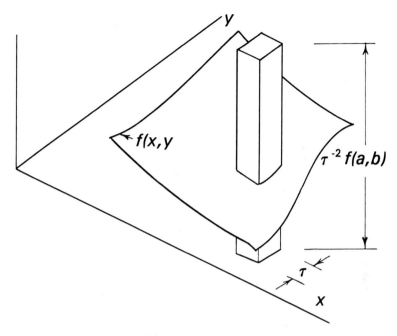

Figure 3-18   The function $\tau^{-2}\ ^2\text{rect}[(x-a)/\tau,\ (y-b)/\tau]f(x,y)$ has a height about $\tau^{-2}f(a,b)$ and a base area τ^2. Consequently, its volume is approximately $f(a,b)$, and the approximation gets better and better as $\tau \to 0$.

DERIVATIVES OF IMPULSES

The concept of the infinitesimal magnetic dipole in physics permits representation in terms of the derivative of a delta function. Since the magnetic dipole is the oldest and most widely familiar of dipoles, it will be used for the following explanation. Suppose that two magnetic poles, one of strength m and one of strength $-m$, are placed at $P(L, 0, 0)$ and $Q(-L, 0, 0)$, respectively. We need to know that a pole of strength m produces a magnetic field F radially outward in all directions of amount m/r^2 at distance r. What is the magnetic field at $(0, y, 0)$? Combining the separate vectors due to the two poles, we find $2mL/(y^2 + L^2)^{3/2}$ in the direction PQ. As $y \to \infty$ the angle between the two vectors diminishes, and we get the well-known inverse cube law $2mL/y^3$. If y is kept fixed and the angle between the two vectors is reduced by bringing the poles closer together, then the magnetic field at $(0, y, 0)$ would cancel, unless the pole strengths were increased in inverse proportion to L. Let the magnetic dipole moment $M = 2mL$ be kept constant as L diminishes; then the limiting field is M/r^3 as before.

The utility of this concept is that if the observing point is far enough away, then it does not matter what the exact configuration of poles is; the constant that counts is the dipole moment M. We understand very well that two opposite poles cannot physically be brought indefinitely close together, particularly if they are made stronger in the process; but the distant effect approaches a limit in the mathematical sense even if the magnetic configuration proposed cannot. Now how is this configuration expressed in delta notation? A pole of strength m, that sets up a field m/r^2 with no restriction mentioned on how small r may be, is conceived of as a point characterized by its location (x_0, y_0, z_0) and strength m, and is therefore representable by $m^3\delta(x - x_0, y - y_0, z - z_0)$. Then the two poles are representable by

$$m^3\delta(x - L, y, z) - m^3\delta(x + L, y, z).$$

The sequence generated as $L \to 0$, while $M = 2mL$ is held constant, defines

$$M\frac{\partial}{\partial x}{}^3\delta(x, y, z).$$

To see this, first reduce the three dimensions to two, since application to plane images is our intent, and discuss $m^2\delta(x - L, y) - m^2\delta(x + L, y)$ as $L \to 0$, with a view to emerging with $M(d/dx)^2\delta(x, y)$. Replacing by rect functions in the spirit of rule 1, we get

$$\frac{m}{\tau^2}\text{rect}\left(\frac{x - L}{\tau}\right)\text{rect}\left(\frac{y}{\tau}\right) - \frac{m}{\tau^2}\text{rect}\left(\frac{x + L}{\tau}\right)\text{rect}\left(\frac{y}{\tau}\right).$$

Each term is constant over a small square of side τ with interior function value $\pm m/\tau^2$ and volume integral $\pm m$ centered at $(\pm L, 0)$. The first moment about the origin is $2mL$.

Rule 2 mentions an operation that is to be carried out. Let us say that the two rectangle function terms are to be convolved with a continuous test function $\phi(x, y)$. The outcome of this operation as $\tau \to 0$ will be $m\phi(x - L, y) - m\phi(x + L, y)$; in short $2mL\,{}^{2L}\Delta_x\phi(x, y)$, the first centered difference of $\phi(x, y)$ over span $2L$ in the x-direction. Now, as $L \to 0$, we find

$$\lim_{L \to 0} 2mL\,{}^{2L}\Delta_x\phi(x, y) = 2m \lim_{L \to 0} \frac{\phi(x - L, y) - \phi(x + L, y)}{2L}$$

$$= M\frac{\partial}{\partial x}\phi(x, y)$$

$$= \left[M\frac{\partial}{\partial x}{}^2\delta(x, y)\right] ** \phi(x, y).$$

(The notation $**$ for two-dimensional convolution, if unfamiliar, can be read about in Chapter 5.) Thus $M(\partial/\partial x)^2\delta(x, y)$ or $M^2\delta'_x(x, y)$ is the delta function representation of a dipole of strength M situated at $(0, 0)$ and oriented along the x-axis. Since ${}^2\delta(x, y)$ was shown to be equivalent to $\delta(x)\delta(y)$, a notational variant is $M\delta'(x)\delta(y)$.

Multipoles are also expressible simply. For example, $\delta'(x)\delta'(y)$ stands for the quadrupole that could be pictured as having positive poles in the first and third quadrants and negative poles in the second and fourth. We can verify that $\delta'(x)\delta'(y)$ is proportional to ${}^2\delta(x, y)\,\text{cas}\,2\theta$, where $\text{cas}\,\theta \equiv \cos\theta + \sin\theta$. Quadrupoles rotated $45°$ are represented by ${}^2\delta(x, y)\,{}^{\cos}_{\sin}\theta$. Linear distributions of dipole moment can also be handled; for example, viewed as a function in the (x, y)-plane, $\delta'(x)$ stands for a situation such as the load distribution on a sheet of metal that is being cut by metal shears on a line along the y-axis.

SUMMARY

Impulse notation in two dimensions is important for digital imaging, which has a discrete basis, because of the analogy betwen the discreteness of a digital image and the compactness of impulsive features, both dots and lines. Impulse notation also bridges the divide between discrete and continuous, because the impulse notation is compatible with analysis of functions of continuous variables. Consequently, much discussion of discrete arrays or images can be conducted in terms of ordinary algebra and calculus with the help of impulse notation. That is why the topic is introduced early.

The three rules for interpreting expressions containing delta notation have been emphasized, because students with prior knowledge of delta functions have a tendency to ignore this unfamiliar doctrine. Such students have trouble with $\delta(r^2 - R^2)$, just discussed, and with impulse functions of two independent variables.

The delta function is more than just a compact notation; it permits algebraic manipulation to be applied to a wider class of entities. The taking of Fourier transforms is also facilitated, which is important because images to be transformed are often impulsive or have an impulsive part. Conversely, a transform may turn out to be impulsive; even such an elementary case as the Fourier transform of a sine wave is an example.

Matrix notation is indispensable in the discussion of images that are describable in terms of two integer variables that specify location together with a third dependent variable that may be binary, integer, or real. For example, using bold face to mark the origin,

$$\begin{bmatrix} 0 & 0 & 0 \\ 0 & \mathbf{1} & 0 \\ 0 & 0 & 0 \end{bmatrix} \quad \text{and} \quad \begin{bmatrix} 0 & 0 & 0 \\ 0 & 0 & 1 \\ \mathbf{0} & 0 & 0 \end{bmatrix}$$

stand for images where pixels are turned on (a) at the origin, (b) at cartesian location (2, 1). One's flexibility in handling such images is much improved by understanding that the matrixes themselves are fully equivalent to (a) $^2\delta(x, y)$ and (b) $^2\delta(x - 2, y - 1)$, expressions that can be subjected to any of the formal operations on functions of two continuous real variables, including integration, as in the Fourier integral.

With impulse notation in our toolbox we can take up the Fourier transform and gain immediate advantage from our ability to discuss discrete entities, hand-in-hand with the continuous, from the beginning.

LITERATURE CITED

The great bulk of the current literature on generalized functions is rather rarefied; the general flavor can be experienced by reference to Schwartz (1950–1951) and Gel'fand and Shilov (1964) and entries in *Mathematical Abstracts* that cite these works or appear under the heading of generalized functions or distributions. Lighthill's book is in the pure mathematical tradition, but is short and sweet, and is recommended.

I. M. GEL'FAND AND G. E. SHILOV (1964), *Generalized Functions*, Academic Press, New York.

M. J. Lighthill (1958), *An Introduction to Fourier Analysis and Generalized Functions*, Cambridge University Press.

L. Schwartz (1950–1951), *Théorie des Distributions*, vols. 1 and 2, Hermann, Paris.

G. Temple (1953), "Theories and applications of generalized functions," *J. Lond. Math. Soc.*, vol. 28, p. 181.

PROBLEMS

3–1. *Two-dimensional impulse.*
(a) A charge Q is located at $x = 0$, $y = 1$ on the (x, y)-plane. How would you write the surface charge density σ in terms of impulse symbol notation?
(b) State the units of every quantity involved. ☞

3–2. *Rectangle function of a function.* On the accompanying graph of $f(x)$ (Fig. 3-19) add a graph of rect$[f(x)]$.

3–3. *Line impulse.* Determine the strength of the line impulse $\delta(3x + 4y + 5)$.

3–4. *Line impulse.* Determine the location and strength of the line impulse $\delta(x/a + y/b - 1)$.

3–5. *Strength of line impulse.* Explain why the linear density, or strength, of the line impulse $\delta(y - mx - c)$ is dependent on the coefficient m describing the slope, whereas the strength of the ring impulse $\delta[(x^2 + y^2)^{1/2} - R]$ is the same everywhere despite the varying slope of the tangent to the circle. ☞

3–6. *Line impulse.*
(a) Make a diagram to illustrate the meaning of $\phi(x, y) = \delta(x \cos \alpha + y \sin \alpha - R)$; what can you say about the strength?
(b) If one thinks of x and y as constants and regards R and α as variables, does it make a difference to the diagram? ☞

3–7. *Row of spikes.* The row of spikes represented by III$(x)\delta(y)$ can also be thought of as being selected from a two-dimensional bed of nails. Is the expression ^{2}III$(x, y)\delta(y)$ an equivalent representation? ☞

3–8. *Impulse function of azimuth.* Let (r, θ) be the polar coordinates of the point (x, y). Explain what $\delta(\theta)$ must mean.

3–9. *Radial impulse.* Use the general formula $[(\partial f/\partial x)^2 + (\partial f/\partial y)^2]^{-1/2}$ for the strength of $\delta[f(x, y)]$ to discuss the strength of $\delta(\theta)$.

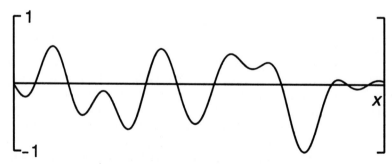

Figure 3-19   Graph of a function.

3–10. *Spoke pattern.* Write an expression for an impulsive function of x and y that has uniform strength along 32 radial spokes equally spaced at $11.25°$ and extending from the origin to unit distance from the origin.

3–11. *Shah function of angle.* What meaning would you attribute to $\text{III}(\theta)$?

3–12. *Evaluation rules.* Evaluate $\int_{-1}^{1} \int_{-1}^{1} \delta(x^2 - y^2)\, dx\, dy.$ ☞

3–13. *Ring impulse.* Show that $\delta(r - a) = 2r\delta(r^2 - a^2)$, where $a > 0$.

3–14. *Mass of nonuniform ring impulse.* Discussing the mass of $\delta(r - a\cos\theta)$, a student first considered $\tau^{-1}\text{rect}[(r - a\cos\theta)/\tau]$ and said, "At $\theta = 0$, $x = a + \tau/2$ or $x = a - \tau/2$, so we are dealing with a crescent-shaped plateau of height τ^{-1} and area $(\pi/4)(a + \tau/2)^2 - (\pi/4)(a - \tau/2)^2$. For small τ, the volume is $\pi a/2$. Therefore the mass is $\pi a/2$." Explain where the error was made.

3–15. *Elliptical impulse.* What is the strength of the impulse $\delta(x^2/a^2 + y^2/b^2 - 1)$?

3–16. *Circular impulse.* What is the strength of the curvilinear impulse $\delta(r^2 - 4R^2)$?

3–17. *Circular impulse.* The equation $y - (1 - x^2)^{1/2} = 0$, with due attention to the allowable signs of the radical, represents a circle of unit radius. Investigate the strength of the impulsive expression $\delta[(y - (1 - x^2)^{1/2}]$ as a function of $\theta = \arctan(y/x)$.

3–18. *Impulse in polar coordinates.* A two-dimensional impulse of unit strength is situated at (R, θ_0). See whether $\delta(r - R)\delta(\theta - \theta_0)$ is a satisfactory representation.

3–19. *Polar coordinates.* Deduce that the strength of $\delta[f(r, \theta)]$ is

$$\left[\left(\frac{\partial f}{\partial r}\right)^2 + \left(\frac{1}{r}\frac{\partial f}{\partial \theta}\right)^2\right]^{-1/2}. \quad ☞$$

3–20. *Triaxial symmetry.*
   (a) Sketch the image represented by $\delta(y^3 - 3x^2 y)$.
   (b) Twelve uniform spokes radiate from a center in the direction of the hours on a clock face. Represent this image as compactly as possible in δ-notation. ☞

3–21. *Impulse function of product.* It has been suggested that

$$\delta[f(x, y)g(x, y)] = \frac{\delta[f(x, y)]}{|g(x, y)|} + \frac{\delta[g(x, y)]}{|f(x, y)|}.$$

Test this relation on the simple case of $\delta(xy)$ and try to establish or disprove the suggestion.

3–22. *Impulse function of product.* Test the formula for $\delta[f(x, y)]$ on the special case $\delta(2x)$, where $f(x, y) = 2$ and $g(x, y) = x$.

3–23. *Line impulse.* Determine the meaning of $\delta(|x| + |y| - 1)$.

3–24. *Rectangle function notation.* Describe in words the meaning of the following expressions interpreted as functions in the (x, y)-plane.
   (a) $\text{rect}(r/2)$,
   (b) $\text{rect}(r - 0.5)$,
   (c) $\text{rect}(r/4 + 3/4)$,
   (d) $\text{rect } r \text{ rect } x$,
   (e) $\text{rect } x + \text{rect } y$, (f) $\text{rect } r^2$. Check your answer to make sure that you have indeed been able to convey your meaning accurately in words without the use of symbols.

3–25. *Spiral slit.* An Archimedean spiral is defined by the polar equation $r = a\theta$, $\theta > 0$. Imagine a narrow slit of uniform width ϵ centered on the spiral $r = \theta/2\pi$. The slit is uniformly illuminated from behind, and we wish to express the brightness distribution $b(r, \theta)$ on the front side of the plane of the slit using δ-function notation, because the slit is too narrow for us to worry about. Express the brightness distribution $b(r, \theta)$ in the form $\delta[f(r, \theta)]$.

3–26. *Slit for optical processor.* A curve $y^2 = 4ax$ is engraved on two identical metal blanks, and a machinist removes the material to the left of the curve from one blank and to the right of the curve from the other, so that the remaining parts fit together with no spaces exceeding 1 μm. The two parts are then separated 100 μm in the x-direction and the slit is illuminated uniformly from behind.
(a) What is the slit width as a function of x?
(b) Naturally the amount of light transmitted per unit length of slit is not constant, because the slit width varies; but let the light per unit length be q_0 at $x = 0$. Let the amount of light transmitted per unit area be $Q(x, y)$. Express $Q(x, y)$ in delta-function notation.

3–27. *Uniform slit.* A slit of uniform width is cut in a flat metal plate with a fine laser beam moving at constant speed along the curve $y = x^2$ over the range $-1 < x < 1$. The slit is uniformly illuminated from behind. Write an expression for the light per unit area emerging from the (x, y)-plane.

3–28. *Notation for fan.* Describe the meaning of $\mathrm{III}(10\theta) \, \mathrm{rect}(\theta/11)$ and make a sketch. ☞

3–29. *Curvilinear sifting.* A student proposed that the integral of the product of a given function $g(x, y)$ with a curvilinear impulse $\delta[f(x, y)]$ must be the area of the vertical cross section of $z = g(x, y)$ along the curve $f(x, y) = 0$. A friend said that this could not be so, because in one dimension $\int_{-\infty}^{\infty} g(x)\delta[f(x)]\,dx$ is a *sum* of terms. Therefore, says the friend, it is not possible that

$$\int_{-\infty}^{\infty} \int_{-\infty}^{\infty} g(x, y)\delta[f(x, y)]\,dx\,dy = \int_{f=0} g(x, y)\,ds, \qquad (1)$$

where ds is differential arc length along the curve $f(x, y) = 0$. Establish a valid equation having the same left-hand side as in the foregoing equation.

3–30. *Curvilinear impulse.* In attempting to give a definition of the strength of a curvilinear impulse $\delta[f(x, y)]$ a student wrote as follows. "Take a point $P(x_0, y_0)$ that is on the line impulse and imagine a circle of radius R centered there. Then the strength of the line impulse at P is, by my definition,

$$\lim_{R \to 0} (2R)^{-1} \int_0^{2\pi} \int_0^R \delta[f(x, y)]r\,dr\,d\theta,$$

where (r, θ) is a polar coordinate system with its origin at (x_0, y_0)." Is this a satisfactory definition?

3–31. *Sifting property.*
(a) In the expression $\int \int {}^2\delta(x, y)f(x, y)\,dx\,dy$ let the integrand be replaced by $(XY)^{-1} \exp[-\pi(x/X)^2 \exp[-\pi(y/Y)^2]$. Evaluate the integral in the limit as X and Y approach zero and verify that the result is the same as when $(XY)^{-1}\mathrm{rect}(x/X)\mathrm{rect}(y/Y)$ is substituted.
(b) Does it make a difference if $(XY)^{-1}\mathrm{sinc}(x/X)\mathrm{sinc}(y/Y)$ is substituted? ☞

3–32. *Line impulse generated by derivatives.* Give an interpretation of

$$\left| \frac{\partial}{\partial x}{}^2\mathrm{rect}(x, y) + \frac{\partial}{\partial y}{}^2\mathrm{rect}(x, y) \right| \qquad \text{and of} \qquad \left| \frac{\partial}{\partial x}\mathrm{rect}\,r \right| + | \frac{\partial}{\partial y}\mathrm{rect}(r)|. \quad ☞$$

3–33. *Impulse on complex plane.* The notation $\delta(z - z_0)$ may be used to represent a unit impulse at $z = z_0$ on the complex plane of z. If $z = x + iy$ and $z_0 = x_0 + iy_0$, compare $\delta(z - z_0)$ with $^2\delta(x - x_0, y - y_0)$. ☞

3–34. *Surface impulse in 3-space.*
  (a) Write the general equation for a surface S in (x, y, z)-space.
  (b) Write an expression for an entity that is impulsive on the surface S.
  (c) What would be meant by the strength of a surface impulse?
  (d) What is the strength at each point of the surface impulse given in (b)? ☞

3–35. *Helicoidal surface.* The surface $z = \theta$ is a helicoid. A surface of a screw with axis in the z-direction and pitch P would be described by the helicoid $z = P\theta/2\pi$. The expression $\delta(z - \theta)$ describes a surface density distribution that is confined to a helicoid and is zero elsewhere. For example, a stretched ribbon uniformly twisted about its axis constitutes a mass distribution sufficiently close to a helicoid that impulsive notation might be appropriate. The durface density of a twisted ribbon would be very nearly uniform (if it was uniform before twisting), whereas the strength of the density distribution $\delta(z - \theta)$ remains to be investigated. What is the strength of $\delta(z - \theta)$ at a point (r, θ, z)?

3–36. *Vector argument.* The notation $\delta(\mathbf{r})$ is sometimes used to represent a unit three-dimensional impulse at the point whose vector displacement from the origin is $\mathbf{r}$. Let the components of $\mathbf{r}$ be (a, b, c); then $\delta(\mathbf{r})$ means the same as $^3\delta(x - a, y - b, z - c)$.
  (a) In the vector argument notation, how would one represent the electric charge density $\rho(x, y, z)$ in C m^{-3} associated with a point charge Q at the origin?
  (b) What would $\delta(0)$ mean?
  (c) How could the sifting theorem be written if we chose to write $f(x, y, z)$ as $f(\mathbf{x})$, where $\mathbf{x} = (x, y, z)$? ☞

3–37. *Intersecting line impulses.* Two unit-strength line impulses intersect at (x_1, y_1), making an angle θ. Show that their product equals $^2\delta(x - x_1, y - y_1)/\sin\theta$, $0 < \theta < 2\pi$.

3–38. *Intersecting sinusoidal impulses.* Interpret the expression $\delta(y - \sin x)\delta(y + \sin x)$.

3–39. *Intersecting curvilinear impulses.* Two curvilinear impulses intersect at (x_1, y_1), making a well-defined angle θ. What is the strength of the point impulse defined by their product?

4

The Two-Dimensional Fourier Transform

In this chapter the two-dimensional Fourier transform is defined mathematically, and then some intuitive feeling for the two-dimensional Fourier component is developed. Just as we understand that a waveform can be broken down into time-varying sinusoids, so also we can acquire a corresponding physical picture of the decomposition of a single-valued surface in space. A set of useful theorems analogous to the more familiar one-dimensional theorems is presented for reference. Since many of these theorems are of enduring value, it is worthwhile choosing and memorizing those that you think may be useful to have at your fingertips in the future.

Analysis of an image into spatial sinusoids running in various directions with different wavelengths offers a divide-and-conquer technique for problem solving—it is often easier to conduct a given operation on a constituent sinusoid than to conduct the same operation on a full image; but linearity is a prerequisite before it is valid to analyze, operate, and resynthesize. A characteristic feature of the world of physics is that a high degree of linearity exists in many of the phenomena of interest. In consequence the diversity of the fields in which Fourier analysis proves useful puts this valuable mathematical tool right at the heart of engineering. There is no tool of similar stature for dealing with nonlinear phenomena. For the time being, then, we concentrate on Fourier technique, bearing in mind that the world of physics is not exclusively linear, visual display and perception being prime examples from the field of imaging.

ONE DIMENSION

In one dimension we define the Fourier transform $F(s)$ of a given function $f(x)$ by

$$F(s) = \int_{-\infty}^{\infty} f(x)e^{-i2\pi sx}\, dx.$$

We may think of the right-hand side as specifying an operation of *analysis* as follows. Multiply the given function $f(x)$ by the factor $\exp(-i2\pi sx)$, a function of x containing a parameter s which is to remain fixed at some constant value for the time being; then integrate the product over all x. The result is no longer a function of x but does depend on the chosen value of the parameter s. The F value so obtained may now be supplemented by other F values corresponding to other choices of s by repeating the multiplication and integration. The computation performed in this way may be thought of as analysis of $f(x)$ into exponential components. The underlying idea is that exponential components of $f(x)$ not having the chosen s value will contribute products that integrate to nothing, whereas the component at the chosen value of s will respond to the analysis because the integral of its product can be nonzero.

The inverse relationship

$$f(x) = \int_{-\infty}^{\infty} F(s)e^{i2\pi xs}\, ds$$

can be thought of as specifying an operation of *synthesis,* in which exponential functions of x with different s values, each having its appropriate amplitude $F(s)\, ds$, are summed. The function $f(x)$ is thus synthesized from exponentials of all frequencies s. Of course, we are talking about exponential functions of imaginary argument, a convenient way of handling sinusoids and cosinusoids simultaneously.

THE FOURIER COMPONENT IN TWO DIMENSIONS

In two dimensions, similar viewpoints are obtained. The two-dimensional Fourier transform $F(u, v)$ of the two-dimensional function $f(x, y)$ is defined by

$$F(u, v) = \int_{-\infty}^{\infty}\int_{-\infty}^{\infty} f(x, y)e^{-i2\pi(ux+vy)}\, dx\, dy.$$

We may view this as an operation of analysis and ask what are we analyzing $f(x, y)$ into. The answer is functions of the form $\exp[-i2\pi(ux + vy)]$ with various amplitudes depending on the choice of u and v. The exponential can be decomposed into two terms

$$\cos[2\pi(ux + vy)] \quad \text{and} \quad \sin[2\pi(ux + vy)]$$

or into four terms

$$\cos 2\pi ux \cos 2\pi vy, \quad \sin 2\pi ux \sin 2\pi vy, \quad \sin 2\pi ux \cos 2\pi vy, \quad \text{and} \quad \cos 2\pi ux \sin 2\pi vy.$$

We see that functions $f(x, y)$ that are symmetrical about both the x- and y-axes may be analyzed into functions of the form $\cos 2\pi ux \cos 2\pi vy$ and that other combinations of

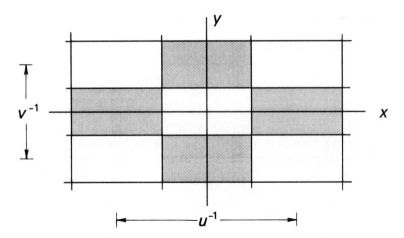

Figure 4-1   The "quilted" surface $\cos 2\pi ux \cos 2\pi vy$. The shaded regions are negative.

even and odd symmetry are provided for. Figure 4-1 shows the character of such a cosine-product function; it is the same function that describes standing waves arising by perfect reflection from a rectangular boundary.

The quantity u is the number of waves per unit length in the x-direction, and the quantity v is the number of waves per unit length in the y-direction. To perform the analysis of a doubly symmetrical function $f(x, y)$ into cosine-product functions, one multiplies $f(x, y)$ by $\cos 2\pi ux \cos 2\pi vy$, where u and v are fixed at certain constant values for the time being, and integrates the product over the whole (x, y)-plane. The result is a function only of the chosen u and v and is the amplitude of the chosen component.

From the synthesis standpoint we may say that a doubly symmetrical function $f(x, y)$ may be synthesized by the superposition, with appropriate amplitudes, of two-dimensional cosine-product functions chosen from the full doubly-infinite range of u and v. The two equations summarizing the above statements are

$$F(u, v) = \int_{-\infty}^{\infty} \int_{-\infty}^{\infty} f(x, y) \cos 2\pi ux \cos 2\pi vy \, dx \, dy$$

$$f(x, y) = \int_{-\infty}^{\infty} \int_{-\infty}^{\infty} F(u, v) \cos 2\pi xu \cos 2\pi yv \, du \, dv.$$

In general, for nonsymmetrical functions, we would need additional terms as expressed compactly in the two standard relations

$$F(u, v) = \int_{-\infty}^{\infty} \int_{-\infty}^{\infty} f(x, y) e^{-i2\pi(ux+vy)} \, dx \, dy$$

$$f(x, y) = \int_{-\infty}^{\infty} \int_{-\infty}^{\infty} F(u, v) e^{i2\pi(xu+yv)} \, du \, dv.$$

The presence of the factor i in front of the term $\sin 2\pi(ux + vy)$ when the second relation is expanded is worthy of comment. Why should a *real* function $f(x, y)$ that is not strictly symmetrical, and therefore requires sine components, require *imaginary* quantities of the sine components? If the function is real, the sine component will have to be real. This means that the coefficient $F(u, v)$ cannot be pure real; it will need to be complex in order to cancel the i that automatically arises from the use of the exponential notation.

We might describe the pattern illustrated as a "quilted" surface, because it resembles the surface of a quilt whose lines of stitching are perpendicular to each other. The Fourier synthesis theorem is like saying that any surface can be built up by superposition of quilted surfaces, all oriented the same way, but having all possible wavelengths in the two perpendicular directions and all possible shifts with appropriate amplitude.

THREE OR MORE DIMENSIONS

Clearly in three dimensions the basic formulas generalize to

$$F(u, v, w) = \int_{-\infty}^{\infty} \int_{-\infty}^{\infty} \int_{-\infty}^{\infty} f(x, y, z)e^{-i2\pi(ux+vy+zw)} \, dx \, dy \, dz$$

$$f(x, y, z) = \int_{-\infty}^{\infty} \int_{-\infty}^{\infty} \int_{-\infty}^{\infty} F(u, v, w)e^{i2\pi(xu+yv+zw)} \, du \, dv \, dw$$

and similarly in four dimensions or more. Fourier analysis in three dimensions is best known historically from x-ray diffraction analysis of crystals, but applications to elasticity, electricity, magnetism and many other fields are numerous for the fundamental reason that space is three dimensional. Four-dimensional transforms arise in plasma dynamics and other branches of fluid dynamics where time variation is inherent. Many authors find it convenient to use compact vector notation that eliminates the multiple integration signs, even if only one integral is saved. In this view $f(x, y, z)$ becomes $f(\mathbf{x})$, where $\mathbf{x}$ is a vector whose components are x, y and z and the transform is $F(\mathbf{u})$.

VECTOR FORM OF TRANSFORM

The point (x, y) on the plane, or (x, y, z) in space, or (x, y, z, t) in the space-time continuum, can be expressed compactly by a multidimensional vector $\mathbf{x}$ having the components stated. For the (u, v)-plane, (u, v, w)-space or the (u, v, w, f) spatiotemporal frequency domain, a multidimensional vector $\mathbf{u}$ can be similarly defined. The Fourier transform relations can then be succinctly stated, for any dimensionality, using vector dot-product notation $\mathbf{u} \cdot \mathbf{x}$ to condense any of the expressions $ux + vy$, $ux + vy + wz$, or $ux + vy + wz + ft$. Thus, for any number of dimensions one can write the one pair of relations

$$F(\mathbf{u}) = \int_{\mathcal{X}} f(\mathbf{x})e^{-i2\pi \mathbf{u}\cdot\mathbf{x}} \, d\mathbf{x}$$

$$f(\mathbf{x}) = \int_{\mathcal{U}} F(\mathbf{u})e^{i2\pi \mathbf{u}\cdot\mathbf{x}} \, d\mathbf{u}.$$

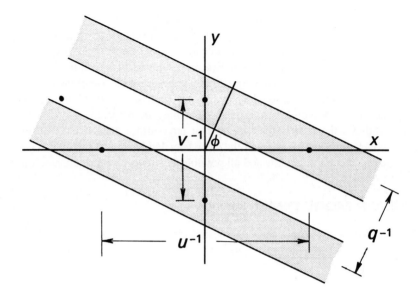

Figure 4-2   A "corrugation" $\cos[2\pi(ux + vy)]$. The shaded zones are negative. The spatial frequency in the x-direction is u (in the y-direction, v); *the* spatial frequency is q.

The differentials $d\mathbf{x}$ and $d\mathbf{u}$, which are not vectors, stand for the area elements $dx\,dy$ and $du\,dv$, volume elements $dx\,dy\,dz$ and $du\,dv\,dw$, and so on. The scalar area element $d\mathbf{x}$ defined as $dx\,dy$, is to be distinguished from the vector differential $\mathbf{dx}$ whose components are (du, dv). The spaces $\mathcal{X}$ and $\mathcal{U}$ are the whole infinite domains of $\mathbf{x}$ and $\mathbf{u}$; consequently one integral sign stands for $\int_{-\infty}^{\infty} \int_{-\infty}^{\infty}$ or $\int_{-\infty}^{\infty} \int_{-\infty}^{\infty} \int_{-\infty}^{\infty}$ or more.

This condensed notation is very handy in contexts such as papers where all the integrals are triple integrals or where extended algebra is needed. However, in a book about two dimensions, where single and double integrals are both frequently encountered, it is better for explanatory purposes to distinguish between one-dimensional and two-dimensional integrals and to write them out in full.

THE CORRUGATION VIEWPOINT

There is another way of viewing the Fourier component in two dimensions. Let us define a "corrugation" (Fig. 4-2) as a surface generated as the locus of a level straight line that passes through a sinusoid perpendicular to the plane containing that sinusoid. Then the two-dimensional Fourier synthesis can also be described in terms of the superposition of corrugations having all possible wavelengths q^{-1} and all possible orientations ϕ, with appropriate amplitudes. The interrelation is

$$q = (u^2 + v^2)^{1/2} \qquad \tan\phi = v/u,$$

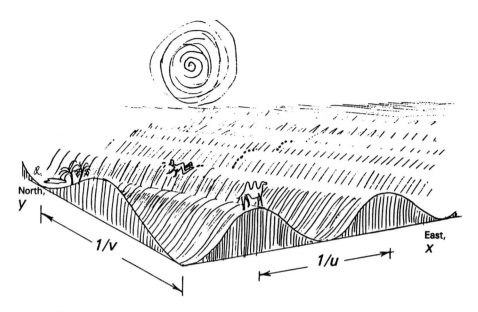

Figure 4-3 The x-component of spatial frequency is u, measured in cycles per unit of x. The spatial period, or crest-to-crest distance traveling east, is u^{-1} (see camel). The spatial frequency q is measured in the direction of hardest going (see man).

where q is the spatial frequency of a corrugation (the number of waves per unit length in the direction normal to the wave crests) and ϕ is the angle between that direction and the x-axis. Alternatively we may write

$$u = q\cos\phi, \qquad v = q\sin\phi.$$

The corrugation viewpoint provides our clearest way of understanding and remembering the basic significance of the variables u and v. Imagine a sandy desert where the sand dunes have a sinusoidal cross section and all run parallel to each other. Then if you ride a camel (Fig. 4-3) from west to east, u is the number of crests per unit horizontal distance traveled. The crest-to-crest distance going west to east is u^{-1}. Likewise, if you ride north, the number of crests per meter is v (assuming that distance is measured in meters), and v^{-1} is the number of meters per crest. The true crest-to-crest distance measured perpendicular to the direction of the crest lines is shorter than either u^{-1} or v^{-1}, in general, and, of course, it is the hardest direction in which to ride. The spatial frequency in that direction is greater than in any other direction and may be calculated by vectorial addition of the spatial frequency components (Fig. 4-4) along any pair of orthogonal axes.

Since any surface can be analyzed into corrugations of appropriate amplitude, direction, and spatial frequency, it follows that a quilted component can be so analyzed. We find that it is just the sum of two corrugations that are equally inclined to the coordinate axes. To prove this we reflect the corrugation $\cos[2\pi(ux + vy)]$ in the y-axis to obtain $\cos[2\pi(ux - vy)]$ and add the two together. Then

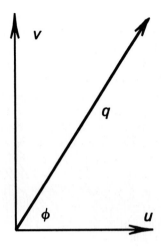

Figure 4-4 Vector addition relates the spatial frequency q to its components u and v.

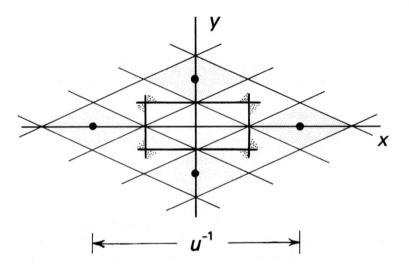

Figure 4-5 Two equally inclined corrugations are shown superimposed. The shaded regions are the intersections of the negative regions of each corrugation considered separately but give only an indication of the whereabouts and shape of the combination. The central positive bulge of the quilt pattern is indicated in heavy outline.

$$\cos[2\pi(ux + vy)] + \cos[2\pi(ux - vy)] = \cos 2\pi ux \cos 2\pi vy - \sin 2\pi ux \sin 2\pi vy$$
$$+ \cos 2\pi ux \cos 2\pi vy + \sin 2\pi ux \sin 2\pi vy$$
$$= 2 \cos 2\pi ux \cos 2\pi vy.$$

We see that the two corrugations into which a unit quilt pattern resolves must be of strength 0.5 each.

Figure 4-5 shows the two corrugations mentioned above together with parts of the four null lines bounding the central maximum of their combination. By fixing attention on a point on one of these null lines, one can see how the contributions from the two corrugations do indeed cancel along noninclined loci.

Here is a physical interpretation of the intimate relation between the quilt and corrugation patterns. Suppose the x-axis is a perfectly reflecting barrier to waves that are incident from the NNE and, after reflection, will be traveling toward the NNW. The previous figure catches these waves at the moment when a crest of the incident wavetrain is at the origin, and the same is true for the reflected wave. Consequently, the superposition of the two waves has its maximum at the origin, and the full pattern representing the interference of the two oblique wavetrains is the quilt pattern. Of course, the lower half of the figure is to be ignored. The two traveling waves interfere to set up a standing wave in the y-direction. For example, there will be nodal lines of zero disturbance running parallel to the x-axis, the first one being at a distance $\frac{1}{4}u^{-1}$ from the reflecting barrier on the x-axis. The standing wave is a consequence of the fact that energy flow to the south is blocked; on the other hand, there is no barrier to energy flow to the west, therefore the whole quilt pattern is to be regarded as moving to the west. Only at the moment previously frozen in time would the positive maximum be over the origin. After the time taken for the incident wavetrain to deliver its next crest to the origin, the quilt pattern will have moved west by a distance u^{-1}. Since this distance, the east-west crest-to-crest distance, is greater than the wavelength q^{-1}, we see that the quilt pattern moves faster than the incident waves. This phenomenon can be confirmed occasionally on beaches where the waves are obliquely incident (usually they are not), and the situation also arises in other cases—for example, in glass, where a ray of light is reflected when it impinges on the glass-to-air boundary. The electromagnetic disturbance then propagates along the glass boundary faster than light. The same happens with microwaves in a waveguide for the same reason.

EXAMPLES OF TRANSFORM PAIRS

A stock-in-trade of two-dimensional Fourier transform pairs is useful for illustrating the analysis and synthesis discussed above and also for illustrating the theorems that follow.

Impulse

As a first example consider the pair

$$\boxed{{}^{2}\delta(x, y) \supset 1,}$$

which is illustrated in Fig. 4-6. Thus, a centrally situated unit two-dimensional impulse transforms into a function that is equal to unity, independent of u and v. To establish this we use the sifting property

$$\int_{-\infty}^{\infty}\int_{-\infty}^{\infty} {}^{2}\delta(x, y)f(x, y)\,dx\,dy = f(0, 0),$$

from which it follows that

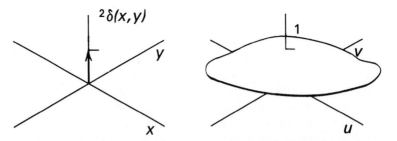

Figure 4-6 The unit two-dimensional impulse and its two-dimensional Fourier transform. The small ticks in this and other illustrations mark unit value along the axes.

$$\int_{-\infty}^{\infty}\int_{-\infty}^{\infty} {}^2\delta(x, y)e^{-i2\pi(ux+vy)}\, dx\, dy = 1.$$

The inverse transform would be

$$\int_{-\infty}^{\infty}\int_{-\infty}^{\infty} e^{i2\pi(ux+vy)}\, du\, dv = {}^2\delta(x, y).$$

To prove this relationship we could, if we wished, take the position that the direct transform has already been established and that, since the impulse notation must conform with ordinary notation, the inverse transform holds by necessity. On the other hand, one could ask for more direct insight. In that case the procedure is to consider a sequence of functions, such as the sequence generated by the expression $\exp[-\pi\tau^2(x^2 + y^2)]$ as $\tau \to 0$, which has the property of approaching unity. Each member of the sequence has a calculable transform. We then look at the sequence of transforms to see whether it is a suitable defining sequence for ${}^2\delta(u, v)$. We will be able to do this below.

As to physical interpretation one could say that sound escaping through a pinhole in a rigid wall spreads uniformly in all directions. This statement will be made rigorous later; meanwhile, it should be noted that further interpretation is needed, because the directions available to the sound reach only to 90° from the normal.

The inverse transform can be illustrated in the same field of acoustics, where it means that a wavefront of constant amplitude and phase over some plane radiates only in the single direction normal to the plane.

This sort of physical interpretation, whether in terms of acoustics, light, electromagnetic waves, or other waves, is of great assistance in thinking about the mathematical material.

Impulse pair

Taking two half-strength impulses at $(a, 0)$ and $(-a, 0)$ and integrating by means of the sifting property as before, we obtain (Fig. 4-7)

$$\boxed{0.5\,{}^2\delta(x + a, y) + 0.5\,{}^2\delta(x - a, y) \supset \cos 2\pi au.}$$

Although the direct transform

$$\int_{-\infty}^{\infty}\int_{-\infty}^{\infty} [0.5\,{}^2\delta(x + a) + 0.5\,{}^2\delta(x - a)]e^{-i2\pi(ux+vy)}\, dx\, dy$$

is very easy to evaluate by splitting into two integrals and using the sifting property, the

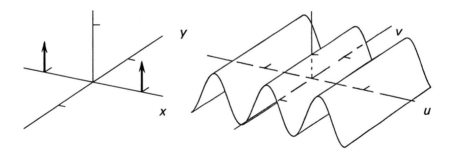

Figure 4-7 Symmetrical impulse pair transforms into a cosinusoidal corrugation. A related example appears in Fig. 4-15.

inverse transform

$$\int_{-\infty}^{\infty} \int_{-\infty}^{\infty} \cos 2\pi au \, e^{i2\pi(ux+vy)} \, du \, dv$$

is, as in the previous example, much more complicated.

One learns that it very often happens that the forward transform and the inverse transform are of unequal difficulty and that it pays to look at both ways of proceeding before forging ahead with what may prove to be the hard way.

As to physical interpretation, here we are dealing with the cosinusoidal interference pattern produced when waves escape through two pinholes, or from two point sources, with the same phase and amplitude at each source point. As before, the inverse transform also has an interpretation and one that is different in character from the interpretation of the direct transform. If we could impose a cosinusoidal amplitude variation, of spatial period a^{-1}, over a plane wavefront, then, according to the transform pair, radiation would be launched in just two directions, equally but oppositely inclined to the normal to the plane. The angle of launch, which is fixed by the spatial period a^{-1}, can easily be deduced.

Gaussian hump

A two-dimensional Gaussian function $\exp[-\pi(x^2 + y^2)]$, which may also be written $e^{-\pi r^2}$ in terms of the polar coordinate r, arises in innumerable connections. It is its own Fourier transform (Fig. 4-8),

$$\boxed{e^{-\pi r^2} \supset e^{-\pi q^2}.}$$

In this statement we use q as the radial polar coordinate in the (u, v)-plane; thus

$$r^2 = x^2 + y^2 \quad \text{and} \quad q^2 = u^2 + v^2.$$

The coefficient π is included in the position shown partly to have the convenience of symmetry, which makes it easier to remember this pair, and partly because $\exp(-\pi x^2)$ has unit area and $\exp(-\pi r^2)$ has unit volume.

To derive the result requires evaluating the integral in a conventional manner or, as is customary with standard integrals, looking it up in a table of integrals.

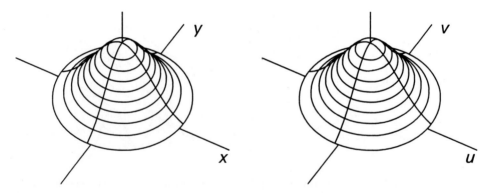

Figure 4-8 The unit Gaussian hump is its own Fourier transform.

Rectangular

A unit rectangle function $^2\text{rect}(x, y) = \text{rect}\,x\,\text{rect}\,y$ transforms into the product of two sinc functions as follows (Fig. 4-9):

$$^2\text{rect}(x, y) \supset \text{sinc}\,u\,\text{sinc}\,v.$$

Since in one dimension rect(·) transforms into sinc (·), one might suspect that there is a very simple direct derivation of this example. It will be given below. Radiation from a square aperture has a directional dependence connected with sinc u sinc v. As we are often concerned with power rather than with amplitude in radiation problems, it will often be $\text{sinc}^2 u\,\text{sinc}^2 v$ that is encountered. Apart from any importance that rectangular structures may have in the world of engineering, the two-dimensional rectangle function, which in speech is more conveniently called rect x rect y, arises in various other ways. For example, it is a two-dimensional gate function which, by multiplication, selects out a portion of a field for retention while putting the surrounding function values to zero.

Pillbox

Corresponding to the square aperture is the equally important circular aperture. Interpreted in two dimensions, the unit rectangle function of radius, rect r or $\Pi(r)$, represents a function that is equal to unity over a central circle of unit diameter and zero elsewhere. Just as the Fourier transform of a rectangle function in one dimension is a sinc function, so the two-dimensional transform of rect r is a jinc function. Naturally, a function as fundamental as the diffraction pattern of a circular aperture is itself intrinsically simple in nature. However, the understanding of the jinc function will be deferred for the moment. Here we content ourselves with introducing the formal definition jinc $q \equiv J_1(\pi q)/2q$, where J_1 is the Bessel function of the first kind of order unity. Then (Fig. 4-10)

$$\text{rect}\,r \supset \text{jinc}\,q.$$

Gaussian ridge

Consider $\exp(-\pi x^2)$ to show what happens when the function $f(x, y)$ is independent of y, i.e., when its representation is a cylinder (in the sense of a surface generated by a straight line

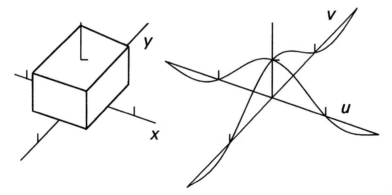

Figure 4-9 Unit two-dimensional rectangle function $^2\mathrm{rect}(x, y) = \mathrm{rect}\,x\,\mathrm{rect}\,y$ transforms into a function sinc u sinc v that is suggested by its two principal cross-sections.

Figure 4-10 Introducing the jinc function, the circular analogue of the sinc function and two-dimensional Fourier transform of the unit pillbox function rect r. Heights are exaggerated by a factor 2.

moving parallel to itself and passing through a fixed curve). The two-dimensional transform is representable by a nonuniform line impulse running along the u-axis. In Fig. 4-11 this is represented by a blade of height equal to the strength of the line impulse at each value of u. The two-dimensional Fourier transform pair is (Fig. 4-11)

$$\exp(-\pi x^2) \,{}^2\!\supset\, \exp(-\pi u^2)\delta(v).$$

Note that this is a case where the sign $^2\!\supset$ is being used for "has two-dimensional Fourier transform" because a possibility of confusion might exist with the symbol $\supset$ "has Fourier transform." To avoid such risk, where it exists, one can write $^2\!\supset$.

Line impulse

We next have a pair that could be generated from the preceding one by considering a sequence of Gaussian cylinders $\tau^{-1}\exp(-\pi x^2/\tau^2)$ which, as $\tau \to 0$, would be a suitable defining sequence for $\delta(x)$. If we regard $\delta(x)$ as a function of two dimensions, but independent of y, we have a unit line impulse running along the y-axis. The transforms pass through a sequence of blades of unit central height whose extent increases as τ^{-1}. In the limit, the two-dimensional

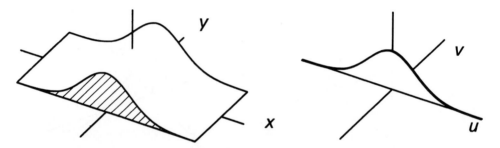

Figure 4-11   The Gaussian cylinder $\exp(-\pi x^2)$ transforms into a Gaussian blade.

Fourier transform of $\delta(x)$ is $\delta(v)$, a unit line impulse running along the u-axis. Thus (Fig. 4-12)

$$\delta(x) \; {}^2\!\supset \delta(v).$$

Two-dimensional signum function

The function $\operatorname{sgn} x$ equals $+1$ for positive x, -1 for negative x, and zero between. Introduce a two-dimensional generalization ${}^2\operatorname{sgn}(x, y) = \operatorname{sgn} x \operatorname{sgn} y$, which equals $+1$ in the first and third quadrants, -1 in the second and fourth, and zero on the axes. Alternatively, ${}^2\operatorname{sgn}(x, y) = \operatorname{sgn}(\sin 2\theta)$. We pronounce $\operatorname{sgn} x$ [sɪgnm̩ ɛks], (using International Phonetic Association symbols), in recognition of the original Latin; this pronunciation avoids a homophone with $\sin x$.

$$ {}^2\operatorname{sgn}(x, y) \; {}^2\!\supset \frac{1}{\pi^2 uv}.$$

Dipole

The following pair appears in the discussion of infinitesimal dipoles.

$$\sin 2\theta \; {}^2\!\supset {}^2\delta(u, v) \sin 2\phi.$$

Gaussian with angular variation

The following self-reciprocal pair has significance as an eigenfunction of the Radon transform as discussed under tomography.

$$e^{-\pi r^2} \sin 2\theta \; {}^2\!\supset e^{-\pi q^2} \sin 2\phi.$$

Bed-of-nails or shah function

Just as in one dimension the shah function is its own transform,

$$\operatorname{III}(x) \supset \operatorname{III}(s),$$

so also in two dimensions (Fig. 4-13)

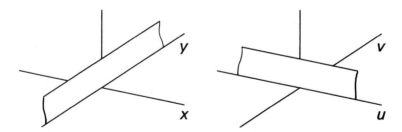

Figure 4-12   A unit line impulse on the y-axis transforms into a unit line impulse on the u-axis.

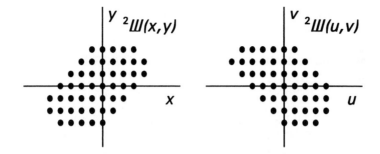

Figure 4-13   The two-dimensional bed-of-nails function $^2\mathrm{III}(x, y)$ is its own transform.

$$^2\mathrm{III}(x, y) \;^2{\supset}\; {}^2\mathrm{III}(u, v).$$

Grid

Regarded as a function of two variables the shah function

$$\mathrm{III}(x) = \sum_{n=-\infty}^{\infty} \delta(x - n)$$

describes a set of vertical unit-strength line impulses uniformly spaced at unit interval. Likewise $\mathrm{III}(y)$ describes a horizontal grid. The product of $\mathrm{III}(y)$ with $f(x, y)$ retains values of $f(x, y)$ along the lines $y = $ integer but abandons values in between. Thus $\mathrm{III}(y)f(x, y)$ contains information that is preserved in a horizontal raster scan of the function $f(x, y)$. Likewise $\mathrm{III}(x)f(x, y)$ would represent the information in a vertical raster scan, which is not seen very often because of the convention that has been adopted with television. The two-dimensional transform of $\mathrm{III}(x)$ is a uniform row of unit impulses $\sum_{m=-\infty}^{\infty} {}^2\delta(u - m, v)$ which can also be written briefly as $\mathrm{III}(u)\delta(v)$. The transform pair is

$$\mathrm{III}(x) \;^2{\supset}\; \mathrm{III}(u)\delta(v).$$

Conversely, $\mathrm{III}(y) \;^2{\supset}\; \mathrm{III}(v)\delta(u)$, a row of spikes on the v-axis, as illustrated in Fig. 4-14.

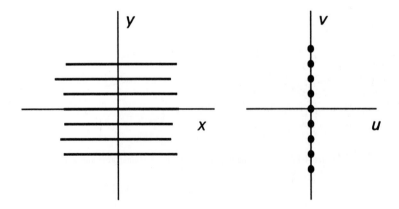

Figure 4-14 The uniform grid III(y) that appears in raster sampling transforms into the vertical impulse row III(v)$\delta(u)$.

THEOREMS FOR TWO-DIMENSIONAL FOURIER TRANSFORMS

Similarity Theorem

All the theorems for the one-dimensional Fourier transform go over into two dimensions. The similarity theorem, which tells us what happens when the abscissa expands or contracts, generalizes in two dimensions to tell us the effect of expansion or contraction of the picture plane (Fig. 4-15). An extra element enters, inasmuch as the expansion need not be uniform but may be unequal along the two axes. The theorem does not tell us what happens if the picture plane is stretched diagonally. Furthermore, it is not possible to combine stretching and compression along the coordinate axes to simulate stretching in the 45° direction. However, stretching in an arbitrary direction is easily handled with the aid of the rotation theorem introduced below.

Similarity Theorem in One Dimension

$$\text{If } f(x) \supset F(s)$$

$$\text{then } f(ax) \supset |a|^{-1} F(s/a).$$

Similarity Theorem in Two Dimensions

$$\text{If } f(x, y) \,^2{\supset}\, F(u, v)$$

$$\text{then } f(ax, by) \,^2{\supset}\, |ab|^{-1} F(u/a, v/b).$$

Shift Theorem

When a one-dimensional function is translated along the axis of abscissas, or, what is equivalent, the origin is shifted, no change occurs in the amplitude of any Fourier component, but the phases will be changed. A given amount of shift will amount to more and

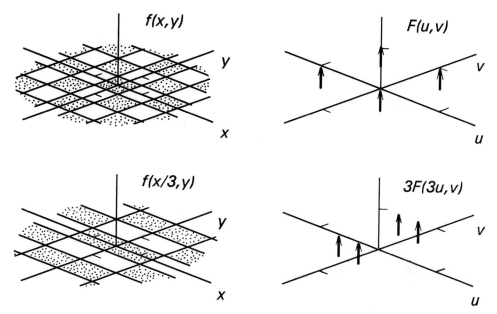

Figure 4-15   The similarity theorem in two dimensions illustrated by stretching a function by a factor of 3 in the x-direction. In this example $f(x, y)$ is the unit amplitude component $\cos \pi x \cos \pi y$. Despite the factor 3, introducing the expression $3F(3u, v)$, the impulses are of equal strength (namely, 0.5) before and after stretching. Representation of the functions on the (x, y)-plane is by a contour map with only one contour (the zero-level contour), supplemented by stippling that identifies the positive regions.

more phase change as components of higher frequency are considered. For example, consider a function of time that is delayed by an amount τ, and think of the effect on the Fourier component $A \cos 2\pi f t$. The period of this component is f^{-1}, and, as a shift of one period would introduce a phase change 2π, we see that a delay τ will change the phase by $-2\pi f \tau$. We see that the phase change of the component not only *depends* on its frequency f but is *proportional* to f (and is also, of course, proportional to the delay τ). A shift in the function domain thus introduces a linear phase gradient in the transform domain.

Shift Theorem in One Dimension

$$\text{If } f(x) \supset F(s)$$

$$\text{then } f(x - a) \supset e^{-i2\pi as} F(s).$$

By *phase gradient* we mean that the phase shift itself is not constant, but that a constant rate of change of phase is introduced along the axis of abscissas of the transform. The phase change at the origin $s = 0$ is zero. In two dimensions the origin can be independently translated along either the x- or y-axes, which leads to a simple generalization.

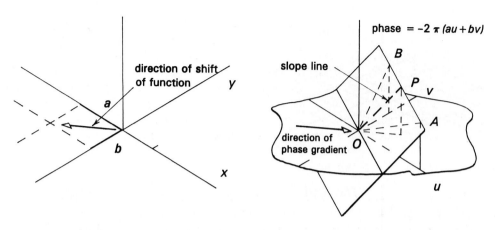

Figure 4-16   A shift of origin in the picture plane introduces a linear phase gradient in the transform domain.

Shift Theorem in Two Dimensions

$$\text{If } f(x, y) \, {}^2\!\!\supset F(u, v)$$

$$\text{then } f(x - a, y - b) \, {}^2\!\!\supset e^{-i2\pi(au+bv)} F(u, v).$$

Again we see that no change in amplitude occurs and that a linear phase gradient is introduced. The phase $-2\pi(au + bv)$ can be contemplated as a function of u and v as in Fig. 4-16, where it appears as a plane surface whose direction of maximum slope is opposite to the direction of shift of the function on the picture plane. In the illustration a and b are negative, the function shifts to the left, and the phase gradient is to the right. In any case the magnitude of the phase gradient is proportional to the magnitude $(a^2 + b^2)^{1/2}$ of the vector shift. Recalling that the gradient of a function of two variables, $\Phi(u, v)$, is defined by

$$\text{grad } \Phi = \mathbf{i} \, \partial\Phi/\partial x + \mathbf{j} \, \partial\Phi/\partial y,$$

where $\mathbf{i}$ and $\mathbf{j}$ are unit vectors in the u and v directions, we see that, when $\Phi(u, v) = -2\pi(au + bv)$,

$$\text{grad } \Phi = -2\pi a \, \mathbf{i} - 2\pi b \, \mathbf{j}.$$

The components are $\text{grad}_u \, \Phi = -2\pi a$ and $\text{grad}_v \, \Phi = -2\pi b$, which gives the slopes of lines OA and OB respectively; and the gradient, or the magnitude of the vector grad Φ, is $2\pi(a^2 + b^2)^{1/2}$ and gives the slope of the line OP.

The converse of the shift theorem states that a linear phase gradient applied to the picture plane, the phase change being zero at the origin, results in a shift of the function, in the (u, v)-plane, without change of form, the direction of shift depending on the direction of the applied gradient, and the amount of shift being proportional to the gradient applied.

Of course, when a phase gradient is applied to a real function, it becomes complex; as a result, the inverse shift theorem might not be expected to arise very often, but in fact it often helps to think in terms of shifting things around on the (u, v)-plane.

Again a suggestion arises of theorems with no one-dimensional counterpart. In this case we can imagine a displacement that comprises not only a two-dimensional translation but also a rotation. While displacement in one-dimension can be described by just one parameter, the most general displacement in two dimensions, without change of form, requires three parameters or can be said to possess three degrees of freedom.

Rotation Theorem

It is physically obvious that rotation of an antenna about its beam axis will result in an equal rotation of the radiation pattern of the antenna, and the same can be said for the diffraction pattern of an optical source. We know that a Fourier transform underlies the relationship in these cases, and so we expect the following theorem.

Rotation Theorem

$$\text{If } f(x, y) \,{}^2\!\supset F(u, v)$$

$$\text{then } f(x \cos \theta - y \sin \theta, x \sin \theta + y \cos \theta) \,{}^2\!\supset F(u \cos \theta - v \sin \theta, u \sin \theta + v \cos \theta).$$

It is not intuitively obvious from the mathematical definition that this theorem follows, and in particular some moments of very attentive mathematical thought are required to assure oneself that clockwise rotation of $f(x, y)$ does *not* result in counterclockwise rotation of $F(u, v)$. This theorem is an example of the kind of power that results from possessing more than one way of interpreting Fourier transforms.

Shear Theorems

The similarity theorem, shift theorem, and rotation theorem are special cases where one distorts a function by shifting points on the (x, y)-plane in accordance with an affine transformation

$$x' = ax + by + c, \qquad y' = dx + ey + f.$$

Pure shear is another special case. For example,

$$x' = x + by, \qquad y' = y$$

describes horizontal shear distortion that would take each square on a sheet of graph paper into an equilateral parallelogram, or rhombus, of the same height and area but sloping sides. A general function $f(x, y)$ is converted into a different function $f(x + by, y)$. The shear theorem tells us that there is a corresponding distortion of the (u, v)-plane which alters the original transform $F(u, v)$, as follows.

Simple Shear Theorem

$$\text{If } f(x, y) \,{}^2\!\!\supset F(u, v)$$

$$\text{then } f(x + by, y) \,{}^2\!\!\supset F(u, v - bu).$$

Thus the corresponding distortion in the (u, v)-plane is also pure shear but in the perpendicular direction. As is known from the theory of elasticity, a state of shear strain is equivalent to a combination of pure compression and extension in diagonal directions. One way of proving the shear theorem is to apply the similarity theorem in two perpendicular diagonal directions in turn.

Vertical shear is described by $x' = x$, $y' = dx + y$, and in this case the transform suffers horizontal shear to become $F(u - dv, v)$.

Compound Shear Theorem

$$\text{If } f(x, y) \,{}^2\!\!\supset F(u, v)$$

$$\text{then } f(x + by, dx + y) \,{}^2\!\!\supset \frac{1}{|\,1 - bd\,|} F\left(\frac{u - dv}{1 - bd}, \frac{-bu + v}{1 - bd}\right).$$

Looking at (x, y) as a vector $\mathbf{x}$ and (x', y') as a vector $\mathbf{x}'$, the compound-shear coordinate transformation can be written concisely as

$$\mathbf{x}' = \begin{bmatrix} 1 & b \\ d & 1 \end{bmatrix} \mathbf{x}.$$

The matrix operator for horizontal shear is $\begin{bmatrix} 1 & b \\ 0 & 1 \end{bmatrix}$ or $\begin{bmatrix} 1 & \tan\beta \\ 0 & 1 \end{bmatrix}$ and for vertical shear $\begin{bmatrix} 1 & 0 \\ d & 1 \end{bmatrix}$ or $\begin{bmatrix} 1 & 0 \\ \tan\delta & 1 \end{bmatrix}$, where β and δ are angles noted on Fig. 4-17. Horizontal shear followed by vertical shear is expressed by

$$\begin{bmatrix} 1 & 0 \\ d & 1 \end{bmatrix}\begin{bmatrix} 1 & b \\ 0 & 1 \end{bmatrix} = \begin{bmatrix} 1 & b \\ d & 1 + bd \end{bmatrix},$$

while vertical shear followed by horizontal shear is expressed by

$$\begin{bmatrix} 1 & b \\ 0 & 1 \end{bmatrix}\begin{bmatrix} 1 & 0 \\ d & 1 \end{bmatrix} = \begin{bmatrix} 1 + bd & b \\ d & 1 \end{bmatrix}.$$

Thus simple shear operations carried out in succession produce different outcomes, depending on the order in which they are applied. In matrix terminology, the matrix factors do not commute. In neither case is the outcome the same as for compound shear as defined. Compound shear results in the same inclination angles β and δ as for simple shear. Figure 4-17 shows that δ is retained in case (d) while a new angle ζ is introduced. In case (e) β is retained and a new angle η comes in. The angles ζ and η are given by $\tan\zeta = b/(1 + bd)$ and $\tan\eta = d/(1 + bd)$. Since shear does not change the area of a figure, all the deformed outlines in Fig. 4-17 retain unit area. The area magnification is equal to $|\,ae - bd\,|$.

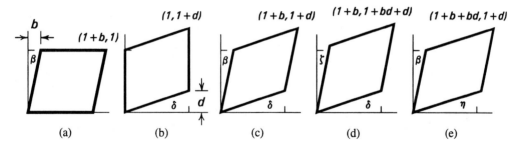

Figure 4-17   A unit square is subjected to (a) horizontal shear, (b) vertical shear, (c) compound shear, (d) horizontal shear followed by vertical shear, (e) vertical shear followed by horizontal shear. Coordinates are shown for the NE vertex.

Table 4-1   Reference Table for Shear

Case	B	C	D	m_{AB}	m_{AC}	m_{AD}
(a)	$(1, 0)$	$(1 + b, 1)$	$(b, 1)$	0	$1/(1 + b)$	$1/b$
(b)	$(1, d)$	$(1, 1 + d)$	$(0, 1)$	d	$1 + d$	∞
(c)	$(1, d)$	$(1 + b, 1 + d)$	$(b, 1)$	d	$(1 + d)/(1 + b)$	$1/b$
(d)	$(1, d)$	$(1 + b, 1 + bd + d)$	$(b, 1 + bd)$	d	$(1 + bd + d)/(1 + b)$	$(1 + bd)/b$
(e)	$(1 + bd, d)$	$(1 + b + bd, 1 + d)$	$(b, 1)$	$d/(1 + bd)$	$(1 + d)/(1 + b + bd)$	$1/b$

A unit square $ABCD$ with A at the origin, B at (1,0), C at (1,1), and D at (0,1), after deformation under the five cases of Fig. 4-17, finds its vertices in the locations listed in the accompanying table. The point A remains at (0,0). The slope of the side AB is listed under m_{AB} and similarly for the diagonal AC and initially vertical side AD.

Other natural parameters are the expansion or compression of the diagonals, which are easily calculable, but, if conversely the extension of the diagonal is specified, it is awkward to extract the parameters a, b, c, d. Bearing in mind that CCW rotation through θ is produced by the operator $\begin{bmatrix} \cos\theta & \sin\theta \\ -\sin\theta & \cos\theta \end{bmatrix}$, while expansion by a factor M in the x-direction is produced by $\begin{bmatrix} M & 0 \\ 0 & 1 \end{bmatrix}$, any desired expansion in a specified direction can be reached by a combination of rotation and expansion and expressed in the form $\begin{bmatrix} a & b \\ c & d \end{bmatrix}$ by a sequence of matrix multiplications.

Affine Theorem

This is a little-known theorem that incorporates several of the foregoing theorems (similarity, shift, compound, shear) as special cases. It is particularly useful in its general form in image processing, where sequences of affine transformations are applied. Derivations have not been given for the simpler theorems, but the derivation of this theorem is instructive (Bracewell et al., 1993). The theorem which we shall derive is:

Affine Theorem

$$\text{If } f(x, y) \text{ has 2-D FT } F(u, v),$$

$$\text{then } g(x, y) = f(ax + by + c, dx + ey + f) \text{ has 2-D FT}$$

$$G(u, v) = \frac{1}{|\Delta|} \exp\left\{ \frac{i2\pi}{\Delta} [(ec - bf)u + (af - cd)v] \right\} F\left(\frac{eu - dv}{\Delta}, \frac{-bu + av}{\Delta} \right),$$

where the determinant Δ is given by

$$\Delta = \begin{vmatrix} a & b \\ d & e \end{vmatrix} = ae - bd.$$

To derive this result, express the affine coordinate transformation

$$x' = ax + by + c, \qquad y' = dx + ey + f$$

in the matrix notation:

$$\begin{bmatrix} x' \\ y' \end{bmatrix} = \begin{bmatrix} a & b \\ d & e \end{bmatrix} \begin{bmatrix} x \\ y \end{bmatrix} + \begin{bmatrix} c \\ f \end{bmatrix}$$

and note the Jacobian relation $dx' \, dy' = |\Delta| \, dx \, dy$. If $\Delta \neq 0$, invert the transformation to get

$$\begin{bmatrix} x \\ y \end{bmatrix} = \begin{bmatrix} a & b \\ d & e \end{bmatrix}^{-1} \begin{bmatrix} x' - c \\ y' - f \end{bmatrix}.$$

In the phase exponent $-i2\pi(ux + vy)$ that occurs in the definition of the two-dimensional Fourier component note that

$$ux + vy = [u \quad v] \begin{bmatrix} x \\ y \end{bmatrix}$$

$$= [u \quad v] \begin{bmatrix} a & b \\ d & e \end{bmatrix}^{-1} \begin{bmatrix} x' - c \\ y' - f \end{bmatrix}.$$

$$= \frac{1}{\Delta} [u \quad v] \begin{bmatrix} e & -b \\ -d & a \end{bmatrix} \begin{bmatrix} x' - c \\ y' - f \end{bmatrix}$$

$$= \frac{1}{\Delta} [eu - dv \quad -bu + av] \begin{bmatrix} x' \\ y' \end{bmatrix} - \frac{1}{\Delta} [eu - dv \quad -bu + av] \begin{bmatrix} c \\ f \end{bmatrix}.$$

Hence

$$G(u, v) = \int_{-\infty}^{\infty} \int_{-\infty}^{\infty} f(ax + by + c, dx + ey + f)e^{-i2\pi(ux+vy)} \, dx \, dy$$

$$= \int_{-\infty}^{\infty} \int_{-\infty}^{\infty} f(x', y')e^{-i\frac{2\pi}{\Delta}[(eu-dv)x'+(-bu+av)y']}e^{i\frac{2\pi}{\Delta}[(ec-bf)u+(af-cd)v]} dx' dy'/|\Delta|.$$

$$= \frac{1}{|\Delta|} \exp\left\{ \frac{i2\pi}{\Delta}[(ec - bf)u + (af - cd)v] \right\} F\left(\frac{eu - dv}{\Delta}, \frac{-bu + av}{\Delta} \right).$$

This completes the derivation. The expression can be condensed by referring to affine transform plane coordinates defined by $u' = (eu - dv)/\Delta$ and $v' = (-bu + av)/\Delta$. Then

$$G(u, v) = \frac{1}{|\Delta|} e^{i2\pi(cu'+fv')} F(u', v').$$

The inverse transformation to (u, v)-coordinates is

$$\begin{bmatrix} u \\ v \end{bmatrix} = \begin{bmatrix} a & d \\ b & e \end{bmatrix} \begin{bmatrix} u' \\ v' \end{bmatrix}.$$

One does not begin with much intuitive feel for the affine coefficients; taken on their own, b and d express x-shear and y-shear, a and e are linear magnifications, while c and f are displacements. It would make sense to introduce mnemonic value by writing

$$x' = M_x x + \sigma_x y + x_0$$

$$y' = \sigma_y x + M_y y + y_0.$$

In this notation some expressions seem to be more intelligible. For example, the area magnification $| ae - bd |$ becomes $| M_x M_y - \sigma_x \sigma_y |$.

Rayleigh's Theorem in Two Dimensions

Although Rayleigh's theorem, first published in 1889, referred to time-dependent waveforms and their spectra, it is convenient to generalize the name to include the corresponding theorem in two dimensions. The theorem is as follows.

Rayleigh's Theorem in Two Dimensions

If $f(x, y) \,^2\!\supset F(u, v)$

then $$\int_{-\infty}^{\infty} \int_{-\infty}^{\infty} | f(x, y) |^2 \, dx \, dy = \int_{-\infty}^{\infty} \int_{-\infty}^{\infty} | F(u, v) |^2 \, du \, dv.$$

In physical situations the relation often represents the equality of two power flows. For example, the left-hand side might represent the power flowing through an aperture as integrated element by element over the aperture, and the right-hand side might be the same power integrated direction by direction over the diffracted beam of the radiated field.

There is a more general theorem, which arises less often. It would be encountered in connection with radiation from an aperture, if the power were to be expressed as the product of an electric with a magnetic field rather than as proportional to the squared magnitude of just one field component. The more general theorem is

$$\int_{-\infty}^{\infty}\int_{-\infty}^{\infty} f(x,y)g^*(x,y)\,dx\,dy = \int_{-\infty}^{\infty}\int_{-\infty}^{\infty} F(u,v)G^*(u,v)\,du\,dv.$$

A derivation of Rayleigh's theorem can be based directly on the autocorrelation theorem, which will be introduced later.

Parseval's Theorem in Two Dimensions

Parseval's theorem, like Rayleigh's, can often be interpreted as a statement of equality between two energies or powers viewed in different ways. For example, in a loss-free electric transmission-line system the stored energy can be expressed as a spatial integral of the squared modulus of the electric field, or it can be expressed as the sum of the energies in the natural modes. In a mechanical system or in an acoustical resonator the sum of the energies in all the modes of vibration is equal to a spatial integral of squared stress or strain or of sound intensity.

If there is a periodic waveform $f(t)$, then $[f(t)]^2$ cannot be integrated to infinity, but it can be integrated over one period; theorems involving $\int_{-\infty}^{\infty}[f(t)]^2\,dt$ do not apply to periodic functions. Correspondingly, if $f \supset F$, the integral of $|F|^2$ cannot be evaluated if F is impulsive, as it will be if f is periodic. Parseval's theorem deals with such situations. Let $f(x,y)$ be periodic in both x and y with unit period. The theorem relates the integral of $|f(x,y)|^2$ over unit cell to the coefficients of the impulses in the (u,v)-plane defined by

$$F(u,v) = \sum\sum a_{mn}[^2\delta(u-m,v-n)].$$

Parseval's Theorem in Two Dimensions

If $f(x+1,y+1) = f(x,y)$ for all x and y

then $\int_{-1/2}^{1/2}\int_{-1/2}^{1/2} |f(x,y)|^2\,dx\,dy = \Sigma\Sigma a_{mn}^2,$

where a_{mn} is the strength of the impulse at $u=m, v=n$.

Marc Antoine Parseval (1755–1836) published his result in 1799 (before Fourier began working on heat diffusion) and also evaluated the spin integral of the cosine function to obtain the series expanson of the zero-order Bessel function.

Derivative Theorem

It is sufficient to discuss differentiation with respect to x, for which the theorem is as follows.

Derivative Theorem

$$\text{If } f(x, y) \,^2\!\!\supset F(u, v)$$

$$\text{then } \left(\frac{\partial}{\partial x}\right) f(x, y) \,^2\!\!\supset i2\pi u F(u, v).$$

We see that all spatial frequency components of $f(x, y)$ are to be increased (or reduced) in proportion to their x-component of spatial frequency and moved 90° in spatial phase. The reason for this is apparent from consideration of a single component $A \sin 2\pi u x \sin 2\pi v y$ whose derivative with respect to x is

$$2\pi u A \sin(2\pi u x + \tfrac{1}{2}\pi) \sin 2\pi v y.$$

The higher the frequency of a given component, the more its amplitude is enhanced in the spectrum of the derivative, because, of course, the maximum slope of a sinusoid of given amplitude is greater in proportion to its frequency.

A list of further results stemming from this theorem follows.

$$\frac{\partial}{\partial y} f(x, y) \,^2\!\!\supset i2\pi v F(u, v)$$

$$\frac{\partial^2}{\partial x^2} f(x, y) \,^2\!\!\supset -4\pi^2 u^2 F(u, v)$$

$$\frac{\partial^2}{\partial y^2} f(x, y) \,^2\!\!\supset -4\pi^2 v^2 F(u, v)$$

$$\frac{\partial^2}{\partial x \partial y} f(x, y) \,^2\!\!\supset -4\pi^2 u v F(u, v)$$

$$\left(\frac{\partial^2}{\partial x^2} + \frac{\partial^2}{\partial y^2}\right) f(x, y) \,^2\!\!\supset -4\pi^2 (u^2 + v^2) F(u, v).$$

Difference Theorems

Define the first difference of $f(x, y)$ in the x-direction over interval a by $^a\Delta_x f(x, y) = f(x + \tfrac{1}{2}a, y) - f(x - \tfrac{1}{2}a, y)$. By the shift theorem, the transform is $\exp i a\pi u F(u, v) - \exp(-i\pi au)F(u, v)$ or $2i \sin \pi au F(u, v)$. Thus

First Difference Theorem

$$\text{If } f(x, y) \,^2\!\!\supset F(u, v)$$

$$\text{then } ^a\Delta_x f(x, y) = f(x + \tfrac{1}{2}a, y) - f(x - \tfrac{1}{2}a, y) \,^2\!\!\supset 2i \sin \pi au F(u, v).$$

Since, by definition,

$$\frac{\partial f(x, y)}{\partial x} = \frac{\lim}{a \to 0} \frac{^a\Delta_x f(x, y)}{a},$$

the derivative theorem follows from the first difference theorem. The second difference $^a\Delta_{xx} f(x, y) = {}^a\Delta_x[^a\Delta_x f(x, y)]$ leads to the

Second Difference Theorem

$$\text{If } f(x, y) \,^2\!\supset F(u, v)$$

$$\text{then } {}^a\!\Delta_{xx} f(x, y) = f(x + a, y) - 2f(x, y) + f(x - a, y) \,^2\!\supset -4\sin^2 \pi au F(u, v).$$

Further results are

$$^b\!\Delta_{yy} \,^2\!\supset -4\sin^2 \pi bv F(u, v)$$

$$^a\!\Delta_x \,^b\!\Delta_y f(x, y) \,^2\!\supset -4\sin \pi au \sin \pi bv F(u, v)$$

$$^a\!\Delta_{xx} + {}^b\!\Delta_{yy} \,^2\!\supset -4[\sin^2 \pi au + \sin^2 \pi bv] F(u, v).$$

Definite Integral Theorem

Very useful for checking calculations, for determining normalization factors, and for evaluating integrals, this apparently trivial theorem is constantly drawn on.

Definite Integral Theorem

$$\text{If } f(x, y) \,^2\!\supset F(u, v)$$

$$\text{then } \int_{-\infty}^{\infty} \int_{-\infty}^{\infty} f(x, y) \, dx \, dy = F(0, 0).$$

First Moment Theorem

Since the operation of taking first moments comes up in a wide variety of circumstances, it is useful to know the corresponding operation in the transform domain. Just as the first moment of $f(x, y)$ is merely one piece of information about $f(x, y)$, so also there is a single property of $F(u, v)$ which reveals the first moment of $f(x, y)$. That property is the central slope of $F(u, v)$ in the u-direction.

First Moment Theorem

$$\text{If } f(x, y) \,^2\!\supset F(u, v)$$

$$\text{then } \int_{-\infty}^{\infty} \int_{-\infty}^{\infty} xf(x, y) \, dx \, dy = \frac{1}{-2\pi i} F'_u(0, 0)$$

where $F'_u(u, v) \equiv (\partial/\partial u) F(u, v)$.

This theorem can be derived from the inverse of the derivative theorem

$$-i2\pi x f(x, y) \,^2\!\supset (\partial/\partial u) F(u, v),$$

where the factor $-i2\pi x$ takes the place of the factor $i2\pi u$ seen in the normal derivative theorem because of the change in sign of i in the inverse Fourier transformation. Now, applying the definite integral theorem to the inverse of the derivative theorem, we get

$$\int_{-\infty}^{\infty} \int_{-\infty}^{\infty} -i2\pi x f(x, y) \, dx \, dy = (\partial/\partial u) F(u, v) \, |_{u=v=0} \, .$$

Second Moment Theorem

Applying the definite integral theorem after taking the second derivative of $F(u, v)$ with respect to u, we obtain immediately the next theorem.

Second Moment Theorem

$$\text{If } f(x, y) \, {}^2\!\supset F(u, v)$$

$$\text{then } \int_{-\infty}^{\infty} \int_{-\infty}^{\infty} x^2 f(x, y) \, dx \, dy = -F_{uu}''(0, 0)/4\pi^2.$$

Other results may be added for reference.

$$\int_{-\infty}^{\infty} \int_{-\infty}^{\infty} y^2 f(x, y) \, dx \, dy = -F_{vv}''(0, 0)/4\pi^2$$

$$\int_{-\infty}^{\infty} \int_{-\infty}^{\infty} xy f(x, y) \, dx \, dy = -F_{uv}''(0, 0)/4\pi^2$$

$$\int_{-\infty}^{\infty} \int_{-\infty}^{\infty} (x^2 + y^2) f(x, y) \, dx \, dy = -[F_{uu}''(0, 0) + F_{vv}''(0, 0)]/4\pi^2.$$

Equivalent Area Theorem

Just as an equivalent width can be defined in one dimension, so in two dimensions there is an equivalent area A_f with the property that

$$(\text{equivalent area}) \times (\text{height of function at origin}) = \text{volume}.$$

Thus

$$A_f \, f(0, 0) = \int_{-\infty}^{\infty} \int_{-\infty}^{\infty} f(x, y) \, dx \, dy.$$

The transform $F(u, v)$ has an equivalent area also. Call it A_F.

Equivalent Area Theorem

$$\text{If } f(x, y) \, {}^2\!\supset F(u, v)$$

$$\text{then } A_f = 1/A_F.$$

Equivalent area is a standard concept in some fields. For example, the concept of beam solid angle Ω of an antenna radiation pattern is the equivalent area of the radiation intensity pattern. Antenna directivity or directive gain D may also be expressed in terms of equivalent area; in fact $D = 4\pi/\Omega$. In other fields equivalent area is never employed but in some cases might be useful. Its simple reciprocal property commends it.

Separable Product Theorem

It may happen that a function of x and y is separable into a product of two functions, one of which is a function of x alone and the other of y alone. In that case there is a very useful relation that enables the two-dimensional Fourier transform to be determined by taking one-dimensional transforms only.

Separable Product Theorem

$$\text{If } f(x) \supset F(u) \quad \text{and} \quad g(x) \supset G(v)$$

$$\text{then } f(x)g(y) \overset{2}{\supset} F(u)G(v).$$

In the important special case where the function of x and y is independent of y, we can think of $g(y)$ as being unity. Then it follows that

$$f(x) \overset{2}{\supset} F(u)\delta(v).$$

THE TWO-DIMENSIONAL HARTLEY TRANSFORM

Given a function $f(x, y)$, and using the abbreviation $\operatorname{cas}\theta = \cos\theta + \sin\theta$, the two-dimensional Hartley transform and its inverse are as follows.

$$H(u, v) = \int_{-\infty}^{\infty} \int_{-\infty}^{\infty} f(x, y)\operatorname{cas}[2\pi(ux + vy)]\,dx\,dy$$

$$f(x, y) = \int_{-\infty}^{\infty} \int_{-\infty}^{\infty} H(u, v)\operatorname{cas}[2\pi(ux + vy)]\,du\,dv.$$

The two-dimensional Hartley transform of a real object has the interesting property of being itself real. With the Fourier transform, which is complex and possesses hermitian redundancy, one-half of the transform plane suffices to determine the original object. Information in the Hartley plane is spread over the whole plane without redundancy or symmetry. The antisymmetry that characterizes the Fourier plane is not possessed by the Hartley plane, every point of which counts. In computing, the property of being real valued is a considerable convenience, as it also is in analogue situations where phase, which is not responded to by optical detectors, has significance.

A fast algorithm for Fourier analysis of images (faster than the FFT) has been discovered and may be useful to those concerned with filtering or other transform domain manipulation of images (Bracewell et al., 1986).

THEOREMS FOR THE HARTLEY TRANSFORM

All the theorems for the Fourier transform have counterparts applying to the Hartley transform. The several theorems that are special cases of the affine theorem (Bracewell, 1994) can be condensed for reference as follows.

Affine Theorem for Hartley Transform

If $f(x, y)$ has 2-D Hartley transform $H(u, v)$,

then $f(ax + by + c, dx + ey + f)$ has 2-D Hartley transform

$| \Delta |^{-1} [H(\alpha, \beta) \cos \Theta - H(-\alpha, -\beta) \sin \Theta]$,

where

$$\Delta = ae - bd$$

$$\alpha = (eu - dv)/\Delta \qquad \beta = (-bu + av)/\Delta$$

$$\Theta = 2\pi \Delta^{-1}[(ec - bf)u + (af - cd)v].$$

Other theorems can be extracted directly from the Fourier version as follows:

Conversion Theorem

If $f(x, y)$ has Fourier transform $R(u, v) + iI(u, v)$,

then $f(x, y)$ has Hartley transform $R(u, v) - I(u, v)$.

DISCRETE TRANSFORMS

The discrete Fourier transform and discrete Hartley transform are well known in one dimension and need only be mentioned here in their two-dimensional form for reference. For the discrete Fourier transform

$$F(\sigma, \tau) = \frac{1}{MN} \sum_{x=0}^{M-1} \sum_{y=0}^{N-1} f(x, y) \exp\left[-i\left(\frac{2\pi\sigma x}{M} + \frac{2\pi\tau y}{N}\right)\right]$$

$$f(x, y) = \sum_{\sigma=0}^{M-1} \sum_{\tau=0}^{N-1} F(\sigma, \tau) \exp\left[i\left(\frac{2\pi\sigma x}{M} + \frac{2\pi\tau y}{N}\right)\right]$$

and for the discrete Hartley transform

$$H(\sigma, \tau) = \frac{1}{MN} \sum_{x=0}^{M-1} \sum_{y=0}^{N-1} f(x, y) \operatorname{cas}\left[\left(\frac{2\pi\sigma x}{M} + \frac{2\pi\tau y}{N}\right)\right]$$

$$f(x, y) = \sum_{\sigma=0}^{M-1} \sum_{\tau=0}^{N-1} H(\sigma, \tau) \operatorname{cas}\left[\left(\frac{2\pi\sigma x}{M} + \frac{2\pi\tau y}{N}\right)\right]$$

The equations refer to a data array of size $M \times N$ with integer indices x and y running from 0 to $M - 1$ and 0 to $N - 1$, respectively, and having the same meaning as on the (x, y)-plane apart from the restriction to integer values. The range restriction, which places the origin at the lower left, is awkward at times when it is more natural to think of a centrally situated origin, but it is firmly fixed by custom. Parts of an image that would be thought of as residing in the NW quadrant with respect to a central origin appear in the SE of the $M \times N$ matrix, SE goes to NW, and SW goes to NE. At the same time reflections are introduced which are correctly implemented when negative values of x and y (relative to a central origin) are replaced by $M - x$ and $N - y$ in the matrix.

Division by the factor MN results in the transform values $F(0, 0)$ and $H(0, 0)$ being equal to the spatial dc level of $f(x, y)$, as is customary for the leading term in the Fourier series expansion of a periodic function. This factor is absent from the inverse formula. In computing practice, it is efficient not to introduce the factor MN at all into the fast algorithm but to combine it subsequently with other normalizing factors, calibration factors, or graphics scale factors that are needed when the results come to be displayed.

The symbols σ and τ, which have to do with spatial frequency, are not customary and are introduced to draw attention to the fact that they differ in meaning from the true spatial frequencies u and v. To get the spatial frequency in cycles per unit of x (and the unit of x is by definition unity in the discrete case) we need σ/M, but that is not all; σ has to be no larger than $\frac{1}{2}M$. In the range $\frac{1}{2}M < \sigma < M$, the spatial frequency represented by σ is $(\sigma - M)/M$. This is because small negative values of u appear as large values of σ in the discrete formulation. No spatial frequency greater then 0.5 thus appears, as is expected for sampling at unit spacing, where the shortest period is 2. For τ the spatial frequency is $(\tau - N)/N$ or τ/N according as τ exceeds $\frac{1}{2}N$ or not.

SUMMARY

Different ways of looking at two-dimensional spectral analysis have been presented. One way is to generalize from what one already knows about time-domain waveforms and their spectra; another is to think pictorially in terms of topographic surfaces such as sinusoidal dunes of all wavelengths and orientations or alternatively of the "quilted surfaces" which, in general, have unequal spatial frequencies in the x- and y-directions. The concept of spatial frequency measured in cycles per unit of distance in the (x, y)-plane needs to become as familiar as the concept of temporal frequency measured in cycles per second; a crucial difference is that spatial frequency has two components and may be thought of as a complex number.

A stock-in-trade of transforms has been recorded compactly for reference and will be found to cover the majority of needs in everyday image engineering, especially when extended to all the variations generated by the use of the reference list of theorems. Many examples of transform pairs so generated are found in the problems; facility at recognizing the applicability of some theorem to a known pair is most useful to cultivate. Only brief mention of the significance or use of the material has been introduced at this stage in order to keep the presentation brief and compact for reference purposes. Still, it is not a good idea to skim this chapter; take a deep breath, accept it as a heavy mathematical pill, swallow it, and make what you think will be useful in the future part of you.

LITERATURE CITED

R. N. BRACEWELL, O. BUNEMAN, H. HAO, AND J. VILLASENOR , "Fast two-dimensional Hartley transforms," *Proc. IEEE*, vol. 74, pp. 1283–1284, September 1986.

R. N. BRACEWELL, K.-Y. CHANG, A. K. JHA, AND Y. H. WANG , "Affine theorem for two-dimensional Fourier transform," *Electronics Letters*, vol. 29, p. 304, February 1993.

R. N. BRACEWELL , "Affine theorem for the Hartley transform of an image," *Proc. IEEE*, vol. 82, pp. 388–390, March 1994.

FURTHER READING

In two dimensions neither Fourier transform pairs nor their theorems have attracted much attention beyond the specific needs of authors dealing with particular problems. The notation and approach adopted here were developed from FTA (1986), which can be recommended for brushing up on one- and two-dimensional aspects. For extensive listing of one-dimensional transforms (from which many two-dimensional ones may be constructed by use of theorems) see Erd (1954) and G. A. Campbell and R. M. Foster, *Fourier Integrals for Practical Applications*, Van Nostrand, New York, 1948. For further information on the Hartley transform see R. N. Bracewell, *The Hartley Transform*, Oxford University Press, 1986, and *Proc. IEEE*, vol. 82, special issue on the Hartley transform, March 1994. For information about the life and times of Fourier and his contemporaries see I. Grattan-Guiness, *Joseph Fourier, 1768–1830*, MIT Press, Cambridge, MA, 1972.

PROBLEMS

4–1. *Synthesis of corrugation.* It has been shown that two corrugations added together can form the quilt pattern $\cos 2\pi ux \cos 2\pi vy$. A corrugation has spatial frequency q_1 and orientation ϕ_1. Show that it can be represented as the sum of quilt patterns, but that in general four quilt patterns are required. ☞

4–2. *Fourier transform of rectangle pair.* What is the Fourier transform of $^2\text{rect}[(x + X)/W, y/H] + {}^2\text{rect}[(x - X)/W, y/H]$?

4–3. *Two-dimensional transform.*
(a) What are the two-dimensional Fourier transforms of the following functions? (i) $\text{rect}(x/a)\,\text{rect}(y/b)$, (ii) $\text{rect}(x/a)\,\text{rect}(y/b - 0.5)$, (iii) $\text{rect}(x/a - 0.5)\,\text{rect}(y/b)$?
(b) Explain why $F(0, 0)$ is the same in all three cases,
(c) Explain why $F(u, 0)$ is the same in the first two cases.

4–4. *Fourier transform of wedge.* The value of a function is unity within the isosceles triangle of height h whose base runs from $(-0.5, 0)$ to $(0.5, 0)$ and is equal to zero elsewhere. Give expressions for the transform cross sections $F(u, 0)$ and $F(0, v)$.

4–5. *A theorem.* Work out the transform of $\cos 2\pi x\, f(x, y) + i \sin 2\pi x\, f(x, y)$.

4–6. *Another theorem.* Work out the transform of $\cos 2\pi x\, f(x, y) + i \sin 2\pi y\, f(x, y)$. ☞

4–7. *Two-dimensional transform.*
(a) Given that $f_1(x, y) = {}^2\text{rect}(x - 3, y) + {}^2\text{rect}(x + 3, y)$, $f_2(x, y) = {}^2\text{rect}(x, y - 5) + {}^2\text{rect}(x, y + 5)$, $f_3(x, y) = {}^2\text{rect}(x - 0.4, y) + {}^2\text{rect}(x + 0.4, y)$, and $f_4(x, y) = {}^2\text{rect}(x - 2, y - 2) + {}^2\text{rect}(x + 2, y + 2)$, work out the two-dimensional Fourier transforms.

(b) Make accurate drawings of the four functions on the (x, y)-plane, designing your drawings for information transfer rather than as personal memoranda.

(c) Devise simple graphical representations of the four transforms that would help someone else to obtain a quick grasp of the character and scale of the transforms. ☞

4–8. *Two-dimensional transform.*

(a) Give the two-dimensional Fourier transforms of the following functions,

$$(i) f_1(x, y) = H \exp\{-\pi[(x/a)^2 + (y/b)^2]\},$$

$$(ii) f_2(x, y) = \exp(-\pi r^2) + R^{-2} \exp[-\pi (r/R)^2],$$

$$(iii) f_3(x, y) = \exp\{-\pi[(x + a)^2 + y^2]\} + R^{-2} \exp\{-\pi\{[(x - a)^2 + y^2]/R^2\}\}.$$

(b) Represent the (x, y)-functions graphically by drawing in the half-peak and tenth-peak contours, using $a = 1$, $b = 2$ and $R = 0.5$.

(c) Verify the transforms by evaluating the central values $F(0, 0)$. ☞

4–9. *Transform of line pattern.* A pattern consists of four unit line impulses which run from $(-1, -1)$ to $(1, 1)$, from $(-1, 1)$ to $(1, -1)$, from $(0, -1)$ to $(0, 1)$, and from $(-1, 0)$ to $(1, 0)$. Experiment with ways of conveying the two-dimensional Fourier transform of this pattern diagrammatically and submit the most effective drawing you can devise.

4–10. *Two-dimensional transforms.* State the two-dimensional Fourier transforms of the following functions:

(a) $^2\text{rect}(x/3, y/2)$,

(b) $^2\text{rect}(x - 3, y)$,

(c) $^2\text{rect}(x - 4, y - 5)$,

(d) $^2\text{sinc}(x - 5, 2y - 7)$,

(e) $3\,^2\text{rect}(x - 8, 3y)$,

(f) $\exp(i\,16\pi x)\,^2\text{sinc}(x, y/3)$. ☞

4–11. *Transform of hole pair.* Give the Fourier transform of

$$f(x, y) = e^{-\pi[x^2 + (y-b)^2]/R^2} + e^{-\pi[x^2 + (y+b)^2]/R^2}.$$

4–12. *Transform with isometric.*

(a) Use the shift, similarity and addition theorems to obtain the two-dimensional Fourier transform of $f(x, y) = 2\,^2\text{rect}[x - 0.5, (y - 3)/4] + 5\,^2\text{rect}[(x - 3.5)/3, 0.5y - 1)]$.

(b) Make an accurate isometric drawing to represent $f(x, y)$ using a scale of 1 cm for the unit of x, y and f. The purpose of this exercise is to gain experience in making choices that best convey your ideas in visual mode.

4–13. *Verbal description.* Consider the function $\cos[2\pi(3u + 4y + 7t)]$. State in words what this function describes. Make your statement the briefest that would permit a reader to reconstitute the function. This is an exercise in information transfer to nonmathematical people but may also give you new insight.

4–14. *Mixed transforms.* In the previous problem,

(a) What is the two-dimensional Fourier transform with respect to x and y?

(b) What is the one-dimensional Fourier transform with respect to t?

(c) Give some thought to a consistent notation for the function and the two transforms, which would be helpful to a third person trying to understand your answers.

4–15. *Hollow square.* A function is equal to zero outside a central unit square and inside a central half-unit square similarly oriented. Elsewhere the function is equal to unity. What is its Fourier transform? ☞

4–16. *Two-dimensional transform.*
   (a) Find the transform of sinc x rect y.
   (b) Make drawings of the function and its transform, selecting a method of presentation so as to be most instructive to a reader. ☞

4–17. *Transform of two squares.* A function is equal to unity on two unit squares, one centered at $(3,0)$ and the other at $(-3,0)$. What is the two-dimensional Fourier transform?

4–18. *Rotation theorem.* Prove the rotation theorem from first principles.

4–19. *Rotation matrix.* An object $g_\theta(x, y)$ is obtained from a given $g(x, y)$ by rotating the latter CCW about the origin through an angle θ. Explain why the rotation matrix

$$\begin{bmatrix} \cos\theta & \sin\theta \\ -\sin\theta & \cos\theta \end{bmatrix}$$

is not the same as the matrix for coordinate transformation between (x, y) and a set of coordinates (x', y') in which the x'-axis points in the direction that lies CCW from the x-axis by an amount θ.

4–20. *Skew rectangle function.* In the rotated coordinate system (x', y'), where the x'-axis lies at an angle θ counterclockwise from the x-axis, consider the function rect x'. What is its two-dimensional Fourier transform?

4–21. *Expert in trouble.* An expert on morphic resonance who is trying to debug an affine transformation program is found studying the following code segment and is heard to say, "It's not doing what I expected."

```
xprime=x
x=a*xprime+b*y+c
y=d*xprime+e*y+f
PRINT x
PRINT y
```

As a consultant, what do you advise? ☞

4–22. *Self-reciprocal transform.* Usually, when a function $f(\ ,\ )$ has a Fourier transform $F(\ ,\ )$ the functions f and F are different. If the transform is the same as the original function we say that the transform is self-reciprocal. Examples include

$$e^{-\pi(x^2+y^2)} \ ^2\!\supset e^{-\pi(u^2+v^2)}$$

$$^2\mathrm{III}(x, y) \ ^2\!\supset \ ^2\mathrm{III}(u, v)$$

$$\mathrm{sech}\, x \ \mathrm{sech}\, y \ ^2\!\supset \mathrm{sech}\, u \ \mathrm{sech}\, v$$

$$^2\mathrm{rect}(x, y) + \ ^2\mathrm{sinc}(x, y) \ ^2\!\supset \ ^2\mathrm{rect}(u, v) + \ ^2\mathrm{sinc}(u, v)$$

$$e^{-\pi(x/5)^2} + 5e^{-\pi(5x)^2} \ ^2\!\supset e^{-\pi(u/5)^2} + 5e^{-\pi(5v)^2}.$$

   (a) Show that if $g(x, y)$ is any real function, even in both x and y, whose transform is $G(u, v)$, then the composite function $g(x, y) + G(x, y)$ is its own transform.
   (b) Show that, in general, if $g(x, y)$ is any complex function with transform $G(u, v)$, then $g(x, y) + g(-x, -y) + G(x, y) + G(-x, -y)$ is its own transform. ☞

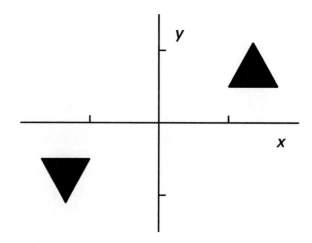

Figure 4-18   A binary function.

4–23. *Fourier kernel.* Endeavoring to form a mental image of the two-dimensional Fourier transform kernel, Šadrak writes

$$e^{-i2\pi(ux+vy)} = (e^{-i2\pi})^{ux+vy} = 1^{ux+vy} = 1.$$

Šadrak's roommate Mišak says, "If that is so, then $\cos[2\pi(ux + vy)] - i\sin[2\pi(ux + vy)] = 1$ and $2\pi(ux + vy) = 0$ or π, which is impossible." Mediate this dispute.

4–24. *Orthogonality.*
(a) Can you prove that $\int_{-\infty}^{\infty}\int_{-\infty}^{\infty}\cos[2\pi(u_1x + v_1y)]\cos[2\pi(u_2x + v_2y)]\,dx\,dy$ is equal to zero unless $u_1 = u_2$ and $v_1 = v_2$?
(b) A sophisticated electrical appliance draws a 50-Hz current when a 60-Hz voltage is applied. Since $\int_{-\infty}^{\infty}\cos\omega_1 t\cos\omega_2 t\,dt$ is zero for $\omega_1 \ne \omega_2$, is it possible that there would be no electricity bill for the use of this appliance?

4–25. *Operator notation.*
(a) Explain the difference between $\mathcal{F}^2$ sinc x and $^2\mathcal{F}$ sinc x.
(b) What does $^2\mathcal{F}^{-1}$ sinc x mean? ☞

4–26. *Axis values of 2-D transform.* A function $f(x, y)$ is equal to unity within the triangular boundaries shown in Fig. 4-18 and is zero elsewhere. It has a two-dimensional Fourier transform $F(u, v)$, but we are interested only in the transform values along the axes. What are $F(u, 0)$ and $F(0, v)$?

4–27. *Laplacian operator.*
(a) A new function $g(x, y)$ is formed from $f(x, y)$ by subtracting the mean of four values at a distance a in the four principal directions; thus $g(x, y) = f(x, y) - \frac{1}{4}[f(x + a, y) + f(x, y + a) + f(x - a, y) + f(x, y - a)]$. If $f(x, y)$ has FT $F(u, v)$, what is $G(u, v)$?
(b) As $a \to 0$, does $G(u, v)$ approach a limit?

4–28. *Shear theorem.* If $f(x, y) \ ^2\supset F(u, v)$, show that the effect of y-shear is that

$$f(x, y + xd) \ ^2\supset F(u - vd, v).$$

4–29. *Double shear theorem.* If $f(x, y) \ ^2\supset F(u, v)$ show that

$$f(x + yb, y + xd)^2 \supset \frac{1}{|\Delta|} F\left(\frac{u - vd}{\Delta}, \frac{v - ub}{\Delta}\right),$$

where $\Delta = |1 - bd|$.

4–30. *Equivalent area theorem.* Derive the equivalent area theorem.

4–31. *Fourier transform of triangle.* The value of a function is unity within the equilateral triangle whose base runs from $-0.5, -0.289$ to $(0.5, -0.289)$ and zero elsewhere. (The origin is at the centroid of the triangle.) What are the cross sections $F(u, 0)$ and $F(0, v)$?

4–32. *Transforms of blades.* Find the two-dimensional Fourier transform of
(a) $\text{rect}(y + \frac{1}{2})\delta(x)$,
(b) $\delta(\theta - \pi/4) \text{rect}(x - 1/2) \text{rect}(y - 1/2)$,
(c) $\delta(\theta - \pi/4) \text{rect}(x + 1/2) \text{rect}(y + \frac{1}{2})$.

4–33. *Textile diffraction.* Textile is to be monitored by illuminating a patch of material with laser light and examining the Fraunhofer diffraction pattern. As a preliminary study for this idea, obtain the diffraction patterns expected for plain, twill, and satin weaves. Hold some different materials up to the sunlight to understand how this proposal would work. ☞

4–34. *Square slit.* A narrow slit 1 μm wide is formed on an opaque overexposed photographic plate by scratching along the edge of a square of side H.
(a) Express the illumination over the plate as seen from the front, when it is evenly illuminated from behind, in terms of delta function notation.
(b) What is the Fourier transform of this impulsive function?

5

Two-Dimensional Convolution

Convolution derives its importance for time-dependent signal processing in a fundamental way. Input and output signals *need* not be related through convolution, but they *are* if (and only if) the processor is linear and time invariant. Now these two conditions are just those that are met very commonly indeed in analogue systems. Therefore, convolution is exactly what we want for the neat expression of the cause-effect relationship in a large fraction of all signal transmission systems, both natural and artificial.

In two dimensions, linearity and *space* invariance are the joint conditions needed to guarantee a convolution relation. The corollary is that sinusoidal input produces sinusoidal output if and only if linearity and shift invariance apply (Fig. 5-1). Of course, the property of sinusoidal response to sinusoidal input is what empowers the Fourier technique of problem solving. Linearity and shift invariance are returned to in the next chapter.

Correlation, or cross-correlation, derives its importance from origins in statistical theory that are quite unrelated to the fundamental reason given above for the importance of convolution, but in mathematical formulation correlation is closely related to convolution. Cross-correlation also arises in signal processing, not only for statistical reasons but from circumstances that have nothing to do with statistics; and it should be added that convolution also can arise in situations having nothing to do with signal theory.

Autocorrelation, a special case of cross-correlation, is extremely important; in fact we are concerned more often with autocorrelation than with cross-correlation. By symmetry one sees that self-convolution should be mentioned to round out these remarks. However, self-convolution is hardly ever encountered.

174

Figure 5-1 The output of a linear and time- or space-invariant system is related to the input by convolution (with the impulse or point response). In the frequency domain, output is related to input by multiplication (by the frequency-dependent transfer function); the result is that sinusoidal input produces sinusoidal output.

    This chapter introduces convolution in terms of continuous independent variables x and y. With this background in place the specialization to images specified discretely comes more easily, but it is desirable to deal with sampling also before moving on to digital images. Nevertheless, it is a good idea to do numerical exercises, as well as algebraic ones, at this stage, so discrete numerical examples are shown and practice is strongly recommended in order to acquire an intuitive feel for the convolution operation. There is nothing like convolving some pairs of small matrix representations of two-dimensional functions by hand in order to appreciate (a) the reversals of data demanded by the minus signs in the convolution integral, (b) exactly how the size of the convolution relates to the size of the factors when these sizes are not infinite, (c) how array dimensioning is affected, (d) how the quantity of arithmetic depends on size. You also get a feel for (e) how far and in which direction a data set is shifted when a convolving operator has a centroid that is offset from its origin, (f) the amplification factor, and (g) whether to expect a result with a zero mean value, a result that has smaller peak values and broader peaks, or the reverse. All these phenomena are inherent in the one definition integral, but it has to be admitted that information of the kinds listed above does not leap out from the integral formulation. Programs for two-dimensional convolution are readily available, but the initial experience recommended here is gained more readily with a hand calculator.

    Time spent becoming familiar with convolution is time well invested. The material has been divided into two chapters.

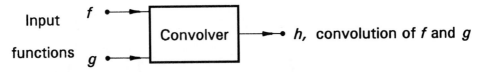

Figure 5-2　A diagrammatic view of convolution showing a "convolver" as a three-port
device. The input ports are interchangeable.

CONVOLUTION DEFINED

Convolution is an operation (Fig. 5-2) that takes two input functions $f(\,)$ and $g(\,)$ and
produces from them a third function $h(\,)$ which is the convolution of $f(\,)$ and $g(\,)$.

Convolution in one dimension is defined by the integral

$$h(x) = \int_{-\infty}^{\infty} f(x - x')g(x')\,dx'$$

and, in abbreviated form when we wish to avoid the repetition of the dummy variable x',
we may write simply

$$h(x) = f(x) * g(x).$$

The asterisk notation is in quite general use and has gathered momentum as its convenience has come to be realized. I first encountered it in a very readable book on integral
equations by Vito Volterra (1860–1940) written in 1930 at a time when the word *convolution* itself was far from generally established (in English the German word *Faltung*
was often encountered). Occasionally authors invent new symbols, such as $\otimes$, which are
doomed to extinction.

In certain cases the functions entering into convolution may not be given in terms of
a continuous independent variable x. For example, fractional values of x may be devoid of
meaning; or, even if fractional values of x have meaning, function values may be available
only at regularly spaced values of x; or even though function values could be obtained for
any value of x, we may choose to work only at regularly spaced values of x for purposes
of computation or to reduce the time needed to acquire data.

One example of a function of an intrinsically discrete variable is the probability $p(j)$
of throwing a total j upon rolling a pair of dice. The independent variable j can assume
only integral values from 2 through 12, and fractional values of the independent variable
have no meaning. An example of a function that is available only at regularly spaced
intervals is the output of a recording thermometer that prints out a permanent record of
temperature on paper tape at fixed intervals of time. Indeed any recording sensor, whether
it is recording on magnetic tape, paper tape, film, or any other medium and whether it is
recording a slowly varying temperature, a much faster signal such as music, or a very fast
television signal can only make data available at spaced intervals, because some time is
necessarily consumed while the sensor responds.

In view of these considerations the *convolution sum* assumes an importance that
is fully equal to that of the convolution integral and that in a deep way may be more
fundamental. We therefore define $\{f_i\}$ and $\{g_i\}$ to be input signals or sequences which

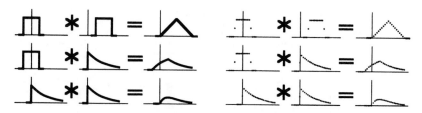

Figure 5-3   Examples of convolution integrals (left) and convolution sums (right).

depend on an independent variable i (shown as a subscript), assuming only integral values. Then the convolution sum $\{h_i\}$ is defined by

$$h_i = \sum_{j=-\infty}^{\infty} f_j g_{i-j}$$

or, expanding the summation shorthand so that we no longer need the dummy variable j,

$$h_i = \cdots + f_2 g_{i-2} + f_1 g_{i-1} + f_0 g_i + f_{-1} g_{i+1} + f_{-2} g_{i+2} + \cdots.$$

Note that, in each term, the sum of the two subscripts is the same and equal to i.

What happens if, as a result of trying to extend the series to infinity in both directions, we arrive at values of the subscript that are impossible? For example, if j represents the total shown on two dice, what do we do about $j = 14$? One way to handle such situations is by proper wording that ensures that the function value is zero for unwanted values of the subscript. Another way is to restrict attention to the acceptable set of j-values when performing the summation, a choice that could be indicated by the notation Σ_j. Elementary examples of convolution are given in Fig. 5-3.

In computational contexts sequences $\{f_i\}$ and $\{g_i\}$ may be of fixed length N, and only N places may be allocated for the convolution, which would normally extend to $2N - 1$ elements. One approach is to limit $\{f_i\}$ and $\{g_i\}$ to $N/2$ nonzero values. Another is to recognize cyclic convolution defined by

$$h_i = \sum_{j=1}^{N} f_i g_{i-j} \quad \text{or} \quad \{h_i\} = \{f_i\} \circledast_N \{g_i\}.$$

The encircled asterisk represents cyclic convolution and the subscript N indicates the period. An analogue calculator made of two concentric discs, each with N places, shows how the cyclic convolution involves overlap when the combined lengths of $\{f_i\}$ and $\{g_i\}$ exceed $N + 1$.

In two dimensions the diagrammatic view of convolution remains as pictured above. The convolution integral becomes

$$h(x, y) = \int_{-\infty}^{\infty} \int_{-\infty}^{\infty} f(x - x', y - y') g(x', y') \, dx' \, dy'$$

and the symbolic abbreviation may be extended to two asterisks,

$$h(x, y) = f(x, y) ** g(x, y).$$

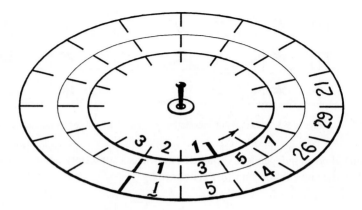

Figure 5-4   A "cyclic convolver" for $N = 16$ set up for convolving $\{\underline{1}\ 2\ 3\}$ with $\{\underline{1}\ 3\ 5\ 7\}$ to give the hand-lettered output $\{\underline{1}\ 5\ 14\ 26\ 29\ 21\}$. If the ordinary convolution has more than 16 elements, then the head and tail overlap in the cyclic convolution.

$$
\begin{bmatrix}
0 & 0 & 0 & 0 & 0 & 0 & 0 \\
0 & 0 & 0 & 0 & 0 & 0 & 0 \\
0 & 0 & 1 & 1 & 1 & 0 & 0 \\
0 & 0 & 1 & 1 & 1 & 0 & 0 \\
0 & 0 & 1 & 2 & 1 & 0 & 0 \\
0 & 0 & 0 & 0 & 0 & 0 & 0 \\
0 & 0 & 0 & 0 & 0 & 0 & 0
\end{bmatrix}
**
\begin{bmatrix}
0 & 0 & 0 & 0 & 0 & 0 & 0 \\
0 & 0 & 0 & 0 & 0 & 0 & 0 \\
0 & 0 & 0 & 0 & 1 & 0 & 0 \\
0 & 0 & 0 & 1 & 0 & 0 & 0 \\
0 & 0 & 1 & 0 & 0 & 0 & 0 \\
0 & 0 & 0 & 0 & 0 & 0 & 0 \\
0 & 0 & 0 & 0 & 0 & 0 & 0
\end{bmatrix}
=
\begin{bmatrix}
0 & 0 & 0 & 0 & 0 & 0 & 0 \\
0 & 0 & 0 & 1 & 1 & 1 & 0 \\
0 & 0 & 1 & 2 & 2 & 1 & 0 \\
0 & 1 & 2 & 3 & 3 & 1 & 0 \\
0 & 1 & 2 & 3 & 1 & 0 & 0 \\
0 & 1 & 2 & 1 & 0 & 0 & 0 \\
0 & 0 & 0 & 0 & 0 & 0 & 0
\end{bmatrix}
$$

Figure 5-5   An example of discrete two-dimensional convolution presented in matrix form.

It might be thought that one asterisk would suffice; would it not be clear from the notation $h(x, y)$ that *two*-dimensional convolution was intended? There are cases where no confusion would result, but quite often convolution occurs in both one and two dimensions in the one document and it is helpful to have the cue provided by the double-asterisk notation. And, of course, $f(x)$, if regarded as a function of two variables, can enter into two-dimensional convolution even though two variables are not explicitly indicated.

In the discrete case the entity corresponding to the convolution integral is the convolution sum, which is defined as

$$
h_{ij} = \sum_{k=-\infty}^{\infty} \sum_{l=-\infty}^{\infty} f_{i-k,j-l}\, g_{k,l}.
$$

Each of the dummy variables k and l now ranges from $-\infty$ to ∞, so the summation covers the whole infinite plane. An example is shown in Fig. 5-5.

Since continuous and discrete variables are not found mixed together very often, there is normally no question whether one is dealing with integrals or sums. For this reason the asterisk notation may freely be used to abbreviate convolution sums. Thus we write

$$\{h_i\} = \{f_i\} * \{g_i\}$$
$$[h_{ij}] = [f_{ij}] ** [g_{ij}].$$

The square-bracket notation reminds us that $[h_{ij}]$ is a two-dimensional matrix. The sequence $\{h_i\}$ may be thought of as a one-dimensional matrix.

CROSS-CORRELATION DEFINED

The cross-correlations $c(x)$ and $c(x, y)$ of f on g, in one and two dimensions, respectively, appear as

$$c(x) = \int_{-\infty}^{\infty} f(x' - x)g(x')\, dx' = \int_{-\infty}^{\infty} f(x')g(x + x')\, dx'$$

$$c(x, y) = \int_{-\infty}^{\infty} \int_{-\infty}^{\infty} f(x' - x, y' - y)g(x', y')\, dx'\, dy'$$

$$= \int_{-\infty}^{\infty} \int_{-\infty}^{\infty} f(x', y')g(x + x', y + y')\, dx'\, dy'.$$

In discrete form we have

$$c_i = \sum_{j=-\infty}^{\infty} f_{j-i}\, g_j = \sum_{j=-\infty}^{\infty} f_j\, g_{i+j}$$

$$c_{ij} = \sum_{k=-\infty}^{\infty} \sum_{l=-\infty}^{\infty} f_{k-i,l-j}\, g_{k,l} = \sum_{k=-\infty}^{\infty} \sum_{l=-\infty}^{\infty} f_{k,l}\, g_{i+k,j+l}.$$

In the abbreviated notation we may distinguish cross-correlation from convolution by use of the pentagram instead of the asterisk. Thus

$$\{c_i\} = \{f_i\} \star \{g_i\}$$
$$[c_{ij}] = [f_{ij}] \star\star [g_{ij}].$$

The pentagram notation is very useful for personal use, but of course in publications needs to be flagged because not everyone distinguishes an asterisk with six spokes $*$ from a pentagram $\star$ with five points. Cross-correlation is not commutative, that is to say $f \star\star g$ is not the same as $g \star\star f$. Therefore the English phrase "cross-correlation of f and g" is misleading by comparison with a phrase such as "product of x and y" where commutativity exists. Terminology such as "cross-correlation of f on g" helps to clarify the nonsymmetrical roles of f and g; sometimes it is appropriate to the subject matter to which f and g refer to read $f \star\star g$ as "f scans g." For example, if g describes a transparency and f describes a small aperture for imaging g (Fig. 5-6), then the image that results when f scans g is indeed $f \star\star g$. The definition integral for $c(x, y)$, in its first form where the integrand is $f(x' - x, y' - y)g(x', y')$, shows that g remains stationary while f scans over it and that the displacement of f relative to g is the same as the argument (x, y) of the cross-correlation integral.

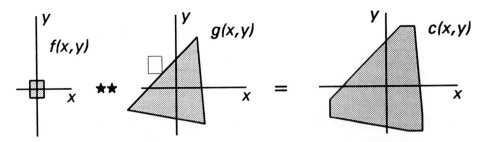

Figure 5-6 Functions f and g are nonzero only within the areas illustrated. When f scans g, the result $f \star\star g$ is contained in the seven-sided polygon. This polygon is the locus of a point fixed to the hollow rectangle (middle) as it translates, without rotation, in contact with the triangle. When g scans f, the result is not the same.

In the variant definition

$$c(x, y) = \int_{-\infty}^{\infty} \int_{-\infty}^{\infty} f(x', y')g(x + x', y + y')\,dx'\,dy'$$

the displacement of f relative to g is exactly the same as before, namely (x, y), even though $g(\ ,\ )$ has shifted. So the absolute displacements are different but when the integral is taken over the configuration, the result is independent of exactly where the configuration is located. When $f(\ ,\ )$ is of relatively small extent, as in this example, $c(x, y)$ has a general resemblance to $g(x, y)$ as regards shape and orientation.

If on the other hand g scans f, then the pattern g is to be rotated half a turn prior to multiplication and integration. The outcome will then resemble $g(-x, -y)$ rather than $g(x, y)$. Indeed, in the case illustrated, where $f(x, y)$ possesses twofold rotational symmetry, the outcome will be precisely the previous $c(x, y)$ rotated half a turn.

If Fig. 5-6 is redrawn using a small **L** for f instead of a small $\square$, the nature of the difference between $f \star\star g$ and $g \star\star f$ will be appreciated. It is also possible to express the distinction mathematically by resolving f and g into their symmetrical and antisymmetrical parts and obtaining the two cross-correlations as a sum of four terms.

FEATURE DETECTION BY MATCHED FILTERING

A graph of received signal strength versus time on a radar display is composed of receiver noise plus, possibly, a faint echo whose location on the time axis will give the range to a reflecting target. If there were no noise, or if the echo were strong, the shape of the echo might be known in advance from previous experience. Assume that the echo waveform $e(t)$ is known as regards shape, but not the amplitude A. Then the signal $s(t)$ seen on the screen in the time taken for one sweep is $s(t) = Ae(t - T_1) + n(t)$, where $n(t)$ is the noise component due to the receiver, and T_1 is the delay associated with the range $\frac{1}{2}cT_1$, c being the velocity of the radar waves. The delay T_1 cannot be found exactly, even if $e(t)$ is known, but a best estimate can be made by cross-correlating $s(t)$ with $e(t)$ to get $e(t) \star s(t) = e(t) \star Ae(t - T_1) + e(t) \star n(t)$. The component $e(t) \star n(t)$ will be more noise of somewhat reduced amplitude depending on the number of noise wiggles spanned by $e(t)$, while the component $e(t) \star Ae(t - T_1)$ will be a symmetrical peak centered at $t = T_1$. The time at which the peak in the sum of the two components occurs is the best estimate of T_1, as is shown in texts on signal processing, under assumptions of simplicity about the statistical behavior of $n(t)$. (A noise component due not to receiver noise but to ground clutter might not meet the theoretical requirements.) This method of echo location is called *matched filtering*, because the impulse response of the filter required to smoothe the signal matches the expected echo.

An analogous situation in two dimensions is to locate the features of known shape $e(x, y)$ in the presence of irregular background $n(x, y)$. An image of the form

$$s(x, y) = \sum_i A_i e(x - X_i, y - Y_i) + n(x, y)$$

can be searched by forming the cross-correlation $e(x, y) \star \star s(x, y)$ and looking for peaks. The rigorous statistical theory behind this is of less relevance in two dimensions, because the assumptions are less likely to be met. To start with, if we were looking for needles spilt on a bed of sand, although the shape of a needle is known perfectly well, its orientation is not. This extra degree of freedom does not arise in the radar-range problem. Also, in that problem, receiver noise really does have a simple statistical description, which an image background rarely has. Nevertheless, feature detection and feature recognition are important in imaging, and matched filtering is one tool to be kept in mind, as illustrated in Chapter 8.

AUTOCORRELATION DEFINED

We may think schematically of cross-correlation as in the upper part of Fig. 5-7. But an autocorrelator would be a device having only one input function rather than two. We could, of course, make an autocorrelator from a cross-correlator, as suggested in the lower part of the figure. The tee-junction T is of the kind that delivers an unreduced copy of f along both arms and not the kind of tee-junction familiar from electric-circuit theory that obeys the Kirchhoff continuity law. From the previous discussion of cross-correlation we may write

One-dimensional autocorrelation:

$$f(x) \star f(x) = \int_{-\infty}^{\infty} f(x' - x) f(x') dx' = \int_{-\infty}^{\infty} f(x') f(x + x') dx'.$$

Two-dimensional autocorrelation:

$$f(x, y) \star \star f(x, y) = \int_{-\infty}^{\infty} \int_{-\infty}^{\infty} f(x', y') f(x + x', y + y') \, dx' \, dy'$$

$$= \int_{-\infty}^{\infty} \int_{-\infty}^{\infty} f(x' - x, y' - y) f(x', y') \, dx' \, dy'.$$

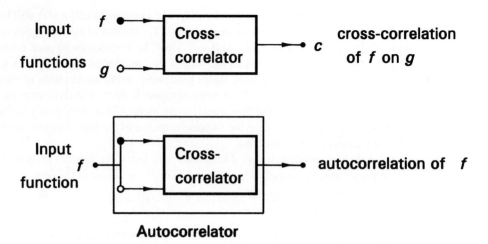

Figure 5-7   Diagramatic views of cross-correlation and autocorrelation. Unlike a convolver, whose input terminals are interchangeable, a cross-correlator has different terminals for f and g. No output can emerge from an autocorrelator until the input is fully inserted.

The discrete expressions are

$$c_i = \sum_{j=-\infty}^{\infty} f_{j-i}\, f_j = \sum_{j=-\infty}^{\infty} f_j\, f_{i+j}$$

$$c_{ij} = \sum_{k=-\infty}^{\infty} \sum_{l=-\infty}^{\infty} f_{k-i,l-j}\, f_{k,l} = \sum_{k=-\infty}^{\infty} \sum_{l=-\infty}^{\infty} f_{k,l}\, f_{i+k,j+l}.$$

Very often indeed, autocorrelation is normalized so as to be unity at the origin. When this is the case, one may emphasize the fact by using the term *normalized autocorrelation*. Using $\gamma(x, y)$ to mean normalized autocorrelation, we have by definition

$$\gamma(x, y) = \frac{\int_{-\infty}^{\infty} \int_{-\infty}^{\infty} f(x' - x, y' - y) f(x', y')\, dx'\, dy'}{\int_{-\infty}^{\infty} \int_{-\infty}^{\infty} [f(x, y)]^2\, dx\, dy}.$$

Of course, there is no great difference in character between the autocorrelation $f \star\star f$ and the normalized autocorrelation $\gamma(x, y)$; it is merely a matter of scale factor. For that reason one often carries calculations through in the nonnormalized form and then, at the end, normalizes if desired. There is never any difficulty in determining the normalizing factor that goes in the denominator; it is just the factor that brings the central value, whatever magnitude it may have, down to unity.

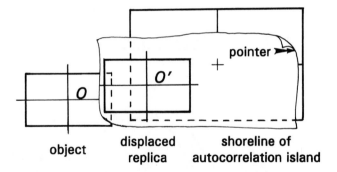

object   displaced   shoreline of
                replica   autocorrelation island

Figure 5-8   Tracing-paper construction for autocorrelation. Displacement of the paper is (x, y), the desired argument.

UNDERSTANDING AUTOCORRELATION

From the expression

$$\int_{-\infty}^{\infty} \int_{-\infty}^{\infty} f(x' - x, y' - y) f(x'f, y')\, dx'\, dy'$$

we see that we start with a given function $f(x, y)$. We relabel the axes x' and y', because we are going to integrate over the plane, whereupon the dummy variables x' and y' will disappear; and we wish to be left with a function of x and y. The given function now appears as $f(x', y')$. The second factor $f(x' - x, y' - y)$ is a replica of $f(x', y')$ but displaced by an amount x to the right and an amount y upward. We may represent this situation by copying the picture $f(x', y')$ on a piece of tracing paper and displacing it as described. If we push a pin through the copy and into the original, then after displacement we have the value $f(x', y')$ at one pinhole and the value $f(x' - x, y' - y)$ at the other. These values are to be multiplied together, and we can imagine the product as a new function covering the (x', y')-plane. According to the defining expression for the autocorrelation we are now to perform the double integral, that is, to find the volume under the product function. That is a lot of work involving multiplications all over the plane followed by summing of products all over the plane. Even so, the result is merely a single value of the autocorrelation, namely for the particular (x, y) describing the displacement. To get another value we have to displace the function to a new position, multiply throughout, and find the volume under the product again. To obtain the whole autocorrelation function we have to repeat over and over until all desired values of x and y are covered.

Figure 5-8 shows an original function and its displaced replica. They overlap, for the chosen displacement OO', only over a thin rectangle. The product is to be integrated over this region of overlap and recorded. Then a new displacement is chosen. One may imagine recording the integral by making a dot on the (x, y)-plane in a location indicated by a pointer carried by the tracing paper and writing the value alongside. The pointer is placed on the paper so that its displacement from the origin of the (x, y)-plane (marked by

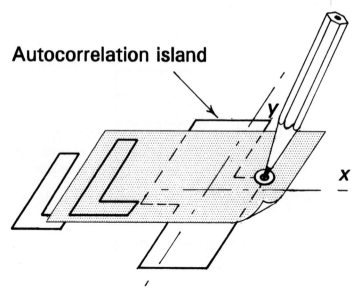

Autocorrelation island

Figure 5-9   Method of tracing shoreline of autocorrelation island by moving the replica (of an ell) to be tangent to the original while maintaining the orientation.

a cross) is equal to the vector OO'. In practice one may mark the pointer in a convenient place and then establish the (x, y)-origin by placing O' on O.

It is a very good idea to get a piece of tracing paper and experiment with this mechanical construction, because it develops certain intuition that remains useful later. A number of general points can be made. For example, if the original function f is zero outside a certain boundary, then $f \star\star f$ will also be zero outside some other boundary. The shoreline of this autocorrelation island may be explored rapidly by translating the replica so as to keep it tangent to the original, while simultaneously drawing the locus traced by the pointer. This can be done by making a reinforced hole in the paper at the tip of the pointer and putting the point of a pencil through, as illustrated in Fig. 5-9.

It is quite possible to dispense with the use of tracing paper and visualize the motion of an imaginary pencil point as it traces out the boundary of the autocorrelation island. However, to gain this ability it is necessary even for people with good powers of visualization to practice with tracing paper. Figure 5-10, which exhibits a number of examples, will enable the reader to determine his current level of ability. For practice, trace the diagrams on the left, verify the autocorrelation islands on the right, and discover which one contains an error, which has been incorporated to test your ability to dispense with the tracing paper.

Considerable practical importance attaches to the autocorrelation boundary in radio interferometry, where it defines the spectral sensitivity islands (Bracewell, 1961), regions in the (u, v)-plane where the instrument is sensitive. In optical interferometry the autocorrelation of the aperture defines the boundary of the optical transfer function. Some of

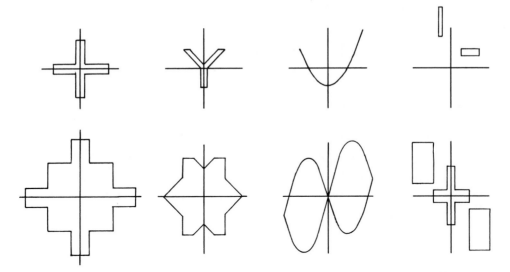

Figure 5-10   Various outlines (above) and their autocorrelation islands (below).

these ideas go back to the time of Rayleigh and Michelson, while others arose in radio astronomy (Bracewell, 1984).

A feature noticeable from the autocorrelation islands shown is a certain symmetry with respect to the axis through the origin perpendicular to the x- and y-axes. If any of the autocorrelation islands are rotated about the perpendicular axis, the boundary shape repeats each half-turn. The islands are said to possess a twofold axis of rotational symmetry. Furthermore, a little thought shows that the autocorrelation function values themselves have the same symmetry; that is, the value of the autocorrelation at any point (x, y), not necessarily on the boundary, is the same as the value at the diametrically opposite point $(-x, -y)$. A formal proof can easily be given by referring to the definition integral, replacing x by $-x$ and y by $-y$ and showing by substitution of variables that the integral is unchanged. Thus $\gamma(-x, -y) = \gamma(x, y)$ as illustrated in Fig. 5-11.

A rotation π about the perpendicular axis can be achieved in another way. Take a sheet of paper printed on one side only. Mark x- and y-axes. Now flip the paper about the x-axis, rotating half a turn about the axis. Then flip about the y-axis half a turn. The net result, as you see, is half a turn about the perpendicular axis. The reason is that reversal of the sign of x followed by reversal of the sign of y takes each point (x, y) to the diametrically opposite point $(-x, -y)$. A further way of looking at this is to understand that it makes no difference whether the replica is displaced a certain distance to the southwest or the same distance to the northeast. Either way, the configuration of both shapes taken together is the same.

A final general feature of the autocorrelation is independence of the origin of $f(x, y)$. That is to say, if you move $f(x, y)$ to another place, it still has the same autocorrelation. This is obvious from the tracing-paper construction and easy to prove mathematically by substituting $f(x + a, y + b)$ into the definition integral and showing that the

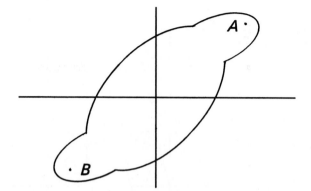

Figure 5-11 Under twofold rotational symmetry, an autocorrelation repeats after half a turn about the origin. Thus the value at A is the same as at the diametrically opposite point B.

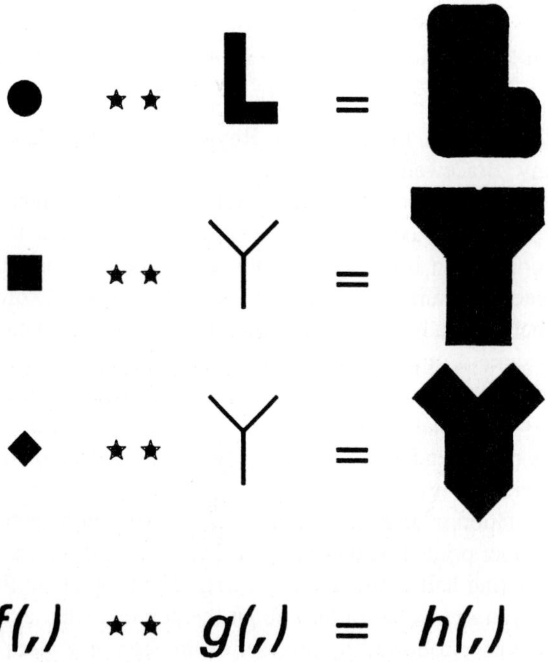

$$f(,) \; \star\star \; g(,) \; = \; h(,)$$

Figure 5-12 Cross-correlating the function $f(\;,\;)$ on $g(\;,\;)$ produces the outline that results from dilating the shape g by the shape f.

integral is unchanged. If you had to design an automatic machine to verify the signatures on checks, part of the problem would be to fix the origin of the signature. This problem would be eliminated by electing to work with the autocorrelation of the signature.

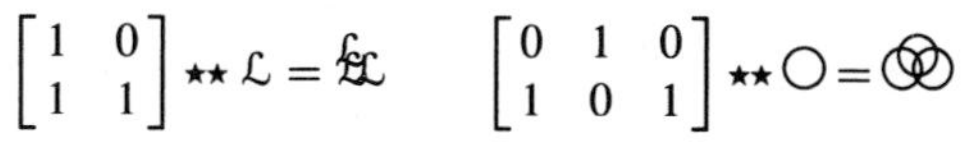

$$\begin{bmatrix} 1 & 0 \\ 1 & 1 \end{bmatrix} \star\star \mathcal{L} = \text{\reflectbox{L}} \qquad \begin{bmatrix} 0 & 1 & 0 \\ 1 & 0 & 1 \end{bmatrix} \star\star \bigcirc = \text{⊗}$$

Figure 5-13   Two continuous shapes $\mathcal{L}$ and $\bigcirc$ which have been dilated by a digital object represented by a matrix.

CROSS-CORRELATION ISLANDS AND DILATION

A good deal of information resides in the autocorrelation island that has just been described, despite the fact that the island boundary itself says nothing about the quantitative values of the autocorrelation inside the boundary. The information has to do with shape. When cross-correlation is considered, we find cross-correlation islands, and again the discussion has to do with morphology, but the development is much richer than in the case of autocorrelation.

Dilation by a circle was introduced in Chapter 2; now an example can be given. Suppose that an ell shape (Fig. 5-12) forms the boundary of a function $g(x, y)$ that is zero outside the boundary. Let $f(x, y)$ be the boundary of a function that is zero outside a circular boundary. Dilating the ell by the circle will produce the plump outline shown on the right, which we see is the same as the boundary of the cross-correlation of f on g. The graphical technique of copying the circle onto a transparency, which is translated without rotation so as to carry the circle around the ell-shape while maintaining contact, can mechanically generate the dilated outline. Putting a pencil through a pinhole anywhere in the moving transparency and writing onto the stationary sheet bearing the ell will trace the outline. The technique also works quite well on a blackboard.

It appears, then, that dilation falls within cross-correlation, and indeed a program for cross-correlation will perform dilation. However, a lot of numbers will be needlessly generated. To be sure, efficiency can be improved by choosing the nonzero function values to be unity, but there is a better way, which will be explained in Chapter 8. The concept of cross-correlation is useful, however, in getting a feel for dilation. For example, one sees at a glance that dilating the circle by the ell will be different, although the handedness will be preserved. The additional examples suggest interesting modifications of given shapes.

Dilation of digital images by shapes defined digitally has become very important in graphics, following a history of growth as a statistical tool in sedimentology. Purely discrete considerations will be taken up later, but it may be remarked here that the continuous methodology can handle the discrete as well as the continuous. This is not merely a convenience; hybrid situations arise where digitization would be inappropriate. Two examples are shown in Fig. 5-13 where objects with a continuous boundary are to be dilated by digital objects represented by matrices.

LAZY PYRAMID AND CHINESE HAT FUNCTION

As examples of autocorrelation that may be worked out analytically from the integral definition we take two cases that are frequently met in a variety of circumstances.

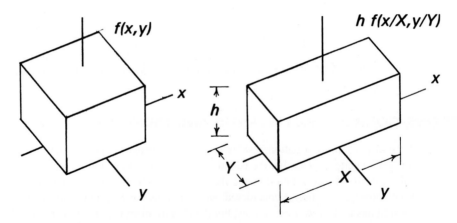

Figure 5-14   The two-dimensional unit rectangle function (left) and an example of how the rectangle function notation may be employed to describe a block of width X, depth Y, and height h (right).

Suppose we have a function $f(x, y)$ that could be described as a square-table function of unit height and unit side. To express this function algebraically we could write

$$f(x, y) = \begin{cases} 1 & |x| < \tfrac{1}{2} \text{ and } |y| < \tfrac{1}{2} \\ 0 & \text{otherwise.} \end{cases}$$

Admittedly this notation is a little awkward, because the inequalities have to be studied rather carefully to see that the function intended is quite simple. Furthermore, the notation is rather lengthy. For these reasons it is convenient to make use of the two-dimensional unit rectangle function ${}^2\Pi(x, y)$ or ${}^2\text{rect}(x, y)$, illustrated in Fig. 5-14, which may be defined by

$$ {}^2\text{rect}(x, y) = \begin{cases} 1, & \text{inside central unit square} \\ 0, & \text{outside.} \end{cases}$$

With this notation agreed upon,[1] suppose we wish to autocorrelate a function $f(x, y)$ such that

$$f(x, y) = {}^2\text{rect}(x, y).$$

From the definition integral for the autocorrelation $c(x, y)$ we have

$$c(x, y) = \int_{-\infty}^{\infty} \int_{-\infty}^{\infty} {}^2\text{rect}(u - x, v - y)\,{}^2\text{rect}(u, v)\, du\, dv.$$

---

[1] In terms of the one-dimensional rectangle function $\Pi(x)$ or $\text{rect}\,x$ we can write ${}^2\Pi(x, y) = \Pi(x)\Pi(y) = \text{rect}\,x\,\text{rect}\,y$. The notation $\Pi(\ )$ is just as good as $\text{rect}(\ )$ but in conversation it is convenient to pronounce "Π" as "rect." The notation ${}^2\Pi$ is valuable for its compactness on occasion. It makes use of a preceding-superscript convention that can be applied systematically to a variety of situations: thus ${}^2\text{sinc}(x, y) = \text{sinc}\,x\,\text{sinc}\,y$, ${}^2\text{III}(x, y) = \text{III}(x)\text{III}(y)$, while ${}^2\mathcal{F}$ means the two-dimensional FT operator.

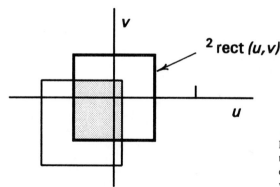

Figure 5-15 The autocorrelation of the two-dimensional unit rectangle function is equal to the area of overlap of $^2\text{rect}(u, v)$ with $^2\text{rect}(u + x, v + y)$.

Assume for the moment that x and y are both positive. Then

$$c(x, y) = \begin{cases} \int_{x-0.5}^{0.5} \int_{y-0.5}^{0.5} dx'\, dy', & 0 \le x < 1 \text{ and } 0 \le y < 1 \\ 0, & \text{otherwise;} \end{cases}$$

$$= \begin{cases} (1 - x)(1 - y), & 0 \le x < 1 \text{ and } 0 \le y < 1 \\ 0, & \text{otherwise.} \end{cases}$$

To see how the limits of integration are arrived at is a topological rather than an algebraic exercise and is most easily carried out with a sliding piece of tracing paper. Figure 5-15 illustrates a fixed situation where $x = \frac{4}{5}$ and $y = \frac{3}{5}$ and will permit the limits of integration to be verified. Clearly the integral is equal to the area of the rectangle of overlap whose dimensions are $1 - x$ and $1 - y$; thus we reach the same conclusion as when we evaluated $\iint dx'\, dy'$ with careful attention to limits.

In order to carry out these discussions we assumed that x and y were both positive. If one or both are negative, the limits change. For example, if x is negative, the integral with respect to x' becomes

$$\int_{-0.5}^{x+0.5} \quad \text{instead of} \quad \int_{x-0.5}^{0.5}.$$

There are four cases to consider altogether, which is not only tedious but fraught with risk of algebraic error if all the changes are reasoned out. Therefore, we apply more fundamental reasoning to argue that the sign of x and y should not affect the autocorrelation, which must be symmetrical with respect to both x and y, and that therefore x and y may be replaced by $\mid x \mid$ and $\mid y \mid$, respectively, in the limited expression first obtained. Hence

$$c(x, y) = (1 - \mid x \mid)(1 - \mid y \mid)\ ^2\text{rect}(x/2, y/2).$$

The resulting function is the *lazy pyramid;* it is like a pyramid, but one which has slumped along its four sloping edges. The lines of steepest descent along the middle of each face are, however, straight (Fig. 5-16).

Our second example is the autocorrelation of a circular pillbox or round-table function, one that is equal to unity inside a central circle of unit diameter and zero outside.

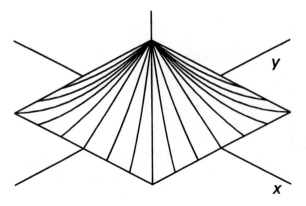

Figure 5-16 The lazy pyramid $(1-|x|)(1-|y|)\ ^2\text{rect}(x/2, y/2)$, the autocorrelation of a unit square-table function $^2\text{rect}(x, y)$. The base is a square of side 2 and the peak height is unity. The parabolic profiles are sections on planes containing the vertical lines.

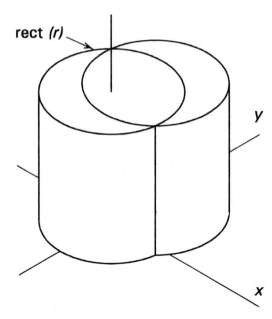

rect (r)

Figure 5-17 The autocorrelation of rect r, the unit round-table function or unit circular pillbox, is determined by integrating the product of rect r with a shifted replica, over the region of overlap.

There is no need[2] to invent new notation to cover this case, because the unit rectangle function of radius rect r suffices. By definition, the rectangle function is zero for $r >$ 0.5 and equal to unity in the range 0 to 0.5. Very often there is no need to talk about

---

[2] In optical circles one may encounter circ r, for describing a two-dimensional function that is equal to unity over a central circle of diameter 2. Goodman's definition is

$$\text{circ}\, r = \begin{cases} 1 & r \le 1 \\ 0 & \text{otherwise.} \end{cases}$$

In the framing of this definition it was clearly not intended to admit negative values of r. For this reason, circ r is not precisely the same thing as rect$(r/2)$.

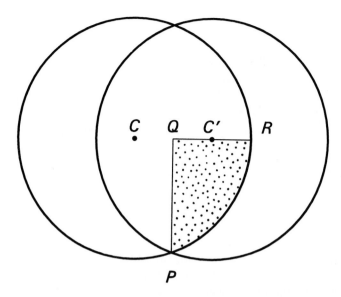

Figure 5-18 The area of overlap of the two unit circles is four times the shaded area PQR. Each circle has radius 0.5, and the separation CC' is r.

negative values of the radial coordinate r, but if negative values of r were introduced, the notation would be unaffected. Although two-dimensional notation could be imagined, one variable suffices because of the circular symmetry that makes it unnecessary to mention the azimuthal coordinate θ.

To picture the geometry involved in calculating the autocorrelation refer to Fig. 5-17. Both the original function and the shifted replica being unity over the region of overlap, their product there is unity. Therefore the autocorrelation will just be equal to the area of overlap, which can be derived simply by reference to Fig. 5-18. The centers of the two circles are at C and C'. The area of overlap may be calculated by noticing that the shaded area, which is one-quarter of the whole, is expressible as the difference between the circular sector CPR and the triangle CPQ.

The separation CC' between the centers of the two circles is the value of r to which the calculation will refer. Thus

$$\text{chat } r = 4(\text{sector } CPR - \text{triangle } CPQ)$$

$$= 2CP^2 \cos^{-1}(CQ/CP) - CC'\sqrt{CP^2 - CQ^2}$$

$$= [\tfrac{1}{2}\cos^{-1} r - \tfrac{1}{2}r\sqrt{1 - r^2} \text{ rect } \tfrac{1}{2}r.$$

Because of the shape of this function (Fig. 5-19) it is referred to as the *Chinese hat* function, after P. Léna. Two simple properties are useful for reference:

$$\text{chat } 0 = \tfrac{1}{4}\pi \quad \text{and} \quad 2\pi \int_0^1 r \text{ chat } r \ dr = (\tfrac{1}{4}\pi)^2.$$

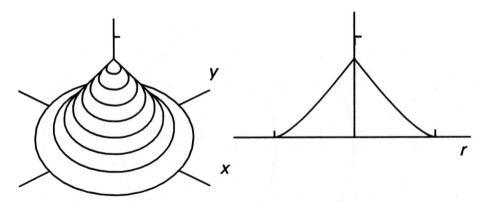

Figure 5-19 A pictorial view of the Chinese hat function (left) and its cross section (right).

The diameter, of course, is 2, being twice that of the original circle.

The chat function is ubiquitous in optical imaging, because it enters into the optical transfer function of a circular lens. Likewise, the lazy pyramid is the transfer function of a uniformly excited square antenna array.

CENTRAL VALUE AND VOLUME OF AUTOCORRELATION

Two simple theorems are of considerable value for checking and for normalizing, the first referring to the central value $c(0, 0)$ and the second to the volume under the autocorrelation function $c(x, y)$. By putting $x = y = 0$ in the definition integral, we obtain

$$c(0, 0) = \int_{-\infty}^{\infty} \int_{-\infty}^{\infty} [f(x, y)]^2 \, dx \, dy.$$

The theorem for volume is

$$\int_{-\infty}^{\infty} \int_{-\infty}^{\infty} c(x, y) \, dx \, dy = \left[\int_{-\infty}^{\infty} \int_{-\infty}^{\infty} f(x, y) \, dx \, dy \right]^2.$$

Thus the central value of the autocorrelation is the volume under the square of the original function. The volume under the autocorrelation is the square of the volume under the original function. Applying these theorems to the chat function, we immediately verify that chat $0 = \pi/4$ and that the volume under chat r is $(\pi/4)^2$.

To derive the volume theorem requires the evaluation of

$$\int \int c(x, y) \, dx \, dy = \int \int \left[\int \int f(x' - x, y' - y) f(x', y') dx' dy' \right] dx \, dy.$$

To perform this forbidding quadruple integration we rearrange things in the form

$$\int \int f(x', y') \left[\int \int f(x' - x, y' - y) \, dx \, dy \right] dx' \, dy'.$$

Since the integral in square brackets is to be performed with respect to x and y, it is permissible to remove the factor $f(x', y')$, which does not depend on x or y, outside the inner integrals. The mathematical condition for changing the order of integration in a case such as this is that the integral of f should exist, which is essentially the only situation we are interested in. Now the integral in square brackets is just the volume under f because the volume is invariant under a shift of origin; that is to say, $f(x, y)$ has the same volume as the shifted version $f(x' - x, y' - y)$. The square bracket, thus being a constant independent of x' and y', may be removed outside the remaining integral, which we see amounts to a second factor that is also equal to the volume of f. Hence the volume of the autocorrelation is the square of the volume of the original function.

The autocorrelation volume theorem is a special case of a more general theorem that is familiar in its one-dimensional form as a relation between the area under a convolution and the areas under the two functions entering into the convolution. Thus, if $h(x) = f(x) * g(x)$, then

$$\int_{-\infty}^{\infty} h(x)dx = \int_{-\infty}^{\infty} f(x)\,dx \int_{-\infty}^{\infty} g(x)\,dx.$$

The area under the convolution $h(x)$ is equal to the product of the areas under the given functions $f(x)$ and $g(x)$. Likewise, for the convolution sum in the discrete case, the sum of the convolution sequence is the product of the sums of the two factors, a relationship that is useful all the time for numerical checking when short sequences or small arrays are being dealt with. The two-dimensional generalization is

$$\int_{-\infty}^{\infty}\int_{-\infty}^{\infty} f(x, y) ** g(x, y)\,dx\,dy = \int_{-\infty}^{\infty}\int_{-\infty}^{\infty} f(x, y)\,dx\,dy \int_{-\infty}^{\infty}\int_{-\infty}^{\infty} g(x, y)\,dx\,dy;$$

the volume under the convolution equals the product of the separate volumes. As autocorrelation is a special case of convolution, the autocorrelation volume theorem follows. It is indispensable for checking calculations and for determining normalizing factors.

THE CONVOLUTION SUM

When convolution of functions of continuous x and y is to be carried out numerically, we deal with two matrices comprising samples that represent $f(x, y)$ and $g(x, y)$. Assume that the samples are taken at unit interval, and thus at integral values of i horizontally and j vertically. The convolution sum

$$h_{ij} = \sum_{k=-\infty}^{\infty} \sum_{l=-\infty}^{\infty} f_{i-k, j-l}\, g_{k,l}$$

will then be an approximation to the integral; if a better approximation is needed and the computing time is available, the samples could be taken at interval 0.5. Since this would now be four times as many samples as before, the new convolution sum would have to be divided by four. The time taken to compute the new values would be four times longer, assuming it was sufficient merely to recompute the convolution sum in the original

$$\begin{bmatrix} 1 & 2 & 3 \\ 4 & 5 & 6 \end{bmatrix} ** \begin{bmatrix} 1 & 0 & 0 \\ 1 & 1 & 1 \\ 2 & 1 & 1 \end{bmatrix} = \begin{bmatrix} 1 & 2 & 3 & 0 & 0 \\ 5 & 8 & 12 & 5 & 3 \\ 8 & 14 & 21 & 11 & 6 \end{bmatrix}$$

Figure 5-20   The convolution of two simple discrete sets presented in the form of small matrices.

$$\frac{1}{36}\begin{bmatrix} 6 & 7 & 8 \\ 3 & 4 & 5 \\ 0 & 1 & 2 \end{bmatrix} ** \begin{bmatrix} 1 & 15 & 14 & 4 \\ 12 & 6 & 7 & 9 \\ 8 & 10 & 11 & 5 \\ 13 & 3 & 2 & 16 \end{bmatrix} = \begin{bmatrix} 0.2 & 2.7 & 5.5 & 6.7 & 3.9 & 0.9 \\ 2.1 & 4.7 & 8.0 & 8.2 & 5.7 & 2.6 \\ 2.3 & 5.1 & 8.9 & 8.8 & 6.3 & 2.6 \\ 2.8 & 5.1 & 7.8 & 7.3 & 6.3 & 4.8 \\ 1.1 & 1.9 & 3.0 & 2.8 & 2.8 & 2.5 \\ 0.0 & 0.4 & 0.8 & 0.2 & 0.6 & 0.9 \end{bmatrix}$$

Figure 5-21   A convolution example in which one of the factors is normalized to unit sum and has a displaced centroid.

locations. However, in cases where more precision is needed, it is commonly desirable to compute the more finely spaced results; the computing time will then have gone up by a factor 16 as a consequence of halving the tabulation interval. Clearly it is wise before heavily increasing the number of samples to consider this nontrivial tradeoff.

Figure 5-20 illustrates a convolution sum. This result should be checked. As described in connection with continuous variables, one can use tracing paper to copy the smaller matrix, rotated through 180°. Then the tracing can be laid over the larger matrix, products of coincident terms can be summed, and the result can be recorded on the lower sheet through a suitably located window in the tracing paper (or against an arrowhead at the edge). Then the tracing is relocated until all relative positions within the boundary of the output have been dealt with.

Fig 5-21 shows a further example; the factor 1/36 has the effect of reducing the mass, or sum, of the smaller matrix to unity. As a consequence the mass of the convolution is 132, the same as the mass of the larger matrix factor, and the centroid has moved off center up to the right by exactly the amount by which the centroid of the smaller matrix is offset. This is because the 4 × 4 matrix is a magic square and therefore has its centroid at the center. The positions of the centroids can be checked numerically and the amount of the discrepancy can be thought about. In both of the above examples cross-correlating instead of convolving gives different results, which you can calculate. Before doing so, reason out some of the properties to be expected, as a basis for checking the output.

Convolution sums do not arise only as discrete samples of some physical quantity; a convolution sum need not be an aproximation at all but rather may correspond to some entity in its own right. In Fig. 5-22, the left-hand matrix represents a set of binomial coefficients chosen for the purpose of smoothing some recorded data and diminishing the magnitude of unwanted measurement noise. Following this step, the data analyst wishes to

$$\begin{bmatrix} 1 & 4 & 6 & 4 & 1 \\ 4 & 16 & 24 & 16 & 4 \\ 6 & 24 & 36 & 24 & 6 \\ 4 & 16 & 24 & 16 & 4 \\ 1 & 4 & 6 & 4 & 1 \end{bmatrix} ** \begin{bmatrix} 1 & -1 \end{bmatrix} = \begin{bmatrix} 1 & 3 & 2 & -2 & -3 & -1 \\ 4 & 12 & 8 & -8 & -12 & -4 \\ 6 & 18 & 12 & -12 & -18 & -6 \\ 4 & 12 & 8 & -8 & -12 & -4 \\ 1 & 3 & 2 & -2 & -3 & -1 \end{bmatrix}$$

Figure 5-22   An example of convolution where one factor has zero sum.

$$\begin{bmatrix} 1 & 1 & 1 \\ 1 & 0 & 0 \\ 1 & 1 & 1 \\ 0 & 0 & 1 \\ 1 & 1 & 1 \end{bmatrix} ** \begin{bmatrix} 1 & 1 & 1 \\ 1 & 0 & 0 \\ 1 & 1 & 1 \\ 0 & 0 & 1 \\ 1 & 1 & 1 \end{bmatrix} = \begin{bmatrix} 1 & 2 & 3 & 2 & 1 \\ 2 & 2 & 2 & 0 & 0 \\ 3 & 4 & 6 & 4 & 2 \\ 2 & 2 & 4 & 2 & 2 \\ 3 & 6 & 11 & 6 & 3 \\ 2 & 2 & 4 & 2 & 2 \\ 2 & 4 & 6 & 4 & 3 \\ 0 & 0 & 2 & 2 & 2 \\ 1 & 2 & 3 & 2 & 1 \end{bmatrix}$$

Figure 5-23   Self-convolution of a function exhibiting twofold rotational symmetry and noticeable chirality.

subtract the resulting image from itself after a right shift of one unit has been introduced. Let us say the purpose is to remove a uniform background that is not relevant, while at the same time emphasizing east-west gradients that have significance for the job at hand. Although two operations are to be carried out, the work can be condensed to one convolution with the data by combining the two operators illustrated. The resulting combination is shown on the right of the figure. This smoothed gradient operator has the following features: its mass is zero, its centroid is offset to the right by half an interval, and it has odd symmetry with respect to its centroid.

Finally a self-convolution example is shown in Fig. 5-23, where we start with a function having twofold rotational symmetry and a sum of 11. The self-convolution shown occupies three times the area, has a mass of 121, its centroid in the center, and exhibits twofold rotational symmetry; in addition it has S-shaped chirality that is noticeable, but less so than on the original.

COMPUTING THE CONVOLUTION

Convolving two matrices is a fundamental operation which applies both to data sets and to situations where one operates on data with a factor that is defined as some function of a continuous variable. The following segment of code convolves two functions $f(\ ,\ )$ and $g(\ ,\ )$ to produce $h(\ ,\ )$. The functionf $f(\ ,\ )$ has M_f columns and N_f rows, $g(\ ,\ )$ has size $M_g \times N_g$, while the size of $h(\ ,\ )$ will be $M_f + M_g - 1 \times N_f + N_g - 1$.

CONVOLUTION OF $f(\ ,\ )$ WITH $g(\ ,\ )$

```
FOR i=1 TO Nf+Ng-1
    FOR j=1 TO Mf+Mg-1
        h(i,j)=0
        FOR k=MAX(1,i+1-Ng) TO MIN(i,Ng)
            FOR l=MAX(1,j+1-Mg) TO MIN(j,Mg)
                P=f(k,l)
                Q=g(i+1-k,j+1-l)
                h(i,j)=h(i,j)+P*Q
            NEXT l
        NEXT k
    NEXT j
NEXT i
```

If both the given matrices are of large size, it may save computing time to use a fast transform method such as described in connection with Table 6-1. However, there is much to be said for a simple program, and, seeing that computers keep getting faster, users may not notice any significant reduction in elapsed time for a particular job if they repeat the computation using transforms. Actual timing will tell. Related programs for autocorrelation and cross-correlation appear in the problems. Naturally, if two mathematically defined functions are to be convolved numerically, the code given here is applicable.

Array operations provided for in some high-level languages do not include convolution; however, some matrix statements are convenient in handling image matrices, for example,

```
DIM h[19,19]
MAT PRINT h prints matrix h in rows and columns
```

Other convenient one-line statements reset the matrix to constant elements, add or multiply element by element, and furnish row or column sums.

DIGITAL SMOOTHING

Two-dimensional data may contain unwanted high spatial frequencies or fine structure that can be reduced or filtered out by convolution with a compact function spread over a suitable area. For numerical purposes the convolving function has to be in the form of a two-dimensional array or matrix; each spatial-frequency component of the data will then be modified by a two-dimensional transfer function. To discuss the spectral aspect of digital smoothing, or filtering in general, we need the two-dimensional convolution theorem, which will be taken up in the next chapter.

MATRIX PRODUCT NOTATION

When two matrices are convolved, one standing for an object (discretely represented) and one standing for a point-spread function, or impulse response, the notation exhibits

$$[f] = \begin{bmatrix} 0 & 0 & 0 & 0 \\ 0 & 1 & 2 & 0 \\ 0 & 3 & 4 & 0 \\ 0 & 0 & 0 & 0 \end{bmatrix}, \quad [g] = \begin{bmatrix} 0 & 1 & 2 & 0 \\ 1 & 7 & 9 & 2 \\ 3 & 11 & 13 & 4 \\ 0 & 3 & 4 & 0 \end{bmatrix},$$

$$[\mathbf{f}] = \begin{bmatrix} 0 \\ 0 \\ 0 \\ 0 \\ 0 \\ 1 \\ 2 \\ 0 \\ 0 \\ 3 \\ 4 \\ 0 \\ 0 \\ 0 \\ 0 \\ 0 \end{bmatrix}, \quad [\mathbf{g}] = \begin{bmatrix} 0 \\ 1 \\ 2 \\ 0 \\ 1 \\ 7 \\ 9 \\ 2 \\ 3 \\ 11 \\ 13 \\ 4 \\ 0 \\ 3 \\ 4 \\ 0 \end{bmatrix}.$$

Figure 5-24   An object $[f]$ and image $[g]$ represented by matrices, and the corresponding column vectors $[\mathbf{f}]$ and $[\mathbf{g}]$.

in a direct, essentially pictorial way, the entities entering into the convolution operation. However, an image matrix is merely an array of coefficients and lacks properties of the matrix of coefficients familiar from linear algebra. For example, it makes no sense to multiply an image matrix by an impulse response in accordance with the rules of matrix multiplication. Nevertheless, conventional matrix manipulations can be brought to bear— not surprisingly, because, after all, convolution is a linear operation. The 4×4 object

$$[f] = \begin{bmatrix} a & b & c & d \\ e & f & g & h \\ i & j & k & l \\ m & n & o & p \end{bmatrix}$$

is equivalent to the 16-element row vector $[a\,b\,c\,d\,e\,f\,g\,h\,i\,j\,k\,l\,m\,n\,o\,p]$, or to the corresponding column vector which we may call $[\mathbf{f}]$. Suppose the matrix $[f]$ is convolved with some point-spread function to give a smoothed image $[g]$; there will be a corresponding column vector $[\mathbf{g}]$. Is there some square matrix $[M]$ which, when multiplied into $[\mathbf{f}]$ will give $[\mathbf{g}]$?

An example will show how the expectation of such a linear relationship works. Fig. 5-24 presents a simple object $[f]$ and an image $[g]$ formed by smoothing (each value of $[f]$ was replaced by twice itself plus the sum of the four adjacent values – those situated one step to the E, N, W and S). The equivalent column vectors $[\mathbf{f}]$ and $[\mathbf{g}]$ are formed by reading off $[f]$ and $[g]$ European style. It is easy to verify that $[M][\mathbf{f}] = [\mathbf{g}]$ written in full is as follows;

$$
\begin{bmatrix}
2 & 1 & 0 & 0 & 1 & 0 & & & & & & 1 & 0 & 0 & 1 \\
1 & 2 & 1 & 0 & 0 & 1 & 0 & & & & & & 1 & 0 & 0 \\
0 & 1 & 2 & 1 & 0 & 0 & 1 & 0 & & & & & & 1 & 0 \\
0 & 0 & 1 & 2 & 1 & 0 & 0 & 1 & 0 & & & & & & 1 \\
1 & 0 & 0 & 1 & 2 & 1 & 0 & 0 & 1 & 0 & & & & & \\
 & 1 & 0 & 0 & 1 & 2 & 1 & 0 & 0 & 1 & 0 & & & & \\
 & & 1 & 0 & 0 & 1 & 2 & 1 & 0 & 0 & 1 & 0 & & & \\
 & & & 1 & 0 & 0 & 1 & 2 & 1 & 0 & 0 & 1 & 0 & & \\
 & & & & 1 & 0 & 0 & 1 & 2 & 1 & 0 & 0 & 1 & 0 & \\
 & & & & & 1 & 0 & 0 & 1 & 2 & 1 & 0 & 0 & 1 & 0 \\
0 & & & & & & 1 & 0 & 0 & 1 & 2 & 1 & 0 & 0 & 1 \\
1 & 0 & & & & & & 1 & 0 & 0 & 1 & 2 & 1 & 0 & 0 \\
0 & 1 & 0 & & & & & & 1 & 0 & 0 & 1 & 2 & 1 & 0 \\
0 & 0 & 1 & 0 & & & & & & 1 & 0 & 0 & 1 & 2 & 1 \\
1 & 0 & 0 & 1 & 0 & & & & & & 1 & 0 & 0 & 1 & 2
\end{bmatrix}
\begin{bmatrix} 0 \\ 0 \\ 0 \\ 0 \\ 0 \\ 1 \\ 2 \\ 0 \\ 0 \\ 3 \\ 4 \\ 0 \\ 0 \\ 0 \\ 0 \\ 0 \end{bmatrix}
=
\begin{bmatrix} 0 \\ 1 \\ 2 \\ 0 \\ 1 \\ 7 \\ 9 \\ 2 \\ 3 \\ 11 \\ 13 \\ 4 \\ 0 \\ 3 \\ 4 \\ 0 \end{bmatrix}
$$

   The argument from linearity tells us that [**f**] and [**g**], if regarded as simple sequences rather than as formal column vectors in matrix terminology, must be related by one-dimensional convolution; in other words $\{g\} = \{h\} * \{f\}$, and in this case

$$\{h\} = \{0\,1\,0\,0 \quad 1\,2\,1\,0 \quad 0\,1\,0\,0 \quad 0\,0\,0\,0\}.$$

   The factor $\{h\}$ is found by performing the inverse convolution $\{f\}^{-1} * \{g\}$. The technique is to fill the array $\{h\}$ with 16 unknowns and to begin to convolve $\{h\}$ with $\{f\}$ by sliding the reverse of $\{h\}$ over $\{f\}$, as explained in FTA (1986, p. 36). When the moving array makes first contact, the two coefficients that overlap are to be multiplied together. It is true that one coefficient is unknown, but the product is known—it is the first element of $\{g\}$. With the first unknown evaluated, the moving array slides one more step, and the second unknown element of $\{h\}$ falls out, and so on.

   Referring back to the 16×16 matrix $[M]$ we see that the rows are progressively displaced copies of the sequence $\{h\}$ written in reverse (the trailing zeros of $\{h\}$ have been omitted to make this clearer). The kinematics of the sliding sequence is caught as a series of still snapshots, one for each location of the moving array. This sort of matrix is called a *circulant matrix*, because the rows circulate progressively from top to bottom. Of course, it is rather extravagant notation when written out numerically in full, but it can be recommended for those fluent in conventional matrix manipulation. The methodology has been systematically exploited, for example by Andrews and Hunt (1977). The approach readily accommodates to space variance, in which case the rows may change as they progress.

   In terms of matrix convolution, introducing the matrix $h(x, y)$ that is formed by stacking the sequence $\{h\}$ in rows of four, the situation would be presented as

$$g(x, y) = h(x, y) ** f(x, y) = \begin{bmatrix} 0 & 1 & 2 & 0 \\ 1 & 7 & 9 & 2 \\ 3 & 11 & 13 & 4 \\ 0 & 3 & 4 & 0 \end{bmatrix}$$

$$= \begin{bmatrix} 0 & 1 & 0 & 0 \\ 1 & \mathbf{2} & 1 & 0 \\ 0 & 1 & 0 & 0 \\ 0 & 0 & 0 & 0 \end{bmatrix} ** \begin{bmatrix} 0 & 0 & 0 & 0 \\ 0 & 1 & 2 & 0 \\ 0 & 3 & 4 & 0 \\ 0 & 0 & 0 & 0 \end{bmatrix}.$$

This notation has the visual appeal of relating to the image plane and shows pretty succinctly what is going on. The boldface **2** in the point-spread function $h(x, y)$ conveys the origin chosen for $h(\ ,\ )$.

As another example, conventional matrix multiplication of coordinates with

$$\begin{bmatrix} \cos\theta & \sin\theta \\ -\sin\theta & \cos\theta \end{bmatrix}$$

can be used for rotating an image, because the matrix product with any vector tells where the value associated with that vector will be moved to. But since the matrix product consists of real rather than integer values, there will not in general be a grid point at the calculated destination. The technique of rotating an image is dealt with in Chapter 7.

So far, the complications of cyclic convolution have not required mention, but the NE and SW corners of the matrix [M] clearly warn of the need to provide a guard zone of zeros around $f(x, y)$, or to make the equivalent adjustments to [$\mathbf{f}$]. A corresponding concern applies to the 4 × 4 matrices. The smoothing matrix $h(x, y)$, when read off European style by raster scanning, yields the one-dimensional sequence $\{h\}$, whose convolution sum with the sequence $\{f\}$ yields the sequence $\{g\}$.

The visual appeal of two-dimensional data arrays and point-spread functions and the relative ease of gaining an intuitive feel for two-dimensional convolution speak in favor of the matrix-convolution equation just given. The formulation in terms of column vectors and circulant matrices is bulkier and less graphic. The bulk of a matrix [M] can become embarrassing in a computer. For example, an image of size 512 × 512 corresponds to a one-dimensional array of size 262,144 elements (and indeed is so stored in memory). The matrix [M] then has 68,719,476,736 elements, which amounts to a nontrivial number of megabytes.

SUMMARY

Convolution is a remarkable operation involving two functions, or discrete sequences, that will remain a useful tool for life. Convolution applies to smoothing and sharpening of images, trend removal, trend detection, duplication, replication, differentiation, integration, differencing, summing, and dilation. Convolution includes cross-correlation and autocorrelation and is useful in subjects where these concepts arise—for example, in statistics, probability, signal analysis, noise, and optical transfer functions. As a thinking tool, convolution is one of the sharpest. Rather simple notation has been introduced that is versatile enough to handle both the continuous and the discrete. Further aspects of convolution come in the next chapter, which deals with the convolution theorem.

LITERATURE CITED

H. C. ANDREWS AND B. R. HUNT (1977), *Digital Image Restoration*, Prentice-Hall, Englewood Cliffs, N.J.

R. N. BRACEWELL (1961), "Interferometry and the spectral sensitivity island diagram,"*IRE Transactions on Antennas and Propagation,* vol. AP9, pp. 59–67, January 1961.

R. N. BRACEWELL (1984), "Early work on imaging theory in radio astronomy," in *Early Radio Astronomy* (W. T. Sullivan, III, ed.), pp. 167–190, Cambridge: University Press, 1984.

PROBLEMS

5–1. *Convolution of line segments.* Evaluate the convolution [rect x $\delta(y) ** $[rect y $\delta(x)$]. ☞

5–2. *Convolution of matrices.* Convolve the following two pairs of data arrays presented in matrix form.

$$
\begin{bmatrix} 1 & 2 & 3 & 4 & 5 \\ 0 & 1 & 2 & 3 & 4 \\ 0 & 0 & 1 & 2 & 3 \\ 0 & 0 & 0 & 1 & 2 \\ 0 & 0 & 0 & 0 & 1 \end{bmatrix} ** \begin{bmatrix} 0 & 0 & 0 & 0 & 0 \\ 0 & 0 & 0 & 0 & 0 \\ -1 & 2 & -1 & 0 & 0 \\ 0 & 0 & 0 & 0 & 0 \\ 0 & 0 & 0 & 0 & 0 \end{bmatrix}, \qquad \begin{bmatrix} 3 & 2 & 1 \\ 2 & 3 & 2 \\ 1 & 2 & 3 \end{bmatrix} ** \begin{bmatrix} 1 & 0 \\ 0 & -1 \end{bmatrix}.
$$

A feature of this mode of presentation is that the origin of indexing is not explicitly stated.
(a) Present the convolutions in matrix form.
(b) Comment on the matter of indexing.

5–3. *Analytic convolution.* What is the two-dimensional convolution of $\exp\{-\pi[(x/a)^2 + (y/b)^2]\}$ with $\exp\{-\pi[(x/c)^2 + (y/d)^2]\}$? ☞

5–4. *Autocorrelation program for serial object.* Consider a rectangular sparse matrix composed of 0s and 1s, such as the bitmap that specifies a letter or numeral for use in a printer or represents a digital signature. Instead of presenting the whole matrix, one can economize by listing coordinates $X(i)$, $Y(i)$ as a function of a serial number i which runs from 1 to N, where N is the number of 1s, or dots. If the bitmap is for a cursive character, it can be drawn on the screen as i runs. Write a program that computes the autocorrelation matrix $A(\ ,\ )$. ☞

5–5. *Self-convolution of half moon.* A function is zero where $x < 0$ and where $x > \sqrt{c^2 - y^2}$ but is otherwise equal to unity. What is its self-convolution?

5–6. *Computed autocorrelation boundary.* Draw two imaginary islands, one convex, the other having a reentrant bay, and determine the autocorrelation boundaries graphically. Think about a computer program for plotting autocorrelation boundaries starting from a digitized list of points on the coastline of the given island. ☞

5–7. *An autocorrelation boundary.* A function $f(x, y)$ is zero where $y < x^2$ and where $y > 1$ but elsewhere is nonzero. Find the autocorrelation boundary. ☞

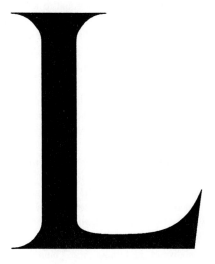

Figure 5-25 The letter ell, with serifs.

5–8. *Analytic autocorrelation boundary.* Think of a closed curve specified analytically where the outline bounding the nonzero values is simple and has continuous slope almost everywhere. How could you determine the autocorrelation boundary analytically? ☞

5–9. *Noncommutativity of cross-correlation.* For the purpose of comparing $f \star\star g$ with $g \star\star f$ let $f(x, y) = S_f(x, y) + A_f(x, y)$, where $S_f(x, y) = \frac{1}{2}[f(x, y) + f(-x, -y)]$ is the symmetrical part of $f(x, y)$ and $A_f(x, y) = \frac{1}{2}[f(x, y) - f(-x, -y)]$ is the antisymmetrical part. Express $f \star\star g$ and $g \star\star f$ in terms of these components and comment on the distinction between the two.

5–10. *Commutativity of cross-correlation.* What is the condition for the cross-correlation of $f(x, y)$ on $g(x, y)$ to be equal to the cross-correlation of $g(x, y)$ on $f(x, y)$? ☞

5–11. *Truncated pyramid.* When the rectangular prism function $^2\text{rect}(x/A, y/B)$ is cross-correlated with $^2\text{rect}(x/C, y/D)$, the result is a sort of pyramid with a flat top, if A, B, C and D are all different. What is the height of that plateau?

5–12. *Central value.* A function $f(x, y)$ is equal to unity inside the two circles $x^2 + (y \pm 1)^2 = 1$ and zero elsewhere. Naturally the autocorrelation function is a little complicated, but what is the central value of the autocorrelation, what is the autocorrelation boundary, and what is the volume under the autocorrelation? ☞

5–13. *Semicircular aperture.* One half of a circular aperture is blocked; the remaining semicircular aperture is describable by

$$f(x, y) = \begin{cases} 1, & x^2 + y^2 < c^2 \text{ and } y > 0, \\ 0, & \text{elsewhere,} \end{cases}$$

where the radius $c = \frac{1}{2}$.
  (a) Determine the boundary of the autocorrelation function $C(x, y) = f(x, y) \star\star f(x, y)$.
  (b) Obtain algebraic expressions for $C(x, 0)$ and $C(0, y)$, the principal cross-sections through $C(x, y)$.
  (c) Represent $C(x, y)$ by a formula or graphically by three or four contours.

5–14. *Autocorrelation of ell.* Find the autocorrelation of the shape shown in Fig. 5-25.

5–15. *Autocorrelation function of elliptical disc.* A function is equal to unity inside the ellipse $(x/a)^2 + (y/b)^2 = 1$ and zero elsewhere. You have to report the autocorrelation function to a client. First decide whether you will report in terms of algebra or graphics or the written word. Then work out the answer and present it.

5–16. *Autocorrelation parameters.* Let $c(x, y)$ be the autocorrelation of $f(x, y) = \Lambda(x)\operatorname{rect}(y/2)$. What is $c(0, 0)$ and what is the volume under $c(x, y)$?

5–17. *Autocorrelator.* Evolve a concept for an optical instrument that would accept a picture in the form of a transparency and would deliver a film bearing the autocorrelation function of the picture.

5–18. *Profiles of lazy pyramid.*
   (a) Show that all cross sections of the lazy pyramid taken parallel to the sides of the base are triangular.
   (b) Someone said the vertical-plane cross section parallel to a diagonal of the base has a rounded top, but a student pointed out that the cross section exactly along the diagonal has a point at the top. In view of the fact that the triangular sections are all pointed at the top, is it possible that the inclined sections could be rounded?

5–19. *Hello.* Obtain the autocorrelation matrix of the bitmap (5×7 binary matrix) representing the following digraph: **H i**. The **H** is five pixels wide and is followed by a space three pixels wide. The **i** is seven pixels high and the space under the dot occupies one pixel.

5–20. *Shape of the lazy pyramid.* A standard method of indicating the shape of a surface is by contours of equal height. If we represent the function $(1 - |x|)(1 - |y|)$ by contours ranging from 0 to 1 at intervals of 0.1 (which would be specified briefly by 0[0.1]1 in the jargon of mathematical tables), the lower contours will be squarish. Will those contours have corners? (A "corner" is a point on a curve where there is a finite discontinuity of slope, as distinct from a cusp, which is more drastic.) Will the higher contours be approximately circular, or will they also be squarish, and, if so, how oriented? Arrive at conclusions by thinking and verify the conclusions by calculation. ☞

5–21. *Apex of a pyramid.*
   (a) The equation $z = \operatorname{pyr}(x, y)$ represents a surface whose height is zero outside a central 2×2 central square, equal to unity at the origin, and is otherwise like an Egyptian pyramid (apart from the fact that the height is half the base instead of about two-thirds). Obtain an expression for $\operatorname{pyr}(x, y)$.
   (b) A lazy pyramid $\mathcal{P}(x, y)$ also has a 2×2 base and a central height of unity. I make clay models of the two sorts of pyramid to a scale modulus of 20 mm. (*Scale modulus* is the number of mm per dimensionless unit, while *scale factor* is the number of dimensionless units per mm, or length unit of choice.) Accuracy of construction is of the order of 0.1 mm. I carefully slice the top millimeter from the apex of each model and give you the two small pieces. Can you tell which is which? If so, say how. If not, say why not.

5–22. *Circularity of binomial array.* The 2D binomial arrays represented concisely as the n-fold self-convolutions $\begin{bmatrix} 1 & 1 \\ 1 & 1 \end{bmatrix}^{**n}$ for $n = 1, 2, \ldots$ look rather too squarish to represent circularly symmetrical Gaussian humps.

   (a) Write out the array $\begin{bmatrix} 1 & 1 \\ 1 & 1 \end{bmatrix}^{**4}$ for study (the central value is 36).
   (b) Let σ^2 be the horizontal variance and τ^2 the diagonal ($45°$) variance. Define the "noncircularity" by $\nu = (\sigma^2 - \tau^2)/(\sigma^2 + \tau^2)$. Estimate the sign and rough magnitude of ν by eye by superimposing graphs of the cross sections taken parallel and at $45°$ to the axes.

(c) Calculate σ^2, τ^2, and ν. ☞

5–23. *Binomial compared with Gaussian.*

   (a) Compare $\begin{bmatrix} 1 & 1 \\ 1 & 1 \end{bmatrix}^{**4}$ with $36\exp(-r^2/2\sigma_0^2)$, where σ_0^2 is the horizontal variance of
   $$\begin{bmatrix} 1 & 1 \\ 1 & 1 \end{bmatrix}.$$

   (b) Draw a conclusion about the interchangeability of the integer binomial-array values and discrete samples of the stated Gaussian expression.

5–24. *Centroid of convolution.*

   (a) If
   $$a(i, j) = \begin{bmatrix} 1 & 4 & 6 \\ 4 & 16 & 4 \\ 1 & 4 & 1 \end{bmatrix}$$

   work out $a(i, j) ** a(i, j)$, the two-dimensional convolution of $a(i, j)$ with itself.

   (b) Determine the centroid of $a(i, j)$.

   (c) Determine the centroid of $a(i, j) ** a(i, j)$.

5–25. *Cross-correlation program for matrix.* Write a simple computer program that accepts a smoothing matrix $S(X, Y)$, $1 \le X \le L_x$, $1 \le Y \le L - y$, composed of zeros and positive and negative integers, and a corresponding data matrix $D(X < Y)$, $1 \le X \le L_x$, $1 \le Y \le L_y$, and delivers a matrix $C(X, Y)$ which represents the cross-correlation function $C(X, Y)$ of $S(X, Y)$ on $D(X, Y)$. ☞

5–26. *Autocorrelation program for matrix.* Write a simple program that accepts a rectangular matrix $M(x, y)$ and computes the nonnormalized autocorrelation $A(x, y)$.

5–27. *Autocorrelation integral by summing.* A lazy person who needs the autocorrelation function of $\text{sinc}^2(x/2)$ decides to calculate an approximation to the value of

$$\text{sinc}^2(x'/2)\,\text{sinc}^2[(x - x')/2],$$

by first selecting the offset x and then summing the product at intervals $\Delta x' = X$, and finally multiplying by a factor X on the grounds that the wider the sampling interval X, the less the sum will be, and so the bigger will be the compensation factor needed. We know that $X < 1$ will satisfy the sampling theorem for sampling the band-limited function $\text{sinc}^2(x/2)$, but for a quick and crude start this fearless numerical analyst tries $X = 2$ and adds up only 21 values symmetrically situated about $x' = x/2$.

   (a) Verify by computation that the rather coarse interval $X = 2$ leads to the correct value of the integral, and

   (b) that it does not matter whether the central sample is situated at $x' = x/2$ or not. ☞

5–28. *Faceted surface for computer graphics.* One computer-graphics technique of giving the appearance of depth to a 2D image is to treat a 3D object as if it were bounded by a faceted surface. For example, a football would be treated as a polyhedron with many plane-triangular faces. Then the projection of each facet on the picture plane is shaded with a grey level determined by some rule. If a curve $z = f(x)$ is sampled at unit intervals of x, then the polygon passing through the samples is the convolution $\Lambda(x) * [f(x)\text{III}(x)]$, where $\Lambda(x)$ is the unit triangle function (unit height, unit area, and base = 2). If a surface $z = f(x, y)$ is sampled at unit intervals of both x and y, can a faceted surface with vertices at the sample points be expressed in terms of convolution? If so, give the formula, and state the shape of the facets. If not, explain. ☞

6

The Two-Dimensional Convolution Theorem

Just as in one dimension, the convolution theorem in two dimensions plays a pervasive role wherever *linearity* and *shift invariance* are simultaneously present. In one dimension shift invariance most commonly means *time invariance*. A time-invariant system has the property that the response to an input impulse is independent of epoch. In other words, if two different input impulses are considered, one shifted in time by any amount with respect to the other, then the responses will be the same, allowing for the time shift. In two dimensions, where the variables represent space, the corresponding attribute of a system is *space invariance*. Suppose that a television camera is pointed at a blackboard. Then a space-invariant system has the property of imaging the blackboard onto the screen of the television display so that two different white dots of chalk, no matter where they are in the plane of the board, produce appropriately shifted identical images (Fig. 6-1).

A *definition* of shift invariance, rather than an illustrative property which is sufficient here, is given with the problems. A definition of linearity is given also.

Time invariance is very common indeed, but of course even the most stable transducer will alter as time elapses, and in any case it is of finite age. It had a beginning, when it was assembled, and will meet with an end. Therefore, even time invariance never occurs with mathematical strictness in the physical world, and we understand that some reasonable judgment has to be made. The idea that the convolution theorem may be inapplicable to a transistor amplifier, on the grounds that transistors slowly age because of diffusion, is one that we reject because our tests will usually be finished long before the aging of a

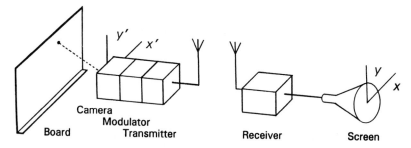

Figure 6-1 Given linearity and space invariance, the image $i(x, y)$ on the screen is derivable from the object distribution $o(x, y)$ by two-dimensional convolution with the point response function $h(x, y)$ of the system.

transistor makes a noticeable difference; and there are other time-varying influences, such as supply voltage and temperature, that take precedence.

By contrast, space invariance is much less common, and one encounters failure of shift invariance much more often than when dealing with time. Thus, a chalk dot on the edge of the blackboard will be less faithfully imaged than one the camera is pointing straight at. And the lens designer will have taken that fact directly into account, because to make sure the deterioration at the edge of the field is tolerable, it is necessary to know the details.

As soon as an image is converted to a time-dependent voltage waveform, familiar circuit conditions govern time invariance. But where we are dealing with optics in space, caution is necessary. Nevertheless, where space invariance exists, to the satisfaction of the engineer, and also linearity, then the convolution theorem applies. That means that the image is representable as the two-dimensional convolution of a point-response, or point-spread function, with a two-dimensional object function. Since the objects pointed at by television cameras are usually three dimensional, some attention is needed to decide just what the two-dimensional object function is. In other words, if x and y are coordinates on a television screen and the response to a point source of light on the camera axis is $h(x, y)$, and if $i(x, y)$ is the image distribution, then the assertion that a convolution relation exists means that:

$$i(x, y) = h(x, y) ** o(x, y),$$

where $o(x, y)$ is the object function. But what do x and y mean in the expression $o(x, y)$? Clearly, (x, y) defines a direction, and $o(x, y)$ is the brightness of light arriving at the camera lens from that direction. It is the *direction* which is imaged onto the *point* (x, y) on the television screen. Very often x and y will be direction cosines of that direction. We now turn to the two-dimensional convolution theorem, which tells what happens in the Fourier transform domain when convolution occurs in the function domain.

When dealing with discrete data, we can apply the discussion given in this chapter without change by drawing on delta-function notation. Therefore the main presentation works in terms of continuous variables.

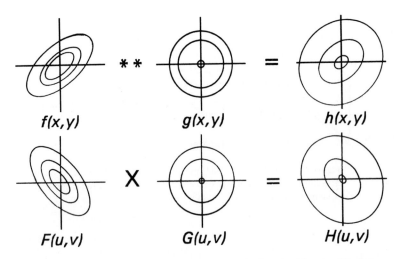

Figure 6-2   Functions $f(x, y)$ and $g(x, y)$ are convolved to give $h(x, y)$, while their Fourier transforms, shown below, combine through multiplication.

CONVOLUTION THEOREM

In the convolution theorem we find the most powerful tool for thinking about two-dimensional matters. It can be expressed in words exactly as for one dimension. "Convolution in the function domain corresponds to multiplication in the transform domain." Just as in one dimension one of the functions must be reversed before the multiplication and integration are performed, so in two dimensions one of the functions must be reversed with respect to both coordinates. As we have seen, this is the same as rotating the function $180°$ about the origin in the picture plane.

Convolution Theorem in 1D

$$\text{If } f(x) \supset F(s) \text{ and } g(x) \supset G(s),$$

$$\text{then } f(x) * g(x) \supset F(s)G(s).$$

Convolution Theorem in 2D

$$\text{If } f(x, y) \,{}^2{\supset}\, F(u, v) \text{ and } g(x, y) \,{}^2{\supset}\, G(u, v),$$

$$\text{then } f(x, y) ** g(x, y) \,{}^2{\supset}\, F(u, v)G(u, v).$$

In the top row of Fig. 6-2 two functions $f(x, y)$ and $g(x, y)$ are convolved to form $h(x, y)$. The example is chosen so that it is qualitatively verified by eye that the some-what elongated distribution $f(x, y)$, after blurring by convolution with the symmetrical function $g(x, y)$, has become broader in general but has also retained a trace of the original anisotropy of $f(x, y)$. In the bottom row we see the Fourier transforms of all three functions: the axis of $F(u, v)$ is at right angles to the axis of $f(x, y)$ as expected, $G(u, v)$

is symmetrical, and $H(u, v)$ is slightly elongated in the direction expected relative to the slight elongation of $h(x, y)$. All these transforms are familiar. What the convolution theorem tells us is that $H(u, v)$ is derivable as the product of $F(u, v)$ with $G(u, v)$. Qualitative examination of the figure shows just how the multiplication leads to the narrowing of $H(u, v)$ and to the reduction of eccentricity.

As we constantly require to use multiplication as well as convolution in the image plane, we often require the convolution theorem in its inverse form

$$f(x, y)g(x, y) \,{}^2{\supset}\, F(u, v) ** G(u, v).$$

AN INSTRUMENTAL CAUTION

Convolution in two dimensions could be prescribed as follows. "Take one of the two given functions, *rotate* it through 180° about the origin of coordinates, *translate* it to a new location by an amount (x, y), *multiply* the two functions together point by point, and *integrate* the product over the whole plane to obtain the value of the convolution for that (x, y). Repeat for other choices (x, y) until the desired region is covered."

It often happens that a function is to be displaced but the displacement is not a pure translation. A searchlight beam scanning the wall of a jail may be raised by elevating the lamp and reflecting mirror about the horizontal axis and may be turned to the left and right about the vertical axis of the mounting. It would be fair to say that the beam is being translated. Let us describe the intensity distribution over the beam by a point spread function, the amount of light received back from a white spot placed at the nearest point on a black test wall. A light meter directed toward the wall of the jail would record one value of the convolution of the point spread function with the distribution of light reflecting power over the wall. The same searchlight scanning the clouds directly overhead would *not* generate a convolution. It is true that the beam still has two degrees of freedom of displacement, and it would be quite possible by carefully manipulating the hand wheels for training and pointing the beam to perform a parallel raster scan exactly as could have been done on the wall. But as the beam is displaced, it also rotates around the beam axis. In this example, the amount of rotation is locked to the direction of pointing. Let $g_\theta(x, y)$ be what $g(x, y)$ becomes when it is rotated through a position angle θ about the origin. Then

$$\int \int f(x', y')g_\theta(x' - x, y' - y)\,dx'\,dy',$$

where $\theta = \arctan(y/x)$, is the kind of integral that would apply. This is not the convolution integral; although, if the searchlight beam had reasonable circular symmetry, it would not make much difference. More complicated situations also arise where it is possible for an instrument to point at a particular location (x, y) and to return to that point on another occasion but oriented with a changed position angle, all three degrees of freedom now being availed of. When astronomical telescopes were on equatorial mounts, exposures on a given star on different occasions would be uniformly oriented. Now that computer drives make it possible to use altazimuth mounts (the same as the searchlight mount) for large telescopes, the telescope can return to a given star in various position angles; even in the course of a single exposure position angle may vary.

POINT RESPONSE AND TRANSFER FUNCTION

In the convolution relation $i(x, y) = h(x, y) ** o(x, y)$, $h(x, y)$ is the point response function. From the convolution theorem we get the following relation between the image transform $I(u, v)$ and the object transform $O(u, v)$:

$$I(u, v) = H(u, v)O(u, v),$$

where $H(u, v)$ is the two-dimensional transfer function, sometimes called the *optical transfer function*. Before the complex entity $H(u, v)$ came into use in optics, it was customary in lens design to talk about the *modulation transfer function,* which, in the present notation, is $| H(u, v) |$. As the underlying unity of the theory in several fields has come to be recognized, the term transfer function has established itself generally.

AUTOCORRELATION THEOREM

Since the autocorrelation $f(x, y) \star\star f(x, y)$ is also expressible as the convolution $f(x, y)$ $** f(-x, -y)$, the convolution theorem may be applied. Let $f(x, y) \ {}^2\!\supset F(u, v)$. Then one can verify from the definition of the Fourier transform that

$$f(-x, -y) \ {}^2\!\supset F(-u, -v).$$

Now $F(-u, -v)$ is just $F(u, v)$ rotated half a turn about the origin. If $f(x, y)$ is a real function, as is common, then the real part of $F(u, v)$ has twofold rotational symmetry and the imaginary part is antisymmetric (i.e., rotation through half a turn results in a change of sign). Consequently, $F(-u, -v)$ is just the complex conjugate $F^*(u, v)$, because the half-turn results in a change in the sign of the term containing i. Hence:

Autocorrelation Theorem in 2D

$$\text{If } f(x, y) \ {}^2\!\supset F(u, v),$$

$$\text{then } f(x, y) \star\star f(x, y) \ {}^2\!\supset F^*(u, v)F(u, v).$$

When autocorrelation is performed in the (x, y)-domain, the corresponding operation in the transform domain can be described as follows: shift every Fourier component to zero phase and then square the resultant amplitude.

The transform $F(u, v)$ need not in general possess rotational symmetry, although its real and imaginary parts do. However $|F(u, v)|^2$ always has twofold rotational symmetry. To see this, split $F(u, v)$ into its real and imaginary parts $\mathcal{R}(u, v)$ and $\mathcal{I}(u, v)$:

$$F(u, v) = \mathcal{R}(u, v) + i\mathcal{I}(u, v).$$

The symmetry properties of the real and imaginary parts are

$$\mathcal{R}(-u, -v) = \mathcal{R}(u, v), \qquad \mathcal{I}(-u, -v) = -\mathcal{I}(u, v).$$

Now

$$|F(u, v)|^2 = [\mathcal{R}(u, v)]^2 + [\mathcal{I}(u, v)]^2.$$

But $[\mathcal{I}(u, v)]^2$ has twofold rotational symmetry, as does $[\mathcal{R}(u, v)]^2$, and therefore so does $|F(u, v)|^2$.

By the rotation theorem we deduce that $f(x, y) \star\star f(x, y)$ also in general possesses twofold rotational symmetry, a property that is well known from examples of autocorrelation.

In some applications $f(x, y)$ is a complex quantity. An example would be the optical excitation at a point (x, y) in an aperture in a screen where both amplitude and phase are to be specified. Exactly the same situation arises with antennas at radio wavelengths and with acoustic radiators. In all these fields we need the complex autocorrelation, which is the cross-correlation of the function on its complex conjugate, $f(x, y) \star\star f^*(-x, -y)$.

When $f(x, y)$ is real—as is often the case, for example in a photographic image, TV display, or light-intensity pattern—it is equal to its own complex conjugate. Therefore, autocorrelation as discussed in connection with real functions is included within the concept of complex autocorrelation. Evidently complex autocorrelation is the fundamental concept, but it would be cumbersome always to carry along the asterisk indicating complex conjugates in subjects where $f(x, y)$ is always real. For this reason we have to be very careful, when $f(x, y)$ is in fact complex, to remember the asterisk. To emphasize this point another way, let us just add that, when $f(x, y)$ is complex, the quantity $f(x, y) \star\star f(x, y)$ never seems to turn up.

A derivation of Rayleigh's theorem can be based directly on the autocorrelation theorem, which will be written here in terms of complex autocorrelation:

$$\int_{-\infty}^{\infty} \int_{-\infty}^{\infty} f(x', y') f^*(x' + x, y' + y) \, dx' \, dy' = \int_{-\infty}^{\infty} \int_{-\infty}^{\infty} |F(u, v)|^2 \, e^{i2\pi(ux+vy)} \, du \, dv.$$

Now put $x = y = 0$ to obtain Rayleigh's theorem.

We see that the theorem is merely an example of the property that the central value of a function is equal to the integral of the transform: if $f(x, y) \;{}^2\!\!\supset F(u, v)$, then $f(0, 0) = \int\int F(u, v) \, du \, dv$. In the present application, ACF of $f \;{}^2\!\!\supset |F|^2$; therefore the central value of the ACF equals $\int |F|^2 \, du \, dv$. But the central value of the autocorrelation is simply the integral $\int\int ff^* \, dx \, dy$.

CROSS-CORRELATION THEOREM

It follows from what has gone before that the theorem applying to the cross-correlation of $f(x, y)$ on $g(x, y)$ is as follows.

Cross-correlation theorem in 2D

$$\text{If } f(x, y) \,{}^2\!\supset F(u, v) \quad \text{and} \quad g(x, y) \,{}^2\!\supset G(u, v),$$

$$\text{then } f(x, y) \star\star g(x, y) \,{}^2\!\supset F^*(u, v)G(u, v).$$

It is important to remember that in this formula it makes a difference which of the factors carries the complex conjugate asterisk.

FACTORIZATION AND SEPARATION

If a given function of two variables is to be transformed, then it is advantageous to be able to recognize factors of the function, because then, if the transforms of the factors are known, it becomes possible to make use of the convolution theorem. In general, one would expect that the transforms of factors would be simpler to arrive at than the transform of a built-up product. In the special case where the factors of a two-dimensional function are one dimensional, then the separable product theorem, which is a special case of the full two-dimensional convolution theorem, simplifies transformation. In the intermediate case where one of the factors is one dimensional, there is also a simplification.

It is not particularly difficult to spot factors of algebraic expressions, because the study of algebra involves practice at factorization. On the other hand, the factorization of data presented digitally or graphically is hardly practiced at all and may be worth thinking about. For example, it may be useful to recognize that a particular data string is modulated by some Gaussian, triangular, or other envelope that enters as a factor. There may also be a rapidly varying factor such as a sinusoidal one. To discover this from a digital printout might be hopeless, but one can cultivate the habit of preparing a graphical representation and examining it for significant features, including factors.

Although one possesses background to help with factorization, it is necessary to cultivate the ability to notice that a given function can be broken down into that other kind of factor which enters into the makeup of a compound function through convolution. The possibility of breaking a function down into components from which the function may be regenerated by convolution is, of course, just as helpful, when one comes to draw on the convolution theorem, as is ordinary factorization. Here is an example of a simple problem that is reducible to an even more simple one by analysis into convolution factors.

A function $f(x, y)$ is zero outside a central 3×3 square. Divide the square into nine unit squares; then the function value is 4 all over the central unit square, unity over the four corner squares, and 2 over the four unit squares remaining. The object is to find the two-dimensional Fourier transform $F(u, v)$. We try to do this by spotting convolution factors.

It is immediately verifiable that ${}^2\text{rect}(x, y)$ is such a factor. The other factor is a set of nine two-dimensional impulses situated at the centers of the nine squares. Therefore it is possible to express $F(u, v)$ as a product of $\text{sinc}\, u \,\text{sinc}\, v$ with the sum of the nine exponentials into which the impulses separately transform. Although this decomposition is verifiable, it may not be immediately apparent the first time a function of the nature of this particular $f(x, y)$ is presented. In order to develop the background that will make such decomposition apparent in the future, it is a good idea to practice with other functions

Figure 6-3 Four convolution factors possessed by the function $f(x, y)$.

that are made up of tabletops in various locations. The tops might be rectangular and still allow decomposition, and some might be square and some might be rectangular without detriment. But not all functions made up of square-tabletop functions are amenable, as a little trial will soon show.

In this example one can go further by noticing that the nine impulses themselves are representable as the convolution $[^2\delta(x, y + 1) + 2^2\delta(x, y) + {}^2\delta(x, y - 1)] ** [^2\delta(x + 1, y) + 2^2\delta(x, y) + {}^2\delta(x - 1, y)]$. It is not at all obvious how one would notice this property from the delta-function representation of the nine impulses, but the result is easily verifiable once it is presented. Here we are interested in how to become aware of such possibilities. I believe that I gained this ability years ago by practising two-dimensional convolution with a movable piece of tracing paper and that I can now think pictorially without the crutch of the piece of paper, which was originally indispensable. To go through the sort of maneuver I have in mind, trace one of the left-hand components in the following pictorial statement and go through the kinematics of scanning the other; then do the opposite to notice that the same result is arrived at (as the commutative property of convolution assures us will indeed happen).

$$\begin{bmatrix} & \cdot & \\ \bullet & & \\ & \cdot & \end{bmatrix} ** [\cdot \quad \bullet \quad \cdot] = \begin{bmatrix} \cdot & \bullet & \cdot \\ \bullet & \bullet & \bullet \\ \cdot & \bullet & \cdot \end{bmatrix}$$

If you follow the advice given above, you will no doubt notice that the procedure can be carried one step further, because each collineation of delta functions can itself be decomposed. Thus

$$^2\delta(x + 1, y) + 2^2\delta(x, y) + {}^2\delta(x - 1, y) = [^2\delta(x + \tfrac{1}{2}, y) + {}^2\delta(x - \tfrac{1}{2}, y)] **$$
$$[^2\delta(x + \tfrac{1}{2}, y) + {}^2\delta(x - \tfrac{1}{2}, y)].$$

Finally we find that the nine impulses have four components, as shown in Fig. 6-3. With this knowledge one can write the Fourier transform by inspection to get

$$F(u, v) = 16\cos^2 \pi u \cos^2 \pi v \operatorname{sinc} u \operatorname{sinc} v.$$

CONVOLUTION WITH THE HARTLEY TRANSFORM

The convolution theorem tells us that the operations implied by the two-dimensional convolution sum, of the order of N^4 in number for $N \times N$ arrays, may be avoided by moving to the transform domain and performing $O(N^2)$ multiplications, in relatively negligible time if N is large. The time taken to compute the transforms, which will go as $N \log N$, is additional. To convolve, one computes the two Fourier transforms, multiplies corresponding elements, and retransforms, for a total of one inverse and two direct transforms. In two dimensions the Fourier transforms comprise four two-dimensional arrays of coefficients, two for real parts and two for imaginary parts. Each complex multiply involves all four arrays and four real multiplies, which is more cumbersome than the notation $F(u, v) \times G(u, v)$ conveys at first glance.

Using the Hartley transform, which is real, improves the situation considerably. Each Hartley transform consists of one real array, and each multiplication is a single real multiply. The product array retransforms into the desired convolution. All this is under the condition that the original data were real, which with images is usual. There are some technical points, such as what to do when the two arrays are not the same size. For an introduction to the Hartley transform, including the application to convolution, see Bracewell (1986). Subsequent applications of the Hartley transform can be traced through the Science Citation Index or in the pages of *Electronics Letters*, the *IEEE Proceedings* (including a special issue on the Hartley transform, 1994), and the *IEEE Transactions on Signal Processing*.

A Hartley Transform Algorithm

An extensive literature developed, much of which was concerned with whittling away the time taken to transform, even if by only a few percent. Remembering that the clock rate of computers regularly improves by factors of two as time passes, we see that fine tuning of algorithms is much less significant than the speed of your central processing unit, and will soon be irrelevant, and probably already is as you read this. Since the rules are now reversing, my recommendation for a general-purpose Hartley program is a radix-4 version which accepts data lengths of 4^P points (16, 64, 256, 1024, . . .). According to this doctrine, one appends zeros to a data set to reach the next value of 4^P. At one time, if you had 365 data points, which is not uncommon, it seemed wasteful to pad the data with zeros simply to get the 512 points that suited a radix-2 algorithm. Under certain circumstances where an algorithm must be optimized without varying the hardware, alternatives to the radix-4 algorithm are appropriate. However, when running-time differences between such algorithms are significant to the user, varying the hardware is worth considering. Lost time due to padding is partly compensated by the speed advantage of the radix-4 algorithm (about 30 percent relative to radix-2); the spacing of the computed transform values is closer than the critical spacing (depending on the fraction of appended zeros), which is sometimes an advantage, for example in graphical presentation.

The following subprogram for the radix-4 fast Hartley transform is based on FHTRX4 of Bracewell (1986). It accepts the data array $f(\ , \ )$ and replaces the contents by the transform values.

Computation of the two-dimensional Hartley transform is implemented by multiple calls to the one-dimensional subprogram, as follows. Starting from an $N \times N$ array, take the Hartley transform of the first row and write the values over the data values. Do this

Table 6-1   A subprogram for the fast Hartley transform.

```
FAST HARTLEY TRANSFORM TO RADIX 4

SUB "FHTradix4" (F(),P)
N=4^P
N4=N/4
R=SQR(2)

                  Permute to radix 4
J=1 @ I=0
a: I=I+1 @ IF I>=J THEN GOTO b
T=F(J-1) @ F(J-1)=F(I-1) @ F(I-1)=T
b: K=N4
c: IF 3*K>=J THEN GOTO d
J=J-3*K @ K=K/4 @ GOTO c
d: J=J+K @ IF I<N-1 THEN GOTO a

                            Get DHT
                            Stage 1
FOR I=0 TO N-1 STEP 4
    T1=F(I)+F(I+1)
    T2=F(I)-F(I+1)
    T3=F(I+2)+F(I+3)
    T4=F(I+2)-F(I+3)
    F(I)=T1+T3
    F(I+1)=T1-T3
    F(I+2)=T2+T4
    F(I+3)=T2-T4
NEXT I

                      Stages 2 to P
FOR L=2 TO P
    E1=2^(L+L-3)
    E2=E1+E1
    E3=E2+E1
    E4=E3+E1
    E5=E4+E1
    E6=E5+E1
    E7=E6+E1
    E8=E7+E1
    FOR J=0 TO N-1 STEP E8
        T1=F(J)+F(J+E2)
        T2=F(J)-F(J+E2)
        T3=F(J+E4)+F(J+E6)
        T4=F(J+E4)-F(J+E6)
        F(J)=T1+T3
        F(J+E2)=T1-T3
        F(J+E4)=T2+T4
        F(J+E6)=T2-T4
```

```
        T1=F(J+E1)
        T2=F(J+E3)*R
        T3=F(J+E5)
        T4=F(J+E7)*R
        F(J+E1)=T1+T2+T3
        F(J+E3)=T1-T3+T4
        F(J+E5)=T1-T2+T3
        F(J+E7)=T1-T3-T4
        FOR K=1 TO E1-1
            L1=J+K
            L2=L1+E2
            L3=L1+E4
            L4=L1+E6
            L5=J+E2-K
            L6=L5+E2
            L7=L5+E4
            L8=L5+E6
            A1=PI*K/E4
            A2=A1+A1 @ A3=A1+A2
            C1=COS(A1) @ S1=SIN(A1)
            C2=COS(A2) @ S2=SIN(A2)
            C3=COS(A3) @ S3=SIN(A3)
            T5=F(L2)*C1+F(L6)*S1
            T6=F(L3)*C2+F(L7)*S2
            T7=F(L4)*C3+F(L8)*S3
            T8=F(L6)*C1-F(L2)*S1
            T9=F(L7)*C2-F(L3)*S2
            T0=F(L8)*C3-F(L4)*S3
            T1=F(L5)-T9
            T2=F(L5)+T9
            T3=-T8-T0
            T4=T5-T7
            F(L5)=T1+T4
            F(L6)=T2+T3
            F(L7)=T1-T4
            F(L8)=T2-T3
            T1=F(L1)+T6
            T2=F(L1)-T6
            T3=T8-T0
            T4=T5+T7
            F(L1)=T1+T4
            F(L2)=T2+T3
            F(L3)=T1-T4
            F(L4)=T2-T3
        NEXT K
    NEXT J
NEXT L
SUBEND
```

for each row until an intermediate array is completed. Now take the transform of each column in turn, writing over the intermediate values and call the result $T(u, v)$. Then the two-dimensional Hartley transform $H(u, v)$ is given by

$$2H(u, v) = T(u, v) + T(N - u, v) + T(u, N - v) + T(N - u, N - v).$$

For further information see Bracewell et al. (1986), Hao and Bracewell (1987) and Buneman (1987).

Computational Complexity in Image Processing

When two $N \times N$ arrays are to be convolved, a number of multiplies proportional in number to N^2 must be performed for each different relative position of the arrays. Thus, for large N, the time taken approaches KN^4, where K is some constant. For values of N in everyday use, the time can be expressed as

$$KN^4 + T_0,$$

where T_0 contains contributions from parts of the program other than the dominant multiplications, as well as such things as disc access time, queuing time, travel time, and programming time which do not ordinarily enter into complexity theory but can be important for a user. The constant K can range in magnitude from milliseconds to hours and defines a characteristic value of N, namely $(T_0/K)^{1/4}$. For values of N smaller than this rough indicator, it makes little difference whether a fast algorithm is used or not.

There may be advantages to slow algorithms. To begin with, they may not be slow, as explained. As a rule the coding is more straightforward and can be adapted under your control in a way that is often infeasible with imported subprograms, which may be inaccessible to you, or, if accessible, not easily understandable by you.

The lapse of decades has changed the balance between sophisticated fast algorithms and naive algorithms as hardware advances have substantially lessened K, with the result that the characteristic values of N have risen in all applications. In some applications opportunities have been opened for working with much larger values of N than were previously imaginable; in other applications there will never be a need for, let us say, $N > 1000$; in these cases there is no significant computer time advantage in a fast algorithm, while the advantage of the naive algorithm may be welcome. The naive convolution and cross-correlation programs mentioned in Chapter 5 can be tried out under reigning circumstances with an eye to identifying your personal break-even value of N.

SUMMARY

Convolution and the convolution theorem are basic to image processing; indeed, convolution goes intimately hand-in-hand with the concept of transfer function. Where there is convolution, there will be a transfer function, and conversely. As a result there are always two options for thinking about certain problems and two ways of composing software, and two ways of designing hardware, according as one thinks in the continuous frequency domain or in the object domain, where things may be continuous but where discontinuity, discreteness, and impulsiveness are also freely entertained. Digital signal processing exemplifies the object-domain approach to handling sampled data streams; delta-function notation is the key to facilitating transfer between this discrete option and the corresponding interpretation in terms of continuous frequency.

The completeness of the duality expressed by the convolution theorem was reached in relatively recent years with acceptance of the delta function, which encountered conceptual difficulties and even some derision. Around the start of the twentieth century Oliver Heaviside had explicitly defined, but not named, the entity that we now call the Dirac delta function $\delta(t)$. Heaviside's shorthand notation was $p1$. The symbol $p \equiv d/dt$ represents the derivative operator, while 1 stands for a function that switches on at $t = 0$, assuming a value of unity. Today we write $\delta(t) = p\mathbf{H}(t)$, where $\mathbf{H}(t)$ is the Heaviside unit step function, thereby bringing the discontinuous into the realm of differential calculus, where continuity was formerly prerequisite.

In two dimensions, the point impulse that is needed to extend the scope of the convolution theorem does not arise in a context of differentiation of a discontinuous function. Instead, a different approach, such as that taken in Chapter 3, provides the basis for completing the range of application of the convolution theorem to images. As this framework becomes more widely known, use of the convolution theorem as a way of relating the discrete option of matrix convolution to the continuous option of the transfer function will gain in prominence.

PROBLEMS

6–1. *Shift invariance.* Let the response of a system to a two-dimensional input $f(x, y)$ be $g(x, y)$. Then the system is space invariant by definition if the response to $f(x - a, y - b)$ is $g(x - a, y - b)$ for all a and b and for all $f(x, y)$.
  (a) A physical system is tested with an input $\text{rect}(x^2 + y^2)$ centered at numerous points (a, b) such that $a^2 + b^2 < 5$ and the response is always $\exp[-(x - a)^2 - (y - b)^2]$ within measurement accuracy. Is there any reason to think that this system might not be space-invariant?
  (b) Another system always gives an output $\exp(-x^2 - y^2)$ regardless of the input. Say why this system is or is not space-invariant. ☞

6–2. *Linearity.* Let the response of a two-dimensional system to an input $f_1(x, y)$ be $g_1(x, y)$ and the response to $f_2(x, y)$ be $g_2(x, y)$. Then the system is linear by definition if the response to $f_1(x, y) + f_2(x, y)$ is $g_1(x, y) + g_2(x, y)$ for all choices of $f_1(x, y)$ and $f_2(x, y)$. It follows that the response of a linear system to $kf_1(x, y)$ is $kg_1(x, y)$ for all $f_1(x, y)$ and all constants k, provided k is a rational fraction. Since irrational numbers are entertained in the mind rather than found in the physical world, seeing whether the response to $kf_1(x, y)$ is $kg_1(x, y)$ for various $f_1(x, y)$ is a common acceptance test for linearity.
  (a) A camera system considered to be linear has a checkerboard pattern of opaque and transparent squares installed over the lens. Is the system still linear?
  (b) What if the checkerboard device is installed well in front of the lens, and is sufficiently large to fill the field of view? ☞

6–3. *Fourier transform pairs.* Give the Fourier transform of the following functions.
  (a) $^2\text{rect}(x - A, y) + {}^2\text{rect}(x + A, y)$,
  (b) $^2\text{rect}(x - A, y/W) + {}^2\text{rect}(x + A, y/W)$,
  (c) $^2\text{rect}(x - A, y) + 2\ {}^2\text{rect}(x, y) + {}^2\text{rect}(x + A, y)$. ☞

6–4. *Fourier transform pairs.* Give the Fourier transform of the following functions.

$$\exp\{-\pi[x^2 + (y - B)^2]/R^2\} + \exp\{-\pi[x^2 + (y + B)^2]/R^2\}$$

$$\exp\{-\pi[(x - A)^2/R^2 + y^2/S^2]\} + \exp\{-\pi[(x + A)^2/R^2 + y^2/S^2]\}$$

$$\exp\{-\pi[(x - A)^2 + (y - B)^2]/R^2\} + \exp\{-\pi[(x + A)^2 + (y + B)^2]/R^2\}$$

$$\exp\{-\pi[(x - A)^2/R^2 + y^2/S^2]\} + \exp\{-\pi[(x + A)^2/S^2 + y^2/R^2]\}$$

6–5. *Inverse convolution theorem.* Given that $f ** g \ ^2\supset FG$, prove that $fg \ ^2\supset F ** G$. ☞

6–6. *Perpendicular exponentials.* Represent the truncated exponential function by $e(x) \equiv \exp(-x)H(x)$. Use the two-dimensional convolution theorem to determine $e(x) ** e(y)$.

☞

6–7. *Cosine hump.* Let $h(x) \equiv \cos \pi x \ \mathrm{rect}\, x$.
   (a) Determine the Fourier transform of $f(x, y) = h(x)h(y)$.
   (b) What is the variance $V_f = \langle x^2 \rangle$?
   (c) Determine the Fourier transform of $\phi(x) = f(x, y) ** f(x, y)$.
   (d) Determine the cross section $C(u')$ of $\Phi(u, v)$ along the $45°$ direction $u' = 0.707(u + v)$.
   (e) Presumably $C(u')$ is approximately Gaussian because of the central limit theorem. What would be the variance $\langle u'^2 \rangle$?
   (f) Show by computing that $C(u')/C(0)$ agrees with $\exp(-u^2/2\langle u'^2 \rangle)$ to within 2 percent of the maximum. ☞

6–8. *Contrast and resolution.* A photographic transparency 128 mm square is to be digitized by raster scanning with a 1-mm-square hole through which light passes to a photodetector. Special transparencies are prepared for test purposes in the form of gratings consisting of alternate opaque and transparent bars of equal width. As the square hole scans across the grating in the direction normal to the bars the digitized output rises and falls between maxima Y_{max} and minima Y_{min}. Define "contrast" in decibels to mean $10 \log_{10}(Y_{max}/Y_{min})$.
   (a) When the opaque bars are 10 mm wide, what is the contrast?
   (b) Define "resolution" as that number of bars per millimeter such that the contrast is 3 db. What is the resolution?
   (c) What output would you expect if the bars were 0.5 mm wide?
   (d) What would the contrast be for bars of width 0.33 mm? ☞

6–9. *Invisible distribution.* A test object whose transparency is $0.5 + 0.5 \sin 2\pi x$, where x is measured in millimeters, is inserted in the digitizing photometer described in the previous problem.
   (a) Explain in physical terms without recourse to mathematics why the photometer cannot see any structure in the object.
   (b) Another transparency distribution $t(x, y) = 0.5 + \cos 2\pi x \, \mathrm{sinc}\, 2y + \cos 2\pi y \, \mathrm{sinc}\, 2x$ is inserted. Show by appeal to the convolution theorem that this is also an invisible distribution.
   (c) Describe what a transparency $t(x, y) = 0.5 + 0.5 \cos 3\pi x$ would yield. ☞

6–10. *Invisible distribution made visible?* Alarmed by the discovery that the newly marketed smart digitizing photometer has blind spots, the section head asks the project engineer to work on it, but the latter is fully engaged on his next project and just comments philosophically, "Once you've squeezed the last drop, why scrape the bottom of the barrel?" So the section head hires a consultant, whose report, reduced to a nutshell, says, "Change the square hole to a triangular one of the same area and that will get rid of invisible distributions." Is that true? ☞

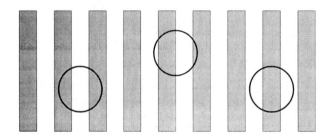

Figure 6-4   Critical locations of a circle on a bar pattern.

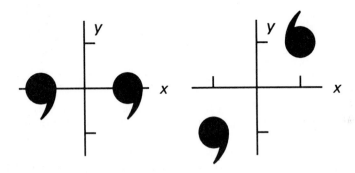

Figure 6-5   The comma-pair on the left is expressible in terms of convolution but the pair on the right is not.

6–11. *Circular aperture.* A densitometer with a 1-mm-diameter circular hole scans a bar pattern of parallel, equally wide opaque and transparent stripes. What is the "resolution," expressed as the number of stripes per mm for which the contrast falls to 3 db (ratio of maximum to minimum response falls to 2)?

6–12. *Invisible bar pattern?* A bar pattern with 1.238 bars/mm is scanned by a circular aperture 1 mm in diameter (Fig. 6-4). When the circle evenly straddles a bar, which is 0.404 mm wide, the area of bar covered equals half the area of the circle. Noting this, a student reasons, "When the circle straddles a space between bars, the area covered is the same. Therefore, a bar pattern having 1.238 bars per circle diameter is the critical case separating spurious antiphase response from normal response. When the bar frequency is just above 1.238, there will be some response but in reversed spatial phase. At the critical frequency there will be no response at all." What has the student overlooked?

6–13. *Comma.* The FT of a comma $c(x, y)$ is $C(u, v)$. To get the FT of a pair of commas (Fig. 6-5, left) we can use the convolution theorem, once we perceive that the comma-pair is expressible as $c(x, y) ** [{}^2\delta(x + 1, y) + {}^2\delta(x - 1, y)]$. Now obtain $K(u, v)$, the FT of the comma pair on the right. ☞

6–14. *Degree of mirror symmetry.* The letter **H** has mirror symmetry about a vertical axis, while **Q** has not, but **Q** more nearly resembles its mirror image than does **G**. Characterization of an image segment $\phi(x, y)$ of a larger image $f(x, y)$ could benefit from a parameter quantifying degree of mirror symmetry. Investigate the suitability of

$$\rho_M = \frac{\int\int \phi(x, y)\phi(-x, y)}{\int\int [\phi(x, y)]^2} dxdy.$$

6–15. *The two-dimensional signum function.* The function sgn x is unity for positive x, equal to -1 for negative x and zero in between. Introduce a two-dimensional generalization ^{2} sgn$(x, y) = $ sgn x sgn y, which is $+1$ where x and y are positive, -1 when x and y are negative, and zero elsewhere. Alternatively, ^{2} sgn$(x, y) = $ sgn$(\sin 2\theta)$. sgn is pronounced as the original Latin *signum* to avoid the homophone with sin meaning sine. Show that

$$^2 \text{sgn}(x, y) \; ^2\supset \frac{1}{\pi^2 uv}. \quad \text{☞}$$

6–16. *Moiré pattern.* A transparent film is ruled with parallel black lines 0.25 mm wide and spaced 1 mm between centers. When such a film, which is obtainable where drafting supplies are sold, is placed on a photocopy of itself, a moiré pattern is seen, unless the two items are accurately oriented in the same direction. If the orientation difference θ is $1°$, dark bands are seen spaced 57 mm apart.

(a) Perform the demonstration, or make a sketch, with a view to understanding the direction and motion of the bands as $\theta \to 0$.

(b) In the vicinity of a grazing intersection of two black lines there may be an illusion of increased blackness, although there is in fact less black ink exposed to view in the area of overlap. Therefore, although the dark bands have a most definite subjective reality, there is a question whether the apparent low-spatial-frequency bands have mathematical reality. Would the Fourier transform of the object exhibiting a moiré pattern confirm the presence of components at much lower spatial frequencies than the frequency of the ruling? Apply the convolution theorem and report.

(c) An instrument has been proposed by one of your employees to measure microscopic angular deflections in a machine subject to torsion by forming a moiré pattern as above. A ruling is attached to a "fixed" part of the machine, and the transparency overlying it is supported at a relative orientation $+0.1°$ by the part whose suspected angular movement is to be monitored. Two photodetectors are positioned equally on opposite sides of a dark band. The idea is that when a minute deflection in the angular direction takes place, say 1 μm, then the much larger motion of the band (570 μm or more than half a millimeter) should be easily detectable by the imbalance between the photodetector outputs. What do you think about the idea? ☞

6–17. *Degree of symmetry.* A real function $f(x, y)$ is equal to $f_e(x, y) + f_o(x, y)$, the sum of its symmetric and antisymmetric parts.

(a) Show that

$$\int_{-\infty}^{\infty}\int_{-\infty}^{\infty} f(x, y)f(-x, -y)\, dx\, dy = \int_{-\infty}^{\infty}\int_{-\infty}^{\infty} [f_e(x, y)]^2\, dx\, dy -$$

$$\int_{-\infty}^{\infty}\int_{-\infty}^{\infty} [f_o(x, y)]^2\, dx\, dy.$$

(b) Show that

$$S = \int_{-\infty}^{\infty}\int_{-\infty}^{\infty} f(x, y)f(-x, -y)\, dx\, dy \Big/ \int_{-\infty}^{\infty}\int_{-\infty}^{\infty} [f(x, y)]^2\, dx\, dy$$

characterizes the degree of symmetry by evaluating S for functions that are purely symmetrical or purely antisymmetrical.

(c) What does $S = 0$ mean?

6–18. *Convolution after compression.* Let $f_3(x, y) = f_1(x, y) * *f_2(x, y)$. Both f_1 and f_2 are compressed by a factor k in the x-direction only and then convolved. What is the relationship between $f_4(x, y) = f_1(kx, y) ** f_2(kx, y)$ and $f_3(x, y)$? ☞

6–19. *Correcting astigmatism.* An east-west ridge is represented by

$$f_1(x, y) = \exp[-\pi(x^2/W^2 + y^2)].$$

See if you can find another function $f_2(x, y)$ which, when convolved in two dimensions with $f_1(x, y)$, produces a result that is circularly symmetrical. ☞

6–20. *Stellar magnitudes.* The relative magnitudes of N stars in a star field have to be measured from a photographic plate, but the star images are very small, and what is more, the brighter stars are overexposed. These two difficulties are overcome by a *Schraffierkassette*, a photographic plate holder that executes a raster scan over a small square area while the exposure is in progress. The resulting photographic density is described by

$$d(x, y) = \sum_{i=1}^{N} D_i{}^2\delta(x - x_i, y - y_i) ** {}^2\text{rect}(x/a, y/a).$$

(a) Obtain the Fourier transform $F(u, v)$.
(b) What is there about the transform that would tell you that the difficulty of positioning a photometer exactly on a very small star image is relieved?
(c) Knowing that one magnitude equals -4 decibels, show that the magnitude m_i of the ith star exceeds that of the reference star $i = 1$ as given by

$$m_i - m_1 = 2.5 \log_{10} \left[\frac{\int_{x_1-a/2}^{x_1+a/2} \int_{y_1-a/2}^{y_1+a/2} d(x, y)\, dx\, dy}{\int_{x_i-a/2}^{x_i+a/2} \int_{y_i-a/2}^{y_i+a/2} d(x, y)\, dx\, dy} \right]. ☞$$

6–21. *Factorization of transform.* Eight equal point impulses are arranged spatially as shown in Fig. 6-6 to constitute a function $f(x, y)$. Let $q(x, y)$ represent four equal impulses at the corners of a square. Then the set of eight is representable as $q(x - a, y - b) + q(x + a, y + b)$, which, one notices can be expressed as $[{}^2\delta(x - a, y - b) + {}^2\delta(x + a, y + b)] ** q(x, y)$. Therefore it follows that $2 \cos 2\pi au \cos 2\pi bv$ is a factor of $F(u, v)$.
(a) Discover two other ways of expressing $f(x, y)$ as a convolution with a point-impulse pair and thus deduce two more factors of $F(u, v)$.
(b) Find out whether the product of all these factors is equal to $F(u, v)$. ☞

6–22. *Factorization of transform.* Six unit point impulses are situated at the vertices of a regular hexagon at unit distance from the origin.
(a) Give the Fourier transform.
(b) Find factors for the transform, or show why none would be expected.

6–23. *Fuzzy edge.* A straight boundary between black and white areas is blurred by scanning with a Gaussian $h(r) = \exp(-\pi r^2/W^2)$, where W is the equivalent width of any cross section through $h(r)$. In one dimension let us define

$$I(x) = \int_{-\infty}^{x} \exp(-\pi x^2)dx = \exp(-\pi x^2) * H(x)$$

to be the Gaussian integral as shown in Fig. 6-7. In two dimensions the fuzzy edge is

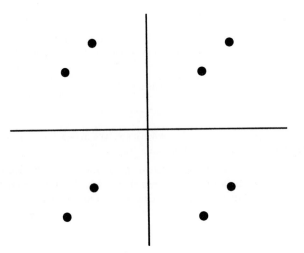

Figure 6-6   A set of eight equal point impulses.

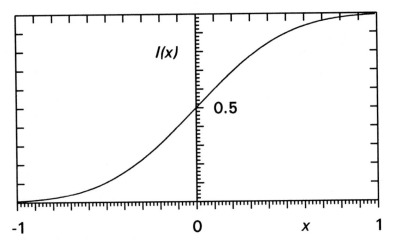

Figure 6-7   The Gaussian integral $I(x)$. Numerical values can be computed from $I(x) = \frac{1}{2}[1 + \mathrm{erf}(\sqrt{\pi}x)]$. The 10 and 90 percent points are at $x = \pm 0.51$.

expressible as

$$\exp(-\pi r^2/W^2) ** \mathbf{H}(x)$$

if the boundary is along the y-axis. Clearly the 50 percent contour lies on the y-axis. Find out how far the 10 percent and 90 percent points are from the y-axis in units of W. ☞

6–24. *Fuzzy corner.* A square corner separating black from white areas is blurred by scanning with a Gaussian $W^{-2}\exp(-\pi r^2/W^2)$. The 50 percent contour departs from the straight edges near the corner.

    (a) How far inside the corner is the point P where the 50 percent contour cuts the bisector of the corner?

(b) In units of W, how far inside and outside P are the points where the 10 and 90 percent contours cross the bisector? ⯈

6–25. *Matched filter*. Construct an image matrix from random values 1 to 10 and add the following numbers to a row: {0 8 4 2 1}. Do this in two or three places, but amplified by different factors k, including $k = 10$. Think about whether artificial signals of known exponential form but arbitrary location would be locatable by eye at a given amplification factor k. Now cross-correlate the matrix with

$$[0\ 0\ 0\ 0/8\ 4\ 2\ 1/0\ 0\ 0\ 0]$$

to see whether the artificial signals have been located and, if so, how well. Experiment with the weakest signal (least k) that is locatable. The signal suggested has the form $8/2^i$, $i = 0$, 1, 2, 3. See what happens if you shift the signal by taking $i = 0.5, 1.5, 2.5, 3.5$. See whether the peak amplitude after cross-correlation agrees with what you expect.

LITERATURE CITED

R. N. BRACEWELL (1986), *The Hartley Transform*, Oxford University Press, New York.

R. N. BRACEWELL, O. BUNEMAN, H. HAO AND J. VILLASENOR (1986), "Fast two-dimensional Hartley transforms," *Proc. IEEE*, vol. 74, pp. 1283-1284.

O. BUNEMAN (1987), "Multidimensional Hartley transforms," *Proc. IEEE*, vol. 75, p. 267.

H. HAO AND R. N. BRACEWELL (1987), "A three-dimensional DFT algorithm using the fast Hartley transform," *Proc. IEEE*, vol. 75, pp. 264-266.

IEEE Proceedings, Special issue on the Hartley transform, March 1994.

7

Sampling and Interpolation in Two Dimensions

Sampling and interpolation in two dimensions is much richer than in one dimension. Not only are there polar coordinates and other coordinate systems in addition to cartesian, but sampling can be done along lines as well as at points. The distinction between point and line sampling will be discussed first. Since sampling at regular intervals plays a major role, we pass on to regular point sampling as expressed in one dimension by the shah function $\text{III}(x)$, an entity that was introduced earlier in connection with delta functions. In two dimensions we find the direct generalization $^2\text{III}(x, y)$, which occupies a square lattice of points, together with a range of other manifestations of the shah function. Since the Fourier transform of a sampling pattern is a transfer function, this treatment brings us to the brink of digital filtering. Before developing this subject in the next chapter, however, we discuss the sampling theorem and interpolation.

Interpolation may be viewed as the inverse of sampling in the sense that interpolation reconstitutes a function from its samples. Thus it starts with the discrete and approaches the continuous. The classical interpolation formulas of Newton, Lagrange and Gauss are of this kind, as well as sinc function interpolation and spline interpolation. In the world of discrete data this sort of interpolation is needed all the time. Another sort of operation, typified by midpoint interpolation, seeks only to provide values midway between pairs of adjacent samples. Starting from a given set of samples in a region of the plane, one can quadruple the number of samples by midpoint interpolation. For many purposes, such a large increase is enough. Examples include the enlargement of images, image representation by grey pixels, and various image-processing operations.

Sampling is inherently discrete. It is noteworthy, though, that the theoretical treatment of this digital topic can draw on the familiar notation for continuous variables. This is because the discrete aspect can be subsumed within the unit impulse function $\delta(x)$ and its replicated form $III(x)$; the essence of discreteness is thus captured by the continuous variable x. The convenience of being able to handle digital data arrays in terms of equivalent continuous notation is especially evident in a topic such as sampling, where continuity and discreteness are both present. This chapter draws on previous material, especially the chapters on the convolution theorem and impulse functions in two dimensions.

WHAT IS A SAMPLE?

A time-dependent signal may be known to us through measurements taken or provided at successive times, usually uniformly spaced, and presented numerically. Time-dependent signals may also be provided in analogue form as a waveform on a screen, recorder chart, or other recording medium, ready for measurement. Spatially dependent "signals," or images, which are not time dependent, may be displayed or recorded similarly. The samples constitute the numerical outcome when the measurements of the record are completed. There will be a finite number of samples, say N, and they will be characterized by limited precision (about one part in 10^4, or 16 bits at the best, if a signal voltage is being measured) and by unavoidable instrumental effects, such as unwanted sensitivity to extraneous physical variables, reduced response to rapidly varying signals, flicker noise at low frequencies, thermal noise at low signal levels, shot noise when quanta are few, and hysteresis. These physical aspects are mentioned here as a reminder of what sampling theory deals with.

The sample number is usually an integer ranging from 1 to N, or from 0 to $N - 1$, but in any case the domain of a sample set is the domain of the ordinal numbers. For programming, a number pair representing position on a coordinate grid is often more convenient than a sample number, though the latter is what the computer itself uses. The domain of the sample values is that of the rational numbers.

There is a risk of confusion between the samples of a measured signal and the regularly spaced function values of a mathematical expression. In practice the main serious consequence of the confusion is that noise may tend to be overlooked. It would be a good idea to call a function value a function value, but in fact the word sample is often substituted by analogy even though no measurement is involved, especially when the function values intended are regularly spaced.

The domain of regularly spaced samples is characterized by a set of integers, but delta-function notation allows us to work with a continuous independent variable, achieving a result that is equivalent to an ordinal-number basis while giving the notational convenience of the continuous set of numbers on the real line.

SAMPLING AT A POINT

When a continuous function $f(x, y)$ is sampled at a point (a, b), the function value, or sample, $f(a, b)$, can be thought of in terms of the sifting integral, where we multiply $f(x, y)$ by a two-dimensional unit impulse at (a, b) and integrate over the whole plane. Thus the sample is expressible as

$$f(a, b) = \int_{-\infty}^{\infty} \int_{-\infty}^{\infty} f(x, y)\,{}^2\delta(x - a, y - b) \, dx \, dy.$$

A feature of this elaborate approach is that the right-hand side is a convolution integral. To see this better, allow the point (a, b) to be variable in position. Then the sample at (x, y) is

$$f(x, y) = \int_{-\infty}^{\infty} \int_{-\infty}^{\infty} f(x', y')\,{}^2\delta(x - x', y - y') \, dx' \, dy'.$$

In order to get agreement with the standard form of the convolution integral we have reversed the arguments of the delta function, having in mind that ${}^2\delta(-x, -y) = {}^2\delta(x, y)$. The integral formulation merely states in extended notation what may be more compactly expressed as

$$f(x, y) = {}^2\delta(x, y) ** f(x, y).$$

So far, the sampling points have not been restricted to some regular lattice but may be anywhere.

SAMPLING ON A POINT PATTERN, AND THE ASSOCIATED TRANSFER FUNCTION

Often we sample an image on a regular array of points and combine clusters of samples; an example of this would be the digitizing of a photograph $f(x, y)$ using a high-resolution scanner followed by the averaging of several samples adjacent to points of a thinned array. This would allow the image to be dumped to an available lower-resolution printer. Suppose that the pattern of points occupied by samples to be combined is describable by N impulses of strength N^{-1} at (a_i, b_i), $i = 1$ to N; then the average of the N sample values would be

$$f(a_i, b_i) = \int_{-\infty}^{\infty} \int_{-\infty}^{\infty} f(x, y) \left[\frac{1}{N} \sum_{i=1}^{N} {}^2\delta(x - a_i, y - b_i) \right] dx \, dy.$$

This is a convolution integral, and we may refer to the impulse set $\sum_i {}^2\delta(x - a_i, y - b_i)$ as the convolving pattern. The integral does reduce to a sum, but for the time being this discrete smoothing operation can be treated, with full equivalence, in terms of functions of the continuous variables x and y.

The operation of averaging N samples naturally gives a result that departs more or less from the function value at the origin of the convolving pattern; some smoothing is to be expected, as well as some displacement if the origin of the pattern is offset from its

Figure 7-1 A drawing (left) to be digitized at a coarse interval d, after which samples situated as on the right will be averaged in accordance with a given convolving pattern (lower left). The design is a medieval ruler-and-compass construction denoting purity. The radii are proportional to small integers, which facilitates layout and reproducibility. An interesting exercise is to attempt to improve the design by relaxing the pure digital tradition.

centroid. Converting to standard form as before, we get

$$f_{av}(x, y) = \int_{-\infty}^{\infty} \int_{-\infty}^{\infty} f(x', y') \left[\frac{1}{N} \sum_{i=1}^{N} {}^{2}\delta(x - x' + a_i, y - y' + b_i) \right] dx' \, dy'$$

$$= \int_{-\infty}^{\infty} \int_{-\infty}^{\infty} f(x - x', y - y') \left[\frac{1}{N} \sum_{i=1}^{N} {}^{2}\delta(x' + a_i, y' + b_i) \right] dx' \, dy'$$

or, in compact form

$$f_{av}(x, y) = \left[\frac{1}{N} \sum_{i=1}^{N} {}^{2}\delta(x + a_i, y + b_i) \right] ** f(x, y).$$

The striking feature of this formulation showing that the N-sample average is expressible as a certain convolution with the original $f(x, y)$ is that the convolution theorem can be brought to bear. The convolution operator, convolving pattern, or

$$h(x, y) = \frac{1}{N} \sum_{i=1}^{N} {}^{2}\delta(x + a_i, y + b_i),$$

tells us that $f_{av}(x, y)$ is related to $f(x, y)$ through a transfer function $H(u, v)$, which specifies how every Fourier component of the original function $f(x, y)$ will be modified in both amplitude and phase. The convolution theorem gives us the $H(u, v)$ from

$$H(u, v) \qquad \text{is FT of} \qquad \frac{1}{N} \sum_{i=1}^{N} {}^{2}\delta(x + a_i, y + b_i).$$

In case the pattern of sample points does not have twofold rotational symmetry, care is needed to rotate the pattern half a turn before transforming, because ${}^{2}\delta(x + a, y + b)$ is on the opposite side of the origin from ${}^{2}\delta(x - a, y - b)$. If this is overlooked, one will get $H(-u, -v)$ instead of $H(u, v)$; but the two are often the same.

Figure 7-1 illustrates a simple case where the point pattern is a 3×3 lattice of points, each occupied by an impulse of unit strength. The set of impulses may be written, taking $a_i = (i - 1) \bmod 3$ and $b_i = (i - 1) \operatorname{div} 3$, as

$$h(x, y) = \sum_{1}^{9} {}^2\delta(x - a_i, y - b_i)$$

$$= {}^2\delta(x, y) + {}^2\delta(x - 1, y) + {}^2\delta(x - 2, y)$$
$$+ {}^2\delta(x, y - 1) + {}^2\delta(x - 1, y - 1 + {}^2\delta(x - 2, y - 1) + {}^2\delta(x, y - 2)$$
$$+ {}^2\delta(x - 1, y - 2) + {}^2\delta(x - 2, y - 2),$$

The nine-point average is then

$$f_{av} = \frac{1}{N}[{}^2\delta(x, y) + \cdots + {}^2\delta(x - 2, y - 2)] ** f(x, y).$$

This lengthy expression is loaded with redundant symbols but lends itself to matrix notation that is more readable, is fully equivalent, and allows the location of the origin to be conveyed clearly (e.g., by boldface type). What is more, when we agree that the matrix of coefficients can stand for the sum of impulse symbols that are functions of a continuous variable, we get the convenience of the indispensable mixed notation

$$f_{av}(x, y) = \begin{bmatrix} 1 & 1 & 1 \\ 1 & 1 & 1 \\ \mathbf{1} & 1 & 1 \end{bmatrix} ** f(x, y).$$

One prepares immediately to be on guard for displacement in a diagonal direction (downward to the left). The convolution shown is itself continuous but will be evaluated only at discrete intervals; if the purpose is to send to the low-resolution printer, evaluation will be on the sparse 3 × 3 lattice.

The transfer function associated with the square array of nine unit impulses is, taking the Fourier transform,

$$H(u, v) = 9 \operatorname{sinc} 3u \operatorname{sinc} 3v \; e^{-i2\pi(u+v)}.$$

Thus very low spatial frequencies will be amplified by 9, while a spatial frequency of $\frac{1}{3}$ (period=3) oriented in either of the principal directions will be filtered out, because one or the other of the two sinc functions will be zero. It is characteristic of the delta-function approach to the handling of discrete patterns that the continuous-variable formulation takes us straight to the answer about the smoothing to be expected from averaging a digital image.

If the sampling pattern has nonuniform weights h_i associated with the points (a_i, b_i), we have a much more general situation, where the transfer function $H(u, v)$ obeys the transform relation

$$\frac{1}{N} \sum_{i=1}^{N} h_i {}^2\delta(x + a_i, y + b_i) \; {}^2\supset H(u, v).$$

This will be developed in the next chapter. For the present it is sufficient to note that the Fourier transform of a sampling pattern is an important entity, and we proceed to exhibit these transforms for several basic cases.

SAMPLING ALONG A LINE

Sampling along a line arises in scanning a transparency with a slit, scanning a microwave emissive source such as the ground with a fan-beam antenna, or irradiating a solid object with a collimated x-ray beam as in tomography. Slits are usually of constant width, and so a uniform line impulse is appropriate; a slit of variable width would be represented by a line impulse of lesser strength where the slit was narrower. A fan-beam antenna normally does not have constant antenna gain and is represented by a line impulse that falls off suitably in strength from the antenna beam axis. An x-ray beam is attenuated by a solid object and is represented by a line impulse whose strength diminishes toward the point of exit.

When a function $f(x, y)$ is to be sampled at a point, we multiply by a unit impulse at that point and integrate; for sampling on a straight line we multiply by a line impulse and integrate. For example, to sample $f(x, y)$ on a unit line impulse along the line $y = x$ we evaluate

$$\int_{-\infty}^{\infty} \int_{-\infty}^{\infty} f(x, y)\delta(y - x)\, dx\, dy,$$

since $\delta(y - x)$ is confined to the line $y - x = 0$ and has unit linear strength. Clearly this is the area of the cross section of $f(x, y)$ along $y = x$, or $\sqrt{2} \int_{-\infty}^{\infty} f(x, x)dx$. The rules for evaluating integrals containing line impulses were given earlier and account in a natural way for the factor $\sqrt{2}$. However, a virtue of delta-function notation is to permit normal algebra, so it will be possible to verify the two-dimensional integral by other methods (try rotation of coordinates).

If $f(x, y)$ is an area density distribution, then the line sample

$$\int \int f(x, y)\delta(x \cos \theta + y \sin \theta + c)\, dx\, dy$$

is the mass associated with a strip of unit width centered on the line $x \cos \theta + y \sin \theta + c = 0$. Thus the line sample is exactly what is wanted in x-ray tomography, under conditions of low attenuation, to represent the absorption that is measured when a narrow parallel beam of given width follows the line mentioned.

CURVILINEAR SAMPLING

Sampling on curves arises in optical and seismic tomography, where ray trajectories are refracted as a result of gradients of wave velocity within the medium. Sampling on circles, or arcs of circles, is characteristic of radar, where echoes from a circle at a given range from the transmitter later arrive simultaneously at the receiver. In general, given a curve $c(x, y) = 0$,

$$\int_{-\infty}^{\infty} \int_{-\infty}^{\infty} f(x, y)\delta[c(x, y)]\, dx\, dy = \int_{c} \alpha(x, y)f(x, y)\, ds,$$

where s is arc length along the curve and the factor $\alpha(x, y)$ is the intrinsic space-dependent strength of $\delta[c(x, y)]$. This integral gives the area of the curvilinear section only if $\alpha(x, y) = 1$.

Curvilinear impulses can be handled numerically by point impulses distributed along the curve. The points may be equally spaced and weighted to suit the assigned strength per unit arc length, or unit impulses may be variably spaced.

It is allowable for the function $f(x, y)$ itself to be impulsive; this would arise with an optical, seismic, or x-ray object that was specified by samples. In that case the numerical procedure would be to interpolate $f(x, y)$.

THE SHAH FUNCTION

In one dimension the shah function performs a dual role—sampling at regular intervals on the one hand, and representing periodic functions on the other. In two dimensions the situation is much richer. To start with, sampling which arises through *multiplication* with the shah function can take place not only on a regular two-dimensional lattice of points but also on a regular grille of straight lines. In addition one can have sampling on spokes regularly spaced in angle or on circles regularly spaced in radius. There is also sampling on a lattice of nonuniformly spaced points, such as sampling points in polar coordinates. The representation of periodic functions, which arises from *convolution* with the shah function, also leads to much variety.

Let us recall that

$$\text{III}(x) = \sum_{n=-\infty}^{\infty} \delta(x - n).$$

We can introduce a two-dimensional shah function, or bed-of-nails function, defined by

$$^2\text{III}(x, y) = \sum_{m=-\infty}^{\infty} \sum_{n=-\infty}^{\infty} {}^2\delta(x - m, y - n).$$

The unit-strength impulses occupy a square lattice of points, one of which is at the origin (see Fig. 7-2).

The product of $f(x, y)$ with $^2\text{III}(x, y)$ is a set of spikes whose strengths are sample values of $f(x, y)$ at unit interval of x and y. If we wish to have samples spaced X and Y, we write

$$\frac{1}{|XY|} {}^2\text{III}(\frac{x}{X}, \frac{y}{Y}) f(x, y).$$

To understand this expression it is necessary to remember that the shah function obeys a scaling rule analogous to $\delta(x/a) = |a| \delta(x)$. Thus

$$^2\text{III}(\frac{x}{X}, \frac{y}{Y}) = |XY| \sum_{m=-\infty}^{\infty} \sum_{n=-\infty}^{\infty} {}^2\delta(x - mX, y - nY).$$

Consequently, in order to represent an array of unit delta functions it is necessary to write

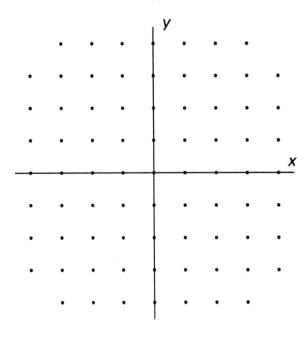

Figure 7-2 The domain occupied by the bed-of-nails function $^2\mathrm{III}(x, y)$ is the square lattice of points (m, n), where m and n are integers. A unit two-dimensional impulse resides at each point. This function became famous as being its own Fourier transform, once the theory of generalized functions showed how to deal with "functions" that severely violate the conditions for existence of the Fourier transform as known from traditional analysis. A different pictorial view was given in Chapter 3.

$|XY|^{-1} \mathrm{III}(x/X, y/Y)$. When we come to take transforms, the factor associated with the scaling will be found to behave exactly as we would wish.

FOURIER TRANSFORM OF THE SHAH FUNCTION

Just as in one dimension the shah function is its own transform

$$\mathrm{III}(x) \supset \mathrm{III}(s)$$

so also in two dimensions

$$^2\mathrm{III}(x, y) \,^2{\supset}^2 \,\mathrm{III}(u, v).$$

When the spacing between impulses is X horizontally and Y vertically, the similarity theorem tells us that

$$^2\mathrm{III}(\frac{x}{X}, \frac{y}{Y}) \,^2{\supset}| XY |^2\mathrm{III}(Xu, Yv)$$

The proof that $\mathrm{III}(x) \supset \mathrm{III}(s)$ and that $^2\mathrm{III}(x, y) \,^2{\supset} \,^2\mathrm{III}(u, v)$ depends upon the idea of generalized functions (Lighthill, 1958). One sets up a transform pair that does not involve nonconvergent integrals and arranges that each side represents a sequence suitable for representing a shah function. The details are shown in the appendix to this chapter.

Although a shah function transforms into a shah function, the individual impulses do not transform into impulses. Each even pair of impulses transforms into a cosine function, and the combination of all the pairs transforms into a combination of cosines of arithmetically increasing frequency that interfere constructively, in the wave-theory sense, to build

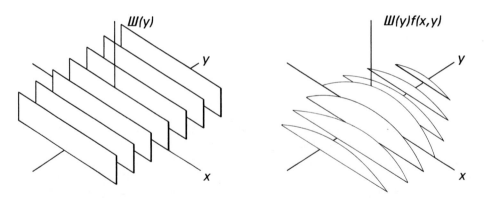

Figure 7-3   An impulse grille and its role in representing raster sampling of a function $f(x, y)$.

up spikes at all the integer locations, including the origin. The central impulse, which was not included with the pairs, contributes the necessary dc bias.

OTHER PATTERNS OF SAMPLING

A wide variety of possibilities arise in two dimensions that have no counterpart in the sampling of time signals.

Impulse Grille

A regular set of horizontal unit-strength line impulses arrayed on a regular grille, or grating, is represented by $III(y)$. As a factor in a product the impulse grille permits the simple representation of raster sampling (Fig. 7-3).

Sampling on a Grille (Raster Sampling)

The product of $III(y)$ with $f(x, y)$ retains values of $f(x, y)$ along the lines $y =$ integer but abandons values in-between. Thus $III(y)f(x, y)$ contains information that is preserved in a horizontal raster scan of the function $f(x, y)$. Likewise $III(x)f(x, y)$ would represent the information in a vertical raster scan, which is not seen very often because of the convention that has been adopted with television.

To take the two-dimensional Fourier transform of $III(x)$ we refer to the separable product theorem and imagine that $III(x)$ is really the product $III(x) \times 1$, where the factor 1 is interpreted as a function of y that just happens to be a constant. Then the two-dimensional transform will be the product of the separate one-dimensional transforms. The transform of 1, as a function of y, is $\delta(v)$. Thus

$$III(x) \; {}^2\!\supset III(u)\delta(v).$$

This expression represents a row of unit spikes spaced regularly at unit interval along the u-axis. Likewise $III(y) \; {}^2\!\supset III(v)\delta(u)$, a row of spikes on the v-axis.

Square Lattice

The bed of nails, which occupies a regular arrangement of points, is complemented by the bed of blades, a regular arrangement of line impulses in the form of a grid, which is represented simply by the sum of two perpendicular grilles:

$$\text{III}(x) + \text{III}(y).$$

A rather interesting question that will immediately occur to you is whether the sum involves an extra, and perhaps unwanted, impulse at each grid intersection. The answer is that a unit line impulse has unit mass per unit length of run and therefore has negligible mass in the infinitesimal length in the intersection. Consequently, there is no effective difference between the simple definition of a grid given above and some more elaborate formulation that seeks to compensate for the overlapping at the intersections.

The term *grid* is adopted here in accordance with other usage, such as a map grid, to describe the combination of two perpendicular grilles. In the next paragraph the term *lattice* continues to be taken in the sense of regularly arranged points, as in a crystal lattice. In other contexts, such as architecture and gardening, lattice means crossed grilles, and in group theory lattice denotes a concept not needed in this book.

Other Square Lattices

A bed-of-nails function whose origin lies between the impulses rather than coinciding with one of the impulses is an even function, and it possesses four-fold rotational symmetry (Fig. 7-4). It may be expressed as ${}^{2}\text{III}(x - \frac{1}{2}, y - \frac{1}{2})$. It will not be its own transform, however. Instead,

$${}^{2}\text{III}(x - \tfrac{1}{2}, y - \tfrac{1}{2}) \;{}^{2}\!\supset\; e^{-i\pi u}\text{III}(u)e^{-i\pi v}\text{III}(v).$$

The imaginary part of the factor $\exp[-i\pi(u + v)]$ is nonzero only where $\text{III}(u)\text{III}(v)$ is zero, and may therefore be ignored. The real part $\cos[\pi(u + v)]$ has values (± 1) only at the points where the impulses of $\text{III}(u)\text{III}(v) = {}^{2}\text{III}(u, v)$ occur. Consequently,

$${}^{2}\text{III}(x - \tfrac{1}{2}, y - \tfrac{1}{2}) \;{}^{2}\!\supset\; (-1)^{u+v}\,{}^{2}\text{III}(u, v).$$

Thus the transform is a two-sided bed of nails, with one nail at the origin and the other nails alternating in sign in the manner of the black and white squares of a chessboard.

There is no square at the center of a chessboard, so impulses distributed on the (u, v)-plane with strength $+1$ on the black squares and -1 on the white are a little different from $(-1)^{u+v}\,{}^{2}\text{III}(u, v)$. Describe the strict chessboard pattern

as $\text{ch}(x, y)\,\text{rect}(x/8)\,\text{rect}(y/8)$, where

$$\text{ch}(x, y) = (-1)^{x+y-1}\,{}^{2}\text{III}(x - \tfrac{1}{2}, y - \tfrac{1}{2}).$$

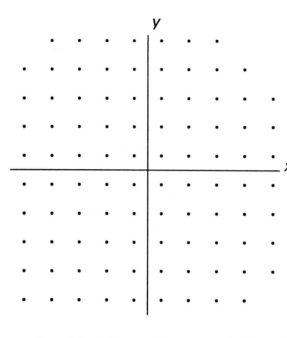

Figure 7-4   An offset bed of nails $^2\text{III}(x - \frac{1}{2}, y - \frac{1}{2})$ whose transform is the double-sided bed-of-nails shown in the next figure. Both of these functions are real and even.

Since $\text{ch}(x, y)$ is even with respect to the diagonal axes, we expect its transform to be real. We already know that

$$(-1)^{x+y}\,{}^2\text{III}(x, y) \,{}^2\!\supset\, {}^2\text{III}(u - \tfrac{1}{2}, v - \tfrac{1}{2})$$

(bearing in mind that this transform pair is reversible, because both left and right sides are even); therefore by the shift theorem

$$(-1)^{x+y-1}\,{}^2\text{III}(x - \tfrac{1}{2}, y - \tfrac{1}{2}) \,{}^2\!\supset\, e^{-\pi(u+v)^2}\text{III}(u, v).$$

Thus

$$\text{ch}(x, y) \,{}^2\!\supset\, \cos[\pi(u + v)]^2 \text{III}(u, v) = \text{ch}(u, v),$$

and so we see that the double-sided chessboard bed-of-nails function $\text{ch}(x, y)$ is its own transform. Fig. 7-5 shows the truncated set of 64 impulses $\text{ch}(x, y)\,\text{rect}(x/8)\,\text{rect}(y/8)$. The transform of this limited set would be an infinite array of narrow sinc-function spikes of the form $\pm 64\,\text{sinc}\,8u\,\text{sinc}\,8v$, distributed on an infinite chessboard with strength $+64$ on the black squares and -64 on the white.

Hexagonal Lattice

Bees make hexagonal cells in their honeycomb, and as a result they achieve a minimum expenditure of wax for a given cross section of cell. Many crystals exhibit hexagonal spacing, also because of a minimum principle. Corresponding virtues have been claimed for two-dimensional data sampling on a hexagonal or triangular basis rather than on a

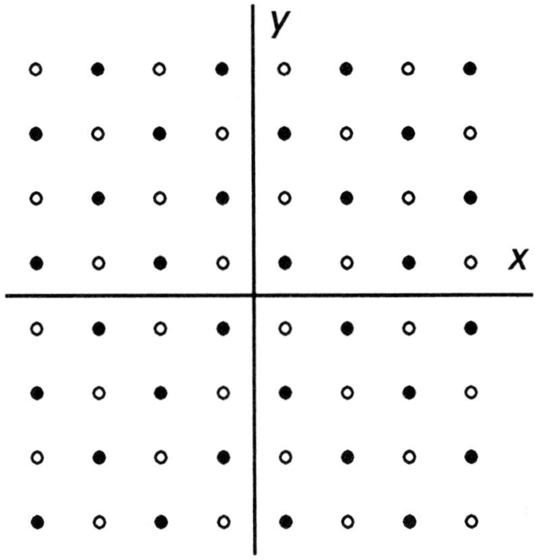

Figure 7-5 A double-sided bed-of-nails function or chessboard function ch(x, y), so called because the sign of the impulses accords with the color of the squares. Negative impulses are shown hollow. Because of the negative nails, this double-sided bed would be just as uncomfortable to lie on if you turned it over.

square basis. Consider the pattern $h(x, y)$ of Fig. 7-6(a). The row of impulses along the x-axis may be expressed as III$(x)\delta(y)$. The horizontal row next above is shifted by an amount $(0.5, \sqrt{3}/2)$. Consequently the whole pattern could be expressed as

$$\sum_{n=-\infty}^{\infty} \text{III}(x - 0.5n)\delta(y - 0.866n).$$

Alternatively, we note the rather simple decomposition into a pair of (nonsquare) bed-of-nails functions

$$h(x, y) = {}^2\text{III}(x, y/\sqrt{3}) + {}^2\text{III}(x - 0.5, y/\sqrt{3} - 0.5).$$

Either expression may be used to obtain the transform. The second form gives

$$H(u, v) = \sqrt{3}(1 + e^{-i2\pi(0.5u+\sqrt{3}v/2)2}\text{III}(u, \sqrt{3}v).$$

The factor in parentheses equals 1 ± 1 at the locations of the impulses comprising ${}^2\text{III}(u, \sqrt{3})$, consequently half of the impulses are suppressed. Hence

$$H(u, v) = (\sqrt{3}/2)h(\sqrt{3}v/2, \sqrt{3}u/2).$$

The pattern of Fig. 7-6(b) can be expressed by subtracting the impulses at the centers of the hexagons; these centers occupy a pattern that is geometrically similar to Fig. 7-6(a) but $\sqrt{3}$ times larger and rotated 90°.

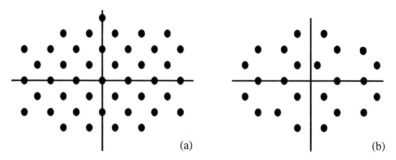

Figure 7-6   Two forms of hexagonal lattice.

The Reciprocal Lattice

Examples have now been seen of arrays of delta functions whose transforms are also arrays of delta functions. Are there other lattices whose transforms are lattices? Clearly, not all arrays of point delta functions have this property. A simple condition must be met, namely that the array be periodic. That means by definition that the array extends to infinity both ways. Periodicity in the plane allows for just two periods (not counting multiples of a fundamental), which will correspond, in the transform domain, to two other periods that are the reciprocals of those in the space domain. Any number of impulse-lattice pairs can be generated. For example, you can experiment with a pentagonal lattice generated by two-dimensional replication of five points at the vertices of a regular pentagon, using periodicities that please you. The transform of your arrangement will be another set of delta functions known as the reciprocal lattice. In theory, fivefold symmetry does not exist in crystals, but x-ray diffraction patterns have shown puzzling phenomena in nature; this development adds topical interest to the experiment proposed above.

FACTORING

It is always worth while to be familiar with ways that may exist of breaking a given function down into factors from which the function can be generated by convolution. In the example of Fig. 7-7 the factors are a row of impulses on the y-axis and a line impulse on the x-axis. Thus

$$\text{III}(y) = \sum {}^2\delta(x, y - n) ** \delta(y)$$
$$= [\text{III}(y)\delta(x)] ** \delta(y).$$

Also note that

$$\delta(x) ** \delta(y) = 1.$$

To confirm this relation, rewrite the right-hand side as $\int \int \delta(x')\delta(y - y') \, dx' \, dy'$ and make the unit-area τ-dependent rectangle function substitutions to obtain

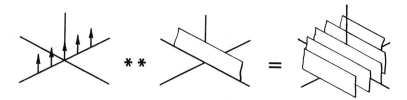

Figure 7-7 A way of generating the impulse grille by two-dimensional convolution between a row of impulses and a line impulse.

$$\tau^{-2} \int \int \text{rect}(x'/\tau)\, \text{rect}[(y - y')/\tau]\, dx'\, dy',$$

an expression that is equal to unity, and in particular in the limit as $\tau \to 0$. The convolution theorem also verifies this result: transforming both sides gives $\delta(v)\delta(u) = {}^2\delta(u, v)$.

It is also worth while to be aware of ways of splitting given functions into multiplicative factors. In terms of the impulse grille we can express the bed-of-nails function as a product of two perpendicular grilles. Thus

$$^2\text{III}(x, y) = \text{III}(x)\text{III}(y).$$

Transforming both sides and using the separable product theorem gives ${}^2\text{III}(u, v) = \text{III}(u)\text{III}(v)$, confirming the factorization.

THE TWO-DIMENSIONAL SAMPLING THEOREM

Consider a continuous function $f(x, y)$ of the continuous variables x and y, and the set of function values belonging to all the integral values of x and y. The sampling theorem says that the exact value of $f(x, y)$ at any point (x, y) lying in-between the square grid of points with integer coordinates can be recovered exactly from the values at the grid points, provided the given function $f(x, y)$ is band-limited.

Band-limited is a convenient term for expressing a property of a function $f(x, y)$ in terms of a property of its transform $F(u, v)$. The term derives from radio usage, where the spectrum has long been divided into bands delimited by upper and lower frequencies. To say that a signal $s(t)$ is band-limited is usually taken to mean that there exist two frequencies f_{min} and f_{max} such that the Fourier transform $S(f)$ is zero wherever $|f| > f_{max}$ or $|f| < f_{min}$. Examples showing some different ways in which signal-frequency bands may be disposed are given in Fig. 7-8. The four on the left meet the condition given, while the one on the right does not.

In two dimensions we continue to apply the term band-limited to a continuous function $f(x, y)$ with the meaning that $F(u, v)$ is nonzero within some finite area of the (u, v)-plane. More often than not we have in mind that this finite area is a central rectangle, $F(u, v)$ being zero wherever $|u| > u_{max}$ or $|v| > v_{max}$, as on the left of Fig. 7-9. But, not uncommonly, we mean that $F(u, v)$ is zero wherever $\sqrt{u^2 + v^2} > q_{max}$, as in (b). More generally, other finite-area arrangements, see (c) and (d), connote band-limited functions.

S(f)

f

Figure 7-8   Some band arrangements meeting the condition $S(f) = 0$ wherever $|f| > f_{max}$ or $|f| < f_{min}$ and one (on the right) that does not.

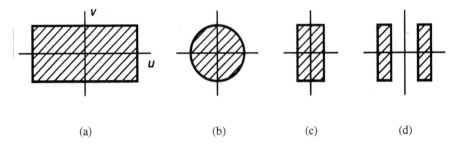

(a)   (b)   (c)   (d)

Figure 7-9   If the nonzero values of a transform $F(u, v)$ are confined to a finite area, as in the examples shown, the function $f(x, y)$ is band-limited.

For the purpose of understanding the sampling theorem it is simplest to start with case (a) of Fig. 7-9, so for the time being assume that "$f(x, y)$ is band-limited" means

$$F(u, v) = \begin{cases} 0, & |u| > u_{max} \text{ or } |v| > v_{max} \\ \text{nonzero}, & \text{elsewhere.} \end{cases}$$

The function $^2\text{sinc}(u/2u_{max}, v/2v_{max})$ is band-limited in this restricted sense, because its transform $(4u_{max}v_{max})^{-1} \, ^2\text{rect}(u/2u_{max}, v/2v_{max})$ cuts off at the specified boundary.

Take the sampling lattice to have unit spacing in both directions. Then the sample set $f(i, j)$, $-\infty < i < \infty$, $-\infty < j < \infty$, can be thought of as a matrix of infinite extent, or we can work with the equivalent impulse signal $s(x, y)$, thought of as a function of continuous x and y. In this case

$$s(x, y) = {}^2\text{III}(x, y)f(x, y) = \sum_i \sum_j f(i, j)^2\delta(x - i, y - j),$$

where x and y are continuous and the matrix values $f(i, j)$ are the strengths of the impulses located at each sample point. This simple change of representation enables us to write the two-dimensional transform of $s(x, y)$ immediately; since $s(x, y)$ is in the form of a product, its Fourier transform $S(u, v)$ is a convolution. We know that the transform of $^2\text{III}(x, y)$ is $^2\text{III}(u, v)$ and the transform of $f(x, y)$ is $F(u, v)$. Hence,

$$S(u, v) = {}^2\text{III}(u, v) ** F(u, v).$$

This completes the first mathematical step; we now examine the result graphically in Fig. 7-10. Let $f(x, y)$ be a band-limited function represented here by a few contours

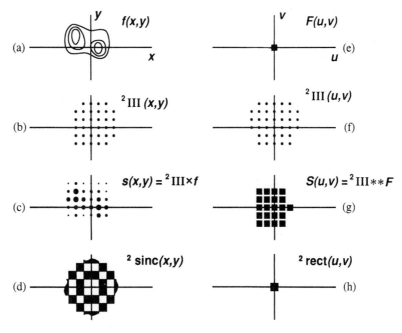

Figure 7-10 (a) A function $f(x, y)$ and its Fourier transform $F(u, v)$, (b) The sampling function $^2\mathrm{III}(x, y)$ and its Fourier transform $^2\mathrm{III}(u, v)$, (c) The sampled function $^2\mathrm{sinc}(x, y)$ and its transform $^2\mathrm{rect}(u, v)$, (d) The interpolating function $^2\mathrm{sinc}(x, y)$ and its transform $^2\mathrm{rect}(u, v)$.

(a) and possessing a Fourier transform $F(u, v)$ whose nonzero values are restricted to the rectangle $|u| \leq u_{max}$, $|v| \leq v_{max}$. For the purposes of illustration, the value 0.4 has been used for both u_{max} and v_{max}. The grid spacing, which is unity throughout, enables the significance of the magnitudes to be followed. Outside the central square island, $F(u, v) = 0$.

The sampling factor $^2\mathrm{III}(x, y)$ shown at (b) consists of unit impulses situated on a square lattice of unit spacing, while its Fourier transform $^2\mathrm{III}(u, v)$ at (f) has the same form. The function $s(x, y)$, representing the information remaining about $f(x, y)$ after sampling, is the product $^2\mathrm{III}(x, y) f(x, y)$, which is shown at (c) on the same convention as for (b), namely dots centered at the sample points and with sizes proportional to the sample values. The Fourier transform of $s(x, y)$ is the convolution of (e) with (f) and consists of islands replicated at unit interval in both directions, as shown at (g).

We are now in a position to appreciate that the sampling theorem is indeed true. Given only the samples, as in (c), can we get back the lost information and resurrect the original $f(x, y)$ in full? The translation of this question into the transform domain is: "Given $S(u, v)$, can we get back $F(u, v)$?" A glance at (g) and (e) shows that evidently we can; we need only disregard, or reject, the outer islands. What was a puzzling proposition in the image domain has become an almost trivial exercise in the transform domain, a characteristic example of the power of transform methods. In addition, not only do we see that the theorem is true, we also see immediately a procedure for implementing the recovery of the full function from the samples, the second mathematical step.

To reject the outer islands of (g), apply the selection operator that retains the central member of the set of islands unchanged, while reducing the remainder to zero. In this case, multiplication by $^2\text{rect}(u, v)$ performs the desired rejection. Of course, if a function is band-limited with nothing beyond a certain cutoff frequency u_{max} it follows that it also has nothing beyond $1.25u_{max}$; conversely, it is allowable for a function band-limited to u_{max} to cut off at any positive frequency less than u_{max}. Consequently, there is some freedom in choosing the selection operator; its boundary need not even be rectangular. The present choice is $^2\text{rect}(u, v)$ as indicated at (h).

The multiplication of $S(u, v)$ by $^2\text{rect}(u, v)$ in the transform domain corresponds to convolution of $s(x, y)$ with $^2\text{sinc}(x, y)$ in the image domain. This convolution operator is indicated at (d) by a schematic rendering in which the negative regions are black; the important thing is the scale of this pattern relative to the sample spacing used in the upper parts of the figure. Thus, finally, the formula for getting $f(x, y)$ back from its samples is expressible as

$$f(x, y) = {}^2\text{sinc}(x, y) ** s(x, y).$$

Continuous-variable notation has been used throughout this discussion, so the answer comes in the same form, but the result may be expressed in terms of the discrete samples themselves. Convolution of the sinc function with any one of the impulses constituting $s(x, y)$ simply replicates the sinc function at the location of that impulse and with an amplitude equal to the strength of that impulse (which is the same as the sample value by definition). Thus to find the value $f(x_1, y_1)$ at some particular point (x_1, y_1) that was not at a sample point and needs to be recovered, we sum all the displaced sinc functions with appropriate amplitudes to obtain

$$f(x_1, y_1) = \sum_i \sum_j s(i, j){}^2\text{sinc}(x_1 - i, y_1 - j).$$

Example

From a given matrix of samples, a new matrix is to be constructed which is four times as dense by inserting new values on the cell sides and at the cell centers. Calculate the value at $x_1 = 0.5$, $y_1 = 0$, which is on a cell side, half a unit to the right of the sample point at $(0, 0)$. ▷The factor $^2\text{sinc}(x_1 - i, y_1 - j)$ becomes $^2\text{sinc}(0.5 - i, j) = \text{sinc}(0.5 - i)\,\text{sinc}\,j$, which is equal to $\text{sinc}(0.5 - i)$, where $j = 0$ and zero elsewhere. Therefore only the samples $s(i, 0)$ along the horizontal line $j = 0$ will contribute, and $f(0.5, 0) = \sum_i s(i, 0)\,\text{sinc}(0.5 - i) = \cdots + s(-2, 0)\,\text{sinc}(-2.5) + s(-1, 0)\,\text{sinc}(-1.5) + s(0, 0)\,\text{sinc}(-0.5) + s(1, 0)\,\text{sinc}(0.5) + s(2, 0)\,\text{sinc}(1.5) + \cdots$. The sinc-function coefficients to be applied to the samples are

$$\cdots \quad 0.1273 \quad -0.2122 \quad 0.6366 \quad 0.6366 \quad -0.2122 \quad 0.1273 \quad \cdots \triangleleft$$

Extra values at other midpoints $(x_1, 0)$, where $x_1 = I + 0.5$ and I is an integer, are obtained by using the same coefficients shifted I places to the right.

The importance of the sampling theorem lies in the method of thinking about discrete samples using the continuous domain of the Fourier transform as a tool. The topics

Figure 7-11 An explanatory diagram by P. Mertz and F. Gray conveying the character of Fourier components having several integer values of u and v. Those shown are sine components; negative values of v and a second family of cosine components would complete the integer set.

of undersampling and aliasing, which are dealt with next, are examples of inherently discrete situations where transform-domain thinking is the customary way of gaining insight even though the conclusions can also be reached by discrete mathematics. As the example above illustrates, the sampling theorem provides a method of interpolation when one has foreknowledge that the samples are taken from a band-limited function (how would one know that?). An infinite set of samples enters into the expression for $f(x, y)$, which therefore is dependent on distant large samples in a way that does not often make practical sense. Interpolation is discussed further below.

Some History

Multidimensional Fourier analysis goes back to the earliest days of exploitation of Fourier's theorem in France with papers by Sophie Germain (1776–1831) and C. L. M. H. Navier (1785–1836). Two-dimensional heat flow in plates and vibrations of membranes provided immediate applications. Two-dimensional images ultimately came to the attention of telephone and television engineers. (See, for example, Mertz and Gray 1934). Fig. 7-11 is an ingenious diagram of theirs.

The sampling theorem is relatively recent, tracing back to a paper by Whittaker (1915) written in a context of interpolation between samples. Whittaker discovered the sinc-function convolution sum but did not use the terms sinc function or convolution. He proved the exactness of the procedure for interpolating between samples from a function expressible as the sum of a Fourier series with a finite number of terms. Then Nyquist

(1928) discussed the subject from the viewpoint of a sampled telephone signal, reaching a readership of electrical engineers and establishing among them the term *Nyquist frequency* for the sampling rate, which is still in use. Time-dependent signals, bandwidths, and sampling rates are terms that herald the adoption by a new constituency of what was previously a topic in computing. Shannon (1948) wrote a culminating work in this vein. Related work was published in Russian by Kotel'nikov in 1933.

The extension from one-dimensional time-dependent telephone signals to two-dimensional space-dependent images occurred in a different venue (Bracewell, 1956, 1958). The (u, v)-plane terminology in radio astronomy originated at that time.

The multidimensional extension was made by Miyakawa (1959) in a paper cited by Peterson and Middleton (1962), who recounted some history and brought the two-dimensional sampling theorem to the attention of a further population of engineers. For other history of these times, relating advances in radio astronomical imaging to television imaging, see Sullivan (1984). It appears that a number of basic ideas—including the two-dimensional sampling theorem, the projection-slice theorem, the proof that $^2\mathrm{III}(\,,\,)$ is its own transform, the rigorous transform relationship between a brightness distribution and measurable fringe visibility, the optical transfer function, and facets of restoration of images in the presence of noise—entered the mainstream of signal theory from a time when radioastronomical imaging was developing under vigorous international competition. Continued activity in this field has given us the best images currently available, both as to resolution and dynamic range.

UNDERSAMPLING

Figure 7-10 also handles the question of what happens when the sampling is too sparse; evidently a function cannot be recovered if there are insufficient samples. To think of this quantitatively before resuming the main discussion, if it takes N pieces of information to specify a function, then N samples will surely be needed if the samples are to suffice. Since there is nothing in the present approach to exclude x and y from running to infinity, we are talking about infinite N, the perils of which will be revisited.

If in Fig. 7-10(c) the samples were spaced 1.5 units, then instead of $s(x, y)$, the sampled function would be $\sigma(x, y) = {}^2\mathrm{III}(x/1.5, y/1.5)f(x, y)/1.5^2$. By the previous reasoning the transform $\Sigma(u, v)$ would consist of seriously overlapping islands. If $\Sigma(u, v)$ were given, or calculated by transforming $\sigma(x, y)$, we would lose some knowledge about $F(u, v)$; only the sum of all the islands would be known, and separation into the component islands would not be possible. There would indeed be a restricted central zone of $F(u, v)$, not subjected to overlap, that could be recovered, but not all of $F(u, v)$. Clearly, full recovery first fails when widening sample spacing, accompanied by narrowing spacing of the islands, brings islands into contact. The condition for this is that the island spacing, center to center, shrinks to $2u_{max}$ horizontally and $2v_{max}$ vertically. The corresponding critical sample spacings on the function to be specified are $X_c = 1/2u_{max}$, $Y_c = 1/2v_{max}$. This third mathematical result tells the sample spacing which suffices for recovery of the band-limited function $f(x, y)$.

In practice, when sampling intervals are being chosen for an application, smaller than critical intervals are chosen. There is no fixed 90 percent rule; the choice depends on the sharpness of cutoff, which often is a statistical parameter. In addition, noise level is a consideration, as well as the spectral character of the noise. Practical tests on ensembles of actual data sets or corresponding simulations may be required.

Even if due caution is exercised, sampling may be too sparse for various reasons. Sometimes the sampling density is set not by a designer but by limitations of current technology or by nature. Also, even if the sampling is adequate on some occasion, there may be other times when the band limits of the function expand. Under conditions of unwanted undersampling the phenomenon of aliasing arises, so frequently in fact that it is a vital topic.

ALIASING

Suppose that a function has been undersampled, but we proceed as if it had not been. The effect will be as if the composite $\Sigma(u, v)$ of overlapping islands were multiplied by $^2\mathrm{rect}(u, v)$ and transformed back to the image domain, yielding a function $\phi(x, y)$ that is not the same as the wanted $f(x, y)$. In what way would $\phi(x, y)$ differ from $f(x, y)$? With the degree of overlap resulting from a sample spacing of 1.5 units, there would be a small central zone correctly representing part of $F(u, v)$; therefore, certain low spatial frequencies originally present in $f(x, y)$ would be preserved. Now divide the higher-spatial-frequency components of $F(u, v)$ into two parts, intermediate and high. The high-frequency part is the margin which is eliminated by multiplication with $^2\mathrm{rect}(1.5u, 1.5v)$; the intermediate part consists of the hollow band between the low and the high. Figure 7-12, which is an enlargement of the central region of Fig. 7-10(g), shows these three parts marked L, I and H. The small shaded area in the upper left of $F(u, v)$ represents a patch of high frequencies with negative u-components, cut off by the multiplication with $^2\mathrm{rect}(1.5u, 1.5v)$. But, since the $F(u, v)$ islands are replicated, the same patch appears again on the lower right, where it is situated in the intermediate zone; it will not be cut off and will appear in $\phi(x, y)$. Thus, the lost high spatial frequencies are not lost forever but are converted with full amplitude to a different frequency in the intermediate range. This frequency shift is called *aliasing*.

When the interpolation procedure is being applied, the sparse samples will be convolved with the same coefficients as would be used for samples that were adequately spaced. But those coefficients are taken from a sinc function scaled in accordance with the wide samples, and so it is clear that frequencies higher than half a cycle per sampling interval cannot emerge. The discussion on the (u, v)-plane explains this cutoff in a complementary way but makes it clear how certain high negative frequencies are translated into intermediate positive frequencies.

It is also possible to understand undersampling and the consequent aliasing without reference to the (u, v)-plane at all. Simply imagine a corrugation of high frequency (i.e., of period less than two sample intervals), take the samples, and replace each sample by a suitably displaced two-dimensional sinc function whose spacing between nulls equals the

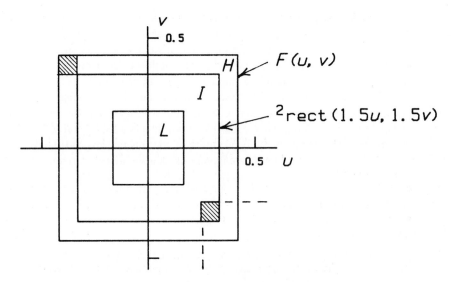

Figure 7-12   The band limit of $F(u, v)$ (outer square) contains an area L of low frequencies that have been preserved, an area H of high frequencies rejected by filtration, and an area I of intermediate frequencies contaminated by contributions from overlapping islands such as the one indicated in broken outline.

sampling interval. The sum of all these sinc functions will be another corrugation, parallel to the first and of the same amplitude but with an intermediate spatial frequency.

Example

A function $f(x, y)$ is sampled at unit interval but inadvertently contains a sinusoidal component $2\sin(2\pi x/1.8)$, whose period 1.8 in the x-direction is shorter than two sample intervals. After the samples are processed with a view to getting $f(x, y)$ back, how will the result differ from the original? $\triangleright$The two-dimensional Fourier transform of $2\sin(2\pi x/1.8)$ is $i^2\delta(u + 0.56, 0) - i^2\delta(u - 0.56, 0)$. After replication with $^2\mathrm{III}(u, v)$, two terms, $i^2\delta(u + 0.44, 0) - i^2\delta(u - 0.44, 0)$, will result from shifting to the right and two others, $i^2\delta(u - 1.56, 0) - i^2\delta(u + 0.44, 0)$, from shifting to the left, plus other terms further out. In the range $-0.5 < u < 0.5$, which is all that will remain after multiplication by the rect function, the only surviving terms are $-i^2\delta(u + 0.44, 0) + i^2\delta(u - 0.44, 0)$, which transform back into $-2\sin(2\pi \times 0.44x) = -2\sin(2\pi x/2.25)$. Thus, the reconstituted function will contain a sinusoidal component of frequency 0.44, aliased down to the lower frequency from the undersampled frequency 0.56. $\triangleleft$

In this example the opportunity was taken to illustrate that high *negative* frequencies alias into intermediate *positive* frequencies, hence the sign reversal of amplitude from 2 to -2. Sometimes this is important, but often the phase relation between sinusoids of different frequencies has no interest. When this is the case, and when it is good enough to say that the undersampled component is a *cosine* wave (whose positive and negative frequency components are of the same sign), then an undersampled component whose frequency u_{high} (0.56 in the example) exceeds the critical frequency $u_c = 0.5$ by δ will alias into a lower frequency u_{alias} that is as far below u_c as u_{high} is above u_c; so

$$u_{\text{alias}} = u_c - (u_{\text{high}} - u_c) = 2u_c - u_{\text{high}}$$
$$= 0.5 - (0.56 - 0.5) = 0.44.$$

Prefiltering

Sometimes, as with signals that arrive by telephone line or by some other channel that is understood, there is a reasonable presumption that the signal has band limits that have been imposed by design, but in other circumstances the band to which the signal is confined may be known only roughly. And yet it may be necessary to sample the signal and thereby risk the generation of aliased components. Assume that the sampling rate is not under the user's control. Then it makes sense to subject the arriving signal to analogue bandpass prefiltering prior to sampling. Just how this filter should be designed, or whether a filter should be introduced at all, depends on the kind of out-of-band noise that is present. Perhaps the signal represents images, and inspection of the images reveals that the kind of noise to be guarded against is in the form of impulses that make white specks on the image. Use a low-pass filter to generate a reference voltage with respect to which positive impulses exceeding normal signal departures can be clipped. This filter is nonlinear. If a linear filter were used with the aim of discriminating against brief impulses, the high peaks would be reduced, but the area under the peak would be conserved and distributed over the neighborhood. A linear analogue filter might be appropriate in some other noise environment. Prefiltering prior to sampling is an unavoidable analogue design exercise (as is sampling) and cannot be performed digitally. The discussion reminds us that an array of numbers is not a real physical filter.

Randomized Sampling

The possibility of finding a low-frequency sinusoid that can be drawn through samples of a sinusoid whose frequency is higher than the critical frequency is hindered if the sample locations depart from strict periodicity. It is true that irregular sampling has deleterious effects, as discussed in a later section, but the benefit of reducing aliasing can outweigh the cost. Experience with the technique has been essentially with time signals, but the principle transfers to image sampling. To follow up this topic see Filicori et al. (1989).

CIRCULAR CUTOFF

We have seen how the sampling theorem applies to full recovery and aliasing when the cutoff boundary is square. Simple scaling extends the conclusions to the rectangular cutoff that arises when the sampling intervals in perpendicular directions are not the same. Unequal sampling is the usual situation in practice, but when the matrix of samples is written down, there does not appear to be any difference, because the rows and columns both advance by serial number, so the different scale factors in the x- and y-directions are absorbed by normalization. It was convenient to adopt the rectangular cutoff boundary for

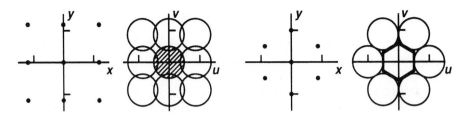

Figure 7-13 A square sampling pattern (left) with spacing 1.128 applied to a function with circular cutoff at spatial frequency 0.5 (grey circle) and the corresponding overlapping islands of spacing 0.886 in the (u, v)-plane. A hexagonal sampling pattern exists (right) that involves no overlap.

introducing the sampling theorem; now we can deal with complications of shape, beginning with circular.

Circular cutoff arises with instruments that have circular symmetry, especially optical instruments. A telescope with a circular aperture, for example, responds only to spatial frequencies not exceeding a critical value that is the same regardless of position angle. A square cutoff, by contrast, implies that structure in the $0°$ direction may cut off at, let us say, 1 cycle per milliradian, and the same in the $90°$ direction, but that structure oriented at $45°$ will be responded to even if the spatial frequency is as much as 1.4 cycles per milliradian. That is not as common as circularity.

Clearly, if $F(u, v) = 0$ where $u^2 + v^2 > q_{max}^2$, then $F(u, v)$ is also zero outside the escribed square, where $|u| > q_{max}$ or $|u| > q_{max}$ and the problem is reduced to the one previously solved as far as choosing a safe sampling interval is concerned. When it comes to interpolation to regain the original function, convolving with the two-dimensional sinc function will likewise work. But convolving with jinc q, the circularly symmetrical bandlimited function with no spatial frequencies greater than 0.5, would also work, in the case of samples at unit spacing, because multiplication by rect q would have the same effect as multiplication by ^{2}rect(u, v).

It appears that some advantage has been lost and indeed this is the case. The area occupied by rect q is less than that occupied by ^{2}rect(u, v) by a factor $\pi/4 = 0.7854$. Therefore, is it not reasonable to ask whether samples based on squares of area $\pi/4$, or a sample spacing of $2/\sqrt{\pi} = 1.128$, might suffice? To answer this question we go to the diagram of overlapping islands (Fig. 7-13) associated with a sample spacing of 1.128 units.

It is clear from the previous line of argument that spatial frequencies in the shaded region will be undisturbed but that other spatial frequencies in the lenticular areas will be half lost and half contaminated by aliasing.

However, the diagram suggests a different sampling pattern, shown on the right, which will allow full recovery. The area of the hexagonal sampling cell is not quite as small as 0.7854 but has an area 0.866, which is noticeably less than unity. Thus the question asked above has led to a demonstration that the density of sampling that allows full recovery of a function band-limited to 0.5 cycles per unit distance in the (x, y)-plane may indeed be less than unity.

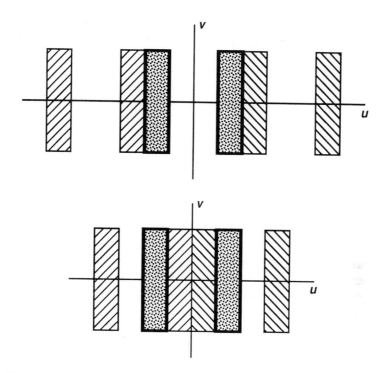

Figure 7-14   A region (shown stippled) lying entirely within unit square.

DOUBLE-RECTANGLE PASS BAND

Fig. 7-14 shows a region in the (u, v)-plane that consists of two stippled rectangles (above) lying within a square and occupying half its area. Spatial spectra fitting into two rectangles (not necessarily having these proportions) arise with two-element interferometry in optics and radio astronomy. Since the two rectangles lie entirely within the square bounded by $u = \pm 0.5$ and $v = \pm 0.5$, it follows that any function whose spatial frequencies are confined to the two rectangles would be adequately sampled on a square grid of unit spacing. But might not a rectangular sampling pattern with the coarser spacing of 2 in the x-direction and 1 in the y-direction also suffice? Translating the island pair to the left (diagonal hatching) and to the right (opposite hatching) as for sampling at unit spacing, we confirm that there is no overlap. Now, translating only half as far as for sampling at spacing 2, we find (below) that there is again no overlapping. It follows that the function can be recovered from the coarse samples—a considerable savings, since the number of samples will have been halved.

Recovery of the function from its coarse samples will not be by simple sinc-function convolution, but it is easy to deduce the sample-domain procedure by noting that the appropriate selection operator in the (u, v)-plane is $^2\mathrm{rect}(u, v) - {}^2\mathrm{rect}(2u, v)$. Therefore, the interpolating function is $^2\mathrm{sinc}(x, y) - 0.5^2\mathrm{sinc}(x/2, y)$.

It is noticeable that the required number of samples is reduced by the ratio of the grey area to unity, a saving that was almost achieved in the circular case. As a further variant, suppose that the two rectangles occupied 60 percent of the unit area so that the space between them was a little too narrow to allow the translated islands to fit in exactly. Would it be sufficient to take samples on a rectangular grid at 60 percent of the unit density? The answer is yes. The replicated islands in the (u, v)-plane will show overlap but of an extricable kind. With these concrete examples before us we can benefit from a general conclusion based on the discrete mathematics that applies as soon as digital computing is involved.

DISCRETE ASPECT OF SAMPLING

In practice there will only be a finite number of samples—for example, $M \times N$ in a rectangular situation—but in any case characterized by a number A that we can think of as the number of unit-area sample cells contained within the boundary of the function $f(x, y)$ to be sampled or as a number of degrees of freedom. The function may have dropped to a negligible level outside its boundary or may be inaccessible. Let the function $f(x, y)$ be band-limited and let the region $\Re$ of the (u, v)-plane within the band limits contain B independent values. If the function is to be recoverable from its A samples, the region $\Re$, whatever its shape, must not be too large, or deleterious overlap will occur. When there is no overlap, all the information in the (u, v)-plane is that which is contained in the central island, because the other islands are only replicas. Applying the sampling theorem in reverse, we see that, since $f(x, y)$ is zero beyond a certain boundary, it follows that $F(u, v)$ is itself fully specified by discrete samples. We can write A equations of the form

$$ f(x_i, y_i) = \sum_k F(u_k, v_k) \exp[i2\pi(u_k x_i + v_k y_i)], $$

where i ranges from 1 to A, and k is a serial number assigned to the discrete samples of $F(u, v)$. If these equations are to be soluble, there must be as many unknowns as equations; therefore B must equal A. (To get this result one allows for the fact that each sample in the (u, v)-plane carries two tags, one real and one imaginary, compensated by the Hermitian property of $F(u, v)$ that halves the number of independent tags.) The simple conclusion that

$$ A = B $$

is consistent with the examples treated above, except for the case of the circular cutoff. The reasoning given here is, however, fundamental, so it follows that sampling at a spacing 1.128 really should suffice to sample a function with a circular cutoff at spatial frequency 0.5. That means that there is some glitch in the customary story about overlapping islands; it arises from the assumed infinite extent of $f(x, y)$.

Other interesting facets emerge when the discrete nature of the practical situation is accepted. For example, the argument that the number of degrees of freedom of the image must equal the number of degrees of freedom of the transform says nothing about the location of the samples. If, because an image is band-limited, it is fully specifiable by 64 samples that are regularly spaced, then is it not fully specified by 64 different samples that are not? The later section on interlaced sampling is relevant.

INTERPOLATING BETWEEN SAMPLES

Between samples one often wishes to know what the measurement might have been, had it been made. Sometimes it is possible to repeat the experimental work with a closer sample spacing until the run of the curve between samples is satisfactorily apparent, but often it is not possible to repeat the measurements because conditions have changed. Working from the original samples, one time-honored method is to draw a smooth curve by hand. A smooth curve can also be calculated by well-tested algorithms bequeathed by Newton, Lagrange, and their successors. Sometimes drawing a polygon through the points is good enough; if not, a cubic spline often is. The element of judgment as to what is good enough makes this a little different from the usual mathematical problem. This is to be expected. Incomplete measurements were made; how can one say exactly what would have been measured in-between when in fact the measurement was not made?

A long history of development lies behind the practice of interpolation. The principal formulas may be referred to in A&S (1964), while computer programs are provided by NumRec (1986) in BASIC, FORTRAN and PASCAL. Here we describe some two-dimensional versions suitable for images.

Nearest-neighbor Interpolation

As mentioned at the beginning of this chapter, one kind of interpolation starts with discrete samples $f(i, j)$ and provides interpolated values at any point (x, y), where x and y are continuous. In one dimension, to find $f(i + p)$, given the samples $f(i)$ and $f(i + 1)$, where p is a fraction from 0 to 1, almost the simplest thing to do is to interpolate linearly; thus $f(i + p) = (1 - p)f(i) + pf(i + 1)$. If p is exactly equal to zero or unity, the formula yields the exact sample value. Interpolation formulas are required to yield exact sample values at the sample points. The simpler procedure, which adopts the sample at the nearest grid point, is to say [using f for $f(i + p)$]

```
IF p<0.5 THEN f=f(i) ELSE f=f(i+1)
```

Since this line of code uses logical comparisons instead of multiplications, it runs faster on some machines (60 percent faster on mine) and may be the best for some purposes. By comparison with linear interpolation, nearest-neighbor interpolation introduces some high-frequency noise, which may or may not be important, depending on the application.

In two dimensions consider the four samples at the corners of a unit square at (i, j), $(i, j + 1)$, $(i + 1, j)$, $(i + 1, j + 1)$, thought of as occupying the SW, SE, NW, and NE respectively (Fig. 7-15). Then one can assign the value $f(i + p, j + q)$ to the point obtained by rounding off $i + p$ and $j + q$ to the nearest integers.

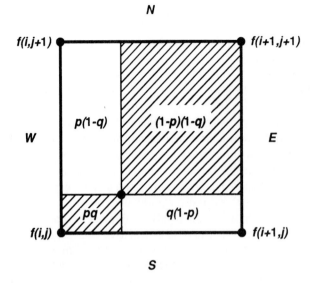

Figure 7-15 The "twisted plane" through the four corner samples has a height at $(i + p, j + q)$ that is a weighted mean of the corner heights; the weight for any corner is equal to the area opposite that corner.

Twisted-plane Interpolation

The two-dimensional analogy with linear interpolation is $f(i + p, j + q) = (1 - p)(1 - q)f(i, j) + p(1 - q)f(i + 1, j) + q(1 - p)f(i, j + 1) + pqf(i + 1, j + 1)$. Just as linear interpolation in one dimension can be thought of in terms of a string stretched between two stakes whose points are at $[i, f(i)]$ and $[i + 1, f(i + 1)]$, so this two-dimensional extension can be thought of in terms of a square tablecloth nailed at the corners to four legs of heights equal to the sample values. When evenly stretched, the tablecloth would be twisted out of the plane into the form of a hyperboloid of one sheet, because a plane cannot in general be passed through four points. However, the straight lines of a checkered cloth would remain straight, because the surface is a ruled surface. One way of thinking about the expression above is to imagine linear interpolation east-west along the straight lines between $f(i, j)$ and $f(i + 1, j)$ and between $f(i + 1, j)$ and $f(i + 1, j + 1)$, using the value of p, followed by north-south interpolation along the straight line connecting the interpolates, using q. Factorizing the expression to express this procedure reduces the number of multiplications. The term *bilinear interpolation* evokes this construction and is used in contexts where no confusion with other meanings of bilinear is likely. The method is also referred to as *area-weighting* and as the *four-point formula*.

Twisted-plane interpolation is a little bit better than might be thought from comparison with linear interpolation in one dimension, where only the first difference, or slope, is taken into account and curvature is ignored. Four sample values provide an element of redundancy, since there are four first differences, two in each direction. In addition, nonplanarity is taken into account through the second difference that corresponds to $\partial^2 f / \partial x\, \partial y$.

The Six-point Formula

If the second differences corresponding to the "curvatures" $\partial^2 f/\partial x^2$ and $\partial^2 f/\partial y^2$ are going to be significant, then the samples must span two units each way. The economical pattern of samples that reaches just this level, responding to both slopes and all three second differences, is provided by the six-point formula. Suppose that the point to be interpolated is in the SW corner of the square sample pattern, where $p < 0.5$ and $q < 0.5$. Then two additional samples, one at $(i - 1, j)$ and one at $(i, j - 1)$, give a bite on the horizontal and vertical second differences centered on (i, j), the closest sample point to $(i + p, j + q)$. The six-point formula is

$$f(i + p, j + q) = \tfrac{1}{2}q(q - 1)f(i, j - 1) + \tfrac{1}{2}p(p - 1)f(i - 1, j)$$
$$+ (1 + pq - p^2 - q^2)f(i, j) + \tfrac{1}{2}p(p - 2q + 1)f(i + 1, j)$$
$$+ \tfrac{1}{2}q(q - 2p + 1)f(i, j + 1) + pqf(i + 1, j + 1).$$

If the point to be interpolated is in another quadrant than the SW, the two additional samples are shifted accordingly. As before, the expression given can be rearranged to reduce the multiplications. Experience indicates that the six-point formula is extremely practical in a wide variety of circumstances. It is dependent on the immediate environment and uninfluenced by nearby image features or edges that are unrelated. The interpolated surface may be any of several different quadrics.

Smooth Plane Curves

When regularly or irregularly spaced samples of a single-valued function are given, the classical theory of interpolation is available for finding intermediate values. The result is that a smooth curve is established between the points of a set defined by the abscissas and ordinates of the samples. The origins of this exercise are in arithmetic, not geometry, even though a geometrical interpretation in terms of points and a curve can be proposed, as has just been done. The inherently geometrical two-dimensional situation into which the curve-plotting exercise generalizes is as follows. A set of ordered points is given for the purpose of indicating a more-or-less complicated curve on the plane; the objective is to interpolate additional points that will improve the appearance of continuity. One can in principle represent the interpolated points by an equation $\phi(x, y) = 0$, understanding that there may be multiple values of y for each x, and conversely. For example, if points happened to be given on the ellipse $x^2/a^2 + y^2/b^2 - 1 = 0$, this equation could be used to generate additional points. Normally, however, the given points are specified in terms of an ordering parameter t by $x(t)$ and $y(t)$, $t = 1$ to N, where there are N given points. The additional points might then be specified for fractional values of t by extended data arrays arrived at by an algorithm, or algebraic expressions might be found for $x(t)$ and $y(t)$ from which additional points could be generated when and where required. An example of algebraic representation is furnished by the cubic spline, which starts from the observation that, with proper choice of the four coefficients a, b, c, and d, the cubic polynomial

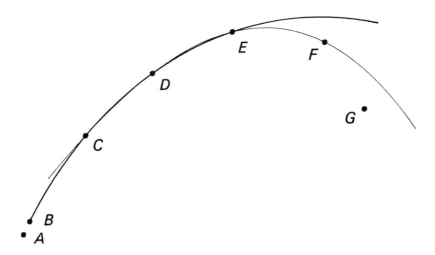

Figure 7-16   Cubic spline interpolation.

$y = ax^3 + bx^2 + cx + d$ can fit any four points in the plane, provided that no two distinct points have the same value of x.

Cubic Spline Interpolation

Classical interpolation, for example of tabulated mathematical functions, aims at optimizing accuracy but in the geometrical problem as it arises in graphics a smooth appearance may be more important than accuracy; indeed there may be no standard against which to quantify accuracy. A satisfactory approach for a large class of cases starts with the ordered set $B, C, D, E, F, \ldots$, corresponding to increasing values of t, not necessarily equally spaced. Use the cubic polynomial through $BCDE$ to draw the arc CD, use the polynomial through $CDEF$ to draw the arc DE, and so on (Fig. 7-16). For convenient programming wisdom see NumRec (1986). The arc BC and the final arc cannot be determined this way; two additional conditions, such as initial and final slope, are required. Sometimes the terminal intervals can be utilized to apply the boundary condition. For example, an additional point A specified close to point B can impose an initial slope. The arc BC would become determinate, while the new initial arc AB would be irrelevant.

     Splines arise in graphics in various ways. If their sides are short enough, polygons appear as smooth curves to the eye; interpolation can provide the additional vertices needed to achieve a cosmetic improvement in appearance as compared with the angular polygon defined by coarsely spaced data. Sometimes data points are deliberately specified at coarse intervals in order to gain the smoothness introduced by a spline.

     A specific application of spline curves to imaging arises in contour plots where the requirement is to put a smooth closed curve through a set of points previously found by interpolation from data on a grid. Descriptions of splines are usually concerned with curves that do not turn around on themselves and close. The way to adapt to closed contours is to extend the data set to $N + 3$ points by appending the first three points at the end.

There is an analogy between cubic splines and the flexible ruler used for drawing smooth curves through given points. When a thin beam is bent, the stored resilient energy per unit arc length is proportional to the square of the local bending moment, which in turn is inversely proportional to the local radius of curvature ρ. Hence the total stored energy is measured by

$$\int \frac{ds}{\rho^2} = \int \frac{d^2y/dx^2}{1 + (dy/dx)^2} \, dx.$$

When a flexible beam such as a wire is bent by constraints that do not apply bending reactions (think of threading an oiled wire through swivels at the given points), then the wire takes up a shape that minimizes the potential energy and in that sense follows the smoothest curve through the points. The mathematical spline does not minimize the energy; instead it minimizes the integral of d^2y/dx^2, which approximates the situation for a weakly flexed beam where $(dy/dx)^2 \ll 1$. In this case, the bending moment, and the second derivative, vary linearly between adjacent constraints, the slope varies quadratically, and the curve is a cubic. In three dimensions there is no analogy with the flexure of thin shells, because, unlike a beam, a shell changes its stiffness as it is bent. One thinks instead in terms of two intersecting systems of simple splines when interpolating on surfaces.

Midpoint Interpolation

When it is not necessary to find interpolates in arbitrary positions, but only at points midway between pairs of samples, interpolation naturally simplifies somewhat. And yet no generality is lost, since repeated midpoint interpolation can be as fine as one wishes. However, the main application is to one stage of midpoint interpolation over the whole of an image (or at least over a substantial part). Discrete-to-continuous interpolation, as discussed above, is more apt to situations where interpolation is required only here and there.

If samples are given on a square grid and a value is to be interpolated midway between two samples on a horizontal line, will the samples on other horizontal lines have an influence, or will the wanted value depend only on samples on that horizontal line? One approach to this simple question is to try to disprove the absence of influence by constructing a numerical counterexample. Here is a line of reasoning based on the sampling theorem. A cross section is taken through a band-limited two-dimensional function to obtain a one-dimensional function, which itself must necessarily be band-limited. The reason is that a cut through a sinusoidal corrugation of wavenumber k at an angle α has wavenumber $k \cos \alpha$, which never exceeds k. To put this another way, if you trudge across sinusoidal dunes, your crest-to-crest distance is never shorter than the spatial period of the dunes. If the samples given on the horizontal row are closely enough spaced to satisfy the sampling theorem, then values on that row are determined by the samples on that row alone. Putting $q = 0$ in the six-point formula will confirm the conclusion for that special case.

Thus, two-dimensional midpoint interpolation reduces to interpolation along the rows followed by midpoint interpolation along columns.

Sinc function interpolation would consist of multiplication of the row samples by the coefficients $\mathrm{sinc}(|n| - 0.5)$, for $n = \pm 1, \pm 2, \pm 3, \ldots$, namely

$$\{\ldots \; -0.091 \; 0.127 \; -0.212 \; 0.637 \; \uparrow \; 0.637 \; -0.212 \; 0.127 \; -0.091 \; \ldots\}$$

$$= \frac{2}{\pi}\{\ldots \; -\frac{1}{7} \; \frac{1}{5} \; -\frac{1}{3} \; 1 \; \uparrow \; 1 \; -\frac{1}{3} \; \frac{1}{5} \; -\frac{1}{7} \; \ldots\}.$$

The sum of the products is the interpolated value at the arrow. As we have seen, there is danger of contamination of an interpolate by unrelated features that are nearby, so for this reason an infinite series does not recommend itself, especially when the coefficients diminish slowly (as n^{-1}). The linear case, taking account of slope only, would be represented by the coefficients

$$\{0.5 \; \uparrow \; 0.5\}$$

For more attention to higher differences we have

$$\frac{1}{16}\{-1 \; 9 \; \uparrow \; 9 \; -1\}$$

and

$$\frac{1}{256}\{3 \; -25 \; 150 \; \uparrow \; 150 \; -25 \; 3\}.$$

These sequences have unit sum, and their central moments are zero up to the first, second, and third, respectively.

Sinc-function Interpolation

If the samples are $s(i, j)$, the interpolated value at (x_1, y_1) is given by

$$f(x_1, y_1) = \sum_i \sum_j s(i, j)^2 \mathrm{sinc}(x_1 - i, y_1 - j);$$

for midpoint interpolation $x_1 = i$ or $i + 0.5$ and $y_1 = j$ or $j + 0.5$. A numerical example will be worked out.

Example

Samples of a band-limited function, taken sufficiently closely, are given by the following matrix. Surrounding values are all zero.

$$\begin{bmatrix} 0 & 0 & 0 & 0 & 0 & 0 \\ 0 & 1 & 2 & 3 & 4 & 0 \\ 0 & 5 & 6 & 7 & 8 & 0 \\ 0 & 9 & 10 & 11 & 12 & 0 \\ 0 & 13 & 14 & 15 & 16 & 0 \\ 0 & 0 & 0 & 0 & 0 & 0 \end{bmatrix}$$

Find the value of the sampled function at the central point. ▷Inspection suggests that the central value will be approximately $\frac{1}{4}(6 + 7 + 10 + 11) = 8.5$. Sinc function interpolation will require adding 16 terms of the form i sinc l sinc m, where $l = 2.5 - (i \bmod 4)$ and $m = 2.5 - (3 + i) \div 4$. A short program for this single value is provided because once it is translated and running it can easily be extended to sinc function interpolation at any point in any finite matrix.

SINC FUNCTION

```
DEF FNsinc(x)=(SIN(PI*x)+(x=0))/(PI*x+(x=0))
f=0
FOR j=1 TO 4
   FOR i=1 TO 4
      s=i+4*(j-1)
      f=f+s*FNsinc(2.5-i)*FNsinc(2.5-j)
   NEXT i
NEXT j
PRINT f
```

The convenient one-line function definition for sinc x appearing in the first line avoids division by zero when $x = 0$. The terms 2.5 in the sixth line arise from the coordinates $(2.5, 2.5)$ of the point (x_1, y_1) at the center. The result for the central value is 6.12 rather than 8.5. Moving SW for a further example, the interpolate at $(1.5, 3.5)$ midway between the samples 9 and 14 is 16.24. ◁

A peculiar thing about sinc-function interpolation is that it purports to give intermediate values precisely, whereas the variety of numerical methods described above clearly do not agree exactly among themselves, let alone with what the intermediate measurement might have been had it been taken. To understand this consequence of the sampling theorem requires asking what band-limited means, because it is the property of band limiting that the sampling theorem is based on. Obviously, if one's only knowledge of some phenomenon is what is available by measurement, it is not possible to deduce from the samples that there is a band limit. However, there may be additional information; samples of a telephone conversation are compatible with what is known of the transmission characteristics designed into the telephone channel and what is known of speech. If the band limit is pushed—that is, if there is substantial signal power close to the limit—then interpolation may be noticeably influenced by the presumptions that the spectrum has a sharp discontinuity and that the high power near cutoff drops sharply to precisely zero beyond cutoff. Still, if the sharp band limit can be relied on, the interpolation will be exact.

In the numerical example given above the image represented by the numbers looked compatible with a square patch with a linear gradient of grey level surrounded by an extended black border. But the interpolated value 6.12 disagreed sharply with the linear expectation 8.5, while the value calculated between 9 and 14 exceeded all four neighboring values. This is a consequence of asking a bright region to drop abruptly to black but to do so in such a way that no high frequencies are called for. If, indeed, the samples were taken from a patch with brightness varying from one corner to the other, then twisted-plane interpolation would have done much better than sinc-function interpolation. Of course, it would be fair to ask how it happened that the samples presented were contrived to precisely avoid the dappled appearance of lights and darks that must have been clearly evident to the person obtaining the samples. The answer is that the function was a mathematical conception specified by function values and the band-limited property, not by a physical quantity that had been sampled by measurement. The conclusion is that sinc-function interpolation relates to conditions of band limiting whose degree of applicability in any particular case needs scrutiny.

Because of the sampling theorem sinc-function interpolation is often referred to in theoretical contexts, but practical implementation must stop short of the infinite number of coefficients implied. One way of doing this is to frankly limit the number of coefficients. If there are N data values in a row, clearly no more than N different coefficients will be needed. If coefficients less than 1 percent of the central value are ignored, then there will be 64 coefficients (4096 in two dimensions). Clearly, there can be few practical applications for such large numbers of coefficients. On the other hand, to retain only eight coefficients means ignoring others as large as 9 percent. To eliminate this discontinuity authors try multiplying by a tapering factor, usually Gaussian. Here is the eight-element sequence, normalized to a sum of unity to avoid magnification, obtained by multiplying the sinc-function values by the eight-element binimial sequence {1 7 21 35 21 7 1}, which descends to zero more gracefully than the truncated Gaussian:

$$\{-0.0024 \quad +0.0239 \quad -0.1196 \quad +0.5981 \quad -0.1196 \quad +0.0239 \quad -0.0024\}$$

This interpolating sequence has the effect of a transfer function that is flat-topped out to about one-half the cutoff frequency, followed by a more or less linear decline to zero. It is satisfactory for some purposes. Even though derived from sinc-function interpolation, however, its transfer function is far from the ideal rectangle with a perfectly sharp cutoff. There is a case, though, where strict sinc-function interpolation is applicable, and that is where the data are obtained on a ring. In this case the fact that the coefficients run on to infinity does not cause a problem, because the data can be thought of as cyclic. The number of coefficients will be equal to the number N of data points on the ring, and computed from $\mathrm{sinc}\, x * \mathrm{III}(x/N)/N$, which is equal to $\sin(\pi x)/N \sin(\pi x/N)$.

INTERLACED SAMPLING

Consider a band-limited function such that samples at the intersection of the grid of unit-area cells shown in Fig. 7-17 would be sufficient for recovering the function. Recall that the method of recovery is to center a two-dimensional solving function at each sample location with amplitude equal to the sample value, and sum. The solving function is a sinc function scaled so as to pass through zero at every sample point other than the one it is centered on, and, of course, the sinc function meets the band limit.

Let the given function fall off to negligible amplitude outside a certain area A, as suggested in the figure by termination of the grid. As the square cells have been defined as having unit area, the number of significant samples is A. In the illustration this number is only about 50 or 60, but it could be as large as we wish. The Fourier transform will be specifiable, out to the band limit in the (u, v)-plane, by the same number of independent tags as there are samples in the (x, y)-plane. The discussion in an earlier section suggests that the original samples need not have been regularly spaced on the grid points but might

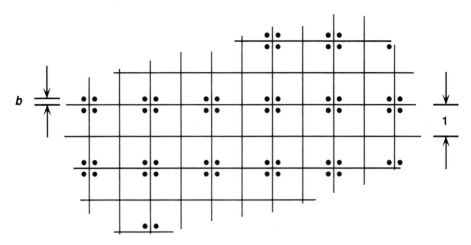

Figure 7-17   Interlaced sampling as represented by $^2\text{III}(x \pm b, y \pm b)$.

have been clustered in some way, as for example in fours as indicated by dots on the figure. Such a plan of sampling can be thought of in terms of four sets of samples at twice the original spacing, but differently staggered by amounts $(\pm b, \pm b)$, and is referred to as *interlaced sampling.* Since there are A interlaced samples, it should be possible to solve for the A unknowns in the (u, v)-plane and hence to interpolate between the new samples to recover $f(x, y)$. However, if we wished to use a function-domain procedure for the interpolation, the method would be to place an as yet unknown solving function at each sampling location, and sum. The required function would need to pass through zero at the three nearby samples of its cluster and at all four members of every other cluster. Clearly, no one function can do this, but if $\phi(x, y)$ has the required properties when centered over some particular sample, then by symmetry $\phi(-x, y), \phi(x, -y)$, and $\phi(-x, -y)$ will work when centered over the others.

The function $\phi(x, y)$ can be discovered (Fig. 7-18) and proves to be

$$\phi(x, y) = (\operatorname{sinc} x - Kx \operatorname{sinc}^2 \tfrac{1}{2}x)(\operatorname{sinc} y - Ky \operatorname{sinc}^2 \tfrac{1}{2}y),$$

where $K = \tfrac{1}{2}\pi \cot \tfrac{1}{2}\pi b$ and $2b$ is the side of the cluster square. A derivation of the one-dimensional result, with and without the effect of errors, can be found in FTA (1985, at p. 201), or we can reason as follows. The functions $\operatorname{sinc} x$ and $x \operatorname{sinc}^2 x$ contain only frequencies from $-\tfrac{1}{2}$ to $\tfrac{1}{2}$. Consequently, by the separable product theorem, $\phi(x, y)$ contains only frequencies in the central square of the (u, v)-plane having unit side. Thus $\phi(x, y)$ has the same band limit as $f(x, y)$. Consequently, the sum of the solving functions with amplitudes given by the sample values has the right band limit and passes through all the sample values as required.

Interlaced sampling has interesting aspects. For example, if some geophysical measurement is to be made on a 1-km grid, then one could consider the alternative of occupying only one-quarter of the sites but deploying four instruments displaced by $(\pm 100, \pm 100)$ meters from a central station.

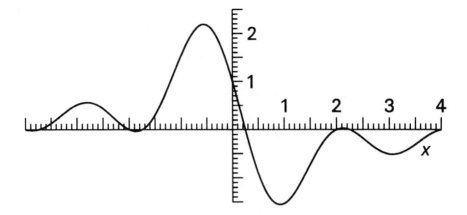

Figure 7-18   Interpolating function for interlaced samples.

Other nonuniform sampling patterns arise, including the irregular ones imposed when samples are taken in villages or along roads. If the samples are irregularly distributed, but are equal in number to those that would specify a band-limited function if uniformly distributed, then different band-limited solving functions can be sought, each equal to unity at one sampling location and equal to zero at the others. Needs of this sort arise in the daily bulk handling of meteorological weather data from places scattered around the country.

Effect of Errors

Suppose that measured samples are subject to errors with zero mean and a standard deviation σ. The error made at one location is very often correlated with the error made at an adjacent location, but it is sufficient here to take a case where it has been ascertained by some means that the errors are uncorrelated. (If the measurements are the source of our knowledge about the entity being measured, how could statistical information about the errors be acquired?) If the measurements are distributed on a uniform grid, then values interpolated by sinc function will also be subject to error, but the mean error will be zero and the standard deviation will have the same σ, which is very reasonable. This striking result is not obvious and is not true for linear interpolation, for example. If the measurements are clustered regularly in fours, the error in the interpolated value at the point most distant from the sampling locations has a standard deviation that exceeds the previous σ, the more so as the cluster is tighter. This makes a lot of sense and is a contributing factor to the assessment of the clustering option.

In the limit, as neighboring samples of a cluster approach each other, the measurement becomes like measuring a derivative, which can be notoriously difficult, and the error in an interpolate will increase indefinitely. On the other hand, samples measured very close together are less likely to be subject to the uncorrelated errors that this discussion started with; consequently, each interlaced sampling proposal needs its own error analysis.

As a general guide, it is not likely to be practical to allow samples to depart locally from the mean area density by a substantial factor. Would the degree of nonuniformity seen in Fig. 7-17 be practical? In this simply structured case the theory can be worked out, but a conclusion can also be reached empirically. Take $f(x, y) = 0$; then allocate the errors with $\sigma = 1$ to the sampling locations, interpolate at a point of interest such as the pole of inaccessibility, and analyze the interpolates statistically after many repeats. This technique extends to irregular structure of the sampling locations and also works when correlation between errors is introduced.

Interpolation from Irregular Samples

Collection of two-dimensional data often takes place irregularly in space. Familiar examples are furnished by topographic surveys, the recording of daily maps of rainfall, temperature, wind velocity, and atmospheric pressure, and observations made from rotating platforms including satellites and the earth itself. The construction of isolines then requires interpolation. Since many different disciplines have generated attention to problems of this type, the literature is extensive and scattered. As a guide consult Okabe et al. (1992) and Ripley (1981). Practice in interferometric imaging is discussed in Chapter 12, and Chapter 18 describes the Dirichlet cells (Voronoi polygons) which are fundamental to the geometry of scattered points.

A common misconception is to treat the data as a set of delta functions to be smoothed by convolution with, for example, a Gaussian function of width related to the mean spacing of sampling locations. It is obvious that this procedure does not interpolate, for the result at a data point does not in general agree with the sample value there.

APPENDIX: THE TWO-DIMENSIONAL FOURIER TRANSFORM OF THE SHAH FUNCTION

The integral $\int_{-\infty}^{\infty} \int_{-\infty}^{\infty} {}^2\mathrm{III}(x, y)\, dx\, dy$ not being convergent, the function ${}^2\mathrm{III}(x, y)$ does not possess a regular two-dimensional Fourier transform. We are concerned here with showing that it has a transform-in-the-limit, and that this transform is identical in form with ${}^2\mathrm{III}$ itself. We prove in full the corresponding result for the one-dimensional function $\mathrm{III}(x)$, and indicate the method of proof for ${}^2\mathrm{III}$.

Consider the function[1]

[1] In following this proof it will be helpful to notice that, in our $*$ and III notation,

$$f(\alpha, x) = \exp(-\pi\alpha^2 x^2)\{\alpha^{-1} \exp(-\pi\alpha^{-2}x^2) * \mathrm{III}(x)\}.$$

Also $\exp(-\pi\alpha^2 s^2)$ is the Fourier transform of $\alpha^{-1} \exp(-\pi\alpha^{-2}x^2)$ the factor α being so chosen that $\int_{-\infty}^{\infty} \alpha^{-1} \exp(-\pi\alpha^{-2}x^2)dx = 1$, irrespective of the value of α, thus ensuring that $\alpha^{-1} \exp(-\pi\alpha^{-2}x^2)$ is a suitable defining sequence for $\delta(x)$, as $f(\alpha, x)$ is for III(x). The line of proof is to deduce, by conventional means, that $f(\alpha, x)$ has a regular transform, viz.

$$F(\alpha, s) = \alpha^{-1} \exp(-\pi\alpha^{-2}x^2) * \{\exp(-\pi\alpha^2 s^2)\mathrm{III}(s)\},$$

and that the sequence $f(\alpha, x)$ and the transform sequence $F(\alpha, s)$ both define III$(\ )$ as $\alpha \to 0$.

$$f(\alpha, x) = \alpha^{-1} \exp(-\pi\alpha^2 x^2) \sum_{n=-\infty}^{\infty} \exp\{-\pi\alpha^{-2}(x-n)^2\}.$$

For each α, $f(\alpha, x)$ represents a row of overlapping strictly Gaussian spikes of width α, their peaks following a Gaussian curve of width α^{-1}. We shall consider the sequence of functions generated as $\alpha \to 0$. It is easy to show that $\lim f(\alpha, x) = 0$ for $x \neq n$, and that, for every n, $\lim \int_{n-\frac{1}{2}}^{n+\frac{1}{2}} f(\alpha, x)dx = 1$. It follows that the sequence $f(\alpha, x)$, as $\alpha \to 0$, defines $III(x)$.

We now consider the transform of $f(\alpha, x)$ and show that, as α tends to zero, this sequence also defines $III(s)$. To find the transform of $f(\alpha, x)$ we first notice that the factor $\alpha^{-1} \sum \exp\{-\pi\alpha^{-2}(x-n)^2\}$ is periodic in x and may therefore be expressed as the Fourier series

$$\sum_{m=-\infty}^{\infty} \exp(-\pi\alpha^2 m^2) \cos 2\pi mx.$$

Hence

$$f(\alpha, x) = \sum_{m=-\infty}^{\infty} \exp(-\pi\alpha^2 m^2) \exp(-\pi\alpha^2 x^2) \cos 2\pi mx.$$

Applying the Fourier shift theorem we find for the transform of $f(\alpha, x)$

$$F(\alpha, s) = \alpha^{-1} \sum_{m=-\infty}^{\infty} \exp(-\pi\alpha^2 m^2) \exp\{-\pi\alpha^{-2}(s-m)^2\}.$$

This function is a row of overlapping, almost Gaussian spikes of width α lying below a nearly Gaussian envelope of width α^{-1}, and clearly, as $\alpha \to 0$, this sequence also defines $III(x)$. Hence the transform-in-the-limit of III is identical with itself. Figure 7-19 illustrates this reasoning; the middle diagram on the right does not mean that the derivation uses delta functions to prove something about delta functions—it is merely graphical shorthand to represent the Fourier series coefficients determined classically for the periodic function (middle left). A similar proof can be given for ^2III by showing that the regular two-dimensional transform of $\gamma(\Gamma * {}^2III)$ is $\Gamma * (\gamma^2 III)$, where

$$\gamma(x, y) = \exp\{-2\pi^2\alpha^2(x^2 + y^2)\},$$

and

$$\Gamma(x, y) = (2\pi\alpha^2)^{-1} \exp\{-(2\alpha^2)^{-1}(x^2 + y^2)\}.$$

Hence γ and Γ are a transform pair, and the sequence Γ as $\alpha \to 0$ has the property of defining $^2\delta$. The doubly periodic function $\Gamma * {}^2III$ can be expressed as a double Fourier series, and the two-dimensional shift theorem can be applied term by term. Both $\gamma(\Gamma * {}^2III)$ and $\Gamma * (\gamma^2 III)$ define ^2III as $\alpha \to 0$. The proof that the shah function is its own Fourier transform first appeared in Bracewell (1956).

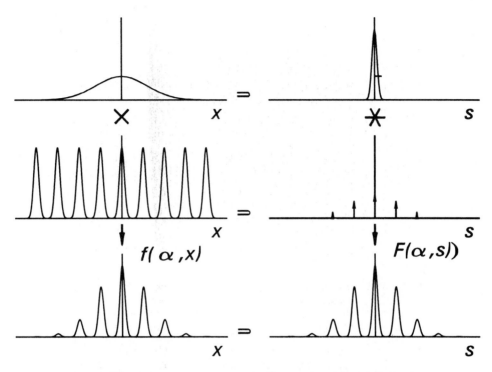

Figure 7-19   The Gaussian envelope $\exp(-3\pi x^2)$ (top left) and the periodic function $\Sigma_n 3 \exp\{-\pi[3(x-n)]^2\}$ (middle left) are multiplied together to give the train of pulses (lower left) which are approximately but not exactly Gaussian. On the right are the Fourier transforms, the top two of which are convolved together to give the train of almost Gaussian pulses at the lower right. As the parameter α (here given the value $\frac{1}{3}$) approaches zero, each of the lower pair of transforms constitutes a sequence defining the shah function.

LITERATURE CITED

R. N. BRACEWELL (1956), "Two dimensional aerial smoothing in radio astronomy," *Aust. J. Phys.*, vol. 9, 297–314.

F. FILICORI, G. LUCULANO, A. MENCHETTI, AND D. MARRI (1989), "Random asynchronous sampling strategy for measurement instruments based on nonlinear signal conversion," *IEE Proc. (London)*, vol. 136, part A, pp. 114–150.

M. J. LIGHTHILL (1958), *Introduction to Fourier Analysis and Generalized Functions*, Cambridge University Press, Cambridge, England.

P. MERTZ AND F. GRAY (1934), "A theory of scanning and its relation to the characteristics of the transmitted signal in telephotography and television," *Bell Syst. Tech. J.*, vol. 13, pp. 464–515.

K. MIYAKAWA (1959), "Sampling theory of stationary stochastic variables in multi-dimensional space," *J. Inst. Elec. Commun. (Japan)*, pp. 421–427.

H. Nyquist (1928), "Certain topics in telegraph transmission theory," *Trans. A.I.E.E.*, vol. 47, pp. 617–644.

D. R. Peterson and D. Middleton (1962), "Sampling and reconstruction of wave-number limited functions in N-dimensional Euclidean spaces," *Information and Control*, vol. 5, pp. 279-323.

B. D. Ripley (1981), *Spatial Statistics*, John Wiley & Sons, New York.

C. E. Shannon (1948), "A mathematical theory of communication," *Bell Syst. Tech. J.*, vol. 27, pp. 379–423, pp. 623–656.

W. T. Sullivan, III, ed. (1984), *The Early Years of Radio Astronomy*, Cambridge University Press, Cambridge, England.

E. T. Whittaker (1915), *Proc. Roy. Soc. Edinburgh.*, vol. 35, pp. 181–194.

PROBLEMS

7–1. *Sampling on a line.* Evaluate $\int_{-\infty}^{\infty}\int_{-\infty}^{\infty} f(x,y)\delta(y-x)\,dx\,dy$ by converting to coordinates (x',y') that are rotated $45°$. ☞

7–2. *Integral evaluated exactly by sum.* A function $f(x)$ is band limited with critical sampling interval X.
   (a) Show that its infinite integral equals X times the sum of any set of critically spaced samples.
   (b) Show that if you take only every second sample, you will still get exactly the correct value if you multiply the sum by $2X$.
   (c) Show that the convolution integral $f*g$ is exactly equal to $\sum_n f(x-nX)g(nX)$, provided only that $f(x)$ is band-limited with critical sampling interval X.
   (d) What is the generalization to two dimensions?

7–3. *Coarse summing for autocorrelation.* Show that the autocorrelation of a band-limited function can be obtained exactly by summing samples of the product of the function with a displaced replica of itself and that the sampling interval may be greater than the critical sampling interval corresponding to the band limit.

7–4. *Bessel function samples.*
   (a) The Bessel function $J_0(\pi t)$ is band-limited; what is the critical sampling interval?
   (b) Explain why in two dimensions $J_0(\pi r)$ must also be band-limited. Does it follow that this circularly symmetrical function is fully defined by the function values on a square lattice?

7–5. *Sinc function expression for Bessel functions.*
   (a) Show by sampling-theorem considerations that

$$\operatorname{jinc} x = \sum_{n=-\infty}^{\infty} \operatorname{jinc} n\,\operatorname{sinc}(x-n)$$

$$= \tfrac{1}{4}\pi\operatorname{sinc} x + \sum_{1}^{\infty}\operatorname{jinc} n[\operatorname{sinc}(x+n)+\operatorname{sinc}(x-n)],$$

where $\operatorname{jinc} x = J_1(\pi x)/2x$.
   (b) Also show that

$$J_0(\pi x) = \sum_{n=-\infty}^{\infty} J_0(\pi n) \operatorname{sinc}(x - n).$$

(c) See whether the expansion for jinc x would constitute a practical way of computing the jinc function to four decimals on a hand calculator, assuming that a modest number of coefficients could be stored for the purpose.

7–6. *Product of impulsive grilles.*
   (a) Derive the relation ${}^2\mathrm{III}(x, y) = \mathrm{III}(x)\mathrm{III}(y)$.
   (b) A grille inclined at $45°$, and represented by $\mathrm{III}(y - x)$, intersects the grille $\mathrm{III}(x)$ at the points where ${}^2\mathrm{III}(x, y)$ is impulsive. Does it follow that $\mathrm{III}(x)\mathrm{III}(y - x) = {}^2\mathrm{III}(x, y)$? ☞

7–7. *Rectangular sampling.* A two-dimensional band-limited function is sampled at twice the critical spacing in the x-direction, but this is compensated for in the y-direction by sampling at one-half the critical spacing. Thus the number of samples N is the same as would have been obtained if the sampling had been done at the correct critical spacing in both directions. From samples taken on the critical array it would be possible to determine values of the original function at points between the sampling points. Is this also possible for the compressed sampling described? Explain why it is not possible or explain how to perform the interpolation. ☞

7–8. *Impulse array.* A function $f(x, y)$ consists of nine impulses as follows.

$$f(x, y) = {}^2\delta(x + 1, y + 1) \quad -2\,{}^2\delta(x, y + 1) + {}^2\delta(x - 1, y + 1)$$

$$- 2\,{}^2\delta(x + 1, y) \quad +4\,{}^2\delta(x, y) \quad - 2\,{}^2\delta(x - 1, y)$$

$$+ {}^2\delta(x + 1, y - 1) - 2\,{}^2\delta(x, y - 1) + {}^2\delta(x - 1, y - 1)$$

The two-dimensional Fourier transform of $f(x, y)$ is $F(u, v)$.
   (a) What can you say about $F(u, 0)$?
   (b) What can you say about $F(0, v)$?
   (c) What can you say about $F(u, u)$, the transform along the diagonal?
   (d) An object $O(x, y)$ is of the form $\alpha x + \beta y + g(x, y)$, where α and β are constants. If you form the 2D convolution $f(x, y) ** O(x, y)$, what will be the result?
   (e) Can you express $f(x, y)$ as the two-dimensional convolution of two simpler patterns? ☞

7–9. *Aliasing.*
   (a) An image consists of several Gaussian humps distributed at random, as described by $f(x, y) = \Sigma_i a_i \exp\{-\pi[(x - x_i)^2 + (y - y_i)^2]\}$. You and I know that the Fourier transform of $\exp[-\pi(x^2 + y^2)]$ is $\exp[-\pi(u^2 + v^2)]$ and therefore never cuts off. Still, it gets very small, so our project leader proposes to sample the image using a sampling interval X that is small enough to introduce only negligible erors. Our leader says, "Choose X so that the sum of all the samples, multiplied by X^2, is the same as $\int\int f(x, y)\,dx\,dy$ within 1 percent." What value of X does this lead to?
   (b) Since we have violated the sampling-theorem condition, the samples are contaminated by aliasing. Comment on the damage done.

7–10. *Aliasing.* A corrugation $\cos[2\pi(0.3x + 0.4y)]$ with a spatial frequency 0.5 and crest-to-crest distance 2 is sampled on a unit square grid. Unit sampling interval is sufficient for spatial period 2 in one dimension, but in this case the period is 3.33 in the x-direction and 2.5 in

Figure 7-20   Bar-width modulation with wave noticeable to the eye.

the y-direction. Smith says, "It will be aliased into a longer wave oriented in a different direction." Jones says, "There will be no aliasing." Robinson says, "Yes, but if you sample on a 1.5×1.5 grid, that will be good enough for the x-direction but not for the y-direction; in that case the direction will be changed." What would you really recover from the samples?

7–11. *CCD camera.* The focal plane of a telescope is occupied by an array of 4096×4096 charge-coupled diodes that sample the image $f(x, y)$ produced by the telescope over an area 117×117 mm. Each diode is 24 μm square.

    (a) What is the highest spatial frequency (in cycles per μm in the image plane to which this camera responds?

    (b) What can you say about the transfer function of this imaging system?

7–12. *Bar code spectrum.*

    (a) Determine the spatial power spectrum of a bar-code image taken from something you have bought, and select a method of presentation that enables quantitative values of spatial frequency to be read off.

    (b) A TA in a DSP course says, "Analogue displays such as bar codes would be more resistant to various kinds of noise if the readout device made digital samples of the display at fine enough intervals and converted into a binary ascii code of 0s and 1s." Examine the power spectrum you have prepared, adopt a tentative value for an effective cutoff frequency, and calculate what the corresponding sampling interval in mm would be.   ☞

7–13. *Modulated grating.* The bars of a grating are regularly spaced between centers, but their width is sinusoidally modulated as illustrated (Fig. 7-20). Adapting the terminology of pulse modulation, we could call this effect *bar-width modulation.* Spectral analysis will show the presence of a fundamental frequency corresponding to the bar spacing and harmonics. There is also apparently a low spatial frequency, visible to the eye as a longer wave.

    (a) Would Fourier analysis confirm the presence of this low frequency?

    (b) Unwanted long-period artifacts sometimes afflict images and need to be removed. In this case, would a spatial band-stop filter that rejects the low spatial frequency remove the bar-width modulation?   ☞

7–14. *Pyramid interpolation.* In one dimension linear interpolation is performed by representing each sample $f(i)$ by a triangle function $f(i)\Lambda(x - i)$ of height $f(i)$ and base 2. The sum of all the overlapping triangle functions is the continuous function of x resulting from linear interpolation. "This generalizes to 2D," says Strephon, "if each sample $f(i, j)$ is represented by a pyramid of base 2×2 and the overlapping plane facets are added. This is the correct geometrical way of generalizing to 2D linear interpolation." Explain whether or not this is equivalent to the four-point (twisted-plane) method.

7–15. *Interpolated contour curves.* A data set is interpolated onto such a fine grid before a contour plot is made that the polygons look like curves to the eye. How can you tell whether four- or six-point interpolation was used?

7–16. *Sinc-function interpolation.*
   (a) Work out the six-element midpoint interpolation sequence derived from sinc x, $x = -2.5, -1.5, , -0.5, 0.5, 1.5, 2.5$, by multiplication with the six-term binomial array followed by normalization to a sum of unity. Give the coefficients to four decimals.
   (b) Graph the sum of the three cosine functions associated with these coefficients.

7–17. *Six-element sequence.* Show that the six-element interpolating sequence

$$\{3 \quad -25 \quad 150 \quad 150 \quad -25 \quad 3\}/256$$

is approximated by $\exp[-\pi(x/W)^2]$ sinc x, $x = \pm 0.5, \pm 1.5, \pm 2.5, \dots$, where $W = 3$.

7–18. *Rectifying polar samples.* Samples are measured for $r = 0, 1, 2, \dots, N$ and all values of θ from 0 to 2π, but interpolated values are wanted for integral values of x and y. Give a discussion based on the sampling theorem leading to an exact expression for the wanted values, provided the indications from measurements at large r are that the samples for $r > N$ would have been negligible. (b) If a continuous record was not measured as θ varied, but instead samples were taken at discrete increments $\Delta\theta = 1/N$, show how an exact expression can still be obtained. ☞

7–19. *Comparing interpolation coefficients.* Take the values of $\exp[-\pi(x/4)^2]$ at $x = \pm 0.5, \pm 1.5$, $\dots$ to be artificial data and interpolate to find the value at $x = 0$ (known to be unity) using
   (a) sinc-function interpolation (ignoring the fact that the Gaussian function is not band-limited),
   (b) the four coefficients $\frac{1}{16}\{-1 \ 9 \ 9 \ -1\}$, and
   (c) the six coefficients $\frac{1}{256}\{3 \ -25 \ 150 \ 150 \ -25 \ 3\}$. How do you explain the results?

7–20. *Sample number.*
   (a) Each of MN data values has a serial number, or sample number, n, which ranges from 1 to MN. The values are arranged in M columns and N rows with coordinates i and j, $1 \le i \le M, 1 \le j \le N$. Show that

$$i = 1 + (n - 1) \bmod M$$
$$j = 1 + (n - 1) \operatorname{div} M$$

   and conversely that

$$n = (j - 1)M + i.$$

   (b) What are the corresponding equations when i ranges from 0 to $M - 1$, j ranges from 0 to $N - 1$, and n ranges from 1 to MN as before?

7–21. *Sinc function.* Show that

$$\operatorname{sinc} x = \prod_{n=1}^{\infty} \left(1 - \frac{x^2}{n^2}\right)$$

and

$$\operatorname{sinc} x = \prod_{n=1}^{\infty} \cos\left(\frac{\pi x}{2^n}\right).$$

7–22. *Tapered sinc-function interpolation.* Take the eight sinc-function values at $x = \pm 0.5, \pm 1.5$,

Figure 7-21   Superposed gratings with $w = 4$, $S_1 = 16$, and $S_2 = 15$.

± 2.5, ± 3.5 with a view to performing midpoint interpolation so that dependence on strong remote features is eliminated. To mitigate the discontinuities produced by truncation of the sinc function, taper the eight values by multiplication with the eight binomial coefficients 1, 7, 21, 35, 35, 21, 7, 1. Normalize the resulting products to a sum of unity.

(a) State the eight coefficients to four decimals.

(b) Convolution with these coefficients will smooth as well as interpolate. What is the transfer function $H(\ )$ corresponding to convolution with the discrete set of coefficients? ☞

7-23. *Irregular sampling.* Tissue-density measurements for an important medical image are available at irregularly distributed points. To interpolate onto a square grid for further processing, a graduate student, who is unaware of the meteorological tradition of drawing isobars and isotherms through measurements made at irregularly located stations, makes the following suggestion. "The value at a given grid point is dependent on the measured values at the vertices of the smallest triangle of sample points enclosing the grid point, and the closest sample should have the most influence. Since differently shaped triangles surround each grid point, adopt a Gaussian weighting coefficient $\exp[-\pi (r/S)^2]$ that falls off with distance r from the grid point to each measurement point. The scale distance S is the mean spacing $\sqrt{A}/N$ when N measurements are taken over a total area A. Using these three coefficients, calculate the grid point value as the weighted mean of the three surrounding measurements." Comment on this suggestion. ☞

7-24. *Converting samples from polar coordinates.* A set of data $P(r, \theta)$ acquired in polar coordinates with $\Delta\theta = 2\pi/N$ and $\Delta r = 1$ is to be converted to samples $C(x, y)$ in cartesian coordinates with $\Delta x = \Delta y = 1$.

(a) Propose a method that will not be too time-consuming when $N = 180$ and the radius W of the field of view is 32.

(b) If your method is approximate, discuss ways of testing the accuracy of the results.

(c) The original data were measured subject to a certain noise level. Has your method preserved the original quality? ☞

7-25. *Beat-frequency artifact.* A printed grating consisting of parallel bars of width w having uniform spacing S_1 between centers is representable by $\mathrm{III}(x/S_1) * \mathrm{rect}(x/w)$. Another slightly longer grating of slightly different spacing S_2 but the same bar width is printed over the first, as illustrated (Fig. 7-21), and it is represented by $\mathrm{III}[(x/S_2) * \mathrm{rect}(x/w)$. Beats with a frequency $(S_1^{-1} - S_2^{-1})$ are visible to the eye, but is this frequency really present in the Fourier analysis? ☞

7-26. *Frequency analysis of modulated gratings.* A plain grating whose center-to-center bar spacing is 1 mm has bars of width 1/3 mm. Sketch the spectrum for spatial frequencies from 0 to 6 cycles per mm. Fig. 7-22 shows similar gratings of the same mean spacing but with sinusoidal grey-level modulation (like AM) and sinusoidal position modulation. Sketch the spectrum of each of these over the same range of spatial frequencies.

7-27. *Annular sampling.* An omnidirectional radar is situated at $(0, 0)$ on a uniformly rough plane, any square meter of which returns an echo amplitude proportional to R^{-4} at the receiver, owing to the fact that the inverse-square law operates twice, once at emission and again at reflection.

  (a) Show that the total echo amplitude received at time $2R/c$ after the emission of a pulse is proportional to R^{-3}.

  (b) If the plane is not uniformly rough, and an element $dx\,dy$ returns an echo amplitude $\sigma(x, y)\,dx\,dy$, show that the echo amplitude received from range R is proportional to $\int\int \sigma(x, y)\delta[(x^2 + y^2)^2 - R^4]\,dx\,dy$.

  (c) Explain why $\delta[(x^2 + y^2 - R^2)^2]$ would not be correct, even though $x^2 + y^2 - R^2 = 0$ and $(x^2 + y^2 - R^2)^2 = 0$ are equally good equations for a circle of radius R.

7-28. *Taylor series for two variables.* Examine the Taylor series for $f(x + h, y)$ with a view to seeing whether the claim that interpolation along a row does not depend on samples outside that row is confirmed or refuted by the dependence of $f(x + h, y)$ on derivatives in the y-direction.

Figure 7-22   A plain grating, a grating with amplitude modulation, and one with bar-position modulation.

7–29. *Honeycomb.*

    (a) Prove the minimum property of hexagonal honeycomb cells mentioned in connection with hexagonal lattices.

    (b) Honeycomb is double-sided, the inside ends of the cells on one side abutting against the inside ends of the other side. If you were a bee interested in partitioning the two sets of cells with the least wax (without violating the wall-thickness code) how would you design the party wall?

8

Digital Operations

Many of the topics already taken up have underlying implications of numerical evaluation, even where the treatment is presented in the form of continuous analysis. In some cases the computational evaluation is straightforward, as with many integrals, but in other cases ideas enter in that are associated with discrete mathematics. Many of these ideas come to the fore in the present chapter, which begins with a variety of frequently needed elementary operations, such as smoothing and sharpening digital images, and continues on to introduce morphological operations, such as dilation and erosion, that are prerequisite to handling binary objects and to following the literature of feature recognition in images. Where it helps to clarify numerical procedures, short segments of computer pseudocode have been supplied, and the opportunity has been taken to point out how algebraic notation such as $A \cup B$ and $A \oplus B$ translate directly into the simple logical expressions acceptable to computers.

SMOOTHING

It is often necessary to smooth data. One reason is that measurements are noisy. Another is the presence of unwanted trends in the background. In the first case what is wanted is the smoothed version; in the second what is wanted is the original after the smoothed version has been subtracted. Because of a long tradition of data smoothing, for instance in meteorology and in census statistics, where the two-dimensional aspect has always been present, and because smoothing is perhaps the most commonly practiced form of digital processing on images, the chapter starts with this special topic.

A related topic is trend removal as practiced by stock-market chartists, which is a traditional data operation with two-dimensional embodiments and can be thought of as the opposite of smoothing (seeing that continued smoothing is what exposes the trend). We will show that trend reduction is accomplished by the technique for smoothing.

Sharpening, another opposite of smoothing, attempts to reverse the action of smoothing. Sharpening is given preliminary consideration here in preparation for a later chapter on restoration.

Digital operations on numerical images seem quite different in character from related operations expressed in terms of continuous-variable analysis. For example, convolving $f(x, y)$ with the unit-variance normal distribution $0.4 \exp[-(x^2 + y^2)/2]$ conjures up two-dimensional integrals. A related operation would be to convolve a digital image $[a_{i,j}]$ with the binomial array

$$[b_{i,j}] = \frac{1}{256} \begin{bmatrix} 1 & 4 & 6 & 4 & 1 \\ 4 & 16 & 24 & 16 & 4 \\ 6 & 24 & \mathbf{36} & 24 & 6 \\ 4 & 16 & 24 & 16 & 4 \\ 1 & 4 & 6 & 4 & 1 \end{bmatrix},$$

that has the same unit mass and the same unit variance as for the normal distribution mentioned. In principle, however, we can retain the continuous formalism if we use the two arrays to construct sets of delta functions having the strengths given by the arrays. From the image matrix $[a_{i,j}]$ we get $A(x, y) = \Sigma a_{i,j}{}^2 \delta(x - i, y - j)$, while the binomial array gives $B(x, y) = \Sigma b_{i,j}{}^2 \delta(x - i, y - j)$. These two functions depend on the continuous variables x and y; their convolution integral $A(x, y) ** B(x, y)$ is another set of delta functions. The strengths of these delta functions are exactly the numbers that would be arrived at by evaluating the convolution sum $[a_{i,j}] ** [b_{i,j}]$. This intimate equivalence means that everything we have learned about continuous variables has a meaning for the discrete digital situation. When we are reasoning about a problem, we can choose whether to think in continuum or quantum mode, whichever suits the problem. There is also a hybrid situation of importance: we often need to apply discrete operators to functions of continuous variables, and delta-function notation allows this directly without additional conventions.

The Origin of a Convolution Operator

When we plot a function of x and y, it is standard practice to show the x- and y-axes, but it is not customary to show the indices i and j on a matrix. This is because in much work with matrices the index pair $(1,1)$ refers to the top left element. With convolution operators shown as matrices we are also interested in an origin which corresponds to the intersection of the x- and y-axes in the continuous equivalent, and this origin may fall in the middle of the matrix. It is often convenient to indicate the element at the origin by using boldface, as was done with the element **36** above. Sometimes the origin does not fall on an element, as for example with a 2×2 or a 2×3 matrix. When it is important to indicate the origin in such a case, some symbol such as a plus sign can be introduced; but very often the origin of indexing is clear to the reader and no special indication is necessary. For programming purposes the element (0,0) or (1,1) may still have to be in the top left or bottom left corner; this shift in indexing is left to the reader to cope with.

Noise Reduction by Smoothing

By substituting the mean of several adjacent values for the measured value, we reduce any noise fluctuations that are more or less independent from one value to the next. For example, if nine completely independent random numbers are averaged, the standard deviation of the average will be reduced to about one-third. A convenient way to perform such averaging on a two-dimensional data array is to replace each value f_{ij} by the mean of the nine values centered on the location (i, j). The signal-to-noise ratio would then improve by a factor of three if the signal were unaffected by the averaging, but of course the signal is smoothed to some degree. As with most convolution operations, the peak signal is likely to be reduced, troughs are likely to be filled in, and the widths of peaks are likely to be broadened.

To illustrate the smoothing effect an 8×8 array, part (a) of Fig. 8-1 has been constructed from the expression

$$ f(x, y) = 65G \left(\frac{x - 2.5}{2}, \frac{y - 6.5}{2} \right) $$

$$ + 132G \left(\frac{x - 4.5}{3}, \frac{y - 2.5}{2} \right) + 33G \left(\frac{x - 6}{2.5}, \frac{y - 6}{2.5} \right), $$

where $G(x, y) = \exp[-\pi(x^2 + y^2)]$. With this notation $AG[(x - X)/W_x, (y - Y)/W_y]$ represents a Gaussian hump of peak height A, centered at (X, Y) and having equivalent widths W_x and W_y in the x- and y-directions. Calling this array $[F]$, we now construct the array

$$ [H] = \frac{1}{9} \begin{bmatrix} 1 & 1 & 1 \\ 1 & 1 & 1 \\ 1 & 1 & 1 \end{bmatrix} ** [F], $$

which is shown in Fig. 8-1(b). Thus the value 24 in the top left-hand corner is calculated as $(2 + 9 + 9 + 9 + 44 + 44 + 9 + 44 + 44)/9$. The contours, which are easily sketched by hand with a little practice, show the noticeable widening of the peaks and the reduction of peak height. Since the original widths of the peaks were not all the same, the changes produced vary in the ways that would be expected. If random noise were added to this example, the contours would become wavy but, after smoothing, would be less so.

Clearly, noise could be further reduced. For example, 25-element averaging would give up to a factor-of-5 reduction, but the original signal would suffer more degradation in the form of reduced peak values and expanded widths. The art of digital smoothing for the purpose of signal-to-noise improvement is clearly to find a compromise between noise reduction and signal degradation. There is no unique procedure. The best thing to do depends on the nature of the noise, which is not always uncorrelated from sample to sample, often does not have zero mean, and ought to be carefully studied in each practical case. Nasty sorts of noise, such as the sample errors that arise when the sampling point is

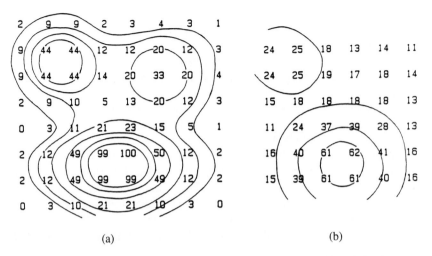

Figure 8-1   (a). An 8 × 8 array created artificially from a formula, with some hand-drawn contours at levels 5, 10, 20, 40, 60, 80, (b) the same array smoothed by nine-element averaging.

displaced spatially from its nominal location, either randomly or systematically, cannot be ignored. A good way to look for surprises of this type is to record measurements on simple known signals such as spatial sinusoids of different periods, strengths, and background levels. In observational contexts where the signal is not under the observer's control, it may still be possible to take measurements in the absence of signal. Sometimes, as with photon noise or shot noise in electronic devices, when there is no signal there is no noise; in other cases when there is no signal there is receiver noise.

Effect of Smoothing

When an isolated hump is smoothed, its peak value goes down and the breadth increases in such a way that the volume is conserved. (That is, of course, provided the smoothing array sums to unity; with a sum Σ the volume of the smoothed hump is magnified by Σ.) Since variance is additive under convolution, the variance of the smoothed hump will increase by $\frac{2}{3}$ under nine-element averaging because the variance of that smoothing operator is $\frac{2}{3}$, both horizontally and vertically. To some degree of approximation, the smoothed hump will be widened horizontally by a factor equal to the ratio of the new to the old standard deviation and vertically by a similar factor. When the volume is conserved, the peak value will go down approximately by the product of these two broadening factors.

Consider an example in one dimension. An isolated hump represented by the binomial sequence $f_i = \{1\ 4\ 6\ 4\ 1\}$ has variance $\sigma_f^2 = (1 \times 2^2 + 4 \times 1^2 + 6 \times 0 + 4 \times 1^2 + 1 \times 2^2)/(1 + 4 + 6 + 4 + 1) = 1$, while the sequence $g_i = \{1\ 1\ 1\}$ has variance $\sigma_g^2 = 2/3$. The sequence $f_i * g_i = \{1\ 5\ 11\ 14\ 11\ 5\ 1\}$ has variance $(1 \times 3^2 + 5 \times 2^2 + 11 \times 1^2 + 14 \times 0 + 11 \times 1^2 + 5 \times 2^2 + 1 \times 3^2)/48 = 1\frac{2}{3}$, which equals $\sigma_f^2 + \sigma_g^2$. Thus, after convolution the standard deviation which was unity has increased to $1.291 = \sqrt{5/3}$. If we were to take running means by convolution with $\frac{1}{3}\{1\ 1\ 1\}$, instead of taking running sums as above, then instead of a central value $14 = 4 + 6 + 4$ we would get 14/3=4.67. This is less than the original peak value of 6 by 1.286, essentially the same as the width-increase factor 1.291. Thus the bulk of the smoothing effect is deducible from the variance σ_g^2 alone. Higher-order moments would only add refinement.

Now compare the effect of plain nine-element averaging with smoothing by means of the nine-element binomial array

$$\frac{1}{16}\begin{bmatrix} 1 & 2 & 1 \\ 2 & 4 & 2 \\ 1 & 2 & 1 \end{bmatrix},$$

whose variance in each direction is $\frac{1}{2}$, somewhat less than the previous value of $\frac{2}{3}$. In this case an isolated hump is not broadened so much by smoothing and the peak height is not reduced so much. The outcome, shown in Fig. 8-2, offers the opportunity to verify the broadening effects numerically. From the equivalent widths W_x and W_y specified for the three humps in the earlier expression for $f(x, y)$ one can estimate the original variances from $\sigma^2 = W^2/2\pi$, which relates standard deviation to equivalent width, and which is exact for a normal distribution. Then the smoothed variances are obtained by adding $\frac{1}{2}$.

The discussion suggests that the peaks will be broadened by a factor $\sqrt{\sigma^2 + \frac{1}{2}}/\sigma$. Test this by taking the ratio of the widths to half level, easily determinable from Figs. 8-1(a) and 8-2. Numerical experiments of this type are to be encouraged for the purpose of developing intuitive feeling for actual data sets whose statistical properties are not necessarily known, and may be unknowable. The acts of judgment that enter into the setting of acceptance criteria for digital smoothing procedures are forced directly on the attention by numerical experiment on actual data.

NONCONVOLUTIONAL SMOOTHING

At this stage the plan is to expound the further riches of convolution, but it is worthwhile to pause and reflect that data may be smoothed in other ways. For example, if f_{av} is the average value of $f(x, y)$ over some region, then $\arctan\{\alpha[f(x, y) - f_{av}]\}$, where α is a choosable constant, will be smoother than $f(x, y)$ in almost any sense of the word smooth. It is easy to compute, too. Of course, the range will be hard-limited to $[-\frac{1}{2}\pi, \frac{1}{2}\pi/2]$, but softer functions such as $[f(x, y) - f_{av}]^{1/4} \operatorname{sgn}[f(x, y) - f_{av}]$ and others with logarithmic growth can be formulated. The theory of such smoothing operations is different, and in particular is not expressible in terms of a transfer function; in other words, a sinusoidal component of the original function to be smoothed is not reduced by a unique factor characteristic of the frequency of that sinusoid. This is simply a consequence of applying an operation that is not linear. The absence of clean theory is apt to remove discussion from the classroom; that does not mean that the medicine is not effective. It can be tried.

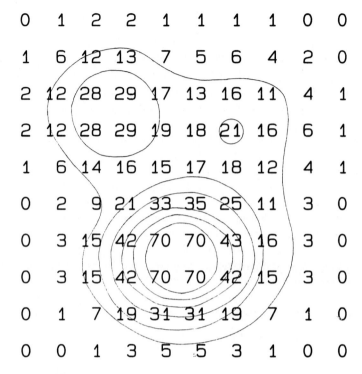

0	1	2	2	1	1	1	1	0	0
1	6	12	13	7	5	6	4	2	0
2	12	28	29	17	13	16	11	4	1
2	12	28	29	19	18	21	16	6	1
1	6	14	16	15	17	18	12	4	1
0	2	9	21	33	35	25	11	3	0
0	3	15	42	70	70	43	16	3	0
0	3	15	42	70	70	42	15	3	0
0	1	7	19	31	31	19	7	1	0
0	0	1	3	5	5	3	1	0	0

Figure 8-2 The previous 8×8 array smoothed by the nine-element binomial array. A single-layer fringe of extrapolated zeros was added before smoothing.

Median Smoothing

Instead of taking the arithmetic mean of the nine values centered at a point, one can take the median. This will consume more time, because the values have to be sorted, but the result of median smoothing is rather interesting. Irregular fluctuations are reduced, much as with ordinary averaging, but sharp edges tend to be preserved rather than converted to ramps. Pixels with erratic values, perhaps due to sporadic noise, are ignored rather than given proportional representation. Therefore there is a possibility that some class of images that you plan to smooth may benefit from median smoothing. On the other hand, some unfamiliar effects set in. Groups of up to four bright pixels on a dark background in a 3×3 block will be eliminated regardless of their brightness. Mass is not conserved; consequently, an area that depends on bright stippling for its tone will change its shade. Corners of bright areas will be cut. The operator is nonlinear, is not describable by a transfer function, and thus does not filter in the sense of passing or discriminating against wanted or unwanted frequencies. In the absence of familiar filter theory, median smoothing is best evaluated by test and judged accordingly. Instruments for general use by customers can be provided with options for the user to choose.

TREND REDUCTION

A satellite photograph of the ground contains a background trend resulting from the slow variation in angle of incidence of sunlight across the scene. Similarly, a flash photograph at close quarters contains a centrally symmetric trend resulting from the variation in distance of the flashbulb from different parts of the illuminated object. For these and many other reasons images contain slowly varying trends that are unwanted and may need to be reduced.

Image smoothing provides a technique for trend reduction as follows. First subject the image to heavy smoothing so that mostly the trend, but little of the desired detail, remains. Then subtract this outcome from the original. This technique can be practiced on photographs by analogue techniques and has in fact reached an advanced level. In digital form, starting from an image $[F]$, we smooth with a broad array $[S]$ which sums to unity, to obtain $[S] ** [F]$, and then subtract from $[F]$ to obtain $[F] - [S] ** [F]$. The two operations can be combined into one. For example, suppose that a 25-element array is extensive enough to smooth out most of the informative detail in an image (although usually a larger spread would be needed). Then the one-stage trend-reducing array would be

$$\begin{bmatrix} -0.04 & -0.04 & -0.04 & -0.04 & -0.04 \\ -0.04 & -0.04 & -0.04 & -0.04 & -0.04 \\ -0.04 & -0.04 & \mathbf{0.96} & -0.04 & -0.04 \\ -0.04 & -0.04 & -0.04 & -0.04 & -0.04 \\ -0.04 & -0.04 & -0.04 & -0.04 & -0.04 \end{bmatrix}.$$

Each element is $-\frac{1}{25}$ except the central element, which is $1 - \frac{1}{25}$.

The design of such a trend-reducing array involves judgment as to how large the array should or can be made, and whether it is advantageous to taper the coefficients. Thus, the 25-element binomial array would be implemented with

$$\frac{1}{256} \begin{bmatrix} -1 & -4 & -6 & -4 & -1 \\ -4 & -16 & -24 & -16 & -4 \\ -6 & -24 & \mathbf{220} & -24 & -6 \\ -4 & -16 & -24 & -16 & -4 \\ -1 & -4 & -6 & -4 & -1 \end{bmatrix}.$$

The different impacts of these two arrays would lie mainly in the fact that the first has a larger variance and thus is more effective in establishing a signal-free trend. But they also act differently on the Fourier components of the image and can be compared as to their transfer functions; and yet a choice cannot be made on the basis of transfer functions alone. It is true that the 25-element running average transmits high spatial frequencies that the binomial array cuts down, but the image we are interested in may have little content at those spatial frequencies. Direct numerical tests on a typical image may provide information that is useful in making the judgment.

Since the operations described here are convolution, it may be that computing time will be saved by using transform methods such as the fast Fourier or fast Hartley transforms, but not necessarily. One of the arrays is commonly much smaller than the other.

Thus, 25 multiplies per point may be fine, and, if 25-element averaging is used, the multiplies disappear and only shifting and adding remains.

Sharpening is the other operation mentioned above that is opposed to smoothing.

SHARPENING

Consider the trend-reducing formula $[F] - [S] ** [F]$ under circumstances where the smoothing array $[S]$ is not broad but on the contrary is rather narrow. As an extreme, suppose that

$$[S] = \frac{1}{9} \begin{bmatrix} 1 & 1 & 1 \\ 1 & \mathbf{1} & 1 \\ 1 & 1 & 1 \end{bmatrix}.$$

Then the operator

$$\frac{1}{9} \begin{bmatrix} -1 & -1 & -1 \\ -1 & \mathbf{8} & -1 \\ -1 & -1 & -1 \end{bmatrix}$$

will not remove trends. What will it do? Since the sum of the coefficients is zero, any uniform background will be removed, and broad features will be reduced, as with trend reduction. But compact features will be retained to greater or lesser extent, and surrounded with a negative halo. If this entity $[F] - [S] ** [F]$ (or a fraction of it) is now added back into the original array $[F]$, to form say $2[F] - [S] ** [F]$, then the compact features will be emphasized. The general effect is described as *sharpening*. The very simple one-stage array that does this is

$$\frac{1}{9} \begin{bmatrix} -1 & -1 & -1 \\ -1 & \mathbf{17} & -1 \\ -1 & -1 & -1 \end{bmatrix}.$$

Any image can be subjected to sharpening of this kind, and the desirability of the outcome can be judged; try it, you may or may not like it. As an example, the following matrix, rounded to integers, shows the result of so sharpening the 8 × 8 array of Fig. 8-1(a). Hand contouring as in Fig. 8-3 will show that, compared to Fig. 8-2, the peaks are narrowed and heightened, the opposite of smoothing.

64	63	6	11	26	13
64	63	9	23	48	26
3	2	−8	8	22	11
−5	−2	5	8	2	−3
8	59	137	138	59	8
9	59	137	137	58	8

What if this sharpening operator is applied to the smoothed version; will the original array be recoverable? We are in a good position for empirical investigation, since applica-

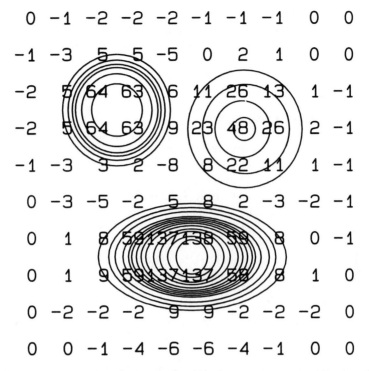

```
  0  −1  −2  −2  −2  −1  −1  −1   0   0

 −1  −3   5   5  −5   0   2   1   0   0

 −2   5  64  63   6  11  26  13   1  −1

 −2   5  64  63   9  23  48  26   2  −1

 −1  −3   3   2  −8   8  22  11   1  −1

  0  −3  −5  −2   5   8   2  −3  −2  −1

  0   1   8  59 137 138  59   8   0  −1

  0   1   9  59 137 137  59   8   1   0

  0  −2  −2  −2   9   9  −2  −2  −2   0

  0   0  −1  −4  −6  −6  −4  −1   0   0
```

Figure 8-3   Partial restoration is achieved by applying the sharpening operator to the
binomial smoothing operator. Alternative proposals may be tested by convolving the two
operators.

tion of two convolution operators in succession reduces to convolution with one equivalent
operator. Thus we find that

$$\frac{1}{16}\begin{bmatrix} 1 & 2 & 1 \\ 2 & 4 & 2 \\ 1 & 2 & 1 \end{bmatrix} ** \frac{1}{9}\begin{bmatrix} -1 & -1 & -1 \\ -1 & 17 & -1 \\ -1 & -1 & -1 \end{bmatrix} ** = \frac{1}{9}\begin{bmatrix} -0.1 & -0.3 & -0.4 & -0.3 & -0.1 \\ -0.3 & 1. & 2.7 & 1. & -0.3 \\ -0.4 & 2.7 & 6.2 & 2.7 & -0.4 \\ -0.3 & 1. & 2.7 & 1. & -0.3 \\ -0.1 & -0.3 & -0.4 & -0.3 & -0.1 \end{bmatrix} **$$

The overall equivalent operator, on the right, is substantially more compact than the binomial smoothing operator, the central value amounting to 39 per cent of the whole, compared to 25 per cent for the smoothing array. Consequently, partial recovery of the original
detail will be achieved by operating on the smoothed version with the simple sharpening
operator whose central value is 17/9.

The purpose of these introductory comments on sharpening is to show that increasing peak height and more or less isotropic narrowing of peaks will result when the convolving operator has unit mass and consists of a positive peak surrounded by a negative ring.

The detailed character of the operator should also be tailored to the nature of the original smoothing, to take account of the sampling interval, and to allow for measurement error. Physical measurement in general necessarily involves smoothing as well as error, consequently sharpening procedures are practiced in a wide variety of disciplines. The topic is returned to later under the title of *restoration.*

WHAT IS A DIGITAL FILTER?

Smoothing, trend reduction, and sharpening, when performed by convolution, are analogous to the operation of an electrical filter that accepts an input waveform and delivers an output waveform, each Fourier component of which has been modified in accordance with some frequency-dependent transfer function. Such filters, which may be constructed of inductances and capacitors and segments of electrical transmission line, or of mechanical resonators such as tuning forks or quartz crystals, are widely used in the telephone system. Acoustic filters (mufflers) and mechanical filters (vibration isolators) are similarly associated with transfer functions. The effect of digital convolution is also describable in terms of a transfer function as discussed below. Thus a matrix, or array of coefficients, can be applied to operate like a filter even though a matrix is not a filter in the physical sense; it has become customary, though, to refer to the matrix operation as a digital filter. The matrix itself is analogous to the impulse response of a filter. Other methods of numerical filtering, which are nonlinear (median smoothing for example), are not equivalent to convolution with any matrix and do not have an associated transfer function; in such cases the set of operations may still be called a digital filter. Some authors write as if a matrix is a digital filter, but although convolution with respect to continuous independent variables performs filtering, we do not call the convolving function a filter. On the whole, if a thing is a matrix it is better to call it a matrix than a digital filter.

GUARD ZONE

If a 16×16 array is to be smoothed with a 3×3 operator, then the operator cannot be centered at a point on the edge of the large array without losing contact. One procedure is to settle for a 14×14 output array. For various purposes it may be appropriate to surround the output by a fringe of zeros to bring the output size back to 16×16; this will also restore the indexing of the center. Another approach is to first surround the 16×16 array by zeros; then the output will be 16×16; the fringing values may then sometimes be just what is wanted, but in other circumstances may be meaningless. Yet another idea is to surround the data with artificial values that have more plausibility as data than zero values would have. One might simply repeat the edge values, or extrapolate the rows and columns in some simple way. People rightly feel nervous when making up data, but adoption of zeros explicitly, or invisibly by implication, is even less defensible. Although the fringe region is narrow, and may seem negligible, one often does not wish to throw it away. Sometimes the operating array is not small at all and may occupy a substantial area of the data array. In the case of contour diagrams, one does not want the contours that reach the edges to behave misleadingly. This is a rather interesting practical topic involving subjective judgment according to the case. The contour diagram provides a good medium for judging acceptability.

TRANSFORM ASPECT OF SMOOTHING OPERATOR

Smoothing by convolution is sometimes referred to in its appplication to time series as the *moving-average method* as distinguished from the Fourier method, which transforms, cuts the high frequencies, and retransforms. A disadvantage cited for the moving average, or running mean, of length N is that the first and last of the N values are not handled properly, while various advantages and disadvantages are cited for the Fourier method. But in fact the strict correspondence between operations in the function and transform domains makes it certain that the Fourier method will produce end effects and that they will be the same as for running means when the weights used for the running mean and the shape of the effective transform domain filter correspond. We now consider the filter characteristic, or transfer function, associated with the various impulse responses.

Nonconvolutional methods of smoothing do not share this duality and do indeed have distinguishing advantages and disadvantages. For example, smoothing by least-mean-squares fitting has an advantage in dealing readily with nonuniformly spaced data, while a restrictive feature is that something has to be known about the answer in advance so that goodness of fit can be judged.

The convolution theorem allows any smoothing operator to be described in terms of the transfer function that modifies each Fourier component. Take as an example the nine-element averaging operator. The Fourier transform $T(u, v)$ can be written by inspection as $(\text{sinc}3u \ \text{sinc} 3v) ** {}^2\text{III}(u, v)$. Along the u-axis this expression reduces to $T(u, 0) = (\sin 3\pi u)/3 \sin \pi u$. The shortest period representable by samples at unit spacing is 2. Thus, the highest frequency of interest is 0.5 cycle per sample interval. A graph of $T(u, 0)$ for $|u| < 0.5$ shows that low spatial frequencies are favored, but response falls completely to zero when u reaches 0.33. Response increases in magnitude as u increases further, but with negative sign, reaching -0.33 at $|u| = 0.5$. This alarming behavior for a low-pass filter can be very objectionable because of the sign reversal, but on the other hand may not be harmful if the image content in the spectral range 0.33 to 0.5 is negligible. This can be found out by experiment. In the event that such spurious response is unacceptable, the nine-element binomial array $(1/9)[1 \ 2 \ 2 \ / \ 2 \ 4 \ 2 \ /1 \ 2 \ 1]$ can be considered. In that case $T(u, 0) = 0.5(1 + \cos 2\pi u)$, a characteristic that falls nicely to zero at $|u| = 0.5$. However, this more elaborate characteristic passes high frequencies, if they are present, in greater strength than does simple nine-element averaging and so cannot be recommended as always giving the best smoothing available from a nine-element array.

Every smoothing operator has a spectral transfer function $T(u, v)$ which, if it is not apparent by inspection, may be expressed as follows. If the operator is

$$\frac{1}{\Sigma} \begin{bmatrix} a & b & c \\ d & e & f \\ g & h & j \end{bmatrix},$$

where Σ is the sum of the coefficients, then

$$T(u, v) = [ae^{i2\pi(-u+v)} + be^{i2\pi v} + ce^{i2\pi(u+v)} + de^{-i2\pi u} + e + fe^{i2\pi u}$$
$$+ ge^{i2\pi(-u-v)} + he^{-i2\pi v} + je^{i2\pi(u-v)}]/\Sigma.$$

This formula works for any smoothing array that fits within the 3×3 frame, but of course symmetry considerations usually yield much simplification.

FINITE IMPULSE RESPONSE (FIR)

The image operations under discussion here can be related to terminology in use with the filtering of time series. For example, FIR filter, where FIR stands for finite impulse response, is a term that could be applied to all the two-dimensional operations discussed above, or to any convolutional operation where the impulse response has finite discrete spatial support. Of course, in practice all impulse responses with discrete support are of finite extent.

The term IIR filter, where IIR stands for infinite impulse response, can be applied to an impulse-response matrix that spreads out in space indefinitely (no term of the impulse response has to be infinite). A theoretical IIR example is offered by samples of $^2\text{sinc}(x/X, y/Y)$. Theory may write these samples as a matrix of infinite order, but not practice; consequently, the FIR/IIR dichotomy is not useful for numerical analysis, although the distinction is a familiar one in the theory of control systems.

The theoretical impulse response

$$\sum_{i=-\infty}^{\infty} \sum_{i=-\infty}^{\infty} {}^2\text{sinc}\left(\frac{i}{X}, \frac{j}{Y}\right) {}^2\delta(x - i, y - j)$$

corresponds to a transfer function

$$T(u, v) = XY \, {}^2\text{rect}(Xu, Yv) ** {}^2\text{III}(u, v)$$

which, for $|X| < 1$ and $|Y| < 1$, consists of isolated islands in the (u, v)-plane. The sharpness of cutoff of these bandpass islands in the continuous-frequency domain, unattainable in practice, is the counterpart of the infinite spread of samples in the space domain that gives the IIR filter its name. If, instead of working with continuous independent variables, we take the view that the impulse response is a discrete sample set represented by a matrix $[a_{ij}]$, where $[a_{ij}] = {}^2\text{sinc}(i/X, j/Y)$, using the same example, then the corresponding transfer function is doubly periodic and also representable by a matrix of samples. To compute these samples one is obliged to truncate the impulse-response samples to a finite set and use the discrete Fourier transform, thus reverting to the FIR case.

For theoretical purposes the doubly infinite set of impulse response samples $[a_{ij}]$ can be rewritten as the two-dimensional z-transform

$$H(z_1, z_2) = \sum_{i=-\infty}^{\infty} \sum_{j=-\infty}^{\infty} a_{ij} z_1^{-i} z_2^{-j},$$

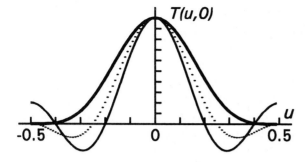

Figure 8-4 Transfer function $T(u, 0)$ for the 5 × 5 binomial array (———) and for the 5 × 5 array of constant coefficients 1/25 (———). Spatial frequencies greater than 0.5 are not represented by samples taken at unit interval in the (x, y)-plane; consequently, the filtering action is fully specified for $|u|$ and $|v| \leq 0.5$.

where the transform variables z_1 and z_2 are continuous variables and complex. Properties of the two-dimensional z-transform, which have been presented by Lim (1990), are interesting when $H(z_1, z_2)$ can be factored. In imaging problems where the coefficients a_{ij} are data samples, factorization never happens, and direct manipulation of the matrix $[a_{ij}]$ is customary, rather than manipulation of a complex two-variable polynomial.

SPECIAL FILTERS

Low-pass Filtering

To illustrate the transfer function $T(u, v)$ associated with a specific filtering operation we consider the 5 × 5 binomial array introduced earlier. The cross section of $T(u, v)$ along the u-axis is given by

$$T(u, 0) = \tfrac{3}{8} + \tfrac{1}{2}\cos 2\pi u + \tfrac{1}{8}\cos 4\pi u$$

as illustrated in Fig. 8-4. This transfer function exhibits a smooth descent to its cutoff at $u = 0.5$. For comparison, the thin line shows the transfer function $\tfrac{1}{5} + \tfrac{2}{5}\cos 2\pi u + \tfrac{2}{5}\cos 4\pi u$ for an impulse response consisting of a 5 × 5 array of 1s, all divided by 25.

For the same span of image elements entering into the convolution the band of low frequencies passed is (in a sense) narrower; this feature is paid for by phase reversal of a substantial band. The choice between these alternatives depends on the spectral content of the image to be filtered. It is possible, of course, that there is little spectral content in the phase-reversed band. Middle ground is provided by the samples of $^2\text{rect}(x/4, y/4)$, a 5 × 5 array with a core of nine 1s surrounded by a margin of values equal to $\tfrac{1}{2}$ on the sides and $\tfrac{1}{4}$ at the corners. The transfer function $\tfrac{1}{4} + \tfrac{1}{2}\cos 2\pi u + \tfrac{1}{4}\cos 2\pi 2u$ is shown dotted in the figure. This transfer function is not identical with sinc $4u$, the cross section of sinc $4u$ sinc $4v$ along the u-axis, because the transform of the *samples* of $^2\text{rect}(x/4, y/4)$ is not the same as the transform of $^2\text{rect}(x/4, y/4)$.

None of the transfer functions of Fig. 8-4 exhibits a very sharp roll-off. Suppose that it is essential to do better and that at the same time the descent to cutoff at $u = 0.5$ is to be smooth and without phase reversal. To find out what we would have to pay for this improved performance we could sketch in a desired transfer function on Fig. 8-4, take its

Fourier transform, and note the unit-interval samples in the object domain. There would be an infinite number of such samples; by truncating and retransforming we could find out whether our desires could reasonably be met.

High-pass Filtering

High-pass filtering (which is often the same as trend removal) is effected by subtracting the low-pass filtered image from the original, and therefore does not introduce any significant new concepts. The step of subtraction need not be carried out separately but may be incorporated into the convolving function by taking the low-pass convolving coefficients (normalized so as to sum to unity) and subtracting unity from the central coefficient. The sum of the coefficients will now be zero, as it should be, since spatial dc (a uniform background) is to be suppressed.

Band-stop Filtering

As an example of the tension between theoretical filtering and reasonable practice suppose that a signal is contaminated with an unwanted addition of known frequency whose amplitude and phase are both subject to slow drift or even jumps. This unwanted noise might be power-supply hum that penetrates the screen of a magnetic-resonance imaging room, where image acquisition is very sensitive to stray magnetic fields, or it might be a 24-hour diurnal variation in a meteorological sensor that is supposed to be temperature independent but is not completely so. What we need is a notch filter that will take out the unwanted frequency and a suitably narrow band around it of width as called for by the statistics of the amplitude and phase drifts. The filter coefficients resulting from the transform approach will alternate in sign with the period of the unwanted signal and will build up and decay in amplitude over a more or less long sequence depending on how narrow a notch is asked for.

An opposite approach starts with the simplest operation that will give some relief, tests it, and if necessary goes a step further. Suppose that interference or some instrumental effect introduces an unwanted spatial frequency into an image and that the defect is manifested by numerous parallel striations. If the striations are not numerous (the unwanted spatial frequency is low), then low-stop filtering may be appropriate (though, of course, wanted frequencies in the neighborhood will also be affected). But if the unwanted frequency is imbedded in the wanted band, band-stop filtering will be needed. The simple operation in this case is to take the data, shift it half a period in the direction perpendicular to the striations, and add it to itself. If the interference were strictly periodic, this would cancel it exactly, except at the edges of the record; in effect we have convolved with two coefficients only (the simplest operation). If the interference drifts slowly in amplitude, then the cancellation is not perfect, but at least the procedure adapts with the amplitude drift rather than invoking a prearranged stop-band whose width is based on previous mean statistics of drift (which may be hard to get).

The effective filter characteristic of the shift-and-add operation is easy to derive; it is simply a cosine function which has a null at the unwanted frequency. It is far from having the appearance of a notch. The simplicity of the operation is paid for by a nonflat response to the wanted frequencies; some damage will have been done at frequencies near the null. At higher frequencies there will be phase reversal of any image frequencies that are present, and the effect of this may be objectionable. Nevertheless the simple shift-and-add technique is very useful; the rule of thumb for judging whether it will be acceptable is to ask whether the wanted image will be significantly degraded if averaged with its shifted copy. If equalization proves necessary after trial with the actual data in question, a further single-stage operation can be considered.

Understanding the shift-and-add operation enables us to see how to avoid the phase reversal—just apply the same operation twice. To take an analogy from a one-dimensional time series sampled at one-second intervals, an unwanted frequency of 0.25 Hz (period 4 s) will be removed by convolution with $\{\frac{1}{2} \ 0 \ \frac{1}{2}\}$. Convolution with $\{\frac{1}{4} \ 0 \ \frac{1}{2} \ 0 \ \frac{1}{4}\}$ will also remove the unwanted 0.25 Hz but will not reverse the phases in the band from 0.25 to 0.5 Hz. The full transfer function will be a squared cosine function with its first null at 0.25 Hz. Instead of squaring the cosine function, one can multiply two somewhat different cosines and place two adjacent nulls in the transfer function, thus hoping to widen the stop-band. The corresponding convolving sequence can be written down. For example, convolution with $\{\frac{1}{2} \ 0 \ 0 \ 0 \ 0 \ \frac{1}{2}\}$ will place a null at $f = \frac{1}{12}$ Hz while $\{\frac{1}{2} \ 0 \ 0 \ 0 \ 0 \ 0 \ \frac{1}{2}\}$ will place a null at $f = \frac{1}{14}$ Hz. The compound operator will be

$$\{\tfrac{1}{4} \ 0 \ 0 \ 0 \ 0 \ \tfrac{1}{4} \ \tfrac{1}{4} \ 0 \ 0 \ 0 \ 0 \ \tfrac{1}{4}\}.$$

This particular numerical example reminds us that the simple shift-and-add cancels not only the frequency whose semiperiod equals the shift but also the frequency three times larger.

Even the most elaborate convolution can be viewed as built up by shifting-and-adding, but with various coefficients applied. Filtering that aims at passing only a narrow band around 0.25 Hz is exemplified by the 21-element array

$$\{-1 \ 0 \ 10 \ 0 \ -45 \ 0 \ 120 \ 0 \ -210 \ 0 \ \mathbf{252} \ 0 \ -210 \ 0 \ 120 \ 0 \ -45 \ 0 \ 10 \ 0 \ -1\}/1024$$

which is derived from the 11-element binomial series $^{10}C_m$ by alternating the signs and interleaving zeros. The associated transfer function is

$$T(u, 0) = (252 - 420\cos 2\pi u + 240\cos 4\pi u - 90\cos 6\pi u + 20\cos 8\pi u - \cos 10\pi u.$$

The central value $T(0.25, 0) = 912/1024$.

Notch filtering that aims at suppresing only a narrow band around $u = 0.25$ would be effected by the corresponding band-stop array, whose central element would be $1024 - 252 = 772$. This array would not completely reject the frequency $u = 0.25$, where the value of the transfer function would be 112/1024. If, however, the central element is changed to $772 - 112 = 660$, the transmission is reduced to zero; the band-stop array becomes

$$\{1 \ 0 \ -10 \ 0 \ 45 \ 0 \ -120 \ 0 \ 210 \ 0 \ \mathbf{660} \ 0 \ 210 \ 0 \ -120 \ 0 \ 45 \ 0 \ -10 \ 0 \ 1\}/1024,$$

for which the transfer function is $(660 + 420\cos 2\pi u - 240\cos 4\pi u + 90\cos 6\pi u - 20\cos 8\pi u + 2\cos 10\pi u)/1024$. Narrower notches, of course, require longer sequences.

The value 660 does result in exact cancellation of any components at $u = 0.25$, but sometimes neighboring frequencies need to be suppressed, too. If reduction by 20 dB is satisfactory, then reducing the factor 660 by another 112 to a value 548 will approximately double the width of the band where the attenuation is 20 dB or more.

In two dimensions the arrays used for representing $T(u, 0)$ may have to be applied to other orientations. The elementary case of $\{\frac{1}{4}\ 0\ \frac{1}{2}\ 0\ \frac{1}{4}\}$ might appear in two dimensions as

$$
\begin{bmatrix} \frac{1}{4} & 0 & \frac{1}{2} & 0 & \frac{1}{4} \end{bmatrix} \quad \text{or} \quad
\begin{bmatrix}
0 & 0 & 0 & 0 & \frac{1}{4} \\
0 & 0 & \frac{1}{2} & 0 & 0 \\
\frac{1}{4} & 0 & 0 & 0 & 0
\end{bmatrix} \quad \text{or} \quad
\begin{bmatrix}
0 & 0 & 0 & 0 & \frac{1}{4} \\
0 & 0 & 0 & 0 & 0 \\
0 & 0 & \frac{1}{2} & 0 & 0 \\
\frac{1}{4} & 0 & 0 & 0 & 0
\end{bmatrix},
$$

matrices which act on orientations of $0°$, $27°$ and $45°$, respectively. To null out a sinusoid of frequency 0.25 oriented at other angles is possible but messy, and this rather simple technique with just three coefficients loses its attraction. However, this question of orientation on a digital map is fundamental; angles $\arctan(m/n)$, where m and n are integers, are the *only* angles at which a pair of coefficients may be placed. This is not to be struggled against. When a discrete map is Fourier analyzed, the only Fourier coefficients found will refer to this same limited set of directions; they will be sufficient to synthesize the map exactly.

In two dimensions image data may be acquired which, on presentation, show an unwanted low-amplitude wavy structure running like parallel dunes across the picture (Bracewell, 1963). If the phenomenon has gone and cannot be reobserved, the picture can be treated by shifting half a period in the appropriate direction and adding. This treatment will largely cancel the wavy interference and, if the period of the interference is short relative to the features of interest, this treatment may be sufficient. On the other hand, if the period is relatively long, the wanted picture will be replicated on top of itself. The appearance will be like the echo seen on television screens in marginal areas, where an indirect ray from the transmitter to the receiver arrives later than the direct ray. Compensation for this kind of disturbance is dealt with systematically later under restoration, but the procedure can be mentioned here: shift the picture again and subtract it from itself. The outcome is that again there is a superposed (negative) replica, but further away and possibly out of the needed field of view; if not, shift and add, shift and subtract, and continue as necessary. An alternative procedure is to begin by shifting a full period and subtracting. As there are some subtle differences, depending on the character of the noise, this alternative should be kept in mind.

Stage-by-stage algorithms like these can be deduced from the transform formalism; they illustrate that there is more to the art of filtering than specifying a filter band shape and implementing it by Fourier transformation.

Digital Filter Design

The many specific digital processing operations introduced above probably cover the majority of digital filtering operations carried out in daily practice. With the exception of the briefly mentioned nonlinear smoothing, all represent convolution and therefore, in terms of the convolution theorem, are expressible by a transfer function. The discrete convolution operations are referred to as *digital filters*. In principle, to design a digital filter whose transfer function is specified, take the Fourier transform to arrive at an array of coefficients which, if convolved with the data, will perform the desired modification of the spectral composition. In practice, often the array of coefficients obtained is too large to use and must be truncated or tapered off, after which it is straightforward to find out what filter function is actually going to result. With this second-best filter characteristic in hand, it will be apparent whether aliasing will occur, given the expected frequency content of the data. Digital filter design thus contains a significant infusion of art.

DENSIFYING

In order to draw smoother contours, to get smoother transitions between grey-level pixels, to gain compatibility with other data sets, or for other reasons, we may wish to convert an $N \times N$ data array to an array with a larger number of elements—for example, by supplying an extra value at the center of each existing cell. An obvious way of doing this is to use the average of the four values at the corners of that cell. In effect that means convolution with $\frac{1}{4}\begin{bmatrix} 1 & 1 \\ 1 & 1 \end{bmatrix}$, a very practical procedure that might be adequate in given circumstances. In one dimension that would be like linear midpoint interpolation between the sample points. The effect is to double the density of array elements. To quadruple the number of elements, convolve first with $\frac{1}{2}[1 \ \ 1]$ to provide midpoint values in each row and then convolve with $\frac{1}{2}\begin{bmatrix} 1 \\ 1 \end{bmatrix}$ to provide values on the new interleaved rows. Methods of interpolation where accuracy is a consideration were discussed in the previous chapter. Densifying an array simply to provide more values is often done where no particular importance attaches to accuracy and even elementary interpolation may be overelaborate; the density can be doubled simply by copying each value half a diagonal unit to the northeast. In those areas of an image where there is little spatial gradient, crude densification is perfectly good.

Enlargement

If we wish to enlarge an image $f(x, y)$ by a factor K, we use $f(x/K, y/K)$. Of course, if there is to be no displacement accompanying the enlargement, the origin of (x, y) will be chosen in the middle of the image. If x and y are restricted to integer values, then x/K may no longer be an integer. For example, if $K = 2$, meaning that we wish to enlarge by a factor 2, then at $(10, 10)$ we put the value originally at $(5, 5)$ and at $(8, 10)$ we put the value originally at $(4, 5)$, but what do we put at $(9, 10)$? There was no value originally at $(4.5, 5)$. One way of dealing with this is by prior densification, as just discussed. Another procedure is to leave the openings but compensate for the lower density by multiplying the values by K^2. Later, if some smoothing operation is applied, the openings will be filled up more or less acceptably.

THE ARBITRARY OPERATOR

Sometimes a convolving operator does not smooth, sharpen, or perform any of the simple operations, but we may wish to describe the effect it will have, and conversely, we may wish to tailor an operator to do something of interest. For example, what would be the effect of convolving with

$$\begin{bmatrix} 0 & 1 & 2 \\ 3 & 4 & 5 \\ 5 & 0 & 0 \end{bmatrix}?$$

The first thing we can say is that the mean level of the operand will be amplified by 20. Because the operator does not have its centroid at its origin (the x-moment is -5 and the y-moment is -2), features will be shifted by an amount $(-0.25, -0.1)$. The second moments are 25 and 8; therefore, features will be smoothed somewhat in the x-direction and rather less in the y-direction. The product moment is 7; therefore, round features will be sqeezed a little into the first and third quadrants.

One could say that this operator is sensitive to the derivatives $\partial/\partial x$, $\partial/\partial y$, $\partial^2/\partial x^2$, $\partial^2/\partial y^2$, and $\partial^2/\partial x\,\partial y$ and has taken these properties into account in various degrees. If we are tailoring an operator that is not to shift features, we arrange that the first moments are zero; if we wish to introduce smoothing, we arrange that the principal second moments are negative; for sharpening we make them positive. To locate differential features we make the sum of the elements zero, and to avoid amplification we make the sum unity. Since there are only six nonzero elements in the example, one amplification factor and five derivatives exhaust the descriptive possibilties.

To quantify the situation, consider the general 3×3 array

$$\begin{bmatrix} a_{-1,1} & a_{0,1} & a_{1,1} \\ a_{-1,0} & a_{0,0} & a_{1,0} \\ a_{-1,-1} & a_{0,-1} & a_{1,-1} \end{bmatrix}$$

The Fourier transform of the set of nine delta functions with coefficients as shown is

$$\sum a_{kl} e^{-i2\pi(ku+lv)}.$$

The sum Σa_{kl} of all the coefficients determines the magnification of the mean value of the operand, the "uniform background," or the "spatial dc." The treatment of components of spatial frequencies other than zero is deducible in terms of the coefficients, as follows. Take the component with spatial frequency $u = 0.5$, $v = 0$. It will be multiplied by $(a_{-1,1} + a_{-1,0} + a_{-1,1})e^{i\pi} + (a_{1,1} + a_{1,0} + a_{1,-1})e^{-i\pi} = Ae^{i\pi} + Be^{-i\pi}$. If $B = A$, the factor is $-2A$: the component will be phase inverted and amplified by $2A$. If $B = -A$, the component will be phase shifted by one-quarter of a period also. In the many cases where the array has twofold rotational symmetry, the transform is real and simplifies to

$$a_{0,0} + 2a_{0,1}\cos v + 2a_{1,0}\cos u + 2a_{1,1}\cos(u+v) + 2a_{1,-1}\cos(u-v).$$

DERIVATIVES

The operation $\partial/\partial x$ may be "approximated" by convolution with $[1 \ -1]$, but it is easier to picture the operation as cross-correlation with $[-1\ 1]$, because this is the pattern that can be copied onto a transparency and translated across the matrix to bring corresponding factors into register; the operators that follow are all cross-correlation operators. Both $[-1\ 1]$ and $\frac{1}{2}\begin{bmatrix} -1 & 1 \\ -1 & 1 \end{bmatrix}$ will have the desired effect, but in these two examples the result does not fall on an existing matrix point. By using

$$\tfrac{1}{2}[-1\ \mathbf{0}\ 1] \qquad \text{or} \qquad \tfrac{1}{6}\begin{bmatrix} -1 & 0 & 1 \\ -1 & \mathbf{0} & 1 \\ -1 & 0 & 1 \end{bmatrix} \qquad \text{or} \qquad \tfrac{1}{8}\begin{bmatrix} -1 & 0 & 1 \\ -2 & \mathbf{0} & 2 \\ -1 & 0 & 1 \end{bmatrix}$$

we avoid the indexing problem but introduce smoothing, whether wanted or not, as we see from the fact that $[-1\ 0\ 1] = [-1\ 1] * [1\ 1]$. Any desired smoothing can always be factored in.

Differencing in the 45° direction is produced by $\dfrac{1}{\sqrt{2}}\begin{bmatrix} 0 & 1 \\ -1 & 0 \end{bmatrix}$ and in the direction θ by

$$\frac{1}{2}\begin{bmatrix} -\operatorname{cas}(-\theta) & \operatorname{cas}\theta \\ -\operatorname{cas}\theta & \operatorname{cas}(-\theta) \end{bmatrix},$$

where $\operatorname{cas}\theta = \cos\theta + \sin\theta$.

Previously, the factor outside the matrix was the reciprocal of the matrix sum, because such normalization gave unit amplification. But the factor when the matrix sum is zero is chosen to give consistency with some standard. If the first difference $f(x+\frac{1}{2}, y) - f(x-\frac{1}{2}, y)$ is the standard, then $\frac{1}{2}[f(x+1, y) - f(x-1, y)]$ will need the factor $\frac{1}{2}$ shown, if the second expression is to give the same result as the first on a tilted plane surface such as $f(x, y) = x\cos\theta + y\sin\theta + c$, whose absolute gradient is 1.

Similar considerations apply to the second derivative $\partial^2/\partial x^2$, which is "approximated" by convolution with $[1\ -2\ 1] = [-1\ 1] * [-1\ 1]$. The mixed derivative $\partial^2/\partial x\partial y$ corresponds to convolution with

$$\begin{bmatrix} -1 & 1 \\ 1 & -1 \end{bmatrix} = [-1\ \ 1] ** \begin{bmatrix} 1 \\ -1 \end{bmatrix}$$

It is, of course, only a rough approximation to suppose that a first derivative is accessible via a pair of discretely spaced samples; one is really talking about a first difference. Still, if the spatial frequencies concerned are low enough, it does not make much difference, and it is common to see terms like "gradient detector" used in contexts where "first difference" would be precise.

The Sobel Gradient Operators

Consider the function $f(x, y) = 1 + x^2y$, $0 \le x \le X$, $0 \le y \le Y$, where x and y are integers. The gradient is given by

Table 8-1 True gradients and azimuth (left) as compared with results from Sobel gradient operators (right).

(x, y)	$(1, 1)$	$(2, 2)$	$(3, 3)$
$\partial f/\partial x$	2	8	18
$\partial f/\partial y$	1	4	9
G	2.24	8.94	20.12
$\tan\theta$	0.5	0.5	0.5
θ	26.6°	26.6°	26.6°

(x, y)	$(1, 1)$	$(2, 2)$	$(3, 3)$
G_x	2	8	18
G_y	5/3	14/3	29/3
$\sqrt{G_x^2 + G_y^2}$	2.60	9.26	20.43
G_y/G_x	0.83	0.58	0.54
$\arctan(G_y/G_x)$	40°	30°	28°

$$\mathbf{G} = \check{\mathbf{i}}\frac{\partial f}{\partial x} + \check{\mathbf{j}}\frac{\partial f}{\partial y},$$

where $\check{\mathbf{i}}$ and $\check{\mathbf{j}}$ are unit vectors in the x and y-directions, respectively. The magnitude of the gradient is

$$\sqrt{(\partial f/\partial x)^2 + (\partial f/\partial y)^2} = \sqrt{(2xy)^2 + x^4}$$

and the direction $\theta = \arctan(x/2y)$. For the three points (1, 1), (2, 2), and (3, 3) we have the results shown on the left side of Table 8-1.

The Sobel gradient operators, which are popular in computer graphics,

$$\frac{1}{6}\begin{bmatrix} 1 & 0 & -1 \\ 1 & \mathbf{0} & -1 \\ 1 & 1 & -1 \end{bmatrix} \quad \text{and} \quad \frac{1}{6}\begin{bmatrix} -1 & -1 & -1 \\ 0 & \mathbf{0} & 0 \\ 1 & 1 & 1 \end{bmatrix},$$

produce "gradient" values G_x and G_y, as shown. The corresponding gradient magnitudes $\sqrt{G_x^2 + G_y^2}$ and azimuth $\arctan(G_y/G_x)$ are shown on the right of the table. The poor agreement with the true values on the left, a consequence of anisotropic smoothing, is disquieting and argues in favor of conducting finite differencing on data whose smoothing is controlled separately.

Once a satisfactory smoothing operator is decided upon, there is no reason not to combine with the differencing operator by convolution before operating on the data. Examples of combined smoothing/differencing convolution operators based on equal smoothing in the cardinal directions are:

$$[1\ -1] * \begin{bmatrix} 1 & 1 \\ 1 & 1 \end{bmatrix} = \begin{bmatrix} 1 & 0 & -1 \\ 1 & 0 & -1 \end{bmatrix}$$

$$[1\ -1] * \begin{bmatrix} 1 & 1 & 1 \\ 1 & 1 & 1 \\ 1 & 1 & 1 \end{bmatrix} = \begin{bmatrix} 1 & 0 & 0 & -1 \\ 1 & 0 & 0 & -1 \\ 1 & 0 & 0 & -1 \end{bmatrix}$$

$$[1\ -1] * \begin{bmatrix} 1 & 2 & 1 \\ 2 & 4 & 2 \\ 1 & 2 & 1 \end{bmatrix} = \begin{bmatrix} 1 & 1 & -1 & -1 \\ 2 & 2 & -2 & -2 \\ 1 & 1 & -1 & -1 \end{bmatrix}.$$

These composite matrices, although not square, nevertheless apply even smoothing. Rotate 90° counterclockwise for differencing with respect to y. Diagonal differencing in the direction of the first quadrant (toward the NE) generates the following composite matrices.

$$\begin{bmatrix} 0 & -1 \\ 1 & 0 \end{bmatrix} * \begin{bmatrix} 1 & 1 \\ 1 & 1 \end{bmatrix} = \begin{bmatrix} 0 & 1 & 1 \\ -1 & 0 & 1 \\ -1 & -1 & 0 \end{bmatrix}$$

$$\begin{bmatrix} 0 & -1 \\ 1 & 0 \end{bmatrix} * \begin{bmatrix} 1 & 1 & 1 \\ 1 & 1 & 1 \\ 1 & 1 & 1 \end{bmatrix} = \begin{bmatrix} 0 & 1 & 1 & 1 \\ -1 & 0 & 0 & 1 \\ -1 & 0 & 0 & 1 \\ -1 & -1 & -1 & 0 \end{bmatrix}$$

$$\begin{bmatrix} 0 & -1 \\ 1 & 0 \end{bmatrix} * \begin{bmatrix} 1 & 2 & 1 \\ 2 & 4 & 2 \\ 1 & 2 & 1 \end{bmatrix} = \begin{bmatrix} 0 & 1 & 2 & 1 \\ -1 & 0 & -3 & 2 \\ -2 & -3 & 0 & 1 \\ -1 & -2 & -1 & 0 \end{bmatrix}.$$

Rotate these matrices 90° counterclockwise for differencing toward the NW.

THE LAPLACIAN OPERATOR

The operator $\partial^2/\partial x^2 + \partial^2/\partial y^2$ is of great importance in electricity, magnetism, elasticity, fluid mechanics, and also image processing. Since $\partial^2 f(x, y)/\partial x^2$ is like the upward curvature of the function $f(x, y)$ along a cross section in the x-direction, the Laplacian $\partial^2 f/\partial x^2 + \partial^2 f/\partial y^2$ can be thought of as a combined measure of total upward concavity of the surface $z = f(x, y)$. When a function $f(x, y)$ is available in discrete form for integer values of x and y, the quantity corresponding to the Laplacian is the sum of the second differences $\Delta^2_{xx} f + \Delta^2_{yy} f$. To get $\Delta^2_{xx} f$ we convolve f with the row matrix $[1 \ -2 \ 1]$, and for $\Delta^2_{yy} f$ we use the corresponding column matrix. Summing these tells us that

$$\Delta^2_{xx} f + \Delta^2_{yy} f = \begin{bmatrix} & 1 & \\ 1 & -4 & 1 \\ & 1 & \end{bmatrix} ** f$$

$$= f(x + 1, 0) + f(x, y + 1) + f(x - 1, y) + f(x, y - 1) - 4f(x, y).$$

PROJECTION AS A DIGITAL OPERATION

When a function of two variables is projected onto a line inclined at an angle θ to the x-axis, line integrals are evaluated for each point on the rotated axis. Thus

$$P_\theta(R) = \int_{-\infty}^{\infty} f(x, y)\, dS,$$

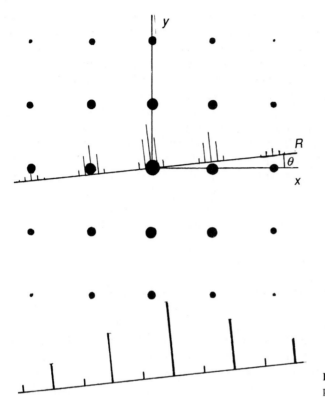

Figure 8-5 A discrete array of impulses projected onto an inclined axis.

where (R, S) is the rotated coordinate system. One can think of discrete data $F_{i,j}$ in exactly the same way in terms of the set of two-dimensional delta functions

$$f(x, y) = \Sigma f_{i,j}{}^2\delta(x - i, y - j).$$

Then $P_\theta(R)$ will be the discrete projection, represented by a set of nonuniformly spaced one-dimensional delta functions distributed along the R-axis. This is illustrated in Fig. 8-5 for the 25-element binomial array projected at an angle $\theta = 12°$.

We see that

$$P_\theta(R) = \Sigma f_{i,j}\delta(R - i \cos\theta - j \sin\theta).$$

The strengths of the one-dimensional impulses give the desired discrete coefficients. In principle the step to the discrete case has now been made, but the result is not acceptable in those situations where discrete samples are to be equispaced. One way of handling this is to subdivide the R-axis into smaller elements so that the impulses can be displaced by up to $\pm\frac{1}{2}$ unit to impose equispaced representation. Of course, there will be a lot of zeroes. But most usually the appropriate sampling interval is the same as that used for the original matrix that was to be projected. To get these sample values one divides the R-axis into unit cells, adds the coefficients within each cell, and assigns the sum to the center of the interval. This is called *cell summing*. Choice of cell phase makes a difference to the values arrived at; in the illustration an interior origin is shown, which was inherited from the central origin of coordinates, but an origin would often be chosen in the lower left corner, and the new sample values would be staggered. This choice of origin may not be trivial when projections at different angles have to be dealt with as a group. The fact that different projections have unequal numbers of elements favors indexing from the center.

Cell summing has the effect of roughening $P_\theta(R)$, as may be seen from the cell-summed projection, and some mitigating step may be desirable. One approach is to smooth $P_\theta(R)$ and resample; another is to assign the coefficient $f_{i,j}$ partly to the location $R = [i \cos \theta + j \sin \theta]_{rounded}$ and partly to $R - 1$ and to $R + 1$. (The subscript means "rounded to the nearest integer.") A third approach is to begin by slightly smoothing the original matrix; a feature of this is that the smoothed image can be judged as to acceptability in terms of the subject matter. The theory of cell summing can be worked out (Thompson and Bracewell 1974); the operation is expressible as convolution with a unit rectangle function followed by resampling at unit interval:

$$[P_\theta(R) * \text{rect } R] \text{ III}(R).$$

The cell-summed sequence has some unpleasant features and should not be acepted incautiously without testing. Two different spatial-frequency components of the data may have their amplitude ratio changed, and one or both may change to a different spatial frequency because of aliasing. The remedy, thinking in terms of continuous R, is to combine the projected impulses in cells of half width (or less), before resampling twice as often. More drastic smoothing than one would have wished is sometimes necessary as a trade-off against unwanted artifacts.

Back Projection

When a projection is given and values have to be assigned to matrix elements by back projection, the essential test is that projection of the matrix should regenerate the projection given at the start. As has been seen, the operation of projection can be messy, but back projection is comparatively straightforward. Here is one way. If the given projection is in the form of discrete samples, arrange to interpolate linearly so that $P_\theta(R)$ is available for any R. Then, at the matrix location (i, j) assign the value $P(i \cos \theta + j \sin \theta)$. Since back projection is often carried out at hundreds of different angles in turn, the matrix values being accumulated, fast arithmetic is a concern. That is why crude linear interpolation was mentioned above; where speed is vital, refined interpolation needs to be justified rather than enjoyed. Even rougher approaches may be effective. For example, interpolation can be avoided entirely by assigning the projection value given at the location that is closest to $i \cos \theta + j \sin \theta$. There is no way of knowing by theory whether the resulting image, after hundreds of stages, will be noticeably degraded in the eyes of the final user; only trial can tell. The art of evolving fast algorithms like this lies in first imagining blunt tools that might just do the job and then refining the best.

Variants of back projection that take into account the size and shape of the region to be back filled were discussed for the continuous case in Chapter 2.

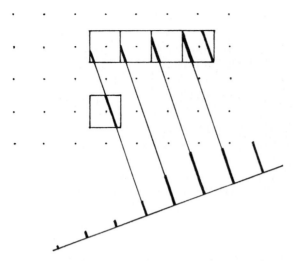

Figure 8-6 Instead of projecting each matrix element unchanged in magnitude, one can use weights proportional to the intercepts (heavy segments). These weighting coefficients are different for every cell and for every angle, which imposes a significant computing delay, but they can be pretabulated.

It is very reasonable to want to know what is the best method of back projection, and so some details may be mentioned. If the mass of one delta function were redistributed uniformly back along a straight line in the S-direction, then the amount of mass falling into the cell centered on a particular matrix location would vary from cell to cell. Therefore, the length of the intercept within each cell could be used to weight the amounts of mass assigned (Fig. 8-6). This comment is especially relevant when the direction of back projection is changed, because the cell spacing is about unity when θ is very small, whereas the average cell spacing in the neighborhood of $\theta = 45°$ is substantially different. One thinks of redistributing along a straight line, because the projection data are discrete, but since each value represents projection along a strip of nonzero width, it also makes sense to redistribute along a strip and to weight in accordance with the overlap of the strip with each matrix cell. The weights are complicated but need not be computed on every run; where much repetitive imaging is done, all such weights can be pretabulated.

MOIRÉ PATTERNS

When a wire screen on a window is viewed from outside, light and dark fringes may sometimes be seen, usually curved but maintaining regular spacing. The fringe separation is several times larger than the wire spacing, and the fringe width is a substantial fraction of the fringe separation, often about one-half. When the eye moves, the fringes move conspicuously, sometimes in the same direction, sometimes not. Other examples can be seen in one's surroundings, but they have to be looked for consciously, because, like shadows and specular reflections, they are dismissed automatically by the brain as having no significant reality.

Figure 8-7 shows an artificial example designed to be sufficiently coarse that the details can be examined. Clearly, the white fringes fall on the locus of intersections of

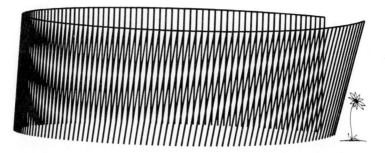

Figure 8-7 A rickety garden fence exhibiting moiré fringes. If you were viewing a real fence, what would the fringes do if you stood on tiptoes or covered your left eye?

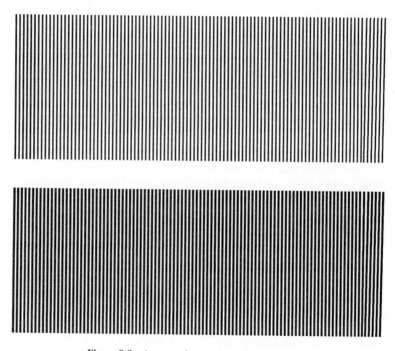

Figure 8-8 A pattern for copying to a transparency.

the black lines. The fringe separation varies regularly, as reported above. The fringe width is fairly sharply defined, an impression that is confirmed by glancing along the fringe with one's eye close to the paper. In this example the fringes are light, and there do not seem to be dark fringes, except by default.

Some intuitive feel for moiré fringes is desirable. For this purpose Fig. 8-8 can be photocopied as a transparency and superposed on the original. If the superposition is at a slight angle of about 5°, parallel east-west fringes will be seen. When the transparency

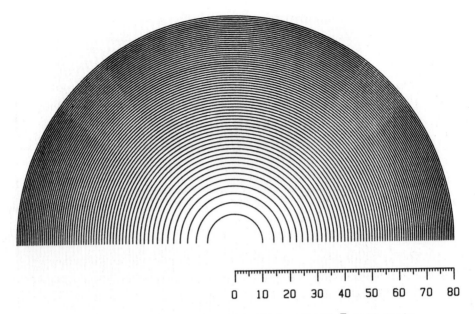

Figure 8-9 Concentric circles, the nth one of which has radius $10\sqrt{n}$. As this drawing
was made with a pen plotter, the phenomena seen are in no way digital or related to pixels.

is moved east-west, the fringes move rapidly north-south, but not vice versa. The grating
illustrated has two deliberate imperfections that can be examined. At one end the lines
depart very slightly from parallelism, while at the other end they depart very slightly from
equal spacing; these imperfections have been made almost too small to see by eye, but
have an easily noticeable effect on the fringes. The behavior of the moiré fringes as one
slides the transparency the full width of the page gives a striking feel for some of the
industrial uses of moiré devices. One is as a translation sensor, another is for sensing and
indicating unwanted irregularities.

Before leaving this diagram, try reversing the transparency and also turning it over.
The reputedly parallel fringes are not in fact straight, but bowed. Investigate that. Other
phenomena that can be investigated include thermal contraction of the transparency as it
cools after emerging from the copier, the distortion of the paper due to humidity, and the
anisotropic properties of recorder paper that comes in rolls.

To discuss moiré fringes in two-dimensional black-and-white images we turn to Fig.
8-9, which presents a set of 64 concentric circles of radius $r = K\sqrt{n}$, the serial number n
ranging from 1 to 64. The spacing between circles is $K(\sqrt{n+1} - \sqrt{n})$ or approximately
$K/2\sqrt{n}$. In this illustration $K = 10$, so the inner circle has radius 10 and the outermost 80.
Figure 8-10 shows what happens when a grille of unit spacing is superposed by means of
a pen plotter. As can be seen, two conspicuous bull's-eyes appear, centered at ± 50, where
the spacing of the circles agrees with the grille spacing. Fainter fringes centered at ± 25
can also be seen and will be discussed later. The shape of the fringes can be investigated
as follows.

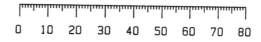

Figure 8-10 The previous family of circles with a superposed grille of unit spacing. The artifacts in the NE and NW quadrants result from an ink line whose width was dependent on the velocity of the pen; the 2° chords of the polygon representing the circles can also be detected.

The first white fringe follows the locus of intersections of circle $n + 1 + m$ with line $x = 50 + m$, where m is an integer. The abscissa of each intersection is $50 + m$, and the ordinate y is derivable from the triangle whose hypotenuse is $r = K\sqrt{n}$, for which $y^2 + (50 + m)^2 = K^2 n$. Replace the value 50 by $K\sqrt{n_0}$, where $n_0 = 25$ is the serial number of the circle of radius 50. The squared distance from $(K\sqrt{n_0}, 0)$ to the intersection will now be shown to be constant independent of the parameter m which indexes the intersections. We have

$$
\begin{aligned}
y^2 &= K^2 n - (K\sqrt{n_0} + m)^2 \\
&= K^2(n - n_0) - 2K\sqrt{n_0}m - m^2.
\end{aligned}
$$

Note that n and m each receive unit increments from one intersection to the next and that $n - n_0 = 1 + m$; also, since the spacing $K/2\sqrt{n_0} = 1$, it follows that $K - 2\sqrt{n_0} = 0$. Then the squared distance mentioned is

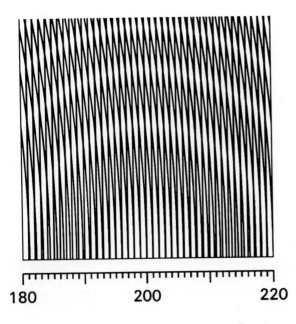

180   200   220

Figure 8-11   Closeup of a bull's-eye showing how light fringes occur where black lines converge. This diagram is for $K = 20$.

$$
\begin{aligned}
y^2 + m^2 &= K^2(n - n_0) - 2K\sqrt{n_0}\,m \\
&= K^2(1 + m) - 2K\sqrt{n_0}\,m \\
&= K^2 + (K^2 - 2K\sqrt{n_0})m \\
&= K^2 + K(K - 2\sqrt{n_0})m \\
&= K^2,
\end{aligned}
$$

which is independent of m as stated. The first white ring thus has radius K. The μth white ring, for which $n - n_0 = \mu + m$, has radius $K\sqrt{\mu}$; consequently the bull's-eye pattern superposes neatly on the original set of circles.

Figure 8-11 shows on an enlarged scale how the impression of fringes arises. If examined from a distance, the white fringes are seen to occupy less than one-half the fringe spacing, while the dark fringes possess a light core. Physiological influences include the perception of lightness where black is expected and simultaneous contrast between adjacent inked and less inked regions. That the perception is not simply associated with the fraction of inked area is shown by the fact that the darkest zone does not have the most ink, while the zone that does have most ink appears less dark (at normal reading distance).

In the case of these bull's-eye patterns an algebraic discussion succeeded because the result was simple. In general, moiré fringes can be traced point by point by simultaneously incrementing the parameters of each family of curves.

A practical application of this particular example is to a halftone screen meter. Make a transparency with 100 semicircles of radius $r = K\sqrt{n}$, where $K = 0.1$. Then the spacing ranges from 0.04 inch between circles $n = 1$ and $n = 2$ to 0.005 inch at the outer radius of 1 inch. This can be done by photographically reducing a larger drawing. When the transparency is dropped onto a halftone illustration in a newspaper small bull's-eyes will appear in locations that enable the number of dots per inch to be read off. As the reciprocal spacing is directly proportional to r, a linear scale of dots per inch can be provided. For example, in Fig. 8-9, the numbers shown will represent dots per inch if the outermost circle is reduced to 80 mm in diameter, or will represent dots per cm if the circle is reduced to 31 mm in diameter.

Another application of the same fringe pattern is to the verification of the pixel spacing on a graphics printer. If you call for concentric circles to appear on a screen, or to be printed, the radii of the circles increasing as $\sqrt{n}$, you can cause the circle spacing to be a small multiple of the pixel spacing. In the case of a graphics screen the pixels are the well-defined elements that light up when addressed. In the case of a laser printer we can speak of virtual pixels; there is a raster defining horizontal rows, and the rows are divided into de facto elements by the digitally modulated voltage waveform that turns the laser beam on and off as it flicks across a rotating light-sensitive drum. Figure 8-12 shows the laser-printer output from such an experiment. The bull's-eyes centered on the extremities of the east-west diameter are positioned where the circle spacing is two virtual pixels. The radius r_0, where the circle spacing would be equal to the pixel spacing, is twice as far out and would print as solid black; the bull's-eyes at the edges, where $r = \frac{1}{2}r_0$, are to be identified with the fainter bull's-eyes at ± 25 previously mentioned in connection with Fig. 8-10. On that figure we can see exactly how the phenomenon arises; an increase of unity in n results in an increase of 2 units in the circle spacing. With the much finer pixel spacing of the laser printer, features are also seen at $\pm\frac{1}{3}r_0$, $\pm\frac{1}{4}r_0$, $\pm\frac{1}{5}r_0$, and $\pm\frac{1}{6}r_0$. Where these bull's-eyes overlap, secondary features appear, the most noticeable being at $0.4r_0$.

Figure 8-13, an enlarged portion of Fig. 8-12, enables the anatomy of the circular structures to be examined. There is no grille of lines for the circles to intersect with; instead, each circle exhibits a step at locations where the abscissa jumps by one pixel unit; this occurs on the locus of the dark moiré fringes of Fig. 8-11. The consequence is that shadows form on the outboard and highlights on the inboard limb of each locus. Exactly the same separation of highlight and shadow is seen when an attempt is made to draw nested circles of diminishing spacing, or a Fermat spiral $r = K\sqrt{\theta/2\pi}$, on a graphics screen. Figure 8-14 shows such a spiral winding from (60, 0) to (127, 0), made with the following program (operating in degree mode for angles).

A TIGHTENING SPIRAL

```
MOVE 60,0
FOR A=3600 to 16129 step 2       Start 10 turns from center
   R=SQR(A)
   X=R*COS(A)
   Y=R*SIN(A)
   CALL "PSET" (X,Y)             Turns on the pixel (x,y)
NEXT A
```

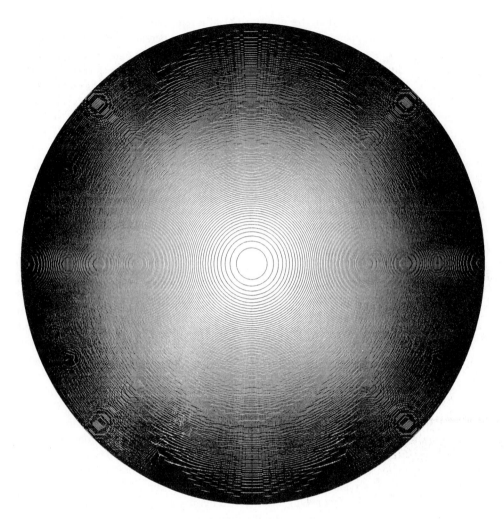

Figure 8-12   A total of 226 circles, the nth having radius $60\sqrt{n}$ in units of the virtual pixel spacing on a laser printer.

If, instead of using a line one pixel wide, one fills alternate annuli with black, as with a Fresnel zone plate, an even more dramatic interaction of the called-for pattern with the graphics screen structure appears, as shown in Fig. 8-15. The program, which builds up the picture row by row, obtains radial square-root dependence without explicitly taking roots. Even so, the program is slow, because every second pixel is visited. The function DIV performs integer division (discarding any fractional part) and is convenient in personal computing (as distinct from community programming) where type declarations are not obligatory.

Figure 8-13 A horizontal sector from the previous figure showing indications of bull's-eyes at radial distances $\frac{1}{2}, \frac{1}{3}, \frac{1}{4}, \frac{1}{5}$ of a diameter on a linearly varying background of grey level.

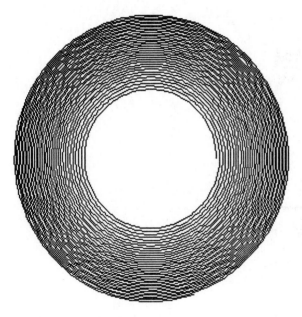

Figure 8-14 A spiral $r = \sqrt{\theta}$ (θ in degrees) drawn on a graphics screen. The spacing between turns passes through 2 pixels exactly at $r = 90$, where shadowed circular features appear.

ZONE PLATE

```
FOR y=-127 TO 127
   FOR x=-128 TO 128
      rr=x^2+y^2
      black=rr DIV 250 MOD 2
      IF NOT black THEN nxt
      CALL "PSET" (x,y)
nxt:  NEXT x
NEXT y
```

Television images of fences may show fringes similar to those discussed. The story of J-B Moiré and the observation of coach wheels turning in front of a picket fence is

Figure 8-15   Concentric dark annuli with radius proportional to the square root of the serial number interact with a graphics screen to generate much fine detail.

told by Weber (1983). Commercial value dates back some centuries to an observation that stacked cloth acquired an attractive "watered" appearance, which was soon reproduced by pressing folded sheets of fine woven cloth, originally mohair. Watered silk and other fabrics are now made by passing material through ribbed rollers. It would be a mistake to suppose that the moiré phenomenon is inherently two-dimensional, as in the imaging examples of this chapter; look into the three-dimensional aspect by obtaining a sample of moiré material and examining it with a lens.

    An inverse problem is to design families of curves to produce specified moiré fringes, or fringes with specified movement. The straightforward approach is to begin with a uniform grille, superpose the specified fringes, and then trace polygons by progressively following diagonals of the resulting network of four-sided cells. There is a good deal of freedom in this construction, since the first family of curves does not have to be a grille. Complete images can be formed as moiré patterns. If made by embossing or layering, such an image has a holographic quality that suits it for use in countries where efforts are made to deter counterfeiting the money.

The Spectrum of Moiré Patterns

Consider two functions $f_1(x, y)$ and $f_2(x, y)$. If they are added together, their spectra add, and clearly no spatial frequencies can appear in $F_1(u, v) + F_2(u, v)$ that are not present in either of the component functions. Let us illustrate with a special case where the first function $f_1(x, y) = 0.5\ {}^2\mathrm{III}(x, y) ** \mathrm{rect}\, 5r$, representable by grey discs on a grid of unit spacing. Its Fourier transform is $F_1(u, v)$. The second function $f_2(x, y)$ is derived by rotating $f_1(x, y)$ through an angle $\arctan(-3/4) = -36.9°$. Let (x', y') be a coordinate system rotated with respect to (x, y) by this angle. Then $f_2(x, y) = f_1(x', y')$. The sum function $f_1(x, y) + f_2(x, y)$ consists of lots of small grey discs but, owing to the special choice of angle (based on a 3, 4, 5 right triangle), occasional discs coincide, as do the two at the origin. The composite pattern is simulated in Fig. 8-16 where black discs stand for coincident pairs of grey discs. It is apparent that the black discs exhibit a low-frequency pattern, of period $\sqrt{5}$, corresponding to a lower frequency than is present in either of the component parts. Although this long period is noticeable to the eye, it could not be discovered by spectral analysis. The low frequency can be made even lower by relative rotation through a smaller angle, behavior reminiscent of moiré fringes produced by small rotations of a transparent copy with respect to an original.

Moiré fringes, however, do not depend on linear superposition, such as is simulated here where two superposed greys add up to black. The moiré effect would be more akin to the superposition of two black patterns, where two blacks would add to make another black, which is a clear nonlinearity. To see that the low frequency would then be present in the spectral analysis of this nonlinear case, notice that the composite picture differs from $f_1(x, y) + f_2(x, y)$ by elements occurring only where overlap takes place. The difference would thus consist of only the black discs of Fig. 8-16. This shows that subjective low-frequency components may indeed exist in the spectral analysis of moiré patterns.

However, that is not the full story. In the figure there are rings of eight grey discs which repeat at the same low spatial frequency as the black discs. Consequently, when there is nonlinear clipping, there may still be low-frequency features that do not show in the spectrum. Furthermore, the appearance of moiré fringes includes psychophysical effects of contrast that are separate from the spectral content.

In x-ray crystallography, spots recorded in the Fraunhofer diffraction plane reveal various spatial frequencies that are present in the crystalline substance under study, which helps to work out the spatial structure. The existence of repetitive structure not revealed by Fourier analysis is naturally of interest to materials science and mineralogy (Pool, 1988).

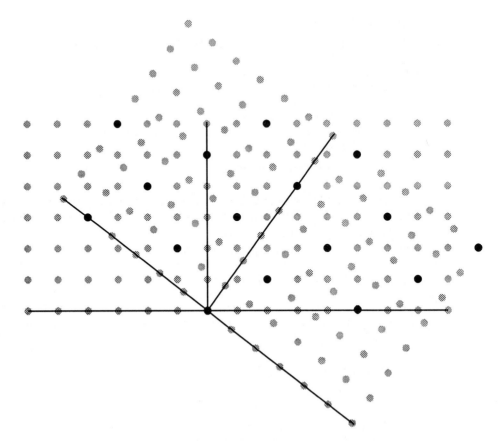

Figure 8-16 Two patterns, superposed with rotation, contain no spatial frequencies not present in one of the original patterns. Nevertheless, a pattern superposed on itself after rotation through certain special angles produces a crystalline appearance, exhibiting low spatial frequencies to the eye which do not appear in the spectral analysis.

FUNCTIONS OF AN IMAGE

Just as one can speak of a function of a function, for example $g[f(x, y)]$, so also one can have a function of the image $f(x, y)$. A common operation on a digital image is to subtract a constant threshold or bias and multiply by a constant, followed if necessary by limiting to the values from zero to unity (or other values by which white and black are denoted). First consider the program for the special case where the median value of an $M \times N$ image $f(x, y)$, x and y running from 1 to M and 1 to N respectively, is first subtracted. An amplification factor is then applied that increases the difference between the maximum and minimum values to unity, and the median is added back in. Any values that are less than 0 or greater than 1 are brought back to those limits. Note that the subprogram "`minmax`" calls a sorting routine.

CONTRAST EXPANSION

```
CALL "minmax" (f(,),Lx,Ly,fmin,fmax,median) Finds extrema and median of f(,)
FOR y=1 to Ly
   FOR x=1 to Lx
      g(x,y)=f(x,y)-median                 Subtract median
      amplif=1/(fmax-fmin)                 Amplify and
      g(x,y)=magnif*f(x,y)+median          restore bias
      g(x,y)=MAX(0,g(x,y))                 Apply upper and lower limits
      g(x,y)=MIN(g(x,y),1)
   NEXT x
NEXT y
```

Contrast expansion is an operation exemplified by photocopying machines, which have the effect of rendering grey levels near 0 as closer to zero and those near 1 as closer to 1. In this way dirty or faded typescript can be much improved. A drastic expression of this operation would be to compute $g(x, y)$ as above and follow with a line

```
g(x,y)=(g(x,y)>0.5)
```

The choice of the threshold 0.5 is dependent on the character of the material in hand: whether fading is to be corrected or whether faint dirty marks are to be removed. A less drastic choice is

```
g(x,y)=g(x,y)-k*SIN(PI*g(x,y))
```

where $k < 1/2\pi$. The line

```
g(x,y)=g(x,y) + k*g(x,y)(1-g(x,y))
```

where $k < 1$, produces a general shift toward dark (positive k) or light (negative k). In the preparation of halftone illustrations from numerical data, contrast expansion or distortion can help to clarify the information to be conveyed. A glance at the technical journals shows that this opportunity is often overlooked.

DIGITAL REPRESENTATION OF OBJECTS

We have been dealing so far with discrete images, where the elements have values that may or may not be integers but in any case a substantial range of values is available. Many important digital operations have to do, however, with binary images that have values chosen from 0 and 1 only. We begin with a simple but important feature, namely the lines or polygons that define the boundaries of discretely represented binary objects.

Straight Boundaries

On the graphics screen straight boundaries can be called for but will in general appear stepped, except for the cardinal directions N, S, E, and W. The angle of inclination of

Figure 8-17   Some of the stepped boundaries associated with discrete pixels.

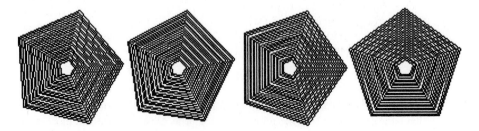

Figure 8-18   Tightly nested pentagons on a graphics screen showing individual pixels. The sets are tilted differently, at 4.5°, 9°, 0°, and 90°.

the line joining any pair of separate pixels will be of the form arctan(m/n), where m and n are integers. Of course, the rational fraction m/n can approximate any real number to any desired accuracy, but not when m and n are limited to a finite range. Consequently, the special angles 45°, 26.6°, 18.4°, 14.0°, 11.3°, 9.5°, 8.1°, 7.1°, 6.3°, . . . , often arise naturally. These nine angles, and their stepped structures, are illustrated in Fig. 8-17 as boundaries in the first octant, and the same forms will be seen in the other octants. If a boundary is called for at some other inclination, say 35°, the best that can be done is to piece together short segments of step structures illustrated for 26.6° and 45°. If 28° is called for, short stretches of the 26.6° structure and occasional single steps at 45° will have to do. The method for constructing these entities is dealt with below under the straight-line algorithm. To understand why this algorithm is worth study in its own right, even though it is built into even the humblest graphics packages, we first look at some of the discrete phenomena associated with higher-level graphics programming.

Figure 8-18 shows attempts to draw nested polygons on a graphics screen. The first set calls for a 4.5° tilt, so that all five sides slope differently with respect to both the horizontal and vertical. Consequently, the expected steps are differently arranged in the five sectors. In addition, within each sector there is irregularity in spacing of the nominally parallel sides, a fact that can be inspected in detail but in any case evidences itself by unruly texture. The second set is tilted 9°, which results in the north-east edges being simplified because they are all inclined at 45°, but the spacing of these strictly parallel edges is still irregular. The textures of the other four sectors have changed. The third and fourth sets of pentagons are tilted 0° and 90°, respectively, to bring one set of sides vertical or horizontal. When the radii of the circumscribing circles are incremented regularly by two pixels at a time, the spacing between the vertical (or horizontal) edges is called on to increase by $2\cos 54° = 1.18$ pixels, which explains the irregular discrete response. Although one can understand this intellectually, the banded appearance of the parallel straight edges is not harmonious. Examination of the third set shows that it is more or less

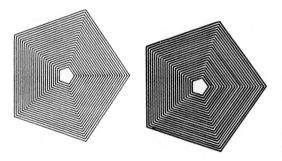

Figure 8-19 Loosely nested pentagons made by a high-resolution laser printer (right) still exhibit unavoidable irregular line spacing caused by discreteness. The version on the left was drawn mechanically with a pen.

symmetrical about a horizontal axis, but not quite; nor is the fourth set exactly symmetrical about a vertical axis. Finally, the third set does not convert into the fourth by rotation through 90°. By nesting the pentagons very tightly we have been able to emphasize the phenomena mentioned, and it is usually possible to work around them, should they appear unexpectedly and be unwanted.

What we have seen is that there are things that are very easy to draw with a pen, or to visualize, that are not suited to discrete representation by pixels. In many cases, the stepped character of simple boundaries becomes imperceptible when the size of the pixels is reduced, and certainly the pixel size is rather coarse in the foregoing figure. However, there are situations where shrinking the pixels does not eliminate the molecular character, and other situations where we do not wish to do so. Figure 8-19 shows the same nested pentagons drawn with an electromechanically driven ink pen (left) compared with a 300-dpi laser-printed version (right) with an area density of pixels 16 times greater than in Fig. 8-18. One of the printed sectors is good, but in general the printer output is not as regular as the pen plotter drawing. It is better than the previous printer versions, because it is a smart printer that tries to maintain a constant line width of 0.35 mm, or 4 dots at 300 dpi, regardless of slope. However, the space between lines suffers accordingly. If tighter nesting is called for than in Fig. 8-18, where the spacing was one pixel, the picture deteriorates again.

Thinking about the nested pentagons merely confirms the obvious, that stepped boundaries will have to be worked around occasionally and that certain images conceived in the continuum cannot be expressed on a discrete medium. On the other hand, extraordinarily coarse discretization can be perfectly acceptable, a good example being provided by the 5 × 7 alphabet widely used for effective electronic displays. The famous moving display in Times Square and similar ones in stockbrokers' offices are of this kind, although it should be mentioned immediately that moving letters are perceived differently. Figure 8-20 shows a pixel dump from a graphics screen to a printer using a 5 × 7 array, together with a normally printed version. If you prop the book up on the table and back away, which one becomes illegible first?

It appears that the brain is quite happy to perform the necessary character recognition subconsciously, and will do very well, provided there is enough information to do so. Experimenting with smaller arrays shows that 5 × 7 is about the minimum effective size, especially if lower-case letters and punctuation are included. Diverse examples show that

ENTIA NON SUNT MULTIPLICANDA PRAETER NECESSITATEM

ENTIA NON SUNT MULTIPLICANDA PRAETER NECESSITATEM

Figure 8-20   Coarsely discretized characters compared with normal printing.

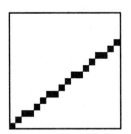 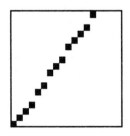 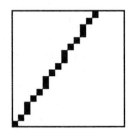

Figure 8-21   Pixel representation of the line segment from $(0, 0)$ to $(17, 13)$ showing some clumping (left), the unacceptably nonuniform result when the same approach (rounding the ordinate) is applied to the line from $(0, 0)$ to $(13, 17)$, whose slope exceeds unity (center), and the conventional approach of rounding the abscissa (right).

the spatial resolution achieved by the display has to be considered in conjunction with the visual interpretation performed by the eye and brain.

We turn now to the discrete representation of a straight line.

The Straight-line Algorithm

How to show a straight line $y = mx + c$ by illuminating a string of pixels is in itself a nontrivial matter and is sufficient to illustrate many of the problems involved on a discretized display medium. Suppose that the line segment is to run from $(0, 0)$ to $(17, 13)$, where the slope $m = 13/17$ and $c = 0$. Then for each integral value of x illuminate the pixel at (x, y_R), where y_R is the value of y rounded to the nearest integer. In every column from $x = 0$ to $x = 17$ there will be one pixel turned on, but in some rows there are two (Fig. 8-21, left). The light from the 18 pixels will be spread over the segment length $\sqrt{13^2 + 17^2} = 21.4$, so the strength, or linear brightness, is $18/21.4 = 0.84$ pixels per unit length. If the slope has the steeper value of $17/13$, then every column still has one illuminated pixel, but some rows have none (center). The mean strength is still 0.84, because the number of pixels is distributed over the same length, but the distribution is more noticeably nonuniform because of the four gaps. Therefore, when a line is closer to the vertical than to the horizontal, the plan is (right) to illuminate one pixel in each row from $y = 0$ to $y = 17$ at location (x_R, y), where in general $x = (y - c)/m$.

Discretizing a straight line, a task that must be done often, was the purpose of a clever presentation by Stockton (1963) in an algorithm that does not use multiplication or division. A similar algorithm by Bresenham (1965) showed that when attention is restricted to inclinations in the range $0°$ to $45°$, the frequently repeated loop contains only a few additions and a logical comparison in addition to the pixel-activation command.

If the straight line from the pixel at (x_a, y_a) to the one at (x_b, y_b) is to be represented by turning on intermediate pixels, the way of doing this will depend on the slope $m = (y_b - y_a)/(x_b - x_a)$. If the slope is gentle, $\mid m \mid \leq 1$, then $x_b - x_a + 1$ pixels will be turned on, one in each column; while if the slope is steep, $\mid m \mid > 1$, a pixel will be turned on in each row. Stockton's algorithm accommodates this distinction.

A different way of dealing with these two cases is to begin by calculating m, and, if $\mid m \mid > 1$, then to swap x_a with y_a and x_b with y_b; the resulting reflection means that only the gentle-slope case, $\mid m \mid \leq 1$, has to be considered in detail. A set of $N = x_b - x_a + 1$ pixels (x_i, y_i) is then determined, where (x_1, y_1) is the same as (x_a, y_a) and (x_N, y_N) is the same as (x_b, y_b). The pixels (x_i, y_i) can then be turned on, or, if the terminal coordinates were originally swapped, then the pixels (y_i, x_i) are turned on.

One additional initial preparation can be made. If $x_a < x_b$, then as i increases, the line will be constructed from left to right, but if $x_a > x_b$, then the pixels will be turned on from right to left. Of course, the Euclidean line joining (x_a, y_a) to (x_b, y_b) is the same as the line joining (x_b, y_b) to (x_a, y_a), but the sets of pixels are likely to be different. The reason for this is seen when the terminal points are a knight's move apart, for example at $(0, 0)$ and $(2, 1)$; three pixels will be turned on: the two terminal pixels and one in-between, let us say $(1, 1)$. However, the intermediate pixel might reasonably fall at $(1, 2)$. In some circumstances it is desirable that the set of pixels for given terminal points should be unique—for example, if a line once plotted is to be erased, as in moving displays. To ensure uniqueness it is only necessary to construct all lines from left to right. Thus, when the terminal points are given, if $x_b < x_a$, then swap x_a with x_b and swap y_a with y_b. This beautiful thought, often mentioned in textbooks, is uniformly ignored by commercial software designers and, having been mentioned, will be ignored here, too.

The degenerate case $m = 0$ $(y_a = y_b)$ needs to be watched but the case of infinite slope $(x_a = x_b)$ is eliminated by the initial plan for restricting attention to the gentle-slope case.

The segment given below, much more compact than in the cited publications, supersedes the Stockton-Bresenham algorithm and works in all four of the gentle-slope octants. It is due to Professor Oscar Buneman. It is worth studying in detail to discover how the multiplication implied by the product mx is avoided.

BUNEMAN'S STRAIGHT-LINE ALGORITHM

```
xa=0 @ ya=0 @ xb=17 @ yb=13        State end points
dx=ABS(xb-xa) @ dy=ABS(yb-ya)
sx=SGN(xb-xa) @ sy=SGN(yb-ya)
dx2=dx+dx @ dy2=dy+dy
d=-dx                              Initial value of discriminant
y=ya                               Start at (xa, ya)
FOR x=xa TO xb STEP sx             The inner loop
   CALL "PSET" (x,y) @ d=d+dy2     Turn on the pixel (x, y) and advance d
IF d>0 THEN d=d-dx2 @ y=y+sy       Discriminant jumps back as the row shifts
NEXT x
```

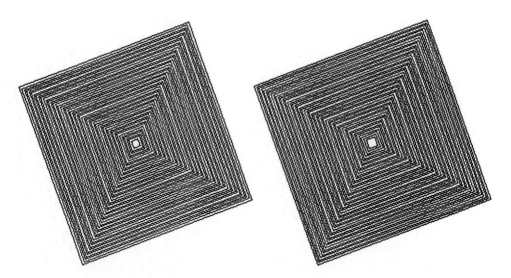

Figure 8-22 Nested squares with sides specified as 8, 12, 16, . . . , 196 (left) and 10, 14, 18, . . . , 198 (right), showing irregularities resulting from discretization.

The variable d is the discriminant whose sign change determines when y is to be incremented. In essence the discriminant starts off negative, follows mx (or $13x/17$ in the example), and, when it exceeds 0.5 plus an integer, shifts y to the next row and drops back to a negative value. But in fact the term mx is amplified by a factor 17 to eliminate the denominator and by a further factor 2 to convert the 0.5 to an integer. Then, when the row shifts, the discriminant drops back by an amount 2×17, after which it accumulates by 2×13 for each further advance in x.

In the repeated loop there are only one addition, one logical comparison, and (on the fraction of occasions when the condition is met) another addition and a subtraction. The user needs to supplement these few lines by reflecting steep into gentle lines, as described above.

If the straight line called for does not run from one integer grid point to another, the coordinates (x_a, y_a) and (x_b, y_b) at the ends have to be adjusted, with consequences that can be displeasing. This is partly because the line of best fit to the pixels is not the line connecting the endpoints; consequently, lines intended to be equispaced and parallel may not appear so when converted to a pixel display. Figure 8-22 presents nominally concentric squares (tilted 20°) to show that the apparent line spacing varies from layer to layer and may even vary along one side, giving an appearance of nonparallelism. Occasionally a particular square is handled in a way that is not at all graceful (Fig. 8-23).

Graphics Commands for Straight Lines

Instead of programming a straight line pixel by pixel, one can call on Graphics Language, which dates back to instruction sets for raising and lowering the pen or pens of XY-plotters

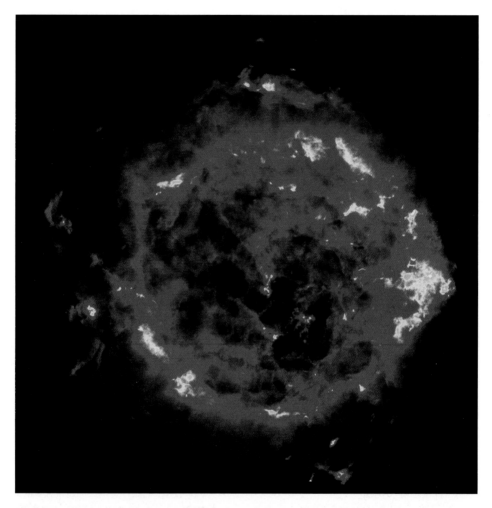

Plate 1 Cassiopeia A. A visual presentation of centimeter-wavelength data obtained by indirect inter-ferometric imaging by R. Braun, S. Gull, and R. Perley using the Very Large Array of the National Radio Astronomy Observatory at Socorro, New Mexico. A seventeenth century supernova explosion left this remnant, which occupies a more-or-less spherical shell twelve light years in diameter, and is still expanding. The angular diameter is four arcminutes. The pixel size of the data is about one-tenth of an arcsecond, while the dynamic range is well beyond the capacity of photography to convey. The National Radio Astronomy Observatory is operated by Associated Universities, Inc. under contract with the National Science Foundation.

Plate 2 San Francisco Bay Area. An image from the Landsat 4 Thematic Mapper based on data acquired in the blue, green, and near infrared (700 nm) bands and processed by William Acevedo. Vegetation, which has a high reflectivity in the infrared is rendered in red. Courtesy NASA Ames Research Center.

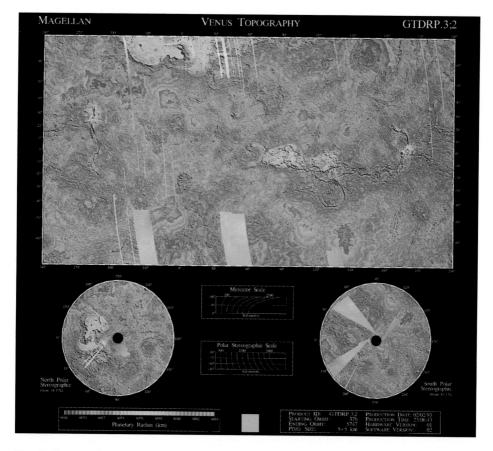

Plate 3 Topography of Venus. A 13-cm radar altimeter carried by NASA's Magellan spacecraft was used to construct a Mercator map in the latitude range from -69° S to 69° N and two polar projections bounded by ±44.1° respectively. The distances from the center of Venus range from 6048 km (blue) to 6064 km (red). Height accuracy is better than 50 m and horizontal resolution at the equator is 10 km or better. The elevated region in the North is Ishtar Terra, dominated by Maxwell Montes. Prepared by the Center for Space Research, Massachusetts Institute of Technology and kindly supplied by Gordon H. Pettengill.

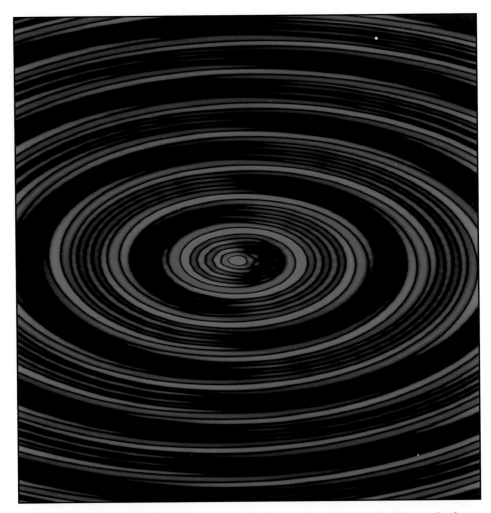

Plate 4 The diffraction field of a 16-turn one-armed spiral slit using color to distinguish negative from positive regions. Both amplitude and phase are full represented, something that is possible with the two-dimensional Hartley transform shown here but would not be convenient for the Fourier transform, which is complex. The diffracted intensity, or Fraunhofer diffraction pattern, which is more usually seen and abandons the phase information, can be constructed as the sum of the squares of the real and imaginary parts of the Fourier transform, or as the sum of the squares of diametrically opposite points of the Hartley transform. Courtesy John Villasenor.

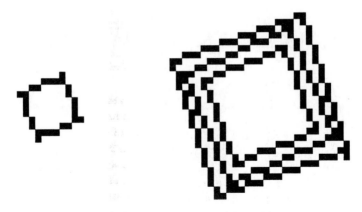

Figure 8-23 The square of side 8 (left) is rather jagged, and the squares 14, 18, 22 do not nest very well.

and drawing straight lines (or vectors). Some higher-level languages provide commands such as MOVE x,y, which raises the pen and sends it to (x, y), and DRAW x,y, which lowers the pen in its current position and draws an ink line to (x, y). These convenient mnemonic commands are invisibly translated into Graphics Language. One pays for this convenience by some loss of freedom, but the higher-level language usually allows imbedding of Graphics Language instructions, if you care to learn them. Graphics software at an even higher level makes drawing accessible to those who do not care to learn any programming language at all; this leaves the user with even less freedom, which may not be noticed much in view of the range of stunning effects that can be called up from a wide repertoire of icons.

To understand the trade-off between freedom and convenience, and the fundamental limitations set by a discrete medium, we limit ourselves to some practical consequences of calling on built-in straight-line algorithms. In Fig. 8-24 a number of small pentagrams have been drawn using a program in which the radius of the escribed circle advances through integral values from 3 to 13.

PENTAGRAMS

```
s=0                              Initial x-shift
FOR R=3 TO 13
   MOVE s,R
   FOR A=90 TO 810 STEP 144      Angle expresed in degrees
      DRAW s+R*COS(A),R*SIN(A)   Draw side of pentagram
   NEXT A
   s=s+3*R                       Shift the origin for the next pentagram
NEXT R
```

Only one of the top row of stars has left-to-right symmetry. Tracing out the polygons in the opposite sense by using   FOR A=810 TO 90 STEP −144 yielded the second

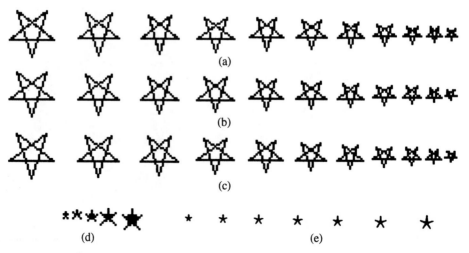

Figure 8-24 From top to bottom: (a) Pentagrams of regularly progressing size drawn CW from a high-level language; the shapes change of their own accord. (b) The same drawn CCW are not exact mirror images. (c) The former two, superposed, are mostly symmetrical, but not entirely. (d) This row shows hand designs that frankly abandon the idea of points spaced 72° and simply try to look like stars. (e) Printed pentagrams from a commercial font with many more dots per inch.

row of stars, nearly every one of which is a mirror image of the one above it. It should be easy to generate symmetrical stars by superposing CW and CCW patterns, as has been done in the third row. Curiously, however, some of the symmetrical stars are not exact superpositions of the two above. This must be connected with the internal roundoff convention that assigns a pixel to a pair of noninteger coordinates.

When these stars are reduced in size so that the individual pixels become hard to see, the phenomena do not go away. One can do better than the programmed outlines by hand-tailoring the pixels for each desired size (row 4), but when printed characters are examined, traces of the digital originals can be seen on the smaller ones (row 5). Row 4 shows some hand-designed stars. The angular spacing between points is noticeably not 72°, but, in the smaller sizes, neither can the patterns based on a regular pentagon achieve equal spacing of the points. The hand designs have a certain character which you can modify, if you wish, so that the effect looks right to you. In the long run a computed character has to meet such an esthetic test, too.

Traverse

Surveying by successive measurements of distance and direction, or *traversing,* is standard practice in recording the boundaries of property and in reckoning positions at sea. In land surveying, a traverse customarily begins and ends at a fixed station, and the closure error is distributed over the intermediate vertices. Here is a standard property specification.

Beginning at the intersection of Arguello Street and Salvatierra Street, thence N. 30° 09′ W. 27.44 feet, thence N. 59° 51′ E. 153 feet, thence S. 30° 02′ E. 129 feet, thence N. 85° 55′ W. 185 feet, thence S. 30° 02′ E. 129 feet, thence N. 85° 55′ W. 185 feet to the beginning.

In navigation, recording courses and distances sailed is called *traverse sailing*. In this case the polygon is not necessarily closed. The sailing record refers to dead water; the dead-reckoning position is to be corrected by giving the polygon an additional leg and using the estimated set of the current as the course and the estimated drift velocity times the time as the distance. The traverse tables formerly used for plane sailing converted course and distance to cartesian increments.

There is an analogy to the traverse in computer graphics where it is often convenient to work with a list of incremental distances and directions rather than with the coordinates of the vertices of an equivalent polygon. In one version all step lengths are unity and there are only the four cardinal directions; in this case a list of directions such as 0, 1, 2, 3 for E, N, W, S specifies a polygon. In another, the eight principal wind directions are listed, and it is understood that the step length is unity or $\sqrt{2}$, according as the direction is cardinal or diagonal. In general, the chain of instructions consists of arbitrary distances and directions. All the above comes in two flavors: direction may be specified absolutely (e.g., with respect to north as in both surveying and navigation), or change of direction may be specified.

Circles

If the line to be discretized is curved, a rather agile program may be needed to switch modes to take account of the local slope as it passes from gentle to steep (at $m = \pm 1$). If the curve is specified by an equation, then solving for y, given x, may require iterative approximation methods; the solution may also permit multiple values for y. Circles and ellipses are by far the most frequently occurring curves and therefore justify attention to fast methods.

If the circle $x^2 + y^2 = R^2$ is to be represented by pixels, one might think of incrementing x, solving for y, and using one's understanding of the twofold ambiguity of a square root by illuminating the two pixels (x, y) and $(x, -y)$. Even if the increment in x is made as small as possible, the result is unsatisfactory. One might also think of the well-known parametric equations $x = a \cos \theta$, $y = b \sin \theta$ for an ellipse of semiaxes a and b, and take $a = b = R$. From a sequence of values of θ pixels can be determined that straddle the circle. An increment $d\theta = \sqrt{2}/R$ should be about right, because at 45°, the arc length between diagonally adjoining pixels is $\sqrt{2}$ pixel units. But because of the vagaries of rounding off, occasional values of R will produce unacceptable outcomes. If computer power permits and an attempt is made to subjugate the problem by use of much smaller increments of angle, other nasty things such as ugly clumping of extra pixels will occur. In surveying, circular curves for roads are laid out without reference to the center by knowing that $d\phi/ds$, the rate of change of direction, is constant, and an algorithm for arcs of extent less than 2π can be based on this property.

When I first had to pixelize some small circles, the parametric method seemed

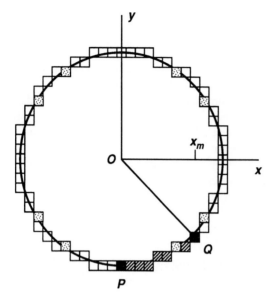

Figure 8-25   The method of pixelizing a circle.

appropriate, but the results were poor. Coloring in squares on graph paper proved to be the best solution and may sometimes still be useful for improving on shapes generated algorithmically. Indeed, the icon-driven drawing packages have raised the technique to a high level of convenience. Nevertheless, the circle algorithm is rather beautiful. Suppose that pixels have to be generated to represent the circle $x^2 + y^2 = R^2$, which passes through the point P in Fig. 8-25. For the time being take the radius R to be an integer. The radius OQ, which is inclined at 45°, defines an abscissa x_m which is the integer closest to $R/\sqrt{2}$. Once pixels whose abscissas range from 0 to x_m have been chosen, then the other seven octants can be filled in by symmetry. The short sector to be handled falls everywhere into the gentle slope category; therefore, only one value of y needs to be chosen for each abscissa from P to Q, though there will be more than one value of x for a given y. The ingenious technique for doing this is exhibited in the following segment. The subroutine labeled octuplicate simply makes eight pixels for each one pixel on the initial octant, as illustrated by the eight light grey pixels. A routine for this was given by Bresenham (1977), written in ALGOL with multistatement lines, as in the pseudocode that follows. The semicolon as a statement separator, despite intentions of international standardization, clashed with use of the semicolon for the suppression of a carriage return; that is the reason for the use of @as a statement separator.

The segment given expresses Bresenham's idea in distinctly less space; the compact subroutine that octuplicates the pixels is not significantly slower than spelling out the eight separate instructions, but may have been so in 1977 when arithmetic was slower.

When R is not an integer, the results are usually satisfactory, but occasional glitches will appear. For example, when R is half an odd integer, roundoff may cause asymmetry. Cure this by a slight change in R.

A CIRCLE ALGORITHM

```
R=7                                      To make a circle of radius 7
xm=INT(0.5+R/sqr(2))                     Largest abscissa
d=1-R                                    Initial value of discriminant
y=-R                                     Start at (0,-R)
FOR x=0 to xm                            The inner loop
    IF d<0 THEN d=d+x+x+1 ELSE y=y+1 @ d=d+x+x+1+y+y
    GOSUB octuplicate
NEXT x

octuplicate:                             Replicates one point into eight
FOR i=-144 TO 1 STEP 2 @ FOR j=-1 TO 1 STEP 2
    u=i*x @ v=j*y @ CALL "PSET" (u,v)    Illuminates pixel at (u,v)
u=i*y @ v=j*x @ CALL "PSET" (u,v)
NEXT j @ NEXT i
RETURN
```

Starting from an initial value $1 - R$, the discriminant d is incremented by an amount $2x + 1$ at each step and therefore generates the parabola $y = x^2 - R$, which guides the pixels along an approximation to the circle $y = -\sqrt{R^2 - x^2}$. However, for $R > 2$ the radial position error is never so great that a better position for any pixel can be found. There are a very few cases ($R = 5, 10, 17$) where the radial displacement of some pixel from the true circle slightly exceeds 0.5, but because of the angular location this is not enough to constitute a position error. Figure 8-26 shows circles for odd radii from 3 to 37 produced by the algorithm. Nesting the circles emphasizes the irregularities; nevertheless, no systematic errors can be seen against the inherently rough structure.

Octuplication results in the pixel at $(0, -R)$ being illuminated twice, which in the world of binary mathematics makes no difference to the appearance. But in the physical world of plotters and printers black on black can look different; an example is the use of overprinting on impact printers to simulate boldface. The effect is also seen with photocopiers and laser printers. A different consequence of addressing the same pixel twice is to eliminate it (as with screen buffers that think $1 + 1 = 0$). If unwanted, the overprinting effect is avoidable by quadruplicating the initial point of the arc and, when necessary, the terminal point.

Ellipses

At the point $(0, -b)$ on the ellipse $x^2/a^2 + y^2/b^2 = 1$ the radius of curvature is a^2/b. Therefore, the arc that sets out to the right from $(0, -b)$ and terminates where the upward slope reaches unity is approximated by a circular arc of radius a^2/b centered at $(0, a^2/b - b)$. A set of pixels may be generated with the circle algorithm and quadruplicated. The same may be done with the arc that starts upward from $(a, 0)$. Efficient algorithms for ellipses in other orientations exist. If the ellipse is specified by an equation or by parameters, however, and an inefficient algorithm will get the job done sooner, it is easy to determine whether a given pixel is centered inside or outside the ellipse. An arc whose

Figure 8-26   Discretized circles for odd radii 3 to 37 nested to permit examination of the shape accuracy.

slope passes from -1 to 1 via 0 can be discretized column by column, and duplicated, and the steep arc whose slope goes from -1 to 1 via ∞ can be discretized row by row.

FILLING A POLYGON

Binary objects represented on a discrete grid are bounded by pixelized line segments which, taken as a complete circuit, can be referred to briefly as closed polygons. The polygon is a binary object in itself and can be specified as a set of 0s and 1s, but to represent the solid object bounded by a closed polygon means assigning 1s to the grid points within the polygon boundary. This is easy enough to do by eye, as we know from coloring books, but instructing a computer to do it is rather interesting. Start with the simple case of Fig. 8-27(a), where a given boundary polygon is indicated. Perform a raster scan starting at the top left of the field until a pixel of the boundary is encountered; then change the 0s of the polygon specification to 1s until the next boundary pixel is encountered, and so on to the end of the row. This procedure will also work on reentrant polygons provided that, in each row, a count is kept of the number of boundary crossings. In an alternative way of looking at this, which is computationally significant, we can determine whether a given pixel is interior to the boundary by counting crossings on the way from that pixel to the edge of the field.

A different way of deciding whether a given pixel is interior is to select a boundary point and, by marching around the boundary, determine whether the angle seen from the given pixel accumulates one or more turns, as happens for interior points of (a) and (b),

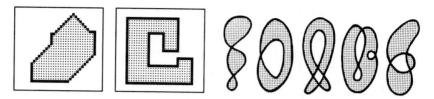

Figure 8-27 Filling a polygon depends on the rule for deciding whether a given point is interior or not. Whether a pixel is inside a given polygon can be determined by counting boundary crossings on a path to the outside, or by a winding rule. The boundary-crossing rule works cheerfully on open polygons, and also leads to contradictions when the way out crosses two boundaries sharing a pixel.

or zero turns, as for an exterior point. It is not necessary to visit all points of the circuit as described; the turns can be counted by starting at the given pixel and proceeding to the edge of the field as before and noting the boundary crossings. But instead of just counting them, combine them with a plus or minus sign according to the direction of march, which can be determined locally, provided the polygon was specified as a sequence of points (x_i, y_i).

Once it is decided that a point is interior, its value can be changed to unity. Then neighboring values can be changed as one works outward, until the boundary is reached; in this way exterior points do not have to be visited at all, except for the initial decision over the starting point. Algorithms for filling are given by Foley et al. (1990).

Turning now to Fig. 8-27(c), we see some of the diverse forms that polygons in reality can exhibit. The counting rules given above can be tested on these cases and will soon show that different filling patterns can result, according to which rule is adopted. The rule adopted for this figure can soon be deduced. Some degree of control is available by cutting and splicing the given polygon boundary to convert self-intersections (such as a figure eight has) to contact points (as with a pair of touching circles). A chessboard coloring problem calling for this kind of choice occurs at the end of the chapter.

Specification of a Polygon

Various ways of specifying a polygon have been mentioned. One is by the matrix having 1s on the boundary lines; another is by a sequence of the same pixels listed by their coordinates. The latter method can be abbreviated to a list of vertex coordinates, since the intervening pixels can be determined from the straight-line algorithm. Instead of absolute coordinates, relative coordinates can be given, as with a traverse; either a chain of octal steps may be specified, each step leading to the next pixel of a sequence, or the abbreviated set of relative vectors leading to the next vertex. For future use the vertex coordinates may be compiled en route as a polygon is traced out from chain code, link by link, or as the vertices are visited as specified by absolute coordinates. Conversely, absolute coordinates may be compiled as a polygon is traced by following relative instructions. Whether a polygon is closed may be determined from its specification by noting whether the total displacements, both east-west and north-south, accumulate to zero.

EDGE DETECTION AND SEGMENTATION

Since much of the information in a scene is conveyed by the edges, our eye-brain system has evolved to extract edges by preprocessing that begins right at the retina. Even the small tremor that the eyeball continually undergoes has the immediate effect of emphasizing sharp spatial variation in the retinal image. Further processing gives us awareness of the collinearity of line segments in the field of view, even though they are interrupted by intervening objects and are subject to nonuniform illumination. Such psychophysical phenomena are of great practical importance, for example in scene analysis performed by advanced robots fitted with cameras. Elaborate thought has been given to this problem (Horn and Brooks, 1989; Pratt, 1978).

Here we are concerned with very elementary preliminaries. For example, we ask whether there is some convolution operator which would detect edges in an image. We wish to be independent of the orientation of the edge; consequently, a circularly symmetrical operator is in question. A little experiment with one-dimensional signals suggests that strong slope is symptomatic of an edge. But if the ordinates are reduced by a factor of 10 to produce weaker slopes, we would not want the edges to alter. Therefore, absolute slope alone is not a sufficient indicator. Using slope/ordinate, which neutralizes the effect of reduction or magnification, will take care of this. In effect we are now finding steep slopes in the logarithm of the image values. If a change in base level is introduced by adding a constant to the whole image, the logarithmic derivative will be weakened. In a photograph, different illumination levels in different areas may mean that a perfectly perceptible edge running across an object subject to space-varying illumination will escape detection in places if a fixed threshold for slope/ordinate is adopted. Clearly, this is not a linear problem likely to respond to a simple convolution operator. The term "edge detector" sometimes used for combinations of smoothing and finite differencing is clearly overambitious.

Here is a proposal. Divide the scene into portions, within which the sort of problem associated with illumination level is not important. Determine the absolute gradient by adding the squares of the first differences in the two directions. Make a histogram of these gradients and determine the gradient that is exceeded by only 10 percent (say) of all values. Then mark the points having these large values. Where these points exhibit some continuity, rather than appearing in isolation or condensations, we have candidates for edges. The missing factor here is judgment; to locate edges by machine requires artificial intelligence. There may, however, be applications for a simple procedure in special circumstances where the same kind of image has to be repeatedly reduced to a line drawing.

Another simple operation that may be useful in special cases is based on the observation that, on a steep slope, the curvature changes sign. Therefore a locus of zero curvature is interesting, especially the one-dimensional curvature in the direction of the local gradient. This locus does not necessarily coincide with a steep slope. Nevertheless, normalization with respect to the ordinate and the effect of an additive term remain as disposable parameters.

Perceptible edges are seen even in the absence of a sharp change in function value. A

Figure 8-28   Perceptible edges need not entail a change of grey level nor an explicit outline.

change in texture with only a small change in grey level, or none at all, may be perceived as an edge or outline. Examples are given in Fig. 8-28. Clearly, any one of the edges could be found by an ad hoc algorithm tailored for the case; equally clearly, there is an endless variety of textures. One tool for discriminating between textures separated by an edge is the grey-level histogram and its statistical parameters (mean, mode, median, variance and kurtosis). Another is the local autocorrelation function, whose widths and orientation would be more effective on Fig. 8-28 than histogram analysis.

Segmentation describes an effort made in machine analysis of images to find the component parts. As a very simple example, on the right-hand side of Fig. 8-28, there are two circular objects which segmentation would hope to label. In a sense this is like edge detection, where the edge closes to define a segment. Other methods of segmentation invoke prior knowledge of shapes and the use of templates, prior knowledge of texture or color, prior knowledge of absolute dimensions, and other descriptors. Some of these topics are returned to below in connection with morphology and random images.

DISCRETE BINARY OBJECTS

Let an integer function $f(i, j)$ be defined on a two-dimensional array of points (i, j), where $0 \leq i \leq M - 1$ and $0 \leq j \leq N - 1$; call this set of points the support $\mathcal{S}$. Suppose that the function $f(i, j)$ has no negative values but does have zero values, which may be numerous. For example, there may be an island of positive values surrounded by a sea of zeros. The function $f(i, j)$ defines a set $\mathcal{F}$ whose members are those points (i, j) such that $f(i, j) > 0$; in the marine metaphor the set $\mathcal{F}$ consists of those array points where the land is above sea level. These entities are illustrated in Fig. 8-29. When the shape of the island is important in its own right, as distinguished from the full topographical height data, it makes sense to introduce the binary function $b(i, j)$, which has values of 0 and 1 only:

$$b(i, j) = \mathbf{u}[f(i, j)],$$

where

$$\mathbf{u}(n) = \begin{cases} 1 & n > 0 \\ 0 & n \leq 0 \end{cases},$$

n being an integer.

The unit step function of integer argument $\mathbf{u}(n)$ shown in Fig. 8-30 may be expressed in terms of the Heaviside unit step function of the continuous real variable x as $\mathbf{u}(n) =$

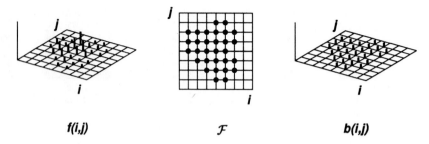

f(i,j) $\mathcal{F}$ **b(i,j)**

Figure 8-29   An integer function $f(i, j)$ defined on the set $\mathcal{S}$; the set $\mathcal{F}$ occupied by positive values of $f(i, j)$; and the binary function $b(i, j)$ which is equal to unity where $f(i, j) > 0$ and elsewhere is zero.

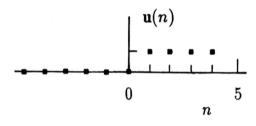

Figure 8-30   The unit step function of integer argument equals unity when the argument is a positive integer.

$\mathbf{H}(n - \frac{1}{2})$. The binary function $b(i, j)$ has much in common with the set $\mathcal{F}$, the difference being that $b(i, j)$ has numerical values and can be represented as a matrix (just as $f(i, j)$ can), while $\mathcal{F}$ is a set. The number of this set is 30 in the case illustrated. Nothing is gained by representing $b(i, j)$ in a 3D view when there is no diversity of function values to be exhibited. Instead it is conventional to use a cartesian plot with black bullets where the value is unity and, for the zero values, open circles, dots as in Fig. 8-31, or no symbol at all. Using some symbol for the zero values has the property that the support $\mathcal{S}$ is exhibited explicitly. Figure 8-31 shows such a representation, together with the conventional and strictly equivalent matrix of values of $b(i, j)$.

When a real function of two continuous variables is discretized by sampling on a lattice or grid, the result is a discrete object; when the sample values (measured to some precision, or specified analytically) are digitized for computer handling, we have a digital object or digital image. A binary object is a digital object having values 0 and 1 only and so is a rather restricted concept. Its point is that the binary object focuses attention on planar shape as distinct from the more general three-dimensional relief. Binary objects form part of the material dealt with by discrete morphology, which is concerned with shapes and their properties and is thus fundamental to image description.

Discrete morphology belongs to finite mathematics as do the theory of graphs and the theory of trees, and since graphs and trees can be given which are fully equivalent to any binary object, some of the established terminology of graph theory is immediately available. A graph is constituted by *vertices*, or *nodes*, and *edges*, as illustrated in Fig. 8-

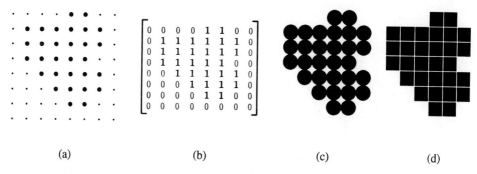

Figure 8-31 Different representations of a discrete binary object. A visual sense of shape is conveyed at (a), where ● stands for 1 and · stands for 0. The matrix formulation (b) is closer to programming needs and retains some visual appeal. Other conventional equivalents are shown at (c) and (d).

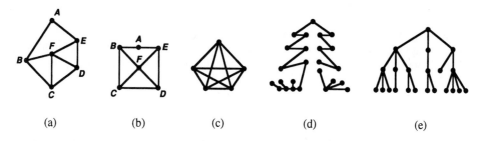

Figure 8-32 (a) A graph, (b) the same graph, (c) a nonplanar graph, (d) a tree, (e) another tree.

32. A *path* is a sequence of edges running from one vertex to another, no edge entering more than once; a path that begins and ends on the same vertex is a *mesh*. The length of a path is the number of edges entering into that path. Since, in general, there is more than one path between two vertices, there is more than one proximity measure associated with a pair of vertices, but there is a well-defined minimum *path length*. Vertices sharing an edge are *adjacent*. The *neighborhood* of a vertex is the set of adjacent vertices. A *graph* is a topological concept which is defined by its set of vertices and the set of vertex-pairs specifying the edges; it may be illustrated by a drawing, but if the drawing is distorted without altering the two defining sets, the graph is unchanged. Consequently, shape and size in the ordinary sense are not present in a graph. The illustration of a graph may be folded over on itself, spindled, crushed, stretched, or even collapsed into one dimension without altering the graph. A *nonplanar* graph is one that cannot be illustrated on the plane without edge-crossings; it can be illustrated in three dimensions without edges making contact except at vertices. A graph is *connected* if at least one path can be found from every vertex to every other, otherwise the graph is *disconnected*.

A *tree* is a connected graph with no meshes. Polygons, important constituents of digital images, specify graphs; an open polygon also defines a tree.

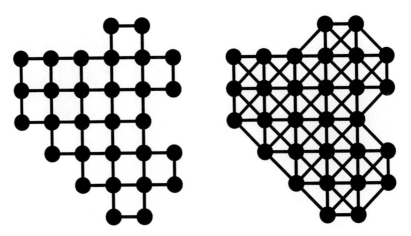

Figure 8-33   Different graphs associated with the same binary object.

By contrast with a graph, a binary object is restricted to a rigid substrate, the array or grid of points (i, j). That does not prevent us from associating one or more graphs with a binary object, and there are two of particular interest, illustrated in Fig. 8-33.

OPERATIONS ON DISCRETE BINARY OBJECTS

The specification of shape and size are fundamental to digital imaging, as well as the determination and manipulation of shape and size. We begin with discrete binary objects and immediately find that intuition based on experience with continuous mathematics does not always generalize readily. Take, for example, a shape with a concavity, such as a single island with a bay. Let C be the closed plane curve bounding the shape and let the set S consist of the points lying inside or on the boundary C. A straight-line segment lies inside C if every point of the segment is a member of S. Then the set S is convex if the line segments defined by every pair of members of S are all inside C. This summary is only a reminder of the set-theory approach and omits the preparatory discussion. Now, shall we call the binary object of Fig. 8-33 convex? There is a notch on the east side that would qualify as nonconvex in a continuous object, but discretization has destroyed the notion of a straight line consisting of members of the set. Some new conventions will be needed. If one is adopted that says the notch on the right nullifies convexity, will the similar notches in the southwest be excepted? If they are not treated differently, then only rectangular objects would qualify as convex. Is the indentation on the northwest to count as a "bay"? These are interesting questions. In many cases the phenomena of discrete mathematics are richer than those of the continuum, just as quantum mechanics is richer than classical physics.

The discussion begins with some simple operations familiar from set theory and moves on to dilation and erosion, extremely powerful tools that operate on (discrete) binary objects and produce useful features.

UNION AND INTERSECTION

Let $A(i, j)$ and $B(i, j)$ be two binary objects that may or may not overlap. They define a new binary object that describes the set of points occupied by $A(i, j)$, by $B(i, j)$, or by both. In set theory the union of two sets $\mathcal{A}$ and $\mathcal{B}$ is written $\mathcal{A} \cup \mathcal{B}$. Now $A(i, j)$ is a binary-valued function, not a set, but we could say that $\mathcal{A}$ is the set of points where $A(i, j) = 1$; then $\mathcal{A} \cup \mathcal{B}$ would be the set of points where either $A(i, j)$ or $B(i, j)$ or both equal 1. I am going to call the binary object $U(i, j)$ that occupies the set $\mathcal{A} \cup \mathcal{B}$ the *union* of the two binary sets, thus modestly extending the meaning of the word union, and the notation will be

$$U(i, j) = A(i, j) \cup B(i, j).$$

To compute $U(i, j)$ is straightforward. For example, one could visit each point and write

```
U(i,j)=MAX(A(i,j),B(i,j))
```

An alternative, which draws on the ability of the computer to evaluate logical relations, is exemplified by the following segment of code, which computes the union of $A(i, j)$ and $B(i, j)$.

UNION OF TWO BINARY OBJECTS

```
FOR i=1 TO M
   FOR j=1 to N
      U(i,j)=A(i,j) OR B(i,j)
   NEXT j
NEXT i
```

A moment's thought will verify that the truth table for OR matches the function MAX(,). Figure 8-34 gives this table and the corresponding ones for AND and EXOR which are referred to below.

Example

Write a program that defines two binary objects $A(i, j)$ and $B(i, j)$ on a grid of size 5×6 and compute their union $U(i, j)\triangleright$

```
INTEGER A(5,6),B(5,6),U(5,6)
M=5
N=6
FOR c=1 TO M
   FOR r=1 TO N
      A(c,r)=(ABS(c-3)<2)      Specify arbitrary objects
      B(c,r)=(r=3)
      U(c,r)=A(c,r) OR B(c,r)   Form the union
   NEXT r
NEXT c ◁
```

The intersection $\mathcal{A} \cap \mathcal{B}$ (Fig. 8-35) represented by $I(c, r)$ is formed using the line

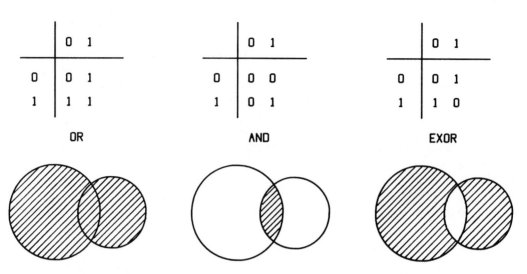

OR	0	1
0	0	1
1	1	1

AND	0	1
0	0	0
1	0	1

EXOR	0	1
0	0	1
1	1	0

Figure 8-34   Truth tables for OR, AND and EXOR and the corresponding circle diagrams introduced by John Venn (1834–1923).

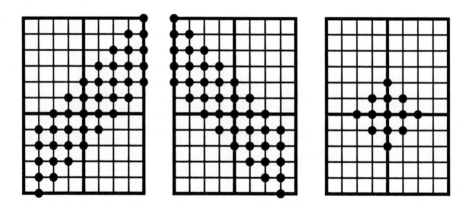

Figure 8-35   Two binary objects A and B and their intersection $A \cap B$.

```
I(c,r)=A(c,r) AND B(c,r)
```

in the program segment above; it is the binary object which is equal to 1 wherever both $A(i, j)$ and $B(i, j)$ are 1 and is zero elsewhere. It can also be calculated from

```
MIN(A(i,j),B(i,j))
```

The binary object which is situated where either $A(i, j) = 1$ or $B(i, j) = 1$, but not both, is computable as A(i,j) EXOR B(i,j) or alternatively from

```
MAX(A(i,j),B(i,j))-MIN(A(i,j),B(i,j))
```

Other operations that are needed include conversion to the complementary object $A^c(i, j)$ (0s and 1s interchanged), expressible as `NOT A(i,j)`, and the form `A(i,j) AND NOT B(i,j)`. The fact that ordinary programming languages accept this notation, as well as long expressions with many terms, means that the explicit logical relations compete with the $\cup$ and $\cap$ notation of Peano. Sometimes $\vee$ is used for the purpose of distinguishing discrete union from continuous; this hardly seems necessary and conflicts with the usage $p \vee q$ introduced into mathematical logic by Whittaker and Russell, where p and q are propositions. These authors said $\vee$ stood for Latin 'vel' meaning 'or.' (Unlike English, Latin had two words for or; 'exclusive or' was rendered by 'aut.')

An extraordinary diversity of mathematical symbols for logical relations can be found in the literature. Cajori (1929) criticized the mathematical world for failure to emulate the high degree of international standardization achieved in electrical engineering, while agreeing that an industry responds to forces that mathematicians do not feel. The convenient AND, OR, EXOR, NOT, from their use in computing, have gained ground in competition with the numerous symbols.

PIXEL MORPHOLOGY

A substantial body of theory has developed under the rubric of *mathematical morphology* from a rigorous statistical monograph by G. Matheron, which has formed a springboard for important developments in computer graphics. Mathematical morphology is too broad a subject to be dealt with here, and even *discrete morphology,* which includes graph theory, trees, and maps, is too broad. Instead, we limit ourselves to *pixel morphology*—and even there confine attention to binary-valued pixel arrays. In effect this means that function values do not count at all, only shapes.

Paths

Consider a square array of points. Each point is surrounded by eight other points, four of which lie in the cardinal directions N, S, E and W, the other four lying in the diagonal directions. A *path* is a sequence of array points, each of which is connected to the next by a cardinal or diagonal link. In general, a path may be self-intersecting.

Path Lengths

Given two separate points (a, b) and (c, d), one can specify many paths that begin at one point and end at the other. Of these, some paths are longer than others, but for distance specification we consider only the minimum such length. The king's-move distance, octal distance, or chessboard distance (a term adopted by Rosenfeld and Kak, 1976) is the number of moves made on a chessboard by a king at (a, b) trying to get to (c, d) in the fewest possible moves. It is easy to see that this distance is equal to the difference in abscissas or the difference in ordinates, whichever is greater. If the points lie on a diagonal, these two measures are the same. Thus,

$$\text{octal distance} = \text{king's move distance} = \max(|c - a|, |d - b|).$$

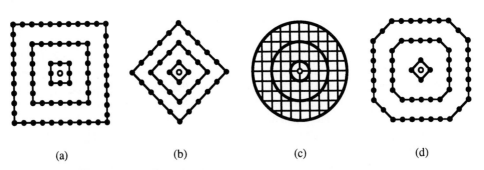

(a)   (b)   (c)   (d)

Figure 8-36   (a) Loci of constant octal (king's-move) distance, (b) of constant cardinal distance, (c) of constant Euclidean distance; (d) pixelized circles of radii 1, 3, and 5.

Figure 8-36(a) illustrates the concept with three loci of constant distance 1, 3, and 5 from a central point. Clearly, these loci are far from the intuitive concept of equal distance; nevertheless, for a king attacking a pawn, that is the reality, both on a chessboard and on a square array.

Some paths are composed of points connected by cardinal links only. Again, there are many such paths between two fixed points, and the minimum length is the cardinal distance between them, given by

$$\text{cardinal distance} = |c - a| + |d - b|.$$

Figure 8-36(b) illustrates three loci of constant cardinal distance. Also shown for comparison are the loci of constant Euclidean distance $\sqrt{(c - a)^2 + (d - b)^2}$, and the pixelized circles from Fig. 8-26, which, of course, follow the circular curves, since that was the intention.

The loci illustrated constitute closed lattice paths whose lengths can be evaluated; the ratios of perimeter to diameter for (a), (b), and (c) are 4, 2, and π, respectively. The same ratio for (d) oscillates weakly in the vicinity of $2\sqrt{2}$.

DILATION

Dilation and erosion, which were introduced in connection with functions of a continuous variable, acquire great practical importance in the world of pixel morphology. Erosion is the opposite of dilation, but if an object is first dilated and then eroded, the original is not in general recovered; it is modified in an interesting way that is called *closing,* because concavities in the boundary tend to be filled in. On the other hand, if an object is first eroded and then dilated, the outcome is equally interesting and is called *opening.*

These fundamental operations are very rich because the object is operated on with a structure function, a small object whose shape can be chosen in many ways. As a result the variety of applications is very broad. What is more, the computations are straightforward. The following sections aim to give a good selection of examples starting with dilation.

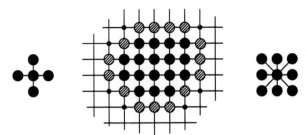

Figure 8-37 An object $A(\ ,\ )$ (large black mass), if dilated with the structure function $B(\ ,\ )$ on the left, accretes the surrounding shaded elements (those linked by edges of the underlying graph to vertices occupied by the mass). This is called *cardinal dilation*. If the structure function on the right is used to dilate, seven extra elements are accreted, as indicated by the small dots.

The treatment concentrates on explaining what the operations do and how to do them in practice, leaving the symbolic algebra to the last.

Figure 8-37 refers to the game of go, which is played on a 19×19 array of points that can be illustrated by a graph with 361 vertices and 684 edges. The two players alternately place black and white (shaded) stones on the points and from time to time remove captured masses. A mass is captured by a move that completes the occupation of all the vertices adjacent to those occupied by the enemy mass. A mass of stones of one color, such as the large group of black stones in the figure, is, of course, equivalent to a binary object. White has just played, has fully surrounded the large black mass, and is about to remove the captured black stones. The score is the difference between the areas occupied by the two colors, and the game terminates when the players agree that the score will not be altered by further play.

Each of the beseiging white stones is connected to a black stone by an edge of the graph ruled on the board. There are four kinds of such links, one in each of the cardinal directions, as illustrated by the link pattern on the left. If now we change the color of the surrounding stones to black, the resulting enlarged black mass is an example of dilation. In this case the rule for dilation is: "Annex all vertices that are connected by an edge to the vertices occupied by the given mass." A different game can be imagined in which the rule of capture requires an impermeable wall comprising the white stones shown plus seven more at the vertices marked by small dots. This would lead to a different example of dilation: "Annex all vertices linked to the given mass by an edge or by a diagonal link." The allowable links in this case are illustrated by the link pattern on the right by showing a central stone and the set of linked stones. The two discrete binary objects on the left and right are the structure functions for the two examples of dilation.

Represent the given mass by the matrix $A(i, j)$, the structure function by $B(i, j)$, and the dilated mass by $D(i, j)$. Then, by analogy with established notation for the direct sum in matrix theory, we write

$$D(i, j) = A(i, j) \oplus B(i, j).$$

The operation of A on B is the same as for B on A; that is, $A \oplus B = B \oplus A$. The reason for this was made clear in Chapter 2, where dilation in the continuous domain was identified in terms of two-dimensional convolution. Computing the dilation in the discrete domain can be done by a two-dimensional convolution algorithm; in other words, using notation that has already become familiar,

$$D(i, j) = \mathbf{u}[A(i, j) ** B(i, j)],$$

where $\mathbf{u}(n)$ is the discrete unit step function that converts positive integers to unity. Thus the operation of dilation falls well within what we already know. In this way of looking at things the structure function is a discrete point response function and the dilation rule can be expressed: "Replace every stone by five stones in the agreed link pattern but take away surplus stones which are piled on top of another stone."

Convolution of matrices results in values being attached to the matrix locations, whereas in morphology only shapes count; therefore, there should be something that is more efficient than accumulating products and throwing away the values. The answer is to take the structure function $B(i, j)$, give it half a turn to get $B(-i, -j)$, and slide it over $A(i, j)$, point by point, noting whether there is overlap or not. This can be decided by a logical OR, as shown in the following snappy segment for performing cardinal dilation.

CARDINAL DILATION

```
FOR i=1 TO M
   FOR j=1 TO N
      D(i,j)=A(i,j) OR A(i+1,j) OR A(i,j+1) OR A(i-1,j) OR A(i,j-1)
   NEXT j
NEXT i
```

As the structure function for cardinal dilation has only five elements, the essential action can be written on one line, and $B(,)$ does not appear explicitly. If $B(,)$ is more complicated, that one line can be expanded to a summation over the field occupied by $B(,)$. However, the explicit form is often appropriate, as with the following exercise, which puts a shadow on the SE side of the given object $A(,)$ by dilating with the simple structure function

$$\begin{bmatrix} 0 & 0 & 0 \\ 0 & 1 & 0 \\ 0 & 0 & 1 \end{bmatrix}.$$

The central **1** replicates the object, and the SE outlier replicates with a SE shift. Indexing from 1 to M and from 1 to N means that the origins of both A and B are lost. The origin of D may be recovered by a subsidiary vector addition $(i_A, j_A) + (i_B, j_B)$ to get (i_D, j_D); however, it is advisable to verify the overall output independently because of possible sign errors. One source of error is inconsistency in treating the indexing as a shift of origin as against a shift of the object; another is confusion between row number (usually measured downward) and ordinate (measured upward). In addition, neither languages nor output devices agree on the meaning of (i, j). However, in some applications the absolute position of the output is not critical, only the shape.

SE SHADOW GENERATED BY DILATION

```
FOR i=1 TO M
   FOR j=1 TO N
      D(i,j)=A(i,j) OR A(i-1,j+1)
   NEXT j
NEXT i
```

Figure 8-38 A binary object (left) subjected to progressive dilation (above) and erosion (below).

In the code for cardinal dilation, if the term $A(i, j)$ is omitted, then, in the example shown, the dilation $D(i, j)$ will be unchanged. That is because there is no member of the mass that is not connected to some other member by an edge. But if there was a member linked diagonally only, the dilation would be changed. Many interesting effects can be found by experimenting with structure functions, including those that are not compact or small.

In the code that adds a shadow in the SE, if one replaces `A(i-1,j+1)` by `A(i-N,j+N)` and N is large enough, then dilation results in duplication of the original object, the extra copy being displaced $(N, -N)$ steps (toward the SE).

Progressive dilation is illustrated in Fig. 8-38 by an original object (left) in the form of a 36×33 representation of Australia. The top row shows a sequence computed by cardinal dilation. The boundaries are loci of constant cardinal distance (see below) from the coast. The bottom row illustrates progressive erosion, where the boundaries are at progressively increasing distance from the original coast. We turn now to erosion.

Erosion

First consider cardinal erosion, the general effect of which is to peel a rind from a given object $A(\ ,\ )$. The structure function chosen is

$$B(\ ,\ ) = \begin{bmatrix} 0 & 1 & 0 \\ 1 & \mathbf{1} & 1 \\ 0 & 1 & 0 \end{bmatrix}$$

as for cardinal dilation, but now a test point qualifies for membership of the erosion of $A(\ ,\ )$ by $B(\ ,\ )$ if and only if the structure function, when centered on that point, fits into $A(\ ,\ )$. In Fig. 8-37, if the black and shaded elements taken together constitute the given object, then the shaded outer rind will be peeled off by cardinal erosion (if the structure function is centered on any one of these shaded points, then one or more of its elements will protrude outside the given object). In another way of looking at this, which enables one to strike out the lost elements by inspection, only those interior members of $A(\ ,\ )$ protected by adjacent members in all four cardinal directions will not be eroded.

It was shown above that dilation with a structure function $\begin{bmatrix} \mathbf{1} & 0 \\ 0 & 1 \end{bmatrix}$ added a fringe on the SE. Erosion with the same structure function removes a fringe from the SE.

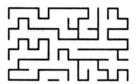

Figure 8-39 A maze whose ⊣ junctions are locatable by erosion.

Erosion can also be presented in terms of convolution just as in the case of dilation, and the alternative mode of approach can be helpful. We use it here to arrive at the method of computing erosion by means of the logical operator AND. Convolving with the structure function $\begin{bmatrix} \mathbf{1} & 0 \\ 0 & 1 \end{bmatrix}$ superposes a shifted copy of a given binary object on itself, the shift being one diagonal step SE. The superposition results in a ternary object, but we are concerned here only with morphology; the eroded set will comprise those elements which are common to the object and its translate—in other words, the intersection. Thus the matrix $E(i, j)$ representing the eroded set can be expressed as

```
E(i,j) = A(i,j) AND A(i-1,j+1)
```

If this line is substituted in the code segment given above for a SE shadow, the effect will be to erode a SE fringe. For a more general structure function with just a few elements, for example

$$\begin{bmatrix} 1 & 0 & 1 \\ 0 & \mathbf{0} & 0 \\ 1 & 0 & 1 \end{bmatrix},$$

the line can be extended to read

```
E(i,j) = A(i+1,j+1) AND A(i-1,j+1) AND A(i+1,j-1) AND A(i-1,j-1)
```

If the elements of the structure function are numerous, a summation loop can be used.

Erosion can be used as a feature locator, as illustrated in Fig. 8-39, which shows a maze $A(\ ,\ )$ containing a variety of junctions. The treasure is reputed to be buried at a ⊣ junction, so we erode with

$$B(i, j) = \begin{bmatrix} 0 & 1 \\ 1 & \mathbf{1} \\ 0 & 1 \end{bmatrix}.$$

The only locations where this structure function can be placed and still fit into $A(\ ,\ )$ are exactly those places where we would wish to dig.

There is a similarity between this technique and the practice of matched filtering, which could be described as follows. Shift a template (a replica of the structure function) to each location of $A(\ ,\ )$ in turn, multiply corresponding elements, and write the sum of the products at that location. Wherever the template agrees with the substrate $A(\ ,\ )$ a large value will be recorded, and the appearance of such peaks will constitute detection of

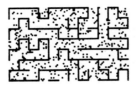

Figure 8-40   The same maze heavily spattered with binary noise.

the looked-for pattern. The values surrounding the peak contain extra information, namely the two-dimensional autocorrelation of the pattern under search, information which could guard against false alarms. However, the simpler multiplication-free arithmetic of the erosion procedure, coupled with the directness of the output map in Fig. 8-39, are impressive as compared with matched filtering. The advantage stemmed from looking for a shape only, which is what morphology is good for. Remembering that the theory of matched filtering produces an optimum signal-to-noise property when a signal of known form is sought in the presence of additive noise with simple statistical properties, we might inquire how resistant erosion would be in the presence of noise. Imagine the maze to be spattered with background noise in the form of randomly placed pixels of value unity. Where such a noise speckle falls on a line of the maze, take the value to remain unity to retain the binary character of the image (Fig. 8-40). One sees immediately that preprocessing by erosion with [1 0 1/0 **0** 0/1 0 1], as just discussed, will remove all isolated speckles as well as any that happen to be diagonally linked. The maze lines themselves will be undamaged but will have acquired a number of warts. All the ⊣ junctions will be found, together with some false responses due to a wart on the left of a vertical line. Foreseeing this, we could use a little more computing time by eroding it with

$$B(i, j) = \begin{bmatrix} 0 & 0 & 1 \\ 1 & 1 & \mathbf{1} \\ 0 & 0 & 1 \end{bmatrix},$$

extending the horizontal stem to the left by one more pixel. Other ways of taking advantage of prior knowledge would occur to an earnest treasure seeker, and it is clear that, even in the absence of a serious theory, feature detection by erosion could be highly resistant to additive noise.

Closing and Opening

It has already been remarked that dilation followed by erosion does not necessarily give the same result as erosion followed by dilation, using the same structure function throughout. The example of dilation/erosion shown in Fig. 8-41 exhibits a character feature: a vacancy in the original gets filled. This property gives dilation/erosion the name of closing. The operation can be used for repairing the boundaries of objects that are notched or have internal holes.

Erosion followed by dilation may restore the original, as would be the case if Fig. 8-41(d) were dilated, but small projections from a boundary may be removed and notches may be deepened. Fig. 8-42 is an example.

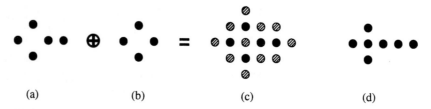

(a)   (b)   (c)   (d)

Figure 8-41   Closing. An object (a) first dilated by the structure function (b) accretes the additional shaded elements as shown at (c). After erosion, the new object (d) retains the original elements plus another which closes the original eye.

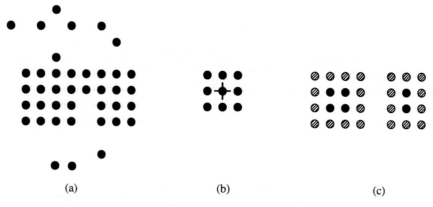

(a)   (b)   (c)

Figure 8-42   Opening. An object (a) after erosion by the structure function (b) consists of the six black elements in (c). Dilation then accretes the shaded elements. The fringing rubbish has been cleaned up, the small projection on the upper boundary has been smoothed out, and the notch on the lower boundary has opened.

Figure 8-43 summarizes the notation for dilation, erosion, closing, and opening.

Fringe Extraction

Closing and opening are only two examples of the wide variety of image operations based on dilation and erosion. Clearly, if an object is subjected to cardinal dilation and then the object itself is subtracted, what remains is an outer fringe. If performed on an island whose boundary was defined by mean sea level, this composite operation would leave what might be described as the low-tide zone. The high-tide zone, or inner fringe, would be represented by the original object from which the cardinal erosion was subtracted. The intertidal zone, a thicker sort of fringe, would be obtained by subtracting the cardinal erosion from the cardinal dilation. Thus,

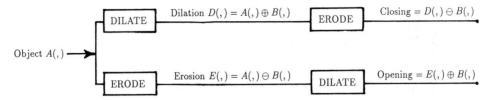

Figure 8-43 Notation for dilation, erosion, closing and opening of a binary $A(\,,\,)$ in terms of a structure function $B(i, j)$.

$$\text{outer fringe} = A \oplus B - A,$$
$$\text{inner fringe} = A - A \ominus B,$$
$$\text{thick fringe} = A \oplus B - A \ominus B.$$

Other structure functions produce other sorts of fringes, or indeed patterns that would not be well described by the term fringe. However, the structure function used to add a SE shadow to an object does produce a sort of fringe, namely the shadow itself. Practical experimentation with a variety of structures functions is the way to gain experience and prepare for future application.

Feature Extraction

It may be important for you to identify the shadow of a hawk or some other pattern in a scene. As a trivial illustration, let us find the places where **L** occurs in object A of Fig. 8-44(a). We can find all the places where an **L** can be fitted into the object (which is a random scatter of nothing but **L**s, added modulo 2) by eroding with **L**. However, the occurrences of a simple uncontaminated **L** are rather fewer. To identify those we have to find occurrences of

$$L(i, j) = \begin{bmatrix} 0 & 0 & 0 & & & & \\ 0 & 1 & 0 & & & & \\ 0 & 1 & 0 & & & & \\ 0 & 1 & 0 & & & & \\ 0 & 1 & 0 & & & & \\ 0 & 1 & 0 & & & & \\ 0 & 1 & 0 & 0 & 0 & 0 & 0 \\ 0 & \mathbf{1} & 1 & 1 & 1 & 1 & 0 \\ 0 & 0 & 0 & 0 & 0 & 0 & 0 \end{bmatrix}.$$

Eroding with this structure function does not allow for our insistence on the guard zone of zeros and will produce the outcome E shown in Fig. 8-44(b). If now we compute the cross-correlation of L_{dilated} on A, where

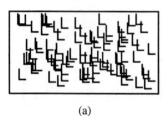
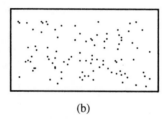
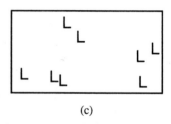

(a)   (b)   (c)

Figure 8-44   (a) A scatter of **L**s constituting an object $A(\,,\,)$, (b) the locations of the **L**s as discovered by eroding A with **L**, (c) the locations of the clean **L**s.

$$
L_{\text{dilated}}(i,\,j) =
\begin{bmatrix}
1 & 1 & 1 & & & & \\
1 & 1 & 1 & & & & \\
1 & 1 & 1 & & & & \\
1 & 1 & 1 & & & & \\
1 & 1 & 1 & & & & \\
1 & 1 & 1 & & & & \\
1 & 1 & 1 & 1 & 1 & 1 & 1 \\
1 & 1 & 1 & 1 & 1 & 1 & 1 \\
1 & 1 & 1 & 1 & 1 & 1 & 1
\end{bmatrix},
$$

the values can range from 0 to 37 (the number of nonzero elements in L_{dilated} but when L_{dilated} is positioned at one of the nonzero elements of E, as shown in Fig. 8-44(b) the value of the cross-correlation will be 11 (the number of nonzero elements in B) only where the **L** is surrounded by the specified guard zone of zeros.

Full cross-correlation is not necessary; only cross-correlation sums at the nonzero values of E need to be evaluated. Consequently, it will be a waste of computer time to call a cross-correlation subprogram; all that is needed is a multiplication-free summation over the domain of L_{dilated} when translated to the locations within A that are specified by E. In many simple cases an ad hoc approach such as described may be suitable, but in complicated cases a general formulation is available.

Suppose that clean features $L(i,\,j)$ have to be extracted from an object A, where L may describe any shape but we adopt the symbol L as a mnemonic. A clean L is one that is surrounded by a guard zone into which other features of the object do not intrude. Introduce a matrix $L_{\text{guard}}(i,\,j)$ defined as the outer fringe of L, or

$$
L_{\text{guard}} = L \oplus
\begin{bmatrix}
1 & 1 & 1 \\
1 & 1 & 1 \\
1 & 1 & 1
\end{bmatrix} - L = L_{\text{dilated}} - L.
$$

The character of L_{guard} can be seen by reference to the earlier example, where we would have had

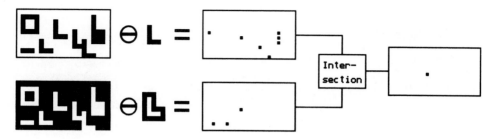

Figure 8-45 Procedure for locating features L in a binary object A by eroding with L, eroding the complement A^c by L_{guard}, and taking the intersection.

$$
L_{\text{guard}} =
\begin{bmatrix}
1 & 1 & 1 & & & \\
1 & & 1 & & & \\
1 & & 1 & & & \\
1 & & 1 & & & \\
1 & & 1 & & & \\
1 & & 1 & & & \\
1 & & 1 & 1 & 1 & 1 & 1 \\
1 & \mathbf{0} & & & & & 1 \\
1 & 1 & 1 & 1 & 1 & 1 & 1
\end{bmatrix}
\quad \text{or} \quad \mathbf{L}.
$$

As was explained in connection with Fig. 8-44, the erosion $E = A \ominus L$ includes the wanted locations of the feature L being looked for, but also additional unwanted locations. To eliminate these, invert the object A to get its complement A^c and find the places where L_{guard}, which can be pictured as a hollow template of the form **L**, can fit over an occurrence of a clean L. The erosion $E_{\text{guard}} = A^c \ominus L_{\text{guard}}$ includes the wanted locations, but also, as before, some unwanted ones, such as places where a fragment of an L may occur fully inside the template. The desired set of locations of a clean L will be the intersection of the two erosions, $E \cap E_{\text{guard}}$. The full procedure is illustrated in Fig. 8-45.

How to compute the dilation and intersection has already been explained. How to dilate L to get L_{guard} depends on how much isolation from the surroundings is thought necessary. Conversely, it might be acceptable if the object was in diagonal contact with a neighboring feature; in that case cardinal dilation instead of dilation with the nine-element matrix would be appropriate. The line of code that forms L_{guard} might read

```
Lguard(i,j)=Ldilated(i,j) AND NOT L(i,j)
```

while the complement of the object is formed by

```
Ac(i,j)=NOT A(i,j)
```

Thus all the symbolic algebra translates readily into computer terms.

One can go a little further than the above example and clean up small amounts of noise that would interfere with the precise matching that the procedure calls for. A charming example of this has been given by Lohman (1991), as shown in Fig. 8-46. Here

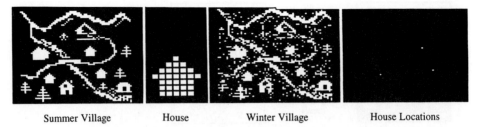

Summer Village	House	Winter Village	House Locations

Figure 8-46   Feature extraction in the presence of binary noise illustrated by a problem of searching for houses in a village.

Figure 8-47   Shadows on the ground may reveal the presence of predators.

a standard house is the feature to be discovered, and in the summer scene the three houses are readily findable, but the snowy winter scene has salt-and-pepper noise that distorts and obscures the outlines. The cleaning up was done by closing the image before eroding to obtain E and by opening the image before eroding to obtain E_{guard}.

The considerable power and versatility of manipulation based on dilation and erosion is well illustrated by this example. Figure 8-47, which shows the shadows of three clearly identifiable hawks, reminds us, however, that the sophistication of the mathematical technique introduced here falls far short of the feature extraction that would be needed by a small rodent. The fact that houses are always vertical and often identical gives the alpine village plausibility for explaining some of the complicated mathematics which will be needed for the even more complicated problem of feature extraction in the real world.

The shape of the feature being sought is subject to tolerances, such as those that result from spatial phase of the discrete elements and from orientation of the object relative to the discretizing axes. To allow for these tolerances, the go no-go templates described above need to be relaxed. The design of such templates then takes on statistical features and requires attention to the cost of false alarms and failures to recognize. Statistically optimal considerations have been studied by Dougherty and Zhao (1992).

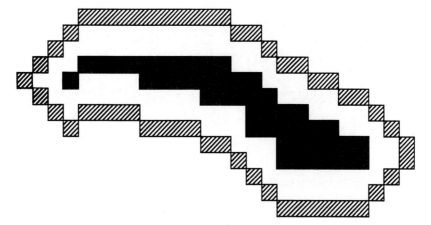

Figure 8-48 An island (dark) and the locus (shaded) of pixels that are at a distance of three cardinal steps from the nearest pixel.

Coastal Distance

Consider the dark mass in Fig. 8-48 representing a binary object and the shaded surround, which has been placed so that the cardinal distance from any member of the surround is three steps from the nearest member of the mass. If a different distance measure had been chosen— for example, if diagonal steps were allowed, as with octal distance—the surround would move outward in places where the gap can now be bridged by a knight's move. If the quasi-euclidean measure based on the pixelized circle of radius 3 were used, the surround would move out a little more, since there is no point on that circle (Fig. 8-26) which is as close to the center as a knight's move. Clearly, the locus of constant offshore distance depends on the distance measure appropriate to the circumstances. Figure 8-48 can easily be constructed by hand. In a more complicated case, given an object $A(i, j)$, how does one determine the object $S(i, j)$ representing the surround at cardinal distance N? If $N = 3$, as in the figure, clearly the answer is to dilate $A(i, j)$ to obtain $D_1(i, j)$, dilate again to obtain $D_2(i, j)$, and again to obtain $D_3(i, j)$. Then $S_3(i, j) = D_3(i, j) - D_2(i, j)$.

The three-step surround is in some ways less complicated than the boundary of the original object and, when the distance is increased becomes simpler still. How the boundary propagates outward can be quickly discovered with pencil and graph paper; in the limit only boundaries in the eight principal directions survive. This propagation effect is illustrated in Fig. 8-49, where the process of conversion of a figure **2** can be followed. Small glitches are progressively rectified, and diagonal sides ultimately dominate.

To determine the locus of constant octal distance one dilates in accordance with the nine-element structure function, and the procedure is otherwise the same. Although distances from a boundary can conveniently be obtained by the dilation technique, and these distances are appropriate to some specific situations, it is noteworthy that the numerical

Figure 8-49   An object in the form of a **2** has been progressively dilated, but alternate stages have been suppressed to bring out the character of the changes.

values are dependent on the orientation of the original object, because the loci of constant distance are square and not circular. The outcome is totally unacceptable for some situations. One approach to defining a locus at integer distance R from the boundary of an object is to replace each member of the original set by the pixelized circle of radius R centered on that member; the fringe of the resulting set, as determinable by erosion, is such a locus.

Since computation of a sum of squares as needed for Euclidean distance is found to be computationally too costly in some applications, a lot of thought has been given to computationally simple ways of approximating Euclidean distance. In a departure from the restriction to binary-valued functions, one can associate numerical values other than 0 and 1 with each pixel. Shih and Wu (1991) apply the concept of a grey structuring function k such as

$$k = \begin{bmatrix} & \frac{1}{4} & \\ \frac{1}{4} & 1 & \frac{1}{4} \\ & \frac{1}{4} & \end{bmatrix},$$

which, on an object A, produces a grey dilation $A \oplus_g k$ defined as

$$A \oplus_g k = \sup\{A(x - m, y - n) + k(m, n)\}$$

with proper attention to edge effects. Grey erosion is defined as

$$A \ominus_g k = \inf\{A(x + m, y + n) - k(m, n)\}.$$

These powerful adjuncts to the toolkit exceed the morphological limitations respected by binary objects but can be used for purely morphological purposes.

Distance inland from the coast can be studied by successive erosions, and the locus of constant interior distance from the boundary can be found in an analogous way. A new feature sets in, because erosion terminates when the last member disappears. That member is, of course, the distinctive element that is farthest from the coast. However, it need not be a single pixel; there may be two different poles of isolation or, indeed, a very elaborate pattern of points that would simultaneously disappear on the next stage of erosion.

Skeleton

Before the last elements erode, there may be earlier disappearances of single isolated elements or groups of elements. The set of all such members of A constitutes its *skeleton*. It has the interesting property that the original object can be restored from knowledge of the skeleton and the distances of the parts from the boundary (Rosenfeld and Kak, 1976; Davies and Plummer, 1981; Jain, 1989). The method is to replace each element of the skeleton by the neighborhood of all elements not further away than that skeletal point is from the boundary. This simple geometrical idea offers a means of image data compression in some applications.

CODING A BINARY MATRIX

A simple binary image is equivalent to a binary matrix, as in the following example.

$$\begin{bmatrix} 0 & 0 & 1 & 1 & 1 & 1 & 0 & 0 \\ 0 & 0 & 1 & 0 & 0 & 1 & 0 & 0 \\ 0 & 0 & 1 & 1 & 1 & 1 & 0 & 0 \\ 0 & 0 & 1 & 0 & 0 & 1 & 0 & 0 \\ 0 & 0 & 1 & 1 & 1 & 1 & 0 & 0 \\ 0 & 0 & 1 & 0 & 0 & 1 & 0 & 0 \\ 0 & 1 & 0 & 0 & 0 & 1 & 0 & 0 \\ 0 & 0 & 0 & 0 & 1 & 0 & 0 & 0 \end{bmatrix}$$

Both the image and the matrix are bulky and clearly susceptible to compression. For example, the first row of the matrix, [0 0 1 1 1 1 0 0], is equivalent to the binary number 00111100 = 60. The square matrix can therefore be condensed to the column matrix

$$\begin{bmatrix} 60 \\ 36 \\ 60 \\ 36 \\ 60 \\ 36 \\ 68 \\ 8 \end{bmatrix}$$

Conversely, we can generate the equivalent row; thus 60 generates the binary number 00111100. The whole binary matrix condenses to the string

```
" 60 36 60 36 60 36 68  8"
```

Three spaces are allowed in the case of an 8×8 matrix, because a row can use three decimal places. The matrix elements $a(r, c)$ can conveniently be recovered from the condensed string using the decimal-to-binary function of a high-level language. A further aspect is that binary image data given in the form of a two-dimensional matrix can be converted to a column matrix automatically in the computer by matrix multiplication. Thus

$$
\begin{bmatrix}
0 & 0 & 1 & 1 & 1 & 1 & 0 & 0 \\
0 & 0 & 1 & 0 & 0 & 1 & 0 & 0 \\
0 & 0 & 1 & 1 & 1 & 1 & 0 & 0 \\
0 & 0 & 1 & 0 & 0 & 1 & 0 & 0 \\
0 & 0 & 1 & 1 & 1 & 1 & 0 & 0 \\
0 & 0 & 1 & 0 & 0 & 1 & 0 & 0 \\
0 & 1 & 0 & 0 & 0 & 1 & 0 & 0 \\
0 & 0 & 0 & 0 & 1 & 0 & 0 & 0
\end{bmatrix}
\times
\begin{bmatrix}
128 \\ 64 \\ 32 \\ 16 \\ 8 \\ 4 \\ 2 \\ 1
\end{bmatrix}
=
\begin{bmatrix}
60 \\ 36 \\ 60 \\ 36 \\ 60 \\ 36 \\ 68 \\ 8
\end{bmatrix}.
$$

GRANULOMETRY

An interesting application of morphological techniques is to granulometric analysis of sand and silt, as practiced in civil engineering since the nineteenth century. A sedimentary rock contains polyhedral particles of size ranging from, let us say, 1 to 5, as measured by the largest dimension seen under the microscope, and in each size range the mass of material is about the same. For example, for each large particle of size 5 there are 125 of size 1. This specification is not representative of sedimentary rocks in general, the particles of which will often have been sorted in various ways by water and wind. Figure 8-50 was constructed by depositing particle images distributed in size as follows:

Size	1	2	3	4	5
Number	375	47	14	6	3

The size distribution obtained by counting through the microscope is, of course, different, because, in the figure, some of the smaller particles are obscured by the larger. If the particles were removed from the rock by dissolving the matrix they are imbedded in and shaken loosely over a microscope slide, the count would be closer to reality. In the figure, the count is further skewed toward larger sizes because two particles that overlap may be counted as one larger particle. An opposing tendency, toward smaller sizes, arises from particle orientation in space. In the specific numerical case mentioned one can estimate the correction to each size range due to the phenomena described; the inverse problem, starting not from a given distribution but from observation, is harder.

Related problems arise when rock is sectioned into slices so thin as to be translucent. Few small particles are then lost by obscuration, but many fail to be counted because they are excluded from the slice; the apparent size of other particles is affected by their positioning within the slice.

The statistical theory required for the industrially important subject of sedimentology was developed largely by G. Matheron (1975) in France and forms the basis for the abundant modern literature of image morphology. Thus granulometric analysis is not merely an interesting application of morphology but is a technological driving force behind new mathematics with relevance to image processing.

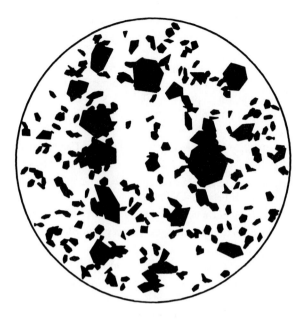

Figure 8-50 Particles whose frequency of occurrence varies as d^{-3}, where d is the major diameter, have been deposited to simulate the view through a microscope. An apparent particle-size distribution over five size ranges can be arrived at by counting.

CONCLUSION

This chapter has shown how many operations applicable to functions of two continuous variables—smoothing and differentiation, for example—can be adapted to functions of discrete variables and how in most cases additional considerations, such as the provision of a guard zone, arise. Many other operations are specific to discrete images. Pixelization of straight lines, dilation, and other discrete morphological operations are of this kind.

Discussion of digital aspects is, however, distributed throughout this book, beginning with convolution. The value of associating the discrete with the continuous in the case of the all-pervasive concept of convolution is apparent. Furthermore, as explained in the Introduction, discretely defined functions can be subsumed under the continuous by use of delta-function notation, and, in the cases of convolution and Fourier transforms, it is advantageous to do so. Other topics, including sampling, are also best treated in company with related continuous-variable material rather than by segregation of the discrete from the continuous.

LITERATURE CITED

R. N. BRACEWELL (1963), "Correction for grating response in radio astronomy," *Astrophysical Journal,* vol. 137, pp. 175–183.

J. C. BRESENHAM (1965), "Algorithm for computer control of digital plotter," *IBM Systems J.*, vol. 4, pp. 25–30.

J. C. BRESENHAM (1977), "A linear algorithm for incremental digital display of circular arcs," *Comm. ACM*, vol. 20, pp. 100–106.

F. CAJORI (1929), *A History of Mathematical Notations*, Open Court Publishing Company, Chicago.

E. R. DAVIES AND A. P. N. PLUMMER (1981), "Thinning algorithms: a critique and a new methodology," *Pattern Recognition*, vol. 14, pp. 53–63.

E. R. DOUGHERTY AND D. ZHAO (1992), "Model-based characterization of statistically optimal design for morphological shape recognition algorithms via the hit-and-miss transform," *J. Visual Communication and Image Representation*, vol. 3, pp. 147–160.

J. D FOLEY, A. VAN DAM, S. K. FEINER, AND J. F. HUGHES (1990), *Computer Graphics Principles and Practice*, Addison-Wesley, Reading, MA.

B. K. P. HORN AND M. J. BROOKS (1989), *Shape from Shading*, The MIT Press, Cambridge , MA.

A. K. JAIN (1989), *Fundamentals of Digital Image Processing*, Prentice-Hall, Englewood Cliffs, NJ.

A. LOHMAN (1991), "Object recognition using image algebra," *Angewandte Optik, Annual Report*, p. 41, Physikalisches Institut der Universität Erlangen, Nürnberg.

G. MATHERON (1975), *Random Sets and Integral Geometry*, John Wiley, New York.

R. POOL (1988), "When crystals collide: grain boundary images," *Science*, vol. 30, pp. 1601–1602.

W. K. PRATT (1978), *Digital Image Processing*, John Wiley, New York.

A. ROSENFELD AND A. C. KAK (1976), *Digital Picture Processing*, Academic Press, New York.

F. Y. SHIH AND HONG WU (1991), "Optimization on Euclidean distance transformation using grayscale morphology," *J. Visual Communication and Image Representation*, vol. 3, pp. 104–114.

F. STOCKTON (1963), "Algorithm 162: xymove plotting," *Communications of the ACM*, vol. 6, p. 161.

A. R. THOMPSON AND R. N. BRACEWELL (1974), "Interpolation and Fourier transformation of fringe visibilities," *Astronomical Journal*, vol. 79, pp. 11–24.

PROBLEMS

8–1. *Smoothing transfer operator.* Show that convolution with the nine-element square matrix summing to unity has the effect in the transform domain of multiplying by

$$\frac{\sin 3\pi u \sin 3\pi v}{9 \sin \pi u \sin \pi v}$$

and that this is the same as $(1 + \cos 2\pi u)(1 + \cos 2\pi v)/9$. ☞

8–2. *Five-point smoothing.* Examine the effect of including the value $f(i, j)$ with its four neighbors and taking the five-point average for both of the cases previously considered for four-point averaging. ☞

8–3. *Skew smoothing.* Construct an artificial image and add a random perturbation to each element. Convolve with

$$\frac{1}{3} \begin{bmatrix} 0 & 0 & 1 \\ 0 & 1 & 0 \\ 1 & 0 & 0 \end{bmatrix}.$$

Choose a means of presentation that makes clear what general effect has been produced.

8–4. *Dot-printer resolution.* Suppose that a dot printer produces "brilliant graphics with 300 dpi resolution." Assuming that a dot is circular and black, what is the transfer function of this printer? ☞

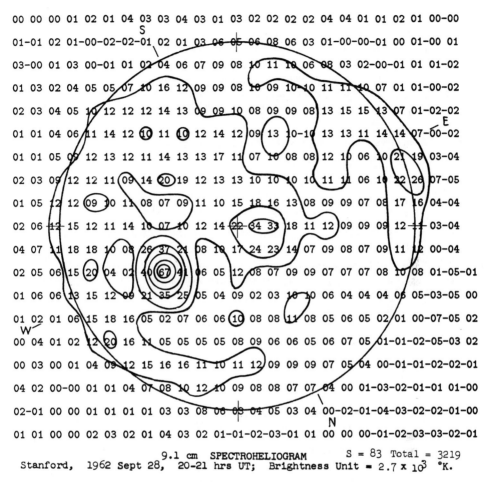

```
00 00 00 01 02 01 04 03 03 04 03 01 03 02 02 02 02 04 04 01 01 02 01 00-00
                 S
01-01 02 01-00-02-02-01 02 01 03 06 05 06 08 06 03 01-00-00-01 00 01-00 01

03-00 01 03 00-01 01 02 04 06 07 09 08 10 11 10 06 08 03 02-00-01 01 01-02

01 03 02 04 05 05 07 10 16 12 09 09 08 10 09 10 10 11 11 10 07 01 01-00-02

02 03 04 05 10 12 12 12 14 13 09 09 10 08 09 09 08 13 15 15 13 07 01-02-02
                                                                    E
01 01 04 06 11 14 12 10 11 10 12 14 12 09 13 10-10 13 13 11 14 14 07-00-02

01 01 05 09 12 13 12 11 14 13 13 17 11 07 10 08 08 12 10 06 10 21 19 03-04

02 03 09 12 12 11 09 14 20 19 12 13 13 10 10 10 10 11 11 06 10 22 26 07-05

01 05 12 12 09 10 11 08 07 09 11 10 15 18 16 13 08 09 09 07 08 17 16 04-04

02 06 12 15 12 11 14 10 07 10 12 14 22 34 33 18 11 12 09 09 09 12 11 03-04

04 07 11 18 18 10 08 26 37 21 08 10 17 24 23 14 07 09 08 07 09 11 12 00-04

02 05 06 15 20 04 02 40 67 41 06 05 12 08 07 09 09 07 07 07 08 10 08 01-05-01

01 06 06 13 15 12 09 21 35 25 05 04 09 02 03 10 10 06 04 04 04 06 05-03-05 00

01 02 01 06 15 18 16 05 02 07 06 06 10 08 08 11 08 05 06 05 02 01 00-07-05 02
  W
00 04 01 02 12 20 16 11 05 05 05 05 08 09 06 06 05 06 07 05 01-01-02-05-03 02

00 03 00 01 04 09 12 15 16 16 11 10 11 12 09 09 09 07 05 04 00-01-01-02-02-01

04 02 00-00 01 01 04 07 08 10 12 10 09 08 08 07 07 04 00 01-03-02-01-01 01-00

02-01 00 00 01 01 01 01 03 03 08 06 03 04 05 03 04 00-02-01-04-03-02-02-01-00
                                                      N
01 01 00 00 02 03 02 01 04 03 02 01-01-02-03-01 01 00 00 00-01-02-03-03-02-01
```

9.1 cm SPECTROHELIOGRAM   S = 83 Total = 3219
Stanford, 1962 Sept 28, 20-21 hrs UT; Brightness Unit = 2.7 x 10³ °K.

Figure 8-51 Example of a 9.1-cm map as published in the years 1962 to 1973. Contour levels are 50,000 and 100,000 K.

8–5. *Zero-phase filter.* The transfer function $H(u, v)$ of an image-plane point-response function $h(x, y)$ is such that pha$[H(u, v)] = 0$. Does it follow that $h(x, y) = h(-x, -y)$? ☞

8–6. *Smoothing sun maps.* In order to give a visual guide to the general temperature distribution over the sun we show contours at 50,000 and 100,000 K (Fig. 8-51). But, for the same reason that meterological data may be smoothed by running means, "smooth" contours were shown. That is, the contours refer to a smoothed map obtained by replacing each observed value by the mean of the 3×3 array of values centered on the observed value. As a result certain spatial frequencies are suppressed. What are the lowest suppressed frequencies
(a) horizontally,
(b) at 45°? ☞

8–7. *Cell-summing.*
(a) Distribute unit masses to 64 random locations on the line from 0 to 16. Produce an

equispaced sample train with values equal to the mass in each unit interval, draw a graph, and satisfy yourself that the condensed sample set reasonably represents the artificial data. The data can possess sinusoidal components with periods ranging from zero to infinity. Select two or three sinusoids in this range and subject them separately to the same cell summing procedure (integrate over each unit interval).

(b) Distribute masses ± 1 alternately to locations spaced regularly 0.8 unit apart and graph the cell-summed values at unit interval. Interpret the results of these numerical experiments. ☞

8–8. *Low-pass filter.* A desired low-pass filter characteristic is designed on the basis of a spline as follows.

$$S(u) = \begin{cases} 2(1 - 8u^2), & |u| < \frac{1}{4} \\ 4(1 - 2u)^2, & \frac{1}{4} < |u| < \frac{1}{2} \\ 0, & \text{elsewhere.} \end{cases}$$

This filter characteristic cuts off at frequency $u = \pm\frac{1}{2}$, but the cutoff is not a sharp discontinuity and in fact is particularly smooth because there is also no discontinuity in slope. This smoothness is desired in order to avoid the appearance of fringes at the edges of images subjected to low-pass filtering. A 2D transfer function with the same smooth properties is defined by $T(u, v) = S(u)S(v)$.

(a) Make a contour plot of $T(u, v)$ and report whether the constructed transfer function clashes with another desideratum, namely that it should be isotropic, treating edges much the same regardless of orientation.

(b) Establish an array of coefficients $a(x, y)$ which, when directly convolved with an image, approximates the filtering effect specified. Remember that there is a premium on the fewness of elements and that small integer values not exceeding 8 or 16 are also a good thing. ☞

8–9. *Nonlinear smoothing.* Devise a nonlinear function $\phi(\ )$ such that $\phi[f(x, y) - f_{av}]$ goes to $\pm\infty$ as $f(x, y) \to \pm\infty$, but does so logarithmically. ☞

8–10. *Four-point smoothing.*

(a) A matrix $f(i, j)$ is smoothed by replacing each element by the average of the four (originally) adjacent values in the row and column passing through the element. Express the degree of smoothing in terms of the variance of the symmetrical Gaussian function that smoothes equivalently. Comment on the isotropy of the four-point procedure.

(b) Repeat the discussion for the four adjacent points situated diagonally. ☞

8–11. *What is a raster graphics line?* Tell your computer, for $i = 0$ to 45 step 1, to draw a line at inclination i degrees to the horizontal starting from the point $(0, i)$ and rising toward the right. Make a note of the pattern of glitches that you see with a view to understanding how the computer decided which screen elements to illuminate. Perhaps this computer evaluates $y = x \tan i + i$, rounds off the result to the nearest integer Y, and illuminates the element at (X, Y), where X is integral. Test this reasonable assumption as follows. Compute the points (X, Y) yourself and erase the computer-drawn picture point by point by inverting the screen (interchanging black and white). If the guess is right, the whole pattern will be progressively erased. If a residual picture of scattered dots remains, use it to investigate how your computer converts the algebraic straight line, or vector, to a set of more or less aligned screen elements.

8–12. *Expanded character.* A bit-mapped character is exploded by multiplying both coordinates of every black pixel on the map by 2. Is the outcome the same (apart from position of the exploded character on the page) regardless of choice of the center of explosion?

8–13. *Numerical intuition.* Let us reason as follows. The five-element binomial series {1 4 6 4 1} has a mass of 16 and a variance $[1 \times 2^2 + 4 \times 1^2 + 6 \times 0^2 + 4 \times 1^2 + 1 \times 2^2]/16 = 1$. Therefore it corresponds with the Gaussian function $g(x) = (16/\sqrt{2\pi}) \exp(-\frac{1}{2}x^2)$ with the same mass and variance.

   (a) Calculate values of $g(0)$, $g(1)$, $g(2)$, and $g(3)$ and compare with the binomial coefficients to get a feel for how close the agreement is.

   (b) Since each coefficient represents the center of its range, compare also against the area under $g(x)$ in the range $x - 0.5, x + 0.5$. Your purpose in doing this exercise, and similar ones that you devise, is to accumulate personal quantitative experience that will guide intuitive judgment.

   (c) The 25-element binomial array can be compared with $256 \exp[-\frac{1}{2}(x^2 + y^2)]$ in both of the above ways. Judge how the comparison will turn out before doing the arithmetic.

8–14. *Mystery operator.* A large digital image is convolved with the following 7×7 matrix. What would be the purpose of doing this?

$$
\begin{bmatrix}
0 & 1 & 1 & 0 & -1 & -1 & 0 \\
1 & 5 & 6 & 0 & -6 & -5 & -1 \\
3 & 12 & 15 & 0 & -15 & -12 & -3 \\
4 & 16 & 20 & 0 & -20 & -16 & -4 \\
3 & 12 & 15 & 0 & -15 & -12 & -3 \\
1 & 5 & 6 & 0 & -6 & -5 & -1 \\
0 & 1 & 1 & 0 & -1 & -1 & 0
\end{bmatrix}.
$$

8–15. *Compass gradient mask.* Convolution with the array

$$
\begin{bmatrix}
1 & 1 & 1 \\
-1 & -2 & 1 \\
-1 & -1 & 1
\end{bmatrix},
$$

known in the literature as a compass gradient mask, is reputed to give the gradient toward the northeast. Test this property on $f(x, y) = x^2 + y^2$, whose northeast gradient at $(1, 1)$ is known to be $+2\sqrt{2}$.

8–16. *Diagonal operator.* A data matrix is convolved with $\dfrac{1}{\sqrt{2}}\begin{bmatrix} 0 & -1 \\ 1 & 0 \end{bmatrix}$ to form a new matrix. What property of the original does the new matrix present?

8–17. *Second cross difference.* The function $f(x, y) = xy$ has a saddle point at the origin, the "strength" of which can be obtained from $\partial^2 f(x, y)/\partial x\, \partial y$.

   (a) Show that convolution of any smooth function $f(x, y)$ with $\frac{1}{4}\begin{bmatrix} 1 & -1 \\ -1 & 1 \end{bmatrix}$ will yield the second difference $\Delta^2_{xy} f(x, y)$.

(b) Explain why $f(x, y) = x^2 - y^2$, which also has a saddle point at the origin, has a second cross difference at the origin that is zero. ☞

8–18. *Gradient operators.* The following matrices are in common use as convolution operators for evaluating the "gradient" of an image matrix in the positive x-direction.

(a) Show that they all evaluate the first difference of a smoothed version of the image and differ only in the choice of how the image is to be smoothed in advance. Show what each smoothing matrix is by factoring the gradient operators into two factors, one of which is $[1 \ -1]$.

$$\begin{bmatrix} 1 & 0 & -1 \\ 1 & 0 & -1 \\ 1 & 0 & -1 \end{bmatrix} \quad \begin{bmatrix} 1 & 0 & -1 \\ 2 & 0 & -2 \\ 1 & 0 & -1 \end{bmatrix} \quad \begin{bmatrix} 1 & 0 & -1 \\ \sqrt{2} & 0 & -\sqrt{2} \\ 1 & 0 & -1 \end{bmatrix}.$$

(b) Is there any advantage in having a repertoire of "gradient detectors" rather than a repertoire of smoothing operators?

(c) Explain why the operators given in (a) are sometimes published with the negative columns on the left.

(d) Characterize the implied smoothing in each case by a verbal description. ☞

8–19. *Gradient direction.* I obtain the first differences of a function $f(i, j)$ by convolution with $[1 \ -1]$ and $\begin{bmatrix} -1 \\ 1 \end{bmatrix}$ to obtain $f_x(i, j)$ and $f_y(i, j)$, while you use the Sobel gradient operators

$$\begin{bmatrix} -1 & 0 & 1 \\ -2 & 0 & 2 \\ -1 & 0 & 1 \end{bmatrix} \quad \text{and} \quad \begin{bmatrix} 1 & 2 & 1 \\ 0 & 0 & 0 \\ -1 & -2 & -1 \end{bmatrix}$$

to obtain "gradients" $f_x(i, j)$ and $f_y(i, j)$. For which of us, if either, is the direction of the (uphill) gradient given by $\theta = \arctan[f_y(i, j)/f_x(i, j)]$? ☞

8–20. *Moiré fringe.* Generate a pattern of bright pixels on a computer screen such that a pixel in row r and column c is illuminated if a line belonging to the family $c = r/10 + 6i$, where $i = 0, 1, 2, \ldots$, passes within a horizontal distance ± 1.5 of the center of that pixel. Inclined bands three pixels wide will be generated, but with toothed edges. Cover an area about 128×128. Now superimpose a set of vertical bands three pixels wide, with the same horizontal spacing, on an OR basis; i.e., a pixel is illuminated if it belongs to one set of bands or to both. Horizontal moiré fringes will be seen.

(a) If the vertical bands are displaced one pixel, left or right, through how many pixels does the moiré fringe move vertically?

(b) This phenomenon is to be the basis of a sensitive displacement sensor. Propose a detection matrix which, scanned over the pattern vertically, will locate the fringes. ☞

8–21. *Nonlinear phenomena.* To gain experience with nonlinear phenomena that pervade digital images, construct the family of parabolas $y = 14\sqrt{x - 10i}$ for $i = 1$ to 30 and superpose a second family $y = 16\sqrt{x - 10i}$. To avoid complications, make the quantity $10i$ about ten times the increment of x appropriate to the medium.

(a) Explain the appearance.

(b) Are spatial frequencies seen that are not present in the component families separately?

(c) Compose a presentation that would convince a skeptic that a nonlinear phenomenon is or is not involved.

(d) Explain why it is or is not correct to say that the phenomenon is digital.

(e) How would you go about deducing the appearance of this display without previously seeing it?

8–22. *Laser-printer accuracy.* The light beam inside a laser printer is swept from side to side by a very rapidly rotating mirror and falls on the light-sensitive surface at a distance proportional to $\tan\theta$, where θ is the angle of incidence of the beam. If the beam is pulsed on and off at uniform time intervals, the spacing between the dots that will print will be proportional to $d(\tan\theta)/dt$.

(a) If θ_m is the maximum value of θ, show that the dot spacing at the ends of a line will exceed the spacing in the middle by a factor $\sec^2\theta_m$.

(b) In a certain laser engine θ_m is $10°$; devise a moiré experiment to discover whether or not the manufacturer compensates for this inaccuracy by correcting the pulse rate in accordance with the mirror position. ☞

8–23. *Grille with frequency modulation.* A grille appears uniform, but in fact the line spacing is faintly modulated at a low spatial frequency. If a transparency is made and slid over the original, what will be seen?

8–24. *Line brightness.* Draw a rasterized straight line between $(0, 0)$ and $(\pi,\ e)$ on a large field with pixel size much less than 1. What is the linear brightness of this line relative to a horizontal line?

8–25. *Arrow of time.* Draw a few rasterized straight lines between chosen endpoints at the edges of the screen. Then alter the program to draw lines between the same endpoints but in the opposite direction and turning the pixels off. Find out whether the computer completely erases its own product. ☞

8–26. *Computer graphics lingo.* The staircase effect seen on "straight" lines drawn on a pixellated screen is referred to as *aliasing* by some authors. Do they mean aliasing in the established sense where a new frequency is generated from an old as a consequence of sampling? If so, explain; otherwise explain physically where the step frequency comes from. ☞

8–27. *Line on grille.* A straight line of slope 3 in 4 is drawn across a horizontal grille of unit spacing to create a binary image $f(x, y)$ equal to unity where there is ink and zero elsewhere. Intersections occur at an interval of five units; does this mean that there will be a feature with spatial frequency 0.2 in the Fourier transform $F(u, v)$?

8–28. *Filling a chessboard.* Define the rulings on a chessboard as a single closed polygon with 34 vertices so that when the "inside" of the polygon is filled, the squares will be shaded in the customary manner. ☞

8–29. *Hatching.* Parallel lines sloping at $60°$ to the horizontal are to be drawn with a vertical spacing of 12 pixels to cover the space with opposite corners at $(0, 0)$ and $(100, 100)$. Draw the set of lines $y = \tan(\pi/3)x + 12k$ for $k = 100/14\tan(\pi/3)$ to 7 using a straight-line algorithm. Your computer will do this for you if you specify the endpoints of each line segment.

(a) Describe the unpleasant artifact that is seen.

(b) Suggest a way of obtaining more acceptable hatching and demonstrate it.

8–30. *Interior of a polygon.* According to Rule 1, a pixel is inside a polygon if a line from that pixel to the edge of the field is crossed by the polygon an odd number of times. Rule 2 depends on the number of times the polygon winds around the pixel and says that the pixel is inside if the algebraic sum of the crossings is not zero. The sign of a crossing is positive if the polygon crosses the line from left to right.

(a) Explain which if either of the rules enables you to tag the whole field as inside or outside, even if the polygon is not closed.

(b) How are the rules affected if the polygon is specified not as an ordered sequence of bounding pixels but as a set of pixels?

(c) What other differences might the two rules exhibit?

8–31. *Perimeter of triangle.* Take the "distance" between two points (x_1, y_1) and (x_2, y_2) to be max $(x_1 - x_2, y_1 - y_2)$.

(a) Give an expression for the sum of the three distances between the vertices $(0, 0)$, (x_2, y_2), (x_3, y_3) of a triangle.

(b) How does this sum compare with the Euclidean perimeter?

8–32. *Dilation in the continuous world.* Let A and B be two subspaces of a vector D. Show that it is reasonable to write $D = A \oplus B$ if any vector $\mathbf{d} = \mathbf{a} + \mathbf{b}$, where $\mathbf{a}$ belongs to A and $\mathbf{b}$ belongs to B. ☞

8–33. *External shadow.* An old machine drawing, which is to be used in a sales brochure, is to be jazzed up as follows. All lettering, broken lines, interior outlines, centerlines, dimension lines, and lettering are erased. The remaining line drawing is digitized into polygons, and a binary object $B(i, j)$ is created by filling in the outline with 1s. The object $B(i, j)$ is then printed in light grey ink. To make up for the disappearance of the outline, the grey mass is provided with a thin edge $E(i, j)$ computed from

```
E(i,j)=B(i,j)   OR   B(i,j+1)   AND NOT   B(i,j)
```

which is overprinted in solid black. On looking at a draft, the editor says the drawing would look better if the top pixel were removed from vertical stretches of the edge, and the left pixel from horizontal stretches.

(a) How should the programmer rewrite the line defining $E(i, j)$?

(b) In your opinion, was the editor right?

(c) Would the editor be happy with the revised graphics in cases where the machine part was perforated? ☞

8–34. *Repeated dilation and erosion.*

(a) An object A is dilated by B_1, and then again by B_2. Express the outcome D as a single dilation with a different structure function B_3.

(b) An object is eroded by B_1 and then again by B_2. Express the outcome E as a single erosion with a different structure function B_3. ☞

8–35. *Speech versus algebra.* "The outer fringe of an object can be obtained by dilating with B and then subtracting the original A," explains a teaching assistant. A class member who prefers algebra to English says, "Do it in one step," and writes on the board

$$A \oplus B - A = A \oplus B - A \oplus I = A \oplus (B - I),$$

where $I = [\mathbf{1}]$. The TA says, "You mean, get the fringe in one go by dilating with the modified structure matrix $B - I$?" The student nods. Prove this conclusion without glossing over the matter of where the origin is placed, or disprove it.

8–36. *Tee junctions.* Eroding an object with $[1 \quad 1 \quad 1/0 \quad 1 \quad 0/0 \quad 1 \quad 0]$ will exhibit all the places where a five-element tee pattern can be found, but would one describe all the places found as tee junctions? How would you find the upright tee junctions only? ☞

8–37. *Odd facts about Pascal.* Imagine Pascal's triangle of binomial coefficients to be continued for about 32 rows and used as the basis for a binary image consisting of black squares wherever the entry in the triangle is odd.

(a) Construct the image.

(b) In the limit, as the number of rows approaches infinity, what fraction of the area is black?

(c) If the horizontal and vertical spacings of the squares are equal, what is the correct slope of the side of the triangle?

9

Rotational Symmetry

A television image is rectangular and uses two independent variables in its representation of a three-dimensional scene. However, the television camera has an optical system with component parts, such as the aperture containing the objective lens, which, while distinctly multidimensional, possess circular symmetry. Only one independent variable, namely radius, may be needed to specify the instrumental properties across such an aperture, because the properties may be independent of the second variable, in this case angle. Circular symmetry also occurs in objects that are under study (some astronomical objects, for example) or in objects that influence imaging (rain drops), and is a property of many artifacts. Because of the prevalence of circular symmetry, particularly in instruments, special attention to circularly symmetrical transforms is warranted. This chapter also deals briefly with objects having rotational symmetry, which are of less frequent occurrence, but related.

Bessel functions play a basic role in the following discussion, especially the zero-order Bessel function $J_0(x)$, or more aptly $J_0(r)$, since the independent variable is ordinarily radial. For reference, Fig. 9-1 shows $J_0(r)$ and $J_1(r)$. For convenience, another function introduced later is included, and, for comparison, the well-known sinc function.

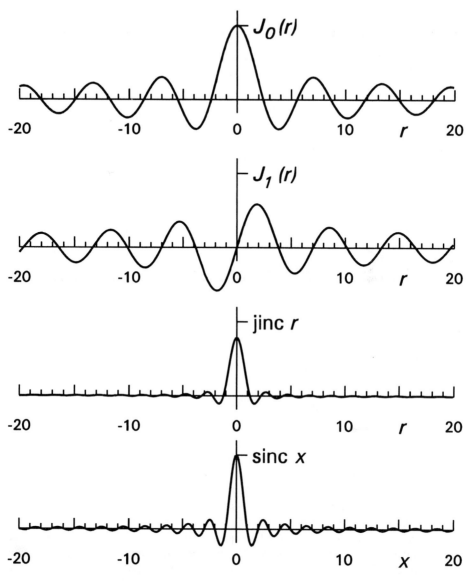

Figure 9-1 The functions $J_0(r)$, $J_1(r)$, jinc r, and sinc x.

WHAT IS A BESSEL FUNCTION?

If you allow a hanging chain to swing, it takes on the shape of a Bessel function, if you bang a drum in the center and wait for the overtones to die out, the fundamental mode of vibration follows a Bessel function of radius; and if you look at the sidebands of an FM station transmitting a tone, their amplitudes are distributed according to a Bessel function.

The chain problem is of particular interest historically, because that is where F. W. Bessel (1784–1846), a great astronomer who first measured the distance to a star and discovered the dark companion of Sirius, found his function.

We may also take a mathematical approach, purified of any physics, and define the Bessel function $J_0(x)$ as the sum of the series

$$J_0(x) = 1 - \frac{x^2}{4} + \frac{x^4}{64} - \frac{x^6}{2304} + \cdots$$

or as a solution of the second-order differential equation

$$\frac{d^2y}{dx^2} + \frac{1}{x}\frac{dy}{dx} + y = 0.$$

Of course we know that a second-order differential equation may possess two independent solutions, and we will automatically generate the second kind of Bessel function $Y_0(x)$ as well as $J_0(x)$ if we take the differential equation as a basis.

By taking x to be large, we find the very useful asymptotic expression

$$J_0(x) \sim \sqrt{\frac{2}{\pi x}} \cos(x - \tfrac{1}{4}\pi), \quad x \gg 1.$$

This expression, together with the series expansion, enables us to calculate with Bessel functions without recourse to tables or to special computer subprograms. Numerical computing itself leads to fascinating viewpoints, exemplified by the following polynomial approximation (A&S 1963, p. 356) for $-3 \le x \le 3$:

$$J_0(x) = 1 - 2.24999\,97(x/3)^2 + 1.26562\,08(x/3)^4 - .31638\,66(x/3)^6 + .04444\,79(x/3)^8$$
$$- .00394\,44(x/3)^{10} + .00021\,00(x/3)^{12} + \epsilon, \quad |\,\epsilon\,| < 5 \times 10^{-8}.$$

A quite different integral definition may also be encountered, namely.

$$J_0(x) = \frac{1}{2\pi} \int_0^{2\pi} \cos(x \cos \phi)d\phi.$$

It is by no means immediately obvious how the three mathematical definitions interrelate, and one would need to exercise some care before choosing a definition as a point of departure into a particular problem.

As the chain, drum, and FM station show, Bessel functions are not at all restricted as to where they may show up, but here we will emphasize situations with uniform circular geometry.

The water in a circular tank (Fig. 9-2) may slosh so that its height is $J_0(kr) \cos \omega t$; the longitudinal electric field in a circular waveguide may follow exactly the same expression, and so may the sound pressure in a circular tunnel. There are many other cases. The light intensity on the photographic film of a circular telescope pointed at a star is described by another Bessel function (Fig. 9-3) and has the appearance of a spot surrounded by rings of decreasing brightness.

These Airy rings were first analyzed by former Astronomer Royal George B. Airy (1801–1892), who is well known for many other things including the Airy functions of

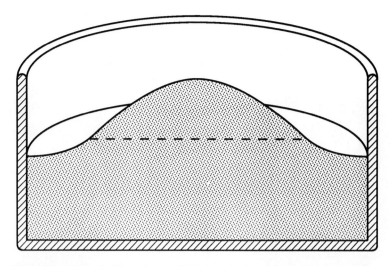

Figure 9-2   Water sloshing at low amplitude in a circular tank may take up a Bessel-function form. This tells us something about the volume under a Bessel function, when we remember that water is rather incompressible.

electromagnetic theory, which are also Bessel functions, and for having turned away J. C. Adams (1819–1892) when the latter appeared at his door at dinner time to talk about the discovery of Neptune.

As groundwork for the discussion we look at the familiar cosinusoidal corrugation as it appears in the polar coordinates which are appropriate to uniform circular circumstances. Consider the corrugation

$$f(x, y) = A \cos 2\pi ux$$

illustrated in Fig. 9-4. At a point (r, θ) the value of the function is $A \cos(2\pi ur \cos \theta)$. Hence in polar coordinates

$$f(x, y) = A \cos(2\pi ur \cos \theta).$$

As a mathematical expression, a cosine of a cosine may appear unfamiliar, but it describes a rather simple entity. Now, if we wish to describe the rotated corrugation $A \cos 2\pi ux'$, where the axis x' makes an angle ϕ with the x-axis, we may write

$$A \cos[2\pi ur \cos(\theta - \phi)].$$

Of course, the rotated corrugation may also be expressed in cartesian coordinates as

$$A \cos[2\pi u(x \cos \phi + y \sin \phi)],$$

which is just as complicated as the form taken in polar coordinates.

Imagine that the full circle is divided into a large number of equispaced directions and that for each of the directions ϕ there is a corrugation $A d\phi \cos[2\pi ur \cos(\theta - \phi)]$, all

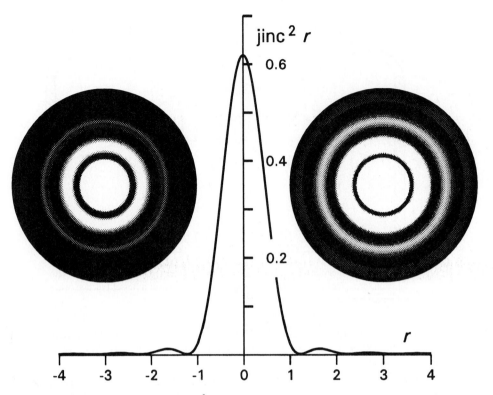

Figure 9-3   The Airy pattern jinc^2 r and two halftone renderings. The three rings, at their brightest, are respectively 18, 24, and 28 dB down relative to the central brightness maximum. The thresholds for black are set at 27 dB (left) and 31 dB (right). Comparison of the widths of the light and dark annuli with the graph will confirm that faithful halftone reproduction of the Airy pattern is impossible in a printed book.

of which corrugations are to be superimposed by addition. The sum will be

$$\int_0^{2\pi} A\cos[2\pi ur\cos(\theta - \phi)]d\phi.$$

After this integral has been evaluated, we will have a new function of r and θ; the variable of integration ϕ will have disappeared. However, if we fix attention on a particular value of r, it is clear from the circular uniformity with which the superposition was carried out that the result will not depend on θ; the integral must be independent of θ. Therefore, it makes no difference for what particular value of θ we evaluate it. Choose $\theta = 0$. Then the circular superposition integral becomes

$$\int_0^{2\pi} \cos(2\pi ur\cos\phi)d\phi.$$

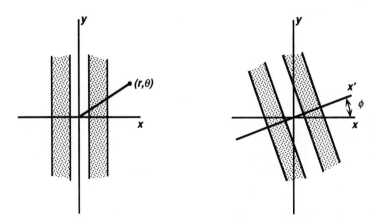

Figure 9-4 The simple corrugation $A \cos 2\pi ux$ (left) becomes $A \cos(2\pi ur \cos\theta)$ in polar coordinates, and the rotated corrugation (right) becomes $A \cos[2\pi ur \cos(\theta - \phi)]$.

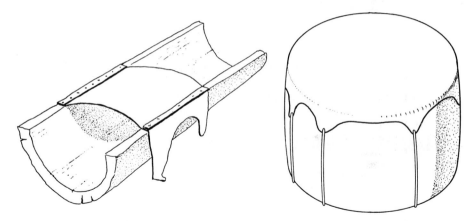

Figure 9-5 The fundamental mode of a laterally stretched membrane is cosinusoidal (left), while the corresponding mode of a circular drum is a zero-order Bessel function.

If we divide by 2π to obtain the angular average rather than the sum, we recognize this as one of the mathematical definitions of $J_0(x)$ mentioned earlier. What have we discovered?

We have found that $J_0(r)$ is to circular situations what cosine is to one-dimensional problems. A laterally stretched square membrane clamped along opposite sides can vibrate cosinusoidally like a clamped string (Fig. 9-5). Clamp the membrane on a circular rim, as with a drum, and all the possible cosinusoidal corrugations differently oriented add up to a Bessel function of radius only (Fig. 9-6).

The same can be said of water waves in square tanks and round tanks, microwaves in rectangular and round waveguides, acoustic waves in bars and rods. Just as the sinusoidal oscillation with respect to one-dimensional time derives its importance from the ubiquity

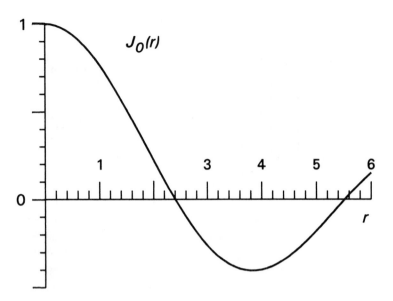

Figure 9-6   The zero-order Bessel function $J_0(r)$.

of electrical and mechanical systems that are both linear and time invariant, so the Bessel function will be found in higher dimensionalities where linearity and space invariance exist and where circular boundary conditions are imposed.

THE HANKEL TRANSFORM

To obtain the Fourier transform of a function that is constant over a central circle in the (x, y)-plane and zero elsewhere, or which, while not being constant, is a function of $r = (x^2 + y^2)^{1/2}$ only, one may of course use the standard transform definition in cartesian coordinates. Let the function $f(x, y)$ be $\mathbf{f}(r)$. Then

$$F(u, v) = \int_{-\infty}^{\infty} \int_{-\infty}^{\infty} \mathbf{f}(r) e^{-i2\pi(ux+vy)} \, dx \, dy.$$

Some comments should be made about the limits of integration. They may certainly be written as above, provided it is understood that $\mathbf{f}(r)$ covers the whole plane $0 < r < \infty$. In the example mentioned, where the function is a constant K over a central circular region C of radius a and zero elsewhere, there would be a simplified alternative

$$F(u, v) = K \int \int_C e^{-i2\pi(ux+vy)} \, dx \, dy.$$

Of course, the boundary of a circle does not lend itself to the simplest kind of representation in rectangular coordinates. Nevertheless, it is possible to do so, in which case we could write

$$F(u, v) = K \int_{-a}^{a} dy \int_{-(a^2-y^2)^{1/2}}^{(a^2-y^2)^{1/2}} e^{-i2\pi(ux+vy)} dx.$$

Sometimes this kind of integration over elaborate boundaries works out painlessly; let us try to proceed in this case.

$$F(u, v) = K \int_{-a}^{a} e^{-i2\pi vy} dy \int_{-(a^2-y^2)^{1/2}}^{(a^2-y^2)^{1/2}} e^{-i2\pi ux} dx$$

$$= K \int_{-a}^{a} e^{-i2\pi vy} \left[\frac{e^{-i2\pi ux}}{-i2\pi u} \right]_{-(a^2-y^2)^{1/2}}^{(a^2-y^2)^{1/2}} dy$$

$$= K \int_{-a}^{a} e^{-i2\pi vy} \left[\frac{e^{-i2\pi u(a^2-y^2)^{1/2}} - e^{i2\pi u(a^2-y^2)^{1/2}}}{-i2\pi u} \right] dy$$

$$= \frac{K}{\pi u} \int_{-a}^{a} e^{-i2\pi vy} \sin\left[2\pi u(a^2 - y^2)^{1/2} \right] dy$$

$$= 2\frac{K}{\pi u} \int_{0}^{a} \cos(2\pi vy) \sin\left[2\pi u(a^2 - y^2)^{1/2} \right] dy.$$

This is the maximum simplification that we can hope to make before turning to lists of integrals for help. We find the integral on p. 399 as entry 3.711 in GR (1965). The answer is

$$F(u, v) = Ka(u^2 + v^2)^{-1/2} J_1 \left[2\pi a(u^2 + v^2)^{1/2} \right] = Ka\frac{J_1(2\pi aq)}{q},$$

where J_1 is the first-order Bessel function of the first kind and $q = \sqrt{u^2 + v^2}$.

As an alternative approach, consider making use of the circular symmetry of the exercise rather than forcing rectangular coordinates upon it. Let (r, θ) be the polar coordinates of (x, y). Then

$$F(u, v) = \int \int_C f(r) e^{-i2\pi(ux+vy)} dx \, dy = \int_0^{\infty} \int_0^{2\pi} f(r) e^{-i2\pi qr \cos(\theta-\phi)} r \, dr \, d\theta.$$

We are now integrating from 0 to ∞ radially and through 0 to 2π in azimuth, and, because of the independence of azimuth, the latter integral should drop out. The new symbols q and ϕ are polar coordinates in the (u, v)-plane. Thus

$$q^2 = u^2 + v^2 \quad \text{and} \quad \tan\phi = v/u.$$

Then the new kernel $\exp[-i2\pi qr \cos(\theta - \phi)]$ arises from recognizing $ux + vy$ as the scalar product of two two-dimensional vectors. Thus

$$ux + vy = \Re[(x + iy)(u - iv)] = \Re[re^{i\theta} qe^{-i\phi}] = qr \cos(\theta - \phi).$$

Removing the integration with respect to θ,

$$F(u, v) = \int_0^\infty \mathbf{f}(r) \left[\int_0^{2\pi} e^{-i2\pi qr \cos(\theta - \phi)} d\theta \right] r \, dr$$

$$= \int_0^\infty \mathbf{f}(r) \left[\int_0^{2\pi} \cos(2\pi qr \cos \theta) d\theta \right] r \, dr.$$

Since u and v, and therefore q and ϕ, are fixed during the integration over the (x, y)-plane, we may drop ϕ, because it merely represents an initial angle in an integration that will run over one full rotation from 0 to 2π. Therefore, the result of the integration will be the same regardless of the value of ϕ; take it to be zero. Again, the sine component of the imaginary exponential may be dropped, because it will integrate to zero; to see this, make sketches of $\cos(\cos \theta)$ and $\sin(\cos \theta)$ for $0 < \theta < 2\pi$.

We now make use of the fundamental integral representation for the zero-order Bessel function $J_0(z)$, namely,

$$J_0(z) = \frac{1}{2\pi} \int_0^{2\pi} \cos(z \cos \theta) d\theta.$$

This basic relation will be returned to later. Meanwhile, incorporating it into the development, we have finally

$$F(u, v) = 2\pi \int_0^\infty \mathbf{f}(r) J_0(2\pi qr) r \, dr.$$

Here we have the statement of the integral transform that takes the place of the two-dimensional Fourier transform when $f(x, y)$ possesses circular symmetry and is representable by $\mathbf{f}(r)$. It follows that $F(u, v)$ also possesses circular symmetry (the rotation theorem is an expression of this) and may be written $\mathbf{F}(q)$, depending only on the radial coordinate $q = (u^2 + v^2)^{1/2}$ in the (u, v)-plane and not on azimuth ϕ. The transform

Hankel transform.

$$\mathbf{F}(q) = 2\pi \int_0^\infty \mathbf{f}(r) J_0(2\pi qr) r \, dr$$

is known as the Hankel transform. It is a one-dimensional transform. The functions, $\mathbf{f}$ and $\mathbf{F}$ are functions of one variable. They are not a Fourier transform pair, but a Hankel transform pair. As functions of *one* variable they may be used to represent two-dimensional functions, which will be two-dimensional Fourier transform pairs.

Hankel Transform of a Disk

Our first example of a Hankel transform pair, obtained directly by integration, was

$$\text{rect}\left(\frac{r}{2a}\right) \quad \text{has Hankel transform} \quad \frac{a J_1(2\pi aq)}{q}.$$

This is such an important pair that we adopt the special name jinc q for the Hankel transform of rect r. Thus

$$\text{rect } r \quad \text{has Hankel transform} \quad \text{jinc } q \equiv \frac{J_1(\pi q)}{2q}.$$

From the integral transform formulation it follows that

$$\text{jinc } q = 2\pi \int_0^\infty \text{rect } r \, J_0(2\pi qr) r \, dr.$$

Likewise, from the reversibility of the two-dimensional Fourier transform, it follows that

$$\text{rect } r = 2\pi \int_0^\infty \text{jinc } q \; J_0(2\pi qr) \, q \, dq.$$

Frequently needed properties of the jinc function are collected below.

Hankel Transform of the Ring Impulse

As an example of the Hankel transform we used the rectangle function of radius 0.5 and found by direct integration in two dimensions that the Hankel transform was jinc q. Now we make use of the Hankel transform formula to obtain another important transform pair. Let

$$\mathbf{f}(r) = \delta(r - a)$$

which describes a unit-strength ring impulse. Its Hankel transform $\mathbf{F}(q)$ is given by

$$\mathbf{F}(q) = 2\pi \int_0^\infty \mathbf{f}(r) J_0(2\pi qr) r \, dr$$

$$= 2\pi \int_0^\infty \delta(r - a) J_0(2\pi qr) r \, dr.$$

Apply the sifting property to obtain immediately

$$\mathbf{F}(q) = 2\pi a J_0(2\pi aq).$$

From the reciprocal property of the Hankel transform it also follows that

$$\delta(r - a) = 2\pi \int_0^\infty 2\pi a J_0(2\pi aq) J_0(2\pi rq) q \, dq$$

a relationship that can be recognized as expressing an orthogonality relationship between zero-order Bessel functions of different "frequencies." Unless the two Bessel functions have the same "frequency," the infinite integral of their product is zero, just as with sines and cosines. Although $J_0(\omega t)$ is not a monochromatic waveform, nevertheless as t elapses, the waveform decays away rather slowly in amplitude and settles down more and more closely to a definite angular frequency ω and a fixed phase, as may be seen from the asymptotic expression

$$J_0(\omega t) \sim \sqrt{\frac{2}{\pi \omega t}} \cos(\omega t - \tfrac{1}{4}\pi).$$

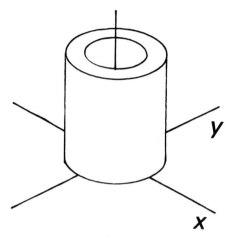

Figure 9-7 A circularly symmetrical function suitable for representing the distribution of light over a uniformly illuminated annular slit. By an extraordinary visual illusion connected with the evolution of vision, the height and outside diameter appear unequal.

Just as jinc r can be described as a circularly symmetrical two-dimensional function that contains all spatial frequencies up to a certain cutoff, in uniform amount for all frequencies and orientations, so also $J_0(r)$ can be seen as a circularly symmetrical two-dimensional function that contains only one spatial frequency but equally in all orientations. The jinc function is thus like the sinc function, and J_0 is like the cosine function.

Annular Slit

We have established the following two Hankel transform pairs:

$$\text{rect } r \supset \text{jinc } q$$
$$\delta(r - a) \supset 2\pi a J_0(2\pi aq),$$

and in what follows we need to recall the similarity theorem in its form applicable to circular symmetry:

$$\text{If } \mathbf{f}(r) \supset \mathbf{F}(q) \quad \text{then} \quad \mathbf{f}(r/a) \supset a^2 \mathbf{F}(aq).$$

Note that the sign $\supset$ may be read "has Hankel transform" if you picture $\mathbf{f}$ and $\mathbf{F}$ as one dimensional, but may alternatively be read "has two-dimensional Fourier transform" if you prefer to think of $\mathbf{f}$ and $\mathbf{F}$ as functions of radius representing two-dimensional entities.

Both the unit circular patch represented by rect r and the ring impulse $\delta(r - a)$ are constantly needed. A third important circularly symmetrical function (Fig. 9-7) is unity over an annulus. For concreteness of description it may be referred to as an *annular slit*, but of course the function is of wider significance.

A narrow circular slit could be cut in an opaque sheet and, if uniformly illuminated from behind, would be reasonably represented by a ring impulse. If the mean radius of the annulus were a, the slit width w, and the amplitude of illumination A, then the distribution of light would be expressible as

$$A \operatorname{rect}\left(\frac{r}{2a + w}\right) - A \operatorname{rect}\left(\frac{r}{2a - w}\right).$$

The "quantity" of light would be $2\pi a w A$, and the quantity per unit arc length would be wA. Therefore, since the ring impulse $\delta(r - a)$ has unit weight per unit arc length, the appropriate impulse representation would be $wA\delta(r - a)$.

We know the Hankel transforms of both the annulus and the ring delta, and the transforms will approach equality as the slit width w approaches zero, provided at the same time the amplitude of illumination A is increased so as to maintain constant the integrated amplitude over the slit. We are saying that as the slit width w goes to zero while wA remains constant, then

$$A(2a + w)^2 \text{jinc}[(2a + w)q] - A(2a - w)^2 \text{jinc}[(2a - w)q] \rightarrow (wA)2\pi a J_0(2\pi aq).$$

It follows that

$$\lim_{w \to 0} w^{-1}\left[(2a + w)^2 \text{jinc}[(2a + w)q] - (2a - w)^2 \text{jinc}[(2a - w)q]\right] = 2\pi a J_0(2\pi aq).$$

The left-hand side is recognizable as a derivative, and therefore the conclusion implies the identity

$$\frac{\partial}{\partial a}(4a^2 \text{jinc } 2qa) = 2\pi a J_0(2\pi aq),$$

a result that can be deduced independently from properties of Bessel functions.

For computing purposes we may sometimes wish to represent a ring impulse by an annulus of small but nonzero width, and we may also wish to do the reverse for purposes of theory—namely, to represent an annular slit by a ring impulse. A slit width equal to 10 percent of the mean radius may seem a rather crude example to take, but with $a = 1$ and $w = 0.1$ let us compare $10[(2.1)^2 \text{jinc}(2.1q) - (1.9)^2 \text{jinc}(1.9q)]$ with $2\pi J_0(2\pi q)$.

We quickly find from a few test points that the agreement is good.

q	0	0.38277	-0.5	1.0
LHS	6.28318	-0.004	-1.9061	1.365
RHS	6.28318	0	-1.9116	1.384

Thus, as far as the transform is concerned, a 10 percent slit width, which seems far from a slit of zero width, gives results within 1 percent or so.

It is worthwhile doing numerical calculations of this sort from time to time to develop a sense of how crude an approximation may be and still be useful. Over the whole range $0 < q < 1$ the discrepancy ranges between limits of 0.0192 and -0.0197 or just under 2 percent of the central value. For less crude approximations the results would, of course, be even more accurate. An approximate solution to an urgent problem is most welcome, provided you have the experience to feel confidence in the quality of the approximation. You gain this feeling for magnitudes by making a habit of comparing rough approximations with correct solutions. Reference lists of Hankel transforms can be found in FTA (1986), in Erd (1954), and others occur in GR (1965).

Table 9-1   Table of Hankel transforms.

$f(r)$	$F(q) = 2\pi \int_0^\infty f(r) J_0(2\pi qr) r\,dr$
$f(ar)$	$a^{-2} f(q/a)$
$f \ast\ast g$	FG
$r^2 f(r)$	$-\nabla^2 F$
rect r	jinc q
$\delta(r-a)$	$2\pi a\, J_0(2\pi aq)$
$e^{-\pi r^2}$	$e^{-\pi q^2}$
$r^2 e^{-\pi r^2}$	$\pi^{-1}(\pi^{-1} - q^2) e^{-\pi q^2}$
$(1+r^2)^{-1/2}$	$q^{-1} e^{-2\pi q}$
$(1+r^2)^{-3/2}$	$2\pi e^{-2\pi q}$
$(1-4r^2)\,$rect r	$J_2(\pi q)/\pi q^2$
$(1-4r^2)^\nu\,$rect r	$2^{\nu-1} \nu!\, J_{\nu+1}(\pi q)/\pi^\nu q^{\nu+1}$
r^{-1}	q^{-1}
e^{-r}	$2\pi (4\pi^2 q^2 + 1)^{-3/2}$
$r^{-1} e^{-r}$	$2\pi (4\pi^2 q^2 + 1)^{-1/2}$
$^2\delta(x,y)$	1

Theorems for the Hankel Transform

Theorems for the Hankel transform are deducible from those for the two-dimensional Fourier transform, with appropriate change of notation. For example, the similarity theorem $f(ax, by) \,^2\!\supset |ab|^{-1} F(u/a, v/b)$ will apply, provided $a = b$, a condition that is necessary to preserve circular symmetry. Thus $f(ar)$ has Hankel transform $a^{-2} F(q/a)$. The shift theorem does not have any meaning for the Hankel transform, since shift of origin destroys circular symmetry. The convolution theorem, $f \ast\ast g \,^2\!\supset FG$, retains meaning for the Hankel transform on the understanding that $f(r)$ and its Hankel transform $F(q)$ are both taken as representing two-dimensional functions on the (x, y)-plane. Then $f \ast\ast g$ has Hankel transform FG. Some theorems have been incorporated in Table 9-1.

Computing the Hankel Transform

It is perfectly feasible to compute the Hankel transform from the integral definition. The infinite upper limit causes no trouble in practice when the given function either cuts off or dies away rapidly. To evaluate the Bessel functions needed for all the q values one uses the series approximation given above for arguments less than 3 and an asymptotic expansion otherwise. In the following sample program the given function $f(r)$ is defined to be $\exp[-\pi (r/7)^2]$, which falls to 3×10^{-6} at $r = 14$, and is integrated from 0 to 14. The Bessel function appears in the inner loop with three explicit multiplies, but at least ten more occur in the function definition for $J_0(x)$. Consequently this program is not fast.

A faster method starts by taking the Abel transform of $\mathbf{f}(r)$ (see below) followed by a standard fast Fourier transform; or, since only the real part of the complex output will be utilized, some may prefer to call a standard fast Hartley transform, which will give exactly the same result faster.

HANKEL TRANSFORM

```
DEF FNf(r)=EXP(-PI*(r/7)^2      Function definition
dr=0.1                          Step in r
FOR q=0 TO 0.25 STEP 0.05
   k=2*PI*q
   s=0
   FOR r=dr/2 TO 14 STEP dr
      s=s+FNf(r)*FNJ0(k*r)* r
   NEXT r
   PRINT q;2*PI*s*dr
NEXT q
END
```

THE JINC FUNCTION

Just as in one dimension there is a sinc function which contains all frequencies equally up to a cutoff, and no higher frequencies, so in two dimensions there is a jinc function (Figs. 9-1, 9-8 and Table 9-1) that has already been referred to. The following material, which is collected in one place for reference, mentions the Abel transform, and the Struve function of order unity, which are discussed later. A table of the jinc function is given as Table 9-4 at the end of the chapter.

Properties of the jinc Function

Definition.

$$\text{jinc } x = \frac{J_1(\pi x)}{2x}.$$

Series Expansion.

$$\text{jinc } x = \frac{\pi}{4} - \frac{\pi^3}{2^5}x^2 + \frac{\pi^5}{2^8.3}x^4 - \frac{\pi^7}{2^{12}.3}x^6 + \frac{\pi^9}{2^{16}.3^2.5}x^8 - \cdots$$

$$= .785398 - .968946x^2 + .398463x^4 - .245792x^6 + .010108x^8 + \cdots.$$

Asymptotic Expression, x > 3.

$$\text{jinc } x \sim \frac{\cos[\pi(x - 3/4)]}{\sqrt{2\pi^2 x^3}}.$$

Asymptotic Behavior.   The slow decay of $J_0(r)$ with r is connected with the fact that its Hankel transform is impulsive, while the relatively rapid decay of jinc r to small values

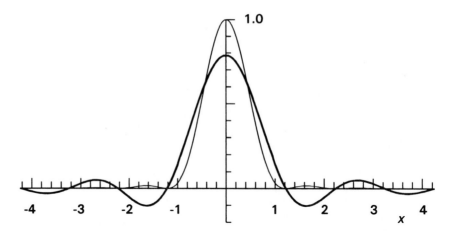

Figure 9-8　The jinc function, Hankel transform of the unit rectangle function (heavy line). The first null is at 1.22, the constant that is familiar in the expression $1.22\lambda/D$ for the angular resolution of a telescope of diameter D. See Table 9-4 for tabulated values. The jinc^2 function normalized to unity at the origin (light line) describes the intensity in the Airy disc, the diffraction pattern of a circular aperture.

occurs because its transform has only a finite discontinuity. The even greater compactness of $\text{jinc}^2(r)$ is associated with the even smoother form of its transform, the chat function. Thus $J_0(r) \sim r^{-1/2}$, $\text{jinc}\, r \sim r^{-3/2}$, $\text{jinc}^2 r \sim r^{-3}$. It appears that, if n derivatives of $\mathbf{F}(q)$ have to be taken to make the result impulsive, then $\mathbf{f}(r) \sim r^{-(n+1/2)}$. The similar theorem for the Fourier transform is that if the nth derivative of $f(x)$ is impulsive, then $F(s) \sim s^{-n}$. However, in the presence of circular symmetry a qualification is necessary, because there cannot be a finite discontinuity at the origin, and a discontinuity in slope at the origin, such as chat r exhibits, counts for less.

Zeros.　$\text{jinc}\, x_n = 0$

n	1	2	3	4	5	6	$\ldots$	n
x_n	1.2197	2.2331	3.2383	4.2411	5.2428	6.2439	$\ldots$	$\sim n + 1/4$

Derivative.

$$\text{jinc}'\, x = \frac{\pi}{2x} J_0(\pi x) - \frac{1}{x^2} J_1(\pi x) = -\frac{\pi}{2x} J_2(\pi x).$$

Maxima and Minima.

Location	1.6347	2.6793	3.6987	4.7097	5.7168	6.7217
Value	−0.1039	0.0506	−0.0314	0.0219	−0.016	0.013

Integral.　The jinc function has unit area under it:

$$\int_{-\infty}^{\infty} \mathrm{jinc}\, x\, dx = 1.$$

Half Peak and 3 dB Point.

$$\mathrm{jinc}(0.70576) = 0.5\,\mathrm{jinc}\,0 = \pi/8 = 0.39270.$$

$$\mathrm{jinc}\,\theta_{3\mathrm{dB}} = \frac{1}{\sqrt{2}}\,\mathrm{jinc}\,0 = 0.55536,$$

where $\theta_{3\mathrm{dB}} = 0.51456$.

Fourier Transform. The one-dimensional Fourier transform of the jinc function is semi-elliptical with unit height and unit base.

$$\int_{-\infty}^{\infty} \mathrm{jinc}\, x\, e^{-i2\pi s x}\, dx = \sqrt{1 - (2s)^2}\ \mathrm{rect}\, s.$$

Hankel Transform. The Hankel transform of the jinc function is the unit rectangle function

$$\int_0^{\infty} \mathrm{jinc}\, r\, J_0(2\pi q r)\, 2\pi r\, dr = \mathrm{rect}\, q.$$

Abel Transform. The Abel transform (line integral) of the jinc function is the sinc function

$$2\int_x^{\infty} \frac{\mathrm{jinc}\, r\, r\, dr}{\sqrt{r^2 - x^2}} = \mathrm{sinc}\, x.$$

Two-dimensional Aspect. Regarded as a function of two variables x and y, jinc r (where $r^2 = x^2 + y^2$) describes a circularly symmetrical hump surrounded by null circles separating positive and negative annuli.

The Null Circles. Nulls occur at radii 1.220, 2.233, 3.239, etc. As the radii approach values of 0.25 + integer, their spacing approaches unity.

Two-dimensional Integral. The volume under jinc r is unity:

$$\int_{-\infty}^{\infty}\int_{-\infty}^{\infty} \mathrm{jinc}\,\sqrt{x^2 + y^2}\, dx\, dy = \int_0^{\infty} \mathrm{jinc}\, r\, 2\pi r\, dr = 1.$$

Two-dimensional Fourier Transform. A disc function of unit height and diameter:

$$\int_{-\infty}^{\infty}\int_{-\infty}^{\infty} \mathrm{jinc}\,\sqrt{x^2 + y^2}\, e^{-i2\pi(ux+vy)}dx\, dy = \mathrm{rect}(\sqrt{u^2 + v^2}).$$

Two-dimensional Autocorrelation Function of the jinc function is the jinc function

$$\int_{-\infty}^{\infty}\int_{-\infty}^{\infty} \mathrm{jinc}\,\sqrt{\xi^2 + \eta^2}\ \mathrm{jinc}\,\sqrt{(\xi + x)^2 + (\eta + y)^2}\, d\xi\, d\eta = \mathrm{jinc}\,\sqrt{x^2 + y^2}.$$

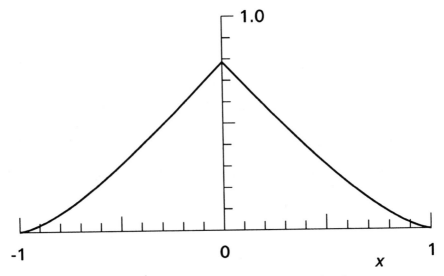

Figure 9-9   The Chinese hat function, autocorrelation function of rect r.

The Chinese Hat Function (Fig. 9-9) is the Hankel transform of $jinc^2 r$.

$$\text{chat } q \equiv \tfrac{1}{2}\left(\cos^{-1}|q| - |q|\sqrt{1-q^2}\right) \text{rect } \tfrac{1}{2}q = \int_0^\infty jinc^2 r \, J_0(2\pi q r) \, 2\pi r \, dr.$$

q	0	0.1	0.2	0.3	0.4	0.5	0.6	0.7	0.8	0.9	1.0
chat q	0.7854	0.6856	0.5867	0.4900	0.3963	0.3071	0.2236	0.1477	0.0817	0.0294	0.0

Autocorrelation of the Unit Disk Function is the Chinese hat function of radius 1.

$$\int_{-\infty}^\infty \int_{-\infty}^\infty \text{rect}(\sqrt{\alpha^2 + \beta^2}) \, \text{rect}(\sqrt{(\alpha+u)^2 + (\beta+v)^2}) \, d\alpha \, d\beta = \text{chat } \sqrt{u^2+v^2}.$$

Two-dimensional Fourier Transform of the $jinc^2$ function is the Chinese hat function

$$\int_{-\infty}^\infty \int_{-\infty}^\infty jinc^2 \sqrt{x^2+y^2} \, e^{-i2\pi(ux+vy)} dx \, dy = \text{chat } \sqrt{u^2+v^2}.$$

See Fig. 9-9.

Abel Transform of $jinc^2$ is (See Fig. 9-10.)

$$2\int_x^\infty \frac{jinc^2 r \, r \, dr}{\sqrt{r^2-x^2}} = \frac{H_1(2\pi x)}{4\pi x^2}.$$

Fourier Transform of Chat.

$$\int_{-\infty}^\infty \text{chat } x e^{-i2\pi s x} dx = \frac{H_1(2\pi s)}{4\pi s^2}.$$

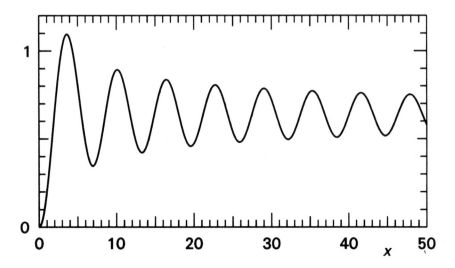

Figure 9-10   The Struve function $\mathbf{H}_1(x)$.

Abel Transform of Chat.

$$2 \int_x^\infty \frac{\text{chat } r \; r \; dr}{\sqrt{r^2 - x^2}} = \text{See Fig. 9-12.}$$

Fourier Transform of jinc^2 **is**

$$\int_{-\infty}^\infty \text{jinc}^2 x \; e^{-i2\pi s x} \, dx =$$

See Fig. 9-12.

Integrals and Central Values.

$$\int_0^\infty \text{jinc}^2 r \; 2\pi r \; dr = \text{chat } 0 = \pi/4,$$

$$\int_0^\infty \text{chat } r \; 2\pi r \; dr = \text{jinc}^2 0 = \pi^2/16.$$

To summarize, we arrange the various functions in groups of four to display their relationships according to the pattern shown in Fig. 9-11. Each algebraic quartet in the table can also be illustrated graphically. We have been accustomed to arranging functions on the left and transforms on the right. A certain convenience accrues from the adoption of conventions of this sort, which provide a constant framework within which different cases may be considered. In the graphical version of the new organization proposed, the

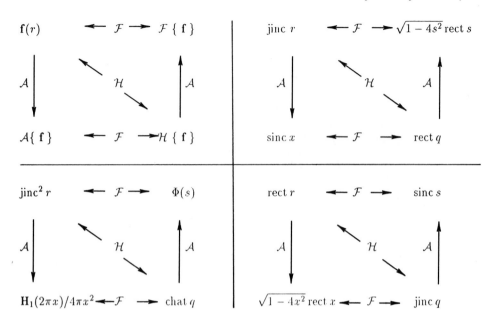

Figure 9-11   The jinc function and its relatives arranged in quartets obeying the relationships specified in the upper left. The function $\Phi(s)$ is both the Abel transform of the chat function and the Fourier transform of the jinc^2 function.

function $f(x, y)$ (or $\mathbf{f}(r)$ in the case of circular symmetry) goes in the northwest and its two-dimensional Fourier transform $F(u, v)$ [or $\mathbf{F}(q)$] goes in the southeast, giving a diagonal arrangement on the page, because in this way the left-right juxtaposition of one-dimensional Fourier transforms can be preserved. In the example where jinc r is in the top left-hand corner and rect q in the bottom right-hand corner, the cross section of each two-dimensional function along the east-west axis can be shown rabatted into the plane. These one-dimensional functions of x and u, respectively, constitute a Hankel transform pair. Where circular symmetry happens to exist and $f(x, y) \overset{2}{\supset} F(u, v)$, then $f(x, 0)$ has Hankel transform $F(u, 0)$. The general situation of no symmetry has more to do with data than with properties of instruments, which can often be designed with cylindrical symmetry, and is taken up later in connection with the projection-slice theorem.

The cross section of $f(x, y)$ along a line $x = $ const has an area which is the ordinate of the Abel transform of $f(x, y)$, viz., sinc x. We see that sinc x and its Fourier transform rect q are arranged left-to-right as planned.

The whole story can now be repeated, since the Hankel transform is reciprocal, starting in the bottom right-hand corner. Thus the cross section of rect q has an area equal to the ordinate of $(1 - 4u^2)^{1/2}$ rect u, which in turn is the one-dimensional Fourier transform of jinc r, the function we began with.

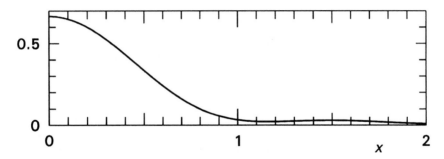

Figure 9-12   The Abel transform of jinc$^2\, r$. This function is also the one-dimensional
Fourier transform of the chat function.

THE STRUVE FUNCTION

The Fourier transform of the chat function, which is the same as the Abel transform of
the Airy diffraction pattern, or jinc^2 function, both arise naturally in optical systems and
may be expressed in terms of the Struve function $\mathbf{H}_1(x)$ as $\mathbf{H}_1(2\pi x)/4\pi x^2$. For values
of x up to about 5 the Struve function can be calculated from the Taylor series $(2/\pi)(1 +
x^2/3 - x^4/3^2.5 + x^6/3^2 \cdot 5^2 \cdot 7 - \cdots)$. For larger values of x use the asymptotic expansion
12.1.31 given in A&S (1964). A graph is shown in Fig. 9-10; the function oscillates with
a period close to 2π about a limiting value of $2/\pi$. The oscillations decay rather slowly in
amplitude, inversely as the square root of x; the only null is the one at $x = 0$.

THE ABEL TRANSFORM

A two-dimensional function $\mathbf{f}(r)$ that has circular symmetry possesses a line integral, or
projection, that is the same in all directions. Call this function $f_A(x)$. The subscript A
refers to Abel and the variable x can be thought of as being the abscissa in the (x, y)-
coordinate system to which the radial coordinate r belongs. Thus $f_A(x)$ is the projection
in the y-direction, or the line integral in the y-direction. As an example, if $\mathbf{f}(r) = \text{rect}(r)$,
then $f_A(x) = (1 - 4x^2)^{1/2} \text{rect}\, x$ (Fig. 9-13). This is because a disk function of unit height
and unit diameter has a cross-section area $(1 - 4x^2)^{1/2}$ on the line $x = \text{const}$, provided
$|x| < \frac{1}{2}$. Where $|x| \geq \frac{1}{2}$, the cross-section area is zero, a fact that the factor rect x reminds us of. The shape of the Abel transform in this example is semi-elliptical, which
is connected with the fact that the given outline was circular. If you wanted to know what
function of r has a semicircular Abel transform, the answer would be $\frac{1}{2}$ rect r.

In lieu of these explanatory remarks it would be sufficient simply to introduce the
Abel transform $f_A(\ )$ of a function $\mathbf{f}(\ )$ by this definition:

$$f_A(x) \triangleq \int_{-\infty}^{\infty} \mathbf{f}(\sqrt{x^2 + y^2})\, dy.$$

Then if the question arose as to the Abel transform of rect( ), we would evaluate it as
follows:

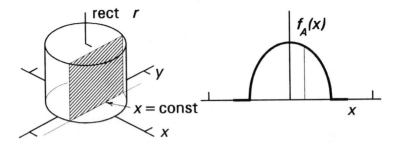

Figure 9-13   The area of the shaded cross section is the Abel transform of the function of r for the particular value of x chosen. In the case of rect r the Abel transform $f_A(x)$ is the semi-ellipse $\sqrt{1-4x^2}$, $|x| < \frac{1}{2}$.

$$f_A(x) = \int_{-\infty}^{\infty} \text{rect}(\sqrt{x^2+y^2})\,dy$$

$$= \int_{-(1/4-x^2)^{1/2}}^{(1/4-x^2)^{1/2}} dy\ \text{rect}\,x$$

$$= 2\sqrt{\tfrac{1}{4}-x^2}\ \text{rect}\,x$$

$$= \sqrt{1-4x^2}\ \text{rect}\,x.$$

The limits of integration were arrived at by noting that points on a line parallel to the y-axis at abscissa x must lie within $y = \pm(\frac{1}{4} - x^2)^{1/2}$ in order for $\text{rect}(\sqrt{x^2+y^2})$ to be unity rather than zero. The integration involved is not difficult, but it is obvious that some geometrical reasoning based on the explanatory introduction is helpful in arriving at the limits of integration.

   An alternative form of the definition can be given in terms of r, which is the natural variable to think of as underlying a circularly symmetrical function $\mathbf{f}(r) = f(x, y)$. Thus

Abel transform definition.

$$f_A(x) \triangleq 2\int_x^{\infty} \frac{\mathbf{f}(r)r\,dr}{\sqrt{r^2-x^2}}.$$

To convert from dy to dr write $dr/dy = \sin\theta = \sqrt{r^2-x^2}/r$. Thus $dy = r\,dr/\sqrt{r^2-x^2}$. To relate this definition to the previous one, note that the minimum value of r is the given value of x. Thus $\int_{-\infty}^{\infty}\ldots dy$ may be replaced by $2\int_x^{\infty}\ldots\dfrac{dy}{dr}\,dr$, the factor 2 arising from the equal contributions from above and below the x-axis. From $y^2 = r^2 - x^2$ we deduce that $dy/dr = r/y$ at constant x. Alternatively, from Fig. 9-14, we can see from similar triangles that $dy/dr = r/y$.

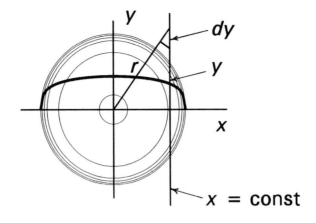

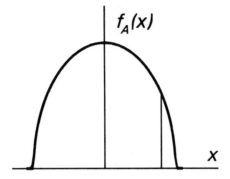

Figure 9-14 A contour map of the plane of $\mathbf{f}(r)$ (above) with a diametral cross section and a graph (below) of the line integral $\int \mathbf{f}(r)dy$ along the line $x = \text{const}$.

Some Theorems for the Abel Transform

Many theorems for the Fourier transform do not have a counterpart when circular symmetry is imposed, but a small number of interesting theorems for the Abel theorem can be mentioned.

Similarity Theorem.   If $\mathbf{f}(r)$ is contracted by a factor a to $\mathbf{f}(ar)$, then clearly $f_A(x)$ will be contracted in the same proportion and, in addition, the values of $f_A()$ will be reduced in magnitude by the same factor. Thus

$$\mathbf{f}(ar) \quad \text{has Abel transform} \quad a^{-1}f_A(ax).$$

This result is verifiable immediately by substituting ar for r in the definition.

Linear Superposition. If $\mathbf{f}(r)$ has Abel transform $f_A(ax)$, and $\mathbf{g}(r)$ has Abel transform $g_A(ax)$, then

$$\mathbf{f}(r) + \mathbf{g}(r) \quad \text{has Abel transform} \quad f_A(x) + g_A(x),$$

for any choice of $\mathbf{f}$ and $\mathbf{g}$.

Convolution Theorem. If $\mathbf{f}(r)$ and $\mathbf{g}(r)$ are convolved in two dimensions, then the Abel transform of the result can be obtained as follows. From each of the Abel transforms $f_A(x)$ and $g_A(x)$ construct a circularly symmetrical function with the same radial section. These two functions are correctly written $f_A(r)$ and $g_A(r)$. After f_A and g_A are convolved two-dimensionally, the radial section in any direction θ is the desired Abel transform. The theorem is:

$$\mathbf{f}(r) ** \mathbf{g}(r) \quad \text{has Abel transform} \quad \left[f_A(r) ** g_A(r) \right]_{\theta=\text{const}}.$$

The derivation of this theorem can be written down starting from the definition integrals, but such a simple result must have a simple explanation, and it is given in Chapter 14.

Conservation Theorem. As the Abel transform is a simple projection of a two-dimensional function, the area integral of $\mathbf{f}(r)$ equals the integral of $f_A(x)$:

$$2\pi \int_0^\infty \mathbf{f}(r) r \, dr = \int_{-\infty}^\infty f_A(x) \, dx.$$

Central Value Theorem. Putting $x = 0$ in the defining integral, we see that

$$f_A(0) = 2 \int_0^\infty \mathbf{f}(r) dr,$$

a relation that is useful for normalizing at the end of a computation in which unnecessary multiplications by constants are dropped.

Table 9-2 lists a variety of Abel transforms for ready reference.

Inverting the Abel Transform

Inversion of the Abel transform is performed by

$$\mathbf{f}(r) = -\frac{1}{\pi} \int_r^\infty \frac{f_A'(x) \, dx}{\sqrt{x^2 - r^2}}.$$

An important special case is where $f_A(x)$ is zero for x greater than some cutoff value r_0. Then

$$\mathbf{f}(r) = -\frac{1}{\pi} \int_r^{r_0} \frac{f_A'(x) \, dx}{\sqrt{x^2 - r^2}} + \frac{f_A(r_0-)}{\pi \sqrt{r_0^2 - r^2}}.$$

The final term, which might be overlooked, arises from the possibility of $f_A(x)$ being discontinuous at $x = r_0$. Numerical inversion of the Abel transform is important, because

Table 9-2 Table of Abel transforms. For compactness rect x is written $\Pi(x)$.

$f(r)$		$f_A(x) = 2\int_x^\infty (r^2-x^2)^{-1/2}f(r)r\,dr$	
$\Pi(r/2a)$	Disk	$2(a^2-x^2)^{1/2}\Pi(x/2a)$	Semiellipse
$(a^2-r^2)^{-1/2}\Pi(r/2a)$		$\pi\,\Pi(x/2a)$	Rectangle
$(a^2-r^2)^{1/2}\Pi(r/2a)$	Hemisphere	$\frac{1}{2}\pi(a^2-x^2)\Pi(x/2a)$	Parabola
$(a^2-r^2)\Pi(r/2a)$	Paraboloid	$\frac{4}{3}(a^2-x^2)^{3/2}\Pi(x/2a)$	
$(a^2-r^2)^{3/2}\Pi(r/2a)$		$\frac{3\pi}{8}(a^2-x^2)^2\Pi(x/2a)$	
$(1-|r|)\Pi(r/2)$	Cone	$[(a^2-x^2)^{1/2} - (x^2/a)\cosh^{-1}(a/x)]\Pi(x/2a)$	
$\cosh^{-1}(a/r)\Pi(r/2a)$		$\pi a(1-|r/a|)\Pi(r/2a)$	Triangle
$\delta(r-a)$	Ring impulse	$2a(a^2-x^2)^{-1/2}\Pi(x/2a)$	
$\exp(-\pi r^2)$	Gaussian	$W\exp(-\pi x^2/W^2)$	Gaussian
$\exp(-r^2/2\sigma^2)$	Normal	$\sqrt{2\pi}\sigma\exp(-x^2/2\sigma^2)$	Normal
$r^2\exp(-r^2/2\sigma^2)$		$\sqrt{2\pi}\sigma(x^2+\sigma^2)\exp(-x^2/2\sigma^2)$	
$(r^2-\sigma^2)\exp(-r^2/2\sigma^2)$		$\sqrt{2\pi}\sigma x^2\exp(-x^2/2\sigma^2)$	
r^{-2}		π/x	
$(a^2+r^2)^{-1}$		$\pi(a^2+x^2)^{-1/2}$	
$J_0(2\pi ar)$	Bessel	$(\pi a)^{-1}\cos 2\pi ax$	Cosine
$2\pi[r^{-3}\int_0^r J_0(r)dr - r^{-2}J_0(r)]$		$\operatorname{sinc}^2 x$	
$\delta(r)/\pi|r|$		$\delta(x)$	Impulse
$2a\operatorname{sinc}(2ar)$		$J_0(2\pi ax)$	Bessel
$\frac{1}{2}r^{-1}J_1(2\pi ar)$		$\operatorname{sinc} 2ax$	
$\operatorname{jinc} r$		$\operatorname{sinc} x$	

line-integrated data can often be obtained in situations where values of $f(r)$ itself are inaccessible. In such circumstances a formula containing a derivative looks unattractive if the derivative $f_A'(x)$ must be formed by differencing, because measurement error is unfavorable. A numerical inversion procedure is described in FTA (1986) which avoids the derivative. A different method of inversion is to take advantage of the Fourier-Abel-Hankel cycle: take the one-dimensional Fourier transform and then take the Hankel transform.

One can take the Hankel transform without the need to invoke Bessel functions, by first taking the Abel transform and then taking the one-dimensional Fourier transform, as displayed in the following diagram.

$$
\begin{array}{ccc}
f(r) & \overset{\mathcal{F}}{\longleftrightarrow} & F(\cdot)\\
{\scriptstyle A}\downarrow & & \uparrow{\scriptstyle A}\\
f_A(x) & \overset{\mathcal{F}}{\longleftrightarrow} & H(q)
\end{array}
$$

It follows that $f(r)$ can be recovered from $f_A(x)$ as indicated by

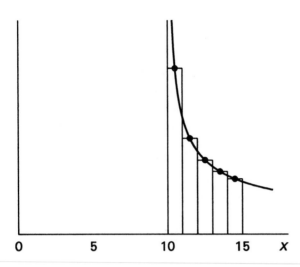

Figure 9-15   Staggering the samples of an integrand to avoid the infinite sample at the discontinuity.

$$f = \mathcal{F}\mathcal{A}\mathcal{F} f_A.$$

Implementation of this sequence of operations numerically is straightforward.

Computing the Abel Transform

A rather interesting item of numerical analysis arises when one is called on to evaluate

$$\int_{r}^{r_0} \frac{dx}{\sqrt{x^2 - r^2}}$$

since the integrand is infinite at the left edge. For example, if we wanted $I = \int_{10}^{15}(x^2 - 10^2)^{-1/2}\,dx$, there would be trouble at $x = 10$, even though the full integral is finite. A way around this would be to stagger the samples (Fig. 9-15). Let $(x^2 - 10^2)^{-1/2} = \phi(x)$; then if one computes $\sum_{j=10.5}^{14.5} \phi(j)$, where $j = 10.5, 11.5, \ldots 14.5$, the result is 0.827, but the exact integral is $I = 0.962$. Obviously the approximation is crude, and fine subdivision of the interval might be needed to achieve desired accuracy. The correct approach is to note that $\int_{10}^{10+w} \phi(x)\,dx = C_1\phi(10 + 0.5w)w$ and that $\int_{10+w}^{10+2w} = C_2\phi(10 + 1.5w)w$, where C_1 and C_2 are coefficients and w is the sampling interval. As $w \to 0$, C_1 and C_2 assume definite values 1.414 and 1.015 that may be used for general-purpose integration in cases such as this, where the integrand diverges inversely as the square root of distance from the pole. Thus

$$\int_{r} \frac{\mathbf{f}(r)\,dx}{\sqrt{x^2 - r^2}} \approx \left[1.414 f(r + 0.5w) + 1.015 f(r + 1.5w) + \sum_{j=2\frac{1}{2}} f(r + jw) \right] w.$$

With $w = 1$, which is rather coarse, the approximation to the correct value 0.962 is 0.959, which is already better than 1 percent. A similar approach works with integrands that go infinite as the inverse three-halves power.

With this useful background, the reader may enjoy the following complete program for the Abel transform in which I avoid the pole at $x = r$. This application assumes that the function of radius is expressible in algebraic form and that the abscissa is scaled so that the function is zero where $r > 1$. The example applies to $f(r) = 1 - r$, whose Abel transform is known to be $\sqrt{1 - x^2} - x^2 \cosh^{-1}(1/x)$. The program can readily be modified for *data* given at equal intervals.

ABEL TRANSFORM

```
DEF FNf(r)=1-r                    Define given function as cone
d=0.1                             Step in x
dy=0.01                           Step in y
FOR x=d/2 TO 1 STEP d
    s=0.5*FNf(x)
    FOR y=dy TO SQR(1-x²) STEP dy
        s=s+FNf(SQR(x²+y²))
    NEXT y
    PRINT x;2*s*dy
NEXT x
```

The results are as follows:

r	.05	.15	.25	.35	.45	.55	.65	.75	.85	.95
$f(r)$	0.98952	0.93053	0.83928	0.72718	0.60210	0.47067	0.33909	0.21404	0.10363	0.02071

Comparison with the theoretical expression shows that the largest error is one digit in the fifth decimal place. Whether the value of dy is too coarse can be checked empirically.

SPIN AVERAGING

A function $f(x)$ may be spin averaged to obtain a new function $f_S(r)$ defined by

$$f_S(r) = \frac{1}{2\pi} \int_0^{2\pi} f(r \cos \alpha) \, d\alpha.$$

If $f(x)$ is an even function, as in all that follows here, we may evaluate $f_S(r)$ from

$$f_S(r) = \frac{2}{\pi} \int_0^{\pi/2} f(r \cos \alpha) \, d\alpha.$$

Two ways of viewing spin averaging will now be described.

Imagine a function defined on the (x, y)-plane so that at any point (x, y) the value is $f(x)$, i.e., independent of y. If the function value represented the height of a surface above the (x, y)-plane, the surface would be a cylindrical ridge running in the y-direction. Then if we traveled on the surface of this ridge so that our track projected onto the (x, y)-plane was a circle of radius r, then our average height would be $f_S(r)$, as given by either of the

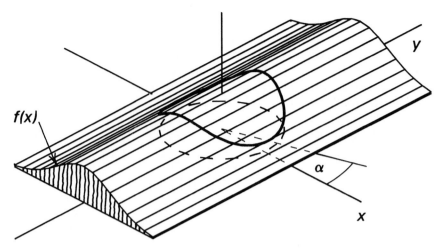

Figure 9-16   The saddle-shaped curve is a circle of radius r in plan view. Its average height at given r is the spin average of the profile $f(x)$.

above integrals (Fig. 9-16). Alternatively, if we were to sit at the point $(r, 0)$ in the (x, y)-plane and spin the distribution at uniform speed, then $f_S(r)$ would be the average of the values passing by in the course of one rotation. If a linear source of light with brightness distributed as $f(x)$ were spun about its center, then a long exposure on a photographic plate would show a radial dependence of photographic density depending on $f_S(r)$. This is the reason for adopting the term *spin-averaging*.

Example

What is the spin average of a cosine function?▷The corrugated ridge pattern $\cos x$ spin-averages into $J_0(r)$. This is an expression of the relation

$$J_0(r) = \frac{2}{\pi} \int_0^{\pi/2} \cos(r \cos \alpha) \, d\alpha,$$

which is sometimes expressed in words by saying that a disturbance $J_0(r)$ (for example, on a drum membrane) can be regarded as made up of cosinusoidal corrugations, all of the same wavelength, superimposed uniformly in all azimuths. ◁

The spin integral of the cosine function was first evaluated in 1805 by M. A. Parseval as follows:

$$1 - \frac{r^2}{2^2} + \frac{r^4}{2^2 \cdot 4^2} + \frac{r^6}{2^2 \cdot 4^2 . 6^2} + \cdots = \frac{1}{\pi} \int_0^\pi \cos(r \sin \alpha) \, d\alpha.$$

A traveler moving steadily around a circular track lingers longer at the larger x-values, the time spent in the vicinity of any x being proportional to the secant of the inclination of the track to the x-axis. Near $x = 0$ each element dx is passed through in about the same time. As $|x|$ approaches r, it takes longer and longer to pass through a given increment dx. Therefore, a weighting function can be imagined that tells how to average the function values $f(x)$ with extra emphasis on the larger x-values.

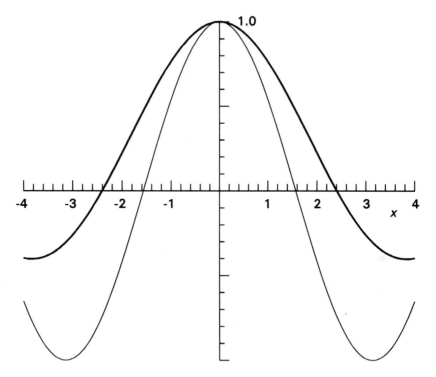

Figure 9-17   Showing how $J_0(2.405) = 0$ from the spin-average viewpoint.

Since the weighting function has to depend on r as well as x, write it $w(x, r)$. It has to be proportional to sec (inclination) = sec $| \arcsin(x/r) |= r(r^2 - x^2)^{-1/2}$. We also want the weighting function to have unit area, as is usual with taking weighted averages, because when r is very small and $f(x)$ is therefore essentially constant around the small circular track, the spin average must approach $f(0)$ in value. Noting that the integral of $r(r^2 - x^2)^{-1/2}$ from 0 to r is $\pi r/2$, (or one-quarter of the perimeter of the circle of radius r), we find

$$w(x, r) = \frac{2}{\pi} \frac{1}{\sqrt{r^2 - x^2}}.$$

Thus the alternative expression for the spin average is

$$f_S(r) = \frac{2}{\pi} \int_0^r \frac{f(x)\, dx}{\sqrt{r^2 - x^2}}.$$

Theorems for Spin Averaging

Denoting spin average by $\mathcal{S}$, the Fourier transform by $\mathcal{F}$, and the Hankel transform by $\mathcal{H}$, we can enunciate the following theorems.

$$f_S(0) = f(0)$$

$$\mathcal{S}(f + g) = \mathcal{S}(f) + \mathcal{S}(g)$$

If $\mathcal{S}[f(x)] = f_S(r)$ then $\mathcal{S}[bf(ax)] = bf_S(ar)$

$$\pi \mid q \mid \mathcal{H}[f_S(r)] = \mathcal{F}[f(x)].$$

$$\lim_{r \to \infty} f_S(r) = \frac{\int_{-\infty}^{\infty} f(x)\,dx}{\pi r}$$

Example

Use the fourth theorem to calculate the spin integral of $f(x) = \cos x$. ▷First rewrite the theorem as

$$\mathcal{H}\left\{\frac{1}{\pi \mid q \mid}\mathcal{F}[f(x)]\right\} = f_S(r).$$

Then

$$\mathcal{F}[\cos x] = \tfrac{1}{2}\delta(q + 1/2\pi) + \tfrac{1}{2}\delta(q - 1/2\pi)$$

and

$$\frac{1}{\pi \mid q \mid}\mathcal{F}[\cos x] = \delta(q + 1/2\pi) + \delta(q - 1/2\pi),$$

because $\pi \mid q \mid \delta(q \pm 1/2\pi)$ is a half-strength impulse. Finally,

$$f_S(r) = 2\pi \int_0^{\infty} \frac{1}{\pi \mid q \mid}\mathcal{F}[\cos x]J_0(2\pi rq)q\,dq = J_0(r).◁$$

Table 9-3 presents a list of spin average functions. The functions of x are shown as even functions of x. However, any entry $f(x)$ may be replaced by $f(x)\mathbf{H}(x)$, which is zero for $x < 0$, and $f_S(r)$ will remain the same because the averaging is done over only one quadrant.

Computing the Spin Average

The spin average is convenient to compute, as shown in the following program.

```
SPIN AVERAGE
DEF FNf(x)=x^2           Define given function as parabola
d=0.1                    Step in radius
N=20                     Number of sample points per quadrant
FOR r=0 TO 1 STEP d
   da=PI/2/N             Step in angle
   s=0
   FOR a=da/2 TO PI/2 STEP da
      s=s+FNf(r*COS(a))
   NEXT a
   spinav=s/N            Take the average
   PRINT r;FNf(r);spinav
NEXT r
END
```

Table 9-3   Spin-average functions

$f(x)$	$f_S(r) = \frac{2}{\pi} \int_0^{\pi/2} f(r\cos\alpha)\,d\alpha$
$\cos x$	$J_0(r)$
$\operatorname{sinc} x - \frac{1}{2}\operatorname{sinc}^2(x/2)$	$\operatorname{jinc} r$
$\operatorname{sinc} x - \frac{1}{2}\operatorname{sinc}^2(x/2)x\mathrm{III}(x)$	$\sum_1^\infty 2\pi n J_0(2\pi n r)$
$\delta(x+a) + \delta(x-a)$	$(2r/\pi)(r^2 - a^2)^{-1/2}H(r-a)$
$\delta(x)$	$1/\pi r$
$\operatorname{rect}(x/2a)$	$1 - (2/\pi)\cos^{-1}(a/r)H(r-a)$
$1 - \operatorname{rect}(x/2b)$	$(2/\pi)\cos^{-1}(b/r)H(r-b)$
$\operatorname{rect}(x/2a) - \operatorname{rect}(x/2b),\ b < a$	$(2/\pi)[\cos^{-1}(b/r)H(r-b) - \cos^{-1}(a/r)H(r-a)]$
$J_1(|x|)$	$2\pi r(1 - \cos r)$
$J_0(x)$	$[J_0(\tfrac{1}{2}r)]^2$
$\cos^2 x$	$\frac{1}{2} + \frac{1}{2}J_0(2(r))$
$(1 - x^2)^{-1/2}$	$(2/\pi)E(r^2) = (2/\pi)\int_0^{\pi/2}(1 - r^2\cos^2\alpha)^{1/2}d\alpha$
$\Lambda(x/2w)$	$(2/\pi)(\pi/2 - \alpha_1) - (2/\pi)(2r/w)(1 - \sin\alpha_1)$ where $\cos\alpha_1 = H(w/2r) + (w/2r)H(r - w/2)$
1	1
$|x|$	$2r/\pi$
x^2	$r^2/2$
$|x|^3$	$4r^3/3\pi$
x^4	$3r^4/8$
$|x|^5$	$16r^5/15\pi$
x^6	$15r^6/16$
$|x|^7$	$52r^7/35\pi$
$\sin x$	$(2/\pi)\sum_0^\infty(-1)^k r^{2k+1}/[(2k+1)!!]^2$†
$x\sin x$	$rJ_1(r)$
$\cosh x$	$I_0(r)$
$J_2(x)$	$J_1^2(r/2)$
$N_0(x)$	$J_0(r/2)N_0(r/2)$
$|x|\,J_0(x)$	$(2/\pi)\sin r$
$J_{2\nu(x)}$	$J_\nu^2(r/2),\ \operatorname{Re} 2\nu > 1$

† $(2k+1)!! = 1\cdot 3\cdots (2k+1)$.

ANGULAR VARIATION AND CHEBYSHEV POLYNOMIALS

When $n = 0$, a function of the form $f(r)\cos n\theta$ is circularly symmetrical. In a typical application it might describe a wave disturbance inside a circular pipe when there is no dependence on azimuth. When $n = 1$, the disturbance would be zero where $\theta = \pm\pi/2$. It could describe a higher-order mode with a null line on one diameter and antiphased disturbances in the two halves. For $n = 2, 3, 4, \ldots$, we say that the function has n-

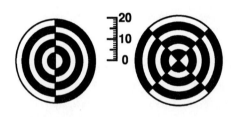

Figure 9-18   Modulated ring impulses $\delta(r - 0.5)\cos n\theta$ have transforms $J_n(\pi q)\cos n\phi$ represented by nodal lines delineating antiphase regions (shaded) for $n = 0, 1, 2, 3$.

fold rotational symmetry. Many functions $f(r)\cos n\theta$ arising from wave propagation in circular structures and from diffraction have transforms that are simply generalizations of cases discussed under circular symmetry, and some will be mentioned here.

The simplest example is a ring impulse of radius a modulated by a cosine function of azimuth, namely

$$\delta(r - a)\cos n\theta.$$

The two-dimensional Fourier transform is

$$(-i)^n 2\pi a J_n(2\pi aq)\cos n\phi.$$

We verify that this reduces to $2\pi a J_0(2\pi aq)$ when $n = 0$. Figure 9-18 shows, for the first four modulated ring impulses $\delta(r - 1)\cos n\theta$, the configuration of the nodal lines of the transforms $J_n(\pi q)\cos n\phi$.

Because $\delta(r - a)\cos n\theta$ is not circularly symmetrical, it does not have an Abel transform, but it has a projection $f_0(x)$ onto the x-axis given by

$$\int_{-\infty}^{\infty} \delta(r - a)\cos n\theta \, dy$$

$$= \int_{-\infty}^{\infty} \delta(r - a)\frac{\cos n\theta}{\sin\theta}dr = 2\frac{\cos n\theta}{\sin\theta}.$$

The integral is taken in the y-direction along a line $x = $ const which intersects the ring impulse where $\theta = \pm\cos^{-1}(x/a)$ and at an angle θ. Consequently the integral is the product of the local strength $\cos n\theta$ with the factor $1/\sin\theta$, all doubled because there are two intersections.

Since $\cos\theta = x/a$ and $\sin\theta = \sqrt{1 - x^2/a^2}$,

$$f_0(x) = \frac{2\cos(n\cos^{-1}x/a)}{\sqrt{1 - x^2/a^2}}$$

$$= \frac{2T_n(x/a)}{\sqrt{1 - x^2/a^2}},$$

where $T_n(x)$ is a Chebyshev polynomial ($T_1(x) = x$, $T_2(x) = 2x^2 - 1$, $T_3(x) = 4x^3 - 3x$, $T_4(x) = 8x^4 - 8x^2 + 1$, ...). This result allows us to understand the Chebyshev polynomials in terms of the simple projections of $\delta(r-1)\cos n\theta$ onto the x-axis. It follows that the one-dimensional Fourier transform of $2/\sqrt{(1-x^2/a^2}T_n(x/a)$ is $(-i)^n 2\pi a J_n(2\pi aq)$.

Any distribution of strength around a circular ring impulse can now be handled readily by one-dimensional Fourier analysis on the circle. Because such a distribution is of necessity periodic, discrete coefficients will result, and the 2D Fourier transform of the general nonuniform ring impulse will take the form

$$\sum J_n(2\pi aq)(a_n\cos n\phi + b_n\sin n\phi).$$

We shall also be able to get the projection of the nonuniform ring impulse. Clearly, this is of general value, since any two-dimensional function can be regarded as built up of concentric nonuniform annuli of infinitesimal width, and in addition to the analytic aspect there is the very practical feature that decomposition into fine annuli may be appropriate in numerical computation with data acquired within a broad annular zone.

As an analytic application consider the 2D Fourier transform of rect $r\cos n\theta$. Regard this as a superposition of ring impulses $\delta(r-a)$, where a ranges from 0 to 0.5. Then the desired transform should be

$$\int_0^{1/2} (-i)^n 2\pi a J_n(2\pi qa)\cos n\phi\, da = (-i)^n 2\pi a\cos n\phi \int_0^{1/2} J_n(2\pi qa)\, da.$$

Integrals of Bessel functions are computable. In the particular case of $n=1$ we find that the 2D Fourier transform of rect $r\cos\theta$ is

$$-i2\pi a\cos\phi \int_0^{1/2} J_1(2\pi qa)\, da = i2\pi a\cos\phi(2\pi q)^{-1}J_0(2\pi qa)\Big]_0^{1/2}$$

$$= i(a/q)\cos\phi[J_0(\pi q) - 1].$$

The projections of $\exp(-\pi r^2)\cos n\theta$ will play a role in the discussion of the Radon transform.

Modulated Ring Delta

Just as a function of the single variable x can be interpreted as a two-dimensional surface $z = f(x)$ above the (x, y)-plane, so also can functions of the single variables r and θ. For example, $z = r^2/4a$, which is a function of r but not of θ, describes a paraboloidal dish with focal length a that is concave upward. The function $\cos n\theta$ which arises in the discussion of rotational symmetry, can be thought of correspondingly. The surface is like a magic carpet with azimuthal undulations. The contour lines are all level, and at the origin, where the contours converge, there is a nasty singularity. The illustration (Fig. 9-19) terminates on a cylinder of unit radius. The curve of intersection with the cylinder of unit radius describes the strength of the modulated ring delta $\delta(r-1)\cos n\theta$, whose character is conveyed in Fig. 9-20 for $n = 0$, 1, and 2, left to right.

Multiplication of a function $f(x, y)$ by $\delta(r-a)\,{}^{\cos}_{\sin}\, n\theta$ has the effect of singling out the infinitesimal annulus of radius a and taking the first step toward finding the Fourier series coefficients for the annular section. We now show how the modulated ring delta $\delta(r-a)\cos\theta$ can arise from differentiation of the familiar function $f(x, y) = \text{rect}(r/2a)$

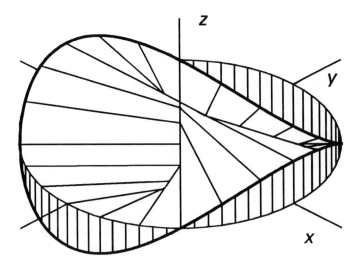

Figure 9-19   The function $\cos n\theta$ exhibited by contours of the surface $z = \cos n\theta$ above the (x, y)-plane for the case $n = 2$.

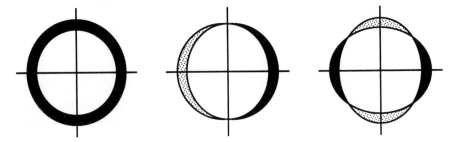

Figure 9-20   Modulated ring deltas $\delta(r - 1)$, $\delta(r - 1)\cos\theta$, and $\delta(r - 1)\cos 2\theta$. Strength is represented by radial line width, and by shading where negative.

and thereby gain a useful algebraic way of relating rotational symmetry to circular symmetry. The proposition is that

$$\frac{\partial}{\partial x}\,\text{rect}\left(\frac{r}{2a}\right) = -\delta(r - a)\cos\theta.$$

We know that differentiation of $f(x, y)$ with respect to x multiplies the Fourier transform $F(u, v)$ by $i2\pi u$. This example, where $F(u, v)$ is known to be the jinc function $4a^2$ jinc $2aq$, implies that $\delta(r - a)\cos n\theta$ has a Fourier transform $i8\pi a^2 u$ jinc $2aq$, which simplifies to $i2\pi a J_1(2\pi aq)\cos\theta$. This relation has in fact already been derived by manipulation of integrals. Knowing the simple proposition stated above provides another way of thinking with a fundamental entity such as the modulated ring delta.

Since $\text{rect}(r/2a)$ is discontinuous on the circle $r = a$, differentiation is not in ordinary circumstances defined on the very locus where the ring delta is situated. Thus, to set

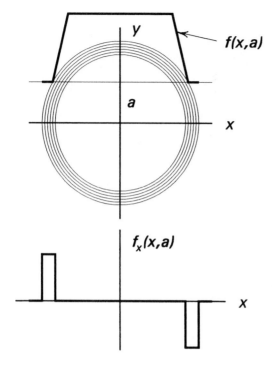

Figure 9-21 Contour representation of an upside-down bucket of top and bottom radii $a - \tau$ and a, a cross section at constant y, and the derivative of this cross section with respect to x (below).

up a consistent relation we think of a sequence of differentiable functions that approach $\text{rect}(r/2a)$ in the limit. The simplest such function is equal to zero outside the circle $r = a$, equal to unity inside the circle of radius $a - \tau$, and equal to the height of the cone forming a continuous surface between the two circular rims of radius a and $a - \tau$. We now differentiate with respect to x and consider the sequence of outcomes as $\tau \to 0$. Along each line $y = \text{const}$ the derivative produces two unit-area rectangle functions of width τ and heights $1/\tau$ and $-1/\tau$, respectively, sequences representing delta functions of strength 1 and -1, situated on the locus $r = a$. How then do we arrive at a strength modulated in accordance with $\cos \theta$? The answer is that the strength of a curvilinear impulse is measured by the cross section area in the direction normal to the curve. What we have done is to measure in the x-direction, thus amplifying the normal cross section by $\cos \theta$. Allowing for this, it is confirmed that $(\partial/\partial x) \text{rect}(r/2a) = -\delta(r - a) \cos \theta$. This rather elegant technique is original and is applicable to other situations where partial differentiation in two or three dimensions is probably the last thing that would occur to you.

SUMMARY

Although circular symmetry is a special case, a rich set of useful tools reveals itself when we restrict attention to radial dependence only. Some important transforms emerge, especially the Hankel transform, many of whose particular instances are in everyday use for computation and for thinking about symmetrical situations. Outstanding examples are

Table 9-4   The jinc function.

	.00	.01	.02	.03	.04	.05	.06	.07	.08	.09
0.0	0.7854	0.7853	0.7850	0.7845	0.7838	0.7830	0.7819	0.7807	0.7792	0.7776
0.1	0.7757	0.7737	0.7715	0.7691	0.7666	0.7638	0.7609	0.7577	0.7544	0.7509
0.2	0.7473	0.7434	0.7394	0.7352	0.7309	0.7264	0.7217	0.7168	0.7118	0.7067
0.3	0.7014	0.6959	0.6903	0.6845	0.6786	0.6725	0.6663	0.6600	0.6536	0.6470
0.4	0.6402	0.6334	0.6264	0.6194	0.6122	0.6049	0.5975	0.5900	0.5824	0.5747
0.5	0.5669	0.5590	0.5510	0.5430	0.5348	0.5267	0.5184	0.5101	0.5017	0.4932
0.6	0.4847	0.4762	0.4676	0.459	0.4503	0.4416	0.4329	0.4241	0.4153	0.4066
0.7	0.3978	0.3890	0.3802	0.3714	0.3626	0.3538	0.3450	0.3363	0.3276	0.3189
0.8	0.3102	0.3016	0.2930	0.2845	0.2760	0.2676	0.2593	0.2510	0.2428	0.2346
0.9	0.2266	0.2186	0.2107	0.2030	0.1953	0.1877	0.1735	0.1656	0.1577	0.1500
1.0	0.1423	0.1347	0.1272	0.1198	0.1125	0.1053	0.0982	0.0912	0.0843	0.0776
1.1	0.0709	0.0643	0.0578	0.0515	0.0453	0.0392	0.0332	0.0273	0.0216	0.0159
1.2	0.0104	0.0051	−0.0002	−0.0053	−0.0103	−0.0151	−0.0199	−0.0245	−0.0289	−0.0333
1.3	−0.0375	−0.0416	−0.0455	−0.0493	−0.0530	−0.0565	−0.0599	−0.0632	−0.0663	−0.0694
1.4	−0.0722	−0.0750	−0.0776	−0.0801	−0.0824	−0.0847	−0.0868	−0.0887	−0.0906	−0.0923
1.5	−0.0939	−0.0954	−0.0967	−0.0979	−0.0990	−0.1000	−0.1009	−0.1017	−0.1023	−0.1028
1.6	−0.1033	−0.1036	−0.1038	−0.1039	−0.1039	−0.1038	−0.1036	−0.1033	−0.1029	−0.1024
1.7	−0.1018	−0.1011	−0.1004	−0.0995	−0.0986	−0.0976	−0.0965	−0.0953	−0.0941	−0.0928
1.8	−0.0914	−0.0900	−0.0885	−0.0869	−0.0853	−0.0836	−0.0818	−0.0801	−0.0782	−0.0763
1.9	−0.0744	−0.0724	−0.0704	−0.0683	−0.0662	−0.0641	−0.0620	−0.0598	−0.0576	−0.0553
2.0	−0.0531	−0.0508	−0.0485	−0.0462	−0.0439	−0.0416	−0.0393	−0.0369	−0.0346	−0.0323
2.1	−0.0299	−0.0276	−0.0253	−0.0230	−0.0207	−0.0184	−0.0161	−0.0138	−0.0116	−0.0093
2.2	−0.0071	−0.0050	−0.0028	−0.0007	0.0014	0.0035	0.0056	0.0076	0.0096	0.0115
2.3	0.0134	0.0153	0.0171	0.0189	0.0206	0.0224	0.0240	0.0256	0.0272	0.0287
2.4	0.0302	0.0316	0.0330	0.0344	0.0356	0.0369	0.0380	0.0392	0.0403	0.0413
2.5	0.0423	0.0432	0.0440	0.0448	0.0456	0.0463	0.0470	0.0476	0.0481	0.0486
2.6	0.0490	0.0494	0.0497	0.0500	0.0503	0.0504	0.0506	0.0506	0.0506	0.0506
2.7	0.0505	0.0504	0.0502	0.0500	0.0497	0.0494	0.0491	0.0487	0.0482	0.0477
2.8	0.0472	0.0466	0.0460	0.0454	0.0447	0.0440	0.0432	0.0424	0.0416	0.0407
2.9	0.0398	0.0389	0.0379	0.0370	0.0360	0.0349	0.0339	0.0328	0.0317	0.0306
3.0	0.0295	0.0283	0.0271	0.0259	0.0247	0.0235	0.0223	0.0211	0.0198	0.0186
3.1	0.0173	0.0161	0.0148	0.0135	0.0123	0.0110	0.0097	0.0085	0.0072	0.0060
3.2	0.0047	0.0035	0.0022	0.0010	−0.0002	−0.0014	−0.0026	−0.0038	−0.0049	−0.0061
3.3	−0.0072	−0.0083	−0.0094	−0.0105	−0.0116	−0.0126	−0.0136	−0.0146	−0.0156	−0.0165
3.4	−0.0174	−0.0183	−0.0192	−0.0200	−0.0209	−0.0216	−0.0224	−0.0231	−0.0238	−0.0245
3.5	−0.0252	−0.0258	−0.0264	−0.0269	−0.0274	−0.0279	−0.0284	−0.0288	−0.0292	−0.0296
3.6	−0.0299	−0.0302	−0.0304	−0.0307	−0.0309	−0.0311	−0.0312	−0.0313	−0.0314	−0.0314
3.7	−0.0314	−0.0314	−0.0314	−0.0313	−0.0312	−0.0310	−0.0309	−0.0307	−0.0304	−0.0302
3.8	−0.0299	−0.0296	−0.0292	−0.0289	−0.0285	−0.0281	−0.0276	−0.0272	−0.0267	−0.0268
3.9	−0.0257	−0.0251	−0.0245	−0.0239	−0.0233	−0.0227	−0.0221	−0.0214	−0.0207	−0.0200
4.0	−0.0193	−0.0186	−0.0179	−0.0171	−0.0164	−0.0156	−0.0148	−0.0140	−0.0132	−0.0124
4.1	−0.0116	−0.0108	−0.0100	−0.0092	−0.0083	−0.0075	−0.0067	−0.0058	−0.0050	−0.0042
4.2	−0.0034	−0.0025	−0.0017	−0.0009	−0.0001	0.0007	0.0015	0.0023	0.0031	0.0039
4.3	0.0046	0.0054	0.0061	0.0069	0.0076	0.0083	0.0090	0.0097	0.0104	0.0110
4.4	0.0117	0.0123	0.0129	0.0135	0.0141	0.0146	0.0152	0.0157	0.0162	0.0167
4.5	0.0171	0.0176	0.0180	0.0184	0.0188	0.0191	0.0195	0.0198	0.0201	0.0204
4.6	0.0206	0.0208	0.0210	0.0212	0.0214	0.0215	0.0217	0.0218	0.0218	0.0219
4.7	0.0219	0.0219	0.0219	0.0219	0.0218	0.0218	0.0217	0.0215	0.0214	0.0212
4.8	0.0211	0.0209	0.0207	0.0204	0.0202	0.0199	0.0196	0.0193	0.0190	0.0186
4.9	0.0183	0.0179	0.0175	0.0171	0.0167	0.0163	0.0158	0.0153	0.0149	0.0144
5.0	0.0139	0.0134	0.0129	0.0124	0.0118	0.0113	0.0107	0.0102	0.0096	0.0090

the relation of the annular slit to the J_0 Bessel function and the corresponding relation between the circular disc and the jinc function. In the world of symmetry which our machines manufacture, these two relationships are as common as the familiar relationships of the cosine function and the rectangle function to their one-dimensional Fourier transforms. Of course, the Hankel transform is one-dimensional, but we use it for handling two-dimensional things. Special instances of the Abel transform are of similar importance—the annular slit and circular disk are again examples. The spin average transform is closely related, but less heard of, not being associated with a great name. It can be recommended for attention. Circular symmetry is a special case of rotational symmetry. Superposition of rotationally symmetrical components permits synthesis of general two-dimensional functions and is important in other ways. It is appropriate to mention the topic in this chapter and to introduce the Chebyshev polynomial, which will be found indispensable if rotational symmetry is pursued.

Professor Pafnutii L'vovich Chebyshev (1821–1894) was a distinguished mathematician of St. Petersburg who worked on the kinematics of machines, the distribution of prime numbers, and many other subjects. He is known as Tchebicheff in French, Tschebischew in German, and Cebiscev in Italian.

Niels Henrik Abel (1802–1829) was a brilliant Norwegian mathematician who solved the spin integral equation in 1823 (*J. für Math.*, vol. 1, p. 153, 1826). Herman Hankel (1839–1873) was Professor of Mathematics at Leipzig. Hermann Ottovich Struve (1854–1920), working at the Pulkovo Observatory near St. Petersburg, introduced his special functions in connection with studies of the diffraction pattern in a telescope.

TABLE OF THE JINC FUNCTION

Table 9-4 (on page 380) is provided to facilitate use of the jinc function. In the event that values are needed for machine use, the series expansion and asymptotic expression given earlier are available. For greater precision, refer to A&S (1964) for computation of $J_1(x)$.

PROBLEMS

9–1. *First null of J_0.* Evaluate the two expressions (i) $\cos[r\cos(15°)] + \cos[r\cos(45°)] + \cos[r\cos(75°)]$, (ii) $0.5\cos r + \cos[r\cos(30°)] + \cos[r\cos(60°)] + 0.5\cos[r\cos(90°)]$ for $r = 2.4048255577$ and comment on the results. ☞

9–2. *Approximation to Bessel function.* A student claims that you can evaluate $J_0(r)$ with an error less than 0.001 from $f_1(r) = \frac{1}{2}(\cos 0.9239r + \cos 0.3827r)$, where $r < 3$. Obviously this claim is easy to check if you have a table of Bessel functions, and you might wish to do so.

   (a) Explain how to test the claim without needing a table.

   (b) Devise a comparable formula that is good to 0.00002 where $r < 3$. [*Hint.* If you do not recognize the source of the formula, identify the coefficients by looking at their arccosines.]

9–3. *Hankel transform theorem.* Observing that rect r and jinc q, which form a Hankel transform pair, both have unit area, a student conceives the following conjectural theorem.

"If $\mathbf{f}(r)$ has Hankel transform $\mathbf{F}(q)$, then $\int_0^\infty \mathbf{f}(r)\,dr = \int_0^\infty \mathbf{F}(q)\,dq$."

A skeptical friend says that one swallow does not make a summer and that the thoughtful normalization build into the definitions of the two functions explains the student's observation. Write a short report on the above material. ☞

9–4. *Annular slit approximation.* We expect that the Hankel transform of $\text{rect}[r/(1+\epsilon)] - \text{rect}[r/(1-\epsilon)]$ is closely approximated by $\mathbf{F}(q) = \pi\epsilon J_0(\pi q)$ when ϵ is small. For example, when $\epsilon = 0.05$, the error is less than 0.02 over the range $0 < q < 1$. Explain why the approximation would not be expected to work for larger values of q.

9–5. *Smooth function.* The circularly symmetrical function $(1-4r^2)^2\,\text{rect}\,r$ falls to zero at $r = \frac{1}{2}$ continuously, and in addition the slope at the cutoff boundary is continuous. What is the Hankel transform and how does it differ from that of $(1-4r^2)\,\text{rect}\,r$, which also falls to zero continuously?

9–6. *Orthogonality.*
(a) Deduce the orthogonality relation $\int_{-\infty}^\infty J_0(2\pi ar)J_0(2\pi br)r\,dr = 0$ when $b \neq a$ directly from the Hankel transform pair $\delta(r-a) \supset 2\pi a J_0(2\pi aq)$.
(b) What can you say about $\int_{-\infty}^\infty J_0(2\pi ar)J_0(2\pi br)\,dr$?
(c) When $a = b$, what useful quantitative statement can you make?

9–7. *Asymptotic variation.* The Hankel transform of a ring impulse consists of a central hump surrounded by fringes whose amplitude decays as $q^{-3/2}$, where q is the radial coordinate in the transform plane, but is the same true of a circular slit? Consider an equivalent narrow annulus of width ϵ centered on a ring impulse, equivalent in the sense that the integral over the (x, y)-plane is the same. The annulus can be represented as the difference between almost equal circular apertures each having a transform that dies out as $q^{-3/2}$.
(a) Explain why the fringes will be of less amplitude than before.
(b) Well away from the origin, how will the fringe amplitude vary with q? ☞

9–8. *Hankel transform of annular dipole.* Show that the Hankel transform of $\delta'(r-a)$ is $2\pi[2\pi aq J_1(2\pi aq) - J_0(2\pi aq)]$. ☞

9–9. *Hankel transform notation.* The Hankel transform of $f(x)$ is defined in the extensive "Tables of Integral Transforms" [Erd (1954), vol. 2] as

$$g(y) = \int_0^\infty f(x)J_0(xy)\sqrt{xy}\,dx, \quad y > 0.$$

Show that if $f(x)$ and $g(y)$ appear as a transform pair in these tables, then $f(x)/\sqrt{x}$ and $2\pi g(y)/\sqrt{y}$ will be a Hankel transform pair under the definition of this book when x is replaced by r and y is replaced by $2\pi q$.

9–10. *Bessel function identity.* Show that

$$\frac{d}{dx}(x^2\,\text{jinc}\,x) = \frac{1}{2}\pi x J_0(\pi x). \quad ☞$$

9–11. *Property of jinc function.* Show that $\text{jinc}\,x = \frac{1}{4}\pi[J_0(\pi x) + J_2(\pi x)]$.

9–12. *Identify diffraction pattern.* State whether the function illustrated in Fig. 9-22 represents a $J_0\ \square$ sinc $\square$ jinc $\square$ $[J_0]^2\ \square$ sinc$^2\ \square$ jinc$^2\ \square$ or other $\square$ function of radius r, giving the reason for your opinion.

9–13. *Power spectrum of zone plate.* A glass plate is alternately opaque and transparent on annuli

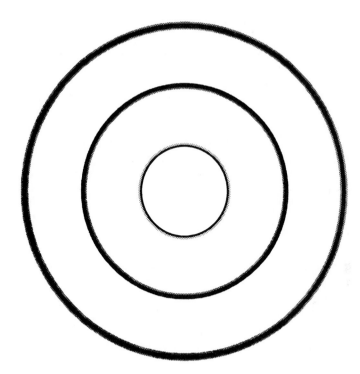

Figure 9-22   A function to be identified.

bounded by radii $1, \sqrt{2}, \sqrt{3}, \ldots, \sqrt{8}, \sqrt{9}$ (the unit-radius disc at the center being opaque). Present the power spectrum graphically, giving numerical values for spatial frequency. ☞

9–14. *Central value of Abel transform.* Prove that $f_A(0) = 2 \int_0^\infty f(r)\, dr$.

9–15. *Integral of Abel transform.* Prove that $2 \int_0^\infty f_A(x)\, dx = 2\pi \int_0^\infty f(r) r\, dr$, a useful result for numerical checking.

9–16. *Repeated Abel transformation.* Show that $f_{AA}(x)$, the Abel transform of the Abel transform, is equal to π times the volume under that part of $f(r)$ where $r > x$.

9–17. *Annulus seen as a convolution.* Is there a function $\phi(r)$ with the following property that

$$\phi(r) ** \delta(r - R) = \begin{cases} 1, & |r - R| < \frac{1}{2} \\ 0, & \text{elsewhere.} \end{cases}$$

In other words, if $\phi(r)$ were convolved with a ring impulse of radius R, the result would have to be unity over an annulus of unit width straddling the circle of radius R.

9–18. *Abel transform by convolution.* A function $f(r)$ whose Abel transform is desired can be expressed as $\hat{f}(\rho)$, where $\rho = r^2$. Consider a kernel $K(\rho)$ which is zero where $\rho \geq 0$ but for negative ρ is given by $1/\sqrt{-\rho}$. Show that by convolving $K(\rho)$ with $\hat{f}(\rho)$ we obtain a function $\hat{F}_A(x^2)$ which, for any x, is the desired Abel transform $f_A(x)$. ☞

9–19. *Accuracy of Abel transform.* Take $f(r) = 64(1 - r/64)\,\text{rect}(r/128)$, a function with a known Abel transform, and form a set of samples at $r = 0, 1, 2, \ldots, 63$. Compute the

Abel transform from the samples and compare with the known transform. Start again with a set of samples at $r = 0.5, 1.5, 2.5, \ldots, 63.5$ and find out whether better accuracy results.

9-20. *Computed Abel transform.* A function of radius is measured as follows.

r	.05	.15	.25	.35	.45	.55	.65	.75	.85	.95
$f(r)$	1.98	1.39	1.04	0.91	0.78	0.64	0.48	0.32	0.16	0.03

Compute and plot the Abel transform.

9-21. *Abel transform of conical crater.* Plot the Abel transform of $f(r) = |r|\,\text{rect}(r/2)$ and compute values for $x = 0[0.2]1$ to an accuracy of ± 0.01. ☞

9-22. *Abel transform to be guessed.*
   (a) Compute the Abel transform of $f(r) = (1 - r^2)^2\,\text{rect}(r/2)$ for $x = 0[0.1]1$.
   (b) The analytic form is likely to be fairly simple in form; try to guess the formula, and verify.

9-23. *Spin average.* What is the spin average of the parabolic function $f(x) = 1 - x^2/a^2$?

9-24. *Sinusoidal spin integral.* Show that $\frac{1}{2}\pi|x|J_0(x)$ is the function whose spin integral is $\sin r$. ☞

9-25. *Spin averaging over the full circle.*
   (a) Show that $f_1(x) = \delta(x + a) + \delta(x - a)$ has the same spin average as $f_2(x) = \delta(x - a)$ under the standard definition.
   (b) What would the situation be if the integration limits were changed to $(1/2\pi)\int_0^{2\pi} \ldots$?
   (c) How would the case of $\delta(x)$ be affected? ☞

9-26. *Spin averaging theorems for initial derivatives.* Show that if $f_s(r)$ is the spin average of $f(x)$, then $f_s(0) = f(0)$, $f_s'(0) = (2/pi)f$, $f_s''(0) = (1/2)f''(0) = 4/3\pi)f'''(0)$, $f_s^{(4)}(0) = (3/8)f^{(4)}$, $f_s^{(5)}(0) = (16/15\pi)f^{(5)}(0)$, $f_s^{(6)}(0) = (15/16)f^{(6)}(0)$, $f_s^{(7)}(0) = 52/35\pi)f^{(7)}(0), \ldots$. ☞

9-27. *Chebyshev polynomials.* A 2D function is a nonuniform ring impulse of unit radius and strength $\cos n\theta$ as represented by $f(x, y) = \delta(r - 1)\cos n\theta$. To get the projection $g_n(x)$ on the x-axis we take the strength at the point $P(x, \sqrt{1 - x^2})$, double it because there are two points on the circle with the same abscissa, and multiply by $(1 - x^2)^{-1/2} = 1/\sin\theta$ to allow for the angle between the line of projection and the tangent at P. The result is $2\cos n\theta/\sin\theta$, or, as a function of x,

$$g_n(x) = 2\cos(n\,\text{arccos}\,x)(1 - x^2)^{-1/2} = T_n(x)2(1 - x^2)^{-1/2}.$$

The factor $2(1 - x^2)^{-1/2}$, which is just the projection of $\delta(r - 1)$, represents the envelope under which $g_n(x)$ oscillates. Show that for $n = 1$ to 4, the oscillatory expressions $\cos(n\,\text{arccos}\,x)$, are

$$T_1(x) = x, \qquad T_2(x) = 2x^2 - 1, \qquad T_3(x) = 4x^3 - 3x, \qquad T_4(x) = 8x^4 - 8x^2 + 1,$$

and verify that these functions are the Chebyshev polynomials defined by

$$T_n(x) = \frac{n}{2}\sum_{m=0}^{\lfloor n/2 \rfloor}(-1)^m\frac{(n - m - 1)!}{m!(n - 2m)!}(2x)^{n-2m}.$$

Compute and plot $T_{40}(x)$ for $0 < x < 1$ using the expression $\cos(40\,\text{arccos}\,x)$. ☞

9–28. *A function with rotational symmetry.* Show that $\operatorname{rect} r \cos \theta$ has the 2D Fourier transform $i(a/q) \cos \phi [J_0(\pi q) - 1]$.

9–29. *Analysis into rings.*

(a) Confirm the following transform pair for a ring impulse whose strength varies cosinusoidally in azimuth:

$$\delta(r - 1) \cos n\theta \;{}^2\!\supset\; (-1)^n 2\pi J_n(2\pi q) \cos n\phi.$$

(b) Show that any nonuniform ring impulse $\delta(r - 1)\Theta(\theta)$ can be transformed in two dimensions by analyzing $\Theta(\theta)$ into a Fourier series $\Theta(\theta) = c_n \exp in\theta$ to find the coefficients c_n and then superposing patterns of the form $J_n(2\pi q) \exp in\theta$ to get

$$\delta(r - 1)\Theta(\theta) \;{}^2\!\supset\; \sum(-1)^n c_n 2\pi J_n(2\pi q) e^{in\theta}.$$

(c) Show further that any function $f(r, \theta)$ can be decomposed into concentric annuli of infinitesimal radial width, each of which can be Fourier analyzed azimuthally, and that the transform $F(q, \phi)$ can therefore be expressed as

$$F(q, \phi) = \int_0^\infty \sum (-1)^n c_{n,a} 2\pi a J_n(2\pi aq) e^{in\phi}\, da,$$

where the coefficients $c_{n,a}$ relate to the azimuthal cut $\Theta(\theta) = f(a, \theta)$ whose Fourier series is $\sum (-1)^n c_{n,a} \exp in\theta$.

(d) Explain the special circumstances under which it might be advantageous to express a two-dimensional Fourier transform as a relatively complicated sum of Bessel functions. ☞

9–30. *Modulated ring impulse.* Establish the following two transform pairs:

$$\delta(r - a) \cos \theta \;{}^2\!\supset\; -i2\pi a J_1(2\pi aq) \cos \phi,$$
$$\delta(r - a) \cos 2\theta \;{}^2\!\supset\; 2\pi a J_2(2\pi aq) \cos 2\phi. \quad ☞$$

9–31. *Chebyshev polynomial.* By considering the Abel transform of the cosinusoidally modulated ring impulse, show that the Fourier transform of a Bessel function is related to the Chebyshev polynomial T_n as follows (where n is an even integer):

$$J_n(2\pi x) \supset (-1)^{n/2}(1 - s^2)^{-1/2} T_{n/2}(s) \operatorname{rect}(s/2). \quad ☞$$

10

Imaging by Convolution

Many images are constructed by active convolution arising from scanning motion. In others, even though motion is absent, the dependence of the image on the object can be expressed as a convolution integral, at least approximately. Examples of the former arise when the ground is scanned by the antenna of a microwave radiometer carried in an airplane or satellite for remote sensing purposes, when the sky is scanned by a radio telescope as in radio astronomy, or when a photographic transparency is scanned by passing light through a small orifice to a photodetector as in microdensitometry. When an object is photographed, the image is produced simultaneously rather than sequentially, but the intervention of the camera lens introduces degradation which also involves two-dimensional convolution. Image construction for the most part is linear, but even with linearity the convolution relation may be perturbed where there is partial failure of space invariance, as with lenses. Not uncommonly (as with overexposure or underexposure) there may also be some degree of nonlinearity. Both linearity and space invariance are prerequisites to the following discussion, which treats cases where there is a simple convolution relation between object and image. First we consider methods such as antenna scanning and photography, where the convolving function is a compact but fuzzy diffraction pattern, and then microdensitometry and other types of imaging, where the convolving function is more or less sharply bounded.

In different fields the convolving function goes by different names: *point spread function (PSF)* and *point response* are common in optics, while antenna practice has established terms such as *antenna pattern* and *directional diagram*. In a context where a range of applications is to be discussed the general term *impulse response* is available.

There is a loss of resolution associated with scanning because those short spatial periods are discriminated against which fit within the impulse response. The relative response to the range of spatial frequencies present in the object is specified by a transfer function. The various categories of transfer function determine the prospects for restoration, a procedure that aims to reduce resolution loss and is taken up in a later chapter.

MAPPING BY ANTENNA BEAM

When an antenna is pointed at the sky with a view to mapping the natural wideband radiation, or conversely is pointed at the land or sea looking down from above, the response is expressible as an integral, over the antenna beam, of the product of the antenna gain in each direction with the radiation brightness in that direction. The operation resembles the taking of two-dimensional weighted running means. When such sky mapping began in the 1940s, it was apparent that the map obtained was a degraded version of what was really radiating. The situation was not at first treated as a problem of convolution, because the sky is curved, and antenna beams were wide enough to make sky curvature a concern. Improved directivity of antennas has changed this, but it is interesting to note that the spherical mapping problem still has unsolved features; for example, the eigenfunctions are not known. The idea that scanning with an antenna is analogous to the operation of a linear filter on a waveform appeared in the 1950s (Pawsey and Bracewell, 1955; Sullivan, 1984, pp. 167–190) when the concept of the spectral sensitivity function of an antenna was introduced. The optical transfer function that is now familiar is strictly comparable and arose somewhat later.

A sequence of temperature maps of the sun made by raster scanning with an antenna beam is shown in Fig. 10-1. The antenna, operating at a wavelength of 10.7 cm, has dimensions of 1000×1000 wavelengths and forms a 3-arcminute beam. A single map shows only the near side of the sun and is subject to perspective foreshortening. A synoptic map covering 360° of solar longitude can be prepared for each solar rotation, as in Fig. 10-2, which presents rotation no. 1455 (counting from November 9, 1853). The presentation is analogous to the photofinish of a horse race, which records with a slit positioned at the finishing line; in the solar case the reference line is the central meridian of the sun as seen from the earth. The synodic period of rotation (averaging 27.2753 days) was first determined by Galileo in 1612, as soon as he had four sunspot drawings in hand. This was pretty good work, seeing that he first had to determine the orientation of the rotation axis from the same data.

Going immediately to the one-dimensional simplification, let us say that $a(x)$ is the response of an antenna scanning over a point source. The sky brightness is periodic in x, with period 2π, and is independent of y; so the following theory would be exact if the sky were cylindrical rather than spherical. Obviously, when we go to two dimensions, some

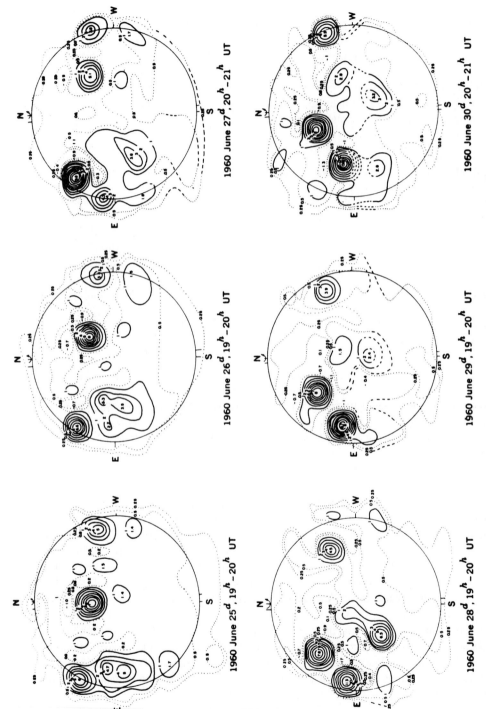

Figure 10-1 The microwave sun, unlike the visible sun which shines at 6000 K and has small dark spots, has large million-degree plages and a disc 10 percent larger than the visible disc. The contour unit is 70,000 K approximately. These maps form part of an 11-year daily series obtained at Stanford (Bracewell and Swarup, 1961; Graf and Bracewell, 1975).

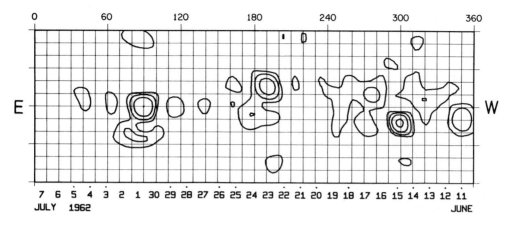

Figure 10-2   An image covering all longitudes of the solar disc computed from a sequence of daily maps. Time runs from right to left in order that East should be on the left and West on the right as with the sun as seen in the sky.

rethinking of fundamentals will be necessary. It is sufficient for the moment to be reminded that x is no longer measured in meters, as on a film sound track scanned by a slit, but has become position along a circle on the sky measured in radians.

If $f(x)$ is the brightness to be found, then the observable function $g(x)$ is given by the convolution relation

$$g(x) = a(x) * f(x) + n(x),$$

where $n(x)$ is unwanted noise. Both $f(x)$ and $g(x)$ are of the nature of power, and for physical reasons cannot go negative, although the noise can go negative. It should be remembered that the quantity actually measured always contains various sorts of unwanted noise that for the moment is left out. Taking the Fourier transform,

$$G(s) = A(s)F(s),$$

where $A(s)$ is the spectral sensitivity function, now usually simply called the transfer function because of convergence of this subject with signal analysis and optics. The idea is that a sinusoidal brightness variation $\sin 2\pi s x$, if scanned with the antenna, will be mapped with amplitude that is reduced by a factor $|A(s)|$ and shifted in (spatial) phase by an amount pha$[A(s)]$. Of course, since $\sin 2\pi s x$ goes negative, there would have to be components present at other spatial frequencies to keep the total brightness nonnegative. The spatial frequency s is measured in cycles per radian. Strictly speaking, s has a minimum value of $1/2\pi$ cycles per radian and in fact all values of s must be integral multiples of this fundamental. The reason is that, if you rotate your antenna through 2π, you will inevitably be pointing at the same place as before and receiving the same signal, unless the brightness is time varying. Consequently, $g(x)$ is periodic in x and must be representable by a Fourier *series*. However, the reason that it is practical to deal with a cylindrical sky at all, when the reality is spherical, is that attention will be limited to mapping a limited region of sky which, to many intents and purposes, is flat. If the field of view is small enough compared to the whole sphere, then the fundamental spatial frequency $1/2\pi$ is negligibly small compared with the spatial frequencies of interest. Consequently, in what follows, we think of s as continuous. This approximation is also inherently present in optical mapping instruments, such as telescopes, but is never mentioned, because the fields of view are so limited. However, in radio astronomy, where fields of view are comparable with those in optics but angular resolution is higher, imaging practitioners are acutely aware of the nonflatness of the sky (Roberts, 1984).

Because an antenna beam pattern $a(x)$, the impulse response, is the diffraction pattern of an aperture distribution over an antenna structure of finite dimensions, it possesses special characteristics that can be analyzed directly in terms of the known aperture distribution. Let the aperture distribution of electric field be the complex phasor $E(\xi)$, where ξ is a dimensionless distance variable defined as distance from the origin of the antenna aperture (usually the center) divided by the mid-wavelength of the radiation to be received. We know that the angular spectrum $P(\ )$ of the aperture distribution $E(\ )$ is its Fourier transform (Booker and Clemmow, 1950), so

$$P(x) = \int_{-\infty}^{\infty} E(\xi)e^{-i2\pi x\xi}\, d\xi.$$

In this equation, we may think of x measured in radians to be the same as the angular variable in $a(x)$, which is satisfactory as long as the angular diameter $2x_{\text{max}}$ of the source distribution is small enough that $\sin x_{\text{max}} \approx x_{\text{max}}$. This approximation will be explained in the two-dimensional formulation.

The angular spectrum $P(x)$ as defined above is not normalized, whereas we want the impulse response $a(x)$ to be normalized so that $A(0) = 1$ or $\int_{-\infty}^{\infty} a(x)\, dx = 1$. Of course, x does not extend to infinity. Our excuse for infinite limits in this integral is that the integrand becomes negligible outside a limited field of view and that therefore the exact limits of the integral do not matter. Therefore, we take them to be infinite in order to have simple and familiar theoretical relations.

As $a(x)$ is of the nature of power, whereas $P(x)$ is of the nature of electric field, we know that $a(x) \propto |P(x)|^2$. Consequently, $A(s)$ is proportional to the autocorrelation function of the known aperture distribution $E(\xi)$ because of the autocorrelation theorem that states, "If $f \supset F$, then $f \star f \supset |F|^2$."

Thus there is a simple relation that goes from specification of the antenna aperture distribution $E(\xi)$ directly to the desired tranfer function $A(s)$. Allowing for normalization,

$$A(s) = \frac{E^*(\xi) \star E(\xi)}{\text{N.F.}}.$$

The normalizing factor N.F. is whatever is necessary to make $A(0) = 1$ after the complex autocorrelation in the numerator has been performed. If needed explicitly, it is given by $\int_{-\infty}^{\infty} E(\xi)E^*(\xi)d\xi$.

As an instructive example, consider a uniformly excited antenna aperture of width W. As we are dealing in distances normalized with respect to wavelength λ we need the normalized width W_λ, which is equal to W/λ. Experience shows that there is danger of

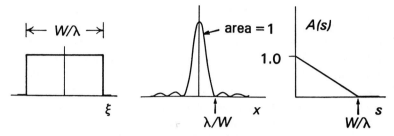

Figure 10-3   A distribution of electric field $E(\xi)$ over an antenna (left), the associated transfer function (center), and the impulse response of the antenna when used to scan a brightness distribution. The system acts as a low-pass filter that selectively operates on the constituent freqiencies of the radiating object. See Bracewell and Roberts (1954).

error at this point and that it is desirable in private work to carry the subscript in W_λ as a reminder. Thus the rectangular aperture distribution adopted is $\mathrm{rect}(\xi/W_\lambda)$. The amplitude does not matter, since it will be normalized out, and so a unit rectangle function suffices. With

$$E(\xi) = \mathrm{rect}(\xi/W_\lambda)$$

it follows that

$$A(s) = \frac{E^*(\cdot) \star E(\cdot)}{\text{N.F.}} = \frac{1}{\text{N.F.}} \Pi\left(\frac{\cdot}{W_\lambda}\right) \star \Pi\left(\frac{\cdot}{W_\lambda}\right) = \Lambda\left(\frac{s}{W_\lambda}\right).$$

This transfer function $A(s)$ is a triangle function that descends linearly from a central value of unity, to zero at $s = s_c = W_\lambda$. For $s > W_\lambda$ it is zero. Instead of using $\mathrm{rect}(\cdot)$, the preceding equation uses the quasipictorial $\Pi(\cdot)$, in terms of which $\Pi \star \Pi = \Lambda$, emphasizing how the shape of the transfer function depends on the aperture distribution (Fig. 10-3).

All antennas share the cutoff feature, because all antennas are finite. If the overall width of the antenna is W_λ wavelengths, then the cutoff spatial frequency s_c will be W_λ; an antenna 1000 wavelengths across will have a cutoff frequency of 1000 cycles per radian. That means that sinusoidal striations in a brightness distribution on the sky, having a period shorter than 1 milliradian, or about 3 minutes of arc, will not be detected. A striated distribution with period 6 minutes of arc will not be responded to in full, because it is halfway down the linear descent of the transfer function, but will be seen at half strength. A striation with period 30 minutes of arc ($s = 100$ cycles per radian) will be seen almost at full strength because

$$A(100) = \Lambda(100/1000) = 0.9.$$

While all antennas exhibit the cutoff at spatial frequency $s_c = W_\lambda$, the shapes of their transfer functions are different and are to be determined by autocorrelating the aperture distribution.

Antenna aperture distributions may be Fourier analyzed into spatial aperture-plane components which are associated with radiation in particular directions, provided the spatial period exceeds the wavelength of the radiation. Components shorter in period than a wavelength are associated with evanescent fields and may be ignored as regards reception from distant sources. When these short spatial components are filtered out, an aperture distribution that cuts off at its extremities acquires a spatially oscillatory extension; consequently, there may be a nonzero response beyond the nominal cutoff. For a discussion of the magnitude, see Bracewell (1962). Students are often bothered by the fact that the aperture distribution exists in the terrestrial domain that they themselves inhabit, where distances are measured in terms of ξ, and yet after autocorrelation, when one would expect to be in the same domain, mysteriously the independent variable has become s, which pertains to spatial frequency in the sky. To feel more at ease with this situation, firmly adopt ξ as the terrestrial or function domain; then x, which is measured on the sky, describes your transform domain. Now s, which has to do with the spectrum of functions of x, must be back in the original domain. The conceptual difficulty arises when attention is transferred to the sky domain x, which you may then tend to think of as your function domain, while s becomes your transform domain. Just remember that this is opposite to the adoption you made when you were dealing with antennas and before you switched your attention to mapping.

The linear filter analogy brings familiar background to bear on the antenna mapping operation. We imagine the object distribution $f(x)$ to be analyzed into superposed sinusoidal components. We remove the components whose spatial frequencies in cycles per radian exceed the cutoff value $s_c = W_\lambda$, and we rearrange the relative amplitudes of the retained components rather as if a waveform were being passed through a low-pass filter. Any uniform background (spatial dc) is kept at full strength; otherwise each sinusoidal component is cut down in strength in accordance with the attention paid to that component by the scanning beam. Finally, the adjusted components are reassembled to synthesize the observed distribution $g(x)$.

In two dimensions the morphology of the cutoff boundary in the (u, v)-plane is important, especially when the antenna aperture comprises well-separated parts, as in interferometry (Bracewell, 1961). The spectral sensitivity boundary is discussed in the next chapter.

SCANNING THE SPHERICAL SKY

To generalize the foregoing theory to two dimensions we have to deal in spherical trigonometry. Let the response of the antenna to a point source be $a(\alpha, \beta)$, where α and β are spherical polar coordinates with respect to axes fixed in the antenna. The polar angle α is measured from the beam axis, and the other angle β is measured from a reference meridian plane containing the beam axis. Take $a(\alpha, \beta)$ to be normalized so that

$$\int_0^{2\pi} \int_0^{\pi} a(\alpha, \beta) \sin \alpha \, d\alpha \, d\beta = 1.$$

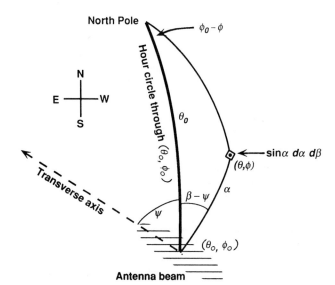

Figure 10-4 A coordinate system for a steerable telescope.

Let $f(\theta, \phi)$, the brightness distribution to be found, be specified in terms of codeclination θ and right ascension ϕ, a different set of spherical polar coordinates which are fixed on the sky, and are the celestial analogues of geographic colatitude (angular distance from the north pole) and geographic longitude (Fig. 10-4). The response to the source necessarily refers to that frequency band and polarization accepted by the antenna and associated transmission lines and electronics.

To specify the orientation of the antenna it is necessary to give not only (θ_0, ϕ_0), the direction on the sky toward which the main axis is pointed, but also a position angle ψ which states the angle through which a reference axis fixed perpendicular to the beam axis is turned eastward from north about the beam axis.

When the antenna is pointed toward (θ_0, ϕ_0) with position angle ψ, the response measured is

$$g(\theta_0, \phi_o, \psi) = \int_{-\pi/2}^{\pi/2} \int_0^{2\pi} a(\alpha, \beta) f(\theta, \phi) \sin\alpha d\alpha d\beta.$$

The values of θ and ϕ appearing in the integral are related to $\alpha, \beta, \theta_0, \phi_0,$ and ψ by

$$\cos\theta = \cos\theta_0 + \sin\theta_0 \sin\alpha \cos(\beta - \psi),$$
$$\cot(\phi_0 - \phi) = -\cos\theta_0 \cot(\beta - \psi) + \sin\theta_0 \cot\alpha \cos(\beta - \psi).$$

This derivation assumes that powers from different directions add, which would not be the case in the presence of coherence such as arises from sea reflections and atmospheric scintillations. Another assumption is that the antenna is rigid, a condition that is violated when antennas of an array are independently mounted.

Note that $\theta - \theta_0$ and $\phi - \phi_0$ constitute approximately rectangular coordinates if attention is limited to a small fraction of the celestial sphere not near the poles. Let x and y be suitably scaled rectangular coordinates and let ψ be constant over the patch of

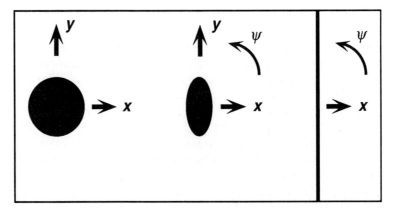

Figure 10-5 The pointing of a beam allows three degrees of freedom. One limiting case
is scanning with a circular beam with two translational motions; the other is scanning in
one dimension with a strip beam that can be rotated.

interest. With a sufficiently directional antenna, contributions to the integral beyond the
patch of interest are small enough that the limits of integration may be taken infinite. With
this train of explicit assumptions we recover the two-dimensional convolution relation

$$g(x, y) = \int_{-\infty}^{\infty} \int_{-\infty}^{\infty} a(x - x', y - y') f(x', y') \, dx' \, dy',$$

where x' and y' are the coordinates of the element $dx' \, dy'$ and $f(x, y)$ and $g(x, y)$ are,
respectively, the brightness to be measured and the instrument response. The symbols $f(\ )$
and $g(\ )$, originally introduced with variables θ and ϕ, have been reused for simplicity.

Scanning with a circular beam (left of Fig. 10-5) allows two translational degrees
of freedom, but if the beam is elliptical (center), there are three degrees of freedom,
one of which is associated with the position angle ψ. A strip beam (right), typical of
tomography, loses the degree of freedom associated with lengthwise translation, but two
degrees of freedom remain, one translation and one rotation. The presence of polarization,
unavoidable when the sensor includes an antenna, adds another degree of freedom.

The foregoing discussion applies not only to the celestial sphere as envisaged in
astronomy but to space in general, whether centered on an observer on the ground, an
aircraft, or a spacecraft. Some of the assumptions mentioned are specific to radiometry but
remind us that, wherever a simple convolution is asserted to apply to data, corresponding
qualifications can be looked for.

PHOTOGRAPHY

Photography differs from antenna scanning because the wavelength is so much shorter.
One consequence is that a cone of directions can be recorded simultaneously rather than

by sequential scanning, but the feature of convolution with a diffraction pattern remains the same. The conditions under which a photograph exhibits reasonably exact convolution are, however, different, freedom from lens aberration and nonlinearity being prominent. Antenna beams have aberration, but the practice of scanning avoids the distortion exhibited by a nonscanning camera lens; in principle a camera could be made to perform a raster scan to get the same advantage. In practice one-dimensional scanning is performed when a photoheliograph records the solar disc by moving a slit across the focal plane and when the finish of a horse race is photographed through a slit. A strict photographic analog with antenna scanning is found in precision astrometry, where a single photodetector at the focus of a telescope records the output from a raster scan performed mechanically by a whole telescope. A variety of more recent instruments use scanning instead of simultaneous image formation, an example being the scanning optical microscope, in which the stage supporting the specimen is moved.

Remarkable photographs of objects in interplanetary space published by NASA are of such high quality that another type of raster scanning is easily overlooked. The optical image formed conventionally by a lens on the spacecraft is converted to digital form by raster scanning and transmitted by radio to an antenna situated on earth in California, Australia, or Spain. The received radio signal is used to reconstitute a photographic print. The transmission method is not the same as for television, which, as the second millennium drew to a close, had not converted to digital modulation. Figure 10-6 shows the quality of halftone achieved.

MICRODENSITOMETRY

A two-dimensional image recorded on a transparency may be read off by scanning with a read head, and the analogue signal so obtained may be passed through an analogue-to-digital converter for further treatment by computer. This procedure originated in astronomy for processing photographic plates. The principle is to pass light through a small hole in close proximity to the emulsion and to measure the amount of transmitted light with a photodetector. Optical sound tracks on movie film are read off in the same way, but the light passes through a narrow slit and the motion is one dimensional.

If the original transparency is represented as an object $f(x, y)$ and the scanning orifice is represented by $a(x, y)$, a binary function that is equal to unity over the orifice and zero elsewhere, then $a(x, y) ** f(x, y)$ is the two-dimensional convolution. This entity is not exactly what is measured, however, because at the end of one traverse in the x-direction the transparency is translated through a distance d in the y-direction. Thus what is read out, call it $g(x, y)$, is given by

$$g(x, y) = [a(x, y) ** f(x, y)]\text{III}(y/d).$$

If it were not for the raster character, the operation could be summarized in terms of a transfer function $A(u, v)$, the Fourier transform of $a(x, y)$. We therefore pay some preliminary attention to the nature of $A(u, v)$.

Figure 10-6   An optical image of the asteroid Gaspra taken in space, digitally telemetered to earth, converted to photographic form, and screened for reproduction in ink. The halftone screen produces a grid of width-modulated parallel black lines. Note how the screen is less obtrusive when sloping at 45°. NASA photo, kindly screened by *USRA Quarterly*.

In the antenna mapping operation, where a raster also exists but was not mentioned, $A(u, v)$ always cuts out to zero on some boundary in the (u, v)-plane, but with microdensitometry the opposite happens: it is $a(x, y)$ which cuts out to zero on a boundary of some shape, taken here as circular with radius R. Consequently, the transfer function $A(u, v)$ dies away but does not have the sharp cutoff characteristic of antenna mapping. In a sense, then, there is a better prospect for restoration, or equalization of the Fourier components whose relative strengths have been disturbed. The effect of the raster is to make each scan line equivalent to a one-dimensional filtering operation.

Let a band of $f(x, y)$ of width D in the y-direction be scanned left to right by a circular hole for which $a(x, y) = \text{rect}(r/2D)$; then the one-dimensional entity denoted by $\int_{-D/2}^{D/2} f(x, y) \, dy$ has been scanned by a function of x which is the Abel transform of $a(x, y)$. This is known to be $D\sqrt{1 - (2x/D)^2}$, which has a semi-elliptical profile. The Fourier transform, which is $D^2 \text{jinc} \, Dq$, is therefore the relevant two-dimensional transfer function.

With advances in charge-coupled device technology, CCD arrays can eliminate the photographic recording stage and the subsequent microdensitometry. There is still a resolution limit set by the size of a single CCD. In addition, a sampling effect in the x-direction is introduced; if the spacing of the CCDs in both directions is d, a factor $\text{III}(x/d)\text{III}(y/d)$ enters into the response. Aliasing in the x-direction thus becomes a consideration.

A machine for digitizing photographic plates is not trivial. A photographic record made under the best circumstances could have a resolution of about one light wavelength, around 1 μm, but take 10μm as a suitable digitizing interval for a photographic plate 100 mm square. Then there would be 10^8 resolution elements and if the photographic density was to be quantized to eight bits, then the amount of information would approach 100 megabytes. The photographic plate would have to be rigid, supported in a strain-free manner, and temperature controlled. The scanning mechanism would need two mechanical degrees of freedom, exhibiting reproducibility to one part in 10^4. To a certain extent nonlinear motion can be calibrated out, but the instrument design must avoid backlash, stiction, hysteresis, vibration and wear. Since not much light can be put through a 10-μm hole, a sensitive and stable detector would be needed, and the dwell time on each of the 10^8 elements to be visited would need to be short. A trade-off against detector noise would have to be accepted. The instrument technology to meet these needs evolved in astronomical observatories as improvements on earlier densitometers. A microdensitometer occupies a small room and weighs a ton or more. After the analogue device does its work, the subsequent digital operations seem simple.

The digital-to-analogue laser printer, which contains equivalent components, weighs practically nothing and is an equal marvel of mechanical optoelectronic design. The analogue-to-digital laser scanner performs the same function as the microdensitometer, but as a business machine takes every possible advantage of the tolerance of the human eye to defects.

VIDEO RECORDING

Television recording on magnetic tape is a sequence of images quite analogous to microdensitometer data and may be discussed in the same terms of the response of the magnetic read head to a point source of light in front of the television camera. Because of the successive stages a Gaussian tendency is to be expected, but the magnetic phenomenon is bipolar, injecting its own characteristic flavor. In addition, spiral recording optically on a disc is a distinctive feature.

ECLIPSOMETRY

Instead of scanning an object with an aperture that selects a small neighborhood, one could scan with an opaque disc that eclipses a part of the object. This procedure might sound perilous from the point of view of signal-to-noise ratio, since fluctuations of the extended uneclipsed portion could drown the signal. Nevertheless, solar eclipses played an important part in studying the solar image at meter wavelengths, and the technique is applicable to some stars which have a dark orbiting companion. Eclipsometry using the moon, a technique known as lunar occultation, was the means by which the first evidence of quasars was obtained (Hazard et al., 1963) (see Fig. 10-7). Later, Pluto was mapped by photometry during the four-hour eclipses of Pluto by Charon that occurred in the years 1985–1990. A striking feature of eclipsometry is the angular resolution obtainable. For example, Pluto subtends only one-tenth of an arcsecond, an angular diameter that is far less than earthbound optical telescopes can reach, and yet within this tiny disc a respectable

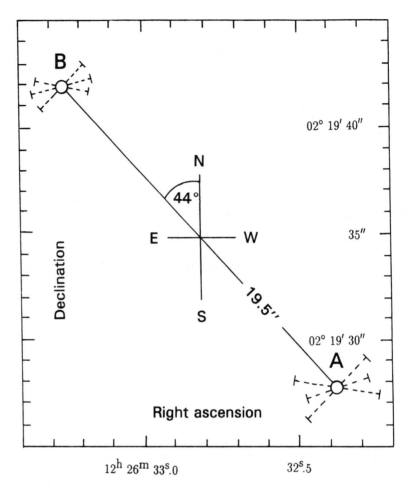

Figure 10-7   The discovery map of the first quasar by Hazard et al. (1963) using the 64-m radio telescope at Parkes, N.S.W.

map could be made. In the case of the quasar, two components spaced 1.11 s in right ascension and 14.2 arcsec in declination could be discerned, and their dimensions, around 2 arcseconds, were measured, together with some indication of orientation.

THE SCANNING ACOUSTIC MICROSCOPE

A monochromatic sound pulse can be focused to a point on the solid surface of an object by a lens, and the reflection will return to the lens to be gathered and detected. The strength of the reflection depends on the acoustic impedance looking into the solid surface relative to the impedance of the propagating medium. If the focal point performs a raster scan over the object, a picture of the surface impedance can be built up. The interesting

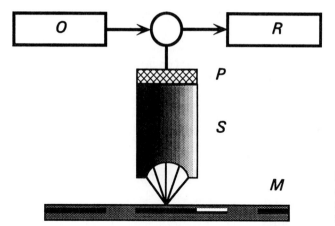

Figure 10-8 Parts of an acoustic microscope: oscillator O, receiver R, piezoelectric transducer P, sapphire rod S, movable specimen M.

things about this method of imaging are that the resolution of an optical microscope is attainable and that the picture looks different. The reason for the difference is that the acoustic impedance of a medium depends on density and elastic rigidity, qualities that do not directly control the reflection of light. In addition, acoustic energy that is not reflected at the surface but enters the solid may be only lightly attenuated and then reflect from subsurface discontinuities to reveal an image of the invisible interior. The scanning acoustic microscope is thus a tool for mapping invisible properties. A major application is in the semiconductor industry for inspecting integrated circuits and detecting subsurface defects, such as air films, where adhesion of metal deposition has failed.

A scanning optical microscope can be made on the same principle to produce an image that is in register with the acoustic image and permits subsurface features to be located with respect to the visible surface. The scanning optical microscope also has value as a means of imaging an extended field without the aberration associated with a lens.

As with all instruments with moving parts, the designer can choose which link of the mechanism will be stationary. For example, the specimen can perform a raster scan in front of a fixed acoustic source, or vice versa, or the frame can be stationary while the specimen performs a sawtooth motion in one direction and the acoustic beam translates slowly in the perpendicular direction. Further possibilities arise when the beam itself can be scanned electronically without the sound source moving.

The acoustic lens is a simple depression (is it paraboloidal?) in one end of a sapphire (Al_2O_3) rod on whose other end longitudinal vibrations are imposed by a piezoelectric transducer made of niobium titanate (Fig. 10-8). Traveling in sapphire at a velocity of 10,000 m s^{-1} (compare 300 ms^{-1} in air), the wave excites the concave surface so as to focus the wave launched into the propagating medium, usually water, in which the specimen is immersed. If the medium were air, the acoustic impedance, the product of density (1 kg m^{-3}) and velocity (300 m s^{-1}), would be 30 kg m^{-2} s^{-1}. For sapphire, the acoustic impedance is $2,500 \times 10,000 = 2,500,000$. With an impedance ratio $\zeta = 8000$, the reflection coefficient at the sapphire-air interface, given by $(\zeta - 1)/(\zeta + 1)$, is essentially unity; in other words sapphire backed by air is an acoustic mirror. Thus a matching transducer is needed, such as a one-quarter-wavelength layer of material of acoustic impedance ratio $\sqrt{8000}$.

Other features of scanning acoustic microscopy, which cannot be discussed here, arise from the creation of elastic waves traveling outward on the surface of the specimen and interior waves guided on velocity interfaces. On return to the detector these paths can be analyzed to enrich the image. An essential technique for disentangling the returns is to vary the distance of the surface of the specimen from the focal surface.

The idea of focusing an acoustic beam was originally suggested by Rayleigh, and the application as a scanning acoustic microscope goes back to 1950. The first implementation was made at Stanford in 1975 by C. Quate and R. Lemmons, on whose work the modern understanding of the complex wave phenomena involved now rests.

FOCUSING UNDERWATER SOUND

A solid parabolic reflector can reflect underwater sound and bring plane waves to a focus where a hydrophone can detect the energy coming from the axial direction. In sonar, where an intense pulse is emitted, omnidirectional sensors are used. In a passive system, where there is only a receiver and no transmitter, the energy received will depend on the level of ambient noise and the capacity of the surroundings, solid and liquid, to reflect or scatter the natural ambient sound. Generators of underwater sound include waves, rain, bubbles in a state of oscillation, shrimp, and other sea creatures. Focal-plane images can be formed in the ultrasonic range, while scanning with a single hydrophone, though much slower, is simpler and is applicable to detection of objects at ranges of hundreds of meters.

LITERATURE CITED

H. G. BOOKER AND P. C. CLEMMOW, "The concept of an angular spectrum of plane waves, and its relation to that of polar diagram and aperture distribution," *J. IEE*, vol. 94, pt. III, pp. 11–19, 1950.)

R. N. BRACEWELL (1961), "Interferometry and the spectral sensitivity island diagram," *IRE Transactions on Antennas and Propagation,* vol. AP-9, pp. 59–67.

R. N. BRACEWELL (1962), "A corrected formula for the spectral sensitivity function in radio astronomy," *Aust. J. Phys.*, vol. 15, pp. 445–446.

R. N. BRACEWELL AND J. A. ROBERTS (1954), "Aerial smoothing in radio astronomy," *Aust. J. Phys.* vol. 7, pp. 615–640.

R. N. BRACEWELL AND G. SWARUP (1961),"The Stanford microwave spectroheliograph antenna, a microsteradian pencil beam antenna," *IRE Transactions on Antennas and Propagation,* vol. AP-9, pp. 22–30.

W. GRAF AND R. N. BRACEWELL (1975), "Synoptic maps of solar 9.1 cm microwave emission from June 1962 to August 1973," *UAG Report No. 44,* National Oceanic and Atmospheric Administration, Boulder, CO.

C. HAZARD, M. B. MACKEY, AND A. J. SHIMMINS (1963), "Investigation of the radio source 3C 273 by the method of lunar occultations," *Nature*, vol. 197, pp. 1037–1039.

J. L. PAWSEY AND R. N. BRACEWELL (1955), *Radio Astronomy*, Clarendon Press, Oxford.

J. A. ROBERTS (1984), *Indirect Imaging*, Cambridge University Press, pp. 177–183.

W. T. SULLIVAN III (1984), *Early Radio Astronomy*, Cambridge University Press, United Kingdom.

PROBLEMS

10–1. *Transfer function of microwave horn.* A microwave horn antenna, such as might be used for scanning the ground from a satellite, has an aperture distribution of electric field that is a rectangle function in one direction and half a period of a cosine function in the other.

    (a) Calculate the two-dimensional autocorrelation $C(x, y)$ of $\cos \pi x \, \Pi(x)\Pi(y)$.

    (b) A particular horn antenna is 10 wavelengths wide and 8 wavelengths high. Show that its transfer function is $T(u, v) = 2C(u/10, v/8)$.

    (c) The antenna beam scans over a spatially sinusoidal microwave source whose spatial frequency components are $u = 3$ and $v = 2.4$ cycles/radian. What is the magnitude of the transfer function at those values of u and v? ☞

10–2. *Contrast ratio.* Define contrast ratio C for a periodic pattern as the ratio of maximum to minimum response.

    (a) What is the general relationship between C and modulation coefficient m, which is defined as (max − min)/(max + min)?

    (b) Illustrate, for an instrument whose transfer function is $\operatorname{sinc}^2 s$, by plotting both C and m versus s over the range $0 < s < 3$ when the instrument is used to scan over a periodic pattern $1 + \sin 2\pi s x$. ☞

10–3. *Semicircular aperture.* Obtain the OTF of a semicircular aperture of unit diameter. ☞

10–4. *Split lens.* The objective lens of an optical system is sawn in two and the equal halves are separated so as to leave a space one radius wide. Compare the new OTF with that of the original.

    (a) Show the cross sections through the OTF along the principal axes.

    (b) Determine the spectral sensitivity island diagram. One way to obtain this informative diagram is graphically with tracing paper (in fact this is the fastest way). It is also possible to write an equivalent program, but not easy. A different way of using the computer is by means of a program that autocorrelates a matrix; in this application all the matrix elements would be 0 or 1. The printout displays the island diagram boundary and also provides the basis of a contour plot. ☞

10–5. *Sky survey.* A relatively small radio telescope (34 m in diameter) operating at 1.5 GHz with a bandwidth of 40 MHz is to map the whole sky, dwelling 5 s on each direction.

    (a) Approximately how long will this survey take?

    (b) A spectrum analyzer will divide the incoming signal into 20-Hz bins. When the final image is composed, how many megabytes will it represent?

10–6. *Microdensitometer.* An object is scanned with a circular aperture 10μm in diameter.

    (a) Graph the absolute value of the transfer function, normalizing to unity at zero spatial frequency.

    (b) Does the graph show that a fine grille with line spacing of 3.732 μm will give a 6.448 percent response relative to zero spatial frequency?

11

Diffraction Theory of Sensors and Radiators

The word diffraction was invented by Francesco Maria Grimaldi (1618–1663) to describe the tendency of light to deviate from propagation in straight lines, for example at a shadow edge, and to break up into dark and light bands. Grimaldi's work was done at about the same time that Isaac Newton (1642–1727) was studying refraction by a prism.

We use the theory of diffraction to calculate the response of an optical system such as a television camera or a telescope to an object in the form of a point source. That response is important because the image—the entity that is accessible to us—relates to what is actually there in object space through a superposition of point-source responses. In the case where linearity and space invariance apply, this image is expressible as a convolution containing the impulse response. Happily for those who are prepared in Fourier analysis, the directions into which light redirects itself after passing through an aperture in an opaque sheet are described by the two-dimensional Fourier transform of the optical field distribution across the aperture.

When these ideas are repeated as they apply to radio wavelengths, *directional diagram of an antenna* is the term used for point-source response. The antenna diagram can be thought of in two distinct ways. First, as in optics, it can be thought of as the response to a point source; it might then be referred to as the *reception pattern*. But it is also customary to use antennas with radio transmitters. In that case the diffraction pattern produced by the distribution of field over the antenna aperture is the *radiation pattern*. Traditionally, telescopes have not been thought of as optical transmitters, although it is perfectly possible to

402

mount a laser at the focus of a telescope, as in fact is done in laser ranging of the moon's surface; ranging on the moon has been possible since retroflectors were installed there. In any case, as the reciprocity theorem asserts, the reception pattern and the radiation pattern are identical, with appropriate provisos about polarization.

What follows also applies to other sensors such as geophones, hydrophones, microphones, loudspeakers, ultrasonic transducers, and arrays of these elements.

The field distribution over an aperture and the directional pattern, or Fraunhofer diffraction pattern, form a two-dimensional Fourier transform pair. Before formulating this statement with more precision, we examine two elementary situations, the source pair and wave pair, that provide a physically intuitive basis of great value.

The chapter contains a discussion showing how convergence toward a focus as produced by a thin lens can be thought of in terms of diffraction by taking the focal plane to be an aperture plane that launches a conical beam.

The Fourier transform relation for the directional pattern breaks down in the immediate vicinity of a diffracting aperture. In the far field, or Fraunhofer diffraction region, the field pattern, as a function of direction, does not change as the distance from the aperture increases; the outward flow of power from the telescope or antenna used for transmitting has become essentially radial. Near the aperture, on the other hand, the mean flow of power starts by being perpendicular to the aperture (under the common condition of isophase excitation) and progressively diverges. This is the Fresnel diffraction region, in which transverse power flow is significant. The transition to the Fraunhofer region is not sharply defined, but we can say that, for an aperture N wavelengths across, the Fraunhofer pattern sets in at a distance of about N aperture widths from the aperture. This distance can be different in different planes if the aperture is elongated. In the immediate vicinity of the aperture is the reactive near field, so called by analogy with electric-circuit reactance, where power flows not only transversely but in and out of the aperture. The time-averaged reactive power flow is zero, but the field will make itself felt if objects are introduced. The following simple special case is used to introduce Fresnel diffraction.

When a plane wave is interrupted by a semi-infinite opaque screen with a straight edge, the unobstructed aperture is an infinite number of wavelengths in extent; consequently no, Fraunhofer region is ever reached—only Fresnel diffraction is seen. Diffraction at a straight edge has an important application to mapping by occultation, surface microscopy, binary optics, and optical lithography, as well as playing a fundamental role in diffraction by polygonal apertures.

THE CONCEPT OF APERTURE DISTRIBUTION

Apertures, in optics, are openings in flat screens dating back to the "round Hole, about one third Part of an Inch broad, made in the Shut of a Window" that Newton pierced in his shutter at Trinity College when he was studying the composition of sunlight. The slit and double slit are other examples of apertures in screens. The *aperture distribution* is the electromagnetic field maintained in the plane of the opening by light that is incident from one side of the screen. When one takes into account the thickness of the screen and its electrical conductivity, it may not be easy to say exactly what the aperture distribution is, so an element of judgment is necessary. However, once that is settled, the problem of what comes out from the other side of the aperture has been converted to the simpler problem of what radiation is launched by the adopted aperture distribution.

The concept of aperture distribution carries over into antenna theory, even though most antennas are not apertures at all, and those that are are mostly not excited by free-space radiation as in optics. Aperture distribution has thus come to mean a distribution of electromagnetic field over some plane. The concept transfers readily to acoustics (where there often is an actual aperture), to ultrasound, underwater sound, and seismic waves. Of course, the leap of judgment in going from a physical arrangement to an adopted aperture distribution varies from case to case. For general references to antennas and their literature see Kraus (1988) and Clarke and Brown (1980); for radio interferometers see Thompson et al. (1986) and Christiansen and Högbom (1969); and for optical aspects see Born and Wolf (1969), Steel (1983), Marathay (1982), and Goodman (1968).

Once an aperture distribution is established, there is more than one way of proceeding. Huygens' principle, according to which the field at a downstream point is the sum of the contributions from elements of the aperture, leads conveniently to an expression for the Fraunhofer diffraction pattern and in the Fresnel region provides a series approximation whose first term often suffices in practice. A less known procedure, which will be followed here, is to make a spatial Fourier analysis of the aperture distribution. An aperture distribution consisting of a single Fourier component is not an everyday object; nevertheless, one can imagine constructing the distribution as follows. Make a photographic transparency whose transmission factor varies between 0 and 1 as $|\cos(2\pi x/L)|$ over an optically parallel glass plate much larger than the characteristic length L; this would look something like a roller blind made of grey bands. Photographic density cannot go negative; so cement strips of glass of refractive index μ, width $\frac{1}{2}L$, and thickness $\frac{1}{2}\lambda/(\mu - 1)$, to every second grey band. Finally, illuminate this plate with a plane wave incident normally on the glass. The extra strips, acting as half-wave plates, will reverse the phase on alternate bands and give rise to an aperture distribution proportional to $\cos(2\pi x/L)$.

A single spatial Fourier component can also be constructed, over a substantial area, by interference between inclined plane waves and so is not a purely theoretical idea. However, the theoretical implication is obvious: if we can handle the case of a sinusoidal aperture distribution, we can handle any aperture distribution by Fourier synthesis.

SOURCE PAIR AND WAVE PAIR

The well-known relationship between a pair of point sources of monochromatic radiation and the familiar sinusoidal interference fringes that are produced does not require advanced mathematics to explain. However, the Fourier transform is well suited to treat the situation and provides a basis for the rigorous development of the Fresnel and Fraunhofer fields produced by any aperture distribution.

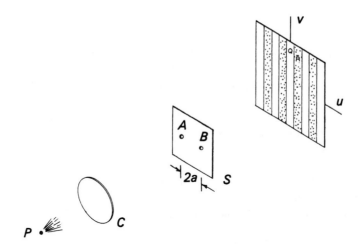

Figure 11-1   Light from a point source P is collimated by lens C to form a plane wave. The pattern of light and dark striations in the (u, v)-plane is produced by diffraction of the plane wave through a pair of pinholes A and B in an opaque screen S.

A Pair of Pinholes

Two pinholes illuminated by a plane wave (as in Fig. 11-1) cause the light to diffract into a pattern of light and dark bands on a distant plane. We understand very well that the bright band along the v-axis results from the fact that the contributions from the pinholes A and B reinforce on the v-axis, because, to take a representative point Q, the distances AQ and BQ are equal. Therefore, because the illuminating wavefronts are plane, and parallel to the diffracting screen S, the two contributions to the field at Q arrive in the same phase. The electric field there is therefore twice as strong as one pinhole would deliver (and of course the power flux density is four times greater). By symmetry, the locus of Q in the (u, v)-plane is a straight line perpendicular to the line AB joining the pinholes.

Likewise, the dark band centered on R is due to cancellation resulting from the fact that $AR - BR$ is half a wavelength. The contributions from A and B arrive at R in antiphase, and the optical intensity is zero along lines that are closely parallel to the v-axis. Along the u-axis the field varies cosinusoidally.

Thus we see very intimately how it is that an aperture distribution

$$^2\delta(x + a, y) +^2 \delta(x - a, y)$$

gives rise to a distant field distribution proportional to

$$2\cos 2\pi au.$$

The greater the spacing $2a$, the higher the spatial frequency of the striations. We recognize in the two expressions a two-dimensional Fourier transform pair:

$$^2\delta(x + a, y) + {}^2\delta(x - a, y) \supset 2\cos 2\pi au,$$

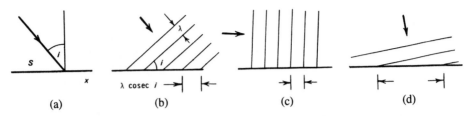

Figure 11-2   Plane wave of wavelength incident on screen S. (a) Ray representation. (b) Incident wavefronts. (c) Grazing incidence. (d) Steep incidence.

and we understand physically how the two sides relate.

The bench setup demonstrating the optical fringes from a pair of pinholes is known as Young's interferometer after Thomas Young (1773–1829). He is also remembered for his modulus of elasticity, for proposing the three-color theory of vision, for progress in deciphering the Rosetta stone, and for choosing the word "energy" to stand for the idea familiar to us all today under that name. If this Fourier transform relation is more than an accident, then, by the reciprocal property of Fourier pairs, a cosinusoidal field distribution across an aperture should give rise to two spots of lights on the distant plane. Such behavior is not immediately apparent from the concept of wavelets, due to Christiaan Huygens (1629–1695), that we used in discussing the pair of pinholes, but it becomes just as clear from a different approach.

Traveling Excitation on a Plane

When a traveling plane wave of wavelength λ impinges obliquely on a plane screen S, the spatial period in that plane exceeds λ by a factor cosec i, where i is the angle of incidence as in Fig. 11-2(a). In the wavefront picture of Fig. 11-2(b), where the incident ray is in the plane containing the normal and the x-axis, the spatial period in the screen plane is $\lambda_x = \lambda$ cosec i.

When the incidence is very oblique, or grazing, the wavecrest separation measured along the screen is not much greater than λ as in Fig. 11-2(c), and when the incidence is very near normal incidence, as in Fig. 11-2(d), the spatial period is very much greater than λ, but in all cases the spatial period is greater than, or in the limit equal to, λ. The periodicity on the screen cannot be shorter than the wavelength of the incident traveling wave.

Now we are ready to address the question of what happens when there is a sinusoidal aperture distribution. First we have to recognize that, except at normal incidence, the field produced on a screen by an incident plane wave is a traveling field. As a familiar example, think of waves breaking on a beach (Fig. 11-3). If the wavecrests are not parallel to the beach but are somewhat inclined, the breaking waves will be seen at intervals CD along the beach that are distinctly greater than the crest-to-crest distance λ, and we see that $CD = \lambda$ cosec i.

In addition, a point of breaking on the beach, such as C, will move along the beach in the direction of D, covering the distance CD in one wave period. Since $CD > \lambda$, the velocity of C will exceed the wave velocity, and if the angle of incidence is near zero,

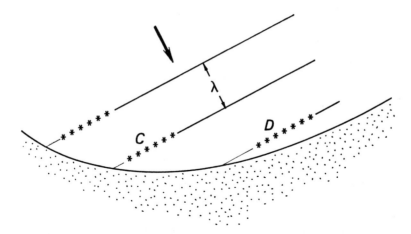

Figure 11-3 Wave crests breaking on the beach. One wave period later, the white water at C will have moved to D, which is more than a wavelength away. So the speed of the phenomenon C exceeds the wave velocity.

as is usually the case, the velocity of the point of breaking may be very high. Such a phenomenon is always noticeable on beaches.

Relation to Wavelength

We have seen in an elementary way that each spatial frequency in the aperture plane is associated with a direction. In other words, a single component of unit amplitude and spatial period λ_x, expressible as $\exp(-i2\pi x/\lambda_x)$, gives rise to a concentrated directional distribution expressible as an impulse function ${}^2\delta(u - \lambda/\lambda_x, v)$. This fundamental result does not involve small-angle approximations and therefore leads to a rigorous Fourier transform relationship between aperture plane distribution and directional pattern of the radiation. Where polarization exists, as in optics and radiophysics, there may be a cosine factor $\cos\theta$ that falls from unity in the forward direction to zero in the direction parallel to the aperture electric field. This factor will be present for one polarization but will be unity for the other polarization. In acoustics, where polarization may be ignored, the factor is unity. With light which is randomly polarized, there is an intermediate factor. We will deal simply with the acoustical case or with that polarization of electromagnetic radiation for which there is no cosine factor.

It is quite possible, and indeed not unusual, to make small-angle approximations when deriving the Fourier transform relationship for diffraction. Consequently, you might then think that the Fourier relationship is an approximation. However, as can be seen from the two-pinhole situation, which gives a cosinusoidal diffraction pattern, and from the cosinusoidal aperture distribution, which launches into just two directions, the transform relation is an exact one in these special cases. By superposition, which is what the Fourier integral describes, we see that the exactitude extends to the general case also.

It now remains to make the connection between the spatial period λ_x and the direction described by $^2\delta(u - \lambda/\lambda_x, v)$. This is directly apparent from Fig. 11-2(b), where we see that

$$\lambda_x = \lambda \operatorname{cosec} i.$$

Hence the direction, as expressed in terms of angle of incidence i, is given by

$$\sin i = \lambda/\lambda_x.$$

Wave Pair

To construct a laterally stationary sinusoidal distribution we superpose two incident waves with equal and opposite angles of incidence. The conventional interpretation of the complex phasor $\exp(-i2\pi x/\lambda_x)$ is that it represents a real wave whose field is obtained by multiplying the phasor by $\exp i\omega t$ and taking the real part of the product. This gives $\cos(\omega t - 2\pi x/\lambda_x)$, which we see is a wave traveling in the positive direction of x. A wave traveling in the opposite direction will be represented by $\exp(i2\pi x/\lambda_x)$ and, when added on, will produce a nontraveling cosinusoidal wave

$$\cos(2\pi x/\lambda_x) = \tfrac{1}{2}e^{i2\pi x/\lambda_x} + \tfrac{1}{2}e^{-i2\pi x/\lambda_x},$$

whose Fourier transform is

$$\tfrac{1}{2}{}^2\delta(u - \lambda/\lambda_x, v) + \tfrac{1}{2}{}^2\delta(u - \lambda/\lambda_x, v).$$

This expression quantifies the expectation that a cosinusoidal aperture distribution will launch plane waves into just two directions; the shifts from the normal direction are $\pm\lambda/\lambda_x$. By inserting a lens, whose property is to bring rays going in one direction to a focus at one point, we would see the two spots of light, confirming that a periodic cosine wave imposed in the screen plane sends out waves in just two directions.

The quantity λ_x is spatial period, λ_x/λ is the spatial period normalized with respect to the free-space wavelength, and λ/λ_x is therefore the normalized spatial frequency of the cosunusoidal variation. When the spatial frequency is low, the two angular deflections are low; as the spatial frequency is raised, the deflections increase toward the limiting values $\pm\tfrac{1}{2}\pi$, which are reached when $\lambda/\lambda_x = 1$. Recalling that $\lambda/\lambda_x = \sin i$ confirms that the angular deflection is $\pm\arcsin(\lambda/\lambda_x)$.

In this section we have connected the Fourier transform relation

$$^2\delta(x + a, y) + {}^2\delta(x - a, y) \supset 2\cos 2\pi au$$

with the cosinusoidal interference fringes of Young's interferometer, and conversely we have connected the mathematically required reciprocal relation

$$2\cos 2\pi ax \supset {}^2\delta(u + a, v) + {}^2\delta(u - a, v)$$

with the launching of plane waves into just two directions by a cosinusoidal distribution of field imposed in the plane of the screen previously containing Young's pinholes. The two physical arrangements, which are quite different, originate from the same Fourier transform pair interpreted in different ways.

TWO-DIMENSIONAL APERTURES

By choosing pinholes oriented along the x-axis, we have avoided the full generality of two dimensions in order to concentrate on the essentials. However, we have been using two-dimensional notation. Before proceeding, we can make a simplification by introducing the normalized distances

$$\widehat{x} = x/\lambda \quad \text{and} \quad \widehat{y} = y/\lambda.$$

The reason is that the relation

$$^2\delta(x + a, y) + {}^2\delta(x - a, y) \supset 2\cos 2\pi au$$

does not contain the wavelength, whereas we know that dependence on wavelength is important. Mathematically it must be true that

$$^2\delta(\widehat{x} + a/\lambda, \widehat{y}) + {}^2\delta(\widehat{x} - a/\lambda, y) \supset 2\cos[2\pi(a/\lambda)u].$$

In the two-dimensional case take the aperture field distribution $f(x/\lambda, y/\lambda)$, now written $f(\widehat{x}, \widehat{y})$, to be the complex phasor representation of the y-component of electric field. Then $F(u, v)$ is defined by

$$F(u, v) = \int_{-\infty}^{\infty} \int_{-\infty}^{\infty} f(\widehat{x}, \widehat{y}) e^{-i2\pi(u\widehat{x} + v\widehat{y})} \, d\widehat{x} \, d\widehat{y}.$$

The quantity $F(u, v)$ is called the *angular spectrum* following radiophysicists Booker and Clemmow who introduced the terminology (Booker and Clemmow, 1950; Clarke and Brown, 1980), and will give the electric-field component of the Fraunhofer diffraction pattern in the direction (u, v), where u and v are the direction cosines relative to the coordinate system of the (x, y)-plane. The electric field will be polarized in the plane containing the y-axis and the ray. When the obliquity factor K that arises in Huygens' theory is to be taken into account, the field will be $KF(u, v)$, and K will be equal to $\sqrt{1 - v^2}$, which is the cosine of the angle between the direction of the ray and the $v = 0$ plane.

RECTANGULAR APERTURE

A uniformly illuminated rectangular aperture described by $\text{rect}(x/a)\text{rect}(y/b)$ would give rise to an angular spectrum $ab \operatorname{sinc} au \operatorname{sinc} bv$, a result that is verifiable in optical experiments. Figure 11-4 presents as an example the angular spectrum associated with an aperture distribution which is taken to be uniform over an aperture 4×3 wavelengths in size. In the (u, v) plane there is a central beam bounded by a rectangular null line whose aspect ratio is the reciprocal of that of the aperture, surrounded by a mosaic of variously

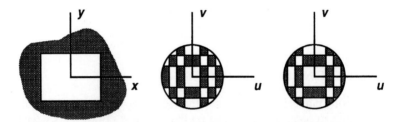

Figure 11-4   An aperture 4×3 wavelengths in size (left) and its angular spectrum (center), assuming the field within the aperture to be $\mathrm{rect}(\widehat{x}/4)\,\mathrm{rect}(\widehat{y}/3)$. If the aperture distribution is taken to be $\cos(\pi\widehat{x}/4)\,\mathrm{rect}(\widehat{y}/3)$, as would be the case with a microwave waveguide horn radiator made of copper, the angular spectrum changes (right); the beamwidths in perpendicular directions are then nearly equal (which is why waveguide horns are proportioned as shown on the left).

shaped regions, which are shaded where they are negative. The angular spectrum by definition continues out to infinity; however, in the figure, only the radiating fields associated with (u, v)-values inside the circle $u^2 + v^2 = 1$ are considered. A total of 32 lobes can be counted surrounding the central beam.

If the aperture were bounded by an electrical conductor, the assumed uniform radiation would violate the boundary condition for the electric field at two of the edges—for example, at the left and right edges if the polarization were in the y-direction—because electric fields cannot be nonzero at perfectly conducting boundaries. The correction may be unimportant if the aperture is thousands of wavelengths in extent. A microwave aperture, on the other hand, may be comparable in dimensions with the wavelength. Thus, at the mouth of a rectangular waveguide, or rectangular horn antenna, if the polarization is in the y-direction, it will not be feasible to maintain uniform illumination; the aperture distribution of electric field will be $\mathrm{rect}(x/a)\,\cos(\pi x/a)\,\mathrm{rect}(y/b)$, in the usual TE_{01} mode. In this case the diffraction pattern will be $\frac{1}{2}ab[\mathrm{sinc}(au + \frac{1}{2}) + \mathrm{sinc}(au - \frac{1}{2})]\,\mathrm{sinc}\,bv$.

EXAMPLE OF CIRCULAR APERTURE

A uniformly illuminated circular aperture of diameter D will, we know, give rise to a directional pattern that is a jinc function (where $\mathrm{jinc}\,q = J_1(\pi q)/2q$), because rect and jinc are a Hankel transform pair. Now we are able to relate the angular scale of the jinc function to the aperture diameter. Of course, the relation depends on wavelength. Working in terms of r/λ in the aperture plane, we rewrite the aperture distribution $\mathrm{rect}(r/D)$ in the form $\mathrm{rect}[(r/\lambda)/(D/\lambda)]$. Then we take the transform, utilizing the similarity theorem to obtain

$$\mathrm{rect}\left(\frac{r/\lambda}{D/\lambda}\right) \supset (D/\lambda)^2\,\mathrm{jinc}[(D/\lambda)q].$$

The angular scale is now fixed. Since we know that $\mathrm{jinc}\,1.22 = 0$, we can say that in the direction q_0 of the first null of the directional pattern we must have

$$(D/\lambda)q_0 = 1.22$$

or

$$q_0 = 1.22\lambda/D,$$

a formula familiar as the resolving power of a telescope of aperture D. For example, if $D = 10$ cm and $\lambda = 600$ nm, then $q_0 = 7 \times 10^{-6}$ radians $= 1.5$ arcseconds.

As pointed out above, the variable q must be understood as the sine of the angle away from the forward direction in order for the transform relation to be rigorous, but with angles as small as seconds of arc the distinction is immaterial.

DUALITY

The analogy between diffraction by a pair of pinholes and diffraction by a cosine grating is a special case of a more general proposition that enables new instrumental configurations to be generated from old. Following Lohmann (1992), let th($\mathbf{x}$) be the theory of an instrument in $\mathbf{x}$-space, which in general is three dimensional. Now take the Fourier transform of the whole theory, obtaining TH($\mathbf{u}$):

$$\mathcal{F}[\text{th}(\mathbf{x})] = \text{TH}(\mathbf{u}).$$

Reinterpreting in $\mathbf{x}$-space gives a new theory TH($\mathbf{x}$) that describes a different but equally valid instrument or experiment. In a Fourier-transforming optical system, a plane-wave source gives rise to a point output. To take the Fourier transform of this whole statement, which may be only part of a fuller theory or description, means to replace the statement by "a point source gives rise to a plane-wave output." The phrase "tilting the source plane shifts the point output" transforms into "shifting the point source tilts the output plane wave."

An analogy with duality in the differential equations of wave propagation or of electric circuits may be noted. Whatever description of the electromagnetic behavior of a circuit may be quoted, a new description of a different situation can be generated by interchanging the words capacitance and inductance, voltage and current, impedance and admittance. Similarly, in geometry, interchanging the words point and plane will produce dual statements. For example, four *points* can be found, one on each of four arbitrary straight lines in space, such that the four *points* have a (different) straight line in common. Therefore, four *planes* can be found, one through each of four arbitrary straight lines, such that the four *planes* have a (different) straight line in common. The first statement may be obvious; the second, obtained merely by changing *point* to *plane,* is clearly different and may be less obvious. The possibility of converting between situations of unequal obviousness permeates Fourier theory and provides a useful thinking tool. A related maneuver exists in the design of mechanisms, where different configurations with different applications can be generated by fixing a moving link to the ground and allowing the frame, which was originally thought of as kinematically identical with the ground, to become a moving part.

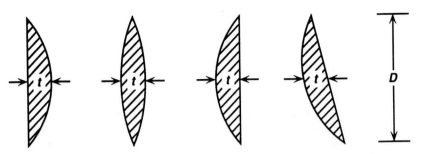

Figure 11-5   Equivalent thin lenses whose thickness varies as $t[1 - r^2/(0.5D)^2]$.

THE THIN LENS

Lenses have endless complexity. Even the simplest lens is rather a marvelous thing and not very easy to understand, although of course nearly everyone knows the lens formula $1/f = 1/v - 1/u$ and can use it. Here we are concerned with gaining some understanding rather than with practical calculations. For this purpose it is sufficient to confine attention for the moment to the "thin lens," one that is thin in relation to its diameter. The thin lens, or alternatively a lightly curved paraboloidal mirror, does not bring about very strong convergence of parallel beams incident upon it. We may suppose that it comprises a lenticular volume filled with a loss-free dielectric whose refractive index n is taken to be 1.5, which is characteristic of glass. We may also suppose that n is independent of wavelength, a supposition that we should recognize as being a physical impossibility. After all, the fact that glass refracts at all is due to the reduced velocity of light in glass, and that is due to the out-of-phase fields produced by electrons vibrating at the optical frequency ω. Unavoidably, the amplitude of oscillation of an electron of mass m must depend on its mechanical impedance $m\omega$, and so the refractive property must change as ω changes.

It does not make much difference whether the thin lens is double convex or planoconvex as long as the thicknesses are the same. Likewise it makes little difference whether the curved surface is spherical or paraboloidal (Fig. 11-5). In fact, if D is the diameter of the lens and t is the thickness of the lens at its center, then $t[1 - r^2/(0.5D)^2]$, which is the thickness at radius r, may be regarded either as describing a paraboloid or as representing the first two terms in the Taylor series for a sphere. The next term, in r^4, will be negligible as long as $t \ll D$, which is the thin lens condition.

Thus the object to be discussed is the lens shown in Fig. 11-6. This is not a particularly thin lens as drawn—indeed, t/D is 0.1—but it is useful for a quantitative example, because the focal point would fall off the page if the lens were thinner.

A bundle of rays arriving from the left parallel to the lens axis will be refracted so as to converge on the focal point F. The focal length f is the distance of F from the lens. The focal property can be verified numerically with little trouble because of the simple geometry. There is no refraction at the plane surface. Let the equation of the curved surface be $z = t[1 - r^2/(0.5D)^2]$. At A, the angle of incidence i_1 is given by $i_1 = \arctan(-dz/dr) = \arctan(8rt/D^2)$. The angle i_2 at emergence is fixed by Snell's law

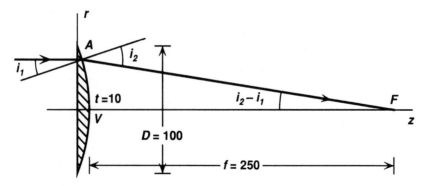

Figure 11-6 A thin lens 100 mm in diameter, 10 mm thick. With refractive index $n = 1.5$ the focal length for paraxial rays is 250 mm.

Table 11-1 Distance VF in mm from the lens to the axis-crossing.

r (mm)	$t = 2\,\text{mm}$	$t = 5\,\text{mm}$	$t = 10\,\text{mm}$
10	1249.8	499.5	249.0
20	1249.2	498.0	245.98
30	1248.2	495.49	240.9
40	1246.8	491.96	233.67
50	1245.0	487.4	224.16
$D^2/8t\,(n-1)$	1250	500	250

$$\sin i_2 = n \sin i_1.$$

Consequently $i_2 = \arcsin\{n \sin[\arctan(8rt/D^2)]\}$. The angle i_3 is given by

$$i_3 = i_2 - i_1 = \arcsin\{n \sin[\arctan \frac{8rt}{D^2}]\} - i_1.$$

Thus the distance VF to the point F where the emerging ray cuts the axis is given by

$$VF = r \cot i_3 - t + z$$

or

$$VF = r \cot[-i_1 + \arcsin\{n \sin[\arctan(8rt/D^2)]\}] - t + t[1 - r^2/(0.5D^2)].$$

This sort of expression may look unfamiliar and in earlier times would have been avoided by introduction of appropriate approximations at an earlier stage but it is perfectly amenable to computation. When the calculations are carried out (Table11-1), we see that VF does not depend much on the "height" r of a ray, that the paraxial rays (those closest to the axis) agree best among themselves, and that the agreement is even better as the lens becomes thinner. As $r \to 0$ the paraxial approximation is

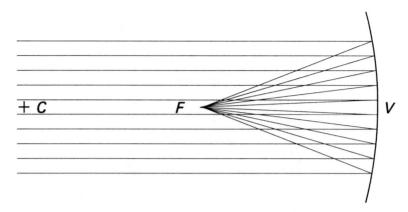

Figure 11-7 The paraxial focal length VF of a spherical reflector (center C) for paraxial rays, is one-half the radius of curvature; other rays pass close to F but cut the axis VF on the reflector side of F.

$$VF = D^2/8t(n-1),$$

a value that is included in the table for comparison. After this, F will denote the crossover point in the paraxial limit and VF will be referred to as the paraxial focal length. A way of looking at this formula is suggested by the property that the paraxial focal length of a spherical reflector is one-half the radius of curvature of the sphere (Fig. 11-7). The radius of curvature of the lens surface is $1/ \mid d^2z/dr^2 \mid = D^2/8t$. Thus the focal length of the lens is proportional to the radius of curvature of the lens and inversely proportional to $n-1$, the refractivity (or departure of the refractive index from that of free space). If the lens has two curved surfaces, then we add the two curvatures (the curvature in radians per meter being the reciprocal of the radius of curvature in meters). After reflection from a spherical surface the rays fail to converge to a single point, and the deficiency is exactly the same as with the thin lens; the rays incident farther from the axis converge more strongly than the paraxial rays. Figure 11-8 shows the focal region of the thin lens. All the light crosses the axis between E and F. A screen placed at F would collect a sizable disc of light. The smallest disc would be seen on a screen between E and F.

This phenomenon is known as spherical aberration because it is exhibited by a spherical reflector. Each annulus of the lens has a definite focus, but the focal points are spread over an axial line segment that stretches from the paraxial focus to a distance dependent on the lens aperture. Refer to Goodman (1968) for a general reference to optics viewed from the Fourier standpoint.

Inclined Rays

When the incident beam is inclined to the lens axis, the focus moves off axis, and one would expect further aberrations to set in with increasing inclination. Here we are going to look into the more elementary question of where the focus is. Consider the following argument, tending to suggest that the off-axis focus falls on a spherical surface. One

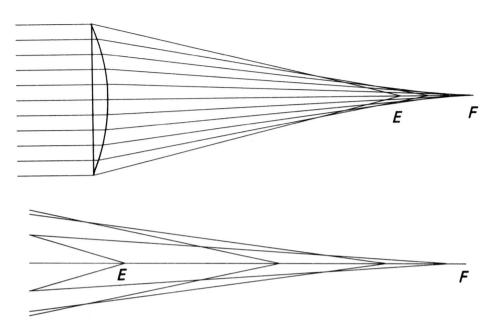

Figure 11-8 Ray paths through a thin lens (above). Detail of the focal region to the left of the paraxial focus F (below) shows spherical aberration.

way of describing the focusing action of a lens is in terms of the net deflection $i_2 - i_1$ undergone by each ray as it passes through the lens. As long as i_1 is small, both i_1 and i_2 increase linearly with the height of the ray, and therefore the angular deflection $i_2 - i_1$ is in proportion to the distance of the incident ray from the axis. Consequently, the rays are brought to a focus, because such proportionality is the property of convergent rays. Of course, approximations involving sines and tangents of small angles are understood.

Now for an inclined beam the situation is the same. The deflection is zero for the ray that passes through V, the vertex of the paraboloid, and is proportional for each other incident ray to its distance from that central ray. Consequently the rays converge at a certain distance from V fixed by the constant of proportionality. Thus the locus of focal points is a certain surface passing through the paraxial focus F. If the focal point F' of a narrow bundle of inclined rays was at the same distance from the vertex V of the lens (Fig. 11-9) as F is then the focal surface would be the sphere Sp centered at V. But we expect VF' to be less than VF because when the lens is viewed at an inclination θ it appears to be thicker by a factor $\sec\theta$ and for this reason alone should be a slightly stronger lens. In addition, however, its vertical-plane diameter as seen from the direction of the source is reduced by a factor $\cos\theta$. From the formula $f = D^2/8t\,(n-1)$ we expect a further factor $\cos^2\theta$ and therefore conclude that $f' = f\cos^3\theta$, writing $f' = VF'$ for the distance to the focus F' of the inclined beam. The curvature of the focal surface S' is deduced to be four times that of the sphere centered at V'; thus the focal surface approximates a sphere centered one-quarter of a focal length from F. As illustrated, the focal surface is four times

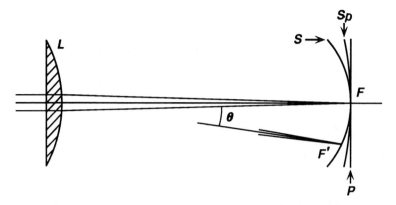

Figure 11-9 F is the paraxial focus of the thin lens L, and F' is the focal point of a fine inclined beam. Since VF' is less than VF, the focal surface S lies inside the sphere centered on the axis at V.

farther from the plane P than is the sphere Sp centered at V. Fig. 11-10 confirms these conclusions by direct ray tracing.

For the purpose of the diagram the rays shown are necessarily taken in the vertical plane. Now imagine a different set of rays belonging to the same cylindrical beam but cutting the lens on its horizontal diameter. For these rays, which do not see a reduction in lens diameter, there is a point F'' lying on a surface S'' such that $VF'' = f \cos \theta$. This second surface approximates a sphere centered midway between V and F. We conclude that the light in a narrow inclined beam is focused onto a line segment extending from F' to F''.

WHAT HAPPENS AT A FOCUS?

Suppose there were a lens of such perfection that, at a certain wavelength, the light emerging from the lens possessed absolutely spherical and concentric wavefronts. Then as the wavefronts shrank about their common center and the rays of light normal to the wavefronts converged, the energy density would build up. If, indeed, this state of affairs could continue, and all the rays could pass through the center and all the wavefronts could shrink to zero radius, then infinite energy density would result. This does not happen. Instead, within a small volume centered on the focus the flow of power deviates from being purely radial. Just what happens there is of considerable interest. As we shall see, the rays do not pass through the focus at an angle at all. On the contrary, the light bends away from the radial direction, passes through the focal volume more or less parallel to the axis of the original cone of rays, then bends back toward the side it approached from (Fig. 11-11).

The dimensions of the focal volume are of the order of the wavelength, generally speaking, but may be quite a few wavelengths. The shape of the focal volume depends on the solid angle of the cone from within which the waves arrive.

We are now in a good position for a rigorous discussion of the field in the focal

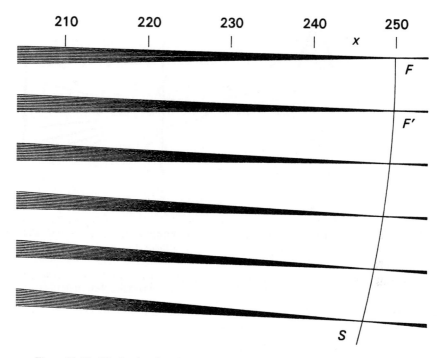

Figure 11-10 The focal surface S (exaggerated vertically) defined by a set of six 10-mm diameter beams leaving a thin lens (100 mm diameter and 10 mm thick), situated to the left at $x = 0$, at progressive angles of inclination. Paraxial focal length measured from the vertex of the lens is 260 mm.

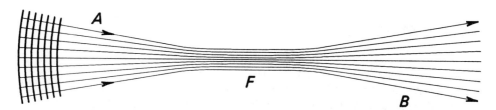

Figure 11-11 Illustrating the general concept of power flow through a focus. Shrinking spherical wavefronts (left) become more and more curved as the rays converge, but the rays do not actually pass through the focus F, and the wavefronts near the focal plane lose their curvature to become more or less plane parallel. The incident ray A does not emerge at B.

region. Let us regard the focal plane (the plane through F normal to the axis) as an "aperture" plane from which radiation is to be launched (Fig. 11-12). The focal-plane excitation is provided by the incident field from the left; the field launched by the "aperture" is simply the continuation of the incident field toward the right. Since the power launched by the focal-plane aperture will, well beyond the focal volume, be essentially of uniform den-

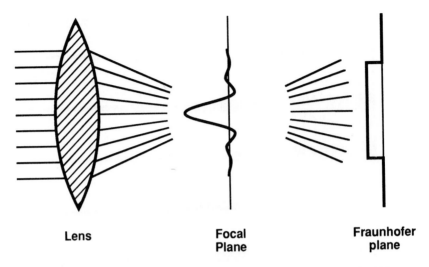

Lens   **Focal**   **Fraunhofer**
                    Plane   **plane**

Figure 11-12 The focal plane of a lens regarded as an aperture plane (center) which produces its own far-field diffraction pattern in a Fraunhofer plane (right).

sity within a certain solid angle and essentially zero outside that cone, let us work out the consequences of having a jinc function distribution in the focal plane. Let the focal-plane distribution be jinc (r_f/R), where r_f is the radial coordinate in the focal plane and R, which sets the scale of the jinc function, is such that $r_f = 1.22R$ at the first null circle.

The Hankel transform of the focal-plane distribution describes the Fraunhofer diffraction pattern or, in other words, the outgoing radiation far from the focal volume. The transform in question is

$$\text{jinc}(r_f/R) \supset (R/\lambda)^2 \ \text{rect}(Rq/\lambda).$$

In this relation, q or $\sqrt{l^2 + m^2}$ is the sine of the angle i between an outgoing direction and the axis. As we see, the outgoing beam has a sharp edge, which of course contains the same solid angle as the incident converging beam. The edges of the beam are at $Rq/\lambda = 1/2$ or at $\sin i = \lambda/2R$.

In terms of the lens diameter D, $\tan i = D/2f$. Now we can solve for the scale R to obtain

$$R = \frac{\lambda}{2 \sin i} = \frac{\lambda\sqrt{1 + \tan^2 i}}{2\tan i} = \frac{\lambda\sqrt{1 + (D/2f)^2}}{D/f}$$

$$= \lambda\sqrt{f^2/D^2 + 0.25}.$$

To an approximation, for $f/D \gg 1$,

$$R = (f/D)\lambda.$$

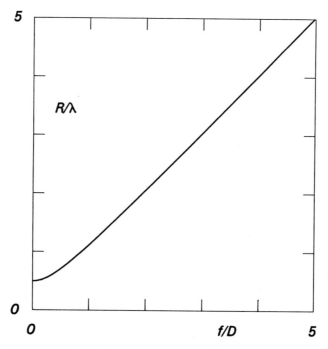

Figure 11-13 Radius of focal spot to first null (in wavelengths) as a function of focal ratio f/D.

As can be seen from Fig. 11-13, the linear approximation covers a wide range of practical cases and will be adopted for the time being.

Evidently the f/D ratio of the lens plays a key role in determining the transverse resolution. With an f/D ratio of ten, the focal spot size is ten wavelengths. In a microscope, where resolution is of the essence, the focal length must therefore be as small as possible with respect to the lens diameter. However, the closer the focal point has to be to the lens, the more serious the aberrations become. Hence, in practice, the design of a microscope involves not only the diffraction geometry but also aberration.

The longitudinal dimension of the focal volume is now easily deducible in terms of the transverse dimension, because we see that the calculation required is the familiar one of determining the distance to the far field or Fraunhofer region. The focal region is clearly the Fresnel region of the field distribution in the focal plane.

Of course, there is no sharp boundary to the Fresnel region, but it may conventionally be taken to extend N aperture widths from the aperture, where N is the number of wavelengths in the aperture; in this case $N = f/D$. Thus the length of the focal region is $2(f/D)^2\lambda$. These results are summarized in Fig. 11-14.

Some remarks may now be added about the approximation $f/D \gg 1$. It is true that $\text{rect}(Rq/\lambda)$ is a uniform beam, but uniform in q, or $\sin i$. If such a beam were intercepted on a distant plane (in the Fraunhofer region), the illumination on the plane would droop toward the edges, instead of being uniform as required for a uniformly illuminated lens.

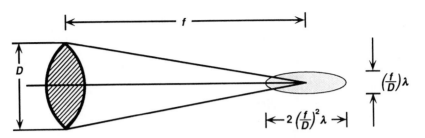

Figure 11-14   Dimension of the focal region (shaded and enlarged) of a lens, illustrated for $f/D = 3$.

The illumination would also droop for another reason—namely, the cosine factor that appears in one polarization. These factors, if allowed for, would adjust the jinc function assumed in the focal plane by convolution with the Hankel transform of the illumination correction factor.

SHADOW OF A STRAIGHT EDGE

We have seen how the action of even a simple lens requires explanation by Fresnel diffraction; now we examine the simplest geometry in which Fresnel diffraction occurs. A semi-infinite opaque plane screen bounded by a straight edge casts a shadow from an infinitely remote point source. As shown in Fig. 11-15, the intensity along the plane of observation AA oscillates about the unperturbed value in the unshadowed zone; and the intensity within the shadow is not zero, but dies away with a certain depth of penetration. The intensity on the geometrical shadow edge is 6 dB down (one-quarter of the unperturbed value). For a full description we need to know the strength and location of the maxima and minima and the depth of penetration for any distance D of AA from the screen. We can say immediately that the strength of the peaks and troughs is approximately independent of D:

Peak strengths are	1.370 1.199 1.151 1.126 1.110 1.090
Trough strengths are	0.778 0.843 0.872 0.889 0.901 0.910
Peak locations are at	$\hat{x} = 1.22$ 2.35 3.1 3.7 4.2 4.7

Approximate values for the kth maximum are $[1 + 1/\pi\sqrt{8k-5}]^2$ and for the kth minimum $[1 - 1/\pi\sqrt{8k-1}/]^2$.

As for the distance of the first maximum X_1 from the geometrical shadow edge, its locus is approximately a parabola with focus at the edge of the screen and axis along the shadow edge. The loci of all the maxima and minima and of given fractional levels within the geometrical shadow are all likewise parabolic.

Thus the perturbation of the geometrical shadow extends further from the geometrical shadow edge as D increases, but not linearly; the penetration goes as $\sqrt{D}$.

The extent of the perturbation also depends on wavelength, being approximately proportional to $\sqrt{\lambda}$. As $\lambda \to 0$, the diffraction phenomena become confined to a narrower

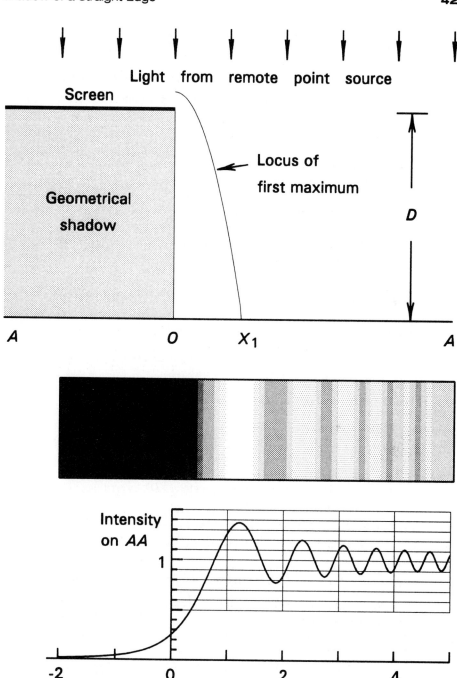

Figure 11-15 Universal curve for illumination on a plane of observation AA. The "geometrical shadow," corresponding to infinite wavelength, cuts AA at O, while maximum intensity, 1.370 times brighter than the incident light, occurs at X_1. The abscissa is the normalized distance $\hat{x} = (D\lambda/2)^{-1/2}x$.

and narrower zone about the geometrical shadow edge. The term *geometrical shadow* is borrowed from geometrical optics, the theory that deals with the limiting situation of zero wavelength.

With the one universal curve of Fig. 11-15, using as abcissa the dimensionless quantity $\hat{x} = x/\sqrt{\frac{1}{2}D\lambda}$, together with the foregoing explanation, we have a detailed description of the shadow of a straight edge, complete for most engineering purposes. The applications go beyond diffraction of light to situations as different as determining noise levels on the far side of masonry walls shielding communities from traffic noise to correction of antennas for adjacent obstructions, such as aircraft present to fuselage-mounted GPS satellite receivers. The high-resolution technique of lunar occultation of radio sources comes under the theory given, as would the scanning light microscope if an edge were used instead of a small aperture.

The word "approximate" may be ignored for most optical situations and for many radio applications also. The nature of the approximations emerges from the derivation of the results.

Derivation

The field at a point (x, D) is given approximately by

$$\int_{-x}^{\infty} e^{-i(2\pi/\lambda)(D+\xi^2/2D)}d\xi$$

and the normalized power is $p(x)$, where

$$p(x) = \frac{\mid \int_{-x}^{\infty} e^{-i\pi\xi^2/D\lambda}d\xi \mid^2}{\mid \int_{-\infty}^{\infty} e^{-i\pi\xi^2/D\lambda}d\xi \mid^2}.$$

We now normalize distance x with respect to $\sqrt{\frac{1}{2}D\lambda}$ by writing

$$\hat{x} = \frac{x}{\sqrt{\frac{1}{2}D\lambda}}.$$

(The distance $\Delta = \sqrt{\frac{1}{2}D/\lambda}$ is related to the first Fresnel zone on AA; the distance from the edge of the screen to the point at $x = \sqrt{\frac{1}{2}D/\lambda}$ on AA is $D + \frac{1}{4}\lambda$.)

Now

$$p(\hat{x}) = 0.5 \mid 0.5 + \int_0^{\hat{x}} \cos(\pi\xi^2/D\lambda)d\xi - i[0.5 + \int_0^{\hat{x}} \sin(\pi\xi^2/D\lambda d\xi] \mid^2$$

$$= [0.5 + C(\hat{x})]^2 + [0.5 + S(\hat{x})]^2$$

where $C(z)$ and $S(z)$ are the Fresnel integrals defined by

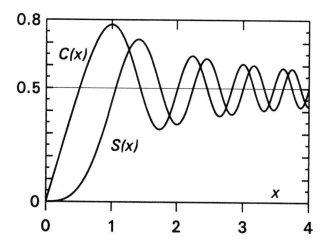

Figure 11-16 The Fresnel integrals $C(x)$ and $S(x)$.

$$C(z) = \int_0^z \cos(0.5\pi t^2)\, dt$$

$$S(z) = \int_0^z \sin(0.5\pi t^2)\, dt$$

(For tabulation see A&S. p. 321.)

Figure 11-17 shows the Cornu spiral, a graph of $S(x)$ versus $C(x)$. The graph of Fig. 11-15 can be obtained from the Cornu spiral as the distance from the vanishing point J' to points on the spiral. The parametric arc distances marked on the spiral are values of $\widehat{x}$. This diagram can be used not only for simple edge diffraction but also for diffraction by an aperture defined by parallel edges.

Example

A slit 10 cm wide is illuminated uniformly at a wavelength of 1 cm. Find the field on a plane at distance 25 cm from the plane of the slit. ▷ The characteristic distance $\Delta = \sqrt{\frac{1}{2}D\lambda} = 3.53$ cm. Since the slit is 10 wavelengths wide, the far field will begin at about 10 aperture widths or 100 cm from the plane of the slit; consequently, the plane at $D = 25$ cm is in the Fresnel region.◁

The slit can be thought of as two semi-infinite opaque screens, and the total field can be thought of as the out-of-shadow field in the presence of one screen, as discussed above, minus the in-shadow field in the presence of one screen only. Representing the slit illumination by $\mathrm{rect}(x/10)$, and the illumination of the first screen by $\mathbf{H}(5 - x)$, we are simply asserting the linear superposition $\mathrm{rect}(x/10) = \mathbf{H}(5 - x) - \mathbf{H}(-5 - x)$, where $\mathbf{H}(\ )$ is the Heaviside unit step function. In terms of the dimensionless parameter $\widehat{x}$, the slit field is

$$\mathbf{H}(5/\Delta - \widehat{x}) - \mathbf{H}(-5/\Delta - \widehat{x}) = \mathbf{H}(1.42 - \widehat{x}) - \mathbf{H}(-1.42 - \widehat{x}).$$

On the median plane, at $x = \widehat{x} = 0$, the first term, as read off from the Cornu spiral at A_0,

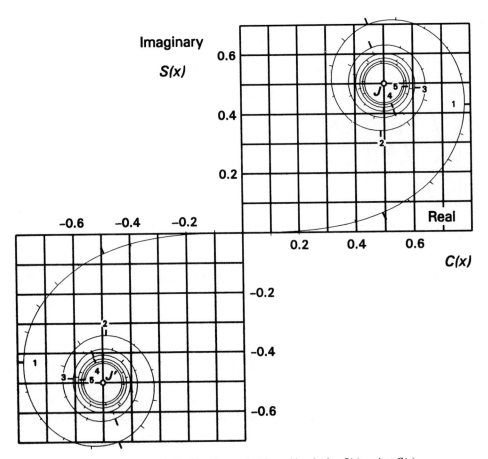

Figure 11-17   The Cornu spiral formed by plotting $S(z)$ against $C(z)$.

is 0.54 + i0.72. The second term (at B_0) is equal but opposite in sign; the difference is thus 1.08 + i1.44, and the intensity is $1.08^2 + 1.44^2 = 3.24$. On the plane of one of the geometrical shadow edges, at $x = 5$ cm, $\hat{x} = 5/3.53 = 1.42$, the first term (at A_5) is zero and the second (at B_5) is $\mathbf{H}(2.83) = 0.5 + i0.39$ (intensity = 0.4).

The unobstructed field, represented by the phasor $JJ' = 1 + i$ on the diagram, has intensity 2. Thus the intensity calculated as 3.24 is 1.61 times greater than the intensity the field would have in the absence of an obstructing slit. The whole field intensity pattern on the chosen plane at $D = 25$ cm can be worked out in this way by sliding the endpoints of the phasor BA to appropriate values of the arc parameter. If you have followed this description, however, you will realize that the value of the graphical analysis lies in showing in advance the general oscillatory character of the field, how strong the fluctuations are, at what distances enhanced or reduced intensities are found, and other qualitative features. Detailed numerical design based on conceptual analysis of this kind can advantageously then turn to Fresnel integrals.

FRESNEL DIFFRACTION IN GENERAL

Having studied two situations of fundamental importance, the focal region of a lens and the shadow of a straight edge, we turn to the Fresnel diffraction produced by apertures in general. Fresnel diffraction occurring on successive planes in front of an aperture can be understood rather simply in terms of spatial Fourier analysis. The idea is to break the aperture distribution down into spatially sinusoidal components. Once we understand the field launched by a sinusoidal aperture distribution, then on any fixed plane downstream the Fresnel field can be expressed as the sum of the effects produced by each Fourier component of the imposed aperture distribution. The procedure will be more general than permitted by the Cornu spiral, which is limited to apertures with uniform illumination.

A cosinusoidal field distribution, one whose amplitude varies as $\cos(2\pi x/\lambda_x)$ on the aperture plane of (x, y), where λ_x is the spatial period in the x-direction, behaves in two different ways. If $\lambda_x < \lambda$, then no net power is launched; the field merely decays with increasing distance z from the aperture plane. This is called an *evanescent wave;* it is familiar in optics as the wave set up in the air outside a block of glass when a wave inside the glass is totally reflected internally. If the interior angle of incidence is i, the field variation along the glass surface has a spatial period λ_x which exceeds the wavelength λ_m in the glass medium by a factor cosec i, but $\lambda_x + \lambda$ cosec i can still be less than the wavelength λ in air, because $\lambda_m < \lambda$. Under these circumstances the field in the air drops off rapidly with distance from the glass surface and no power is radiated normal to the glass—there is total internal reflection. Another well-known evanescent wave exists in a hollow rectangular waveguide whose width a is too narrow to permit microwave propagation. The waveguide field varies with the transverse coordinate as $\cos(\pi x/a)$; thus the spatial period is $2a$ and the waveguide cuts off where $\lambda_x < \lambda$ or $a < \lambda/2$. Under these conditions an excited waveguide presents a reactive impedance and fully reflects incident power, while within the waveguide the field decays exponentially with distance from the plane of excitation.

Although evanescent fields carry away no power, they are important to allow for. When a given aperture distribution is Fourier analyzed, the first step is to omit the Fourier components which are cut off. The remaining components, which will contribute to the radiated field, behave differently in accordance with their spatial transverse wavelengths λ_x and are handled one by one.

On the aperture plane $z = 0$ consider an electric field

$$E_y(0) = E_0 \cos(2\pi x/\lambda_x)e^{i\omega t},$$

which is independent of y. What is radiated across the planes of increasing z? The cosinusoidal pattern, which is fixed in space but oscillating in time, could be set up by the interference of two plane waves of infinite transverse extent impinging with equal but opposite angles of incidence $\pm i$ on the plane $z = 0$. The angle of incidence would be such that $\lambda_x = \lambda$ cosec i. Consequently the field existing where $z > 0$ is just the field of the two incident waves as they continue on across the nominal aperture plane.

It is true that apertures of finite extent are what we have in mind, but each spatial Fourier component of such a finite aperture distribution has infinite extent. We now understand in a simple way what each component launches into $z > 0$ space. On any plane parallel to the aperture plane at distance z the field will still be cosinusoidal in x, with a maximum at $x = 0$, but the amplitude will depend on z as $\cos(2\pi x/\lambda_z)$, where $\lambda_z = \lambda/\sin i = \lambda/\sqrt{1 - (\lambda/\lambda_x)^2}$. The phase will be that associated with a phase velocity $v_z = \lambda_z/\text{period}$ or $f\lambda_z$. Thus on the plane at distance z the field is

$$E_y(z) = E_0(\frac{x}{\lambda}) \cos \frac{2\pi x}{\lambda_x} e^{i(\omega t - 2\pi z/\lambda_z)}.$$

The total Fresnel field at distance z will be the resultant of all such contributions from the constituent Fourier components of the aperture distribution.

With a general aperture distribution $E_y(0) = E_0(x/\lambda) \exp i\omega t$ the complex amplitude $P(s)$ of the Fourier component of spatial frequency $s = \lambda/\lambda_x$ cycles per free-space wavelength is the Fourier transform of the aperture distribution $E_0(x/\lambda)$, expressed as a function of the dimensionless distance $\widehat{x} = x/\lambda$. Thus

$$P(s) = \int_{-\infty}^{\infty} E_0(x/\lambda) e^{-i2\pi s(x/\lambda)} d(x/\lambda).$$

Integrating over the various spatial components of the aperture distribution, but dropping the evanescent components ($|s| > 1$), we get for the total Fresnel field at distance z,

$$\int_{-1}^{1} P(s) \cos 2\pi \widehat{x} \, e^{-i2\pi(z/\lambda)\sqrt{1-s^2}} ds.$$

The factor $\exp i\omega t$ has been omitted as is customary.

It is often satisfactory to drop the evanescent waves from the Fresnel field, because they die out in a short distance, and it helps to do so because the integration from -1 to 1 is simpler numerically. However, the Gibbs phenomenon will appear in the evanescent zone if the integral is truncated in this way; to take the evanescent component correctly into account the limits of the integral must be extended. The factor $\exp(-i2\pi z/\lambda_z)$ will change to a decaying exponential where $|s| > 1$.

A previous example considered a slit describable by an aperture distribution $E_0(\widehat{x}) = \text{rect}(\widehat{x}/10)$ and looked at the Fresnel diffraction field at $z/\lambda = 25$. Unlike the method using Fresnel integrals, the present method can handle aperture distributions of nonuniform amplitude, but let us repeat the example. As the Fourier transform of $\text{rect}(\widehat{x}/10)$ we find the angular spectrum

$$P(s) = 10 \, \text{sinc} \, 10s.$$

The integral for the Fresnel field is therefore proportional to

$$\int_{-1}^{1} \text{sinc} \, 10s \cos 2\pi \widehat{x} s \, e^{-i2\pi 25\sqrt{1-s^2}} \, ds.$$

Because of symmetry it will be sufficient to integrate from 0 to 1. The real and imaginary parts need to be evaluated separately.

LITERATURE CITED

H. G. Booker and P. C. Clemmow (1950), "The concept of an angular spectrum of plane waves, and its relation to that of polar diagram and aperture distribution," *Proc. IEE*, vol. 97, Pt. III, p. 11–17.

M. Born and E. Wolf (1965), *Principles of Optics*, 3d ed., Pergamon Press, New York.

W. N. Christiansen and J. A. Högbom (1969), *Radiotelescopes*, Cambridge University Press, Cambridge, England.

R. H. Clarke and J. Brown (1980), *Diffraction Theory and Antennas*, Ellis Horwood Limited, Chichester, United Kingdom.

J. W. Goodman (1968), *Introduction to Fourier Optics*, McGraw-Hill, New York.

J. Kraus (1988), *Antennas*, 2d ed., McGraw-Hill, New York.

A. Lohmann (1992), "Duality in optics," *Optik*, vol. 89, pp. 93–97.

A. S. Marathay (1982), *Elements of Optical Coherence Theory*, John Wiley & Sons, New York.

W. H. Steel (1983) *Interferometers*, 2d ed., Cambridge University Press, Cambridge, England.

A. R. Thompson, J. M. Moran, and G. W. Swenson, Jr. (1986), *Interferometry and Synthesis in Radio Astronomy*, John Wiley & Sons, New York.

PROBLEMS

11–1. *Apodized apertures.* A circular aperture of unit diameter has a Hankel transform that is a jinc function, and the first sidelobe, with a value -0.104, is 20 dB down. Investigate how the strength of this sidelobe diminishes with n as a sequence of unit-diameter aperture distributions of the form $(1 - 4r^2)^n \operatorname{rect} r$ is considered, taking $n = 0, 1, 2, 3$. ☞

11–2. *Duality.* Three points define a plane (unless the three points lie on a straight line). What is the dual of this statement? ☞

11–3. *Young's interferometer and duality.* We have said that diffraction by a pair of pinholes and the launching of a wave pair by a cosinusoidal field distribution over a plane involve distinct physical arrangements. But does not one case convert into the other if the direction of time is reversed? ☞

11–4. *Binocular vision.* The giant Polyphemus, who lived between Catania and Messina in the time of Ulysses, had only one eye, but it was 10 cm in diameter. Was he deprived of the depth perception enjoyed by two-eyed people, or is it possible that his one giant eye afforded equivalent depth perception?

11–5. *Microlaser beam.* A round sandwich of indium gallium arsenide between indium gallium arsenide phosphide, about 400 atoms thick and 2 μm in diameter, emits laser light at 1.4-μm wavelength in a beam which starts off being 2 μm in diameter. How far does this beam extend in collimated form before it begins to diverge?

11–6. *Iris transfer function.*
    (a) The aperture of a camera is controlled by an iris diaphragm with a pupil that is hexagonal, because it is mechanically more convenient to enlarge a hexagonal opening than to enlarge a circular one. The transfer function of a circular opening is a chat function.

Compute and plot the principal cross sections of the transfer function of the hexagonal aperture and superimpose the chat function pertaining to the circular aperture of the same area.

(b) Explain the deviations noticed between the three curves.

(c) In what way would a photograph of a bright point source reveal whether the camera had a circular or hexagonal aperture?

11–7. *Maladjusted photocopier.* A photocopying machine has the annoying defect of making copies 2 percent smaller than the original, which makes the machine unsuitable for some purposes. An electrical engineering student reasons as follows. "The focal length of the lens is 10 inches. Therefore, if the original is supported 0.2 inch above the top window, which is 2 per cent of 10 inches, a copy of the correct size will result." A professor waiting to use the machine says, "Don't you think 0.1 inch would be the right amount?" What is your opinion? ☞

11–8. *Inclined beam through thin lens.* The axis of a cylindrical beam passes through the center of a thin lens whose diameter is 10 mm and focal length 250 mm, making an angle of $10°$ with the lens axis.

(a) Give the coordinates of the ends of the line segment through which all rays of the beam would pass if followed by the ray tracing formula.

(b) By considering rays of different position angle within the beam, determine the density of rays on the line segment with a view to finding out whether the concentration is dominated by the depth of focus expected at the wavelength in use, let us say 0.5 μm, rather than by ray geometry. Look into this.

11–9. *Rainbow.* Are the colors of the rainbow due to diffraction or refraction or both? The word diffraction derives from Latin *diffrango* = "I break up"; what is it that is broken up? ☞

11–10. *Focal spot.* The 200-inch mirror on Palomar mountain (now referred to as the 5-meter) has a focal length of 666 inches.

(a) In the focal plane of the mirror, what is the theoretical diameter d_0 in μm of the image of a star, measured across the first null circle of the Airy pattern at wavelengths of 400 and 800 nm?

(b) What is the depth of focus in mm? ☞

11–11. *Satellite camera.* A white-light camera looking vertically downward from a height of 275 km has a resolution of 10 cm; i.e., a horizontal object 10 cm long subtends an angle $1.22\lambda/D$ at the camera, where D is the diameter of the camera aperture.

(a) What is the value of D?

(b) If the 10 cm object registers on the film as a spot 1 μm in diameter, what is the focal length f?

11–12. *Spy in the sky.*

(a) It is claimed that a certain satellite-borne camera orbiting 300 km above the ground can read the license number of a parked automobile. Assuming the owner had carelessly left his license plate on the roof, what would be the approximate diameter of the camera aperture? (Assume a wavelength of 0.7 μm.)

11–13. *Edge diffraction.* After passing by a straight edge, a uniform plane wave is no longer uniform but exhibits maxima and minima in any wavefront. Count the nearest maximum to the geometrical shadow boundary as the first extremum, the first minimum as the second extremum, and so on. Show that the intensity at the *l*th extremum is $[1 \pm \pi^{-1}(4l - 1)^{-1/2}]^2$.

11–14. *Transparent straight edge.* A quarter-wave plate is not one-quarter of a wavelength thick;

it is a plate which injects one-quarter of a wavelength extra path when a normally incident wave passes through. A semi-infinite quarter-wave plate abuts a semi-infinite three-quarter wave plate along a straight edge. When a plane wave falls normally on this nonopaque screen, what emerges? ☞

11–15. *Acoustic diffraction.* Tire noise originates on a freeway along a line which is 300 m from a row of bedroom windows. A solid concrete wall is to be constructed 50 m from the noise source with a view to throwing the windows into the sound shadow. How tall will the wall need to be if a 10-dB reduction is to be hoped for at frequencies of 1000 Hz and higher? Take the windows to be 3 m above road level. ☞

11–16. *Experiment.* Empty an aluminum beverage can, break off the tab, and pierce a pinhole about half a millimeter or less in diameter in the bottom. A sewing needle may help. Peep at the pinhole through the top of the can, holding the can up to the blue sky or a bright indoor light. Do not point directly at the sun.

(a) Make a drawing of what you see with each eye separately and keep for future reference. Estimate the angular diameter of the bright patch. (You can calibrate the angular scale by piercing a second hole 10 mm from the first and noting the length of the can.) Explain what you see.

(b) Look for a striated arc of light on the inside wall of the can. Ascertain whether the striations are surface structure on the wall of the can, are associated with the rough edge of the pinhole, are due to the imperfections of the eye, or are caused by some other phenomenon. ☞

12

Aperture Synthesis and Interferometry

To record the image of a sunset we point the camera toward the setting sun, and the rays diverging from the various elements of the scene fall on the lens and are focused, element by element, on the film. Some people undoubtedly conceive of a camera as a device that reaches out and captures the distant scene, and years ago it was vigorously contended that vision reached out from the eye much as a hand reaches out in a dark room to explore the surroundings. It is, of course, perfectly all right to describe photography as in the introductory sentence above, but with the caveat that the description involves information about certain entities (the source elements and the propagation medium) that are unascertainable by the camera alone. In many branches of science the output of the observing instruments is the *only* information—astronomy is one example, and seismic exploration of the earth's interior is another. Both of these endeavors produce images of inaccessible regions of physical space, and they suggest another, more operational, way of describing image formation.

IMAGE EXTRACTION FROM A FIELD

The objective lens of a refracting telescope is immersed in an electromagnetic field of optical frequency that exists at the telescope. The telescope does not reach out into celestial space at all. On the contrary, the lens is in contact with a terrestrially situated field, and the image on the film is formed by electric fields produced at the film when the image-forming instrument, in this case a telescope, operates on the field available to it at the big lens. The same is true of a seismic geophone array, which, far from reaching into the interior of the earth, is merely in contact with the quaking surface of the earth.

By discussing the entities that are available to an instrument to operate upon, rather than emissive or scattering elements in inaccessible regions, we get a tighter, if less familiar, theory.

Now consider the following question. One can see that the sun is a bright circular object half a degree in diameter. Therefore, at the surface of the lens of the eye, or on any other surface on earth, such as the surface of paper exposed to sunlight, the optical electromagnetic field contains the information about the circular form and the diameter. Where in the electromagnetic field on the intercepting surface does this information reside?

If you could record the electric field at a point in full and precise detail as a function of time, you would know everything that could be known about the field at that point, and you could ascertain the spectrum of sunlight and the solar constant, or how many kilowatts fall on a square meter. But you could not deduce the shape of the sun. If you recorded the field at any other point on earth, the recording would be different in detail, but the intensity and the spectrum would be the same. So how does an instrument such as the eye or a camera operate on such data to extract an image?

The answer is that the fields at adjacent points are different but not independent; the way in which the dependence varies with the spacing of the points contains the information about the image. To specify the image-forming characteristic of the field we must deal in terms of *coherence* as a function of vector spacing on the intercepting surface. To understand the coherence of the field is to lay the foundation for an appreciation of a broad class of image-forming instruments.

INCOHERENT RADIATION SOURCE

We say that the field variations at two points in space are *coherent* if they possess such phase relationships as to permit interference. Starting from this basis, we can expand the concept of coherence. For example, a single waveform, such as the output from a signal generator, or the carrier wave from a radio transmitter, or the waveform supplied by the power distribution network, is sometimes said to be coherent. This statement can be given meaning in terms of our point of departure if two waveforms, generated from the one waveform by counting from two different instants of time, can interfere.

Of course, one sees immediately that the question whether or not a waveform is coherent cannot be answered without reference to the time interval between the two selected instants, because it is quite conceivable that a waveform may be coherent over a short time interval but not over a long one. This thought underlies the notion of *coherence interval,* a time interval characterizing the fall-off in coherence as the time interval is increased. In the case of a traveling wave the corresponding term *coherence length* may be applied. Thus, an optical waveform might be split into two wavetrains by a transparent mirror and reunited after one has traveled farther than the other, and it might be found that interference occurs, but not if the extra distance of travel exceeds one meter. Then we would say that the wave train has a coherence length of one meter.

Clearly, to quantify the concept of coherence along the above lines it would be necessary to establish a criterion for deciding in a consistent manner whether interference was present or not. One possibility would be to follow the example set by the definition of time constant, taking an arbitrary but nonzero amount of interference as the boundary between coherent and not-coherent fields. But no such standard is generally current. Another possibility would be to set the boundary where interference becomes undetectable. One might argue that the result would be instrument dependent and not intrinsic to the radiation field itself. Nevertheless, people have done precisely this in optics, where the ability of the eye to perceive interference fringes or not has been a successful basis for the measurement of the spacing between double stars, the angular diameter of stars, and the widths of spectral emission lines. A superior treatment, however, can be given by the introduction of a quantitative measure of coherence, which will be defined below.

It helps the discussion if one expands the terminology to include the idea of an incoherent radiation source. A thermal source of radiation such as a white-hot tungsten filament might be a typical example, but many wideband sources, regardless of emission mechanism, would qualify. A sunlit piece of paper might be an incoherent source. A little care is needed in framing a definition of an incoherent source. For example, you might say that an incoherent source shall be one such that the fields at any two points outside the source are not coherent. However, no such source exists, as will become clear. It is necessary to take the two points on the surface of the source. Furthermore, if the source is translucent, as with incandescent gas, the opaque surface must be chosen, if there is one. A nonopaque incandescent source—red-hot glass, for example—even though its radiation comes from microscopic random events that are essentially independent of each other, will exhibit coherence between adjacent surface points. Our present purpose is not to construct an incoherent source; if you want one, you will have to do the best you can to build one— then try it and see. Our purpose is to make use of the idea of an incoherent source, one whose surface fields exhibit no coherence between any pairs of separate points. As we have stated, the test will lie in the capacity to allow interference, not in the emission mechanism.

FIELD OF INCOHERENT SOURCE

Generally speaking, the field of an incoherent source has a wide spectrum, yet even a narrow spectral line, such as the emission line of cesium-13, on which the definition of the international second is based, has a nonzero bandwidth. The wavetrain of a narrow spectral line merely has a long coherence length. Before we can discuss the field of an incoherent source, we need to agree on a bandwidth. If we are discussing a spectral line, we may use the natural bandwidth of the line, but in other cases the appropriate bandwidth is an instrumental one. For example, we may be studying the infrared radiation from the moon. Unlike white moonlight, the moon's infrared radiation is thermally generated by the moon itself. But it does not have a blackbody spectrum; consequently, measurement of the spectrum can provide information about the moon's surface and the underlying

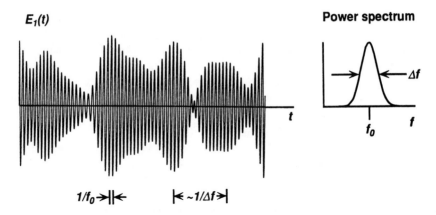

Figure 12-1 A quasi-monochromatic waveform. The amplitude distribution is Gaussian and the envelope has a Rayleigh distribution.

material down to the depths from which the thermal radiation can escape. To determine the spectrum necessarily requires a receiving instrument that limits the spectral band. Therefore the mathematical discussion involves an electric field $E(t)$, which is not the total field of the source but that part of it which is about to be responded to by the receiver. This apparently clumsy notion seems unavoidable. It is, however, in keeping with the operational attitude which refuses to discuss quantities that cannot be measured. The fractional field $E(t)$ is by its definition measurable.

Interestingly enough, even electric field is treated as a second-class citizen by Born and Wolf (1959), who regard intensity as the measurable optical quantity and who prefer theoretical formulations in terms of measurables rather than of ideal concepts such as field. However, we know from experience with radio frequencies that the difficulty with optical field is not fundamental but merely a result of historical limitations to nonlinear detectors (eye and film) for measuring light. The development of instruments such as lasers and optical mixers has made radio techniques available to the world of optics.

Figure 12-1 illustrates a waveform $E_1(t)$, one of the three components of the vector $E(t)$, at a point labeled by subscript 1. It is characterized by a frequency f_0, a bandwidth Δf, and an intensity $\langle E_1^2 \rangle$. It could be described as an oscillation at frequency f_0 modulated by an envelope which rises and falls. In addition, the phase of the oscillation at frequency f_0, as revealed by the time of zero-crossings, drifts back and forth at a rate characteristic of the fading of the envelope. If the spectral shape were specified in more detail than by a single bandwidth parameter, the character of the envelope variations could be better defined, but without more than a statement of Δf we can say the following. The values of $E_1(t)$ are distributed about a mean of zero in a Gaussian manner with a standard deviation $\sigma = \langle E_1^2 \rangle^{1/2}$. That is, the probability of finding E_1 in the interval $E_1 \pm 0.5\,dE$, is $p(E_1)\,dE_1$, where

$$p(E_1) = \frac{1}{\sqrt{2\pi}\,\sigma}\exp[-E_1^2/2\sigma^2].$$

The reason that the probability distribution is Gaussian, without regard to the shape of the spectrum, is that the field is the sum of a large number of independent contributions. Such a sum has a Gaussian distribution under the central limit theorem (FTA 1986, pp. 168–172).

We can also say that the envelope $R(t)$ has a Rayleigh distribution. That is, the probability of finding the envelope in the interval $R \pm 0.5\,dR$ is $p(R)$, where

$$p(R) = \sigma^{-2} R e^{-R^2/2\sigma^2}.$$

This distribution is arrived at by thinking in the $(E_1, \widehat{E}_1)$ plane, where $\widehat{E}_1(t)$ is the Hilbert transform of $E_1(t)$ defined by

$$\widehat{E}_1(t) = \frac{1}{\pi} \int_{-\infty}^{\infty} \frac{E_1(t')\,dt'}{t' - t}.$$

In this plane, $R^2 = E_1^2 + \widehat{E}_1^2$ and the interval $R \pm 0.5dR$ represents an annulus of area $2\pi R\,dR$. The two-dimensional probability density per unit area in the $(E_1, \widehat{E}_1)$-plane is $p(E_1, \widehat{E}_1)$ where

$$P(E_1, \widehat{E}_1) = \frac{1}{\sqrt{2\pi}\sigma} \exp(-E_1^2/\sigma^2) \frac{1}{\sqrt{2\pi}\sigma} \exp(\widehat{E}_1^2/2\sigma^2).$$

Hence $p(R)\,dR = 2\pi R\,dR\,p(E_1, \widehat{E}_1)$; which yields the Rayleigh distribution.

Fading Rate

The two statistical properties just stated do not depend on the spectrum, nor does the phase, which is distributed uniformly over the range 0 to 2π. The details of the envelope fading do depend on the spectrum, but the characteristic fading time is of the order of $1/\Delta f$. One way of looking at this is to subdivide the spectrum into monochromatic components and to note that any pair of such components would produce an envelope fluctuating with a beat frequency equal to the frequency difference between the pair. It is true that frequency differences as great as $2\Delta f$ may be found within the spectrum, but not in as many ways as smaller frequency differences and not with as large amplitudes. Hence we expect the fading envelope to contain frequencies from zero up to Δf and beyond, but not much beyond. (The envelope spectrum can be calculated. The simplest way to state the result is to say that the spectrum of the squared envelope is obtained by autocorrelating the power spectrum.)

Evidently, when the spectrum is narrow, the fading rate is slow, and vice versa. We remember the order of magnitude by saying that if $\Delta f/f_0 = 100$, i.e., if the bandwidth is 1 percent of the mid-frequency, then there will be 100 cycles in one fading cycle. This could be taken as a definition of fading cycle. However, people have an intuitive feeling for fading cycles and have often published "fading rates" in peaks per unit time obtained by counting peaks in records of measured physical quantities. The number of peaks per second is deducible from the spectrum (FTA 1986, pp. 340–341). The fading cycle is clearly connected with coherence interval, because the drift rate of phase is linked to the fading rate.

Field in Randomly Polarized Wavefront.

The electric field in a plane perpendicular to the direction of propagation from a thermal source possesses both linear polarizations simultaneously, but the amplitude and phase of each polarized component are independent of the amplitude and phase of the other. Consequently, the tip of the vector describing the electric field sometimes rotates clockwise and sometimes counterclockwise and occasionally changes sense of rotation by passing through a state of linear polarization. Since the amplitudes are independent, the locus will in general be elliptical and almost never circular; a circular locus would require that the two amplitudes be equal at the same time that the phases are in quadrature. While the mean value of either linearly polarized component of field is zero, the resultant field has a nonzero mean in accordance with the Rayleigh distribution that describes its statistics. This distribution was first worked out by Lord Rayleigh in connection with the drunkard's walk, the locus that consists of a sequence of discrete steps after each one of which the drunkard falls down. When he or she gets up, the new direction is completely independent of the direction of the previous steps; consequently, the mean distance from the point of origin has a nonzero value, averaged over many drunkards, while at the same time the most probable place to find such a person is precisely at the origin, which is where the centroid of the pattern is. However, the mean distance from the origin increases with time in proportion to the square root of the number of steps.

CORRELATION IN THE FIELD OF AN INCOHERENT SOURCE

Figure 12-2 shows an incoherent source of finite extent which produces a field $E_1(t)$ at point P and $E_2(t)$ at Q. Consider just two elements A and B of the source. Element A contributes the same waveform to P as to Q, except for a delay associated with the path difference $AP - AQ$. The field $E_1(t)$ is the sum of the contributions from all of the independent elements of the source, of which just two are illustrated. Now the field $E_2(t)$ is the sum of the same contributions but with relative delays that are different. If Q is close to P, the path difference $AP - AQ$ to element A is much the same as for B, the relative time shift is small, and waveform $E_2(t)$ is not much different from $E_1(t)$. The same applies if A is close to B or if the source is of small angular diameter. Consequently, we may expect correlation between the fields $E_1(t)$ and $E_2(t)$, even though the source is incoherent, but the correlation will fall off as the spacing PQ increases.

The spacing PQ that characterizes a substantial fall-off in correlation will be related to wavelength and to the angular diameter of the source and can be estimated as follows, from fundamental principles. Suppose the source is remote, the element A being in a direction making an angle i_A with the plane normal to the baseline PQ, and similarly for i_B. Then the path difference $AP - AQ$ is $PQ \sin i_A$, and the path difference from B is $PQ \sin i_B$. The relative shift is $PQ(\sin i_A - \sin i_B)$. When this reaches one wavelength, we may expect substantial falling off in correlation, i.e., when

$$PQ(\sin i_A - \sin i_B) = \lambda.$$

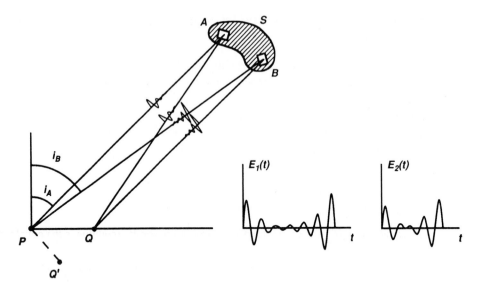

Figure 12-2   The fields $E_1(t)$ and $E_2(t)$ at the points P and Q, respectively, due to an incoherent source S.

When PQ is approximately at right angles to the rays from the source, $\sin i_A - \sin i_B \approx i_A - i_B$. The angular diameter Θ of the source is the maximum value of $i_A - i_B$, and it is the widely spaced source elements that exhibit the loss of correlation. So the condition for loss of correlation may be expressed as

$$PQ'/\lambda = \Theta^{-1},$$

where we have taken the opportunity to relieve the condition that PQ should be perpendicular to the rays by introducing a point Q', so that PQ' is perpendicular to the rays. Evidently, when the radiation arrives obliquely, larger spacing between P and Q may be reached before the correlation deteriorates.

From the foregoing equation we can see how the angular separation of double stars could be estimated by separating a pair of slits at the entrance to a telescope until interference fringes could not be seen. Then, if the slits were 200,000 wavelengths apart, the angular separation would be 1/200,000 radians or one second of arc. There are some roughnesses in this approach; for instance, the profile of the individual stars does not influence the result. Also, disappearance of fringes could occur at different slit spacings for one and the same angular separation, depending on the faintness of the double star. Then, of course, the one-wavelength relative shift criterion was arbitrary.

Quantifying the Correlation

A way to measure the agreement between $E_1(t)$ and $E_2(t)$ would be to multiply them together and then to average the product over many fading periods. We introduce the notation $\langle \ldots \rangle$ to stand for time averaging over many fading periods. One might normalize the quantity $\langle E_1(t)E_2(t)\rangle$ in order to have a value of unity when P and Q were coincident. Then our measure would be

$$\frac{\langle E_1(t)E_2(t)\rangle}{\langle [E_1(t)]^2\rangle}.$$

A theory could be based on this, but in fact the customary measure is based not on the electric field itself but on the time-varying phasor.

The Time-varying Phasor

Let

$$E_1(t) = \mathcal{R}F_1(t)e^{i2\pi f_0 t}.$$

The quantity $F_1(t)$ is the complex time-varying phasor. Its modulus is the envelope of the quasi-monochromatic waveform, and its phase describes the timing of the zero-crossings. Just as alternating-current theory is facilitated by the introduction of a complex phasor, which itself does not alternate with time, so when we generalize from the situation of a single frequency to consider a frequency band the analogous theory makes use of a complex phasor that does not alternate at the mid-frequency f_0, but does vary slowly with time, at the fading rate, so that at any instant it describes the amplitude and phase of the steady-state oscillation that would match the field variation then prevailing (FTA 1986, p. 270). Multiplication by $\exp(i2\pi f_0 t)$ introduces the frequency of oscillation f_0, and taking the real part $\mathcal{R}$ gives the real electric field $E_1(t)$, which both fades and alternates.

Complex Degree of Coherence

Let

$$\gamma = \frac{\langle F_1(t)F_2^*(t)\rangle}{\langle F_1(t)F_1^*(t)\rangle},$$

where γ is defined (Zernike, 1938; Bracewell, 1958) to be the complex degree of coherence between the fields $E_1(t)$ and $E_2(t)$. To gain familiarity with this quantity we investigate $\langle E_1(t)E_2(t)\rangle$. Thus

$$\langle E_1 E_2\rangle = \langle \mathcal{R}F_1 e^{i2\pi f_0 t} \mathcal{R}F_2 e^{i2\pi f_0 t}\rangle$$

$$= \left\langle \frac{(F_1 e^{i2\pi f_0 t} + F_1^* e^{-i2\pi f_0 t}}{2} \, \frac{F_2 e^{i2\pi f_0 t} + F_2^* e^{-i2\pi f_0 t}}{2}\right\rangle$$

$$= \frac{1}{4}\langle F_1 F_2^* + F_1^* F_2\rangle = \frac{1}{2}\mathcal{R}\langle F_1 F_2^*\rangle.$$

The cross-product terms with frequency $2f_0$ average out to zero because they are oscillatory. The normalized quantity is obtained by dividing by the value that results when P and Q coalesce, and turns out to be the real part of γ. Thus

$$\frac{\langle E_1 E_2 \rangle}{\langle E_1^2 \rangle} = \frac{\mathcal{R}\langle F_1 F_2^* \rangle}{\langle F_1 F_1^* \rangle}$$

$$= \mathcal{R}\gamma.$$

VISIBILITY

A. A. Michelson (1852–1931) introduced a method of determining the shape of a spectral line by splitting the radiation from a source into two beams with a relative time delay δ. If the electric field of the incident beam is $E(t)$, then, after recombining, the field is $\frac{1}{2}E(t) + \frac{1}{2}E(t + \delta)$. The intensity as perceived by the eye, or as registered by a photodetector, is a time average of the square of the field; thus the incident intensity I_{inc} is given by $I_{\text{inc}} = \langle [E(t)]^2 \rangle$. The intensity observed after recombining is $I = \frac{1}{2}I_{\text{inc}} + \frac{1}{2}\langle E(t)E(t + \delta) \rangle$, where the second term is recognizable as the autocorrelation of $E(t)$. As the time lag δ departs from zero, the autocorrelation falls off, slowly if the width of the spectral line is narrow, or rapidly if the bandwidth is broad. In general, the Fourier transform of the autocorrelation will yield the power spectrum of the radiation.

This is the basis of modern instruments using Fourier transform spectroscopy for chemical analysis, the time delay being introduced by a moving mirror. In the absence of a fast photodetector Michelson introduced a special wrinkle. If the beams being recombined are inclined at a very slight angle, interference fringes will be seen in place of a reduced intensity distributed over the whole field of view, and the range of observed intensity will be $\frac{1}{2}I_{\text{inc}} \pm \frac{1}{2}\langle E(t)E(t + \delta) \rangle$. The minus sign will refer to a trough line in the field of view along which the path difference due to inclination adds $n + \frac{1}{2}$ wave periods of delay to the delay δ introduced by design. Now it will be possible with an uncalibrated detector to get the ratio $\langle E(t)E(t + \delta) \rangle / \langle E(t)E(t) \rangle$ from $V = (I_{\text{max}} - I_{\text{min}})/((I_{\text{max}} + I_{\text{min}})$, where I_{max} and I_{min} are the values of I at interference maxima and minima, respectively. Of course, Michelson did not really introduce such fringes deliberately. The fact is, it is difficult to bring the interferometer into adjustment in such a way that the fringes are absent. In the absence of a radiometer it is still possible to estimate V by eye, which is what Michelson did, introducing the term *visibility*. Thus 25 percent visibility, which is "rather poor," means that maximum intensity is only 25 percent above the mean and minimum intensity is 25 percent below. Visibility is analogous to *modulation coefficient* in radio communication.

Complex visibility $\mathcal{V}$, which was introduced much later (Bracewell 1958), associates a phase angle σ with the magnitude V, so that

$$\mathcal{V} = V e^{i\sigma}.$$

The phase angle σ describes the spatial displacement of a fringe maximum from some agreed origin, converted to radians on the basis of 2π radians per fringe spacing.

Although visibility V as measured by Michelson is a positive quantity, theoretical visibilities due to multiple spectral lines of identical shape were treated by Michelson as

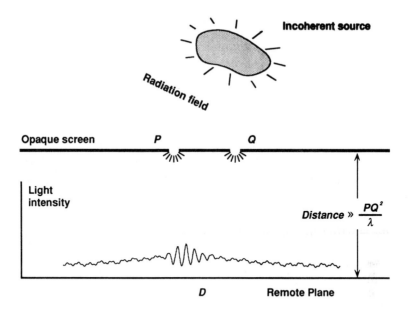

Figure 12-3 Exploring coherence at P and Q optically. The interference is displayed spatially.

going negative, so the idea of phase was partly present. In fact the concept was specifically alluded to by Rayleigh (1892).

MEASUREMENT OF COHERENCE

First we must occupy the two locations P and Q, for example with antennas to extract voltages that may be brought together at a central point and multiplied. In the optical case, which is dealt with first, we may place pinholes at P and Q and allow the light to propagate, screened off from the principal radiation field, and produce interference on a distant plane as shown in Fig. 12-3. The interference pattern may be recorded on film. Or, a single photomultiplier D might be used as a moving detector, or the pattern might be caused to traverse a fixed detector by sliding a phase-changing wedge across one pinhole. One way or another, by phase shifters, bodily rotation of the interferometer (screen and collecting plane), motion of the detector, or use of multiple detectors, a spatial or temporal interference pattern can be revealed. The complex visibility deduced from the interference pattern will yield the complex degree of coherence in both modulus and phase.

In the case of antennas (Fig. 12-4) phase may be steadily added at one antenna and steadily subtracted at the other, as happens in astronomy when a rigid interferometer is rotated by the earth. Interferometer elements situated S wavelengths apart on the terrestrial equator, studying a source on the celestial equator, will experience phase changes of $\frac{1}{2}\phi$

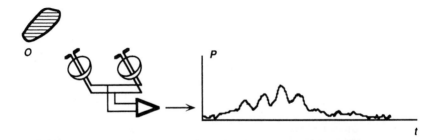

Figure 12-4　Radio measurements of the coherence of the field of a celestial object O by means of an interferometer of spacing S wavelengths whose available power in a band Δf is recorded. The interference is displayed temporally as the earth rotates. The depth of modulation, or fringe visibility V, is about 0.4.

and $-\frac{1}{2}\phi$, where $\phi = 2\pi S \sin \Omega t$ and Ω is the angular velocity of the earth. As a result, a record such as shown in Fig. 12-4 might be obtained. This interferogram slowly rises from and falls back to the receiver noise level as the source moves into and out of the beam of the antennas, but also exhibits the temporal fringes due to interference. The fringe visibility $|\,\mathcal{V}\,|$ is seen to be about 0.4. The quantity pha $\mathcal{V}$ is determined from the offset between the maximum of the central fringe and the maximum of the response $A(t)$ of a single antenna to a point source. In principle some other instant for the origin of phase could be chosen; the moment when the antennas are pointed at the nominal origin of the source is used here. Neglecting the omnipresent noise fluctuations, the record is represented by

$$A(t)[1+\,|\,\mathcal{V}\,|\,\cos(2\pi S \sin \Omega t + \text{pha }\mathcal{V})] + \text{receiver noise level.}$$

Although complex coherence is theoretically defined in terms of microscopic properties of a field, while complex visibility is defined in terms of features measurable from a recording, the two quantities are numerically equal if the equipment is symmetrical (equal losses, phase delay and bandwidth in the two arms). In general, coherence has to be calculated from the measured visibility, taking instrumental parameters into account.

NOTATION

In the following, l, m and n are direction cosines with reference to the coordinate system (x, y, z) as usual. We shall also have occasion to use the off-axis angles α, β, γ defined by $l = \cos\alpha$, $m = \cos\beta$, $n = \cos\gamma$. When the (x, y)-plane is the aperture plane, then the angle γ between the ray direction and the z-axis is the angle of incidence. The symbol γ has already been used for complex degree of coherence, but no confusion is expected.

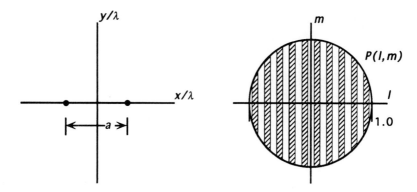

Figure 12-5   The angular spectrum of a pair of pinholes is cosinusoidal with respect to *l*.

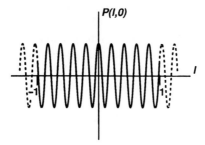

Figure 12-6   The cross section of the angular spectrum along $m = 0$. The broken continuation refers to off-axis angles in the (x, z)-plane whose sine exceeds unity.

INTERFEROMETERS

An opaque screen with two well-separated apertures will produce interference fringes in the Fraunhofer region, if uniformly illuminated.

Pinhole Example

As an example (Fig. 12-5) take two pinholes represented by an aperture field distribution

$$E_y(x/\lambda, y/\lambda) = \left[\tfrac{1}{2}\delta(x/\lambda + \tfrac{1}{2}a/\lambda) + \tfrac{1}{2}\delta(x/\lambda - \tfrac{1}{2}a/\lambda)\right]\delta(y/\lambda).$$

By definition, $E_y(x/\lambda, y/\lambda \supset P(l, m)$, where $P(l, m)$ is the angular spectrum associated with the aperture distribution, and so

$$P(l, m) = \cos[\pi(a/\lambda)l].$$

The stripes so outlined are uniformly wide and continue beyond the unit circle, although the continuation is not shown in the figure. In Fig. 12-6 we have the cross section $P(l, 0)$. In this case the continuation beyond $l = \pm 1$ is shown by a broken line. It simply represents the amplitude of the evanescent wave extending from the (x, y)-plane for each value of l (i.e., of λ/λ_1).

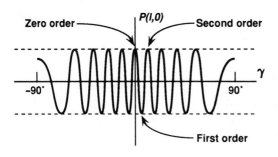

Figure 12-7   The cross section of the angular spectrum plotted against angle of incidence γ.

These diagrams are drawn for $a = 9.7\lambda$, i.e., one-third of a wavelength less than 10 full wavelengths. In an optical application the hole spacing would ordinarily be much greater, and many more stripes would appear within the unit circle.

In Fig. 12-7 we have $P(l, 0)$ plotted against angle of incidence $\gamma = \arccos n$. Here we see that, for small γ, the situation is approximately the same; but as γ increases toward $90°$, the fringes become wider. Finally, for $\gamma > 90°$, $P(l, 0)$ is imaginary and not shown. (Note that $P(l, 0)$ was real for $l > 1$, as shown in the previous figure.)

The widening of the fringes can be understood physically as due to the foreshortening of the aperture distribution as viewed from the direction l. We can work out the angular fringe spacing from the original expression $\cos(\pi a \lambda^{-1} \cos \alpha)$ to be

$$\frac{1}{2} \frac{2\pi}{(d/d\alpha)(\pi a \lambda^{-1} \cos \alpha)}.$$

This gives fringe spacing $= \lambda/a \sin \alpha$, or approximately λ/a in the vicinity of the z-axis. For example, in Fig. 12-8 the first-order fringe forms in the direction where the path difference is λ, i.e., at an angle λ/a to the direction of the zero-order fringe in the direction of the z-axis. The reason for the factor $\frac{1}{2}$ is that two fringes of intensity occur in one period of the fringe pattern.

Example of two rectangular apertures

Consider an aperture distribution

$$E_y(x/\lambda, y/\lambda) = [\mathrm{rect}(x/\alpha) \, \mathrm{rect}(y/\beta)] ** [\tfrac{1}{2}\delta(x/\lambda + \tfrac{1}{2}a/\lambda) + \tfrac{1}{2}\delta(x/\lambda - \tfrac{1}{2}a/\lambda)]\delta(y/\lambda).$$

Again we take the two-dimensional Fourier transform to obtain the angular spectrum automatically. We use the two-dimensional convolution theorem

$$f(x, y) ** g(x, y) \supset F(u, v)G(u, v)$$

and the separable-product theorem

$$f(x)g(y) \supset f(u)G(v)$$

to obtain

$$P(l, m) = [(\alpha\beta/\lambda^2) \, \mathrm{sinc}(\alpha l/\lambda) \, \mathrm{sinc}(\beta m/\lambda)] \cos(\pi a l/\lambda).$$

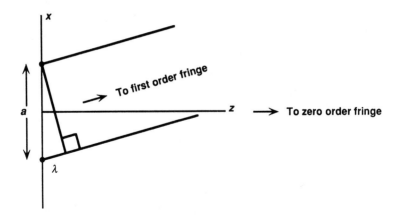

Figure 12-8   The first-order fringe is formed in the direction where the path difference is one wavelength.

RADIO INTERFEROMETERS

Cliff Interferometer

An antenna looking out to sea from the top of a cliff (Fig. 12-9) possesses a deeply dissected directional pattern because of interference between the direct ray and the ray reflected from the sea surface. The antenna can be thought of as possessing an image below the sea and thus, as far as the space above the sea is concerned, the directional pattern is like that of a pair of apertures. In the direction of the broken line (Fig. 12-10) the geometry is such that the two rays arriving at the antenna do so in antiphase, producing a null. Used as a radar, a single antenna elevated above a flat surface, such as an airfield, can provide information about the height of an approaching aircraft, as well as about the range as deduced from echo delay. The height of an aircraft approaching will be characterized by the fading of the echo as a function of range.

Two Steerable Elements

A more controllable radio interferometer consists of a pair of antennas pointed in the same direction. An interference pattern appears within an envelope set by the beam of the incident antennas (Fig. 12-11). As a point source moves through the beam, nulls will occur just as with the two-pinhole interferometer. In this case, however, the antennas are much more directional than a pinhole, and so the interference maxima are constrained to follow the beam pattern as an upper envelope, as shown in Fig. 12-12. An interferometer of this kind permits the angular diameters of small sources to be deduced from the depth of the minima and permits accurate location of a point source in direction to give much better resolution than the beamwidth of a single antenna.

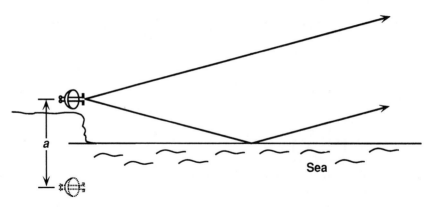

Figure 12-9   An antenna on a cliff and its image in the sea surface constituting a virtual two-element interferometer.

Figure 12-10   Vertical-plane directional pattern of a cliff interferometer comprising a small antenna on a cliff 10 wavelengths high. The first null is at an elevation angle of 2.87 degrees. The intersections of the null cones with a horizontal flight path help fix the altitude of a distant plane (after allowance for atmospheric refraction).

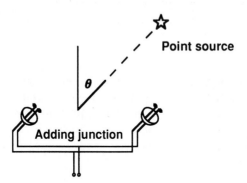

Figure 12-11   A pair of steerable antennas connected by transmission line to a junction where the currents delivered add together.

Multiplying Interferometer

In Fig. 12-13 we see a modification where the adding junction is replaced by a device that multiplies the signals $E_1(t)$ and $E_2(t)$ delivered from the antennas. The output is centered on zero, extended background is suppressed, and undesirable effects of slow-gain drifts are reduced. The simple adding junction, followed by a square-law detector and time averaging produces $\langle E_1^2 + E_2^2 \rangle = \langle E_1^2 \rangle + \langle E_2^2 \rangle + 2\langle E_1 E_2 \rangle$, whereas the multiplier followed by time averaging just produces $\langle E_1 E_2 \rangle$, free from the intensity terms that are devoid of correlation information. Multiplying interferometers were first constructed by

**Response to point
source (power)**

**Response to non-point
source**

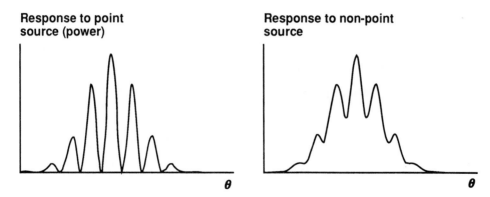

θ θ

Figure 12-12 Power response of a two-element adding interferometer to a point source
(left) and a nonpoint source (right) when the source is moving but the steerable elements
are not tracking. The response to the extended source exhibits minima that are not nulls
and maxima that are weaker than for a point source.

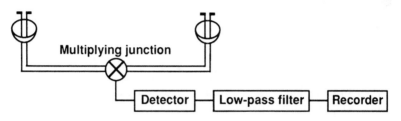

Figure 12-13 A multiplying interferometer, the most usual configuration.

Ryle (1952) and Mills and Little (1953). Ryle's method was to alternate the phase of
$E_2(t)$ by mechanically commutating a half-wavelength of transmission line in and out of
one arm of the interferometer; Mills and Little reversed the phase electronically at the
intermediate frequency. Then, with the previously understood adding junction, square-
law detection, and time averaging, one obtained $\langle (E_1 \pm E_2)^2 \rangle = \langle E_1^2 + E_2^2 \rangle \pm 2\langle E_1 E_2 \rangle$.
A phase-sensitive detector then isolated the product term.

RATIONALE BEHIND TWO-ELEMENT INTERFEROMETER

The virtues of a two-element radio interferometer are that it permits (a) the angular location and (b) the angular diameter of a compact source to be determined with a resolution
comparable to what would be obtained from one large, more expensive, antenna of dimension approximating the spacing of the interferometer.

   In the optical-wavelength range the Michelson stellar interferometer exemplifies
the behavior under discussion. It permits angular-diameter determination but not angular
position. The reason for this deficiency seems to be atmospheric variability, which keeps

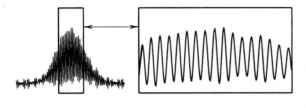

Figure 12-14 An interferometer record (left) and a magnified portion around the interference maximum illustrating how the presence of noise hinders identification of the zero-order fringe. The standard deviation of the noise is only about four per cent of the envelope maximum and of course much fainter sources than shown here can be detected.

the fringes in motion, making them difficult to follow. But, in fact, difficulty with angular location sets in when the size of a single aperture becomes too small compared with the spacing for it to be possible to identify the central fringe. The radio interferometer also breaks down in the same way (Fig. 12-14). On Fig. 12-12 there is no trouble locating the central fringe, but in Fig. 12-14, because of noise, the central fringe is not the largest one and cannot be identified. We are limited in angular-location determination to some fraction of the beamwidth, in practice often of the order of one-tenth. This means that location difficulties set in if the spacing of the elements is not kept down to, say, ten element diameters.

Bandwidth

When the wavelength is changed, all the fringes are moved except the zero-order fringe, which by definition is the one where the path difference is zero. An interferometer pattern such as Fig. 12-6 then becomes a superposition of patterns that are stretched or compressed according as the wavelength is longer or shorter. Since only the zero-order fringe (referred to in optics as the white-light fringe) remains in the same direction, it alone remains at full strength; the other maxima are reduced and the minima are filled in. If the bandwidth is $\Delta\lambda/\lambda$ (and the bandshape is rectangular), then at a certain value of l, the path difference at one band-edge will have slipped a whole wavelength. In this direction, the fringes will be totally washed out. Since the field pattern is described by $\cos[\pi(a/\lambda)l]$, this means

$$\frac{\pi a l}{\lambda - \frac{1}{2}\Delta\lambda} - \frac{\pi a l}{\lambda + \frac{1}{2}\Delta\lambda} = 2\pi.$$

Solving for l,

$$l = \frac{2\lambda}{a}\frac{\lambda}{\Delta\lambda}$$

$$= 2 \times \text{fringe spacing} \times \frac{\lambda}{\Delta\lambda}.$$

Thus, for $\Delta\lambda/\lambda = 1/1000$, the 2000th fringe is washed out. (We are counting fringes according to intensity $\cos^2[\pi(a/\lambda)l]$; thus the first fringe is the antiphase fringe at $l = \lambda/a$ and corresponds to a path difference of one wavelength.)

This phenomenon permits the measurement of the line-width of optical spectral

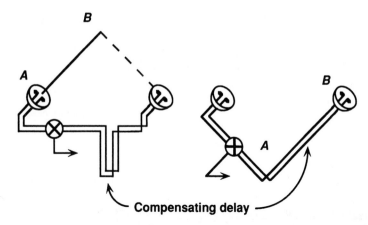

Figure 12-15   The compensating delay necessitated by observations made away from the median plane of an interferometer.

lines and needs to be considered in radio interferometers, where narrow bandwidths are involved.

In Fig. 12-14, if the path difference to the two antennas was 1000 wavelengths, then a (rectangular) bandwidth as wide as $\Delta\lambda/\lambda = 1/1000$ would cause the interference patterns for the two band-edges to separate by one cycle of relative phase, thus destroying the pattern. In order to use such an interferometer, pointed well away from the median plane (plane normal to the baseline joining the antennas), it would be necessary to convert the strongest fringe (the fringe in the center of the beam of a single element) to zero order by inserting a compensating delay. This could be in the form of a piece of air-filled transmission line equal in length to the path difference AB, as in Fig. 12-15. However, as the antennas move across the sky, the compensating delay has to be altered at appropriate intervals.

Since the angular spacing of the fringes becomes finer as the antenna elements are separated, higher precision of angular location of a source becomes possible. However, the precision is limited by noise, which introduces ambiguity into the determination of absolute angular position (Fig. 12-14). Increasing the element width for a given element spacing makes the envelope maximum sharper, thus facilitating identification of the zero-order fringe, and at the same time helps by increasing the signal-to-noise ratio.

APERTURE SYNTHESIS (INDIRECT IMAGING)

Once it is understood that the elementary measurable entity in the radiation field is the complex degree of coherence $\gamma(,)$ between two points, then it becomes conceivable that the high-resolution information extractable by a large antenna could also be extracted by small low-resolution antennas.

The procedure would be to have two small antennas, one fixed at the origin and the other movable, as in Fig. 12-16. At each vector spacing the complex degree of coherence

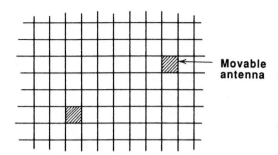

**Movable
antenna**

Figure 12-16 Aperture synthesis by
sequential measurement of coherence values
between a fixed and a movable antenna, a
diagram used by Ryle (1962) when the term
aperture synthesis was introduced.

γ would be measured. Then, by occupying all the vector spacings embraced by a given
large antenna, we would get all the available information out to the boundary defined by
the maximum spacing reached in each direction. The Fourier transform of the coherence
values for the different antenna spacings, obtained sequentially, would yield the source
distribution as a function of direction. The angular resolution would be set by the size of
the boundary reached by the movable antenna.

By contrast, a single large antenna would obtain a reading on one resolution element
at a time. However, all the resolution elements would have to be looked at in turn in order
to build up the source distribution.

This leads to a very interesting question. Does it take as long to map a source with
a single pencil-beam antenna as by aperture synthesis? The mapping time very reasonably
is inversely proportional to the collecting area invested in. Another factor allows for the
fraction of the field of view occupied by the source distribution: a pencil-beam map does
not require dwelling on empty parts of the field, whereas aperture synthesis inevitably
attends to the whole field accepted by its single elements. Other practical considerations in
the comparison concern signal-to-noise ratio and the time taken to alter the configuration
of an interferometer.

The notion that a two-element interferometer responded to one Fourier component is
due to McCready, Pawsey, and Payne-Scott (1947) in Australia, where they made interferometric determinations of the angular diameters and positions of radio sources associated
with sunspots on the face of the sun. The term *aperture synthesis* was introduced and the
technique was first exploited extensively by Ryle (1962) and his associates in Cambridge,
U.K. Intervening history is recounted by Scheuer (1984) and Bracewell (1984).

Large interferometers of a variety of designs that have been set up in many locations
around the world now represent mature technology from which images of extraordinary
quality are flowing. The associated techniques are tailored in accordance with the purposes
of the instruments. Many niches are occupied in the multidimensional space of time, frequency, directional resolution, polarization, and siting. In addition, instrumental design has
often been specialized to suit the endless variety of astronomical and terrestrial sources.

Open worldwide competition moderated by the International Scientific Radio Union
and the International Astronomical Union, beginning in the forties, propelled radio astronomy into the forefront of indirect imaging. However, the principles are fundamental and,

with adaptations, are applicable in optics and acoustics and to other subjects, including seismography and oceanography, where wave propagation is important.

In the radio spectrum intercontinental links were achieved in the sixties with techniques permitting the transfer of timing accurate to a fraction of a microwave period over transoceanic distances. An immediate fruit of the long baselines reached was the discovery of numerous sources too small to be resolved ("point" sources) but still powerful. The angular precision permitted, comparable with the angular diameter of the source, then allowed the detection of baseline changes, over time, associated with continental drift. The importance of geodetic mensuration to such precision led directly to the substitution of strong earth-satellite sources for the exploitation of interferometric technique for everyday application to surveying and navigation. Satellites of the U.S. Global Positioning System and its Russian counterpart have revolutionized these two fields. The remarkable precision achieved is sufficient to determine the height differences between the wingtips and tail of an aircraft, and hence its attitude. This information, together with the altitude difference of the aircraft and the landing field, is adequate to control the attitude and glide angle and thus to replace existing and previously planned blind landing systems. No doubt further everyday applications will stem from indirect imaging, a subject to be watched. The well-documented origins of these developments add a chapter to the stream of discoveries and applications from astronomy, the history of which is well worth contemplating (Bracewell, 1992). An excellent source for following up the topics of this chapter has been provided by Thompson et al. (1986).

LITERATURE CITED

M. Born and E. Wolf (1959), *Principles of Optics*, Pergamon, New York.

R. N. Bracewell (1958), "Radio interferometry of discrete sources," *Proceedings of the Institute of Radio Engineers*, vol. 46, pp. 97–105.

R. N. Bracewell (1961), "Interferometry and the spectral sensitivity island diagram," *IRE Transactions on Antennas and Propagation,* vol. AP-9, pp. 59–67.

R. N. Bracewell (1984), "Early work on imaging theory in radio astronomy," in W. T. Sullivan, III, ed., *The Early Years of Radio Astronomy*, Cambridge University Press.

R. N. Bracewell (1992), "Planetary influences on electrical engineering," *Proc. IEEE*, vol. 80, pp. 230–237.

A. A. Michelson (1902), *Light Waves and Their Uses*, The University of Chicago Press, Chicago.

B. Y. Mills and A. G. Little (1953), "A high resolution aerial system of a new type," *Aust. J. Sci. Res.*, vol. 6, *p.* 272–278.

Lord Rayleigh (1892), "On the interference bands of approximately homogeneous light; in a letter to Prof. A. Michelson," *Philosophical Magazine*, vol. 34, pp. 407–411.

J. A. Roberts (1984), *Indirect Imaging*, Cambridge University Press, pp. 177–183.

M. Ryle (1952), "A new radio interferometer and its application to the observation of weak radio stars," *Proc. Roy. Soc.*, **A**, vol. 211, pp. 351–375.

P. A. G. Scheuer (1984), "The development of aperture synthesis at Cambridge," in W. T. Sullivan, III, ed., *The Early Years of Radio Astronomy*, Cambridge University Press.

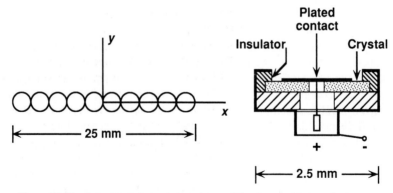

Figure 12-17   An array of ultrasonic transducers of the piezoelectric type shown.

A. R. THOMPSON, J. M. MORAN AND G. W. SWENSON, JR. (1986), *Interferometry and Synthesis in Radio Astronomy*, John Wiley & Sons, New York.

F. ZERNICKE (1938), "Concept of degree of coherence and its applications to optical problems," *Physika*, vol. 5, pp. 785–795.

PROBLEMS

12–1. *Indirect imaging.* To what does the concept of complex degree of coherence apply? (In other words, what is it a property of: a system, a source, an antenna, an observing procedure, or something else, or some combination?). ☞

12–2. *Ultrasonic array.* An ultrasonic array of piezoelectric crystals consists of 10 elements each 2.5 mm in diameter and a total length of 25 mm (Fig. 12-17). The crystals are excited at 3.5 MHz in such a way that the flat circular face of each crystal vibrates in a direction normal to the surface. The velocity of sound waves in air is 330 m s^{-1} and in muscle 1570 m s^{-1}. The attenuation constant in air is 12 db cm^{-1} and in muscle 2 ± 1 db cm^{-1}, depending whether the direction is across or along the fibers.

   The excitation can be pulsed and the crystals exhibit reciprocity; i.e., a 3.5-MHz vibration falling on the flat face will produce a 3.5-MHz terminal voltage. This is all that is known about the device, except that it is in use for imaging the interior of the body. See if you can figure out how it works. For instance, can you estimate
   (a) the dimensions of the volume it can explore,
   (b) what resolution would be achievable in depth and transversely,
   (c) whether the elements are excited coherently in phase and, if so, how? ☞

12–3. *Two-element interferometer.* An opaque plane has two circular holes of diameter 1000 wavelengths separated by 4000 wavelengths as shown in Fig. P12-3. They are uniformly illuminated in amplitude and phase.
   (a) Write down the angular spectrum $P(l, m)$ as a function of the direction cosines l and m.
   (b) Make a sketch on the (l, m)-plane, reaching out to a radius beyond the first and second null circles, that is adequate to convey the chief features (indicate the location of null loci by full lines and crosshatch the antiphase regions. For quantitative purposes you might like to recall that jinc $(1.22) = 0$.

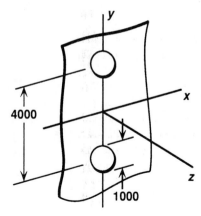

Figure 12-18   Arrangement of interferometer (dimensions in wavelengths).

12–4. *Interferometer design.* A demonstration of Young's interference fringes was to be prepared by allowing light to pass through a plate with two pinholes (0.2 mm in diameter, spaced 1 mm between centers) and placed 1 m from a screen. Unfortunately, no fringes could be seen. Things were rearranged by mounting the plate in an aperture in a garage door so that direct sunlight could fall on the holes. Even when an orange photographic filter was used (peak wavelength about 0.5 μm), no fringes were seen.

    (a) Work out the fringe separation in mm to be expected if a plane wave of $\lambda = 0.5$ μm fell normally on the apparatus.

    (b) Over an area of what size would the fringes be expected?

    (c) If you were Thomas Young, what advice would you give to the experimenter?

12–5. *Infrared interferometer for planet detection.* It would be of great interest if an image could be formed of the planetary system of a nearby star, because so far no planets have ever been seen other than those in our own solar system. As we are now able to place a telescope outside the earth's atmosphere, which has set a limit to angular resolution in the past, let us consider a hypothetical planetary system 10 parsecs away (3.1×10^{17} m) which is otherwise the same as ours. The angular diameter of the star would be $2 \times 6.96 \times 10^8 / 3.1 \times 10^{17} = 4.5 \times 10^{-9}$ radians = 0.0009 arcseconds, while the diameter of the planet would be about 10 percent of that of the star. The planet most likely to be detected, the Jupiter-like planet, could reach an angular separation of $7.78 \times 10^{11} / 3.1 \times 10^{17} = 2.52 \times 10^{-6}$ radians = 0.52 arcseconds or about 500 star diameters.

    An infrared interferometer has been proposed which would have an interference maximum lying on the planet but a null lying on the star. What would the baseline length of the interferometer have to be if the wavelength were chosen at 40 μm?

12–6. *Ideas for more resolution.* Here are two ideas for improving the resolution of a given two-element interferometer.

    (a) Frequency doublers are inserted at the two elements. (Fig. 12-18). (One way of doing this is to use a detector diode and then filter out all but the double-frequency band.) It is claimed that when the signals arrive at the multiplier junction, an interferogram is recorded whose lobes are twice as narrow as before. To obtain such narrow lobes would normally require going to a baseline twice as long.

    (b) The second plan is go to extremely wideband operation, say one octave. Then the element spacing in wavelengths actually covers a 2-to-1 range, thus furnishing all the

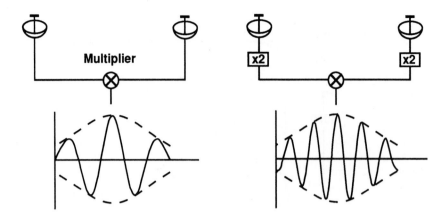

Figure 12-19 Does a two-element interferometer (right) with frequency doubling yield narrower fringes and higher resolution?

information that normally would be obtained by stationing one of the elements at many places in turn, at distances up to twice the minimum spacing. Comment on these plans one by one. ☞

12–7. *Switched interferometer.* An antenna array consists of 16 parabolic reflectors 33.5λ m in diameter spaced 83.68λ in a straight line.

(a) What is the approximate strength of the first-order grating lobe relative to the main lobe?

(b) An effort to reduce the grating lobe is to be made by placing an additional antenna at a distance 41.84λ from an existing end antenna. The original array will become one element of an interferometer, and the additional antenna will be the other element. Will this be effective?

12–8. *Sampling coherence.* The angular diameter of the sun as seen at a wavelength of 600 nm is 0.01 radian. Because the brightness distribution over the sun has a sharply cut off boundary, it would be sufficient to sample the coherence of the solar radiation between points whose spacing D need not be closer together then K wavelengths. What is the value of K, and what is the value of the critical spacing D in millimeters?

12–9. *Unsymmetrical interferometer.* Complex degree of coherence is deducible from an interferogram exhibiting maxima I_{max} and mimima I_{min} by the relation $\gamma = (I_{max} - I_{min})/(I_{max} + I_{min})$, provided that the elements of the interferometer contribute equal power.

(a) If the power available from one element exceeds that from the other, when they are both pointed at the same source, by a ratio R, show that

$$\gamma = \frac{I_{max} - I_{min}}{I_{max} + I_{min}} \frac{\sqrt{R} + \sqrt{1/R}}{2}.$$

(b) If the ratio R is two, show that the correction factor makes a change of less than 6 percent.

13

Restoration

Whenever a time-varying quantity has to be measured, there is an inevitable blurring due to the nonzero time interval necessary to make a single measurement. Consequently, the measurement relating to a given instant always lumps together values that occurred during the measurement interval. If the time-varying quantity is varying slowly, or if the time resolution is short, the measurement may be very good; but some degree of smoothing is in principle always present. That means that a measured waveform never faithfully represents the original quantity. Therefore, one may ask what correction has to be applied to the measurement. In the development given below it is supposed that convolution is involved, but it is understood that convolution is merely an important case of the more general linear functional. Measurements may also involve a little nonlinearity and a little hysteresis. Even so, as will be seen, there is enough complication in the simple presentation below to suffice for a first study. A further limitation may be mentioned. In some subjects, including astronomy, meteorology, and geophysics, the measurements, or observations, may be all we know about the time-varying quantity; in fact the purpose of the observations may be to find out what is there. In such a case the "time-varying quantity" is just a concept without any reality of its own; only the measurements exist and are available to work with. In experimental subjects, as distinct from the observational, the underlying quantity may be verifiable by alternative methods, but in the final analysis measurement is always conducted to a finite resolution. Procedures for combating the smoothing effect of instrumental intervention are known as *restoration*.

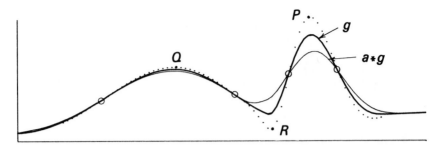

Figure 13-1 A given function g, which is derived from an unknown original f by smoothing, changes to $a * g$ (thin line) if subjected to further smoothing. Reversing the changes leads to a first approximation (dotted line) to the original unsmoothed function.

These remarks are not restricted to time variation, or to one dimension, but one dimension is treated first; the independent variable is taken as x, and the one-dimensional Fourier transform variable is s. We begin with the elegant procedure of successive substitutions, which leads to the useful concepts of the principal solution and invisible distributions and is applicable to a variety of special cases.

RESTORATION BY SUCCESSIVE SUBSTITUTIONS

Let $g(x) = a(x) * f(x)$, where $g(x)$ is given and $f(x)$ is to be found, subject to the caveats mentioned above. We will bear in mind that the assumption of a convolution relation may require verification in any particular application to actual data. The instrumental response $a(x)$, e.g. the response of an antenna, is also taken as given and is normalized so that $\int_{-\infty}^{\infty} a(x)\, dx = 1$. Then $A(0) = 1$, where $A(s)$ is the Fourier transform of $a(x)$.

The method of finding $f(x)$ by successive substitutions starts from a first approximation $f_1(x)$ to $f(x)$. Here is an argument leading to such a first approximation. If we subject g to further smoothing with the antenna pattern, the change produced is $a * g - g$. Thus, where there is an isolated peak in g, the further smoothing will produce a negative change, making $a * g$ fall below the peak. So we may argue that f must have passed *above* the peak in g—for example, at point P in Fig. 13-1. If we place P as far above g as $a * g$ falls below g, we have our approximation $f_1(x)$. The dotted line shows other points arrived at in the same way. Note that the correction is stronger at the sharp peak P than at the blunt peak (point Q). Points of inflection are marked with small circles. At such points the correction $g - a * g$ changes sign, so the dotted line cuts through at the inflections. Thus troughs are deepened (point R) as peaks are sharpened.

We can say that

$$f_1 = g + (g - a * g) = g + \epsilon_1,$$

where $\epsilon_1 = g - a * g$. How can we test the quality of the approximation f_1? Well, if f_1 were exactly right, we could smooth it with $a(x)$, and we would find that $a * f_1$ agreed exactly with g. But if there is a discrepancy $g - a * f_1$ when this test is performed, then

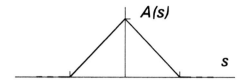

Figure 13-2   A triangular transfer function plotted against spatial frequency s.

the discrepancy can be used as a further correction leading to a second approximation

$$f_2 = f_1 + (g - a * f_1)$$

and in general

$$f_n = f_{n-1} + (g - a * f_{n-1}).$$

We stop when we judge the discrepancy between the given g and the smoothed function $a * f_n$ to have fallen below some tolerable error level. Of course, we do not know yet that the corrections will get smaller. We shall find that sometimes they do, and sometimes not. Even if they do get smaller, the sequence f_n may not converge to a limit. And if there is a limit, we do not know that it will equal f.

As mathematicians we should now investigate whether the sequence $\{f_n\}$ converges and, if so, to what. Taking Fourier transforms gives $G = AF$. Therefore

$$\frac{G}{A} = \frac{G}{1 - (1 - A)}$$

$$= G[1 + (1 - A) + (1 - A)^2 + \cdots]$$

$$= G + (1 - A)G + (1 - A)^2 G + \cdots.$$

Retransform the first two terms of the final expression to get $g + (\delta - a) * g = 2g - a * g$. We recognize this as f_1. Additional terms generate f_2, f_3, etc. We have thus converted the convergence problem in the space domain to one in the transform domain, but the new problem merely involves a geometric series, which we understand. Noting that the geometric ratio is $1 - A$, we know that the sequence of transforms converges, provided

$$|1 - A(s)| < 1.$$

We deduce that the series of approximations may or may not converge, and that whether it does or not depends on a property of the instrumental response $a(x)$. Let us try a particular case such as $a(x) = \text{sinc}^2 x$, where

$$A(s) = (1 - |s|)\,\text{rect}(s/2).$$

We see from Fig. 13-2 that $|1 - A| < 1$ for all s except where $|s| > 1$. But g contains no frequencies $|s| > 1$. Therefore we may expect convergence when applying successive substitutions to data obtained with a sinc^2 instrumental response. An example would be a uniformly excited sensor array.

What does the sequence converge to? We see that

$$F_1(s) = G + (1 - A)G = G + |s|G, \quad |s| < 1.$$

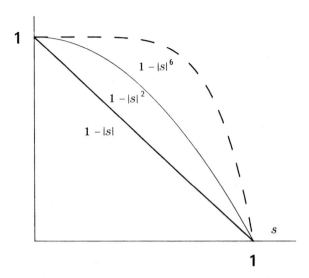

Figure 13-3 Effective transfer functions after one (thin line) and five (broken line) stages of restoration. The limiting effective transfer function is unity where $-1 < s < 1$ and zero elsewhere, perfection for the low frequencies and rejection for the high.

Since $G = AF = (1 - |s|)F$,

$$F_1(s) = (1 - |s|^2)F(s).$$

For the nth stage, $F_n(s) = (1 - |s|^{n+1})F(s)$ as illustrated in Fig. 13-3. The factor $1 - |s|^{n+1}$ describes the way in which $F_n(s)$ falls below the true value of $F(s)$ over the range $|s| < 1$ and this factor gets closer and closer to unity as $n \to \infty$. In the limit $\lim_{n\to\infty}[1 - |s|^{n+1}] = 1$, $|s| < 1$. Thus $\lim F_n(s)$ is the same as $F(s)$ up to the cutoff and zero beyond.

We call this limiting function $P(s)$. Its transform $p(x)$, which is called the *principal solution,* was introduced in connection with radio telescopes (Bracewell and Roberts, 1954) and is discussed further below. Any function $i(x)$ whose Fourier components are all at higher frequencies than the cutoff is an "invisible distribution," because there is no response to convolution with $a(x)$—i.e., $a(x) * i(x) = 0$. It follows that all solutions are expressible in the form $p(x) + i(x)$, including the original function $f(x)$.

RUNNING MEANS

When a sound track on film is scanned by a slit, the response is degraded by the fact that a compromise has to be made between the fidelity achievable with a narrower slit and the speed of scanning achievable with the larger signal from a wider slit. The nature of the degradation is describable, to a good approximation, in terms of convolution with a rectangle function. The read head of a magnetic recording device is similar except that not such a sharp rectangular envelope is realized. The operation performed by a sound track in going from the function $f(x)$ on the record to the "given" function $g(x)$ available as output is known traditionally as taking *running means*. The term *weighted running means*

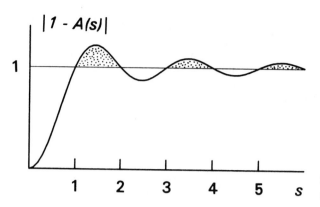

Figure 13-4 Convergence condition is not met in the stippled zones.

refers to weighting which, rather than being uniform as with a strict rectangle function, is nonuniform.

Running means are much used in statistical subjects such as meteorology to remove unwanted fluctuations. For example, if you need to know how much rain will fall on June 15, the 10-day running mean for last year may be more suitable than the average of all previously recorded June 15s. The running mean over a segment of unit length is defined by

$$g(x) = \text{rect } x * f(x).$$

Many other subjects will be found where convolution with a rectangle function arises. For example, a commercial recording voltmeter that prints out a voltage at regular intervals probably contains a voltage-to-frequency converter and a counter that counts the number of cycles fitting within the chosen averaging interval T. Its output therefore consists of samples of the running mean of the voltage to be recorded; the sampling interval is equal to, or perhaps a little shorter than, the averaging interval T. We now consider the problem of recovering $f(x)$ when $g(x)$ is given. The method of successive substitutions yields immediate insight. We take T to be unity.

In this case $a(x) = \text{rect } x$ and $A(s) = \text{sinc } s$. The graph of $|1 - A(s)|$ shown in Fig. 13-4 demonstrates that the convergence condition $|1 - A| < 1$ is not met. Therefore, you might not hope to recover the original data $f(x)$ given the running mean $g(x)$. However, Fig. 13-1 and the verbal argument accompanying it would sound just as convincing in this application as before. And in fact the method works in practice, giving what is technically known as an asymptotic series. What happens is this. The first correction improves the spectrum in the frequency band $0 < |s| < 1$ while making it worse in $1 < |s| < 2$. There will be a net improvement if $G(s)$ has more content in the first band than in the second. So the approximations will get better for a certain number of steps but may ultimately diverge.

Invisible distributions exist whose Fourier transforms are of the form $\Sigma_k a_k \delta(s - k)$, where k is any nonzero integer. Such spectral lines will fall on the nulls of $\text{sinc } s$ and so will not be seen. It follows that all the invisible distributions are periodic with unit period. It is also easy to see directly why such a periodic function when convolved with a rectangle function of unit width produces no response; any periodic function, whether sinusoidal or not, if averaged over one period, yields zero.

There are many situations (for example, daily rainfall in a Mediterranean climate where there are rain-free months) where an additive periodic term can be excluded on prior knowledge. However, in general, it is vital to be aware that the inversion of running means will not recover a unit-period term that was originally present.

EDDINGTON'S FORMULA

In connection with the study of optical spectral lines Eddington (1913) proposed the following formula for $f(x)$ the corrected line profile in terms of the $g(x)$ which is observed when the instrumental profile is proportional to $\exp(-x^2)$:

$$f(x) = g(x) - (1/4)g''(x) + (1/32)g''''(x) - \cdots.$$

This is the earliest example of correction for instrumental broadening. It is particularly important, because instruments often do have a Gaussian response, a fact which can be explained in terms of the central limit theorem. By the use of a narrow spectral line produced in the laboratory one first determined the instrumental profile $a(x)$. A spectrograph (Fig. 13-5) contains several components that contribute to the broadening of the response. For example, the entrance slit S, not being of zero width, guarantees that even strictly monochromatic light will have some spread in the focal plane F due to the fact that elements of the slit are located in different directions. Diffraction makes a contribution. There is some irregular refraction at the prism surfaces because the surfaces are not plane, and there is internal refraction because the prism material is not homogeneous. A strictly parallel beam of light is opened up into a small cone by scattering from discrete irregularities, such as minute bubbles, that afflict optical components. Refraction in the telescope tube due to the temperature gradient in the air adds further to the broadening, as do mechanical vibration and tracking error during the time of exposure if the source of light is changing direction. If the image is caught on a photographic plate, there is halation in the emulsion. Finally, if a chart record is to be produced for numerical analysis, light transmitted through the photographic record must be scanned with a microdensitometer whose aperture is of nonzero width, detected on a photosensitive cell of nonzero dimensions, and the resulting electric signal must be passed through an amplifier with nonzero time constant to drive a pen subject to inertia and friction making an ink line of variable weight on paper advancing at a nominally uniform speed. After the chart is torn off the recorder, it changes dimensions anisotropically, having been previously stored on the roll under longitudinal tension which is now released. The central limit theorem leads us to expect that, to some degree of accuracy, the overall instrumental response will be Gaussian, as is indeed observed.

Spectrographs (Fig. 13-5) may be made with mirrors instead of lenses, gratings instead of prisms, and photomultipliers instead of cameras, so that the phenomena listed above are not of general significance. When a complete instrument is specified, however, it is usual to find many causes contributing to spectral line broadening and tending to give

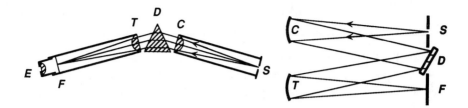

Figure 13-5 The light from slit S after collimation by a lens or mirror C falls as a parallel beam on a prism or grating D, which disperses the light while retaining parallelism. A telescope or camera T focuses any parallel beam to a point on a focal plane F, where images of the slit at those wavelengths present may be viewed through an eyepiece or recorded photographically.

the instrument a Gaussian response to an infinitesimally narrow spectral line. Eddington's procedure was to take the instrumental profile determined in the laboratory and fit it to a Gaussian curve $\exp(-x^2)$ by choosing the unit of x to suit. For normalization to unit area we must deal with $\pi^{-1/2}\exp(-x^2)$.

Probably only one stage of correction was contemplated in practice in which case the correction term $-0.25g''(x)$ would be proportional to the second derivative of the data and yield a positive correction when the curvature was downward. Figure 13-1 indicates that the correction passes through zero at or near points of inflection where the second derivative is zero; has opposite sign to the second derivative, and is greater where the curvature is greater (compare the corrections at the two peaks). Where there is very little curvature, there is very little error committed by the spectrograph.

Eddington's formula may be derived as follows. Let

$$a(x) = w^{-1}\exp(-\pi x^2/w^2) \quad \text{and} \quad A(s) = \exp(-\pi w^2 s^2).$$

Since $A(s) \neq 0$,

$$F(s) = G(s)/A(s) = G(s)[1 + \pi w^2 s^2 + (\pi w^2 s^2)^2/2 + \cdots]$$

and thus

$$f(x) = g(x) - (w^2/4\pi)g''(x) + (w^4/32\pi^2)g''''(x) - \cdots.$$

We have used the derivative theorems $f'(x) \supset i2\pi s F(s)$, $f''(x) \supset (i2\pi s)^2 F(s)$, ... The coefficients reduce to Eddington's 1/4, 1/32, etc., when $w = \pi^{1/2}$, i.e., when $a(x) = \pi^{-1/2}\exp(-x^2)$.

Eddington's formula became well known in spectroscopy (Fig. 13-6), but it is doubtful whether those who used it tried to carry out second-order differentiation, which is particularly susceptible to the effects of noise, even small amplitude noise. They probably used their common sense and determined the derivatives graphically. This suggests a more appropriate theory that is applicable to the actual data manipulation that is to be carried out in practice.

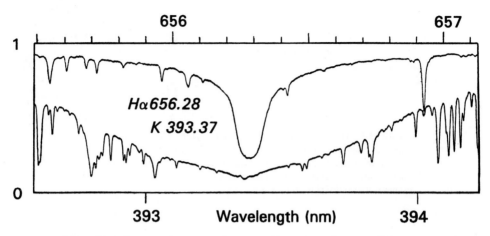

Figure 13-6 Some genuine spectra (the $H\alpha$ and K Fraunhofer lines of sunlight) are presented here to offset any false impressions created by imagined spectra.

FINITE DIFFERENCES

Let the finite difference over interval a be defined by

$$\Delta_a f(x) \equiv f(x + a/2) - f(x - a/2) \supset 2i \sin \pi a s F(s);$$

also,

$$\Delta_a^2 f(x) \supset -4 \sin^2 \pi a s F(s).$$

We see that finite differencing may be regarded as convolution with a pair of unit impulses of opposite sign. Thus

$$\Delta_a f(x) = [\delta(x + \tfrac{1}{2}a) - \delta(x - \tfrac{1}{2}a)] * f(x)$$

and

$$\Delta_a^2 f(x) = [\delta(x + a) - 2\delta(x) + \delta(x - a)] * f(x).$$

If we were carrying out the convolution numerically the operations of differencing would be performed with the sequences $\{1 \;-1\}$ and $\{1 \;-2 \;1\}$. Although the operation of second differencing takes account of highly localized information from $f(x)$, the effect on the Fourier transform is pervasive. The factor $\sin^2 \pi as$ has the effect of reducing almost every Fourier component in magnitude. However, the phase is left unchanged and no component is reversed in phase, whereas phase reversal of some components does take place under first differences. The zero-frequency component is entirely removed, and so are a series of higher frequencies. These properties may be readily understood by considering sine waves of different frequencies and noting directly how convolution with $\{1 \;-2 \;1\}$ gives a result that is sometimes zero, sometimes unchanged in magnitude, sometimes reduced, and never altered in phase.

We may also note that $\{1 \;-2 \;1\} = \{1 \;-1\} * \{1 \;-1\}$. Now, in two dimensions convolution with the same arrays of coefficients may be applied in any direction and the result will be to give first and second differences in those directions. However, more complicated things are conceivable. For example, one may convolve with the array

$$\begin{bmatrix} 1 \\ -1 \end{bmatrix} ** [1 \quad -1] = \begin{bmatrix} 1 & -1 \\ -1 & 1 \end{bmatrix}.$$

The effect of this will be to measure a quantity related to $\partial^2 f/\partial x\, \partial y$. Noting that

$$\begin{bmatrix} 1 \\ -2 \\ 1 \end{bmatrix} ** [1 \quad -2 \quad 1] = \begin{bmatrix} 1 & -2 & 1 \\ -2 & 4 & -2 \\ 1 & -2 & 1 \end{bmatrix}.$$

we arrive at a two-dimensional array that will respond to curvature in both coordinate directions.

FINITE DIFFERENCE FORMULA

Now try for a solution in the form

$$g(x) + \psi_2 \Delta_a^2 g(x) + \psi_4 \Delta_a^4 g(x) + \cdots \supset G(s) - 4\psi_2 \sin^2 \pi a s\ G(s)$$
$$+ 16\psi_4 \sin^4 \pi a s\ G(s) + \cdots.$$

For the Gaussian case with

$$a(x) = w^{-1} \exp(-\pi x^2/w^2) \quad \text{and} \quad A(s) = \exp(-\pi w^2 s^2)$$

and using the expansion

$$\exp \theta^2 = 1 + \sin^2 \theta + (5/6) \sin^4 \theta + (61/90) \sin^6 \theta + \cdots, \quad \theta^2 < (\pi/2)^2$$

we find

$$F(s) = [1 + \sin^2(\pi^{1/2} ws) + (5/6) \sin^4(\pi^{1/2} ws) + \cdots]G(s).$$

Hence the finite-difference formula, deduced by matching transforms, is

$$f(x) = g(x) - (1/4)\Delta_a^2 g(x) + (5/96)\Delta_a^4 g(x) - \cdots$$

The result very closely resembles Eddington's formula but describes explicitly an operation of finite differencing. The differences are to be taken over an interval a that may be related in a definite way to the width of the instrument profile by equating $\pi a s$ to $\pi^{1/2} ws$. Thus we find

$$a = \pi^{-1/2} w.$$

Moderate noise now does not vitiate the procedure, because the finite difference is subject to an error not much greater than the error in $g(x)$ itself. The first correction term produces only minor degradation as compared with differentiation, where noise will be magnified in proportion to the frequency of the noise component and in proportion to frequency squared for the second derivative and so on.

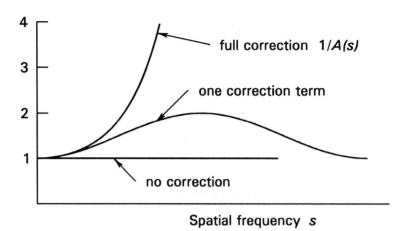

Figure 13-7 For $a(x) = \pi^{-1/2} \exp(-x^2)$ the inversion factor $\exp(\pi^2 s^2)$ for full correction is compared with $1 + \sin^2 \pi s$, which gives the effect of a single correction term based on the second difference. The normal operating region is at small values of s.

Light is shed on the sensitivity to error by considering the correction operation in the frequency domain. The factor $\sin^2 \pi a s$ and its higher powers never rise above unity no matter how large s becomes, whereas the factor $\pi w^2 s^2$ associated with the second derivative amplifies by indefinitely large amounts.

Figure 13-7 compares the inversion factor $[A(s)]^{-1} = \exp(\pi^2 s^2)$ corresponding to $a(x) = \pi^{1/2} \exp(-x^2)$ with the factor $1 + \sin^2 \pi s$ that describes what happens when the finite difference procedure is truncated to just two terms. In this comparison the equivalent width w of the instrumental response is $\pi^{1/2}$ and $a = 1$.

CHORD CONSTRUCTION

From the definition of finite difference it follows that the second finite difference is expressible as

$$\Delta_a^2 f(x) = f(x + a) - 2f(x) + f(x - a).$$

By presenting this in the form $2\{[f(x+a) + f(x-a)]/2 - f(x)\}$ we may give an interpretation in diagrammatic form as follows. In Figure 13-8, the point B represents the ordinate $f(x)$ and the point D represents the ordinate $[f(x+a) + f(x-a)]/2$ of the midpoint of the chord. Thus twice the sagitta BD represents the second finite difference at x. A correction equal to minus one-quarter of the second difference is half of BD in magnitude, and, when applied, leads to the result indicated by the small circle.

By drawing a chord with horizontal span $2a$, one may therefore evaluate the first correction graphically. The graphical procedure is quite practical, but the greatest benefit of the construction is the ability it confers to judge by eye the character and general

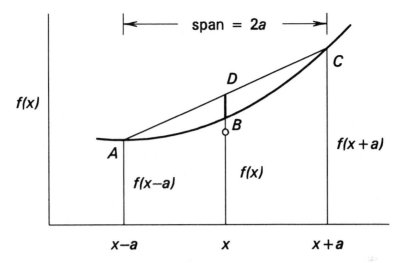

Figure 13-8   Draw a chord AC with horizontal span $2a$. The vertical intercept or sagitta BD measures the second difference at point B.

magnitude of corrections for smoothing. Quantitative work is, of course, best done on any substantial amount of data by finite differencing numerically.

It often happens that the tabulation interval of available data does not agree with the differencing interval indicated by the foregoing theory, according to which the span $2a$ would be $2\pi^{-1/2}$ or 1.13 times the equivalent width of a Gaussian instrumental profile. One can adapt to this by noticing that a 5 percent increase in span produces about a 10 percent increase in the sagitta. Thus the span $2a$ and the coefficient $\psi_2 a^2$ may be varied over a modest range, provided the product $\psi_2 a^2$ is kept constant.

Because this element of choice is available, there is the possibility of optimizing the choice if the spectrum of the data is known, if only to some extent, because the degree of fit between $\exp \pi^2 s^2$ and $1 + 4\psi_2 \sin^2 \pi a s$ depends on the range of s to be considered. Thus, one could weight the fit at each value of s, having in mind the spectral content of the data at that s.

A perfectly practical procedure that responds to the consideration of the two foregoing paragraphs is as follows. Accept a value of span $2a$ compatible with the tabulation or indexing of the data. Then take an artificial waveform typical of the $f(x)$ under consideration. Subject the waveform to numerical convolution with the instrumental profile to obtain artificial data. Determine the second difference at a variety of places, and discover empirically what coefficient gives a satisfactory correction. Even more practical is to select a Gaussian curve of width that one might reasonably aspire to restore, convolve it with the instrumental profile, draw a horizontal chord having the necessary span, and establish the factor needed to convert the sagitta to the known correction. Since the chord construction is a space-invariant operation, one chord suffices. These empirical procedures contain quite explicitly an element of subjective judgment. Such judgment would be present in any case in the optimization exercise referred to above, although the act of judgment might be obscured by the weighty mathematics involved.

The chord construction is equivalent to two or three stages of successive substitutions but involves only one multiplication per point rather than the largish number needed for even one convolution. It could be considered for a hybrid attack, whereby a first approximation by the chord construction became the input for one stage of verification by substitution in the convolution integral. The savings in computational effort would be even more important in two dimensions, where convolution can represent a heavy load.

For a discussion of what happens with profiles other than Gaussian see R. N. Bracewell (1955). I have found the chord construction indispensable for hand correction of graphical data as a means of discovering the magnitude, character, and location of the corrections that would be needed if a decision were made to compute those corrections.

THE PRINCIPAL SOLUTION

We have seen that the series of approximations generated by successive substitutions converges if the condition $|1 - A(s)| < 1$ is met; but the limit of the series, where there is a limit, may or may not be the same as the function $f(x)$ that is to be found.

In the common particular case where $A(s) = 1 - |s|$ the limiting function could be characterized as having a transform the same as $F(s)$ where $|s| < 1$, and zero beyond. The general case can be described by introducing a symbol $\mathcal{D}$ to represent the visible domain, or that set of frequencies where $A(s)$ is nonzero (this is not the same as the domain of convergence, which is a subset of $\mathcal{D}$). Now we define the principal solution $p(x)$ to the integral equation $g = a * f$ in terms of its transform as follows:

$$P(s) = \begin{cases} F(s), & s \in \mathcal{D} \\ 0, & \text{elsewhere.} \end{cases}$$

If $F(s)$ is nonzero only where s belongs to $\mathcal{D}$, then $p(x) = f(x)$; but if $F(s)$ spreads outside $\mathcal{D}$, then convolution of $f(x)$ with $a(x)$ will result in multiplication of $F(s)$ by $A(s)$, and some of the content of $F(s)$ will be destroyed, since $A(s)$ is zero outside $\mathcal{D}$.

This is illustrated in Fig. 13-9. Clearly some information is lost when the spectrum $F(s)$ of the wanted function $f(x)$ spreads outside the visible domain $\mathcal{D}$. The quadratic error incurred can be expressed quantitatively by evaluating

$$\int_{-\infty}^{\infty} [f(x) - p(x)]^2 \, dx,$$

the integral of the square of the discrepancy.

From Rayleigh's theorem we see that this measure is equal to

$$\int_{-\infty}^{\infty} |F(s) - P(s)|^2 \, ds = 2 \int_{s_c}^{\infty} |F(s)|^2 \, ds.$$

If $|F(s)|^2$ represents power per unit frequency interval, as is often the case, we can express the agreement between the principal solution $p(x)$ and the wanted function $f(x)$

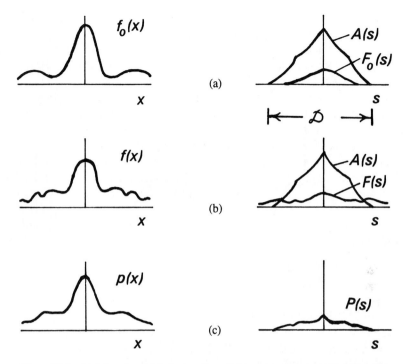

Figure 13-9   (a) A function $f_0(x)$, the transform $F_0(s)$, and a spectral sensitivity function $A(s)$, showing how the instrumental response depends on frequency, being zero outside the visible domain $\mathcal{D}$. The principal solution gives the correct answer. (b) A different function $f(x)$ with more detail, whose transform $F(s)$ extends outside the instrumental domain $\mathcal{D}$. (c) The principal solution $p(x)$ for case (b) approximates $f(x)$, but some detail is lost.

in terms of the power that $f(x)$ contains inside the frequency domain $\mathcal{D}$ to which the instrument responds—namely, where $A(s)$ is nonzero.

Among the infinite set of functions with transform supported on $\mathcal{D}$, the principal solution is optimum in the sense of possessing the least quadratic error.

The principal solution is a genuine solution to $g(x) = a(x) * f(x)$ because it satisfies the integral equation; in other words, $a(x) * p(x)$ does give $g(x)$. In the absence of any knowledge about $f(x)$ other than that contained in the observation $g(x)$ and the instrument specification $a(x)$, the principal solution is thus unique. It is correct in the frequency band where information is available and contains nothing in the band where there is no information. In many cases the principal solution is acceptable even though it may differ from the original $f(x)$. Of course, if computing by series, one has to terminate after a finite number of steps; even so, one or two correction terms may lead to acceptable results.

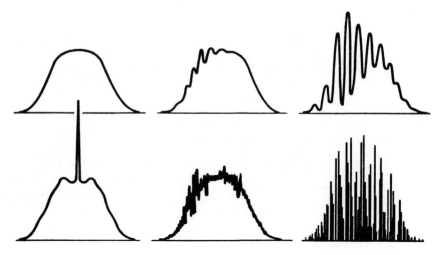

Figure 13-10   Examples containing invisible distributions (Bracewell and Roberts, 1954).

Invisible Distributions.

Not only $f(x)$ and $p(x)$ are solutions of the equation $g = a * f$, but an infinity of other solutions exist. These solutions differ in the frequency band where $A(s) = 0$, i.e., outside $\mathcal{D}$. We conclude that there are functions $i_n(x)$ that are invisible, i.e.,

$$a(x) * i_n(x) = 0.$$

Thus all the solutions are expressible in the form

$$\text{solution} = p(x) + i_n(x),$$

and this includes the true solution $f(x)$, which is expressible by

$$f(x) = p(x) + i_f(x),$$

where $i_f(x)$ is the particular invisible distribution such that

$$I_f(s) = \begin{cases} 0, & s \in \mathcal{D} \\ F(s), & \text{elsewhere.} \end{cases}$$

Examples of invisible distribution are shown in Fig. 13-10. The invisible distributions are characteristic of the operator, such as $a*$, in the integral equation, and one needs to be familiar with the nature of the invisible distributions in order to understand how solutions may differ among themselves and from the wanted solution.

Computing from Series.

When a principal solution exists,

$$P = G + (1 - A)G + (1 - A)^2 G + \cdots,$$

and so

$$p(x) = g(x) + [\delta(x) - a(x)] * g(x) + [\delta(x) - a(x)] * [\delta(x) - a(x)] * g(x) + \cdots.$$

Approximations of various levels may be summarized as follows.

$$f_1(x) = 2g(x) - a(x) * g(x)$$
$$f_2(x) = 3g(x) - 3a(x) * g(x) + a(x) * a(x) * g$$
$$f_3(x) = 4g(x) - 6a(x) * g(x) + 4a(x) * a(x) * g - a(x) * a(x) * a(x) * g(x).$$

One of these relations alone would be sufficient if a fixed number of corrections were adopted. However, one might wish to terminate correction in accordance with the size of the correction. Even more sophisticated procedures might make sense such as terminating correction in one range of x while continuing in another part; this might be valuable on a large image. Therefore, let us express the series as a sum of successive corrections added to $g(x)$; thus for $f_n(x)$, the outcome after n corrections are added, we have

$$f_n(x) = g(x) + c_1(x) + c_2(x) + c_3(x) + \cdots + c_n(x).$$

We see that $c_1(x) = [\delta(x) - a(x)] * g(x)$ and in general that

$$c_1 = (\delta - a) * g, \quad c_2 = (\delta - a) * c_1, \quad c_3 = (\delta - a) * c_2, \quad \ldots, \quad c_n = (\delta - a) * c_{n-1}.$$

In other words, each correction is obtainable by an identical operation on the previous correction, and that operation is the same as the operation carried out on $g(x)$ to get the first correction. These and other neat results were first given by Bracewell and Roberts (1954).

Degradation and Extrapolation of the Principal Solution.

The principal solution has the distinction of being the only solution of the integral equation $g = a * f$ whose Fourier transform agrees exactly with the transform of the desired distribution f and which contains no Fourier components whatever at frequencies to which the instrument did not respond. All other solutions can be obtained by adding to the principal solution invisible distributions $i(x)$ whose transforms $I(s)$ meet the condition

$$I(s) = 0, \quad s \in \mathcal{D}.$$

In spite of this unique character, the principal solution may be unacceptable for some purposes. For one thing, it may be physically unacceptable. Suppose that $f(x)$ is a temperature distribution and therefore may not go negative. As an extreme case take $f(x) = T_0 + Q\delta(x)$, which describes the temperature distribution on an infinite bar, mainly at ambient temperature T_0, when an additional quantity of heat proportional to Q resides at $x = 0$. If this distribution is measured with a radiometer whose instrumental response is $\text{sinc}^2 x$, then we observe $g(x) = T_0 + Q \text{sinc}^2 x$, and the principal solution $p(x)$ is given by $p(x) = T_0 + 2Q \text{sinc} 2x$. If Q is sufficiently large, or T_0 sufficiently small, the negative lobes of the sinc function may cause $p(x)$ to go negative. This is physically impossible.

From the point of view of the magnitude of the error, which is $2Q \text{sinc} 2x -$

$Q \operatorname{sinc}^2 x$, there are errors just as serious, or more so, in the neighborhood of $x = 0$ without violation of positivity. Furthermore, the width estimate is seriously in error. These errors may be just as unacceptable as the negative-going overshoot but do not suffer from the same stigma.

If T_0, or perhaps some other space-variable but positive additive background, is sufficiently large to overcome the negative temperature in the principal solution, the negativity objection can no longer be applied, even though the error is just as great. It appears then that negativity may be less important than error magnitude and in any case may be counterbalanced by the need to avoid degradation of peak intensities, width of peaks, or counts of peaks within clusters. For instance, it may be more important in some context to know that a target in the visual field is single or multiple than to eliminate a negative halo. If further degradation of resolution is acceptable, then negative overshoot may always be removed by further smoothing; this will appear as tapering in the transform domain that reduces the discontinuity at the limit of detectable frequency. Alternatively, extrapolation of the transform into the frequency region beyond the limit of detectability may be attempted; this is the province of the popular methods known as CLEAN and MEM that are discussed below under nonlinear methods.

The idea of the principal solution and its corollary the invisible distribution were launched in connection with the processing of radioastronomical images by Bracewell and Roberts (1954) in a paper that also contains the interpretation of antenna action as the spatial analogue of time-domain filtering, and a presentation of the method of successive substitutions in a context of convolving patterns arising from antenna-aperture distributions of finite extent.

FINITE DIFFERENCING IN TWO DIMENSIONS

A philosophy of limiting numerical operations to the minimum that will produce *some* restoration lay behind the original introduction of the finite-differencing procedure, and in two dimensions any argument favoring numerical simplicity is even more cogent. The idea then will be to apply at each point a correction that is proportional to the "convexity" there. The convexity of a continuous surface could be measured by the amount by which the value at the point lies above a plane that is parallel to the tangent plane but cuts the surface, in some way analogous to the "span" on which the sagitta in one dimension is based.

To formulate a definition of convexity at a point in a two-dimensional array one must first decide which data points should enter into the definition. This is a matter of choice. It might seem that any three points cardinally adjacent to the sample point under consideration would do. The three points would define a plane and the plane would pass below the sample point at a certain distance that could be determined (Fig. 13-11). However, if one starts with a rectangular data array, this vertical distance is independent of the value at one of the adjacent points (the one that is out of line). Therefore, three points so placed are not in fact sufficient to sense the two-dimensional curvature; four adjacent points are the minimum. Of course, a plane cannot in general be passed through four points, but a

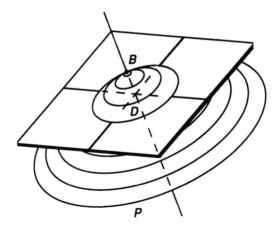

Figure 13-11   Generalization of idea of convexity measure based on vertical offset DB of sample B at point P from the twisted plane through the four adjacent samples.

"twisted plane" such as the hyperbolic paraboloid [1] represented by

$$f(x, y) = a + bx + cy + dxy$$

has four adjustable parameters and meets the requirement in a practical way. As may easily be proved, the height of the twisted plane at a sample point is the mean of the four adjacent values through which it passes.

We now consider two-dimensional Gaussian smoothing and how to correct for it by the two-dimensional analogue of the chord construction. We allow for the usual circumstance that the sampling interval in one direction is not the same as in the other by taking the Gaussian function to have different variances σ_x^2 and σ_y^2 in the x- and y-directions, respectively. Let

$$g(x, y) = \frac{1}{2\pi \sigma_x \sigma_y} \exp\left[-\left(\frac{x^2}{2\sigma_x^2} + \frac{y^2}{2\sigma_y^2}\right)\right] ** f(x, y).$$

Taking Fourier transforms,

$$G(u, v) = \exp[-2\pi^2(\sigma_x^2 u^2 + \sigma_y^2 v^2)]F(u, v).$$

Using the formula

$$\exp \theta^2 = 1 + \sin^2 \theta + \frac{5}{6} \sin^4 \theta + \cdots, \qquad \theta^2 < \tfrac{1}{4}\pi^2,$$

we find

$$G(u, v) = (1 - \sin^2 \pi \alpha u - \sin^2 \pi \beta v + \cdots)F(u, v),$$

---

[1] A quadric surface containing two families of straight lines; in this example all intersections of the curved surface with planes $x = $ const and $y = $ const are straight lines.

or

$$F(u, v) \approx (1 + \sin^2 \pi \alpha u + \sin^2 \pi \beta v) G(u, v),$$

where $\alpha = \sqrt{2}\sigma_x$ and $\beta = \sqrt{2}\sigma_y$. Taking inverse transforms, and provided that $G(u, v)$ does not extend outside the central rectangle of width α^{-1} and height β^{-1},

$$f(x, y) = g(x, y) - \tfrac{1}{4}{}^\alpha \Delta_x^2 g(x, y) - \tfrac{1}{4}{}^\beta \Delta_y^2 g(x, y) + \cdots,$$

where the operator ${}^\alpha \Delta_x^2$ stands for the second finite difference taken over intervals α in the x-direction. It is thus possible, under the conditions stated, to estimate $f(x, y)$ from $g(x, y)$ by differencing the data, just as with the chord construction. As a practical approach we propose to use only the first two difference terms as an estimate f_{est} for $f(x, y)$. The approximate solution, expressed in full, is

$$f_{\text{est}}(x, y) = g(x, y) - \tfrac{1}{4}[g(x + \alpha, y) - 2g(x, y) + g(x - \alpha, y)]$$

$$- \tfrac{1}{4}[g(x, y + \beta) - 2g(x, y) + g(x, y - \beta)]$$

$$= 2g(x, y) - \tfrac{1}{4}[g(x + \alpha, y) + g(x - \alpha, y) + g(x, y + \beta) + g(x, y - \beta)].$$

The operations arrived at by this discussion are indeed simple, requiring only a correction based on the mean of the four values adjacent to (x, y) at intervals α and β in the x- and y-directions.

It is instructive to consider the preceding material from other points of view. We found an exact connection

$$f(x, y) = k(x, y) ** g(x, y)$$

between $f(x, y)$ and $g(x, y)$ and made an approximation that led to an estimate of $f(x, y)$ at any point by adding a correction to $g(x, y)$. The correction at (x, y) is the excess of $g(x, y)$ over the simple mean of the four values adjacent to the g-value in question. We now ask how this proposed estimate compares with the value of f by comparing k with k', where

$$f_{\text{est}}(x, y) = k'(x, y) ** g(x, y),$$

and

$$k'(x, y) = 2^2 \delta(x, y) - \tfrac{1}{4}[{}^2\delta(x + \alpha, y) + {}^2\delta(x - \alpha, y) + {}^2\delta(x, y + \beta) + {}^2\delta(x, y - \beta)],$$

where ${}^2\delta(x, y) = \delta(x)\delta(y)$. A convenient notation results if k' is represented by the two-dimensional array of the coefficients of the impulse functions at the intersections of the (α, β)-lattice. Thus

$$f_{\text{est}}(x, y) \approx 2g(x, y) - \begin{bmatrix} 0 & \tfrac{1}{4} & 0 \\ \tfrac{1}{4} & 0 & \tfrac{1}{4} \\ 0 & \tfrac{1}{4} & 0 \end{bmatrix} ** g(x, y).$$

Note that the sum of the coefficients of the array is unity. The meaning of this fact is that if $g(x, y)$ is constant, then the indicated operation will leave $g(x, y)$ unchanged.

In terms of the two-dimensional Fourier spectrum this means that the zero-frequency component is left unaltered when the sum of the coefficients is unity.

Now that the estimate of the solution is in the same form as the exact solution, assessment of the accuracy obtained should be possible by comparison. At first sight it may seem that a set of four impulses is a rather drastic substitution for the single smooth Gaussian peak that we know would work in the corresponding algorithm based on the method of successive substitutions. But it must be remembered that $g(x, y)$ contains no spatial frequencies beyond certain limits; consequently, any spatial frequencies composing the impulse tetrad, which are not present in $g(x, y)$, will produce no effect. Therefore

$$\begin{bmatrix} 0 & \frac{1}{4} & 0 \\ \frac{1}{4} & 0 & \frac{1}{4} \\ 0 & \frac{1}{4} & 0 \end{bmatrix} ** g(x, y) = m(x, y) ** \begin{bmatrix} 0 & \frac{1}{4} & 0 \\ \frac{1}{4} & 0 & \frac{1}{4} \\ 0 & \frac{1}{4} & 0 \end{bmatrix} ** g(x, y)$$

$$= \tfrac{1}{4}[m(x + \alpha, y) + m(x - \alpha, y) + m(x, y + \beta) + m(x, y - \beta)] ** g(x, y),$$

where $m(x, y)$ is the function whose Fourier transform is constant at all the frequencies contained in the given function $g(x, y)$ and zero elsewhere. As an example, if $G(u, v)$ occupies a rectangular area with cutoff frequencies u_c and v_c, then

$$m(x, y) = \operatorname{sinc} 2u_c x \operatorname{sinc} 2v_c y;$$

and in general $m(x, y)$ is the principal solution associated with system response to a point source. Hence the function with which $k(x, y)$ may be compared is the single smooth-peaked function of bounded spectral extent, of which the quoted array is a set of defining samples at the critical spacings set by the sampling theorem.

A practical method for improving the approximation is immediately suggested by this approach. If a five-point representation is based on the array

$$\begin{bmatrix} 0 & h & 0 \\ h & 1 - 4h & h \\ 0 & h & 0 \end{bmatrix},$$

a degree of freedom is available in the choice of the factor h for a best fit with $k(x, y)$.

The inverse operation to smoothing may be approached in the same spirit of looking for the simplest discrete operation that will give a reasonable degree of improvement. Arguing as before, let $g(x, y) = a(x, y) ** f(x, y)$. If we smooth the given function $g(x, y)$ once more, using the same $a(x, y)$, to get $a(x, y) ** g(x, y)$, the result will be even smoother than before, and the difference $g(x, y) - a(x, y) ** g(x, y)$ is an approximate estimate of $f(x, y) - g(x, y)$. Therefore, as a first approximation,

$$f_1(x, y) = 2g(x, y) - a(x, y) ** g(x, y).$$

We stop at this first approximation and substitute for $a(x, y)$ the discrete array a' already discussed. Then

$$f_1(x, y) = [2\,^2\delta(x, y) - a'] ** g(x, y),$$

or, if the smoothing array a' is

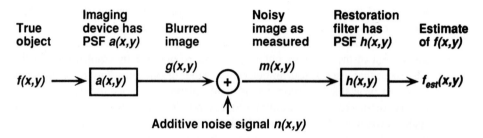

Figure 13-12   Model for restoration of a noisy image

$$\begin{bmatrix} 0 & \frac{1}{4} & 0 \\ \frac{1}{4} & 0 & \frac{1}{4} \\ 0 & \frac{1}{4} & 0 \end{bmatrix},$$

then the sharpening array is

$$\begin{bmatrix} 0 & -\frac{1}{4} & 0 \\ -\frac{1}{4} & 2 & -\frac{1}{4} \\ 0 & -\frac{1}{4} & 0 \end{bmatrix}$$

and adjustments for scale factor are to be made as before.

Two-dimensional sharpening arrays analogous to chord correction in one dimension thus consist of a positive central element flanked by negative elements so as to have a sum equal to unity. If a fixed digitizing interval has to be accommodated, then a variable parameter is needed. In the following simple sharpening array the quantity h is the adjustable parameter.

$$\begin{bmatrix} 0 & -\frac{1}{4} - h & 0 \\ -\frac{1}{4} - h & 2 & -\frac{1}{4} - h \\ 0 & -\frac{1}{4} - h & 0 \end{bmatrix}.$$

As h is changed, the variance of the array changes and can be chosen as wished. For a discussion of this type of simple sharpening array see Menzel (1960).

RESTORATION IN THE PRESENCE OF ERRORS

Hitherto we have discussed deconvolution, or the solution of the convolution-integral equation $g(x) = a(x) * f(x)$, and concluded that the possibilities for finding $f(x)$ when $g(x)$ is given depend sharply on the character of the known factor $a(x)$. In the absence of any information other than $a(x)$ and $g(x)$ the best estimate of $f(x)$ is the famous *principal solution*. This conclusion is helpful in many practical situations, but in others we would wish to take into account prior information about $f(x)$. Also, in many cases it is essential to allow for the presence of measurement error or noise which prevents precise knowledge of $a(x) * f(x)$. This latter complication is taken first.

Consider the linear, space-invariant imaging system shown in Fig. 13-12. A two-dimensional object $f(x, y)$ lying within a region $\mathcal{R}$ in the (x, y)-plane is imaged by a device that has a known point-spread function $a(x, y)$. A blurred image $g(x, y)$ can be imagined, and it has been discussed above, but it is not in fact accessible to observation because in practice some errors, or noise, are inevitably introduced. Naturally we prefer to present a theory that deals in observable quantities. Assume that the noisy blurred image, which is the only one accessible to us, is expressible as the sum of $g(x, y)$ and a two-dimensional noise signal $n(x, y)$. Then the accessible, or measured, quantity $m(x, y)$ is given by

$$m(x, y) = g(x, y) + n(x, y) = a(x, y) ** f(x, y) + n(x, y).$$

From the known quantities $a(x, y)$ and $m(x, y)$ we wish to find out about the true distribution $f(x, y)$.

As a caution, we should remember that we have made an assumption that will need to be verified at a later stage—namely, that the measured image is expressible as a convolution followed by an addition of noise. Just how to verify this assumption is an interesting question in itself, when we bear in mind that in many cases $f(x, y)$ really is not known and that $m(x, y)$ represents the only kind of information we can get about it. In other cases, however—for example when we are not pushing the frontiers of resolution, sensitivity, stability, etc.—it is possible to calibrate by the use of a known $f(x, y)$. How to find out about $n(x, y)$ is a separate matter to be taken up later.

We assume that we know the instrumental response $a(x, y)$, which is defined as the point-spread function in the absence of noise. Since noise is never absent in the situations we are now electing to discuss, one may very well ask how $a(x, y)$ is to be ascertained. There may, of course, be a class of theoretical problems where $a(x, y)$ is deducible from the construction of the system. For example, $a(x, y)$ might be a calculable property of a known lens system. Or we may wish to *assume* an $a(x, y)$ for exploratory calculation. But if we deal with an actual instrument, which is what many people do, we must ascertain $a(x)$ by experiment. That will entail experimental error. Experimental error cannot be eliminated but it can be reduced. One approach is to apply a very large impulse so that the response is very large and the noise by comparison becomes negligible. Sometimes, as with an amplifier, unwanted nonlinearity sets in if too strong a pulse is applied; in that case, sinusoidal signals of moderate strength may be applied at different frequencies sequentially. In other cases, where noise is not merely an unwanted artefact of the system, but is an attribute of the signal itself, it may be possible to repeat measurements N times and take an average. That will often reduce the noise by a factor $N^{1/2}$, but of course noise will always remain. Many different circumstances can arise in practice. Sometimes averaging repeated measurements does not reduce noise much, and sometimes measurements cannot be repeated, as with seismograms of major earthquakes.

What follows is therefore a special example chosen from a wide range of practical problems. It constitutes, however, an important illustration of what may happen where noise is an essential element of the restoration problem. For simplicity we will deal with functions of one variable.

Let us assume that we have access to only the measured signal $m(x)$, which is an additive combination of what the instrumental output would be in the absence of noise, plus a noise signal $n(x)$ of as yet unspecified character, i.e.,

$$m(x) = g(x) + n(x) = a(x) * f(x) + n(x).$$

We would like to design a restoration filter $h(x)$ which does the "best" job of estimating $f(x)$ from the measurements. Call the estimate $f_{est}(x)$.

We define the error in the estimate as

$$e(x) = f_{est}(x) - f(x)$$

and the optimal filter as the one which minimizes

$$\int_{-\infty}^{\infty} |e(x)|^2 \, dx.$$

In order to use infinite limits of integration, which is a convenience, we remember that the object is confined to a finite region $\mathcal{R}$. Outside that region we adopt zero as the value of both the object and the noise. By taking $\mathcal{R}$ to be somewhat larger than the object, we can accommodate the image, which because of smearing by $a(x)$ extends beyond the boundaries of the object. By this choice of procedure we part company with the type of theory that considers quantities such as

$$\lim_{A \to \infty} A^{-1} \int \int_{\mathcal{A}} [e(x, y)]^2 \backslash dx \, dy,$$

where A is the area of some suitably defined region $\mathcal{A}$. The reason for our approach is that it is hard to translate such limiting procedures into operations carried out on actually available quantities.

By Rayleigh's theorem, we know that it would be equally good to minimize

$$I = \int_{-\infty}^{\infty} |E(s)|^2 ds,$$

where $E(s)$ is the Fourier transform of $e(x)$. Now

$$f_{est}(x) = h(x) * [a(x) * f(x) + n(x)]$$
$$e(x) = h(x) * [a(x) * f(x) + n(x)] - f(x)$$

and the Fourier transform of the error estimate is given by

$$E(s) = H(s)[A(s)F(s) + N(s)] - F(s),$$

where capital letters denote the Fourier transforms of the appropriate functions. Assuming we can choose every value of $H(s)$, we have complete control over the integrand I, and we can attack our problem by attempting to minimize $|E(s)|^2$ separately for every value of s.

THE ADDITIVE NOISE SIGNAL

We now must pay some attention to the noise signal $n(x)$, which of course is not known. It presumably has a statistical component, for if it were completely deterministic, we could simply subtract it from the measurements and proceed with restoration of $g(x)$ without

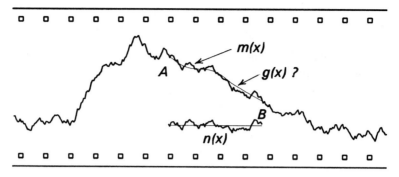

Figure 13-13   A record $m(x)$.

concern for noise. If $n(x)$ is partly statistical and partly deterministic, we would be advised to investigate the latter component and subtract it prior to restoration (the linear filter $h(x)$ cannot perform simple subtraction).

To deal with the statistical component we might try to minimize $\langle |E(s)|^2 \rangle$ where $\langle \ \dots \ \rangle$ denotes an ensemble average over all possible noise signals. An *ensemble average* is a mysterious entity, arising in the study of systems that are defined by explicit differential equations and are excited by sample functions taken from an ensemble described as the set of all possible excitations. When one is dealing with measurements made on exciting functions $f(x)$ which it is the object to estimate, it is putting the cart before the horse to specify in advance details about that which is to be found. However, where random additive noise signals $n(x)$ are concerned, one can attempt to verify separately that the noise is additive, and has zero mean, a constant variance, and a well-defined power spectrum. To do this, lengthy measurements may be possible in the absence of input $f(x)$, from which the mean, variance, and power spectrum can be determined. To check additivity it may be possible to repeat the study in the presence of a steady input signal. Alternatively, it may be possible to analyze the source of the noise and judge that the noise is likely to be additive random noise. Either way, an act of judgment is involved. Thereafter one can imagine an ensemble of sample functions possessing the spectrum and strength adopted and take ensemble averages.

As a practical example consider Fig. 13-13 which presents a record of a measurement $m(x)$. If we knew $g(x)$, we could perform the subtraction to get $n(x) = m(x) - g(x)$. Between A and B, if $g(x)$ were as indicated by the thin line, then $n(x)$ would be as shown directly below. But $g(x)$ might not have been guessed correctly, in which case $n(x)$ would be different. If $n(x)$ is random as believed, there is no general way of apportioning $m(x)$ between the systematic part $g(x)$ and the random part $n(x)$.

However, a moment's contemplation suggests how the mean square value of $n(x)$ might be estimated by paying attention to the record in an area where the signal is absent. If we imagine the continuation of the record to be available, and if the character of the record enables the observer to judge that no signal is present, then it would be quite possible to determine statistical parameters of $n(x)$ in the signal-free region. Thus, given

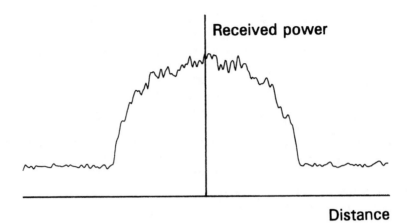

Figure 13-14   A record of thermal emission from the exhaust of a jet engine as a function of distance from the jet axis. A good part of the noise is not additive but results from the fact that the signal is itself thermal noise.

$n(x)$ in the range $-X/2 < x < X/2$, we can take the mean-square value

$$s^2 = X^{-1} \int_{-X/2}^{X/2} [n(x)]^2 \, dx$$

for the variance of $n(x)$. As the power spectrum $N(s)$ we can take the Fourier transform of the autocorrelation $n(x) \star n(x)$.

But would such results derived from the presumed signal-free zone be applicable in the place where the signal existed? We might respond to this question by examining the preceding part of the record. If the noise statistics varied with x, then that behavior would evidence itself. On the other hand, stationarity might be noted. In order to have reasonable assurance about the character of the noise and the presence or absence of signal, it is clearly desirable to have a lengthy record, much exceeding the length of the record of interest. In spite of this obvious conclusion, ordinary human impatience works against the thought of recording reams of uninteresting noise; when the event of interest concludes, the temptation is to stop the recording and go home. The next day it might seem that the noise level was perhaps higher after the event than before, but it would be too late to tell for sure.

Even if a good long record were available which showed the noise to be the same at the beginning as at the end, there would be no guarantee that the noise was the same in the presence as in the absence of signal. Some sorts of noise are independent of signal level (receiver noise often is), but other sorts of noise are inherently signal dependent—for example, shot noise and photon noise (Fig. 13-14). Usually a knowledge of the instrument and the quantity being measured tell what may be expected. Clearly, however, a general theory cannot be given. Therefore it must not be forgotten that in this present discussion the noise is additive by assumption.

To represent the statistical component of the noise spectrum, since we do not know the value of $|N(s)|^2$ at each value of s, we would be content with knowing the ensemble average $\langle|N(s)|^2\rangle$, where $\langle\ldots\rangle$ denotes an average over all possible noise signals. But we must be prepared to recognize at a later point in the discussion that the ensemble average may be unknowable.

DETERMINATION OF THE REAL RESTORING FUNCTION

We will assume that $a(x)$ is even (as well as real) so that the instrumental transfer function $A(s)$ is real. For reasons to be explained later, the optimal restoring function $H(s)$ will then be real, and the algebra is simplified.

For notational convenience, we will drop the variable s from our equations, remembering that small letters are used for functions of x while capitals are used for functions of s. Since minimization is to be performed for all s, the suppressed variable s is treated as a constant for purposes of the differentiation needed for the minimization that determines H. From the equation for $E(s)$ we have (letting asterisks represent complex conjugates)

$$\langle|E|^2\rangle = \langle[H(AF + N) - F][H(AF^* + N^*) - F^*]\rangle$$

$$= H^2\langle(AF + N)(AF^* + N^*)\rangle - H\langle F^*(AF + N) + F(AF^* + N^*)\rangle - \langle FF\rangle^*.$$

To minimize this function $\alpha H^2 + \beta H + \gamma$ we differentiate with respect to H, finding a turning point at $-\beta/2\alpha$. Thus

$$H_{\text{opt}} = \frac{2\langle AF^*F\rangle + \langle F^*N\rangle + \langle FN^*\rangle}{2[\langle AFAF^*\rangle + \langle AF^*N\rangle + \langle AFN^*\rangle + \langle NN^*\rangle]}.$$

For those values of s where $A(s) \neq 0$ we can substitute $F = G/A$ and $F^* = G^*/A$. For s beyond cutoff we adopt a principal-solution philosophy and set $H(s) = 0$, so that high-frequency noise does not corrupt our estimate. Since we know $A(s)$, it can be removed from the averaging operation to yield a general solution

$$H_{\text{opt}} = \begin{cases} \dfrac{1}{A}\dfrac{\langle GG^*\rangle + \frac{1}{2}\langle G^*N\rangle + \frac{1}{2}\langle GN^*\rangle}{\langle GG^*\rangle + \langle G^*N\rangle + \langle GN^*\rangle + \langle NN^*\rangle}, & s < \text{cutoff} \\ 0, & s > \text{cutoff}. \end{cases}$$

We may describe $\langle NN^*\rangle$ as the noise power spectrum and $\langle GG^*\rangle$ as the signal power spectrum. The word *signal* is used in the sense in which it occurs in the phrase *signal-to-noise*, where signal means that which would have been observed had there been no noise present. Strictly speaking, of course, this sort of noise-free signal is not accessible, but the terminology is customary. We may describe $\frac{1}{2}\langle G^*N\rangle + \frac{1}{2}\langle GN^*\rangle$ as the cross power spectrum; the cross power spectrum vanishes for the case of zero-mean noise which is uncorrelated with the signal. The term "power" spectrum is purely conventional and may be undesirable in cases where f and g themselves represent power in the physical sense; "quadratic" spectrum might then be preferable.

To design a restoration filter which works optimally on a class of input distributions, the general solution for H_{opt} can be used directly, once we define the class and calculate the various averages involving the signal spectrum G. If we are willing to tailor-make a filter for a particular distribution, G becomes deterministic and can be removed from the averaging operation. We take the latter viewpoint as we work out some special cases. A practical note on this point will be added later.

Case 1: No Noise

$$H_{\text{opt}}(s) = \frac{1}{A(s)}, \quad s < \text{cutoff}.$$

Full restoration of all components below cutoff gives the best estimate in the squared-error sense.

Case 2: Signal and (Zero-mean) Noise Uncorrelated

$$H_{\text{opt}}(s) = \frac{1}{A(s)} \frac{|G(s)|^2}{|G(s)|^2 + \langle |N(s)|^2 \rangle}, \quad s < \text{cutoff}.$$

This result should be familiar to those who have studied Wiener smoothing of time signals. At frequencies where the signal power is much greater than the noise power, full restoration is called for. Near cutoff, where $|G(s)| \simeq 0$,

$$H_{\text{opt}}(s) \simeq \frac{|G(s)|^2}{A(s)\langle |N(s)|^2 \rangle}.$$

Case 3: Noise Proportional to Signal

$$\langle |N|^2 \rangle = K|G|^2 \quad \text{and} \quad \langle GN^* \rangle = \langle G^*N \rangle = 0$$

$$H_{\text{opt}}(s) = \frac{1}{A(s)(1 + K)}, \quad s < \text{cutoff}.$$

Full restoration differs from optimal restoration by only a constant multiplier, so the estimated shape is unaffected by consideration of errors.

Case 4: Triangular Transfer Function

$$A(s) = \Lambda(s); \quad F(s) = 1; \quad \langle |N(s)|^2 \rangle = K; \quad \langle GN^* \rangle = \langle G^*N \rangle = 0$$

$$H_{\text{opt}}(s) = \frac{1}{\Lambda(s)} \frac{\Lambda^2(s)}{\Lambda^2(s) + K}, \quad s < \text{cutoff}.$$

The second factor, which modifies the full restoration that would be expressed by $1/\Lambda(s)$, is plotted in Fig. 13-15 for various values of noise power K.

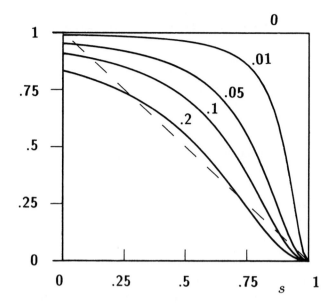

Figure 13-15 The modifying factor $\Lambda^2/(\Lambda^2 + K)$ for various strengths of noise power K with reference to a triangular transfer function $A(s)$ (broken line).

DETERMINATION OF THE COMPLEX RESTORING FUNCTION

We now remove the restriction that $A(s)$ be even, and we solve for the optimal complex restoring values for each s by minimizing $|E(s)|^2$ with respect to both the real and imaginary parts of $H(s)$. The straightforward manipulations reveal that

$$\Re H_{\text{opt}} = \frac{\Re\langle F^*(AF + N)\rangle}{\langle |Q|^2\rangle}$$

$$\Im H_{\text{opt}} = \frac{-\Im\langle F^*(AF + N)\rangle}{\langle |Q|^2\rangle},$$

where $|Q|^2 = (AF + N)(A^*F^* + N^*)$. We find immediately that the restoring function phase is given by

$$\text{pha } H_{\text{opt}} = -\text{pha}\langle F^*(AF + N)\rangle$$

$$= -\text{pha}(\langle A|F|^2\rangle + \langle F^*N\rangle).$$

If the noise has zero mean and is uncorrelated with F, we have pha $H_{\text{opt}} = -$ pha A, which is an obviously appealing result: the restoration filter merely removes the phase shift introduced by the instrument.

SOME PRACTICAL REMARKS

The proposed design of an optimal restoration filter by means of a formula containing G is open to the criticism that we may have access only to the measurements $m(x)$. If we did

in fact know G and thus g, then the restoration would reduce to the noise-free problem!
If we do not know g, how is it possible to use the general solution for H_{opt} to design a
filter based on a full statistical treatment of the problem? This difficult question has no
universally accepted answer. We now explore some considerations on both sides.

A problem frequently posed in texts on statistical estimation theory reads: "A class
of random signals describable by the following statistics . . . is corrupted by additive noise
described statistically by . . . . Design a restoration filter which has optimal average performance in the squared-error sense."

There is no doubt that the student who has mastered the mathematical techniques of
this specialty could complete the assignment and that the filter, if subjected to extensive
testing with input waveforms constructed according to the specified statistics, would behave optimally on the average. There are various methods of obtaining the answer, which
is determinate. But the problem statement has generously supplied sufficient information
to render the solution into an exercise (not necessarily trivial) merely in mathematical manipulation. Nature is rarely so generous in problems encountered in practice. While there
are physical models which can be analyzed to yield statistics for some well-studied noise
processes (thermal noise from resistors, shot noise, etc.), it is not so common to be able to
derive useful statistics for the types of signals that are to be processed. As an extreme example, imagine trying to devise probability densities for the class of passport photographs.
We know that such photographs have a lot in common, but it would be difficult to arrive
at a probabilistic description and even more so to gain general assent. At another extreme,
in observational science, a single record may be all the information that is available. The
approach through time-averaging, which avoids the shaky step of asserting that which we
are trying to find out, is presented by Gardner (1988).

Thus, our mathematical solution to the restoration problem is not really the complete
solution to the practical restoration problem. In fact it is overambitious.

It may be that we must be content with finding an estimate of the original distribution
which is only modestly better than the noisy measurement we started with, rather than
optimal. The notion of an optimal estimate had the virtue of yielding a mathematical
result; however, there is no disgrace in suboptimal performance. This is particularly true
in practice, where a person who provides a timely suboptimal estimate will be held in
higher esteem than another who produces a "better" estimate that is too late. In any case,
an optimum based on minimum squared error is not unique; other error measures may
yield better optima.

In the diffraction-limited scanning problem (Case 4), an essential feature of the
restoring function H is the reduction to zero near the cutoff frequency, which ensures that
the noise is not multiplied by large factors. We might hope that the details of just how the
restoring function falls to zero are of secondary importance and that the performance of
a particular restoration filter does not deteriorate too rapidly as one moves away from the
optimum. This expectation can be investigated in practice by evaluating $\int |e(x)|^2 \, dx$ for
different artificial signal and noise inputs and a variety of restoring functions.

When one is working with an actual record, which is to be interpreted as well as
possible, one cannot take an "ensemble average over all possible signals"; $\langle GG^* \rangle$ is thus
not available for use in the general solution for H_{opt}. Correspondingly, we don't know g,

only $g + n$. But from a single record, $g + n$ is a better estimate of g than an ensemble average of the class of "all possible gs."

For application to an actual record, $\langle GG^* \rangle$ may be replaced by GG^*. The general solution can also be modified as follows, by putting $G = M - N$ to obtain the simplified expression:

$$\frac{1}{A} \frac{MM^* - \langle NN^* \rangle - \frac{1}{2}\langle GN^* \rangle - \frac{1}{2}\langle G^*N \rangle}{MM^*}.$$

If there is no correlation between G and N, which is common, then MM^* is determinable from the measured record, and $\langle NN^* \rangle$ may be estimated by calculating NN^* from a stretch of record where signal is known or judged to be absent. When $\langle GN^* \rangle + \langle G^*N \rangle$ is zero, the simplified expression leads to an operational solution involving available quantities.

But situations where $\langle GN^* \rangle \neq 0$ exist and can be investigated as follows. Let f_{AV} be the average of several values of f_{est} determined by repeated observations $m_i(x)$ of the same f. Subtract $a(x) * f_{AV}(x)$ from the $m_i(x)$ and note whether the residuals appear to be connected with $f_{AV}(x)$. Probably correction for such effects has never been carried out, but it would be quite feasible (Bracewell, 1958). In fact, interesting opportunities exist for image restoration in a range of practical situations where Wiener filter theory has been inapplicable.

Restoration of radio astronomical images was practiced empirically at an early stage but the practitioners soon became cautious, because the modest gain in resolution from restoration was soon overtaken by new instruments under construction. A school of thought developed that it was safer to publish what was observed, without correction for the known degradation due to the antenna. If no account is taken of noise, restoration leading to the principal solution may produce negative brightnesses, which are physically unrealizable but will produce source widths and peak brightnesses that are closer to the truth. The astronomer thus has to tread delicately on the horns of a trilemma. In interferometry, where restoration for the reduction of Fourier components is not called for, the problem does not arise. However, the fact that the transfer function cuts off sharply at a boundary suggests extrapolation into the part of the (u, v)-plane that has not been observed. The relative conservatism over restoration has been matched by enthusiasm for extrapolation, though not under that name. The maximum-entropy method and CLEAN, discussed below, are of this kind.

ARTIFICIAL SHARPENING

Restoration aims to compensate for the blurring due to a known instrument in the presence of specified noise. However, the outcome is generally in the form of an additive correction proportional to curvature, in some sense. Suppose that a digital image is available, but nothing is known about any deterioration that it may previously have suffered. What would happen if a small correction were computed from the curvature? We shall call this *artificial sharpening*. Experience shows that the result may be pleasing. In place of the theoretical optimum concept that guides mathematically developed restoration we have an optimum based on judgment by the user, depending on the amount of artificial sharpening applied. There is a second parameter to be varied, namely the scale of the patch whose curvature is assessed. The finest scale is represented by the curvature operator proportional to [0 1 0/1 −4 1/0 1 0], but this may be too fine, either because it merely amplifies fine-scale noise or because the sampling is finer than the scale of features that could favorably be sharpened. In principle one could smooth and even resample such an image on a coarser grid. Alternatively, the same sample patch could be retained by use of a broadened curvature operator constructed by convolving the fine-scale operator with a set of two-dimensional binomial coefficients spanning a suitable patch size. One could work out the best degree of broadening in terms of the autocorrelation function of the image, but the spirit of the present approach is to vary both the magnitude of the correction and the patch size and to assess the optimum subjectively. This may sound to some people as a rather low-brow admission of defeat, but the counterbalancing consideration is that mathematical optima are sometimes based on mathematical assumptions about the data that have only flimsy applicability. There is certainly room for an empirical tool in one's kit, especially in the absence of demonstrable linearity, Gaussian statistics, additivity of noise, isoplanaticity, space invariance, and other familiar givens of statistical signal processing. The moral here is: Think of a variety of convenient numerical operations that can be promptly applied and try them; you may like one.

ANTIDIFFUSION

The two-dimensional diffusion equation

$$\frac{\partial^2 f}{\partial x^2} + \frac{\partial^2 f}{\partial y^2} - \frac{1}{D}\frac{\partial f}{\partial t} = 0$$

describes how a temperature distribution over a flat conducting plate develops as time elapses and the heat rearranges itself over the plate under the influence of temperature gradients. The diffusion coefficient D, whose S.I. unit is $\mathrm{m^2\ s^{-1}}$, depends on the thermal resistance r and the thermal capacitance c, both per unit area, through the relation $D = 1/rc$. In terms of thermal resistivity ρ, specific heat s, density δ, and plate thickness t, we have $r = \rho/t$ and $c = \delta s/t$.

Figure 13-16 shows a one-dimensional temperature distribution at $t = 0$ and later at $t = 1$, by which time the peak temperatures have diminished and the trough has filled in a little. Since heat is driven in proportion to the temperature gradient the high peak on the left has moved rapidly toward equilibrium because of the steep nearby gradients. The equally hot plateau on the right has not noticeably cooled—the action is taking place in the area of the sharp bordering gradients. Some fine ripples have died out; the temperature differences were not great, but the gradient was strong and the heat did not have to move far to fill in the troughs.

The outcome conspicuously resembles image smoothing and suggests the following sharpening procedure. Take a given image $f(x, y)$ that is to be sharpened and subject it

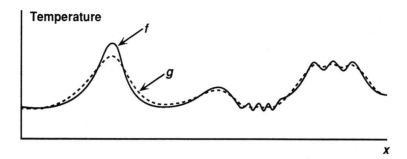

Figure 13-16 Temperature distributions at $t = 0$ (heavy) and $t = 1$ (broken).

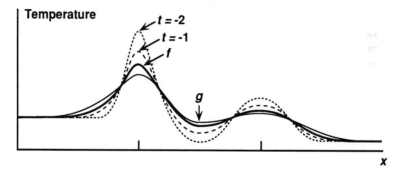

Figure 13-17 Temperature distributions at $t = 0$ and $t = 1$ as before, and prior
distributions at $t = -1$ and $t = -2$.

to the diffusion equation, but let time run backward. A sequence of prior sharp distributions will be generated, any of which could have diffused down into the given $f(x, y)$. An interesting feature of this suggestion is that antidiffusion does not produce negative temperatures; in the restoration of images, avoidance of negatives is often a required constraint. A curve resembling part of Fig. 13-16 has been subjected to inverse diffusion to show two prior temperature distributions (Fig. 13-17) at $t = -1$ and $t = -2$.

Clearly the prior distributions are plausible candidates for image sharpening. Curiously, however, no specification of the smoothing mechanism, such as knowledge of its transfer function, has been invoked. There is only one disposable parameter, namely the time lead. It can be shown that this parameter is equivalent to the central second derivative of the transfer function or to the second moment of the corresponding spatial impulse response. This theory need not be given here, because the story is identical to Eddington's method.

The two-dimensional Laplacian $\partial^2 f/\partial x^2 + \partial^2 f/\partial y^2$ is convenient to translate into its discrete counterpart; all that is required is convolution with

$$\begin{bmatrix} 1 & -2 & 1 \\ -2 & 4 & -2 \\ 1 & -2 & 1 \end{bmatrix}$$

Selection of a diffusion constant amounts to choosing the amount of time advance, provided it is small. In practice, the sharpened image is

$$f(x, y) - \alpha \begin{bmatrix} 1 & -2 & 1 \\ -2 & 4 & -2 \\ 1 & -2 & 1 \end{bmatrix} ** f(x, y),$$

where α is chosen so that the correction is not too large. To go back further in time, repeat the procedure. Combining two stages to go back two steps in one operation will require a larger array (5×5) in conformity with the fact that heat will flow in from further away in the longer time allowed.

So far this discussion has simply recovered some previous practice, though giving it a physical interpretation that is intuitively helpful. But here is a possibly fruitful, hitherto unexplored avenue. As we go back further in time with the example illustrated, we will find out possible ways in which the present temperature distribution may have originated. Any one of the prior temperature distributions could have been suddenly imposed over the whole plane by external intervention a finite time ago; but what if time is simply allowed to recede indefinitely?

What will happen is that heat will flow backward up the temperature gradient, producing higher and higher temperatures, until the gradient somewhere becomes infinite. There is no physically allowable state to precede this. In the example of Fig. 13-17, the sharper peak will reveal an origin as a spatial delta function, or a sudden point deposit of heat at $t = -A$. The process does not have to stop here. We can remove that point peak and run time backward further still to discover that the second peak originated at time $t = -B$. After that point peak has been removed, proceed similarly until what remains is negligible.

This line of reasoning leads to the replacement of a complete image by a list of locations of delta functions, their strengths, and their times of introduction. In other words a means of data compression has been arrived at that may be very economical. Figure 13-17, in its one-dimensional form, used 128 pixels. Now we see that it can be specified by seven numbers only: two locations, two strengths, two times, and one diffusion constant. In general, the compaction factor cannot exceed the area of the image divided by the effectively independent point-spread area.

One can contrive images that could not arise by diffusion from isolated hot points. Possibilities include distributed line deposits of heat, area deposits of heat, and heat infusions that are not sudden. All of these can be expressed discretely. Thinking about restoration of images in this way can be recommended as a way of gaining insight—into the role of spatial noise, for instance.

NONLINEAR METHODS

As we have seen, the transfer function of antennas and other sensors, including optical and ultrasonic, falls off to zero at a cutoff spatial frequency. If the object spectrum has noticeable content around this cutoff, the discontinuity in the spectrum of the principal solution causes undesirable oscillations in the image domain. Restoration can mitigate the effect of the discontinuity by tapering at frequencies near to but less than the cutoff. Additional tapering may be resorted to if there is not much noise.

Information is lost if additional tapering is applied, because signal-to-noise ratio is lost. On the other hand, extrapolation of the visibility to (u, v)-values beyond the resolution boundary reached by observation may improve the signal-to-noise ratio, but the enhanced signal may be erroneous. Presumably, what was not observed has had to be guessed; nevertheless, there may be prior information that helps. Determination of model parameters, if a correct model is chosen, is an application which seems reasonable and can at the same time be described as extrapolation. Two other methods of extrapolation beyond the cutoff are practiced which are both nonlinear.

RESTORING BINARY IMAGES

Many objects are black and white and thus representable by a binary value, zero or one. The image resulting from some sort of reproduction may, however, contain intermediate grey levels. A blurry photograph of a printed page would be an example, and the problem might be to restore the blurry image to legibility. Another application is to the reading of bar codes on railway rolling stock which are both moving and dirty. A simple idea is to apply the method of successive substitutions and then to clip, or hard-limit, the result, converting any values above a threshold to unity. At the same time rectify the result by replacing any negative values by zero. This scheme has been studied for the case of a low-resolution printer alphabet degraded by a point response which is itself a sharp-edged black-on-white element (Bracewell, 1985). Favorable results were obtained. As with many nonlinear procedures, the theory cannot be worked out with generality, but specific proposals can be attacked. Successive substitutions with clipping can be recommended as a numerically simple approach that may be modifiable to suit a variety of situations where image values assume only a limited number of levels.

CLEAN

Högbom (1974, 1979) introduced the method known as CLEAN, which makes use of knowledge of unwanted sidelobes of an antenna beam in order to produce a radio image that is free from the identifiable defects due to those sidelobes.

CLEAN is a procedure that does not make use of prior knowledge about the scene being mapped. The instrument response $a(x, y)$ is picturesquely described as the "dirty beam" in reference to its possession of unwanted sidelobes. The observed distribution $g(x, y)$ is described as the "dirty map." The entity to be found does not have a name in this phenomenological approach, which is strictly operational. Suppose the object observed consists of a finite number of discrete objects of various sizes and shapes but more or less compact, or discrete. Then a very compact object would be represented in the dirty map by a reproduction of the dirty beam centered on the position of that object, complete with the characteristic sidelobes. The dirty map as a whole could be regarded as a superposition of more-or-less smoothed versions of the dirty beam centered on the various objects. The procedure is to locate the strongest peak in the dirty map and subtract a dirty beam with an amplitude big enough to reduce the peak severely. About 50 percent cancellation is enough, because only part of the peak is in fact due to the object presumed to be located there; other objects will be making a contribution at that place through sidelobes.

Start a table consisting of the coordinates and strength S of the subtracted distributions. Now move to the second largest peak and repeat. Construct a table of x_i, y_i, and S_i, stopping when a residual map $r(x, y)$ has been reached where no further outstanding features can be seen. This decision may be based on experience with the class of maps in hand, which recognizes when no further peaks meet the reliability criterion that the user adopts. A "clean beam" is now constructed which possesses a beamwidth appropriate to the resolution achieved but is free of sidelobes; Gaussian is the ordinary choice. Make a superposition of clean beams, using the table of coordinates and strengths, and finally add in the residual map $r(x, y)$. The result is the "clean map," from which negative-going sidelobes will have been eliminated. It is possible to strain a little harder by adopting a clean beam that is a little narrower than you may be entitled to, and the result may look good, but you will not know how reliable the resultant map is. This is where the wisdom of retaining the residual term $r(x, y)$ comes in, because it provides a sort of stabilizing noise level that helps guard against overinterpretation. CLEAN has been widely used in the construction of radio astronomical images, which have the highest resolution and dynamic range of any images today. For further reading see Cornwell and Wilkinson, 1981; Cotton, 1979; Thompson et al. (1986).

MAXIMUM ENTROPY

Restoration techniques known under the name *maximum entropy* (Smyllie et al., 1973, Ables, 1974) can also be interpreted as extrapolation in the Fourier transform domain. The idea is to select among positive images whose Fourier transforms agree with the transform of the principal solution so as to maximize the image entropy. There is a certain air of mystery surrounding the definition of *image entropy*. Two definitions of entropy of an image $b(x, y)$ with positive brightness which are current (Frieden, 1972; Ponsonby, 1973) are as follows:

$$\int \int \log[b(x, y)] \, dx \, dy \quad \text{and} \quad - \int \int b(x, y) \log[b(x, y)] \, dx \, dy.$$

The fainter brightness values make a major contribution to the entropy under the first definition, because the absolute value of the logarithm of small numbers is large. Where the image brightness falls to zero, as it may at the outer edges or even in the interior of the field, the logarithm cannot be taken. Therefore, an arbitrary boundary has to be established at the 5 percent or some other contour. But, of course, this contour is not known in advance, and a lot of entropy lies around the low-brightness fringe, so the choice of the boundary is not a trifle. Likewise, if the map to be improved contains observational errors, as it must, it is quite legitimate for negative brightness to occur where the true brightness is comparable with the error level. To treat such negative values as zero would be philosophically bothersome because of the bias introduced. Maximum-entropy algorithms in action, however, must actually assign an arbitrary positive value in places where negatives occur, because they cannot take the logarithm of zero. These considerations tend to detract from the elegance of the logarithm and in a way have higher priority than the discussions of when to use $\log b$ and when $b \log b$ (Ponsonby, 1978).

The whole philosophical basis of the maximum-entropy method can be put aside in favor of an empirical understanding that is well described by Nityananda and Narayan (1982) and Narayan and Nityananda (1984, 1986). These authors have found satisfactory results in practice with a modified entropy that can have a variety of functional forms including $\sqrt{b}$ and $-\exp(-b)$, provided that, like $\log b$ and $b \log b$, they are convex upward and have a negative third derivative.

Some difficulty was experienced at first in translating the maximum-entropy technique from time signals to two-dimensional images. The work done at the Radio Astronomy Institute, Stanford and the Mullard Radio Astronomy Observatory, Cambridge, in connection with radio astronomical images (Wernecke and D'Addario, 1977; Gull and Daniel, 1978) resolved this difficulty by introduction of the Lagrange multiplier concept. Under this innovation, which explicitly recognizes the presence of errors, interpretation of the term "agreement" with the transform of the principal solution is relaxed to mean compatibility rather than identity.

Unlike the linear methods, the maximum-entropy method introduces spatial frequencies that are invisible to the instrument, extrapolating the spectrum beyond the instrumental cutoff. The inherent risk of error can be held in check by involving an acceptable approximate map in the definition of modified entropy. Work in terms of normalized brightness $f = b/b_{\max}$ that ranges from 0 to 1 (as reflectance does), and let $f_0(x, y)$ be the observed map so normalized. Then as modified entropy H of an approximation $f_n(x, y)$ adopt

$$H = \int \int f_n(x, y) \, dx \, dy - \int \int f_n(x, y) \log[f_n(x, y)/f_0(x, y)] \, dx \, dy.$$

The implication that the observed map is acceptable is disturbing, but for a favorable opinion of the outcome based on experience see Cornwell (1988).

LITERATURE CITED

J. G. ABLES (1974), "Maximum entropy spectral analysis," *Astron. Astrophys. Suppl.*, vol. 15, pp. 383–393.

R. N. BRACEWELL (1955), "A simple graphical method of correcting for instrumental broadening," *J. Opt. Soc. Amer.*, vol. 45, pp. 873–876.

R. N. BRACEWELL (1958), "Restoration in the presence of errors," *Proc. I.R.E.*, vol. 46, pp. 106–111.

R. N. BRACEWELL (1985), "An imaging problem: restoration of blurred digital characters," *Computer Vision, Graphics and Image Processing*, vol. 29, pp. 329–335.

R. N. BRACEWELL (1963), "Correction for grating response in radio astronomy," *Astrophysical Journal*, vol. 137, pp. 175–183.

R. N. BRACEWELL AND J. A. ROBERTS (1954), "Aerial smoothing in radio astronomy,"*Aust. J. Phys.*, vol. 7, pp. 615–640.

T. J. CORNWELL (1988), "Radio-interferometric imaging of very large objects," *Astron. Astrophys.*, vol. 202, pp. 316–321.

T. J. CORNWELL and P.N. Wilkinson (1981), "A new method for making maps with unstable radio interferometers," *Mon. Not. Roy. Astronom. Soc.*, vol. 196, pp. 1067–1086.

W. D. COTTON (1979), "A method of mapping compact structure in radio sources using VLBI observations," *Astrophys. J.*, vol. 84, p. 1122–1128.

A. S. EDDINGTON (1913), *Mon. Not. Roy. Astr. Soc.*, vol. 73, p. 359.

B. R. FRIEDEN (1972), "Restoring with maximum likelihood and maximum entropy," *J. Opt. Soc. Amer.*, vol. 62, pp. 511–518.

W. A. GARDNER (1988), *Statistical Spectral Analysis: A Nonprobabilistic Theory*, Prentice-Hall, Englewood Cliffs, N.J.

S. F. GULL AND G. J. DANIEL (1978), "Image reconstruction from incomplete and noisy data," *Nature*, vol. 272, pp. 686–690.

J. A. HÖGBOM (1974), "Aperture synthesis with a non-regular distribution of interferometer baselines,"*Astron. Astrophys. Suppl.*, vol. 15, pp. 417–426.

J. A. HÖGBOM (1979), "The introduction of a priori knowledge in certain processing algorithms," in C. van Schoonefeld, ed., "Image Formation from Coherence Functions in Astronomy," Reidel, Dordrecht, The Netherlands.

D. H. MENZEL , ed. (1960), *The Radio Noise Spectrum*, Harvard University Press, pp. 142–150.

R. NARAYAN AND R. NITYANANDA (1984) in J.A. Roberts, ed., *Indirect Imaging*, Cambridge University Press, Cambridge, England, pp. 281–290.

R. NARAYAN AND R. NITYANANDA (1986), "Maximum entropy image restoration in Astronomy", *Ann. Rev. Astron. Astrophys.*, vol. 24, pp. 127–170.

R. NITYANANDA AND R. NARAYAN (1982), "Maximum entropy image reconstruction—A practical non-information-theoretic approach," *J. Astrophys. Astron.*, vol. 3, pp. 419–450.

J. E. B. PONSONBY (1973), "An entropy measure for partially polarized radiation and its application to estimating radio sky polarization distributions from incomplete 'aperture synthesis' data by the maximum entropy method," *Mon. Not. Roy. Astron. Soc.*, vol. 163, pp. 369–380.

J. E. B. PONSONBY (1978) in P.L. Baker, ed., *Formation of Images from Spatial Coherence Functions in Astronomy*, IAU Colloquium no. 49, Groningen, The Netherlands, August 1978.

D. E. SMYLLIE, G. K. C. CLARKE, AND T. J. ULRYCH (1973), *Methods Comput. Phys.*, vol. 13, pp. 391–430.

A. R. THOMPSON, J. M. MORAN, AND G. W. SWENSON, JR. (1986), *Interferometry and Synthesis in Radio Astronomy*, John Wiley, & Sons, New York.

S. J. WERNECKE AND L. R. D'ADDARIO (1977), "Maximum entropy image reconstruction," *IEEE Trans. Computers*, vol. C-26, pp. 351–364.

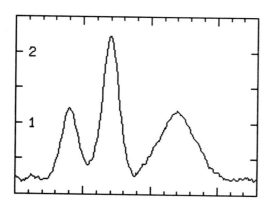
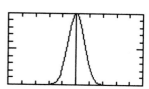

Figure 13-18 An artificial record (left) to be corrected for impulse response (right).

PROBLEMS

13–1. *Restoring an artificial record.* The record shown in Fig. 13-18 was computed artificially to simulate the performance of an instrument whose impulse response is shown on the right. By any means, correct for instrumental smoothing and show your result superposed on the given record. Why do you think the record appears more wiggly at the beginning and end?

13–2. *Restoration.* Explain how, by comparing $a(x, y)$ with $a(x, y) ** a(x, y)$, you would calculate a correction to be applied to $g(x, y)$ from knowledge of $g(x, y + 1)$, $g(x - 1, y)$, $g(x + 1, y)$, and $g(x, y - 1)$. ☞

13–3. *Derivatives cancel running means?* There is a rumor that the integral equation $g(x) = \Pi(x) * f(x)$ for unweighted running means has a solution

$$f(x) = g'(x - 0.5) + g'(x - 1.5) + g'(x - 2.5) + \cdots.$$

Mr. Wu says, "If there is a true formula of this type, it must contain terms with shifts to both left and right, because the inverse operation to a symmetrical operation must be symmetrical." Mr. Cho says, "I remember using a formula like that and it worked very well. Of course, I had to use finite differences instead of derivatives." What do you say? ☞

13–4. *Running means corrected by successive substitution.* Suppose we are given $g(x)$, where

$$g(x) = \Pi(x) * f(x)$$

and it is required to find $f(x)$. It is proposed to apply the method of successive substitutions.
(a) Is the convergence criterion satisfied?
(b) If you make a numerical trial, is your answer above confirmed?
(c) Make any wise comments or further investigations that seem appropriate. ☞

13–5. *Correction for running means.* The method of successive substitutions is subject to the criticism that the convergence criterion is not met in the special case of restoration for smoothing by running means. Here is an alternative approach. Take the given smooth record and subject it to a further stage of smoothing by running means over the same interval. The new record can be regarded as the data for a new problem, where a certain *weighted* running mean now takes the place of the original running mean. Perhaps in this new problem the condition for convergence will be met. Give your opinion on this idea.

13–6. *Restoration.*

(a) Construct a 20 × 20 matrix by summing four or five terms of the form

$$A_i \exp\{-\pi[(\frac{u - U_i}{X_i})^2 + (\frac{v - V_i}{Y_i})^2]\},$$

where $u = x \cos \theta_i + y \sin \theta_i$ and $v = y \cos \theta_i - x \sin \theta_i$. Restrict the amplitudes A_i to a 2-to-1 range—for example, by taking random numbers between 10 and 20. Choose the horizontal and vertical widths X_i and Y_i to be $\ll$ 20 but $>$ 2. Choose the coordinate rotation angles θ_i at random and the associated locations U_i and V_i also at random. Find the largest matrix element, and rescale the values so that the largest value becomes 99 exactly.

(b) Print out the matrix on the international standard with a horizontal spacing between elements of 0.4 inch and a vertical spacing of one-third inch. (These spacings are between the centers of the elements.) This matrix represents the brightness of a true incoherently radiating object, which is unknowable to us except through observation. So we are just pretending that we know it. Sketch in the 20, 40, 60, 80 contours by eye.

(c) Now slip into the matrix one or two weak features in places where you think they might be hard to detect. For example, you might think you can hide a circular Gaussian hump of amplitude 10 and equivalent width 2 in a saddle between the main peaks.

(d) Having prepared a true distribution, now simulate the observational step by convolving the whole matrix with a 5 × 5 binomial array (central row is {6 24 36 24 6}/5 × 2⁵.) One has to do something explicit at the boundaries. In this case, since we have an algebraic expression for the true object, we can use this expression for values outside the 20 × 20 field.

(e) Prepare a careful contour map of the "observed" data. This is the usual starting point of image analysis. Purge your memory banks and look at it with new eyes. You ought to see a certain number of peaks with certain strengths and locations and perhaps noticeable orientations, and there may be suggestions of one or two extra features.

(f) Subject the observed map to one stage of restoration by successive substitutions, print out the restored matrix (rounded to the nearest integer), and redo the contour map.

(g) Make a catalog of the peaks in descending order of amplitude, with their locations and widths, alongside the "true" catalog.

(h) Write up comments on what you saw from this exercise. Possible topics: Accuracy achieved in the catalog entries, how accuracy varies from peak to peak, whether the catalog is constructed more readily from the printed restoration or from the restored contours, would a staggered-profile representation help, whether restoration made a significant improvement, etc.

(i) We saw a simple nine-element sharpening procedure in Chapter 8. Explain why it would not be as good as the restoration method that you have just applied.

(j) Now change hats. How do you know it is not as good? Figure out how to force the simple procedure to be more cost-effective than the longer procedure. Be prepared to accept a tolerable reduction in accuracy in exchange for a desirable reduction in processing time. Since the simple procedure had nothing in it to characterize the width of the instrumental response, the sharpening correction should not, in general, have the correct magnitude. Can you make it correct by introducing a space-independent magnification M to get a sharpening matrix

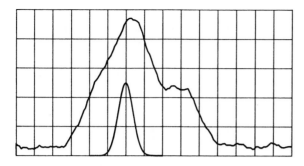

Figure 13-19 A record obtained with the point-response function shown.

$$\frac{M}{9}\begin{bmatrix} -1 & -1 & -1 \\ -1 & 17 & -1 \\ -1 & -1 & -1 \end{bmatrix}?$$

13–7. *Noisy record.* The record shown in Fig. 13-19 was obtained with an instrument whose point response is given.

    (a) Use any method to obtain a restoration that should agree better than the record does with the original quantity measured.

    (b) A colleague asks you what is your peak-signal-to-r.m.s.-noise ratio. What would you say?

    (c) Does the noise seem to be (i) stationary (statistical parameters independent of abscissa), (ii) additive (parameters independent of signal strength)? ☞

13–8. *Circular aperture.* Suitable units are chosen for q so that the modulation transfer function of a certain instrument is chat q.

    (a) Make a graph of the restoring function $H(q)$, $0 < q < 1$, that would apply in the absence of noise.

    (b) Show how you would modify this graph for restoration of images of objects said to have a flat power spectrum, in the presence of noise which also has a flat spectrum one-fifth as strong as that of the object.

    (c) If it was necessary to restore a single image of particular importance and you wished to err on the side of caution, so as to reduce the risk of introducing spurious fine detail, what simple further modification might you make? ☞

13–9. *Deblurring.* A film negative found in a spy's room is digitized on the spot by scanning with a square aperture, 3 mm square, moving in contact with the film. The digitizing interval is 0.1 mm in both directions, which is rather fine. The relatively large scanning area is used so as to get more light to the photometer, and a shorter turn-around time between finding the film and putting it back in place, apparently untouched.

    (a) What is the transfer function $H(u, v)$ as a function of the spatial frequencies u and v in cycles per mm?

    (b) On examination, the digitized material is found to be very important but in need of restoration so that fine detail can be interpreted. Explain why the method of successive substitutions does not converge in this case.

    (c) A consultant makes the following suggestion: "Process the digital data by replacing each value with the mean of all the values in a 3-mm square centered on that value. It's true that the result is blurrier than before, but we now have a new transfer function to which the nonconvergence objection does not apply. Therefore we can restore in the usual way by the method of successive substitutions." Give your opinion on this suggestion.   ☞.

13–10. *Invisible distribution.* A student says that an invisible distribution is one whose transform is zero where the transform of the principal solution is nonzero, and is nonzero elsewhere. Is that equivalent to saying that an invisible distribution is one that elicits no response?

14

The Projection-Slice Theorem

When a function has circular symmetry, its two-dimensional Fourier transform can be expressed as a Hankel transform in terms of the single radial variable $q = \sqrt{u^2 + v^2}$ in the transform plane, as noted earlier. In the notation previously used, the Hankel transform $\mathbf{F}(q)$ of $\mathbf{f}(r)$ is given by

$$\mathbf{F}(q) = 2\pi \int_0^\infty \mathbf{f}(r) J_0(2\pi q r) r \, dr.$$

The Abel transform of $\mathbf{f}(r)$, which was also introduced in connection with circular symmetry, is given by

$$f_A(x) = \int_x^\infty \frac{\mathbf{f}(r) r \, dr}{\sqrt{r^2 - x^2}}.$$

This transform is simply the projection of a circularly symmetrical function $f(x, y) = \mathbf{f}(r)$ onto the x-axis (of course, the projection onto any other line through the origin is the same).

Projection of a two-dimensional function onto a line is the basic operation which if carried out with respect to lines oriented in various directions, provides the data on which tomographic reconstruction is based. However, the material in this chapter is more generally applicable.

While the projection-slice theorem introduced below is not restricted to circular

493

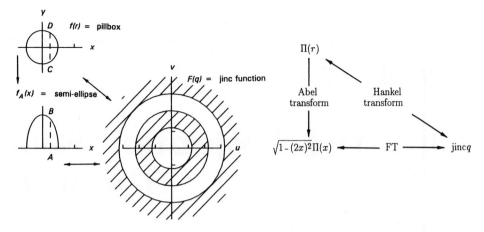

Figure 14-1   The Abel-Fourier-Hankel cycle for rect r or $\Pi(r)$. For a pillbox of diameter D the triplet of functions are $f(r) = \text{rect}(r/D)$, $f_A(x) = D\sqrt{1 - (2x/D)^2}\,\text{rect}(x/D)$ (semi-ellipse), and $\mathbf{F}(q) = D^2\,\text{jinc}\,Dq$. The double-ended arrows refer to reversible transformations (Fourier and Hankel) while the single-ended arrow refers to the unidirectional Abel transformation.

symmetry, it can conveniently be approached and illustrated by making use of background from Chapter 9.

CIRCULAR SYMMETRY REVIEWED

First we show that an interesting relationship holds between the Abel, Fourier, and Hankel transforms. We give those examples which arise frequently in dealing with circularly symmetrical functions, such functions as are met in optical instruments, other symmetrical artifacts of technology, and symmetrical natural objects encountered in geophysics and astrophysics. Symmetry is also injected into certain observations by the rotation of the earth.

Then we take up the projection-slice theorem, which follows easily from the circular-symmetry material and which supplies a useful tool in the innumerable situations where symmetry does not exist. This interesting theorem is the basis for understanding the reconstruction techniques underlying the modern techniques of tomography, which, from beginnings in radiology antedating electronic computers, has expanded into many fields of image formation and is continuing to find new applications. The subject of tomography is taken up in the next chapter.

It was previously noted that the unit-diameter pillbox function rect r has as its Abel transform the semi-ellipse $\sqrt{1 - (2x)^2}\,\text{rect}\,x$, as can be verified by inspection. We also know that the Hankel transform of rect r is jinc q. Interestingly, the semi-ellipse and the one-dimensional jinc function that is the cross-section of the Hankel transform, form a one-dimensional Fourier transform pair. Figure 14-1 records these relationships.

The Abel transform $\int \mathbf{f}(r)\,dy$ is the set of line integrals through $\mathbf{f}(r)$ taken along

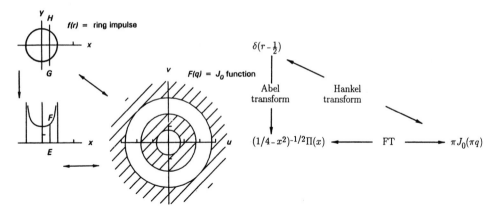

Figure 14-2 The Abel-Fourier-Hankel cycle for $\delta(r - \frac{1}{2})$.

lines parallel to the y-axis. Thus, the ordinate AB is equal to the area of the cross section whose base is on CD, namely $\sqrt{1 - (2x)^2}$. Since the pillbox or disc function rect r has unit height, AB is of course equal to CD, and so the locus of B is an ellipse. To show that the Fourier relation between the Abel and Hankel transforms is not an isolated case, we have Fig. 14-2 for the ring delta of unit diameter. In this figure, the ordinate EF equals the integral through the ring impulse along the line GH. The contribution from each cut exceeds unity by the factor $\sec \alpha = [1 - (2x)^2]^{-1/2}$. This function is familiar in connection with the probability of finding the bob of a long pendulum at position x.

Several features that are useful for checking can also be verified. The volume, for example, under $\mathbf{f}(r)$ equals the area under the Abel transform, which in turn equals the central value $F(0, 0)$ of the Hankel transform of $\mathbf{f}(r)$. This value is $\frac{1}{4}\pi$ for the first illustration and 2π for the second. The volume under $F(u, v)$ equals $f(0, 0)$. The central value of the Abel transform equals the cross-section area of $\mathbf{f}(r)$ along the y-axis, and of $F(u, v)$ along the u-axis.

THE ABEL-FOURIER-HANKEL CYCLE

Using script capitals to refer to the three transform operators, if $\mathbf{f}(r)$ is a function of r, then

$$\mathcal{H}\mathcal{F}\mathcal{A}\,\mathbf{f} = \mathbf{f}.$$

Since the sequence of operations is cyclic, one may start with any one of the operators. Thus

$$\mathcal{H}\mathcal{F}\mathcal{A} = \mathcal{A}\mathcal{H}\mathcal{F} = \mathcal{F}\mathcal{A}\mathcal{H} = \mathcal{I},$$

where $\mathcal{I}$ is the identity operator that converts an operand into itself.

If one starts with $\mathbf{f}(r) = \text{jinc}\,r$, instead of starting with rect r as in the example above, a new triangular diagram can be formed in which the Hankel transform is rect q. It follows from the Fourier transform pair that the Abel transform of jinc r is sinc x, as

Table 14-1   Table of Abel-Fourier-Hankel quartets.

f(r)	Fourier transform $\mathcal{F}f$
Abel transform $\mathcal{A}f$	Hankel transform $\mathcal{H}f$

rect r	sinc u
$\sqrt{1-(2x)^2}\,\text{rect}\,x$	jinc q
jinc r	$\sqrt{1-(2u)^2}\,\text{rect}\,u$
sinc x	rect q
jinc$^2\,r$	See Fig. 9-12
$(4x)^{-2}H_1(4\pi x)$	chat q
sinc r	rect u
$J_0(\pi x)$	$\pi^{-1}(\frac{1}{4}-q^2)^{-1/2}\,\text{rect}\,q$
$\delta(r-a)$	$2\cos(2\pi au)$
$2a(a^2-x^2)^{-1/2}\Pi(x/2a)$	$2\pi a J_0(2\pi aq)$
$\Lambda(r)$	sinc$^2\,u$
$(1-x^2)^{1/2}-x^2\cosh^{-1}x^{-1}$	$2\pi\left[q^{-3}\int_0^q J_0(x)dx - q^{-2}J_0(q)\right]$
$\exp(-\pi r^2)$	$\exp(-\pi u^2)$
$\exp(-\pi x^2)$	$\exp(-\pi q^2)$

illustrated in Fig. 14-3. It is evident that two triangular cycles, such as this one and the one in the first example, can be condensed into a single rectangular diagram, as in Fig. 14-4 and the first four-function entry in Table 14-1.

Many examples of sets of functions that are related by the three operators, and which tend to arise in circularly symmetrical systems, are exhibited for reference in Table 14-1, which adopts the rectangular organization throughout. This interesting mathematical phenomenon was first presented in a paper (Bracewell, 1956) which also contained direct and iterative mathematical solutions for the inversion problem of tomography. For the inversion technique that later came into universal use (the modified back-projection algorithm), see Bracewell and Riddle (1967); for other aspects, see Herman (1979) and Bracewell (1979); for historical context see Sullivan (1984) and Deans (1983).

Writing $\mathcal{H}[\mathcal{H}\mathcal{F}\mathcal{A}f]=\mathcal{H}f$ and noting that $\mathcal{H}\mathcal{H}=\mathcal{I}$, leads to $\mathcal{H}f=\mathcal{F}\mathcal{A}f$. There are three substitution relations of this kind: $\mathcal{H}=\mathcal{F}\mathcal{A}$, $\mathcal{A}=\mathcal{F}\mathcal{H}$, $\mathcal{F}=\mathcal{A}\mathcal{H}$, all of which are deducible directly from the cyclic result by taking appropriate transforms of both sides and noting that $\mathcal{H}\mathcal{H}=\mathcal{F}\mathcal{F}=\mathcal{I}$. (For circular symmetry, cross sections are even functions,

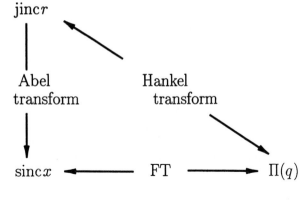

Figure 14-3 The Abel-Fourier-Hankel cycle for jinc r.

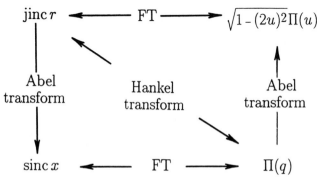

Figure 14-4 A way of condensing the relations of two previous diagrams (Figs. 14-2 and 14-3) into a compact form for reference.

so $\mathcal{F} = \mathcal{F}^{-1}$ and $\mathcal{F}\mathcal{F} = \mathcal{I}$.) Operational algebra leads to these several results painlessly; nevertheless, the full integral expressions look formidable. What we have deduced is that

$$\int_{-\infty}^{\infty} \mathbf{f}(r) J_0(2\pi qr)2\pi r \, dr = \int_{-\infty}^{\infty} e^{-i2\pi qx} \left[\int_{x}^{\infty} \frac{\mathbf{f}(r)r \, dr}{\sqrt{r^2 - x^2}} \right] dx$$

$$\int_{x}^{\infty} \frac{\mathbf{f}(r)r \, dr}{\sqrt{r^2 - x^2}} = \int_{-\infty}^{\infty} e^{-i2\pi xq} \left[\int_{0}^{\infty} \mathbf{f}(r) J_0(2\pi qr)2\pi r \, dr \right] dq$$

$$\int_{-\infty}^{\infty} e^{-i2\pi xr} \mathbf{f}(r) dr = \int_{x}^{\infty} \frac{r}{\sqrt{r^2 - x^2}} \left[\int_{0}^{\infty} \mathbf{f}(r) J_0(2\pi rq)2\pi q \, dq \right] dr.$$

These useful integral relations are all special cases of the projection-slice theorem.

The Inverse Cycle

By pursuing the cycle of transforms in the opposite direction, one finds three further cyclic statements:

$$\mathcal{F}\mathcal{H}\mathcal{A}^{-1} = \mathcal{H}\mathcal{A}^{-1}\mathcal{F} = \mathcal{A}^{-1}\mathcal{F}\mathcal{H} = \mathcal{I}$$

and three further substitution relationships:

$$\mathcal{H} = \mathcal{A}^{-1}\mathcal{F}, \qquad \mathcal{A}^{-1} = \mathcal{H}\mathcal{F}, \qquad \mathcal{F} = \mathcal{H}\mathcal{A}^{-1}.$$

Of these, the second relationship expands into an integral formulation expressing a procedure leading to the inverse Abel transform of a given function $f_A(x)$:

$$\mathbf{f}(r) = \mathcal{A}^{-1}f_A(x) = \int_0^\infty J_0(2\pi rq)\left[\int_{-\infty}^\infty f_A(x)e^{-i2\pi qx}dx\right]2\pi q\,dq.$$

This form of solution for Abel's integral equation is quite different in appearance from the standard solutions in terms of a single integral and looks more difficult. In fact, though, the right-hand side is composed of familiar Fourier and Hankel transforms, which have been thoroughly explored. Consequently, this inversion of the Abel transformation is useful in locating an exact integral inverse, if there is one, for mixed analytic/computational solutions, and for inverting functions defined piecewise.

The Four-stage Cycle

Figure 14-4 makes it clear without explanation that

$$\mathcal{F}\mathcal{A}\mathcal{F}\mathcal{A} = \mathcal{A}\mathcal{F}\mathcal{A}\mathcal{F} = \mathcal{I}$$

and, by proceeding around the rectangular outline in the opposite direction, that

$$\mathcal{F}\mathcal{A}^{-1}\mathcal{F}\mathcal{A}^{-1} = \mathcal{A}^{-1}\mathcal{F}\mathcal{A}^{-1}\mathcal{F} = \mathcal{I}.$$

By symbolic manipulation, or directly from the circuit of Fig. 14-4, we obtain the substitution relationships

$$\mathcal{A} = \mathcal{F}\mathcal{A}^{-1}\mathcal{F} \quad \text{and} \quad \mathcal{A}^{-1} = \mathcal{F}\mathcal{A}\mathcal{F}.$$

The second of these, expressed in full, offers a further inversion of the Abel transformation, this time not involving a Bessel function kernel but instead two ordinary one-dimensional Fourier transforms:

$$\mathbf{f}(r) = \mathcal{A}^{-1}f_A(x) = \int_{-\infty}^\infty e^{-i2\pi ru}\left[\int_u^\infty \frac{1}{\sqrt{q^2 - u^2}}\left[\int_{-\infty}^\infty f_A(x)e^{-i2\pi qx}dx\right]q\,dq\right]du.$$

In diagrammatic terms, in order to travel north, from bottom left to top left, we have taken the route via the other three sides of the rectangle, going east, north, and west.

THE PROJECTION-SLICE THEOREM

It is clear that more general theorems must exist which are not limited by circular symmetry. Modifying the initial cyclic result of the previous section by taking Hankel transforms of both sides to get

$$\mathcal{H}\mathbf{f}(r) = \mathcal{F}\{\mathcal{A}\mathbf{f}(r)\},$$

we see that one can obtain the Hankel transform of $\mathbf{f}(r)$ by first taking the Abel transform and then taking the one-dimensional Fourier transform. In other words, the "projection"

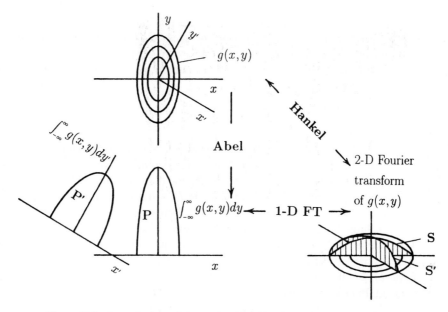

Figure 14-5 Synoptic chart of the relationship of line integration to one- and two-dimensional Fourier transforms (see Bracewell, 1956).

$\mathcal{A}\mathbf{f}(r)$ and the "slice" $F(u, 0)$ through the two-dimensional Fourier transform $F(u, v)$ are a one-dimensional transform pair.

In the latter way of phrasing the connection, the words apply even when circular symmetry is not present.

Projection-slice theorem   The projection of $f(x, y)$ in the direction θ is the one-dimensional Fourier transform of the slice through $F(u, v)$ in the corresponding direction.

To prove that the slice $F(u, 0)$ along the u-axis is the one-dimensional Fourier transform of the x-axis projection $\int_{-\infty}^{\infty} f(x, y)\, dy$ of $f(x, y)$, we start with the two-dimensional Fourier transform relation

$$F(u, v) = \int_{-\infty}^{\infty} \int_{-\infty}^{\infty} f(x, y) e^{-i2\pi(ux+vy)}\, dx\, dy.$$

Now put $v = 0$, because it is the slice along $v = 0$ that we want, to get

$$\int_{-\infty}^{\infty} \left[\int_{-\infty}^{\infty} f(x, y)\, dy \right] e^{-i2\pi(ux)}\, dx = F(u, 0).$$

This completes the proof for the particular direction $\theta = 0$. We see that the quantity in square brackets is the desired projection, or line integral, and that the whole expression is the one-dimensional Fourier transform of this projection. The general situation is summarized in Fig. 14-5, a diagrammatic synopsis of the above relationships reproduced from the original paper. The projection shown as the line integral $f_L(x)$ is the Fourier transform of the slice $F(u, 0)$, shown in the original notation as $\bar{f}(u, 0)$. The spatial organization, with function domains on the left and Fourier domains on the right, is the same as before.

In terms of the projection operator $\mathcal{P}_\theta$ introduced in Chapter 2, the previous equation could be written

$$^1\mathcal{F}\{\mathcal{P}_0 f(x, y)\} = \left[{}^2\mathcal{F} f(x, y) \right]_{\theta=0}.$$

A projection taken in a second direction is shown in the figure; the Fourier transform of that projection is the slice through the two-dimensional transform along the line QR. To prove this assertion we call on the rotation theorem of Chapter 4, according to which rotating a function $f(x, y)$ rotates its two-dimensional Fourier transform through the same angle without other change. Instead of projecting in a second direction, rotate $f(x, y)$ through an angle θ and call the modified version $f_\theta(x, y)$. Then the projection $\int f_\theta(x, y)\, dy$ onto the x-axis is the same as the former oblique projection. The one-dimensional Fourier transform of this projection is the slice $F_\theta(u, 0)$, where $f_\theta(x, y) \,{}^2\!\supset F_\theta(u, v)$. But the slice $F_\theta(u, 0)$ is the same as the slice along QR because of the rotation theorem. In projection-operator notation the projection-slice theorem may be summarized as

$$^1\mathcal{F}\{\mathcal{P}_\theta f(x, y)\} = \left[F(u, v) \right]_{\theta=\theta}.$$

The projection-slice theorem appears almost trivial when derived in this mathematical way, which simply substitutes zero for an existing variable, but the theorem provides a powerful thinking tool. It was first presented in the context of radio astronomy in the 1956 paper mentioned above, where it was given in the form

$$\mathcal{F} f_A = \int_{-\infty}^{\infty} e^{-i2\pi ux}\, dx \int_{-\infty}^{\infty} f(x, y)\, dy = \int_{-\infty}^{\infty}\int_{-\infty}^{\infty} e^{-i2\pi ux} f(x, y)\, dx\, dy = F(u, 0),$$

followed by copious particular examples. The diagram of Fig. 14-5 possesses the systematic spatial arrangement of the functions that is the recognizable ancestor of many modern-day descendants. Adoption of the term (u, v)-plane in radio astronomy dates from this time. Some early history may be found in Sullivan (1984) and Deans (1983). The subsequent history of the diffusion into radiology and neurology and thence into geophysics and various technologies is also interesting and can be traced through the "Science Citation Index."

Particular Examples

Figure 14-6 illustrates a function $f(x, y)$ that could be described as a skew elliptic disc. The zero-degree projection is the same semi-ellipse that is the projection of the elliptical disc derived from $f(x, y)$ by Steiner symmetrization. The latter operation would slice $f(x, y)$ into thin vertical slices and slide each one in the y-direction until centered on the x-axis. From the rotation and similarity theorems we know that $F(u, v)$, the 2D Fourier transform of $f(x, y)$, has radial sections that are jinc functions. The projection of $F(u, v)$

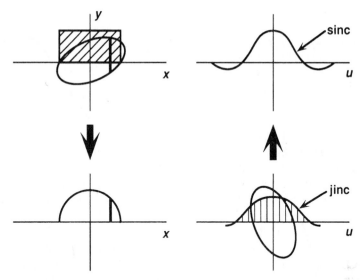

Figure 14-6 A quartet of functions on which the results of the projection-slice theorem can be tested.

would be hard to work out directly, but from the diagrammatic arrangement we know that it must be a sinc function, whose scale is fixed by the length of the segment of the x-axis within the elliptic disc.

Figure 14-7 is a case where the top right-hand corner is again a sinc function. The 2D Fourier transform is a rotated ^{2}sinc function, whose projection would again be hard to work out by integration. The zero-degree projection of $f(x, y)$ is a familiar profile; we see that the Fourier transform of such a profile is an oblique cut through a ^{2}sinc function.

In each of these examples some of the transformation relationships are obvious and others less so. An interesting feature of the diagrammatic presentation is that it offers options as to which way to go when the relationships are being worked out; sometimes the full set of functions can be ascertained by taking easy steps only. Sometimes only two of the functions enter into the original problem, and the way to find the relationship between them is to invoke one or both of the other functions that the diagram provides for.

Many of these ideas can be expressed formally as integral relationships, just as was done at some length for the Abel transform and its inverse, for there are several forms of the projection-slice theorem. This chapter has shown that the diagrammatic arrangement of operations is a basic thinking tool that substitutes for an extended formulary.

LITERATURE CITED

R. N. BRACEWELL (1956), "Strip integration in radio astronomy," *Aust. J. Phys.*, vol. 9, pp. 198–217.

R. N. BRACEWELL (1979), "Computer image processing,"*Annual Reviews of Astronomy and Astrophysics,* vol. 17, pp. 113–134.

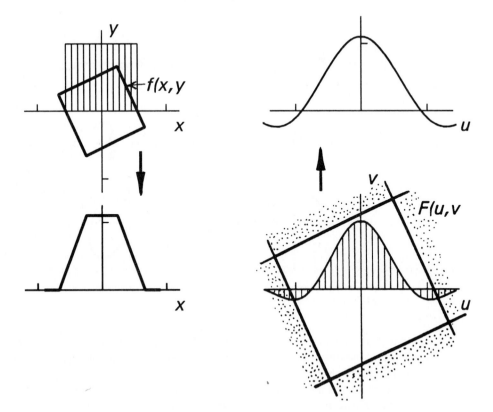

Figure 14-7   Another quartet of functions, consisting of a skew rectangle function
$f(x, y)$ (top left), its projection (bottom left), and its two-dimensional transform $F(u, v)$
(bottom right), which is a skew ^{2}sinc function. The sinc function (top right) is both the
projection of the ^{2}sinc function and the Fourier transform of the cross section of the
rectangle function along $y = 0$. Rabatted cross sections of the functions of two variables
are shown hatched.

R. N. BRACEWELL AND A. C. RIDDLE (1967), "Inversion of fan beam scans in radio astronomy,"
   Astrophysical Journal, vol. 150, pp. 427–434.

S. R. DEANS (1983), *The Radon Transform and Some of Its Applications*, John Wiley & Sons, New
   York.

G. T. HERMAN, ed. (1979), "Image reconstruction in radio astromony," chap. 3 of *Image
   Reconstruction from Projections: Implementation and Applications, Topics in Applied Physics*,
   vol. 32, pp. 81–104, Springer, Berlin.

W. T. SULLIVAN, III, ed. (1984), "Early work on imaging theory in radio astronomy," in *The Early
   Years of Radio Astronomy*, pp. 167–190, Cambridge University Press.

PROBLEMS

14–1. *Abel transform.* A circular image of radius a gets brighter toward the edge, as described by $f(r) = \text{rect}(r/2a)/\sqrt{a^2 - r^2}$. Show that the light collected through a straight slit placed on the image is independent of the position of the slit.

14–2. *Concentric annular slits.*
  (a) Show that N concentric, coplanar, annular slits of radii $R, 2R, \ldots, NR$ have a diffraction pattern given by Hankel transform $\sum_{n=1}^{N} 2\pi n R J_0(2\pi n R q)$.
  (b) Explain why this expression must represent a central Airy diffraction pattern corresponding to a circular aperture of radius $(N + \frac{1}{2})R$ surrounded by an indefinite number of ring lobes spaced R^{-1}.
  (c) Show that a radial section through the kth ring lobe is expressible in terms of the half-order derivative of $\text{sinc}[(2N + 1)(Rq - k)]$. ☞

14–3. *An Abel-Fourier-Hankel quartet.* Set up an Abel-Fourier-Hankel quartet with $f(r) = \text{rect}(r/2a)/\sqrt{a^2 - r^2}$ in the top left corner. In the bottom left corner put $f_A(x)$. Fill in the two spaces on the right.

14–4. *Abel and Hankel transforms.* A function of radius falls off quadratically from a central value H to zero on a circle of diameter D, i.e., $f(r) = H[1 - (2r/D)^2]$. What are its Abel and Hankel transforms?

14–5. *Limb-darkened disk.* The sun is not quite as bright toward the limb as in the center, but the limb darkening differs over the spectrum. Consequently, white light comes from a sun that is reddish at the limb and bluish at the center of the disk. In the red, the brightness is representable by $I_0[1 - \frac{1}{4}(r/R)^2]$, where $R = 960''$. Calculate how the red sunlight falling on an exposed surface of a space vehicle will diminish from 100 percent to zero as the sun is eclipsed by the straight edge of a solar panel mounted on the vehicle, as the vehicle slowly rotates in the sunlight. ☞

14–6. *Chain of AFH cycles.* The quartet in problem 14-3 can be rescaled so that its bottom members agree exactly with the top two members of the first entry in the table of Abel-Fourier-Hankel quartets. Thus a table of six entries can be constructed in which going down the left-hand side corresponds to successive Abel transforms, while the same is true going up the right-hand side. Extend this chain of transforms as far as possible.

14–7. *Discrete projection.* A 5×5 matrix is filled by writing in integers 1 to 25 left to right, starting at the top (the central element is 13).
  (a) Work out the sequence of projected values at $\theta = 20°$, arranging for these values to have unit spacing and to be phased so that there is a value at the projected location of the central element. Round the results to one decimal. Verify that the sum of the sequence equals the sum of the matrix values.
  (b) Explain why the projection is or is not symmetrical.

14–8. *Discrete projection.* A 5×5 matrix consists of all 1s.
  (a) Project it at $20°$ and $45°$ under the conventions of the previous exercise.
  (b) Change the phasing convention so that the lower left corner element projects into an element of the projected sequence.
  (c) Explain why the results are or are not symmetrical.

14–9. *Five discrete projections.* The elements of a 5×5 matrix have values given by $100/[1 + (x - \frac{1}{4})^2 + (y - \frac{3}{4})^2]$, rounded to the nearest integer, where x and y are integers ranging from -2 to 2.

(a) Determine the five projections at $\theta = 9°, 27°, 45°, 63°, 81°$.

(b) From these projections alone, would it be possible to reconstruct the original matrix exactly?

14–10. *Four discrete projections.* The elements of a 5×5 matrix are calculated from $99/[1 + \sqrt{5 \mid x \mid} + \sqrt{9 \mid y \mid}]$), rounded to the nearest integer, and x and y range from -2 to 2. Find the projections for $\theta = 0°, 30°, 60°, 90°$. ☞

14–11. *Projection into half-intervals.* Sometimes quantity of computation is not limiting; therefore consider projecting a matrix onto an inclined line and collecting the projected values into bins of width 0.5 with a view to getting a smoother result than with unit spacing.

(a) Test both methods by projecting a 63×63 matrix of 1s at $45°$.

(b) Comment on the comparison.

14–12. *Repeated projection.* A circularly symmetrical function $f(r)$ is projected to form a new function $g(x)$. The new circularly symmetrical function $g(r)$ is now projected to form $h(x)$. Show that $h'(x) = -2\pi x f(x)$.

14–13.

(a) Construct the horizontal and vertical projections of the capital letters as represented on a 5×7 matrix.

(b) Which letters are not distinguishable by their horizontal projections alone?

(c) Which letters have the same vertical projections?

14–14. *Rotated quartet.* Show that an $\mathcal{AFH}$ quartet, if rotated half a turn in the plane of the paper, retains the original transform relations.

15

Computed Tomography

When a two-dimensional function $f(x, y)$ is line-integrated in the y-direction or, as we say, is projected onto the x-axis, the resulting one-dimensional function $\int f(x, y)\, dy$ does not contain as much information as the original. But other projections, such as the projection $\int f(x, y)\, dy'$ onto the x'-axis, where (x', y') is a rotated coordinate system, contain different information. From a set of such projections one might hope to be able to reconstruct the original function. The best-known example comes from x-ray computed tomography, where Hounsfield's brain scanner has had a dramatic effect in radiology and dependent fields such as neurology, but reconstruction of projections was already known in more than one branch of astronomy and has continued to arise in many different contexts. Some indispensable preliminary theory for following the reconstruction techniques is given in the preceding chapter. In this chapter the theory is developed in terms of the x-ray application in order to get the benefit of being able to interpret the equations in physical terms.

The attenuation of x-rays by a particular tissue may result mainly from interaction with orbital electrons and thus depend in a simple way on electron content. But in addition to this photoelectric effect there may be other mechanisms at work, especially Compton scattering. To appreciate the realities of x-ray imaging and nuclear medicine as distinct from basic tomographic theory, it will be necessary to refer to a specialized text such as Macovski (1983) or to sources such as Deans (1983). In the following discussion we speak as if x-ray attenuation depended on mass alone, glossing over the fact that different tissues

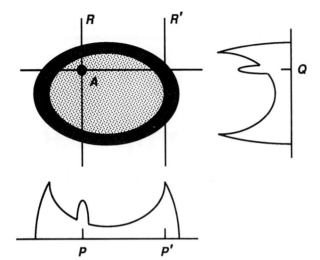

Figure 15-1   A density distribution and two projections.

have different attenuation coefficients, depending upon the mix of atomic weights of the nuclei present. In addition it is supposed that attenuation is small; then the power lost in an element of mass is the same no matter where the element is in the x-ray beam. In reality, an element of mass at the exit end of the beam causes less power loss, because the incident power is already slightly reduced.

Greek *tomos* meaning slice is familiar to biologists in the word microtome, which denotes a precision instrument for preparing thin slices of tissue for microscopy; tomography, which is the representation of a slice of a solid object under study, was in use in radiology to describe an x-ray technique in which an x-ray source moved in the opposite direction relative to the film during a time exposure. When reconstruction from projections entered radiography the term *computed tomography* was coined to distinguish the new technique.

WORKING FROM PROJECTIONS

Consider a density distribution composed of a uniform substance surrounded by a dense shell and containing a small dense object A (Fig. 15-1). The whole density distribution is to be mapped, starting from projections, of which two examples are shown. The value of the projection at point P equals the line integral of density along the straight ray R. The larger values at the ends of the projection, for example at P′, arise because the ray, in this case R′, passes through a greater distance within the shell. The second projection has larger peaks, but is less extended; however, the two projections have equal areas, because each area represents the total mass of the original distribution. If the area density is measured in kg m^{-2}, then the unit of the projected values is kg m^{-1} and the area under the projection is measured in kg.

A peak P due to the compact object tells us that the object lies on a certain ray, and

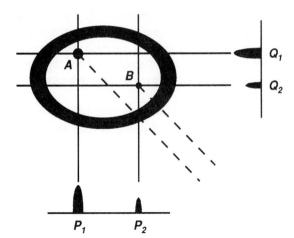

Figure 15-2 A density distribution containing two compact objects.

the corresponding peak Q places the object at the intersection. Two projections would thus suffice to locate an interior object. Of course, the two bumps at P and Q have equal areas, thus telling about the mass of the object.

How many projections would be needed to map the density in full? That is the first key question. After that is attended to, we move on to the question of constructing the map from those projections.

In a sense, any density distribution can be thought of as made up of numerous compact objects of different mass; therefore, it makes sense to think about two objects, three objects, and so on. If there are two different interior point masses, at A and B, bumps will be seen (Fig. 15-2) at P_1 and P_2 in the first projection and at Q_1 and Q_2 in the second. Two projections will still suffice to locate them, but there may be an ambiguity if the masses are comparable, because there are now four intersections at the vertices of a rectangle. The objects might be at the unmarked vertices; this concern could be settled by a third projection, whose back-projected peaks are suggested by the broken lines in Fig. 15-2.

Two projections would be enough if the masses of the compact objects were different. Two projections would also suffice for three unequal masses, and indeed for any small number of masses. With limited precision of measurement as to location or amplitude, the third projection would then clinch the identification of each peak. Of course, as the number of objects grows larger, accidental three-way intersections rapidly become more numerous, and measurement accuracy, or noise, steadily introduces uncertainty. It is therefore necessary to look at the problem another way to get an answer for the full mapping problem, but it is encouraging to know that in special circumstances a very small number of projections may yield a lot of information.

A single projection determines the location of a compact object with a certain accuracy in the transverse direction, but in depth all objects collinear with the ray are integrated together, and there is no resolution in depth at all. This is taken care of by two projections at right angles. But it often happens that some directions are not accessible. If only two projections can be obtained, and if they are separated by only a small angle, even then resolution in depth improves dramatically. For example, on a distribution 200 mm in diameter, where transverse resolution of 3 mm is achievable, suppose that a second projection can be made, but only 5° away (about one-tenth of a radian). Then the resolution in depth becomes 30 mm, which divides the 200-mm-long ray into seven bins and represents respectable resolution.

These elementary considerations have been introduced to intimate that spatial resolution, usually set by the width of the strip integral that in practice replaces the simple integral, is of the essence, and that so too is measurement accuracy.

AN X-RAY SCANNER

Let a source of x-rays S move along the line AA' in the plane $z = $ const (Fig. 15-3) emitting a collimated needle-beam into a detector D that moves in unison along BB' in the same plane. Source and detector are mounted together on a rigid yoke that is not shown. A graph of the departure of the detector output from the value obtained when only free space intervenes between S and D represents the integrated density along the beam. One such curve is merely a silhouette but, from scans versus R taken at other angles θ in addition to that illustrated, one can reconstruct the full two-dimensional density distribution in the plane $z = $ const; and then by repetition in neighboring parallel planes a three-dimensional image of density distribution may be arrived at. This is the basis of the x-ray tomograph or scanner described by Allan Cormack (1963) and Gordon Hounsfield (1972), who later shared a Nobel prize.

Any tool for nondestructive exploration of internal organs of the human body from outside is of great interest, and in the case of the heart the possibility of observing motion adds another dimension. The resolution achieved is around 1 mm, and the time taken to form a single two-dimensional image is around 1 second. Even better resolution and speed can be obtained if one accepts conditions such as limited field of view and increased x-ray dose.

Other fields such as electron micrography of viruses (DeRosier and Klug, 1968; Crowther, 1970) and ultrasonic exploration of objects and possibly the reconstruction of visual space by the striate cortex (Pollen et al., 1971) also offer biomedical applications of the scanning principle, and these applications are merely part of a broader context of nondestructive inspection which in medicine includes the use of gamma-ray cameras, positron emission, digital radiography, and dichromography using synchrotron radiation. Magnetic-resonance imaging, which is discussed further below, is a different and more complicated application of computed tomography. Other applications range from geophysical mapping of the space between boreholes to radar mapping of planetary surfaces. The principle of computed tomography is quite deep and may be expected to continue to find new applications. The problem of reconstruction of a map from a set of scans arose long ago in fields apparently having nothing to do with x-rays or medicine, for example in astronomy (Bracewell, 1956; Bracewell and Riddle, 1967; Taylor, 1967). The modified backprojection algorithm that arose from radio astronomy is in universal use in commercial

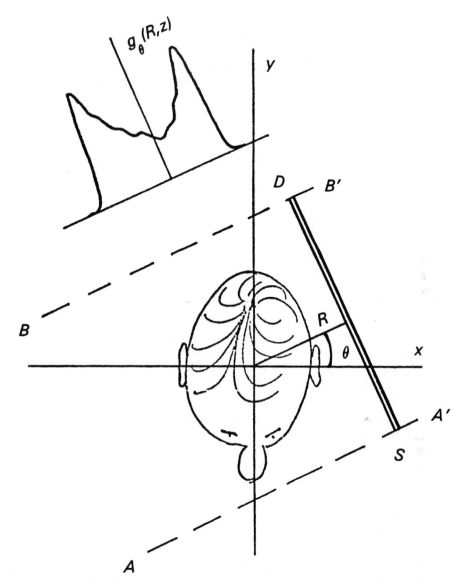

Figure 15-3   Scan $g_\theta(R, z)$ in plane $z = $ const with moving source S and detector D.
The x-ray beam lies along the line $x \cos \theta + y \sin \theta - R = 0$.

x-ray scanners. The following explanation of the algorithm is based on the Fourier transform, which furnishes an excellent tool for understanding the nature of line integration and
reconstruction, and which leads to two distinct classes of reconstruction procedure. The
first procedure involves the taking of Fourier transforms either numerically, as is usual, or
by analogue methods. The second procedure involves manipulation of the data in the data

domain, without the need to take transforms; but we do use transforms to arrive at and understand this method of reconstruction.

Our starting point is a finite set of projections presented as functions of a continuous variable. However, the notion of finite resolution, which is always in our minds when we deal with physically recorded data, is incorporated from the beginning in the transform theory given here. The discussion also benefits from the concept of impulses on the plane introduced in Chapter 3.

FOURIER APPROACH TO COMPUTED TOMOGRAPHY

The basic relation for the response $g_\theta(R, z)$ is

$$g_\theta(R, z) = \int_{-\infty}^{\infty} \int_{-\infty}^{\infty} \int_{-\infty}^{\infty} \rho(x', y', z') A(R - x'\cos\theta - y'\sin\theta, z - z') dx' dy' dz',$$

where $\rho(x, y, z)$ is the density, $A(R, z)$ is the response to unit mass at the origin $(0, 0, 0)$ and R and θ have the meaning indicated by Fig. 15-3. The function $A(R, z)$ may closely approximate the cross section of the beam but depends also on details of the detector, such as its physical size and screening from small-angle scattering, and on the degree of collimation at the source. It is clear that the image distribution can represent the object with a resolution only as faithful as the fineness of $A(R, z)$ will allow. It may be possible to compensate to some extent for beam smearing, depending on the magnitude and character of the errors (Bracewell, 1958). Following the terminology of this paper, we may refer to such compensation as *restoration* to distinguish it from *reconstruction* (Bracewell, 1956), which is the process of combining the silhouettes. There is a sharp difference between the two operations, only a modest amount of restoration being advisable as a rule, whereas full reconstruction is possible in principle. To focus clearly on reconstruction, let us deal from now on only with the smeared density distribution $\widehat{\rho}(x, y, z)$, which is the three-dimensional convolution of the true density $\rho(x, y, z)$ with a small volume distribution $B(x, y, z)$ having dimensions characteristic of the beam cross section. Thus the integral equation to be solved is

$$\widehat{\rho}(x, y, z) = \int_{-\infty}^{\infty} \int_{-\infty}^{\infty} \int_{-\infty}^{\infty} \rho(x', y', z') B(x - x', y - y', z - z') dx' dy' dz'.$$

In any plane $z = $ const, the two-dimensional Fourier transform of $B(x, y, z)$ sets a limit to the highest spatial frequency in $\widehat{\rho}(x, y, z)$, a frequency that is needed in fixing the increment in viewing angle θ. By scanning a point scatterer suspended in the beam, one can determine $A(R, z)$, and then $B(x, y, z)$ is obtainable from

$$B(x, y, z) = \frac{-1}{\pi} \int_{(x^2+y^2)^{1/2}}^{\infty} (R^2 - x^2 - y^2)^{-1/2} \frac{\partial}{\partial R} A(R, z) dR,$$

which may be recognized as the inverse Abel transform (FTA, 1986) or may be expressed by saying that $B(x, y, z)$ is that distribution (Bracewell, 1956), having circular symmetry about the z-axis, whose line integral in any plane $z=$const is $A(R, z)$.

Solving the integral equation is the restoration problem. The pure reconstruction problem, where the beam is a geometrical line of zero width, and the response is $f_\theta(R, z)$, is then to obtain $\hat{\rho}(x, y, z)$ by solving

$$f_\theta(R, z) = \int_{-\infty}^{\infty} \int_{-\infty}^{\infty} \int_{-\infty}^{\infty} \hat{\rho}(x', y', z')\delta(R - x'\cos\theta - y'\sin\theta)\delta(z - z')\,dx'\,dy'\,dz'.$$

A good deal is known about this integral equation (Bracewell, 1956); it possesses a unique solution, and the solution is known for many particular functions, especially ones with circular symmetry. A direct inversion process is known in the case of circular symmetry (the equation for $B(x, y, z)$ is an illustration of this); and the corresponding direct process for the general case was described and experimented with in the days of desk calculators and perhaps should be reexamined. There are various Fourier transform methods based on the fact that the two-dimensional Fourier transform

$$F(u, v, z) = \int_{-\infty}^{\infty} \int_{-\infty}^{\infty} \hat{\rho}(x, y, z)e^{-i2\pi(ux+vy)}\,dx\,dy$$

is obtainable in a way that is best explained by going to cylindrical coordinates (q, θ, z) in terms of which

$$F(u, v, z) \equiv F_\theta(q, z).$$

Then, in any plane $z = $ const and for a given θ, values of $F(u, v, z)$ may be calculated from the fact that $F_\theta(q, z)$ is the one-dimensional Fourier transform of $f_\theta(R, z)$, i.e.,

$$F_\theta(q, z) = \int_{-\infty}^{\infty} f_\theta(R, z)e^{-i2\pi qR}\,dR. \tag{3}$$

To make use of the fast algorithm for the two-dimensional Fourier transform it is necessary to have values on a rectangular lattice. For each lattice point (u, v, z) one can determine $\theta = \arctan(v/u)$ and $q = (u^2 + v^2)^{1/2}$ and evaluate the right-hand side of this equation. It is not obvious that the full power of the fast two-dimensional Fourier transform will make itself felt, because the subsidiary calculations per lattice point are substantial and because the values of θ called for may not be ones for which data exist, and so interpolation calculations may be needed.

A different Fourier scheme is to evaluate the equation for $F_\theta(q, z)$ for equis-paced values of θ corresponding to the plan of data collection and then to perform one-dimensional Fourier transformation of the corresponding narrow sectors, including a weight factor $|q|$ before transformation to take account of the taper of the sectors. The resulting one-dimensional transforms have to be distributed all over the image plane for each value of θ in turn, an operation which again is not attractive digitally, as it involves interpolation, but is very attractive for certain analogue presentations, such as accumulation on a television tube or on a photographic plate.

Other Fourier methods can be contemplated that are appropriate to various procedures for data collection and image presentation. For example, θ and R may be varied either discretely or continously, and, instead of fixing θ while R is varied, it is possible to fix R and vary θ. Another possibility is to vary R and θ simultaneously in such a way that the beam passes through a fixed point and then moves on to a neighboring point, a procedure that contains redundancy but may have an application. For speed of image formation a large number of detectors are used to deal with a large number of values of θ at the same time.

BACK-PROJECTION METHODS

Since the fast Fourier transform is encumbered by the awkward subsidiary calculations mentioned above, it could be that direct numerical operations on the data in the object plane have an advantage.

The Layergram

A solution of this kind begins by assembling the observed scans $f_{\theta_i}(R, z)$ into a "layergram" $l(x, y, z)$ by the formula

$$l(x, y, z) = N^{-1} \sum_{i=1}^{N} f_{\theta_i}(x \cos \theta_i + y \sin \theta_i, z),$$

where θ_i is the angle of the ith scan and it is assumed that a range π is covered at equal intervals. Each of the summands may be pictured, at $z = $ const, as a ridge on the (x, y)-plane described by straight and parallel contours in the θ direction and having a cross-section profile f_θ. In other words, each scan f_{θ_i} is back-projected across the plane, and all the back projections are summed. The layergram so obtained is not the same as the original object but may be usable for some purposes. Such a superposition was described by Vainshtein (1971).

To understand the nature of a layergram it is sufficient to picture the point response. Each scan of a point mass is a simple delta function, and the back projection of each scan is a line impulse passing through the position of the point mass. Thus, for N equispaced scans the layergram for unit mass at the origin is

$$l_0(x, y, z) = \delta(z)N^{-1} \sum_{i=1}^{N} \delta[y \cos(\pi i/N) - x \sin(\pi i/N)]$$

as illustrated in Fig. 15-4.

While it is true that there is a heavy concentration near the origin, which is where the mass is, the layergram extends out in all directions. In any annulus of unit width centered on the origin, the mass contained is inversely proportional to radius; in a sense, the distribution $l_0(x, y, z)$ resembles the function r^{-1}, and for this reason it is sometimes said that the layergram for an infinite number of angles is the two-dimensional convolution between the desired distribution $\widehat{\rho}(x, y, z)$ and r^{-1}. Thus

$$l(x, y, z) = r^{-1} ** \widehat{\rho}(x, y, z).$$

The proof of this relation involves some niceties that will be relegated to the problems. Since r^{-1} is not very compact, possessing a broad skirt of substantial strength (the volume

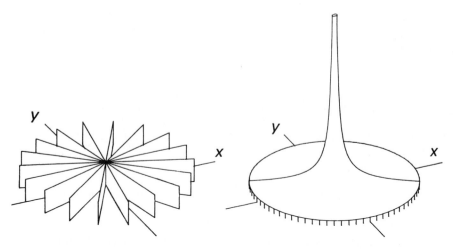

Figure 15-4 The point-mass layergram $l_0(x, y, z)$ formed from 18 equispaced scans (left) and the inverse dependence on radius to which it is likened (right).

under the skirt beyond any given radius is infinite), the layergram leaves something to be desired.

In three dimensions the corresponding superposition $r^{-2} *** \hat{\rho}(x, y, z)$, which arises in magnetic-resonance imaging, may be more useful because r^{-2} is in a way more compact than r^{-1}. However, the triple integral of r^{-2} beyond any given radius is still infinite.

Rigorous Function-domain Solution

One can arrive at a correct reconstruction procedure by noticing where the layergram is at fault. The Fourier transform of the layergram is, taking $P(u, v, z)$ to be the two-dimensional transform of $\hat{\rho}(x, y, z)$,

$$L(u, v, z) = P(u, v, z)\left\{N^{-1}\sum_{i=1}^{N}\delta[u\cos(\pi i/N) + v\sin(\pi i/N)]\right\},$$

which can be understood as $P(u, v, z)$ sampled on a spoke pattern. The density of sampling is high near the origin. Consequently, the layergram is relatively strong in low frequencies and deficient in high frequencies. In terms of the radial variable q in the (u, v)-plane we can say that, since the sampling density goes inversely as the spacing between adjacent spokes, or as q^{-1}, the equalizing factor is simply q. Therefore it makes sense to ask, "What operation in the function domain corresponds to multiplication by q in the transform domain?" Then, if this operation were carried out on the layergram, the layergram would be corrected as desired. Still, multiplication by q is a potentially dangerous operation, because the required amplification increases steadily without limit as spatial frequency increases.

Since we do not expect spatial frequencies higher than the resolution justifies, we

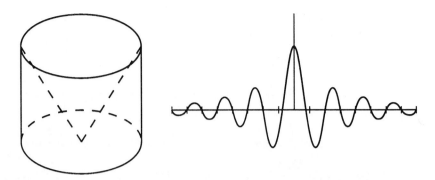

Figure 15-5 The conical crater $\text{rect}(q/2M) - \Lambda(q/M)$ and a radial section through its two-dimensional Fourier transform.

now recognize this practical fact by introducing a parameter M, which is a certain high spatial frequency beyond which no content is expected to be present. One way of arriving at a value of M is through knowledge of the beam pattern $A(R, z)$, which enables M to be fixed by the condition that

$$\int_{-\infty}^{\infty} A(R, z)e^{-i2\pi qx}\, dx$$

should be negligible for all $q > M$. Choice of M in this way will automatically ensure that two-dimensional components of $\widehat{\rho}(x, y, z)$ are negligible for spatial frequencies greater than M.

Then for multiplication by q we substitute multiplication by the product of q and $\text{rect}(q/2M)$. This factor, which is illustrated in Fig. 15-5, may be expressed as $\text{rect}(q/2M) - \Lambda(q/M)$, because it consists of a disk function with a conical crater, rather like the funnel of a rain gauge. Therefore the reconstructed distribution is obtainable from two-dimensional convolution with a circularly symmetrical function of radius only, which in the limit of closely spaced scans is

$$4M^2 \text{jinc}(2Mr) - M^2 \overline{\Lambda}(Mr),$$

where $\overline{\Lambda}(r)$ is the Hankel transform of $\Lambda(q)$ and is equal to $2\pi[x^{-3}\int_0^x J_0(x)\, dx - x^{-2}J_0(x)]$. This solution has not been published before. It says: First form the layergram from the scans. Then convolve in two dimensions with a certain circularly symmetrical function as illustrated. A feature of this solution is that all the operations are carried out in the function domain—that is, the domain in which the scans are taken and in which the reconstructed solution is desired.

Modified Back-projection Method

Starting from the ideas above, Bracewell and Riddle (1967) showed how to reduce the operations to a number of one-dimensional convolutions on the scans themselves, which is of direct practical significance, because that is the form in which the data come when scans are made in successive orientations. The solution in this form gives the answer as a superposition of back-projected profiles h_θ over all angles. Thus

$$\widehat{\rho}(x, y, z) = \int_o^\pi h_\theta(x \cos\theta + y \sin\theta, z)\, d\theta.$$

The function h_θ is derived from the corresponding line scan f_θ by modifying it with a correction term, which is derived by convolving f_θ with a certain sinc^2 function. Thus

$$h_\theta(R, z) = f_\theta(R, z) - \int_{-\infty}^\infty f_\theta(R - R', z) M \, \text{sinc}^2(MR')dR'.$$

The sampling theorem tells us that it is sufficient to measure $f_\theta(r, z)$ at discrete intervals $1/2M$, because $f_\theta(r, z)$ has an upper cutoff frequency M. This fact bears on instrument design either by setting the step length if the source is to move in steps or setting the product of velocity with integrating time if the motion is continuous. Since $f_\theta(R, z)$ is band-limited, the foregoing equation reduces exactly to a sum

$$h_\theta(R, z) = \sum_{n=-\infty}^\infty a_n f_\theta\left(R - \frac{n}{2M}, z\right) = \{a_n\} * \{f_\theta\},$$

where the coefficients a_n are given by

$$a_n = \begin{cases} 0.5, & n = 0 \\ -0.5 \, \text{sinc}^2(n/2), & n \neq 0 \end{cases}$$

and $\{f_\theta\}$ is a sequence of samples of $f_\theta(R, z)$ spaced at $1/2M$. Substituting numerical values for the coefficients, we find the indicated operations are: form a string of values of f_θ at intervals $1/2M$, modify them by forming the convolution sum with a set of coefficients

$$\left\{ \cdots \quad \frac{-2}{25\pi^2} \quad 0 \quad \frac{-2}{9\pi^2} \quad 0 \quad \frac{-2}{\pi^2} \quad \frac{1}{2} \quad \frac{-2}{\pi^2} \quad 0 \quad \frac{-2}{9\pi^2} \quad 0 \quad \frac{-2}{25\pi^2} \quad \cdots \right\},$$

and then assemble the image distribution by the discrete equivalent of the equation for $\widehat{\rho}(x, y, z)$. The sequence of coefficients can be verified to add up to zero. The related sequence $\{1\} - \{a_n\}$ was given by Ramachandran and Lakshminarayanan (1971) (to within a numerical factor).

The modification procedure does not make a profound change. An example presented by Vainshtein (1971), who omitted the modification, shows perfectly legible reproduction of a simple black-and-white test pattern. But where precision is required, as in the rendering of the very low contrast objects constituted by many biological tissues, the rigorous theory cannot be sidestepped.

Number of Scans Needed

Just as there is an upper limit to the step length in R if the potential resolution of the beam is to be realized, so also there is an upper limit to the step length in θ.

Let N scans be taken at an interval π/N in θ. Then, if the object diameter is D, it was shown by Bracewell and Riddle (1967) that a useful choice is

$$N = \pi M D.$$

The reason is that D^{-1} is the critical sampling interval in the (u, v)-plane. Along the semi-circumference of length πM, the sampling interval will equal D^{-1} if there are $N = \pi M D$ scans. A small error is involved whose magnitude can best be studied, not by comparing various artificial objects with their reconstructions, but in terms of a point-response pattern $b(x, y)$, which is the image produced when the object is a point mass. Measurements are made at intervals $1/2M$ in R and at intervals π/N in θ, and reconstruction proceeds by the discrete approximation to the equation for $\widehat{\rho}(x, y, z)$. This is given by

$$b(x, y) = \sum_{i=1}^{N} \frac{\pi M}{N} k(x \cos \theta_i + y \sin \theta_i),$$

where $k(R) = 2M \operatorname{sinc} 2MR - M \operatorname{sinc}^2 MR$. This equivalent pattern, or response of the overall system, consists of a central spike of width of the order of $1/M$ accompanied by characteristically shaped sidelobes, including remote haloes of radius D or more. The tolerable smearing, sidelobe level, and closeness of haloes is a matter for each application. A maximum discrepancy of 1 percent of maximum between $b(x, y)$ and the ideal was reported by Bracewell and Riddle for $N = \pi M D$, but reducing N by a factor of two, which would double the speed of operation, led to quite small discrepancies of the order of $130/(2N + 1)$ percent. Thus the choice of N is one involving experience, not merely theory. There are theoretical reasons why halving N should work. Requiring the arc interval $\pi M/N$ at the circumference of radius M to equal D^{-1} means that all interior arc intervals are shorter than critical, less on the average by a factor of two. The original discussion of the problem of the number of scans (Bracewell, 1956) led to $\frac{1}{2}\pi M D$, and in the absence of noise this would be applicable. However, as explained in Chapter 7, nonuniform sampling leads to degradation, even though the mean density is the same as for samples spaced uniformly at the critical interval.

Assessment of Algorithms

The relative merits of convolution and Fourier transformation, and the various combinations of these, will vary with each case. But in assessing different methods of reconstruction by the use of an artificial test object, specified numerically or measured using an artifact or "phantom," it is important to avoid objects with edges or other sharp features. Otherwise the comparison of results achieved will be contaminated by questions of restoration for degraded resolution. This is important, because blurring can be only partially compensated (Bracewell and Roberts, 1954), whereas reconstruction deals with information that has merely been scrambled rather then partly removed. Thus one should first fix on a resolution limit (e.g., as expressed here simply by a value of M, the highest spatial frequency), and then set up an artificial test object $\widehat{\rho}(x, y, z)$ not containing frequencies that will be filtered out. Otherwise the comparison of different reconstructions against the test object will be invalid.

Other practical aspects of this subject include the problem of scans that are available over less than the full 180 degrees (the missing-sector problem), the problem of a limited number of scans at unequal spacing, and the problem of a beam whose cross section is not constant along its length. The question of restoration has already been mentioned. In acoustic, seismic, and underwater applications, refraction of the beam is an essential feature. Refraction depends on the spatial distribution of the quantity that is to be found; therefore iteration is necessary because the ray paths are initially only approximations. Completely different techniques, such as analysis into cylindrical harmonics (Klug and Crowther, 1972) or eigenfunctions (M. Ein-Gal, 1974) and iterative algorithms (Crowther and Klug, 1974), may all have a place under various practical conditions. For an overview see Herman and Lewitt (1979), Brooks and DiChiro (1976). Presentation of density distributions in space is important in fields, especially medicine, where volumetric data sets can be acquired. Techniques for conveying three-dimensional images to the eye include holography, stereography, and volume rendering on plane images by various depth cues (Totsuka and Levoy, 1993).

THE RADON TRANSFORM

Limiting attention frankly to the two dimensions of the (x, y)-plane, we can say that a density distribution $\rho(x, y)$ has a Radon transform which may be abbreviated by use of the Radon transform operator $\mathcal{R}$ to read

$$\mathcal{R}\rho(x, y) = \int_{-\infty}^{\infty} \int_{-\infty}^{\infty} \rho(x, y)\delta(R - x\cos\theta - y\sin\theta)\, dx\, dy.$$

The operand ρ may be thought of as analogous to the input or cause in other subjects, and the result of the integral operation or Radon transformation as the output or response.

Vector notation is convenient for personal use and in multidimensional theory. Let $\mathbf{t}$ be the unit vector in the scanning direction θ, having components $(\cos\theta, \sin\theta)$ in the case of two dimensions. Let the vector $\mathbf{x}$ have components (x, y), and let the scalar $d\mathbf{x}$ stand for $dx\, dy$. Then one can write

$$\mathcal{R}\rho(x, y) = \int \rho(\mathbf{x})\delta(R - \mathbf{x} \cdot \mathbf{t})\, d\mathbf{x}.$$

The condensed notation does not save enough space to be worthwhile in this chapter, but one needs to be prepared to encounter this notation in the literature.

Johann Radon (1887-1956) was a Bohemian whose paper (Radon 1917) initiated the literature of the Radon transformation. The German text, which was written in Vienna, has been reproduced by Helgason (1980), and an English translation by R. Lohner appears in Deans (1983).

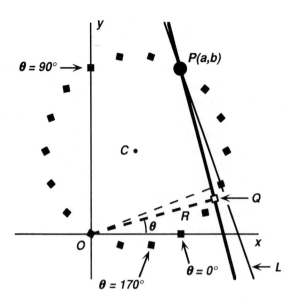

Figure 15-6 A point mass P scanned by a line L (shown in two successive positions) at discrete values of angle.

THE IMPULSE RESPONSE OF THE RADON TRANSFORMATION

For a geometrical interpretation of the operator $\mathcal{R}$ acting on a density distribution $\rho(x, y)$ it is convenient to imagine the scans arranged radially to constitute an output function, which, like the input distribution or operand, is also a two-dimensional function. Of course, the notation $g_\theta(R)$ suggests how the set of scans may be assembled in a plane so that an output value is placed at an angle θ and distance R. In this way of looking at things we may write

$$g_\theta(R) = \mathcal{R}\rho(x, y).$$

As with any operator, one's understanding is deepened by knowing its impulse response and its eigenfunctions. Here we consider the impulse response.

Let there be a unit point mass at $P(a, b)$ describable in terms of an area density distribution $\rho(x, y)$ by

$$\rho(x, y) = {}^2\delta(x - a, y - b).$$

The Radon transform will naturally depend on a and b, and we may write

$$\mathcal{R}^2\delta(x - a, y - b) = I(a, b, R, \theta).$$

This equation defines $I(a, b, R, \theta)$, the impulse response of the Radon transform, i.e., the Radon transform of a unit point mass at (a, b). If we knew the form of the two-dimensional impulse response $I(a, b, R, \theta)$, or point-spread function, we could picture the Radon transform as a superposition of the separate contributions from all the elements of the density distribution and write $\mathcal{R}\rho(x, y) = \int_{-\infty}^{\infty} \int_{-\infty}^{\infty} \rho(x, y) I(x, y, R, \theta) \, dx \, dy$.

Fix attention on a particular value of θ and picture a line L scanning across the (x, y)-plane, maintaining an inclination θ to the x-axis as in Fig. 15-6. The integral of $\rho(x, y)$ along the line will be zero except when the line passes through the point mass at P as shown. Thus $g_\theta(R)$ is nonzero, for the selected value of θ, only where R equals OQ, the perpendicular distance from the origin to the line L when this line passes through P. We place a square dot at Q to remind ourselves that $g_\theta(R)$ is nonzero there. To explore the full behavior of $g_\theta(R)$ we must examine other values of θ over the range from 0 to π; we see that contributions to $g_\theta(R)$ will be confined to the locus of Q, the foot of the perpendicular from the origin O to lines passing through P. Since angle OQP is a right angle, a well-known theorem of geometry tells us that the locus of Q is a circle of which OP is a diameter.

This circular locus has been picked out by a number of dots corresponding to equal intervals of θ. The dots themselves are equidistant, a fact suggesting that $I(a, b, R, \theta)$, the two-dimensional impulse response that we seek, may be a circular line impulse, or ring delta, of the character of

$$\delta(r' - \alpha),$$

where r' is a radial coordinate measured from the center C of the circle and α is the radius of the circle.

The Radon Transform as a Superposition Integral

Since $I(a, b, R, \theta)$ is by definition the Radon transform of $^2\delta(x - a, y - b)$, we may write formally

$$I(a, b, R, \theta) = \int_{-\infty}^{\infty} \int_{-\infty}^{\infty} {}^2\delta(x - a, y - b)\delta(R - x \cos - y \sin \theta) \, dx \, dy.$$

The trouble with this integral is that it does not come within the scope of the theory of generalized functions. Indeed, that theory explicitly refrains (Lighthill, 1958, p. 19) from assigning any meaning to the product of generalized functions, which is what we appear to be dealing with here. However, in two dimensions the subject has features that do not arise in one dimension. Without going into a general discussion, we can solve our present dilemma as follows.

Consider a density distribution which is nonzero only on a narrow straight strip of width τ, within which the density $\rho(x, y)$ has a value τ^{-1} (Fig. 15-7). The area of a cross section perpendicular to the direction of the strip is unity. The sequence of density distributions, generated as $\tau \to 0$ and as the load concentrates toward a line which we may call the centerline of the strip, defines a linear mass distribution that is uniform. The linear density is equal to the area of a perpendicular section, as may easily be verified, and is therefore unity. Now, a cross section taken obliquely along a line making an angle i with the centerline of the strip has a cross section area that is greater by a factor $\sec i$, regardless of the value of τ when τ is small. Consequently, if it were a question of determining the linear density of a line distribution at a point where an oblique cross section area was

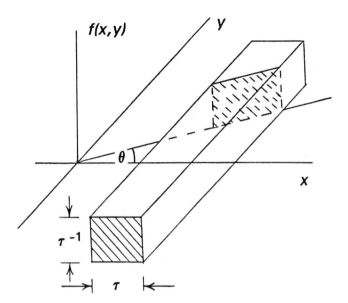

Figure 15-7   Oblique section through a wall.

known, it would be necessary to reduce the area by a factor cos i. This procedure may be applied to situations where the density varies along the line. It is true that, given τ, the cross sections will in general reflect an average density taken over some length of the line distribution. But as $\tau \to 0$, the limit of the cross section area, multiplied by cos i, will represent the local linear density.

The equation for $I(a, b, R, \theta)$ will now be discussed with the aid of Fig. 15-8. Replace the unit point mass at P by a unit mass spread over a small circular area of diameter τ surrounding P. The line integral $g_\theta(R)$, displayed as a function of R for fixed θ, will then be a narrow hump of unit area and width τ as shown shaded. Along a line $\theta =$ const in the (R, θ)-plane the line integral $g_\theta(R)$ will be nonzero only on a short segment of length τ, such as the one indicated at Q. The nonzero part of $g_\theta(R)$ will be confined within the crescent-shaped strip whose width as measured along any line through O has a constant value τ. As $\tau \to 0$, $g_\theta(R)$ remains nonzero only on the circle of which OP is a diameter, but the linear density along this circle will be nonuniform and given by cos i. Indeed, the linear density is proportional to the strip width normal to the centerline of the strip (the line toward which the strip shrinks as $\tau \to 0$). As a further aid to visualization of the nonuniform line impulse $I(a, b, R, \theta)$ we have Fig. 15-9, which invokes the convention of a vertical blade whose height at any point measures the line density at that point. The line density falls off cosinusoidally, being equal to the cosine of half the angle PCQ, or alternatively it is proportional to the radial distance OQ.

We now have a clear understanding of the character of $I(a, b, R, \theta)$. It remains to express it mathematically, but we now understand that our earlier surmise about $\delta(r' - \alpha)$ is wrong, because that expression represents a uniform ring delta of unit line density. The

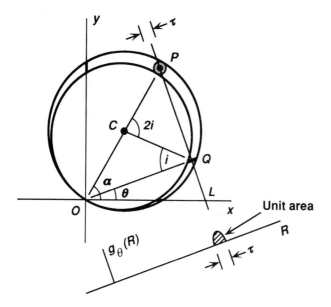

Figure 15-8 Boundary of Radon transform of a small disc centered at P.

circle [1] $r - (a^2 + b^2)^{1/2} \cos(\theta - \beta) = 0$, where $\beta = \arctan(b/a)$, is the one where the two-dimensional impulse response is concentrated. Perhaps

$$I(a, b, R, \theta) = \delta[r - (a^2 + b^2)^{1/2} \cos(\theta - \beta)] = \delta(r - a \cos\theta - b \sin\theta).$$

As no generality is lost by taking the point mass P to be on the x-axis, let us put $b = 0$. We are now supposing that

$$\mathcal{R}\{^2\delta(x - a, y)\} = \delta(r - a \cos\theta).$$

Since $r - a \cos\theta$ is zero on the circle whose diameter is OP (Fig. 15-10), the right-hand side represents a curvilinear line impulse in the correct location, but its strength has to be examined. Replacing $\delta(\cdot)$ by $\tau^1 \text{rect}(\cdot/\tau)$ converts the right-hand side to

$$\tau^{-1} \text{rect}\left(\frac{r - a \cos\theta}{\tau}\right).$$

Now rect x falls to 0 where $x = \pm 1/2$, so this expression represents a function having a value τ^{-1} on a region bounded by

$$\frac{r - a \cos\theta}{\tau} = \pm 1/2.$$

The polar equation

$$r = a \cos\theta \pm \tau/2$$

[1] It is customary with line integrals to limit θ to a range of 180° and to think of R as ranging over negative and positive values. Use of R rather than r helps to remind us that R is really a rotated cartesian coordinate. But for other purposes there is no reason to depart from the usual polar coordinates (r, θ).

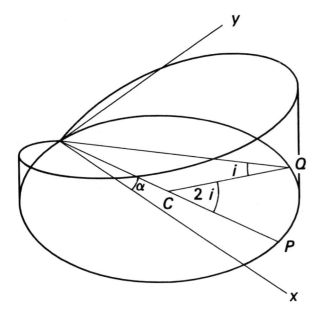

Figure 15-9   Representation of the Radon transform of a point mass at P by means of a blade of variable height.

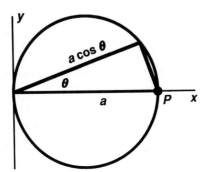

Figure 15-10   Polar equation of the circle shown is $r = a \cos \theta$.

represents precisely the kind of curve which bounds the crescent-shaped region in Fig. 15-8.

In general, the equation $r = a \cos \theta + b$ describes curves known as the *limaçons of Pascal.* They are also *epitrochoids,* being traced by a point fixed on the rim of a disk of radius a that rolls on a disk of radius $b - a$.

We know that the cross section area of $\tau^{-1} \mathrm{rect}[(r - a \cos \theta)/\tau]$, on a radial cut through the origin, is unity, because the height is τ^{-1} and the width is τ. Consequently, the sequence generated as $\tau \to 0$ correctly describes a circular line impulse whose linear density equals $\cos i$. Thus the previous conclusion is confirmed. The linear density ranges between 0 and 1, as given by $\cos(\theta - \beta)$.

Historical Note

The epitrochoid played an important role in the development of mechanics and the laws of motion, and thus of Western culture, because planetary orbits, whose study ultimately led Newton to his laws of motion, are describable to a first order as epitrochoids in the Ptolemaic earth-centered system. For one and a half millennia, from the time of Ptolemy until after the time of Copernicus, astronomers computed planetary positions using epitrochoids. We now understand that the radius vector to a point moving on an elliptical orbit may be Fourier analyzed into a series of harmonic terms and that if we limit ourselves to the fundamental and second harmonic we get the epitrochoid.

Analytic Interpretation of $\delta[f(x, y)]$

As shown in an earlier chapter, the linear density of the line impulse $\delta[f(x, y)]$ is given, at points on $f(x, y) = 0$, by [2]

$$\left[\left(\frac{\partial f}{\partial x}\right)^2 + \left(\frac{\partial f}{\partial y}\right)^2\right]^{-1/2}$$

or, if f is given in polar coordinates, by

$$\left[\left(\frac{\partial f}{\partial r}\right)^2 + \left(\frac{1}{r}\frac{\partial f}{\partial \theta}\right)^2\right]^{-1/2}.$$

The reason is that the local strength at any point on $f(x, y) = 0$ is determined solely by the magnitude of the steepest slope of $f(x, y)$ at that point, just as in one dimension the strength of $\delta[f(x)]$ is determined by the magnitude of $f'(x)$, where $f(x) = 0$. The strength is $\mid f'(x) \mid^{-1}$, as in the special case $\delta(mx) = \mid m \mid^{-1} \delta(x)$.

Example

▷With $f = r - a \cos\theta$, $\partial f/\partial r = 1$, $r^{-1}\partial f/\partial \theta = \tan\theta$, and the strength is $(1 + \tan^2\theta)^{-1/2} = \cos\theta$, as already deduced.◁

Point Response of a Strip Beam

An x-ray beam is not of infinitesimal width, nor is it uniform in cross section when formed by a simple collimator which makes use of two spaced slits. The intensity distribution over the cross section is then trapezoidal in form, the shape of the trapezoid varying with position along the beam; the sloping sides correspond to the penumbra of a beam of sunlight. The elementary discussion to be given here aims only at taking account of what happens when an ideal ray is spread out into a strip with constant trapezoidal density (Bracewell, 1977). The response to unit mass at the origin when a uniform beam parallel to the y-axis is scanned in the x-direction was previously defined as $A(x, z)$, which was then replaced by a delta function in the subsequent theory; thus attention was restricted

---

[2] Or we could say linear density = $|1/\operatorname{grad} f|$ where $f = 0$ and zero elsewhere.

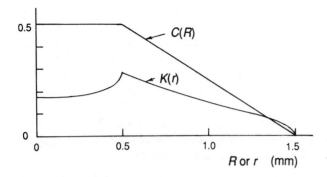

Figure 15-11   The one-dimensional beam cross section $C(R)$ corresponds in to dimensions to a circularly symmetrical cratered cone whose cross section is $K(r)$.

to reconstruction, on the understanding that questions of restoration for blurring could be treated separately. Dropping the coordinate z, the cross section to be considered is

$$A(x) = \begin{cases} 0, & |x| > 1.5 \\ 0.75 + 0.5x \operatorname{sgn} x, & 0.5 < |x| < 1.5 \\ 0.5, & |x| < 0.5. \end{cases}$$

If results obtained with such a beam are reconstructed as if the beam were a ray of infinitesimal width, then in the absence of other defects the image obtained will be the blurred version of the object that would be generated by two-dimensional convolution with a circularly symmetrical function $K(r)$. The function $K(r)$ has the property that its line-integrated profile, or Abel transform, is $A(x)$. Referring to a table of Abel transforms gives

$$K(r) = \begin{cases} [\operatorname{arcosh}(1.5r^{-1})]/2\pi & 0.5 < |r| < 1.5 \\ [\operatorname{arcosh}(1.5r^{-1}) - \operatorname{arcosh}(1.5r^{-1})]/2\pi, & |r| < 0.5 \end{cases}$$

This blurring function has the appearance of a cratered cone, but is shown in Fig. 15-11 only in cross section, together with the adopted trapezoid.

When the digitizing interval used is not much smaller than the width of the beam, it is practical to represent the blurring in terms of one width parameter (such as the equivalent width of $K(r)$). With finer digitization, or under circumstances where an instrument detects the penumbra when scanning a fine filament, more attention can be given to the phenomena associated with strip beams (Verly and Bracewell 1979).

Numerical Computation

Suppose that a density distribution is represented by an array of point masses distributed on a rectangular coordinate system, as would be natural in computing practice. Then the mass at $P(a, b)$ is to be distributed nonuniformly over the perimeter of the circle whose diameter is OP. At a point Q on this circle the density is proportional to OQ, and the mass at P is conserved on distribution. For computing purposes the mass is subdivided into a set of smaller point masses, either uniformly spaced around the circle and varying in mass, or equal in mass and nonuniformly spaced. In either case the set does not fall on the array points. Each small mass may then be allocated to the nearest array point, or it may be split between a group of array points. Allocations from all the points (a, b) are accumulated to yield the Radon transform. This method will work for any arbitrary given distribution and allows for ingenuity in numerical programming.

A different procedure is to form discrete equivalents of the line integrals at progressively spaced angles and to allocate the small masses to the array points adjacent to the points $(a \cos^2 \theta + b \sin \theta \cos \theta, a \cos \theta \sin \theta + b \sin^2 \theta)$ on the radial line for each angle. This amounts to a lot of arithmetic and can lead to unwanted artifacts reminiscent of moiré effects.

Inverting the Radon transform by modified back projection was explained above. For numerical purposes, each projection needs to be digitized and then modified by convolution. The coefficients for doing this were deduced. A numerical procedure for back projection was not part of this discussion, but back projection onto a discrete lattice was treated in Chapter 2 and was the subject of problems. A variety of possibilities can be made to work, each with its own degree of accuracy. How to balance the computing time spent on back projection against the suitability of the final image for its required purpose is the real heart of the matter. The only way to optimize the back-projection algorithm is to assess the trade off between image quality and the time delay injected by computing—a balance that changes with time and other variables. Consequently, the range of techniques described in Chapter 2 needs to be kept under review.

Separability of Restoration and Reconstruction

Strip integration produces two effects; one is the blurring associated with the lumping together of the values of $\rho(x, y)$ that lie within the width of the strip, and the other is the confusion, or scrambling, resulting from integrating the values distributed along the line of integration. So far we have considered line integration rather than strip integration and found that little information is lost if the number of scanning directions is sufficiently great. As a result, not only can the problem of reconstruction from line integrals be solved theoretically, but also practical numerical reconstruction is achievable. There are no distributions that are invisible to line scanning.

By contrast, blurring by convolution with a beam involves a loss of information as well as perturbation of the information that is not lost. Restoration, the art of doing the best one can in the circumstances, was the subject of Chapter 13. We found that there are invisible distributions—patterns on the (x, y)-plane to which there is no response anywhere.

Line integration is the limiting case of strip integration as the width of the strip approaches zero, and it is clear that the strip integral scan at angle θ is related to the corresponding line integral scan by one-dimensional convolution with the profile $A(x, z)$, which represents the combined effects of the beam profile and other instrumental parameters. Thus

$$g_\theta(R) = \int_{-\infty}^{\infty} \int_{-\infty}^{\infty} \rho(x', y') A(R - x' \cos \theta - y' \sin \theta) \, dx' y'$$

$$= A(R) * \int_{-\infty}^{\infty} \int_{-\infty}^{\infty} \rho(x', y') \delta(R - x' \cos \theta - y' \sin \theta) \, dx' y' = A(R) * f_\theta(R),$$

where $g_\theta(R)$ is the strip-integral scan that is measurable (if one discounts noise), while $f_\theta(R)$ is the theoretical line-integral scan.

We have seen how to work from a set of theoretical line scans $f_\theta(R)$. How should one proceed from the given strip scans $g_\theta(R)$? References to several methods are given in the literature (Bracewell, 1956).

> *Method I:* Restore the strip scans and reconstruct.
>
> *Method II:* Reconstruct as though the strip scans were line scans, then restore.
>
> *Method III:* Compute Fourier transforms of all the strip scans, correct each transform for blurring, assemble into a two-dimensional array, and compute the two-dimensional Fourier transform.

The third method was the first to be practiced, but is onerous; for an example that starts by deriving the scans from lunar occultation data see Taylor (1967). The second method raises an interesting question: after reconstruction as prescribed, how does the result relate to the original $\rho(x, y)$? Answer: it is a blurred version, derivable from $\rho(x, y)$ by two-dimensional convolution with the circularly symmetrical pattern whose Abel transform is the profile $A(\ )$. Restoration for two-dimensional smoothing seems less convenient than for one dimension but needs to be evaluated in detail in a given computational environment. Current practice in x-ray tomography is partial implementation of Method II. Of early historical interest (O'Brien, 1953) is the solving path (Fig. 15-12) going up from $^2\mathcal{F}\{\rho\}$ and turning left to $\rho(x, y)$, which applies not to strip-integral data but to two-element interferometry with a small movable antenna.

The interrelationships of line integration and convolutional smoothing are displayed in Fig. 15-12, where the space domain is kept on the left and the transform domain on the right, as is customary. Many different results can be seen by studying this diagram. For example, the question of commutativity between line integration and smoothing is examined by considering the left-hand wall of the cubical arrangement. The conclusion that one-dimensional convolution of the scans with $A(R)$ is equivalent to two-dimensional smoothing with the inverse Abel transform of $A(R)$ is reached by replacing $\rho(x, y)$ by $^2\delta(x, y)$. These ideas and other paths through the diagram were discussed in the original paper. They formed the basis of transform reasoning that led to reconstruction technique purely in the data domain without numerical recourse to transforms at all.

SOME RADON TRANSFORMS

The selection of transforms below exemplifies some procedures and provides material for numerical checking.

Offset Point Mass

As previously deduced, the Radon transform of $^2\delta(x - a, y - b)$ is $\delta(r - a\cos\theta - b\sin\theta)$.

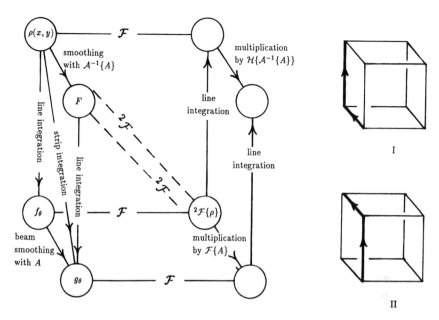

Figure 15-12   Relationships between line and strip integration and two methods of treating data (Bracewell, 1956).

Central Disc

Let $\rho(x, y) = \text{rect}(r/2a)$. Then

$$g_\theta(R) = \mathcal{R}\,\text{rect}(r/2a) = 2a\sqrt{1 - R^2/a^2}.$$

Where circular symmetry exists, a whole range of Radon transforms is directly available from a list of Abel transforms, as with the following example.

Central Ring Impulse

The ring impulse $\rho(x, y) = \delta(r - a)$, of radius a, has the known Abel transform $2a/\sqrt{a^2 - x^2}\,\text{rect}(x/2a)$; consequently

$$\mathcal{R}\delta(r - a) = \frac{2a}{\sqrt{1 - R^2/a^2}}.$$

To deduce the result directly from the definition integral,

$$g_\theta(R) = \mathcal{R}\delta(r - a) = \int_{-\infty}^{\infty}\int_{-\infty}^{\infty} \delta(\sqrt{x'^2 + y'^2} - a)\delta(R - x'\cos\theta - y'\sin\theta)\,dx'\,dy'.$$

Note that the result will, by circular symmetry, be independent of θ, and put θ to zero to get

$$g_\theta(R) = \int_{-\infty}^{\infty} \delta(\sqrt{R^2 + y'^2} - a)\, dy'.$$

The argument $z = \sqrt{R^2 + y'^2} - a$ of the delta function is zero at $\pm\sqrt{a^2 - R^2}$, and so $\delta(z)$ consists of an impulse at each of those zeros, with strength equal to the reciprocal of the absolute slope $|\, dz/dy' \,| = \sqrt{1 - R^2/a^2}$. Hence

$$g_\theta(R) = \frac{2}{\sqrt{1 - R^2/a^2}}.$$

Central Gaussian

If $\rho(x, y) = \exp(-\pi r^2/W^2)$ then we know from the Abel transform that

$$g_\theta(R) = \mathcal{R}e^{-\pi r^2/W^2} = We^{-\pi R^2/W^2}.$$

Consequently, this function is proportional to its own Radon transform; it is one of the eigenfunctions of the transformation and has eigenvalue W. As an exercise in working from the definition,

$$g_\theta(R) = \int_{-\infty}^{\infty}\int_{-\infty}^{\infty} e^{-\pi(x'^2+y'^2)/W^2}\delta(R - x'\cos\theta - y'\sin\theta)\, dx'\, dy'$$

$$= \int_{-\infty}^{\infty}\int_{-\infty}^{\infty} e^{-\pi(x'^2+y'^2)/W^2}\delta(R - x')\, dx'\, dy'$$

$$= \int_{-\infty}^{\infty} e^{-\pi(R^2+y'^2)/W^2}\, dy'$$

$$= e^{-\pi R^2/W^2}\int_{-\infty}^{\infty} e^{-piy'^2/W^2}\, dy'$$

$$= We^{-\pi R^2/W^2}.$$

Square in the First Quadrant

The Radon transform

$$\mathcal{R}\rho(x, y) = \mathcal{R}\,\text{rect}\left(\frac{x - \frac{1}{2}}{a}\right)\text{rect}\left(\frac{y - \frac{1}{2}}{a}\right),$$

when $a < 0.25$, is given by

$$g_\theta(R) = \begin{cases} aR/\sin\beta\cos\beta, & 0 \le R \le \sin\beta \\ a/\cos\beta, & \sin\beta \le R \le \cos\beta \\ a\dfrac{\sin\beta + \cos\beta - R}{\sin\beta\cos\beta}, & \cos\beta \le R \le \sin\beta + \cos\beta \\ 0, & \text{elsewhere,} \end{cases}$$

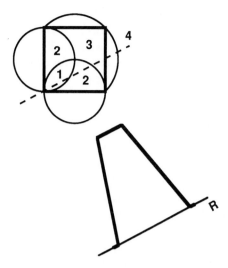

Figure 15-13   Deducing the Radon transform of a square: $g_\theta(R)$ in region 1 rises linearly with R, in region 2 is constant along any radial direction, in region 3 descends linearly, and in region 4 is zero. A cross section along the broken line is shown below.

where $\beta = \frac{1}{4}\pi - | \theta - \frac{1}{4}\pi |$. The angle β is the same as θ where $\theta < \frac{1}{4}\pi$, otherwise it is the complement of θ. The result is easily seen from the construction in Fig. 15-13, where it is understood that $\theta < \frac{1}{4}\pi$.

Straight Line

Consider the straight line $x/a + y/b = 1$. A projection onto the rotated coordinate x'-axis will in general yield a constant independent of x, but the constant will be greater the closer the direction of projection is to the given line. However, when the direction of projection is exactly along the line, the result is zero everywhere except at one point, where the result is infinite. Since the mass of a projection equals the mass of the object projected, the projection in this case has infinite mass; therefore we are not dealing with an ordinary delta function but with one of infinite strength. When the scanning direction makes an angle η (not equal to $90°$) with the straight line, the projected value is $| \sec \eta |$ along the scan. Thus the Radon transform is $\sec[\arctan(b/a) + \theta]$, where θ is the scanning direction, as in previous figures. When $\eta = 90°$, the projected value is infinite at $R = R_0$, where $R_0 = ab/\sqrt{a^2 + b^2}$. Any straight feature present in an arbitrary density distribution should therefore reveal itself by a great peak at (R_0, θ) in the Radon transform plane. One can picture the Radon transform of a line segment as a superposition of circular point responses, all intersecting at the foot of the perpendicular from the origin to the straight feature. In discrete computing, the superposition can be accumulated in a matrix representing grey levels. A heavy deposit of ink in a pixel at (X, Y) then implies a straight feature in an object known only from its projections. The intercepts of the straight line will be $a = (X^2 + Y^2)/X$ and $b = (X^2 + Y^2)/Y$. The occupied portion of the line is deducible from the sector structure at P or from the pedal curve (where it is apparent). The presentation of the Radon transform as a halftone image is not a widely used tool and here is an interesting practical application. The method of reciprocal polars described elsewhere offers an alternative direct method for detecting collineation of elements in an object.

THE EIGENFUNCTIONS

We have seen that

$$\mathcal{R}e^{-\pi r^2} = e^{-\pi R^2},$$

an observation that raises the question whether any other functions are proportional to their own Radon transforms. Some others are $\sqrt{\pi}r \exp(-\pi r^2)\cos\theta$, $\sqrt{\pi}r \exp(-\pi r^2)\sin\theta$, and $\sqrt{\pi}r \exp(-\pi r^2)\exp i\theta$. A whole series $\gamma(r)\exp in\theta$ was derived by Ein-Gal (1974). Starting from the radial functions $\gamma_0(r) = \exp(-\pi r^2)$, $\gamma_1(r) = \sqrt{\pi}r \exp(-\pi r^2)$, remaining functions are obtainable from the recursion relation

$$\gamma_{n+1}(r) = r^n \frac{d^{2n}}{dr^{2n}}[r^n \gamma_{n-1}(r)].$$

THEOREMS FOR THE RADON TRANSFORM

Linear superposition applies to the Radon transform, i.e.,

$$\mathcal{R}[\rho_1(x, y) + \rho_2(x, y)] = \mathcal{R}\rho_1(x, y) + \mathcal{R}\rho_2(x, y)$$

for all choices of $\rho_1(x, y)$ and $\rho_2(x, y)$. It follows that $\mathcal{R}a\rho(x, y) = a\mathcal{R}\rho(x, y)$, where a is a rational number.

Similarity theorem.   If the given distribution $\rho(x, y)$ is expanded or contracted into a similar distribution $\rho(ax, ay)$, then the Radon transform is expanded or contracted in the same ratio:

$$\mathcal{R}\rho(ax, ay) = a^{-1}g_\theta(aR).$$

We can call this the similarity theorem. The more general theorem for $\rho(ax, by)$ can be viewed as a special case of the affine theorem given below.

Shift theorem.   If the given distribution $\rho(x, y)$ is translated to become $\rho(x - a, y - b)$, then

$$\mathcal{R}\rho(x - a, y - b) = g_\theta(R - a\cos\theta - b\sin\theta).$$

According to this theorem, the transform values are not changed, but merely reallocated to new positions on the plane; the value at (R, θ) moves radially to $(R + a\cos\theta + b\sin\theta, \theta)$. The coordinate transformation is such that a point (x, y) moves to (x', y'), a point on the same radial line, but the motion is not radially outward, it is in the positive direction of

R. This is a reminder that R is more like a rotated cartesian coordinate than a radial polar coordinate.

Coordinate Transformation. Suppose that the given distribution $\rho(x, y)$ is distorted by a simple transformation of coordinates such as could produce stretching, shear, or rotation, and we wish to know the new Radon transform

$$\mathcal{R}\rho(ax + by, dx + ey).$$

Let

$$u = ax + by, \qquad v = dx + ey,$$

where u and v are the modified coordinates. We know that the element of area $du\,dv$ equals $|\Delta|\,dx\,dy$, where $\Delta = ae - bd$ is the determinant of the matrix $\begin{bmatrix} a & b \\ d & e \end{bmatrix}$, a special case of the Jacobian of a transformation. We also know that $x = (eu - dv)/\Delta$ and $y = (-bu + av)/\Delta$. Thus

$$\mathcal{R}\rho(x, y) = \int_{-\infty}^{\infty} \int_{-\infty}^{\infty} \rho(u, v)\delta\left(R - \frac{eu - dv}{\Delta}\cos\theta - \frac{-bu + av}{\Delta}\right) du\,dv$$

$$= \frac{1}{|\Delta|} \int_{-\infty}^{\infty} \int_{-\infty}^{\infty} \rho(u, v)\delta\left(R - \frac{e\cos\theta - b\sin\theta}{\Delta}u - \frac{-d\cos\theta + a\sin\theta}{\Delta}v\right) du\,dv$$

$$= \frac{1}{|\Delta|} \int_{-\infty}^{\infty} \int_{-\infty}^{\infty} \rho(u, v)\delta(R - u\cos k - v\sin k)\,du\,dv,$$

where $\mathbf{k}$ is a new vector whose components are $(e\cos\theta - b\sin\theta)/\Delta$ and $(-d\cos\theta + a\sin\theta)/\Delta$. Let the magnitude of $\mathbf{k}$ be $|\mathbf{k}|$. Then, to reduce the foregoing expression to the standard form for a Radon transform, a scale factor $|\mathbf{k}|$ must be introduced so that $\cos k/|\mathbf{k}|$ and $\sin k/|\mathbf{k}|$ can play the role of components of the unit vector $\mathbf{t}$.

$$\mathcal{R}\rho(u, v) = \frac{1}{|\Delta|} \int_{-\infty}^{\infty} \int_{-\infty}^{\infty} \rho(u, v)\delta[|\mathbf{k}|\left(\frac{R}{|\mathbf{k}|} - u\frac{\cos k}{|\mathbf{k}|} - v\frac{\sin k}{|\mathbf{k}|}\right) du\,dv$$

$$= \frac{1}{|\mathbf{k}||\Delta|} \int_{-\infty}^{\infty} \int_{-\infty}^{\infty} \rho(u, v)\delta(\frac{R}{|\mathbf{k}|} - u\cos\theta - v\sin\theta)\,du\,dv$$

$$= \frac{1}{|\mathbf{k}||\Delta|} g_\theta\left(\frac{R}{|\mathbf{k}|}\right).$$

This result is simple in the sense that values of the new Radon transform are proportional to old values but rotated in position to a new angle. However, both the proportionality factor and the angular displacement depend on θ and cannot readily be visualized except for special cases. From the general result the more general similarity theorem for $\rho(ax, by)$ can be deduced, as well as the simple shear theorem for $\rho(x + by, y)$, and so on. A more general theorem for a full affine transformation $u = ax + by + c$, $v = dx + ey + f$ can be deduced by applying the shift theorem for a displacement (c, f).

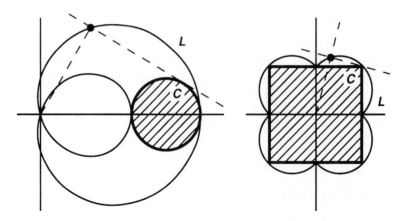

Figure 15-14 Object boundaries C and the corresponding Radon boundaries L.

The symbolic presentation is not very enlightening; when computing is involved, a fully general transformation is rarely required.

Example

Obtain the Radon transform of $\rho(x, y) = \exp[-\pi(\alpha x^2 + \beta y^2)]$. ▷Start from the known $\mathcal{R}\exp[-\pi(x^2 + y^2)] = g_\theta(R) = \exp(-\pi R^2)$. Make the coordinate transformation $u = \alpha x$, $v = \beta y$, with determinant $\Delta = \alpha\beta$. The vector $\mathbf{k}$ has components $(\beta\cos\theta/\Delta, \alpha\sin\theta/\Delta)$ and $|\mathbf{k}| = \sqrt{\beta^2\cos^2\theta + \alpha^2\sin^2\theta}$. Hence,

$$\mathcal{R}e^{-\pi(\alpha^2 x^2 + \beta^2 y^2)} = \frac{1}{|\mathbf{k}||\Delta|}g_\theta\left(\frac{R}{|\mathbf{k}|}\right)$$

$$= \frac{\exp[-\pi R^2\alpha^2\beta^2/(\beta^2\cos^2\theta + \alpha^2\sin^2\theta)]}{\sqrt{\beta^2\cos^2\theta + \alpha^2\sin^2\theta}}. ◁$$

THE RADON BOUNDARY

If a density distribution is confined within a closed contour C, which is always the case in practice, then the Radon transform will have a boundary curve L that defines where the Radon transform is zero. The curve L is pedal to C, a relationship from the theory of plane curves according to which L is the locus of the foot of the perpendicular from the origin to the tangent to C. The pedal curve L to the circle C of radius b centered at $(a, 0)$ in Fig. 15-14 is the epitrochoid $r = a\cos\theta + b$. The pedal curve to the square in the same figure has the four-leaved outline shown.

APPLICATIONS

Among applications that have developed for reconstruction of images from projections, only a few can be mentioned. Of course, there are industrial applications of the x-ray scanning technique and of corresponding techniques using γ-rays. In geophysical exploration efforts have been made to map the vertical plane between two boreholes by lowering a source down one hole to a succession of stations while, for each station, a receiver is scanned from top to bottom of the second hole. A feature of this technique is that there will be sectors missing from the coverage in θ (Bracewell and Wernecke, 1975).

Instead of measuring the line integral of absorption, one can measure transit time. This is possible with acoustic arrangements but is also possible with a light beam passing through a transparent biological tissue, and it permits mapping the velocity distribution in the plane explored. Different states of tissue can be seen in this way. Naturally, the transit-time differences are very brief, but they are readily detectable by causing the exit beam to interfere with a reference beam arriving by a fixed path. A variety of applications, including astronomical, seismic, and positron emission, are described by Deans (1983) with references to the literature.

A fascinating oceanographic development is the demonstration that underwater sound generated in the Southern Ocean at Heard Island can be detected as far away as California and other remote coasts. Transit time is a curvilinear line integral of the reciprocal of sound velocity, which is dependent on water temperature. Hence, it will be possible to make global maps of ocean temperature. An immediate return will be the ability to measure minute changes in the average temperature of a vast volume of sea water. The concern about whether the globe is warming would benefit, possibly in time to avert unwarranted expenditures.

The development of magnetic-resonance imaging has rapidly become of such importance as to warrant special mention.

Magnetic-Resonance Imaging

A bar magnet suspended freely in a magnetic field H tends to align itself and, if disaligned, will oscillate with an angular frequency $\sqrt{MH/I}$, where M is the magnetic moment and I is the moment of inertia about the axis of rotation. The nucleus of a hydrogen atom also has a magnetic moment, and tends to align itself in a magnetic field, and to recover if disaligned. However, a nucleus also has angular momentum (spin); consequently, the tendency to recover, although oscillatory, has the nature of gyroscopic precession. The angular frequency of precession, or Larmor frequency ω_0, is directly proportional to the magnetic field, the constant of proportionality being characteristic of the nucleus. For the hydrogen nucleus the Larmor frequency is 4.26 MHz per kilogauss (42.6 MHz per tesla). Thus, nuclear resonance is excitable by Hertzian waves and detectable by radio receivers, since the natural nuclear emission is in the radiofrequency band.

An incident pulse of circularly polarized radiofrequency energy at the Larmor frequency will disalign hydrogen nuclei, by an angular amount depending on pulse duration.

After the cessation of the pulse, a sensitive radio receiver can detect faint emitted radiation at the same frequency while the precession damps out exponentially with a relaxation time T_1 of about 1 s. The strength of the received signal is proportional to the number of hydrogen atoms engaged.

Suppose that an object is situated in a field that is almost uniform but in fact varies in strength from plane to plane. The Larmor frequency will change slightly from plane to plane. All the hydrogen nuclei in the object can be excited by a short pulse, but the faint radiated fields will be frequency coded as to the plane occupied by the emitting nuclei. Those nuclei in any one plane will emit on the same frequency. Thus by spectral analysis of the decay signal following pulse excitation one can count the hydrogen atoms plane by plane. This is analogous to integrating along a line with x-rays, but in a higher dimensionality; and taking account of a full set of planes is analogous to one complete scan or projection in x-ray tomography.

The three-dimensional projection-slice theorem for projection onto the (x, y)-plane reads

$$^2\mathcal{F} \int_{-\infty}^{\infty} f(x, y, z)dz = F(u, v, 0)$$

and is helpful in understanding how to explore the full three-dimensional transform $F(u, v, w)$ of a density distribution of nuclei. To do this will require reorienting the magnetic field gradient in a sequence of all necessary directions. Various geometries and time schedules are available, from which not only the density distribution but also the distribution of $T_1(x, y, z)$ can be reconstructed. For biological specimens hydrogen is particularly important, because of its abundance in water molecules and organic tissues, but other nuclei are also important. Injections of elements such as gadolinium, which are taken up preferentially by different tissues, allow further richness in imaging methods.

Three-dimensional reconstruction was pioneered experimentally by Lauterbur (1973). See also Lauterbur and Lai (1980) and Andrew (1980). Some of the history, and access to current literature, can be gleaned from special issues of various journals, including *IEEE Trans. Nuclear Science*, *Computers in Biology and Medicine*, *Physics in Biology and Medicine*, and *J. Computer Assisted Tomography*, and later developments can be followed through the *Science Citation Index*.

The time taken to acquire an image has always been a concern. Examination times as long as an hour are expensive, while times as short as only a few seconds affect chest images. Much faster methods (Stehling et al., 1991) are, however, under development.

Curved Line Integrals

Because x-rays are not refracted, line integrals are appropriate to the discussion of computed x-ray tomography, but in other applications of tomography the measurable integrals may pertain to curved rays. For example, the phase of a light ray emerging from transparent biological tissue is the integral of the phase constant $\beta = 2\pi/\lambda$, measured in radians per meter, along the ray. Of course, the phase constant need not be spatially constant. If it is not, then the light ray will be refracted. It is perfectly possible to measure the optical phase by interferometry, using a reference beam that passes through a uniform medium in

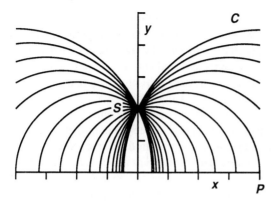

Figure 15-15 A scatterer at S detected by a radar at P could be anywhere on the range circle C.

which the wavelength is λ_0, and to make scans in all directions. From all the phase integrals (one hesitates to call the phase integrals projections, because the term *projection* is associated with straight rays) one might hope to get the refractive index λ_0/λ throughout the object. Such a map of transparent tissue would be an interesting adjunct to the conventional microscope image of transmitted light intensity. However, the problem is clearly difficult, because until the refractive index distribution is known, the ray paths cannot be determined. A special case is tractable where the rays accumulate measurable phase but do not deviate significantly from the straight line. This situation arises both with light and with radio waves in cold plasma. In medical imaging by nuclear magnetic resonance one attempts to eliminate curvature by designing magnets that produce extremely uniform magnetic fields. In geophysics, the curvature of seismic rays has to be faced (Claerbout, 1992), as well as the presence of reflections from sharp boundaries.

Curvilinear tomography will become more important as mathematical methods develop and as more and more measurement techniques are found to be interpretable as basically tomographic. Two related examples will be described. A radar transmitter is transported on a ground vehicle with velocity v along the x-axis (Fig. 15-15) and transmits more or less isotropically into the half-plane in the general direction of the y-axis. The received power, which is displayed as a function of range R and track position x, represents the integral of the energy scattered back from the semi-annulus bounded by ranges R and $R + dR$. Although this radar measures range only, and does not have a directive antenna, it is nevertheless clear that the terrain adjacent to the vehicle track can be mapped two-dimensionally, and this is interesting because an isotropic antenna is small and cheap.

To process the data one could get an immediate rough indication of strongly scattering features by making a layergram of superposed circular arcs. For each vehicle position, the intensity measured at range R would be written along a circular locus into a matrix in the (x, y)-plane and accumulated with the contributions from successive positions. If there was one strongly scattering object, say at $(0, d)$, then its range would have been $\sqrt{d^2 + v^2 t^2}$ at time t, and we would write measured values into the matrix along circles $y^2 + (x - vt)^2 = R^2$. Since these circles intersect at $(0, d)$, there would be a large accumulation there. In the vicinity there would be a general smear, falling off with distance in a way reminiscent of the $1/r$ variation of the layergram formed by unmodified back projection as described earlier. One notable difference is two missing sectors that fill in only as

$x \to \pm\infty$. Other differences are the correction needed to compensate for the divergence of radiation (inverse fourth power of range) and the correction for the radiation pattern of the antenna. This example is a special case of synthetic aperture radar (Chapter 16), which, despite the missing sectors, permits the reconstruction of excellent images of terrain.

In the second example, a finite two-dimensional object propagates sound internally at constant velocity but scatters isotropically with strength $f(x, y)$. The function $f(x, y)$ is to be mapped. Each of many transducers surrounding the object emits a short pulse and makes a record of the echoes as a function of time. The system is analogous to the moving radar described above, but simultaneous use of a set of fixed transducers will enable time changes in the medium to be tracked. These problems have been discussed by Norton (1977).

Further interesting problems arise where the medium is not an isotropic scatterer, the sound velocity depends on position, and the rays are curved. An example of the latter is furnished by the proposal to map ocean temperatures by timing the arrival of underwater sound from Heard Island at multiple sites on the shores of the Pacific and Atlantic Oceans. This is of great environmental interest, because global temperature changes suspected to be under way could be detectable through the influence of temperature on sound velocity. Temperatures mapped in this way would pertain to a vast volume of water; thus, higher precision might be hoped for, as compared with averaging temperatures recorded traditionally at discrete stations.

LITERATURE CITED

E. R. ANDREW (1980), "Nuclear magnetic resonance imaging: the multiple sensitive point method," *IEEE Trans. Nuclear Science*, vol. NS-27, pp. 1232–1238.

R. H. T. BATES AND M. J. MCDONNELL (1986), *Image Restoration and Reconstruction*, Clarendon Press, Oxford.

R. N. BRACEWELL (1956), "Strip integration in radio astronomy," *Aust. J. Phys.*, vol. 9, pp. 198–217.

R. N. BRACEWELL (1958), "Restoration in the presence of errors," *Proc. Inst. Radio Engrs.*, vol. 46, pp. 106–111.

R. N. BRACEWELL (1977), "Correction for collimator width (restoration) in reconstructive x-ray tomography," *J. Computer Assisted Tomography*, vol. 1, pp. 6–15.

R. N. BRACEWELL AND A. C. RIDDLE (1967), "Inversion of fan beam scans in radio astronomy," *Astrophys. J.*, vol. 150, p. 427–434.

R. N. BRACEWELL AND J. A. ROBERTS (1954), "Aerial smoothing in radio astronomy," *Aust. J. Phys.*, vol. 7, pp. 615–640.

R. N. BRACEWELL AND S. J. WERNECKE (1975), "Image reconstruction over a finite field of view," *Journal of the Optical Society of America*, vol. 65, pp. 1342–1346.

R. A. BROOKS AND G. DICHIRO (1976), "Principles of computer-assisted tomography (CAT) in radiographic and isotropic imaging," *Phys. Med. Biol.*, vol. 21, pp. 689–732.

J. F. CLAERBOUT (1992), *Earth Soundings Analysis*, Blackwell, Boston.

A. CORMACK (1963), "Representation of a function by its line integrals, with some radiological applications," *J. Appl. Phys.*, vol. 34, p. 2722–2727.

T. CROWTHER (1970), *New Scientist*, vol. 47, p. 228.

R. A. CROWTHER AND A. KLUG (1974), "Three-dimensional image reconstruction on an extended field—a fast stable algorithm," *Nature,* vol. 251, p. 490–492.

S. R. DEANS (1983), *The Radon Transform and Some of Its Applications*, John Wiley & Sons, New York.

D. J. DEROSIER AND A. KLUG (1968), "Reconstruction of three-dimensional structures from electron micrographs," *Nature,* vol. 217, p. 130–134.

M. EIN-GAL (1974), "The Shadow Transform: An Approach to Cross-sectional Imaging," Ph.D. thesis, Stanford University.

S. HELGASON (1980), "The Radon Transform," Birkhäuser, Boston, Massachusetts.

G. T. HERMAN AND R. M. LEWITT (1979), "Overview of image reconstruction from projections," in G. T. Herman ed., *Image Reconstruction from Projections*, vol. 32, pp. 1–8, Springer-Verlag, New York.

G. HOUNSFIELD (1972), *EMI-Scanner, A New Perspective on Brain Disease,* EMI Central Research Laboratories, Hayes, Middlesex.

A. KLUG AND R. A. CROWTHER (1972), "Three-dimensional image reconstruction from the viewpoint of information theory," *Nature,* vol. 238, p. 435–440.

P. C. LAUTERBUR (1973), "Image formation by induced local interactions: examples employing nuclear magnetic resonance," *Nature*, vol. 242, pp. 190–191.

P. C. LAUTERBUR AND C-M LAI (1980), "Zeugmatography by reconstruction from projections," *IEEE Trans. Nuclear Science*, vol. NS-27, pp. 1227–1231.

M. J. LIGHTHILL (1958), *Introduction to Fourier Analysis and Generalized Functions*, Cambridge University Press, England.

A. MACOVSKI (1983), *Medical Imaging Systems*, Prentice-Hall, Englewood Cliffs, N.J.

S. J. NORTON (1977), "Theory of Acoustic Imaging," Ph.D. thesis, Stanford University.

P. A. O'BRIEN (1953), "The distribution of radiation across the solar disc at meter wavelengths," *Mon. Not. Roy. Astronom. Soc.*, vol. 113, pp. 597–612.

D. A. POLLEN, J. R. LEE, AND J. H. TAYLOR (1971), "How does the striate cortex begin the reconstruction of the visual world?" *Science,* vol. 173, pp. 74–77.

J. RADON (1917), "Über die Bestimmung von Funktionen durch ihre Integralwerte längs gewisser Mannigfaltigkeiten," *Berichte über die Verhandlungen der Königlichen Sächsischen Gesellschaft der Wissenschaften zu Leipzig. Mathematisch-Physikalische Klasse*, vol. 69, pp. 262–277.

G. N. RAMACHANDRAN AND V. A. LAKSHMINARAYANAN (1971), "Three-dimensional reconstruction from radiographs and electron micrographs; Applications of convolutions instead of Fourier transforms," *Proc. Nat. Acad. Sci. USA,* vol. 68, p. 2236–2240.

S. W. ROWLAND (1979), "Computer implementation of image reconstruction formulas," in G. T. Herman, ed., *Topics in Applied Physics*, vol. 32, pp. 7–79.

R. C. SPINDEL AND P. F. WORCESTER (1990), "Ocean acoustic tomography," *Sci. Amer.*, vol. 263, pp. 94–99.

M. K. STEHLING, R. TURNER, AND P. MANSFIELD (1991), "Echo-planar imaging: magnetic resonance imaging in a fraction of a second," *Science*, vol. 254, pp. 43–50.

J. H. TAYLOR (1967), "Two-dimensional brightness distributions of radio sources from lunar occultation observations." *Astrophys. J.,* vol. 150, p. 421–426.

T. TOTSUKA AND M. LEVOY (1993), "Frequency domain volume rendering," *Proc. SIGGRAPH 93*, Anaheim, Calif., August 1993, pp. 271–278.

B. K. VAINSHTEIN (1971), "Synthesis of projecting functions," *Soviet Physics-Doklady,* vol. 16, p. 66.

J. G. VERLY AND R. N. BRACEWELL (1979), "Blurring in tomograms made with x–ray beams of finite width," *J. Computer Assisted Tomography,* vol. 3, pp. 662—678.

PROBLEMS

15–1. *X-ray beam.* X-rays emanating at a rate M photons m^{-2} steradian^{-1} s^{-1} from a small anode pass through a 3-mm hole in a lead sheet 100 mm away, and those that get through the hole impinge on a second plate 100 mm further away, having a second such hole. The anode is elliptical, but so inclined that if you were to look toward it through the two holes (with the x-rays turned off) it would appear circular with a projected diameter of 3 mm. The escaping beam contains $N(r, z)$ photons m^{-2} s^{-1}, where z is measured from the x-ray source along the beam axis and r is measured transversely. We are interested in the cross section of the emerging beam.
  (a) Find a way of representing $N(r, z)$ in the range 200 mm $< r <$ 300 mm, $-10 < r < 10$.
  (b) Comment on the suitability of the beam for use in an x-ray scanner.

15–2. *Projection of hollow shaft.* A drive shaft is 100 mm in outside diameter and is hollow. The inside diameter is 80 mm where $z < 0$ and is 90 mm where $z > 200$ mm. In the range $0 < z < 200$ the interior surface is conical. A photographic plate is placed against the shaft while it is rotating so that an x-ray picture can be obtained of the tapered section. A point source of x-rays, situated far enough away that the rays within the steel are effectively parallel, is turned on for an integral number of revolutions.
  (a) If there are no flaws in the steel, what would be the length of the ray within the steel for points $0 < y < 100, 0 < z < 200$, on the photographic plate?
  (b) What would you expect for $I(x, y)$, the x-ray intensity on the photographic plate for $0 < y < 100, z = 100$? (c) If the x-ray source is placed in a convenient close-up position, say 100 mm from the center of the shaft, graph the length of the ray within the steel as a function of position y on the photographic plate. Restrict attention to the plane containing the source that is perpendicular to the shaft at $z = 100$. ☞

15–3. *Reconstruction from two projections.* Projections of the same $f(x, y)$ taken in two perpendicular directions are shown in Fig. 15-16.
  (a) Can you construct $f(x, y)$?
  (b) The origins of the two given line integrals are indicated, but did you need that information?

15–4. *Radon point response.* To confirm that the Radon transform of a point mass is a circular line impulse, a student begins by considering a mass at (1,0) that is very small but of nonzero diameter τ and takes projections at regularly advancing angles. For each projection she writes the maximum value g_{max} of the projection at the appropriate point (R, θ) in the Radon transform plane, and soon the circular locus begins to appear. As all the values she has written down along the circle are the same, it seems to her that, contrary to expectation, the point response of the Radon transform must be uniform around the circle. The reason is that, although the values of g_{max} approach infinity as $\tau \to 0$, nevertheless all the values are equal for any given value of τ. Her friend suggests that although the values are equal, they may not be uniformly spaced.

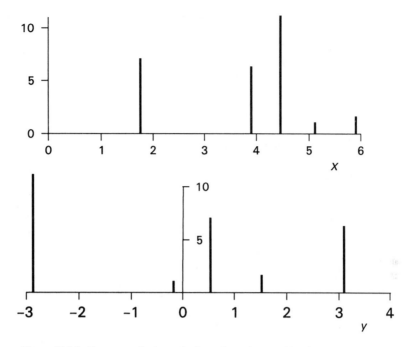

Figure 15-16   Two perpendicular projections of an unknown $f(x, y)$.

(a) If they are not uniformly spaced, does the spacing satisfactorily account for the expected angle-dependent strength?
(b) If they are uniformly spaced, does it mean that the Radon transform of a very small mass is essentially uniform around the circular locus?

15–5. *Invisible distributions.*
   (a) Explain what is meant by an invisible distribution.
   (b) Does the Radon transformation possess invisible distributions? If no, explain. If yes, give an example. ☞

15–6. *Model head.* As a first approximation the following expression is proposed for the density distribution over a cross section of the head

$$f(x, y) = 0.2 \, \delta(x^2/a^2 + y^2/b^2 - 1) + 0.8 \, \mathbf{H}(1 - x^2/a^2 - y^2/b^2),$$

where $\mathbf{H}(\ )$ is the Heaviside unit step function.
   (a) Derive an expression for $g_0(R)$, the line-integrated profile in the $\theta = 0$ direction.
   (b) Make a graph of the result to illustrate the relative magnitude of the contributions made by the two terms.
   (c) Clearly a delta function, which implies a skull of zero thickness, is a coarse approximation. However, does this approximation noticeably affect the relative magnitude of the two terms at, let us say, $R = 0$?

15–7. *An unstable scanner.* A brain scanner obtains 180 line integrals at one-degree intervals, but there are some problems.
   (a) On the fifty-second scan, owing to a temporary gain change, sensitivity change, or

change in x-ray intensity, the whole scan was reported to the computer as 10 percent weaker than it should have been. Is there a way in which the computer algorithm could adapt to such an occurrence without the use of an instrumental monitoring system?

(b) As the rotating yoke turns upside-down, it is quite conceivable that the varying gravity forces on x-ray tube, collimators, detectors, connectors and other components will cause the system gain to depend on position angle. Does your proposal for (a) cope with this more general situation? Explain.

(c) During the forty-fifth scan the patient's head moved. To simplify matters, let us say that during the whole of the scan, but not during any other scans, his head was displaced 3-mm to his left and 3-mm forward (i.e., in the direction of his nose). Clearly, if position sensors were attached to the head, and the motion was reported to the computer, the computations could be modified to correct accordingly. But, is there some way in which the reconstruction algorithm could be refined to adapt automatically to such a head motion without the use of such sensors?

(d) A generalization of situation (c) would arise if head motion were not restricted to scan 45 alone. Can you propose an algorithm to adapt to that?

15–8. *Trapezoidal beam.* An x-ray beam has a trapezoidal intensity distribution across it described by the following formula, where R is measured in millimeters.

$$I(R) = \begin{cases} 0 & R < 1.5 \\ 0.75 + 0.5\,R, & -1.5 < R < -0.5 \\ 0.5, & -0.5 < R < 0.5 \\ 0.75 - 0.5R, & 0.5 < R < 1.5 \\ 0, & 1.5 < R. \end{cases}$$

The intensity distribution does not vary much from the above formula with change of position along the beam (S-direction).

(a) If a two-dimentional reconstruction $f_1(x, y)$ is made, ignoring the beam spread described (the profiles obtained being treated as if they had been obtained with a beam of zero width), show that

$$f_1(x, y) = K(r) ** f(x, y),$$

where $f(x, y)$ is the reconstruction that would have resulted with a zero-width beam used in all position angles and

$$2\pi K(r) = \begin{cases} \cosh(1.5/r), & 0.5 <| r |< 1.5 \\ \cosh(1.5/5) - \cosh(0.5/r), & | r |< 0.5. \end{cases}$$

(b) Give an explanation of why an x-ray beam might have a trapezoidal density distribution.

15–9. *Elliptical object.* An elliptical object 190×210 mm is to be mapped by scanning with an x-ray beam. It has been determined that the beam cross section is such that spatial frequencies only as high as $M = 0.6$ cycles per millimeter are discernible. A total of N scans equally spaced in angle will be taken.

(a) Select a suitable value of N on the basis that gaps between sampled points in the (u, v)-plane should not exceed 2/210 c/m.

(b) Discuss whether you can reduce the number of scans needed by relaxing the management decision on equal angular spacing, while keeping to the spirit of the original basis.

15–10. *Three projections only.* Three projections certainly do not suffice to reconstruct a density distribution $f(x, y)$ but, a student of computer science was heard to say that exact recon-

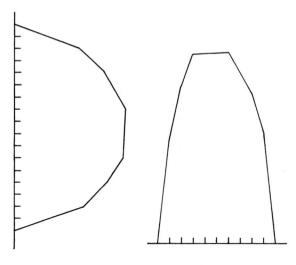

Figure 15-17 Projections of an octagon in its own plane.

struction would be possible from just three projections, provided a priori knowledge were available that $f(x, y) = \Sigma^2 \delta(x - x_i, y - y_i)$. This rather restrictive condition limits us to patterns of point masses all of which are equal and none of which is negative. The student's professor said, "If you can go that far, can't you handle positive impulse patterns of unequal strength? What do you think? ☞

15–11. *Octagonal lamina.* Two projections of a uniform octagonal lamina are shown in Fig. 15-17. Prove that the shape of the uniform lamina is uniquely determined by just the two projections or destroy the possibility of such a proof by constructing a second uniform lamina with the same two projections. ☞

15–12. *Two projections of polygon.* Any density distribution can be approximated by stacking uniform polygons. Sometimes you see three-dimensional "maps" that have been made by stacking laminae cut to the outlines of the contours on the map. Comment on the possibility of reconstructing a density distribution that is known to be stepped in the way that would characterize stacked polygons, from two projections only.

15–13. *Projection of letters.* Printed material in 7×5 dot-matrix capitals (Fig. 15-18) is to be read automatically by scanning with a slit that, in effect, yields the vertical projection of the letters. However, it is noted that several letter pairs have the same projection—for example, **M W** and **S Z**. It is thought that a horizontal projection in addition would eliminate the ambiguity. The virtues of scanning with a slit include speed and tolerance to location.
(a) Is it true that two projections would suffice?
(b) Do you have any suggestions for perfecting the scheme? ☞

15–14. *Mass conservation under scanning.* A two-dimensional density distribution $f(x, y)$ has a mass $\int_{-\infty}^{\infty} \int_{-\infty}^{\infty} f(x, y)\, dx\, dy$.
(a) Show that a projection in any direction has the same mass.
(b) This result is reminiscent of the fact that $g(x, y) ** f(x, y)$ has the same mass as that of $f(x, y)$, provided that the area integral $\int_{-\infty}^{\infty} \int_{-\infty}^{\infty} g(x, y)\, dx\, dy = 1$, and of course we know that projecting, or equivalently scanning with a unit line impulse, is describable in terms of convolution. But the scanning line impulse $\delta(x \cos\theta + y \sin\theta)$, unlike $g(x, y)$, does not have an area integral equal to unity. How do you explain that?

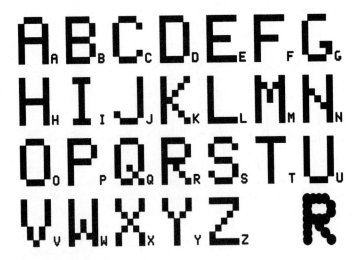

Figure 15-18   The capitals in 7 × 5 dot matrix.

15–15. *Match trick.* Twenty-four matches are arranged in eight piles as shown.

If one match is removed, can you rearrange the remaining matches, keeping to the same eight piles so that there are still nine matches on each side? This problem is clearly an exercise in invisible distributions.

(a) If you can remove one, can you remove more, while maintaining eight piles?

(b) How many matches could be added to the original arrangement while preserving the line sums of the eight piles at nine?

15–16. *Radon transform of normal distribution.* Show that the Radon transform of

$$\frac{1}{\sqrt{2\pi}\sigma} \exp[-(x^2 + y^2)/2\sigma^2] \quad \text{is} \quad \exp(-R^2/2\sigma^2).$$

15–17. *Reconstructing a point pattern.* An object consists of unit point masses situated at (i, j), where both i and j run from 0 to $N - 1$. There are N point masses altogether. Consequently, the area density of masses may be quite low—for example, less than 1 percent if $N > 10$. The projection at $\theta = 10°$ is computed and given in the form $\sum_{n=1}^{N} \delta(p - p_n)$. The coordinate p is measured normal to the direction of projection and has its origin at $i = 0$, $j = 0$. The projection at $\theta = 80°$ is also computed and given as $\sum_{n=1}^{N} \delta(q - q_n)$, where the coordinate q is measured from the same origin. A mathematics student says that, given the list of values p_n and q_n, which will amount to $2N$ values, he ought to be able to solve for the $2N$ coordinates of the N point masses.

(a) Show that the student's opinion is incorrect or

(b) if you find it to be correct, explain why the proposed solution would, or would not be, of use in reconstructing images from two projections only.

15–18. *Consistent projections.* Athos and Porthos were asked to think of complicated 2D functions and calculate the projections g_0 and g_{90} at $0°$ and $90°$, respectively. Athos came up with

$$g_0(R) = 2\operatorname{rect}(\tfrac{1}{4}R) + 2\sqrt{1-R^2}\operatorname{rect}(\tfrac{1}{2}R) \quad \text{and} \quad g_{90}(R) = 8\operatorname{sinc} R + \pi e^{-\pi R^2},$$

while Porthos provided

$$g_0(R) = 4\operatorname{sinc}^2 R \quad \text{and} \quad g_{90}(R) = 7\operatorname{rect} 2R * \frac{1}{\pi}e^{-R^2/4}.$$

(a) Which student, if either, needs help?

(b) Can you supply a 2D function that is consistent with the projections offered by either student?

15–19. *Hole in Radon transform.* The Radon transform of a density distribution is calculated to be zero outside a certain radius and to have a circular hole in the center where the transform is also zero. Does this mean that the density distribution has a hole in the middle?

15–20. *Variance of ring delta.* A density distribution $f(x, y)$ has x-variance defined by

$$\sigma_x^2 = \frac{\int_{-\infty}^{\infty}\int_{-\infty}^{\infty} x^2 f(x, y)\,dx\,dy}{\int_{-\infty}^{\infty}\int_{-\infty}^{\infty} f(x, y)\,dx\,dy}.$$

If $f(x, y) = \delta(r - a)$, show that $\sigma_x^2 = \tfrac{1}{2}a^2$.

15–21. *Layergram.* There is a proposal to make a quick correction to layergrams by using one stage of the method of successive substitutions, on the grounds that a rough approximation to $f(x, y)$ might be useful if it were available almost instantly. To test the proposal numerically, an $N \times N$-point digital representation R_{ij} of r^{-1} is needed, where R_{ij} is the coefficient to be associated with unit square cell centered at $x = i$, $y = j$. Explain how you would determine a value for R_{00}.

15–22. *Zero response of the Radon transformation.* Does the Radon transformation

$$f_\theta(R) = \int_{-\infty}^{\infty}\int_{-\infty}^{\infty} \widehat{\rho}(x', y')\delta(R - x'\cos\theta - y'\sin\theta)\,dx'\,dy'$$

exhibit zero response for any function $\widehat{\rho}(x, y)$ that is itself not zero everywhere? If no, explain. If yes, give an example.

15–23. *Trisectrix.* A computer science student has a solution to a problem inherited from the Greeks, namely how to trisect an angle. "It makes no use of algebra or trigonometry," he says, "but is purely geometrical." To trisect the angle XOA in Fig. 15-19, where $XO = OA$, join X to A. Let XA intersect the curve OBX at B. Then the line OB trisects the angle XOA. When asked how he got the curve OBX, he says he is not at liberty to divulge more because of negotiations with a manufacturer of video games for whom he does software consulting. An electrical engineer says the curve is a limaçon of Pascal, as familiar in x-ray tomography. A mechanical engineer says the curve is an epitrochoid, generated when a circle rolls on another circle. Confirm or refute the statements of the three.

15–24. *Relation between projections.* Authors sometimes write that there is no connection whatever between projections in adjacent directions. If this is so, a digital sundial can be imagined that casts a moving shadow telling the exact time in Arabic numerals (I. Stuart, *The Problems of Mathematics*, Oxford University Press, New York, 1992). Thinking in terms of the projection-slice theorem, explain why adjacent projections are or are not entirely different.

15–25. *Extraterrestrial message.* Problem 2-21 of Chapter 2 reported reception of a binary signal of seemingly extraterrestrial origin. Another radio ham claims to receive the following signal

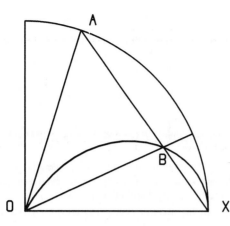

Figure 15-19 Construction for trisecting angle XOA.

from an extraterrestrial source:

35 0 0 0 0 0 0 0 0 0 0 0 0 0 0 0 0 0 0 0 41 19 0 22 25 0 1 0 0 0 0 0 0 0 0 0 0 0 0 0 0 0 0 13

How do you interpret that? ☞

15–26. *Two projections of rod assembly.* A number of circular parallel steel rods of different diameters imbedded in opaque plastic are x-rayed from two directions that are perpendicular to the rods and to each other. In a cross-section plane where there are no defects in the rods the digitized projections of density are as follows;

Horizontal: 1 3 1 0 1 5 5 7 5 5 1 1 3 5 3 1
Vertical: 0 1 4 8 4 1 0 1 5 5 7 5 5 1 0 0

Attempt to determine the locations and diameters of the rods.

15–27. *Convexity.* A function $f(x, y)$ is equal to unity inside a singly-connected region whose boundary curve is C, and zero outside. Two projections $g_0(x)$ and $g_{\pi/2}(y)$ are available. Prove or disprove the following assertions.
(a) If both the projections are convex, then C is convex.
(b) If C possesses a concavity, it will be revealed by a projection discontinuity.

15–28. *Radon transform island.* A function of x and y is nonzero only inside the triangular area bounded by the lines $x = 1$, $y = \pm x$. Construct the area in which the Radon transform is nonzero.

15–29. *Pedal curve of central ellipse.*
(a) Find the pedal curve of $x^2/a^2 + y^2/b^2 = 1$, an ellipse centered at the origin—in other words, of a shape that approximates the transaxial section through the head.
(b) For interest also deal with the case of a highly eccentric ellipse. ☞

15–30. *Pedal curve of focal ellipse.* An object is bounded by a curve which in polar coordinates is given by $r = l/(1 - e \cos \theta)$, where l and e are constants. What is the pedal curve?

15–31. *Pedal curve of circle.* A circle passes through the origin and the point $(1,1)$. What is its pedal curve?

15–32. *Radon transform boundary.* A function of x and y is nonzero only inside the region bounded by $x = 1$ and $y^2 = 2 - x$. Make a drawing on the (x, y)-plane, and crosshatch the region within which the Radon transform is nonzero.

16
Synthetic-Aperture Radar

Some of the most striking images of the earth's surface have been made from aircraft in level flight using radar in a sophisticated modality. The foliage of trees is stripped away and the ridges and gullies of the hard surface are shown in stark relief. For the purpose of topographic survey the technique supersedes airborne stereophotography, which itself had earlier superseded ground survey, in mountainous and forested areas. An example of such an image is shown in Fig. 16-1.

DOPPLER RADAR

When operational radar was introduced in the 1930s, its strong point was accurate range measurement; directional discrimination and height-finding came later. A different sort of development used Doppler effect to give line-of-sight velocity in addition to range. If a source emits radiation of frequency f, and an object returning an echo is in motion, the echo frequency is $f_D = f + \Delta = f - \dot{P} f/c$, where $\dot{P}$ is the rate of change of the round-trip phase path P from the source to the receiving point. If a small signal having the transmitter frequency f is added to the returning echo on arrival, the resultant will rise and fall in amplitude at the frequency of the Doppler shift Δ. In this way both the range and line-of-sight velocity of a discrete target can be measured, in the common case where the transmitter and receiver are at the same location and the propagation medium is effectively free space.

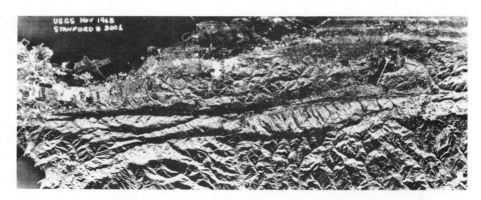

Figure 16-1   Synthetic-aperture radar map of a part of San Francisco Bay made at K band.

The expression $-\dot{P}f/c$ for the Doppler shift is applicable to a moving observer or to a moving source and applies also when both are moving. If neither moves, there will still be Doppler shift if time changes in the medium along the ray cause P to change. In the case of a satellite moving in or above the ionosphere the electron content along a ray varies significantly with time, as does the neutral gas content in the tropospheric portion of the path. In terms of the space-dependent refractive index n, we have $P = \int n\, ds$, where ds is a distance element along the ray. At ground level n is approximately 1.0003, while for ionospheric propagation $0 < n < 1$. In general, the ray will be curved, as calculable if $n(x, y, z)$ is known. Evidently, the phase path P can change if either end of the path moves or if the part of the medium along the ray undergoes a change in refractive index.

The expression $-\dot{P}f/c$ for Doppler shift Δ is general; it applies when both transmitter and target are moving and where the receiver moves relative to the transmitter, as may happen with bistatic radar. If the medium refracts the rays, which both the ionosphere and troposphere do, $\dot{P}$ is measured along the ray; in that case, line-of-sight velocity becomes relative velocity along the ray. If transmitter, receiver, and target are all fixed but the medium is inhomogeneous and moving due to wind or turbulence, there may still be a Doppler shift, and the expression $-\dot{P}f/c$ will give it.

If the radar is in uniform motion, for example if it is carried by an aircraft in level flight (Fig. 16-2), then the echo received at a given time delay will be contributed to by points contained in a circular annulus of width equal to the range resolution. But each component contributing to the echo will be affected by its own Doppler shift. Lines of constant Doppler shift are lines of equal $\dot{P}$. These are almost straight radial lines, but the height of the aircraft makes them more nearly hyperbolas asymptotic to the radial lines; an exact specification must also take into account the curvature of the earth and tropospheric refraction. Therefore, if the echo is sorted into range bins and Doppler-shift bins, an instantaneous retroflectivity map of the terrain can be determined. There is a left-right ambiguity resulting from the fact that, at any given range, there are two points with

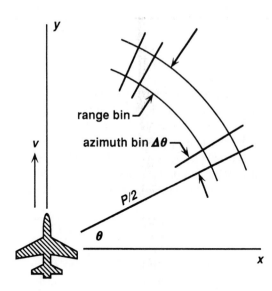

y

range bin

azimuth bin $\boldsymbol{\Delta\theta}$

P/2

v

θ

x

Figure 16-2 The echo received from a
scattering element at range $\frac{1}{2}P$ exhibits
a Doppler shift $\Delta = -\dot{P}f/c$. When $\dot{P}$
is negative, as here, the Doppler shift is
positive.

the same Doppler shift. However, this ambiguity can be suppressed by use of a directive
antenna that looks to only one side of the track.

Resolution

When range is determined by emission of a pulse of radiation at frequency f and measurement of the time delay of the echo, the pulse duration τ determines the range resolution d_r
in meters through the relation

$$d_r = \tfrac{1}{2}c\tau \sec\delta,$$

where δ is the angle of depression given by $\tan\delta = h/\sqrt{x^2 + y^2}$. The slant range $\frac{1}{2}P$ to a
target at $(x, y, 0)$ from a radar at $(0, 0, h)$ flying in the y-direction with velocity v is given
by $\frac{1}{2}P = y/\sin\theta \cos\delta$, where $\tan\theta = y/x$. Consequently $\dot{P} = -2v\sin\theta \cos\delta$ and

$$\Delta = -\frac{\dot{P}f}{c} = 2\frac{v}{c}f\sin\theta \cos\delta.$$

To relate the Doppler-bin width $d\Delta$, which is set as described later, to the azimuthal
resolution angle $d\theta$ we differentiate to obtain $d\Delta/d\theta = 2fv\cos\theta \cos\delta/c$. Thus

$$d\theta = \frac{d\Delta}{2(v/c)f\cos\theta \cos\delta}.$$

In meters the along-track resolution dy is given by

$$dy = \tfrac{1}{2}P\cos\delta \, d\theta = R\,d\theta = \frac{R\lambda_{fs}d\Delta}{2v\cos\theta \cos\delta},$$

where $R = \frac{1}{2} P \cos \delta$ is the ground range and the free-space wavelength is given by $\lambda_{\text{fs}} = c/f$.

Synthetic Aperture

It has not been necessary to allow for the displacement along the y-axis, nor for the fact that the receiver, at the moment of reception, is not in the same location as the transmitter was at the moment of emission of the pulse. The results for resolution in range and azimuth come simply from knowing the instantaneous velocity v. However, there is a way of describing the good resolution achievable by reference to the displacement of the radar in the time between emission of a pulse and its subsequent reception. In fact, this way of looking at things gives rise to the term "synthetic-aperture radar." The idea is that the resolution angle

$$d\theta \approx \frac{\lambda_{\text{fs}} \, d\Delta}{2v}$$

is the beamwidth of a certain hypothetical antenna that lies along the y-axis ahead of and behind the actual radar antenna. Since the beamwidth (measured from peak to first null) of a uniformly excited antenna of length D_{synth} at free-space wavelength λ_{fs} is $\lambda_{\text{fs}}/D_{\text{synth}}$, one can say that the SAR radar is equivalent, as regards along-track resolution, to an antenna of length $D_{\text{synth}} = 2v/d\Delta$ in the y-direction.

In order to obtain the length D_{synth} of the synthetic aperture it is first necessary to establish the width $d\Delta$ of the Doppler bins. Although an expression for this bin width was derived earlier, the numerical magnitude was not examined; the synthetic-aperture approach provides a good physical basis for this. The smaller the bin width $d\Delta$ into which the spectrum of an echo is analyzed, the better the along-track resolution will be; but it is not effective to reduce the bin width indefinitely, even though that may be computationally possible. The practical lower limit to $d\Delta$ arises as follows.

We introduce the transit-time interval T, which is defined as the time interval during which a fixed ground point in the target swath remains in the antenna beam. Let the antenna beamwidth in the horizontal plane be Θ, as shown in Fig. 16-3. A ground point A whose minimum range is R will be within the beam for the time taken by the vehicle to move a distance $2R \tan \frac{1}{2}\Theta$, or approximately $R\Theta$. Thus for the transit-time interval T we have

$$T = \frac{R\Theta}{v}.$$

The echoes received from a particular ground point A as the plane flies past continue for a total time duration T and are sampled at the regular pulse repetition intervals. Nevertheless, as with any signal of duration T, the frequency analysis consists of a fundamental at frequency T^{-1} and harmonics separated by T^{-1}. The sampled nature of the echo sets a high-frequency limit to the useful spectrum but does not affect the separation of the harmonics. Consequently, T^{-1} is the appropriate value for $d\Delta$; use of narrower Doppler bins will not result in extra information. Focusing on the broadside direction $\theta = 0$, approximating $\cos \delta$ by unity, and putting

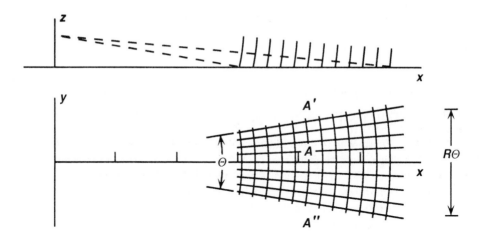

Figure 16-3   View in the (x, z)-plane looking along the flight track (above). View from above showing circles of constant range and the hyperbolas of constant Doppler shift (below).

$$d\Delta = T^{-1} = \frac{v}{R\Theta}$$

gives

$$Rd\theta = \frac{R\lambda_{\mathrm{fs}}d\Delta}{2v} = \frac{1}{2}\frac{\lambda_{\mathrm{fs}}}{\Theta}.$$

Characterize the beamwidth Θ by the angle between the peak and the first null. Since that beamwidth is obtained from an on-board antenna of length $D = \lambda_{\mathrm{fs}}/\Theta$, we arrive at the very simple and interesting result that the along-track resolution is given by

$$R\,d\theta = \tfrac{1}{2}D.$$

Thus, the along-track resolution distance is one-half the length of the on-board antenna. Interestingly, this resolution is independent of the velocity v, even though v appears in the equation $R\,d\theta = R\lambda_{\mathrm{fs}}\Delta/2v \cos\theta \cos\delta$; the reason is that the Doppler shifts observed are proportional to v. Finally, the along-track resolution is independent of frequency. Paradoxically it would seem possible to improve the along-track resolution by using a smaller antenna. The distance $A'A''$ has the practical significance that echoes can be integrated for a time equal to the transit-time interval T in order to build up signal-to-noise ratio. This integrating time is seen to depend upon the velocity v.

Expressing the along-track resolution that the system achieves in terms of D_{synth}, the size of the single antenna that is in effect synthesized, and is defined by $d\theta = \lambda_{\mathrm{fs}}/D_{\mathrm{synth}}$, we have:

$$D_{\text{synth}} = \frac{\lambda_{\text{fs}}}{d\theta} = \frac{2v}{d\Delta} = 2vT.$$

Thus D_{synth} equals twice the distance traveled in the transit interval T or twice the distance $A'A''$ in Fig. 16-3.

SOME HISTORY OF RADIOFREQUENCY DOPPLER

During World War II shore radars could detect ships at meter-wave frequencies and the ship's range could be read from an A-scan, which plots strength of echo against the time that elapses after the emission of the pulse. It is very noticeable on such a display that the sea returns are in a constant state of flux, while a ship's echo, at least if it is strong enough, presents a break in the baseline of the display that an operator learns to recognize, and to use for reading off the range. A second noticeable characteristic is that the ship's echo may exhibit variations over and above those due to the rolling of the ship; in fact the ship may disappear for longish periods. This defect was found to be due to leakage of the transmitter power directly through the transmit-receive switch and into the receiver. Even if this unavoidable leakage is at a very low level, it will still be comparable with the echo level received from some targets. As a result there will be interference between the two radiofrequency signals, and the consequence may be destructive interference and loss of detectable echo. The way to eliminate this is to improve the T-R switch and the oscillator screening; but by going the other way and deliberately introducing a controlled amount of oscillator signal at the receiving antenna terminals, one arrives at the Doppler velocity technique known at one time as Flutter. The effect on the A-scan (Fig. 16-4) is to cause the ship echo to rise and fall, or flutter, at a rate that gives a direct measurement of the radial component of the ship's velocity, and is generally palpably faster than pitching or rolling.

Now, of course, it is quite possible to measure the closing rate of the ship from the rate of change of range. But it will be necessary to wait for lengths of time comparable with the range resolution divided by the radial component of the speed of the ship. In contrast, the Flutter method gives you the speed in the very short time taken for the ship to change its range by a few wavelengths. At first, operators could judge the speed quite effectively by measuring the Flutter frequency with the aid of a comparison audio oscillator, but it is not difficult to automate this system and present the velocity in digital form.

This historical example of an elementary Doppler radar is easy to understand in full detail. Already, however, it contains the striking feature that sometimes causes puzzlement about the SAR system, namely the property of far exceeding the spatial resolution in meters that one might feel entitled to from the apparently relevant system parameter, in this case the pulse width.

As a postscript to this account of how the Doppler radar principle originated from simple beginnings it is interesting to recall that it was Doppler shift at radio wavelengths that first drew attention to the possibility of ship detection by radio. In 1922 it was noticed that the signal received on a point-to-point radio link across the Potomac was fluctuating rhythmically in strength and the happening was attributed, quite correctly, to the reflection

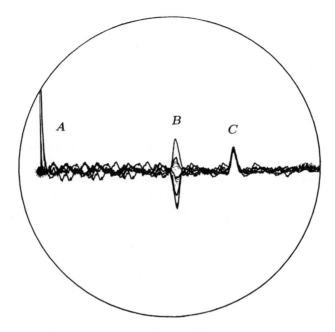

Figure 16-4   The A-scan of a radar equipped with Flutter showing the transmitter pulse
A, the echo from a moving ship *B*, and the echo from a reef *C*.

of a little energy from a moving ship on the river. The appearance was exactly analogous
to the rhythmic fading that is commonly experienced these days on television receivers as
a plane flies nearby. Owing to the Doppler shift of the reflected wave, a beat is produced
at the frequency equal to the difference between the frequency of the transmitted signal
and the Doppler-shifted frequency after reflection. In 1930 such an observation, on a
radio receiver rather than on TV, was made at the Naval Research Laboratory. At the
time it was considered worthwhile following up this accidental observation by conducting
planned experiments for the purpose of investigating the possibility of aircraft detection
(Allison, 1981). While simple target detection is a long way from radio location and
velocity measurement, still it is interesting to recall that the Doppler principle was present
in this precursor of radar and that the concept was proven and available for building upon
from these very early dates. Today we would describe the arrangement as bistatic Doppler;
it is precisely the technique that was used with the Mariner space probe to detect and
measure the density of the atmosphere of Mars (Kliore et al., 1965).

RANGE-DOPPLER RADAR

The development that lead to the application of range-Doppler radar to mapping came
from efforts to map the surface roughness of the moon and Venus. It was possible by the
1950s to build receiving antennas for passive mapping whose beamwidth was less than the

half-degree diameter of the moon, but it was a long time before radar antennas reached that capability. Meanwhile it had been discovered how to make radar maps with angular resolution much exceeding the resolution of the antenna beam (Evans and Hagfors, 1968).

Although the method was applied to studies of the moon, there are some complications in the moon's case that make it preferable for explanatory purposes to describe the principle in connection with Venus as studied from the earth.

Since the period of Venus is 224.70 days while that of the earth is 365.26 days, it is clear that a radar echo from Venus will exhibit Doppler shift, if only because of the changing distance between the two planets. But in addition there will be a further shift due to the fact that the radar is rotating about the earth's axis. This second effect is much smaller, because the rotation speed of the earth's surface is only 0.4 km s^{-1} even at the equator, whereas the orbital velocities of the earth and Venus, 29.79 and 35.03 km s^{-1} respectively, can combine to produce a relative velocity in the line of sight up to about 55 km s^{-1}. Nevertheless, at a frequency of 3×10^9 Hz, a velocity of 0.4 km s^{-1} results in a Doppler shift of 8 kHz, which is quite measurable and in fact can be measured with high precision.

Since the daily rotation of the earth is easily detectable, it follows that the corresponding rotation of Venus might also be detectable. At the time radar studies of Venus began, the rotation of Venus was not correctly understood. Observations of transient features on the cloud surface had been seen in telescopes for years and variously interpreted, but it was not until the 1960s that the radar observations themselves produced the missing knowledge about the rotation of Venus. Unexpectedly, the rotation of Venus is retrograde. The rotation is also very slow—the period is 244.3 days—and has the surprising property of being synchronized with the revolution of the earth. While it is a familiar fact that the rotation of the moon is synchronized with the revolution of the moon, it is also a fact that is understandable in terms of tidal interaction with the earth; but it was a complete surprise to learn by radar that the rotation of Venus was controlled in some way by the revolution of the earth.

With a period of 244.3 days and a radius of 6050 km we find an equatorial velocity of 0.007 km s^{-1}, which corresponds to a Doppler shift of about 140 Hz. This again is a very measurable Doppler shift.

The big radars operating at Jodrell Bank, Arecibo, and Goldstone detected surface features on Venus that gave rise to irregularities in the echo signal strength variation, but it was not at first easy to deduce where these features were located from range observations alone. One range measurement on a detected feature merely localizes that feature to an annulus centered on the subterrestrial point, as indicated on the left of Fig. 16-5.

Since the radius of Venus is 6050 km, rather long pulse lengths may be used and still be effective for preliminary mapping. For example, to achieve a range bin 600 km deep a pulse length of 4 milliseconds would be appropriate. Even so, the echo is so weak after diverging twice over interplanetary distances that long integration times are desirable, and a convenient way of achieving this is by pulse coding. Under this scheme pulses much longer than a few milliseconds are transmitted, and the returns from different range annuli are sorted out on reception by an autocorrelation method.

The use of long pulses makes it convenient to extract the Doppler shifts due to

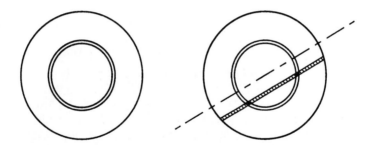

Figure 16-5   An annulus (left) bounded by circles of constant range to which a target on a sphere is confined by one echo-delay measurement. The ambiguity in range-Doppler mapping (right).

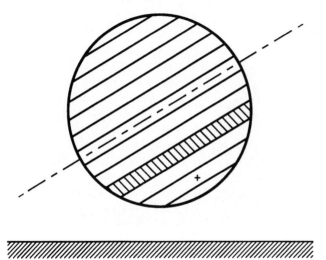

Figure 16-6   Zones bounded by lines of constant Doppler shift are parallel to the rotation axis of a spherical target. In this example, Venus is shown setting on the western horizon. Positive Doppler shift is associated with approach.

planetary rotation. The information resulting from this step enables the surface targets to be assigned on a strip-coordinate system that is complimentary to the annular system furnished by the range data. Thus in Fig. 16-6 we see that the Doppler shifts are distributed over the rotating surface in such a way that a zone of constant Doppler shift is parallel to the instantaneous axis of rotation. To one side of this axis the surface is approaching while to the other side it is receding, and the magnitude of the shift is proportional to distance from the axis.

Consequently, by sorting an echo into range bins one restricts attention to one annulus, and by further analysis into Doppler frequency components one attributes the contents of that annulus to the Doppler bins that intersect it. Although the coordinate system resulting is not orthogonal, it is effective. A noticeable defect, however, is the ambiguity due to the fact that each range annulus cuts each Doppler strip in two places as in Fig. 16-5 (right). This difficulty is rather easy to deal with if the primary resolution of the antenna is

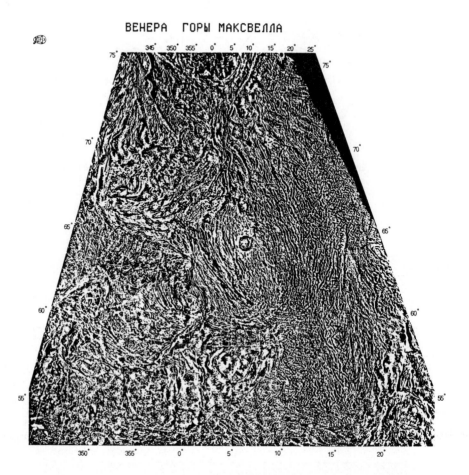

Figure 16-7 Range-Doppler map of the Maxwell Hills of Venus, achieving a resolution of 1 kilometer or better.

sufficient to suppress one of the alternatives, but in the application to Venus, where many of the details were first worked out, the ambiguity caused a delay in progress. A further sorting has to be applied on a time basis, a step that will not be described in detail here but works because the axis of rotation changes with time.

This completes the description of the range-Doppler method of mapping as it was developed in connection with the technically very demanding problem of Venus. The principle is not at all complicated and is easy to follow. Severe technical hurdles were overcome by the use of very large antennas, very sensitive receivers, and sophisticated digital electronic filtering and other data processing. The competitive environment provided by competition between a number of laboratories in open communication led to rather rapid resolution of the difficulties. An example of such mapping is given in Fig. 16-7, which shows the Maxwell Hills of Venus situated in the latitude zone 58.3° to 72.4°, which covers

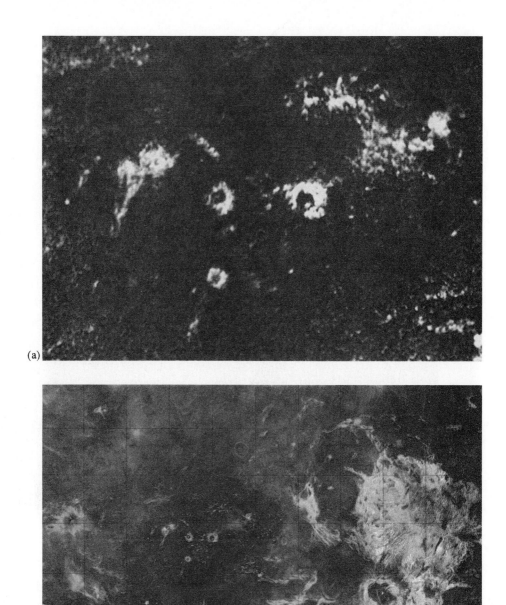

Figure 16-8 Three range-Doppler maps of craters on Venus: (a) made from the Arecibo Observatory in 1977 by Donald B. Campbell, (b) made in 1988 with improved technique, (c) made in 1990 from the Magellan spacecraft orbiting Venus (NASA photo kindly provided by Gordon H. Pettengill). Continues on page 556.

(c)

Figure 16-8   (Concluded)

a north-south distance of approximately 1500 km. This map was distributed by the Soviet delegation to the 1985 meeting of URSI in Tokyo. The remarkable progress over time is illustrated in Fig. 16-8. Frame (a) was made by D. Campbell in 1977 from the ground at Arecibo, Puerto Rico, and shows the "Crater Farm" in the Lavinia region with a resolution between 5 and 10 km. By 1988 Campbell had achieved the improvement shown in frame (b), where the resolution is between 1 and 2 km; this is quite extraordinary relative to the diameter of Venus (6070 km.). One kilometer on Venus subtends an angle of about 0.005 arcsecond. From the Magellan spacecraft orbiting Venus in 1990 even finer detail was discernible (resolution 0.12 km) after the formidable image-processing technique was worked out, as shown in frame (c). By 1991 better maps were obtained from the Magellan mission, as in Fig. 16-9, which shows the crater Cleopatra, about 100 km across, in the Maxwell Hills. The smallest objects have a size of about 120 m.

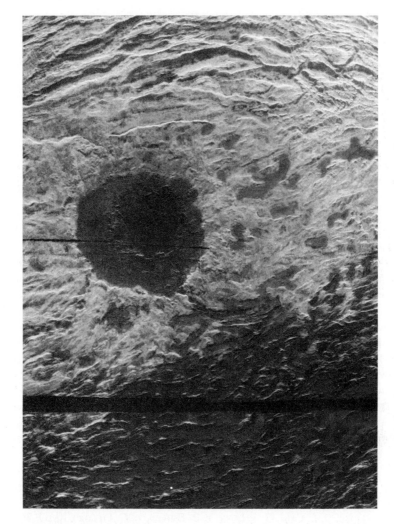

Figure 16-9   Range-Doppler map of the crater Cleopatra on Venus made from the spacecraft Magellan in 1991 (NASA photo).

RADARGRAMMETRY

A radar traveling horizontally above the ground in an orbit through O' perpendicular to the plane of Fig. 16-10 receives a reflection from a scattering element P, which is observed to be at range r'. The element P is not at mean sea level, or other reference level, but at an altitude h above the reference ground-plan position T; consequently, the range measurement alone only confines P to a position line PP', all of whose points are at range r' from O'. On a later orbit passing through O'' the range r'' is measured which

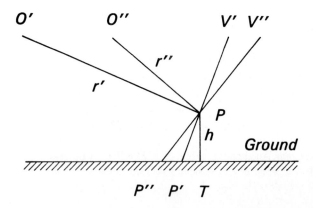

Figure 16-10   Radar range measurements from O' and O'' on a scatterer P imply the same equivalent ground locations P' and P'' as do photographs taken from V' and V''.

locates P on PP'' and fixes the altitude h, provided that the orbits are accurately known and that the reflections from P are correctly identified.

   The situation corresponds to photogrammetry, the precise measurement of photographic stereo pairs obtained by flying over terrain. Stereograms were introduced in Chapter 2 and are discussed in a different context in Chapter 17. A camera positioned at V' would confine P to the same line PP' as does the radar at O'. A second camera at V'' suffices to fix the altitude of P. The photographs refer P to the different ground-plane locations P' and P'', and viewing with a stereoscope elicits the relief. The radar pictures exhibit similar ground displacements and the relief can be recaptured by stereo viewing with the eyes placed in the positions appropriate to photographs made with the corresponding cameras.

LITERATURE CITED

D. K.ALLISON (1981), *New Eye for the Navy: The Origin of Radar at the Naval Research Laboratory*, Naval Research Laboratory.

J. V. EVANS AND T. HAGFORS , eds., (1968), *Radar Astronomy*, McGraw-Hill, New York.

A. KLIORE, D. L. CAIN, G. S. LEVY, V. R. ESHLEMAN, G. FJELDBO, AND F. R. DRAKE (1965), "Occultation experiment: results of the first direct measurement of Mars's atmosphere and ionosphere," *Science*, vol. 149, p. 1243, 10 September 1965.

PROBLEMS

16–1. *Battlefield Doppler.* A pulse radar can detect objects approaching across the ground at night, even very stealthily, by adding the echo to a sine wave from a stable oscillator set at the transmitter frequency and noting the rise and fall in resultant strength (by one cycle for each half-wavelength of range decrease) as the target advances. A smart robot is to be designed that creeps forward between pulses, but arranges to be always stock still while an incident pulse is being reflected. As a result, no reflected pulse exhibits Doppler shift. An expert is recommending that the project be dropped on the grounds that it would not work. Explain why it would not, or confirm that it would. ☞

16–2. *Line-of-sight velocity.*
   (a) Show that the line-of-sight component $\dot{r}$ of the relative velocity between an airplane flying at 300 knots and a fixed ground point at range R is $-vy\cos\delta/r$, where $r = \sqrt{x_0^2 + y^2}$ and x_0 is the distance of the fixed point from the ground track of the aircraft.
   (b) Make a graph of $\dot{r}$ in knots versus y in km for the case $x_0 = 20$ km and $\delta = 0$.
   (c) Show that the curve possesses asymptotes at $\dot{r} = \pm v$ and that $d\dot{r}/dy = -v/x_0$ when $y = 0$.

16–3. *Lines of constant Doppler shift.* An aircraft at height $(0, 0, h)$ is flying with velocity v in the y-direction emitting waves of frequency f. Assuming the earth is flat, show that the ground loci of constant Doppler shift Δ are hyperbolas given by $x^2 - (K^2 - 1)y^2 + h^2 = 0$, where $K = 2vf/c\Delta$.

16–4. *Moving target.* An aircraft flying along a track due north with velocity V detects a ship sailing east with velocity v. Consequently, when the aircraft is exactly astern of the ship, the echo from the ship exhibits a negative Doppler shift. There is a stationary surface point at the same range R as the ship which has just this Doppler shift. Where is this point, and will the ship's image appear at this point?

16–5. *Tropospheric ray deviation.*
   (a) A ray from above the troposphere arrives at the ground, where the refractive index of air is n_0, with an angle of incidence i_0. Show from Snell's law that the ray suffered an angular deviation Δ given approximately by $\Delta = \arcsin(n_0 \sin i_0) - i_0$.
   (b) What are the deviations for $i_0 = 10°$ and $80°$ if $n_0 = 1.0003$?

16–6. *Acceleration.*
   (a) The aircraft carrying a SAR radar always has some forward and backward acceleration, but suppose that the data reduction is carried out as if the velocity were constant. At the moment when the increasing velocity passes through the nominal velocity, there happens to be a point scatterer on the starboard beam. How will it appear on the image?
   (b) Consider the case of lateral acceleration where the speed is constant but the track is slightly circular.

16–7. *Design concept for radar mapper.* A satellite is to be launched eastward from Cape Canaveral (28° 24′ N, 80° 37′ W) into a circular orbit with a 24-hour period, mainly for communications. From its longitude of 97° W the Corn Belt will lie to the north at an angle of depression of just a few degrees off vertical. There is a proposal to place a lightweight radar on board for an experiment to see whether useful data will result from range-Doppler mapping through the seasons. Government experts on remote sensing will review the sections of the proposal dealing with crop damage, crop height, crop disease, soil moisture, flooding, cloud cover, etc., but first they wish to find out rapidly from you as a consultant whether range-Doppler mapping is possible at all from a geostationary satellite under the above conditions.
   (a) Calculate whether usable Doppler shifts at a wavelength of 3 cm could be expected.
   (b) Explain why the proposal should be rejected out of hand or explain why it may merit further consideration.

17

Two-Dimensional
Noise Images

Noise in an image can be defined as the unwanted part of that image. The noise may be random in some way, as is the pepper-and-salt appearance on a television screen when the station goes off the air, or it may be systematic, as with the ghost seen when an echo of the wanted signal arrives with a time delay after reflection from a hill. When the television image responds independently to sparks in a faulty thermostat in the nearby refrigerator or to a faulty ignition system on a passing motorcycle, the noise exhibits both random and systematic features. In other cases, the wanted signal may be random; thermal microwave or infrared radiation used for mapping the ground is of this nature. As a result, one person's noise may be another person's signal and vice versa. Very often it does not matter much what the character of the noise is, only its magnitude is needed, an attitude that is reflected in the term signal-to-noise ratio. As the examples show, the noise in an image need not be independent of, but can be closely connected with, the wanted signal itself. In the latter case if the signal is removed, the noise will change. When the noise is independent, it may be studied on its own in the absence of any wanted signal.

Physical origins of noise are only mentioned incidentally here; the attention is mainly on descriptive features that make two-dimensional noise perhaps a little different from the more familiar one-dimensional noise on time signals. In addition, practical aspects of computing with noise are explained. Only random noise of the commonest types is considered.

Both the spatial spectrum and the autocorrelation of random noise are important but can impose heavy computation. The clipped autocorrelation, the binary object that forms the support of the autocorrelation of a sampled function, is introduced as an explanatory tool that has much of the diagnostic value possessed by the autocorrelation itself but is easier to construct.

Other sorts of noise image are important that do not qualify as unwanted components. There are test images used like the noise-signal generators in electronics. Texture may be random and plays a basic role in computer graphics, both constructive and analytic. Fractal images of material objects and computed fractal images may be random and are related to textural concerns. Finally one should not overlook the commercial use of random images in the printing industry, especially for advertising; the modern applications have roots in halftone lithography, engraving, and oil painting.

SOME TYPES OF RANDOM IMAGE

Figure 17-1 shows some simple random images that may be encountered. The first two consist of unit impulses scattered at random, the third consists of impulses positioned on a regular array that is coarse enough to be apparent, and the fourth is a random function with values distributed uniformly between 0 (white) and 1 (black). In (a) the random scatter diagram exhibits clustering and zones of avoidance possessing a character that one can learn to recognize. The pattern, reminiscent of a field of stars, teaches us that collineations, vacant channels, and apparently significant condensations arise by chance. In reality, star fields do exhibit systematic features that are not stochastic. To be able to recognize the meaningful features, or corresponding ones in other two-dimensional scatters, one needs to study nominally uniform scatter attentively.

In (a) the filling factor (fraction of area covered by ink), set by the dot size, is roughly one-quarter. The filling factor in (b), which uses the same 8192 dots as (a) plus an additional 4000, is about one-tenth. There is a difference in appearance; the zones of avoidance seem to me to be more strikingly void. Clustering seems to favor curved lines. It might be interesting to survey one's friends and ask whether the different distributions appear to them to be random; what we have just seen is that visual impression involves more than the simple statistics. Examination of this image with a lens is also recommended. In (c) the random matrix has approximately 50 percent occupancy but the filling factor is about one-quarter, because the spacing between matrix points is a little over one dot diameter. The regular array is conspicuous in (c) but is not at all apparent in (a), although the minimum dot spacing is less by only a factor two. When the diagrams are viewed from a distance of two or three meters, the top three look much the same. Some of the difference is attributable to the array spacings (respectively about 30, 5, and 60 μm). The random surface shown in (d) is in a different category and will be discussed later.

Binary Random Scatter

If there is a random scatter of N dots in an area A, as in Fig. 17-1(a), the mean density of dots $\bar{n}$ is given by

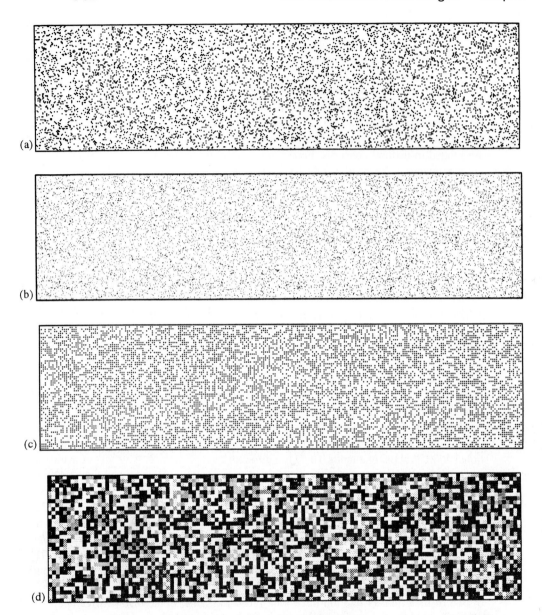

Figure 17-1   From top to bottom: (a) A random scatter of 8192 dots. (b) The effect of decreasing the dot size and increasing the dot density while reducing the filling factor. (c) A randomly occupied matrix. (d) A grey-level representation of a random surface.

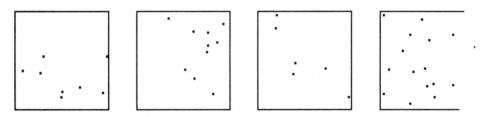

Figure 17-2   Four subfields from a random scatter with 10 dots expected per frame.

$$\bar{n} = N/A.$$

The concept is the same as population density, where we divide the total population by the total area to find the number of people per square kilometer. Just as people are concentrated geographically in accordance with the attractiveness of the locale, so also the random scatter shows places where there are more or fewer dots. However, the specification for construction of the illustration did not contain spatial variation: the same rule of generation was applied uniformly over the whole area. Thus the spatial fluctuations arise purely from statistics and not from variations in the locale. It is of interest to know how great the natural fluctuations can be and to see something about their character when the mechanism of generation is spatially uniform.

Fig. 17-2 shows four subfields selected from an original figure like Fig. 17-1(a), each amounting to 1 percent of the area. Let us say that an original was 10 cm square. It contained 1000 dots with a mean density $\bar{n} = 10$ dots per square cm. The subfields were 1 cm square and would be expected to contain an average of 10 dots each. In fact the numbers of dots are 8, 10, 6, and 17. We would like to know the standard deviation about the expected mean in order to settle design questions such as how long an exposure to use in order to achieve a desired signal-to-noise ratio with an imaging device.

A preliminary thought hints at the answer. Suppose the sample subfield had been 4 cm^2 instead of 1 cm^2; then 40 dots would be expected. In fact, we could take the four subfields already considered to be lumped together as an example of the four-times-larger subfields. The new total is 41 rather than exactly 40, but it is noticeable that the sample total agrees better with 40 than the smaller totals agree with 10. In fact, the sum and the variance for the larger field both tend to be four times greater, which means that the new standard deviation relative to the new sum will be reduced by a factor $\sqrt{4}$, from about 32 percent to about 16 percent. Consequently, we expect the standard deviation of the total number of dots n in a subfield to get smaller relative to the total as the number n increases and to do so in inverse proportion to $\sqrt{\bar{n}}$. Thus, as n varies,

$$\frac{\sigma}{\bar{n}} \propto \frac{1}{\sqrt{\bar{n}}}.$$

Interestingly, the coefficient of proportionality will be close to unity, for large n. Thus with $\bar{n} = 10$ we expect $\sigma = \sqrt{10}$. An experimenter with access to all four frames might report that $\bar{n} = 10.2 \pm 1.6$.

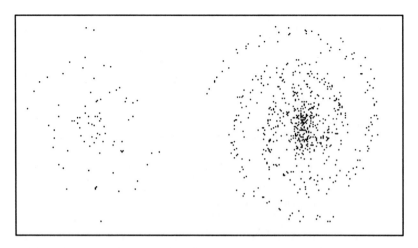

Figure 17-3   An image of an imaginary faint galaxy recording the arrival of 80 photons (left). Even though everything recorded is signal, the record is conceived of as part wanted signal and part noise. A longer exposure recording 670 photons (right) tells more about the galaxy, but separating the "true" structure from the "noise" remains elusive. In a sense the whole image is noise.

If the count n is governed by a Poisson distribution with parameter λ, the probability $p(n)$ that a subfield will contain exactly n dots is

$$p(n) = \frac{\lambda^n}{n!} e^{-\lambda}.$$

The customary symbol λ for the parameter of a Poisson distribution represents a population density like $\bar{n}$. The distinction is that the statement $\lambda = 10 \text{ cm}^{-2}$ is an assertion, while the quantity $\bar{n} = N/A$ is an observable. For large n,

$$p(n) \rightarrow \frac{1}{\sqrt{2\pi\lambda}} \exp\left[\frac{-(n-\lambda)^2}{2\lambda}\right],$$

which we see to be a normal distribution about mean λ with standard derivation $\sigma = \sqrt{\lambda}$.

The sort of random scatter described here may appear as unwanted interference, but it also characterizes the fine structure of many images that are free from interference. For example, the image built upon a photographic plate during the course of an exposure is due to the arrival of photons in locations that are random, even if the scene being photographed is uniformly bright. The telescopic image of a faint galaxy contains spatial fluctuations of this kind; to reduce the standard deviation relative to the mean it would be necessary to extend the exposure time and collect more photons. Although the image of a uniformly bright patch is a legitimate random scatter, the spatially nonfluctuating part constitutes the wanted signal and the fluctuating part is the noise. Of course, when the object photographed has spatial structure of its own, as a galaxy does (Fig. 17-3), determining what part of the spatial variation is signal and what part noise is the interesting

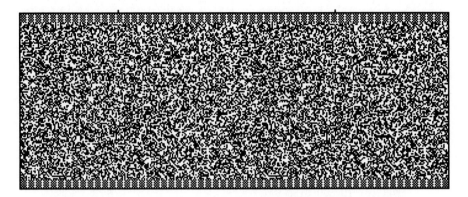

Figure 17-4   A stereogram that conveys nothing to the casual eye but exhibits depth through a stereoscope. Some viewers can dispense with the stereoscope by making the optic axes of their eyes parallel (convergent at infinity), which is facilitated by causing the small ticks on the upper frame to fuse.

problem. In photographs other sorts of noise may dominate. Graininess in the structure of the photographic emulsion is one such mechanism, and distortion due to refraction in the atmosphere is another.

Random Stereograms

Let a rectangular area covered randomly with equal numbers of black and white pixels consist of left and right halves which are identical, the separation of corresponding elements being 60 mm (the normal interpupillary distance). If viewed through a stereoscope, the object will have a sharp appearance, because the superposition of the halves will be exact. A blank patch is prepared on the right and is filled with a copy of a patch of the same size and shape situated at some location on the left. If the borrowed patch comes from exactly 60 mm to the left of the blank opening, the status quo is restored, but if the patch is borrowed from 58 or 59 mm to the left, then three-dimensional relief will be seen through the stereoscope. This is illustrated in Fig. 17-4, where a square patch, shifted five pixels, has been copied as described. To the unaided eye the general appearance of the object is that of an ordinary binary random background, neither the left-right repetition nor the fault lines associated with the shift being apparent without close inspection. The consequence is that the stereoscope shows a square platform distinctly elevated above a floor. This idea was originally described by Julesz (1960); later developments can be traced in "Science Citation Index." A less conspicuous effect also arises with the top and bottom fringes, where the apparent height change is different and can be thought about.

Many people are able to fuse stereographic pairs without an instrument but it can be difficult, especially if the eyes are different. An alternative to focusing on infinity is to make the optical axes of the eyes converge on a point halfway between the object and the eyes, which is helped by placing a finger there. As before, the ticks are to fuse; it will be necessary both to roll the head to bring the line joining the eyes parallel to the line joining the ticks, and to move the finger to and fro. The consequence will be that the platform appears to be below the floor.

A convenient way of making speckle diagrams on a screen uses the byte plot command of high-level graphics languages, which converts a string such as "S11 7B11" to the 64-bit sequence 01010011 00110001 00110001 00100000 00110111 01000010 00110001 00110001 and lights a row of pixels accordingly. One can fill a strip of size 512 × 32 with random speckles as follows.

SPECKLE BY BYTE PLOTTING

```
FOR r=1 TO 32
    A$=""
    FOR c=1 TO 505 STEP 8
        A$=A$&CHR$(255*RND)    concatenate 64 chars
        BPLOT A$,64            byte plot a row of 64 chars (512 pixels)
    NEXT c
NEXT r
```

The function CHR$(N) is the character corresponding to the integer N (from 0 to 255), for example CHR$(65) is A. Your computer knows very well that CHR$(65.1415926535898) is also A, which makes it unnecessary for a private programmer to add a line specifying that N be an integer and to take the extra step of ensuring that it is (unless your language of choice obliges you to perform the keystrokes that your computer will not need).

A shorter way of byte plotting $32 \times 64 = 4096$ characters is as follows.

SHORTER SEGMENT

```
A$=""
FOR i=1 TO 4096
    A$=A$&CHR$(255*RND)    concatenate 4096 chars
    BPLOT A$,64            byte plot 64 chars per row
NEXT i
```

By choosing characters randomly from 0 to 255 one ensures an expectation of 0.5 for black pixels. Choosing the ASCII characters (0 to 127), all of whose binary representations start with 0, will reduce the average grey level to 7/16; other restrictions on the set of characters will allow a full range of greys.

A different kind of stereogram, or autostereogram, can be constructed that is for many people easier to see without the mechanical aid of a stereoscope. The background is randomly speckled as before but is banded periodically, left to right, with a period in the range 10 to 20 mm. The example shown in Fig. 17-5 is a little on the coarse side, in order to allow inspection. A first interesting feature of this object is the presence of seemingly systematic features in the background. These are purely random artifacts, but seem much more emphatic than those mentioned in connection with Fig. 17-1, no doubt gaining some attention from the left-to-right and top-to-bottom symmetry that has been imposed within each band. The design resembles that of some rugs. For a purpose where attention should not be distracted from the stereoscopic features the symmetries

Figure 17-5 Flying saucer hovering above ground with open cargo bay—a banded stereogram that is relatively easy to invoke.

should be dropped. Because the repetition interval (about 25 mm) is much less than 60 mm, it is relatively easier to adjust the eyes until corresponding features of the pattern fuse, whereupon both raised and lowered details will be perceived. Ticks spaced one period are provided in the upper margin. A stereoscope also works, but the perception is different because two or three periods are spanned by the 60-mm spacing of the lenses. With unaided inspection it is possible to see higher-order images up to perhaps the ±3rd.

Construction of an autostereogram $S(x, y)$ of a function $z(x, y)$ on a random banded background $B(x, y)$ can be explained as follows. Both $S(x, y)$ and $B(x, y)$ are binary functions, equal to 0 or 1, of the integer variables x (1 to X) and y (1 to Y). The function $z(x, y)$ to be represented rises above a floor level $z = 0$ with values 1, 2, . . . distributed over some domain—let us call it an island. First fill the background matrix $B(x, y)$ with eight initially identical bands as follows.

BANDED BACKGROUND B(x,y)

```
FOR y=1 TO Y
   FOR x=1 TO 64
      B(x,y)=(RND>0.5)        fill first band
      FOR k=1 TO 7            additional bands
         B(x+64*k,y)=B(x,y)
      NEXT k
   NEXT x
NEXT y
```

At each location (x, y) a value 1 is assigned if a number taken at random between 0 and 1 exceeds 0.5, otherwise 0 is assigned.

The height function $z(x, y)$ is a matrix that may be filled by a variety of methods. For purposes of explanation it is sufficient to take a simple case where $z(x, y)$ is in most places zero but equals unity at grid points within the circle $(x - 200)^2 + (y - 60)^2 = 30^2$, a circle of radius 30 centered at (200, 60). In this case the specification of $z(x, y)$ may be done in the foregoing segment, as each point (x, y) is visited, by inserting a line

```
z(x,y)=((x-200)^2+(y-60)^2<30^2)
```

between NEXT k and NEXT x.

The stereogram $S(x, y)$, which on casual inspection will look much the same as the original background $B(x, y)$, is derived by making small changes to the elements of $B(x, y)$ as determined by the $z(x, y)$ to be exhibited. We visit each point (x, y) again, row by row, to form the stereogram $S(x, y)$. Initialize by copying $B(x, y)$ into $S(x, y)$. In a particular row $y = y_1$, work from left to right until, at $x = x_1$, the first nonzero value of $z(x, y)$ is encountered. In order for this pixel to appear raised as viewed with a stereoscopic shift of 64 pixels, there will need to be a pixel of the same color 63 elements to the right. As things stand, the element $B(x_1 + 63, y_1)$ may be black or white. Change it to agree with $B(x_1, y_1)$, putting

```
S(x1+63,y1)=B(x1,y1)
```

The effect will not be significant to the eye because of the random speckling. With that single change, if the modified pattern is viewed stereoscopically, the pixel at (x_1, y_1) will appear raised above the floor. However, one band to the right there will be a failure of exact superposition at one point because of the change made, and an unwanted depressed point will be seen. Compensate for this by changing not only the element at $(x_1 + 63, y_1)$ but also those at $(x_1 + 63 + 64, y_1)$, $(x_1 + 63 + 128, y_1)$, . . . , to the full width of the picture. The change in appearance will still be insignificant. Continuing to the right in row $y = y_1$, do as above until the right edge of the band is reached. Within the next band, when it is entered, the evolving $S(x, y)$ will already contain elements of $B(x, y)$ inherited from the initialization and elements altered as prescribed; however, proceed inexorably making color changes to $(x_i + 63 + 64 * k, y_1)$, whenever an elevated point (x_i, y_i) has $z(x_i, y_i) = 1$.

In a more general case where $z(x, y)$ is multivalued, then at a point where $z(x_i, y_i) = 2$, make a color change 62 steps to the right and at additional steps 64 across the picture. Thus the segment establishing $S(x, y)$ is

STEREOGRAM S(x,y)

```
FOR y=1 TO Y
   FOR x=1 TO X
      FOR k=1 to 7
         S(x+64-z(x,y)+64*k,y)=B(x,y)
      NEXT k
   NEXT x
NEXT y
```

It is good practice not to create a separate large matrix $S(x, y)$ but to reuse $B(x, y)$. It then becomes unnecessary to range from $(1, 1)$ to (X, Y); the smallest rectangle containing the island suffices. This restriction avoids visiting zero-level pixels where no action is required and which can be a large majority. Finally, within the smallest rectangle there will almost always be further zero-level pixels. Unnecessary assignments and index eval-

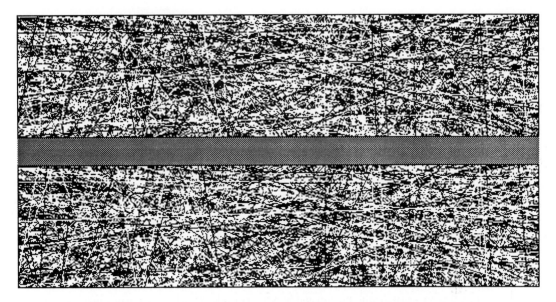

Figure 17-6 Random black and white lines (above) and the negative (below).

uation is avoidable by applying the condition $z(x, y) \neq 0$. The statement in the inner loop above then becomes

```
IF z(x,y)#0 THEN B(x+64-z(x,y)+64*k,y)=B(x,y)
```

For a review of the literature with numerous examples of autostereograms with explanations see Tyler and Clarke (1990). For an astonishing set of autostereograms in color see Thing (1993).

Random Lines

Straight lines may be drawn randomly over a plane area and will exhibit properties analogous to randomly distributed dots. An additional effect results from the intersection of some lines. On a field of N pixels, if $\frac{1}{2}N$ are black, the result is midgrey; but if black lines are drawn until $\frac{1}{2}N$ pixels are visited, the result will be lighter than midgrey because some of the black pixels are shared. Nevertheless, random lines offer a distinctive way of creating greys, so consider a balanced approach in which black and opaque white lines are drawn sequentially. If enough lines are drawn to fully cover the original substrate, the result will presumably represent a grey level of 0.5, because the effect of the shared black pixels will be cancelled by the shared white pixels. This conclusion can be tested by reversing black and white to form a negative, as has been done in the lower part of Fig. 17-6. If the grey level created is not midgrey, then there will not be a grey match with the negative image. The test needs to be carried out by the reader, viewing the figure from a distance of, say, 20 m. You may agree that the two greys are indistinguishable,

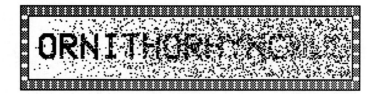

Figure 17-7   A binary object afflicted with salt-and-pepper noise whose density builds up to midgrey at the right.

as predicted, but you may also note blotchiness. The blotchiness will appear complimentary in the two bands, and may show other features. By backing off to a distance (tens of meters) where the blotches are unresolved and you see only uniform grey, you can verify the reasoning above, where it was presumed that the two bands should show equal greys. Equality of greyness implies a grey level of 0.5, exactly what is expected for the chessboard-pattern band that separates the halves. If the separating band, to your eye, does not present a match with the top and bottom, consider whether the positive/negative match is an absolute calibrator for midgrey or whether an unresolved chessboard pattern is. In this discussion midgrey refers to the ink in use; with blacker ink, midgrey in the present sense would be darker.

Additive Noise

It is very common in telecommunications to encounter noise, often with a Gaussian amplitude distribution centered on zero mean, that combines additively with the wanted signal. Additive noise also affects images in a strictly analogous way; however, binary images may be subject to addition modulo 2, or to hard limiting, because black plus black is black and white on white is white. This offers some interesting phenomena. If there is weak non-binary noise $n(x, y)$, then a binary signal $s(x, y)$ may not be much affected. Over much of the frame $[s(x, y) + n(x, y)]$ mod 2 may be exactly equal to the original signal $s(x, y)$. If the noise has a Gaussian amplitude distribution, there will be occasional pixels where $n(x, y) < -1$; if $s(x, y) = 1$, then the effect of the addition of weak noise is to reduce one image value to zero, making a white speck, a 100 percent change. Conversely, if $s(x, y) = 0$, then the noise can change the value from 0 to 1, a black speck. Of course, as the noise is weak, these white and black specks occur only here and there. Addition of this kind is nonlinear, producing white specks on black areas and black specks on white areas, as shown in Fig. 17-7, where the strength of the noise has been raised progressively from left to right. On the left, where the noise is weak, the unwanted specks are isolated and can be removed almost entirely by erosion followed by dilation as discussed in Chapter 8, but this does not repair the damage to the letters. Not unreasonably, it is practically impossible to recover the last two or three letters, where the fraction of damaged pixels approaches 50 percent.

Figure 17-8   Discs whose size and population density are what would be expected from a uniform distribution of equal spheres in space, down to a limiting absolute size.

Scatter of Finite Objects

Clustering, as it occurs in pure random scatter, is emphasized in Fig. 17-8; where discs of increasing size have been successively overprinted by larger discs. The number of discs in each size group has been made inversely proportional to the area of the disc, to match the appearance of a universe randomly occupied by equal spheres. Despite the distribution in depth implied by the obstruction of distant spheres by nearer ones the picture looks rather flat, perhaps more like coins of different denominations scattered on a table. The size range covers a range of only 5:1 in radius, and it is clear that a range of 8:1 or so would result in virtual disappearance of the grey background except possibly for a few chinks. H. W. M. Olbers (1758–1840) argued that a line of sight from the eye out into the night sky would, with a probability that exponentially approaches unity, ultimately intercept a star, or for that matter an external galaxy. Although distant stars are faint, owing to reduced solid angle subtended, they are not less bright, measuring per steradian, than they would be if they were as close as the sun, unless sunlight is absorbed in intergalactic space, which it isn't. The fact that the night sky is dark is Olbers' paradox.

Figure 17-9 suggests a variant that arises in x-ray diffraction analysis of crystals that have been ground to a powder. The purpose is to randomize the orientations and produce diffraction patterns consisting of concentric rings whose radii are diagnostic of lattice spacings in the unfractured material. This important technique obviates the need to rotate a single crystal into different directions. Not only do we see random shapes in random

Figure 17-9   Powdered crystalline material that would require at least half a dozen statistical parameters to describe.

locations in the figure, but also (threefold) randomness in the orientations of the internal crystal planes.

Random superposition of objects arises in various ways. Figure 17-10 shows the effect of depositing black and white squares of different sizes alternately in random locations. The outcome resembles an aerial view of an agricultural area in need of land reform, or a view from the sea of building construction on a Mediterranean hillside. One purpose of studying such patterns is to learn to analyze structures encountered in practice. Clearly the principle of opaque superposition can yield a diverse range of appearances.

Inhibition

Figure 17-11 illustrates the general appearance of vegetation seen from the air over the Sonoran desert. Individual creosote bushes are scattered apparently at random, but they keep their distance from one another. This elbow room gives an advantage to the plant by relieving it of the stress of competition for soil moisture. Inhibition in plant communities

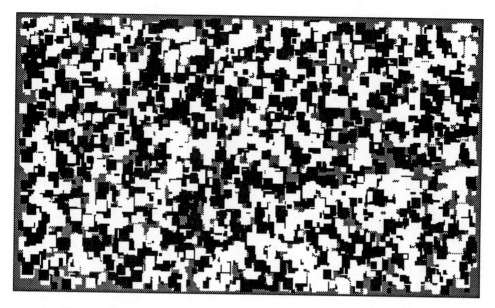

Figure 17-10   Superposed black and white squares of various sizes.

such as a savannah may be brought about by germination control. To germinate a seed requires moisture, light, and warmth, any one of which may be denied in the neighborhood of an existing tree. A living creosote bush efficiently removes surface moisture and possibly produces a chemical that interferes with germination. There is a subtle difference in character between the distributions of sites in Fig. 17-11 as compared with a uniform Poisson scatter. The present illustration was constructed by introducing a sharp minimum spacing between nearest neighbors. To impose this condition rigorously could take a lot of computing; in this illustration 324 displaced sites were first established with reference to a regular 18×18 grid, and only the eight neighbors associated with the eight grid points adjacent to each grid point in turn were tested for Euclidean distance. Two-thirds of the sites were eliminated, and the remainder were furnished with small polygonal bushes. A technique for studying inhibition in data is discussed below in connection with the clipped autocorrelation.

Clustering

A tendency to clump would set in as time elapsed if initially random point masses were subject to a mutual attraction of magnitude r^{-2} between pairs of points separated by a distance r, as with gravitation. The effect would be the opposite of the repulsive force tending to make points maintain their distance. One could parametrize both tendencies by introducing a force of mutual attraction varying as r^{-n}, where n could be positive or negative. Inhibition, which is local, is not the same as global repulsion; local rules of attraction can also be imagined.

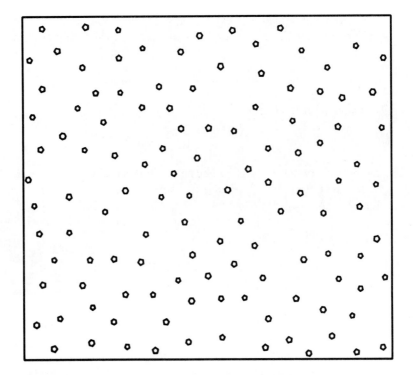

Figure 17-11 Desert scene with self-avoiding creosote bushes.

Statistical studies of spatial populations are important in agronomy, epidemiology, counting blood cells, granulometry, and other subjects where spatial distribution is quasi-random and where it would be useful to characterize the distributions with a view to discovering the influences at work.

Other mechanisms than displacement under a force can give rise to spatial modification. One is diffusion, which tends to reduce density fluctuations, and can be implemented directly, without invoking the diffusion equation, by repeatedly shifting the marked pixels ±1 at random in each direction. Another is contagion, a term applied to the spatial clustering of victims of epidemics and other circumstances where random points appear by creation rather than by migration.

The Nearest Neighbor

Each point in a random scatter possesses a nearest neighbor whose mean distance r_m is inversely proportional to the square root of the population density λ. For pure random scatter

$$r_m = \frac{1}{\sqrt{2\lambda}},$$

which is rather less than the spacing $1/\sqrt{\lambda}$ of points on a uniform grid with the same population density. The probability distribution for the distance r to the nearest neighbor is

$$p(r) = 2\pi\lambda r e^{-\pi\lambda r^2} = \frac{\pi r}{r_m^2} e^{-\pi r^2/2r_m^2}.$$

The cumulative distribution

$$\int_0^R p(r)\,dr = e^{-\pi\lambda R^2}$$

is the probability that there is no neighbor within the circle of radius R. This is recognizable as a special case of the Poisson distribution $(\lambda A)^k e^{-\lambda A}/k!$ for the probability of finding k points in an area A, when $k = 0$ and $A = \pi r^2$. The probability density $p(r)$ follows by differentiating the cumulative distribution. The probability that there are no more than $k - 1$ neighbors within radius R is

$$\sum_{\kappa=0}^{k-1} e^{-\pi\lambda R^2} \frac{(\pi\lambda R^2)^\kappa}{\kappa!},$$

(Ripley, 1981).

The theory of the nearest neighbor has a long history going back to 1909 (Holgate, 1972, Bartlett, 1975) and is significant in subjects as far apart as astronomy and ecology. An interesting practical application which stimulated development of the theory was to forestry and the possibility of estimating the population density of trees from measurements of the distances between nearest neighbors. Statistics of nearest neighbors may be presented as a histogram but also, bearing in mind that each nearest neighbor has a direction, as a scatter diagram which is capable of indicating isotropy as well as magnitude.

Population Density

Before a random scatter can be compared with, let us say, a Poisson distribution, guidance as to the choice of the parameter λ would be helpful. Consider Fig. 17-12. For the sake of concreteness let us suppose that the 100 discs are trees found by a surveyor and that we are interested in studying the population density. If we scan with a window of area A, then the number of trees N in that window will rise and fall in accordance with the position of the window, but a definite value of N/A is determinable as a function of continuous x and y. Naturally, that value is not unique but depends on A. In addition, for any given x and y, N/A depends not only on the magnitude of A but also on the shape of the window. For example, different results will come from square and circular windows of the same area. Furthermore, the value depends on the orientation of the window (unless circular); the identical tree map, scanned with a square window, can give different values of N/A, according as geographic or magnetic north is at the top. A beautiful idea due to P. G. L. Dirichlet [1805–1859] that eliminates arbitrary choices of procedure and enables the spatial variation to be focused on is to delineate domains belonging to each tree. A point belongs to a tree if it is closer to that tree than it is to any other tree. The boundaries

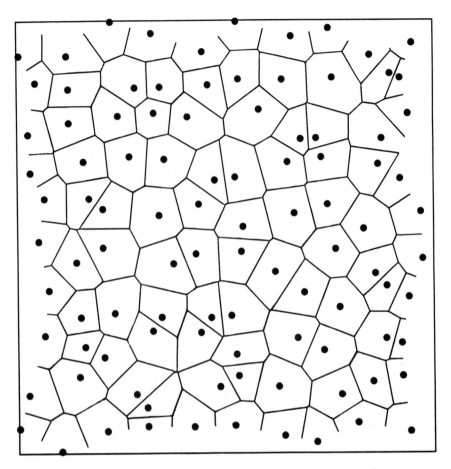

Figure 17-12 One hundred discs representing trees and the polygonal domains, squames, or Dirichlet cells belonging to each tree.

of these domains describe polygons as shown and define a certain area per tree. The reciprocal of the area per tree has the dimensions of population density; it is defined only at the discrete locations of the trees, and converts a list of coordinates to a unique data set relevant to study of density.

Naturally occurring patterns of interlocking polygons occur in wood parenchyma and other plant cells, in the squamous cells of human tissue, mud flats, and other situations where some resource-depleting influence spreads out from randomly situated nuclei.

Analogous polyhedra which exist in association with points scattered in space were introduced by Georgii Fedoseevich Voronoi (1868–1909), who published a substantial treatise (Voronoï, 1908) in the same journal that carried the work of Dirichlet (1850). For three-dimensional applications in various fields, see Stoyan et al. (1987) and for images treated in terms of Voronoi polyhedra, see Icke and van de Weygaert (1991).

Pencil lines connecting close neighbors are seen to be bisected at right angles by

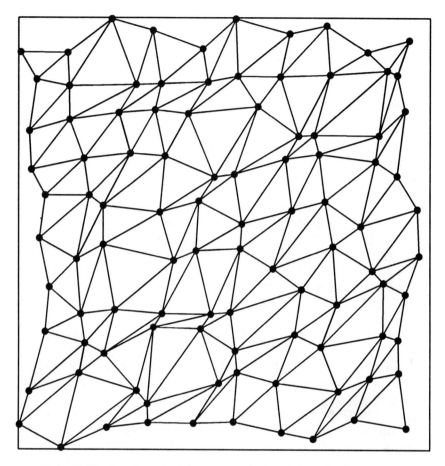

Figure 17-13   A graph constructed as a systematic triangulation taking the trees as nodes.

the domain boundaries; this suggests a construction for the domains starting from a triangulation network. The three perpendicular bisectors of the sides of a triangle intersect at the circumcenter, the center of the circle that passes through the three trees at the vertices of the triangle. The domain boundaries are collinear with such bisectors, and the domain vertices fall at the circumcenters, in many cases. The following segment takes a triangle $(x_i, y_i), i = 1, 2, 3$; it gives the circumcenter (X, Y) and determines whether the triangle is obtuse or not.

CIRCUMCENTER

```
SUB "circumcenter" (x1,y1,x2,y2,x3,y3,X,Y,obtuse)
dx12=x2-x1 @ dy12=y2-y1
dx13=x3-x1 @ dy13=y3-y1
m12=-dx12/dy12                          slope
```

```
m13=-dx13/dy13                          slope
M=(m12-m13)^2
u3=x1+x2 @ v3=y1+y2                      midpoint
u2=x1+x3 @ v3=y1+y2                      midpoint
X=(v3-v2+m13*u2-m12*u3)/M
Y=(m13*v3-m12*v2+m13*m12*(u2-u3))/M
A=dx12^2+dy12^2                          squared side
B=(x3-x2)^2+(y3-y2)^2
C=dx13^2+dy13^2
IF B+C-A<0 OR C+A-B<0 OR A+B-C<0 then obtuse=1 else obtuse=0
SUBEND
```

This subprogram does not guard against division by zero.

However, after a few lines of the triangulation network have been pencilled in, it will be seen that the pencil network, even if known in advance, does not furnish a direct construction for the polygons. The reason is that some of the triangles have one angle greater than a right angle and thus have circumcenters that are outside the triangle. A closed algorithm to draw the polygonal domains is bound to be sophisticated (Skellam, 1952); for a full treatment with exhaustive bibliography see Okabe et al. (1992).

An interactive approach is feasible. The triangulation (Delaunay, 1934), which you have pencilled in, is unique and is itself not easy to construct from the site data. Therefore specify an arbitrary triangulation as in Fig. 17-3, which is systematic and has six branches on all internal nodes. For each triangle in turn draw the perpendicular bisectors by starting from the center of the side and converging at the circumcenter, if it is an acute-angled triangle (no obtuse angle). If the triangle is obtuse, draw from the midpoint of each side only three-quarters of the way to the circumcenter.

The resulting diagram contains most of the domain boundaries. The missing sides, which are defined by intersections, are to be filled in interactively, and superfluous stubs are to be erased. I made Fig. 17-12 this way without much effort.

Visual inspection of the domains is instructive in itself. The domains do not vary much in size or shape and do not reveal noticeable anisotropy or spatial trends. Experts in various fields could use sequences of such diagrams as a diagnostic tool. The polygons themselves have mostly six, five, or seven sides, occasionally four or eight; the breakdown may be telling us something.

From the map of polygonal domains a unique triangulation can be constructed using the perpendiculars from the trees to the sides of the polygons to obtain the graph shown in Fig. 17-14. In order to facilitate comparison with the arbitrary triangulation of Fig. 17-13 alternate bands have been shaded, each containing chains of quadrilaterals as laid out in geodetic surveys. In the systematic triangulation of Fig. 17-13 the diagonals were all chosen so as to make six branches per internal node, whereas in the unique Delaunay triangulation the number of branches equals the number of sides of the polygon surrounding a node. Each node is thus associated with a definite number of other nodes with which it is in contact via a branch. These are the contiguous nodes or, loosely, the neighbors. The nearest neighbor, its distance and direction have been already discussed, but it has been mentioned that it may be computationally expensive to find the nearest neighbors. Now we see that it

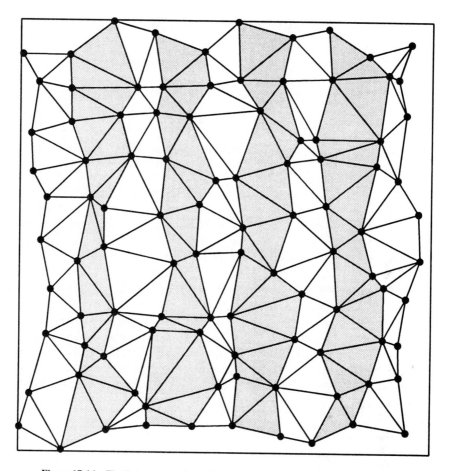

Figure 17-14 The Delaunay graph, a unique triangulation equivalent to the subdivision into domains.

will suffice to compare the distances to the contiguous nodes only. To find the contiguous nodes unequivocally also may involve nontrivial programming, even though by eye it would be easy to identify, let us say, eight candidates, compute the eight Euclidean distances, and sort them. Instead of addressing the problem of finding the neighbors starting from a list of N coordinate pairs (x_i, y_i), it would make sense to provide for an independent way of identifying serial numbers i belonging to nodes in the general neighborhood of a particular node i_k. Systematic renumbering of the coordinates, systematic acquisition of the data in the first place, and interactive screen graphics are ways of avoiding the cost of heavy programming that aims at making no mistakes at all. Data sets acquired observationally are subject to error; to return to the example of a tree survey, there is uncertainty as to what is a tree, and some trees may be excluded because obscured.

Boris Nikolaevich Delone (1890–1980) published in French and it is reasonable

Figure 17-15   A stringy noise image resembling loose weaving.

to suppose that his grandfather or earlier forbear was French, judging by his choice of transliteration as Delaunay.

Random Weave and Stringiness

An interesting random network can be generated in the manner of weaving by allowing a thread to be subject to a diversion of one pixel to its left or right as it proceeds across the frame. Nets loosely woven something like this are incorporated into materials such as paper or resin to add strength. In Fig. 17-15 40 threads start out evenly spaced from each side. The uneven spacing on arrival at the opposite side produces strong spatial nonuniformity, which is more or less compensated by passing the shuttle to and fro (as it were).

Figure 17-16 is introduced as an example of stringy noise as distinct from blobby noise. It has some features in common with the sea surface disturbed by waves from a distant storm that subtends a noticeable spread in azimuth. Autocorrelation of this object would inform us about the azimuth and its spread, the wavelength, and the spread in

Figure 17-16 Simple stringy noise resembling the sea surface in the presence of intersecting waves from a narrow cone of directions.

wavelengths. The illustration was made by copying 20 threads from the random weave of the previous figure with a broad pen.

Figure 17-17 is an example of blobby noise reminiscent of commercially available peel-off patterns as seen in advertisement art. The individual blobs are filled reentrant decagons whose vertices are randomly chosen in the vicinity of a 16×42 grid. Since the 43 rows are closer together than the 17 columns, the polygons tend to be elongated horizontally. Holes in the blobs come from the filling algorithm for self-intersecting polygons, while sharp corners have been rounded off by use of a broad pen.

Both elongated and more or less isotropic elements are commonly seen in random noise, and it is helpful to have some characterization of anisotropy before attempting to extract statistical parameters of a given noise sample. However, the range of possibilities is vast. Figure 17-18 shows a single filled polygon whose 40 vertices are chosen at random. From such a simple prescription, much visual complexity has emerged, and the result defies easy characterization; it would be hard to describe over the telephone.

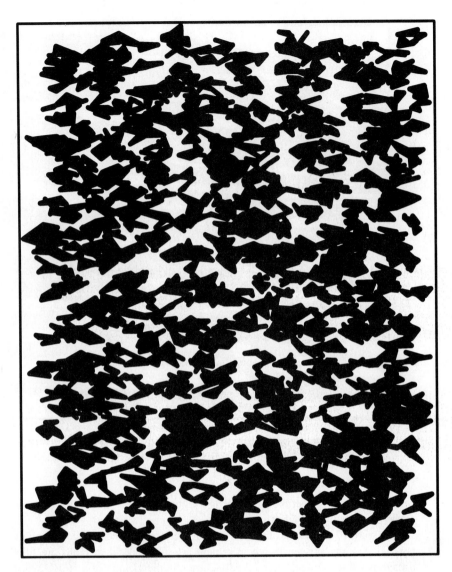

Figure 17-17 Blobby noise constructed from random decagons.

GAUSSIAN NOISE

The sort of noise signal $B(x, y)$ now to be discussed has values that are not limited to 0 or 1. We begin by fixing on a single location (x, y) and taking x and y to be continuous. At that location let the brightness of a given image be $B(x, y)$. Associated with the point is a frequency distribution $p(B)$. We normalize so that $\int_{-\infty}^{\infty} p(B) \, dB = 1$ as if $p(B)$ were a probability distribution. Since brightness does not go negative, nor does it go infinite, some

Figure 17-18   Self-portrait of Dr. Rohrschach, as he imagined himself.

comment on the limits of integration is required. Outside the range of brightnesses that do occur in an image take $p(B)$ to be zero. Then no harm will be done if the normalization is expressed conventionally with infinite limits. As we shall see immediately, negative brightnesses are in fact required all the time, even though they do not occur in nature.

The frequency distribution $p(B)$ possesses a mean $\overline{B}$ and standard deviation σ_B given by

$$\overline{B} = \int_{-\infty}^{\infty} B p(B)\, dB$$

$$\sigma_B^2 = \int_{-\infty}^{\infty} B^2 p(B)\, dB - \overline{B}^2.$$

There is no need for a denominator in these expressions because of the normalization to unity. In general, what is meant by a frequency distribution? If a set of N images were provided, then at the same place (x, y) in each image the brightness $B(x, y)$ could be ascertained and a histogram could be constructed showing the number of occurrences of brightness falling in a specified range ΔB centered on each B. The *frequency* of occurrence would be this number divided by N.

If B is quantized, the frequency distribution is a function of discrete B; the definitions given for $\overline{B}$ and σ_B would be replaced by

$$\overline{B} = \sum_B Bp(B)$$

$$\sigma_B^2 = \sum_B B^2 p(B) - \overline{B}^2.$$

The normalization $\sum_B p(B) = 1$ is ensured implicitly by the definition. For mathematical convenience we often speak as though B were a continuous variable and as though $p(B)$ were not really a discontinuous histogram. However, when it comes to numerical reality, counting supersedes continous analysis.

The set of N images might be 1024 passport photographs, or the images might arise sequentially in time as with successive frames of television. Very often, however, we have to discuss a single image only, or a small group of images not necessarily of the same object. The brightness histogram of a single image exists and is readily ascertainable, but it is not the same as the histogram or frequency distribution $p(B)$ we have been talking about.

It is not necessary that $p(B)$, regarded in terms of continuous B, should be a normal distribution; in fact, that would be unusual. However, in the theoretical case of $B(x, y) = n(x, y) + \overline{B}$, where $n(x, y)$ is a zero-mean Gaussian noise process, the normal distribution for n is

$$\frac{1}{\sqrt{2\pi}\,\sigma_n} e^{-n^2/2\sigma_n^2}.$$

For this normal distribution, which is automatically normalized to unit area, the mean is $\overline{B}$. The only statistical parameter quantifying $n(x, y)$ is its standard deviation σ_n. To estimate the strength of $n(x, y)$ in terms of its r.m.s. value we keep (x, y) fixed and run through the set of N images. If a sufficiently large set of images is not available to verify that the distribution is normal (with zero mean), that fact should obviously be kept in mind.

Clearly $n(x, y)$ can go negative, since it has zero mean. This is understandable in terms of a total image composed of a wanted image plus noise or when the wanted image is superimposed on a uniform background. However, the image of a bounded object, such as an x-ray scan of the brain and skull, cannot be described in terms of additive Gaussian noise, which would require negatives in the surround. This would also be true of a photographic image where the noise took the form of graininess in the film.

A short digression now explains how to set up numerical Gaussian noise for purposes of simulation.

Computed Gaussian Noise

Suppose that your computer provides random numbers ranging from 0 to 1 under the name RND and you would like to have a long sequence of integers in the range 1 to 64 with mean value 32 and standard deviation S. If you call up 12 independent values of RND and add them together, the expected mean will be 6, and the expected standard deviation will be unity. This is because RND comes from a population of mean 0.5 and standard deviation $1/\sqrt{12}$. Addition of independent samples from probability distributions $p_1(x)$

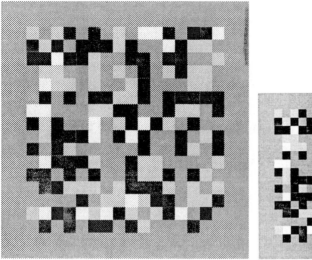

Figure 17-19   A Gaussian noise image on a midgrey background, with a reduced photocopy.

and $p_2(x)$ results in a probability distribution $p_3 = p_1 * p_2$—that is, the convolution of the two parent distributions; and under convolution means and variances add (FTA 1986). If p_1 and p_2 are identical distributions, the mean of p_3 will be twice the mean of p_1, and the variance will be twice the variance of p_1. Consequently, if twelve draws are made from the same distribution $p_1(x) = \text{rect}(x - 0.5)$, for which the mean is 0.5 and the variance $\int_{-\infty}^{\infty} x^2 \, \text{rect} \, x \, dx$ is 1/12, and added together, the resulting mean and variance will be 6 and 1, respectively.

Having added the twelve random numbers, subtract 6, multiply by S, add 32, and round off to the nearest integer. The short program segment that does this assigns the desired integer with mean 32 and standard deviation S, to the integer variable X.

NORMAL VARIATES

```
X=0
FOR I=1 TO 12 @ X=X+RND @ NEXT I
X=(X-6)*S+32
X=INT(0.5+X)
```

The symbol **@** is a statement separator, often represented by a semicolon and sometimes by a colon. This impressively compact maneuver for normally distributed variates is handy for avoiding excursions into packaged subroutines.

An example constructed to illustrate Gaussian image noise is shown in Fig. 17-19. It uses 16×16 pixels and nine grey levels (0 to 8), which is about as many as can be counted on for reproduction in journals. To allow for negatives, a background level of 4, visible in the surrounding frame, has been added. The standard deviation is unity; as a result no pixels have the extreme values 0 or 8. The illustration has been made somewhat tolerant to repeated photocopying by use of a relatively coarse dot pattern; the reduced version shown alongside, which was made on a photocopying machine, illustrates the degree of fidelity that is often accepted as good enough. Had the illustration been prepared to a finer mesh, or with true grey tones, it would not have been possible to make even a rough representation on an office photocopier; instead, resort to photography would have been required.

THE SPATIAL SPECTRUM OF A RANDOM SCATTER

An object $f(x, y)$ consisting of N unit delta functions $^2\delta(x - x_i, y - y_i)$ situated at (x_i, y_i) has the simple Fourier transform

$$F(u, v) = \sum_{i=1}^{N} e^{-i2\pi(x_i u + y_i v)}.$$

To gain some feel for the nature of this expression take a specific example, such as the first frame of Fig. 17-2. Taking the frame to have a side of 100 units, we can measure the positions of the eight points as (8, 39), (28, 37), (31, 53), (49, 12), (51, 18), (70, 23) (95, 17), (98, 54). The Fourier transform is thus

$$F(u, v) = \cos[2\pi(31u + 53v)] + \cos[2\pi(49u + 12v)] + \cdots 8 \text{ terms} \cdots + \cos[2\pi(98u + 54v)]$$
$$+ i \sin[2\pi(31u + 53v)] + i \sin[2\pi(49u + 12v)] + \cdots 8 \text{ terms} \cdots + i \sin[2\pi(98u + 54v)]$$
$$= R + iI.$$

The imaginary part must have exactly the same character as the real part, so consider first the sum of the eight cosine terms. At a given point in the (u, v)-plane we are adding eight samples from corrugations that have various orientations and wavelengths but the same unit amplitude. The sample values are more likely to be close to ±1 than to zero, because the cosine function dwells near its peaks and troughs longer than near its zero crossings. The situation is like drawing samples from a frequency distribution $p_t(t) = \pi^{-1}(1 - t^2)^{-1/2} \text{rect}(t/2)$. The factor π^{-1} ensures that the integral of $p_t(t)$ over its range from -1 to 1 is unity. The variance of this distribution is 0.5. If N samples are taken at random from such a distribution, and added, the mean and variance of the resulting distribution will be close to N times the original mean and variance. In this example the expected mean of the sum of eight cosine samples would be zero, while the expected variance would be $8 \times 0.5 = 4$. In addition, if N is large, the distribution of the samples will be Gaussian in form, approaching $\exp(-R^2/2\sigma^2)/\sqrt{2\pi}\sigma$ with $\sigma^2 = 4$. Eight is not a large number, but the sum of eight cosines is so easy to compute that we can soon verify that the frequency distribution of values of R as specified above can be distinguished only with difficulty from $\exp(-R^2/8)/2\sqrt{2\pi}$. The histogram of computed values of R does not fit the normal probability curve exactly, but neither does the result for the ten points of the second frame of Fig. 17-2. The degree of concordance between different trials, or frames, sets the criterion by which closeness of fit to the theoretical curve is to be judged.

An exact result $p_R(R)$ for the sum of eight cosines can be calculated from the distribution $p_t(t)$. Thus

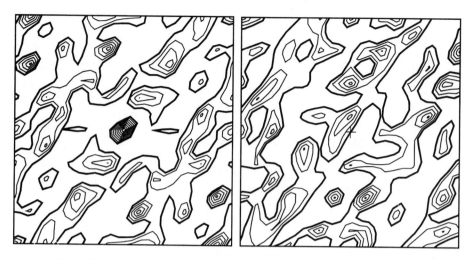

Figure 17-20 The real (left) and imaginary parts of the Fourier transform of Fig. 17-2, shown on the (u, v)-plane. The contour levels are **0**, 1, 2, 3, . . . ; no negative contour levels are included. The left-hand diagram, which rises to a value of 8 at the origin, if given half a turn, is the same as itself. The right-hand diagram is zero-valued at the origin and becomes the negative of itself if rotated; the negative contours are thus deducible from those shown.

$$p_R(R) = p_t(t) * p_t(t) * p_t(t) * \cdots 8 \text{ factors} \cdots * p_t(t) = [p_t(t)]^{*8},$$

an expression that can be evaluated numerically. However, from the Fourier transform pair

$$p_t(t) \supset J_0(2\pi s),$$

it follows by the convolution theorem that

$$[p_t(t)]^{*8} \supset [J_0(2\pi s)]^8.$$

So, as an alternative to comparing $p_R(R)$ with the normal distribution $\exp(-R^2/8)/2\sqrt{2\pi}$ we can move to the transform domain and compare $[J_0(2\pi s)]^8$ with the transform of the normal distribution. From the transform pair

$$\frac{1}{\sqrt{2\pi}\sigma} e^{-t/2\sigma^2} \supset e^{-2\pi^2\sigma^2 s^2}$$

we note that $[J_0(2\pi s)]^{*8}$ is to be compared with $\exp(-8\pi^2 s^2)$, which is itself of Gaussian form. Here is the result.

s	0	.05	.1	.15	.2	.25	.3
$\exp(-8\pi^2 s^2)$	1	.821	.454	.169	.042	.007	.001
$[J_0(2\pi s)]^8$	1	.820	.445	.152	.029	.002	.000

The conclusion is that the randomly oriented corrugations associated with a random scatter of N unit impulses add up in the transform domain to a function with mean close

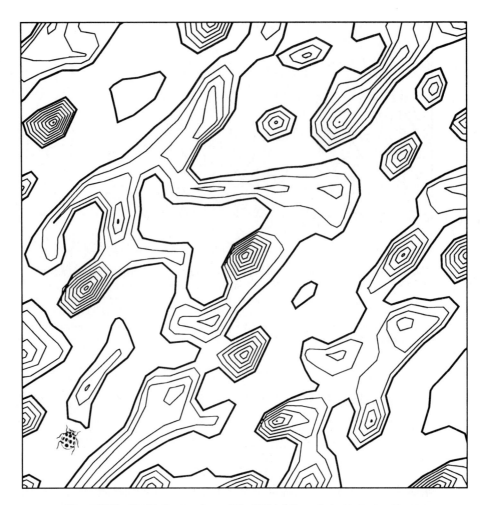

Figure 17-21   The Hartley transform of Fig. 17-2 is fully equivalent to the two diagrams
of Fig. 17-20. The null contour is shown as a heavy line, and the negative contours are not
included. The value at the origin is 8.

to zero and having a normal distribution with variance close to $0.5N$. The fact that even a
few sine waves add up to have a nearly Gaussian amplitude distribution has been utilized
for commercial noise-signal generators and is useful for computational simulation of short
lengths of random noise.

   Figure 17-20 illustrates the Fourier transform of the set of eight impulses in Fig. 17-
2. Two diagrams are needed, one for the real part and one for the imaginary part. Because
of the cumbersome nature of the transform this sort of illustration is not usually presented,
but it is important to see one once in order to be aware of the nature of the entity being

Figure 17-22   The amplitude of the Fourier transform of the eight impulses of Fig. 17-2. Contour levels range from 7 down to 1, which is shown as a heavy line surrounding deep minima.

dealt with when the Fourier transform of an image is manipulated in a computer program. The standard deviation of both R and I is $1/4\pi$.

Figure 17-21, which illustrates the Hartley transform $H(u, v)$ of the same set of impulses, is purely real, and free from the rotational symmetry and associated redundancy of the Fourier transform. It is a more convenient entity to manipulate, since only a single array of real numbers are involved. There is, as well, a certain practical convenience in the fact that a call to the Hartley transform subprogram takes you to the *other* domain, whereas subprogram calls to the Fourier transform need to distinguish the inverse transform from the direct transform.

Figure 17-22 shows how the amplitude $|F(u, v)|$ is distributed over the (u, v)-plane, descending from 8 at the origin to minimum values close to but almost never equal to zero. The lowest contours (value = 1), shown in heavy outline, surround these minima. While the null contours of $R(u, v)$, $I(u, v)$ and $H(u, v)$ wander all over the plane, the deep minima of $|F(u, v)|$ are confined. The amplitude may be computed as $\sqrt{[R(u, v)]^2 + [I(u, v)]^2}$ or directly from the Hartley transform as $\sqrt{\frac{1}{2}\{[H(u, v)]^2 + [H(N - u, N - v)]^2\}}$.

One sees that the amplitude $|F|$ of the Fourier transform cannot have a Gaussian distribution because it does not go negative. Instead, $|F|$ has a Rayleigh distribution

$$p_{|F|}(|F|) = \frac{2|F|}{\sigma_N^2} e^{-|F|^2/\sigma_N^2},$$

for which the expected mean value F_m is $\sqrt{\pi/2}\sigma_N$, or 2.51. For a graph of this distribution normalize the amplitude $|F|$ with respect to its mean and introduce the variable $F_r = |F|/F_m$. Then Fig. 17-23 shows the frequency of occurrence of amplitudes in the range $0 < F_r < 3$. The distribution is distinctly skew, favoring amplitudes less than the mean. Some critical parameters are indicated on the figure; other details can be extracted from the normalized distribution

$$p_{F_r}(F_r) = \tfrac{1}{2}\pi F_r e^{-\pi F_r^2/4}.$$

The effects of this skewness can be examined on Fig. 17-22 by shading areas where the value exceeds the mean.

As values of the two-dimensional power spectrum are simply the squares of the amplitude values, the contour labels are changed. Consequently, Fig. 17-22 shows how the power spectrum would appear if the contour labels were 1, 4, 9, . . . , 49. If, as is more usual, contours are chosen at equal increments of power, then emphasis is given to the peaks by the crowding of contours around them. The general effect can be seen by pencilling in the amplitude contours for $\sqrt{8}$, $\sqrt{16}$, $\sqrt{24}$, . . . , $\sqrt{56}$. The difference in visual impact is striking and offers an option for achieving the information emphasis desired.

There is also a change in the frequency distribution $p_P(P)$, where $P = |F|^2$, which can be understood as follows. In the range $|F|$ to $|F| + \Delta|F|$ the number of amplitude values $p_{|F|}\Delta|F|$ equals the number of power values $p_P\Delta P$, where ΔP corresponds to $\Delta|F|$. But $\Delta P = 2|F|\Delta|F|$. Therefore,

$$p_{|F|}\Delta|F| = p_P^2 |F|\Delta|F|$$

and

$$p_P = \frac{p_{|F|}}{2|F|} = \frac{1}{\sigma_N^2} e^{-P/\sigma_N^2}.$$

Values of the power spectrum are thus distributed exponentially about a mean σ_N^2 and with rms deviation from the mean equal to that mean. We have already seen that the amplitude is not strongly confined about its mean; the tendency of the power spectrum to dwell near zero is much more marked.

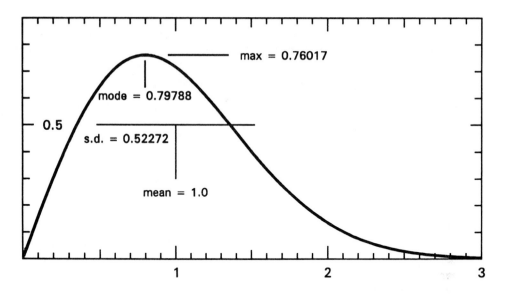

Figure 17-23 The frequency of occurrence of amplitude values of the Fourier transform of a random scatter, relative to the mean, with critical numerical values for reference. The area under the curve is unity. This is the Rayleigh distribution, which governs the envelope of a quasi-monochromatic waveform such as laser light or fading radio waves.

AUTOCORRELATION OF A RANDOM SCATTER

Proceeding on the principle that a thing should be exhibited before its qualities are deduced, let us find the autocorrelation of the eight-point scatter $f(x, y)$ defined previously by

$$f(x, y) = \sum_1^8 {}^2\delta(x - x_i, y - y_i)$$

with the same specific values for the locations (x_i, y_i) as before. The autocorrelation $C(x, y)$ will consist of an impulse at the origin of strength 8 surrounded by 56 unit impulses in the locations $(x_j - x_i, y_j - y_i)$, where i and j run from 1 to 8 and $i \neq j$. It is possible for two of these impulses to coincide and produce an impulse of strength 2, but less likely when the number of points is small. The autocorrelation will have twofold rotational symmetry and occupy a frame of twice the dimensions of the frame specified for $f(x, y)$. The quantity $(x_j - x_i, y_j - y_i)$ can be thought of as the vector displacement of the jth with respect to the ith point; reference to Fig. 17-24 will confirm that each vector starting at the origin and going to a point of $C(x, y)$ can be identified on Fig. 17-2 (a) as the vector displacement of some point-pair. The fact that these displacements can have either of two signs accounts for the symmetry of $C(x, y)$. Formally,

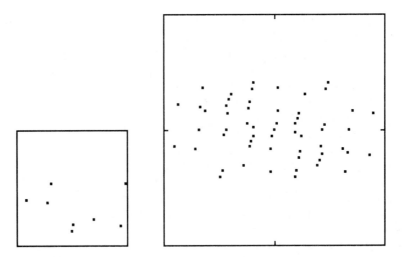

Figure 17-24   Autocorrelation (right) of the eight-point scatter of Fig. 17-2(a). The origin of the autocorrelation is at the center of the frame.

$$C(x, y) = \int_{-\infty}^{\infty} \int_{-\infty}^{\infty} \left[\sum_{1}^{8} {}^2\delta(x' - x_i) \right] \left[\sum_{1}^{8} {}^2\delta(x' + x - x_i, y' + y - y_i) \right] dx' dy'.$$

The Fourier transform of $C(x, y)$ is the power spectrum $P(u, v)$:

$$C(x, y) \, {}^2\supset P(u, v).$$

It is clear by inspection that the points of the autocorrelation are far from appearing to have been taken from a spatially uniform population. The anisotropic outline is peculiar to the particular sample set, but the central concentration is not. If one cares to make a tracing of the given eight points, with their frame, and proceed to establish the autocorrelation graphically, it will be immediately obvious that, in the neighborhood of any given displacement, one is looking for coincidences within an area of frame overlap that diminishes with displacement. This area is described by the lazy pyramid function, which controls the area density of the points of $C(x, y)$. If there are lots of points, area density can be measured in the same way as population density is measured, by specifying an area unit analogous to the square kilometer. If there are not many points but if additional data, as for example the remaining three parts of Fig. 17-2, can be supplied, then for each area unit the count can be averaged over additional cases. Any precision specified in advance can be obtained by invoking the necessary quantity of additional data. Of course, the supplier of the data should be asked to guarantee that the source is uniform and that the the data are statistically representative—something that is often known only to professors of statistics. A third approach to area density avoids the notion of data sequences, asserts statistical properties of the data source, and in this example would directly yield the lazy pyramid exactly. This approach is familiar from statistical signal processing.

The vector plot of nearest neighbors, mentioned earlier, will be found as a subset of eight points in the right frame of Fig. 17-24. Identify these points by reference to the left frame and pencil in the smallest polygon that contains them. There is only one foreign vector included; it represents a second-nearest neighbor. To make the scatter diagram of nearest-neighbor vectors can mean heavy computing; however, in the presence of inhibition, the second-nearest neighbors may be separated enough to leave the nearest neighbors segregated. In these circumstances, the clipped autocorrelation introduced below is an economical way of obtaining a full vector plot for studying nearest-neighbor statistics. The theory of this and related topics is presented by Ripley (1981).

It will have been noticed that the term autocorrelation is used in the nonprobabilistic sense of the experimentalist who plans the taking and analysis of data, and stems from the statistical notion of correlation coefficient for comparing two sets of ordered data. The autocorrelation is then the sequence of correlation coefficients when one set of ordered data is compared with its shifted self. It is the generalization of this statistical tradition to two dimensions that is employed here. A different school of thought postulates a random process $\{X(x, y)\}$ defined mainly by its expectation $E\{X(x, y)\}$, or mean, and its auto-covariance $w(x, y) = E\{X(x' + x, y' + y)X(x', y')\}$. At zero shift, the value of $w(0, 0)$ is $\sigma^2 = E\{[X(x, y)]^2\}$, the variance of the random process. The autocorrelation function, normalized to have unity value for zero shift, then *defines* the power spectral density function $P_{\text{process}}(u, v)$, which agrees with the use of the word spectrum in physics—for example, the Planck black-body spectrum.

Specification by postulate is the preferred approach to random processes, but the topic is also often introduced in terms of ensembles of "sample functions" of the process. The word *ensemble,* is a useful cue to insert when a sentence is framed in random process mode. While the power spectrum $P(u, v)$ is the squared magnitude of the Fourier transform of an actual function of x and y, it has deep fluctuations by nature, while the power spectral density $P_{\text{process}}(u, v)$ is smooth. If the spectrum of sunlight were to be determined experimentally, one might obtain a million or so samples of the optical electric field as a function of time for one picosecond, then Fourier transform. The result would not look smooth, as Planck's spectrum does; numerical smoothing would be applied at the expense of loss of frequency resolution. Nevertheless, one could determine the presence of departures from Planck's formula, such as absorption features due to molecules in the earth's atmosphere and their variation from day to day. Discoveries of this kind cannot be made from the random-process approach, which postulates in advance that which observational science tries to find (Gardner, 1988).

PSEUDORANDOM SCATTER

Spatially random arrangements are needed for various purposes, including some for which purely random noise is not suitable because clustering and zones of avoidance are features of pure randomness (Fig. 17-25) that may be undesired. Figure 17-26 is an example of an apparently random arrangement that looks quite pleasing and might be suitable for dress material or as a test object for a medical imaging system. Such random patterns also arise naturally in biological systems, in mechanical systems subject to thermal or other

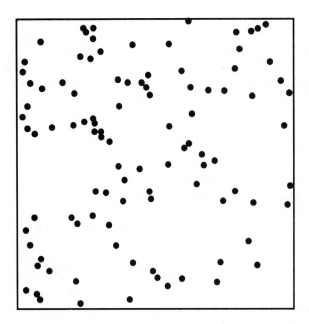

Figure 17-25 One hundred spots whose locations were determined by a random-number generator. Zones of avoidance and clustering are very noticeable concomitants of pure randomness.

agitation, and in convective turbulence. How was the figure constructed? One way would be to place spots at points having coordinates generated at random, rejecting any point closer than one spot diameter to a previous point. That would stop clustering but would not take care of zones of avoidance. The simpler way used was to start with a square array of points. Then each point was shifted in both coordinates by amounts randomly distributed between 0 and $w - d$, where w is the array spacing and d is the spot diameter. The same 100 random numbers were used as for Fig. 17-2(a). A maximum displacement $w - d$ from the regular lattice precludes overlap (but does not exclude contact). Images of this kind have been made for optical purposes in the past by arranging confetti agreeably on the floor and photographing from the top of a ladder, but that is no longer necessary.

If a transparency was made from Fig. 17-26 and subjected to optical diffraction, or if a giant array of dish antennas was constructed as shown, the Fourier transform would differ in some way from that of a more purely random arrangement. As a matter of fact, it is not obvious how the close spacings allowed by pure randomness could be implemented optically or with antennas; a design calling for a random array would probably lead to just the pseudorandom placement that we are talking about.

The Clipped Autocorrelation

Before considering the Fourier transform of Fig. 17-26, look first at the autocorrelation. The autocorrelation of the 100 delta functions located at the spot centers will occupy a space of twice the dimensions and contain 10,000 points. These points are shown in Fig. 17-27, whose frame has been scaled down by a factor of 2. Points which fall on top of other points are not identified in this presentation. The area density is a reasonably

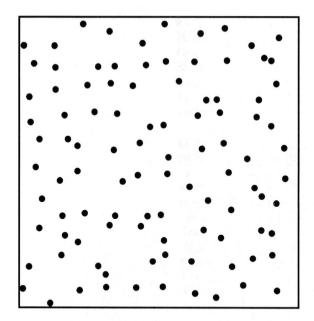

Figure 17-26 A pseudorandom array of 100 spots that does not exhibit the clustering and zones of avoidance that arise with a purely random scatter, but does exhibit inhibition and traces of underlying rows and columns.

smooth approximation to the expected lazy pyramid function; however, some features are noticeable. First of all, the method of construction of Fig. 17-26 from a square grid has left traces of long-range correlation which are especially visible under oblique viewing. Second, the existence of inhibition shows itself by a vacancy of diameter $2d$ surrounding the origin, caused by the absence of neighboring impulses at separations less than d. This phenomenon was first reported by B.M. Oliver. At the very center there is a single dot, but it represents an impulse of strength 100.

Figure 17-27 presents not the autocorrelation $C(x, y)$ itself but its clipped form $C_c(x, y) = \mathbf{u}[C(x, y)]$, in which values exceeding unity are replaced by unity. This is a useful diagnostic diagram, especially for sparsely sampled functions $f(x, y)$, and is easy to program. The samples $f(x, y)$ need not be binary valued. A binary function $C_c(x, y)$ is also formed by the following segment for future use.

CLIPPED AUTOCORRELATION FUNCTION

```
k=0                           serialize the array
FOR x=1 TO LX
    FOR y=1 TO LY
        IF f(x,y)>0 THEN X(k)=x @ Y(k)=y @ k=k+1
    NEXT y
NEXT x
N=k
FOR i=1 TO N                  replicate the N points N times
    FOR j=1 TO N
        x=X(i)-X(j) @ y=Y(i)-Y(j)
        CALL "PSET" (x,y)         turn on a pixel
```

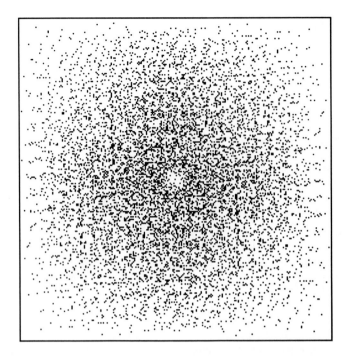

Figure 17-27 The clipped autocorrelation of Fig. 17-26, a binary function that computes easily and clearly shows the dearth of near neighbors due to the mode of construction of Fig. 17-26.

```
    Cc(x+x0,y+y0)=1
  NEXT j
NEXT i
```

The subprogram PSET typesets a pixel at (x, y). The size of the array of samples is LX×LY and of the clipped autocorrelation (2*LX-1)×(2*LY-1). The constants x_0 and y_0 are to prevent negative array indices; use $x_0 = X_{max} - X_{min} + 1$ and $y_0 = Y_{max} - Y_{min} + 1$.

For many practical purposes the displacement method of generating agreeable random spot patterns with collision avoidance is satisfactory. However, if the spot size is increased, the underlying grid becomes apparent, as can be verified by copying Fig. 17-26 and enlarging the spots with a pencil.

A diffraction pattern will also reveal the underlying regularity; in the example of a random antenna array four sidelobes would appear close to the main beam. A way of reducing such artifacts, while still following the idea of limited random displacements from a systematic array, is to base the pattern on a set of points spaced at unit arc length along the spiral $r = \theta/2\pi$. This Archimedean spiral expands unit distance per turn. Consider a point at (r, θ). An arc element $ds = r\, d\theta$, if equal to unity, subtends an angle $d\theta = 1/r$ at the origin. If now the point is displaced to $(r + \Delta r, \theta + \Delta\theta)$, where $0 < \Delta r < 1$ and

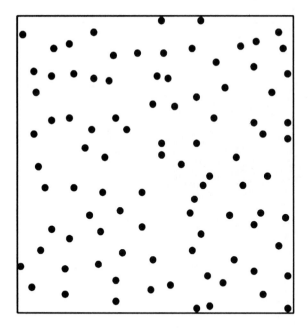

Figure 17-28 A pseudorandom array base on an underlying spiral which is difficult to discern.

$0 < \Delta\theta < 1/r$, the point will remain within a quasi-square cell not obtruding on the similar cell associated with any of the other basis points. If spots of diameter d are to replace points, the displacements can be assigned as follows:

$$\Delta r = (1 - d)R_1, \qquad \Delta\theta = (1/r - d)R_2,$$

where R_1 and R_2 are independent random numbers between 0 and 1. The result of doing this is shown in Fig. 17-28. It is difficult to discern the systematic basis, even knowing about the spiral. If ensembles of such patterns are needed, it is good to put the origin of the spiral at the corner, instead of at the center, because the square-cell idea breaks down at the origin.

Some structure can be detected in the autocorrelation (Fig. 17-29) of the spiral-based pattern of impulses, which may be compared with the structure in Fig. 17-27, and could be further reduced if necessary. Otherwise the principal features are the vacancy around the origin and the very strong central impulse.

Returning to the matter of the Fourier transform of this type of pseudorandom scatter, and bearing in mind that the power spectrum is the Fourier transform of the autocorrelation we see that the departure of the power spectrum of a pure random scatter arises solely from the circular vacancy around the origin. An ensemble average of autocorrelations such as Fig. 17-27 would lead to a lazy pyramid. Picture the autocorrelation being found by superposing a displaced transparent copy of Fig. 17-26 on itself and counting coincidences. While the counts will vary with small changes in displacement, the total will be roughly proportional to the area of overlap of the frames if an ensemble of cases is worked through. The lazy pyramid description can be tested by comparing the number of

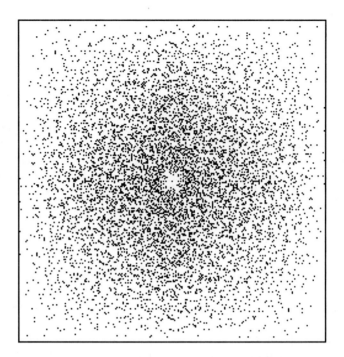

Figure 17-29   The autocorrelation of a pseudorandom array of impulses distributed under guidance from a spiral as in Fig. 17-28.

dots that can be seen through a 3-mm square hole positioned halfway from center to edge versus three-quarters of the way from center to corner. In each case there are about 20, in accordance with the linear edge-to-center variation and the quadratic corner-to-center variation. It follows that the ensemble power spectrum of the pure random scatter is a two-dimensional sinc-squared function added to the constant value, independent of spatial frequency, that is associated with the strong central impulse.

As the frame size expands to include more impulses, the sinc-squared component narrows, so that, at spatial frequencies greater than a few cycles per frame width, the constant component dominates. The effect of removing a small circular patch from the autocorrelation is to subtract a jinc function that is both weak and extended. Of course, the flat spectrum associated with an ensemble becomes, for any particular instance, deeply modulated in accordance with the Rayleigh distribution. Thus the power spectrum of a given pseudorandom scatter is hardly to be differentiated from the power spectrum of a given pure random scatter. The further modification to the Fourier transform when a scatter of spots replaces a scatter of impulses is the modification that takes place to a transform when the original function is convolved with a narrow rect function of radius, namely multiplication by a broad jinc function. The power spectrum is multiplied by the corresponding jinc-squared function.

RANDOM ORIENTATION

Figure 17-30, which resembles a regiment of bacilli, features independently randomly oriented linear elements centered on a regular grid. This is a pattern whose Fourier transform can be approached in terms of the autocorrelation function. An interesting application is in optics, where an aperture is desired whose transparency diminishes toward the rim so that sidelobes, or fringes, in the diffraction pattern can be reduced. Cutting slits in an opaque plate, or making a binary transparency, is more convenient than grading the transparency of an aperture, especially at small sizes. If a plate were machined as in the figure, then the autocorrelation would approach zero at a displacement of one-half the grid spacing, but do so in the gentler way associated with spin integrals because of the superposition of many orientations. The strength of unwanted diffraction orders, especially the third, can be controlled and sidelobes on the diffracted beams can be reduced by reducing the slit length toward the edges of the plate. Since devices built to probabilistic models disagree among themselves, and never agree with the ensemble average, the design of such devices requires trial and selection. In the end, the product is in a sense not random at all.

NONUNIFORM RANDOM SCATTER

Examination of the simple case of a random scatter sampled from a spatially uniform population has allowed some fundamental properties to be discussed. Further ideas can be introduced by considering a spatially nonuniform scatter of impulses such as has just arisen in the discussion of autocorrelation or as illustrated in Fig. 17-31 where 34,000 impulses, represented by dots, have been distributed in accordance with a normal distribution. A certain number of dots have fallen on other dots, and therefore the image has already undergone nonlinear limiting. Hundreds of unoccupied sites still remain, represented by white specks in the interior, so the image cannot be said to be heavily overexposed. Two-dimensional Gaussian distributions are often referred to but not often illustrated; consequently, it is opportune to note some features. The span of corners is 8 standard deviations, and clearly very few dots indeed would fall outside that range; the 5σ rule of thumb for peak-to-peak variation of temporal Gaussian noise can be assessed. Apparently systematic clustering and avoidance can be seen, including some alignment features that appear on oblique viewing. At distances greater than about 6 meters one receives a definite impression of finite size; with the help of a friend moving a pointer you can measure this apparent size. At shorter distances a dark umbra and midgrey penumbra are seen, both of fairly definite size.

Conceive now of an image, not illustrated, that consists of the *linear* superposition of the 34,000 impulses, and suppose that each impulse represents a photon from a luminous celestial object, perhaps a comet. The spatial randomness is associated with the sensitivity of the recording instrument rather than with the spatial form of the celestial object, and it is the latter that we are interested in. We therefore suppose that there is a true brightness distribution, which we would like to determine, contaminated with additive zero-mean spatial noise. The sum of these two hypothetical entities comprises the observed data.

Figure 17-30   Randomly oriented rods on a square grid.

One way of approaching this is by smoothing in the image domain, hoping to reduce the fine structure associated mainly with the noise, while conserving the broad features of the wanted distribution. However, by moving to the transform domain we get an alternative view.

Taking the Fourier transform of the data will give a complex function, which we could display as a power spectrum, abandoning the information contained in the phase. This display might be a useful input to the judgment needed to assess which high frequencies are likely to be noise. After multiplication by the adopted noise-filter function the modified transform will possess the strong fluctuations discussed in connection with Fig. 17-22. In that case the fluctuations were true features of the spectrum of the eight-point object; in the present case the deep fluctuations are unwanted. The fact that a very large number of impulses constitute the image has not smoothed out the fluctuations. This

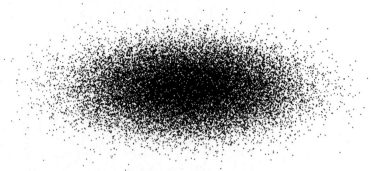

Figure 17-31   Pattern composed of 34,000 dots drawn from a normal distribution.

fluctuating character is a fundamental characteristic of the transform of random images, or of images with a random component. Averaging over a moving patch of restricted area in the spatial-frequency domain will reduce the fluctuations, and the outcome may be a more useful representation of the spectrum. In some applications the spectrum is the final product that is wanted, but in the present case the luminous entity in the image domain is our interest. If we smooth the transform, how will the object be affected? Clearly, fine-scale fluctuations in the transform are associated with the outlying borders in the image domain; smoothing of the fluctuations by convolution with a restricted array will result in multiplying the image by a broad factor that tapers down at the borders. The maneuvers in the transform domain thus correspond first to smoothing in the image domain and then truncating the outlying fringes. Mathematically the approaches in each domain are equivalent, but the judgment involved varies from case to case, especially when information is available in addition to what is contained in the data. If we know that our object is a solid object such as a comet, we will be more confident removing fringes in the image domain than we will be in designing a small array for smoothing in the transform domain.

SPATIALLY CORRELATED NOISE

In Figure 17-1, the noise in any one pixel was independent of the noise in adjacent ones. We now consider a spatial variation that is not so drastic and where there is a correlation patch of a certain size, within which brightness values are related. Let us construct an example where the brightness possesses spatial autocorrelation. Any autocorrelation function could be specified, including ones with negative surrounds, but this example will take Gaussian autocorrelation. The autocorrelation function can be skewed with respect to the axes of x and y, but we will avoid this and assume isotropy. Accordingly, the spatial autocorrelation function $C(r)$ will be taken as

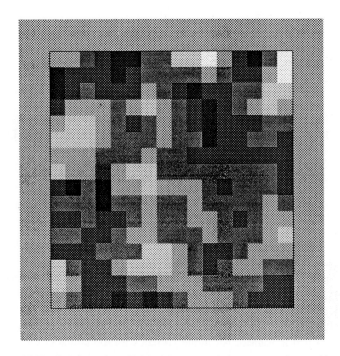

Figure 17-32   Gaussian noise with isotropic autocorrelation superposed on a midgrey
background to avoid negative brightness.

$$C(r) = e^{-\pi r^2 / r_0^2}.$$

The quantity r_0 is the radial correlation distance, and the normalization is such that the
autocorrelation is unity when the radial separation r is zero. The reason for specifying in
terms of a correlation distance r_0 rather than a circular standard deviation $\sigma_r = \sqrt{\sigma_x^2 + \sigma_y^2}$
is partly to emphasize that the autocorrelation function is not necessarily a normal distri-
bution. However, for reference, we note that $\sigma_r^2 = r_0^2 / 2\pi$.

Figure 17-32 was made from an uncorrelated original by convolving with a nine-
element array of coefficients. A result of this is to smooth the brightness distribution.
Clustering and zones of avoidance inherent in spatially uniform sources are still visible,
but now contours of equal brightness are suggested by the boundaries between the discrete
grey levels. The natural shapes of the hills and dales can therefore be studied; ridge
lines, valley lines, saddles, and watercourses can be traced, an exercise that will help
in recognizing spatially uniform Gaussian noise in applications to real data. Naturally,
the topography is unlike the rolling hills of inhabited regions. An extended example of
similar smoothing is given in Fig. 17-33, while Fig. 17-34 was smoothed with staggered
coefficients so that the hills and dales show a general NW-SE trend.

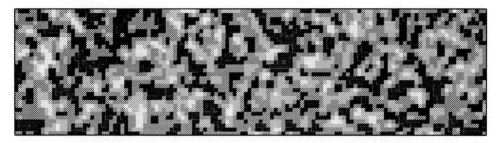

Figure 17-33 A random field smoothed with a nine-element matrix [1 2 1 / 2 4 2 / 1 2 1].

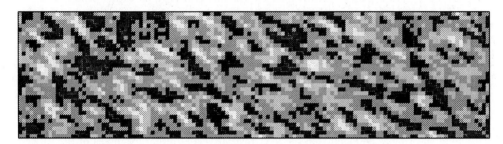

Figure 17-34 Replacing each element of a random field by [1 1 1 0 0 / 0 1 2 1 0 / 0 0 1 1 1] both smoothes and introduces diagonal correlation.

Determining the Associated Smoothing Function

If, instead of constructing Fig. 17-32 we had been presented with the figure, we could try to find out what the array of smoothing coefficients had been, on the assumption that the image was expressible as a convolution with spatially independent random noise. How to do this is easily understood in terms of Campbell's theorem, which, in its original one-dimensional form, goes as follows. Consider a train of impulses occurring at random times which is passed through a filter whose impulse response is $I(t)$. The output waveform is a convolution of the form $\Sigma a_k I(t - t_k)$. Occasionally an isolated replica of $I(t)$ may be seen where a relatively isolated impulse occurs; mostly there will be substantial over-lapping that prevents recognition of the shape $I(t)$. Given the convolution, we cannot in general find out what $I(t)$ was except in special cases, where information about $I(t)$ may be available beyond what is furnished in the form of the output waveform alone. But we can get the autocorrelation function of $I(t)$. In fact, this autocorrelation is the same as the autocorrelation function of the output waveform, in the limit as the number of constituent pulses approaches infinity. Therefore, the procedure is simply to autocorrelate the wave-form; or, in the two-dimensional generalization, the autocorrelation of the image supplies the estimate of the autocorrelation of the smoothing function. It is very reasonable to apply the name *Campbell's theorem* to the two-dimensional version, which then reads as follows: The normalized autocorrelation of an image of the form $\Sigma_{i,j} a_{ij} I(x - i, y - j)$ approaches the autocorrelation of $I(x, y)$ as the number of terms becomes large. It follows, of course, that the power spectrum of the image approaches the power spectrum of the smoothing function represented by the coefficients a_{ij}.

For a derivation of Campbell's theorem based on specific assumptions see Papoulis (1965). The conditions needed for a formal proof are not likely to be encountered in practice, and especially not under numerical circumstances and with images of finite size. Rather, the scientific interest is in the departure of images from spatial uniformity, because that is where the information in an image must reside. A simple example of the sort of problem that is factored out when proofs based on axioms (such as spatial uniformity, or temporal stationarity) are being polished is this. Given the image in Fig. 17-32, and nothing about how it was constructed, how can you tell whether it is or is not likely to have arisen as a convolution? Remember that the model here is an experimental or observational situation where statistical ensembles, or random processes defined by probability distributions given in advance, will be hard to apply. The theory of random signals based on probability distributions specifying at the outset that which we wish to find out has a role, but requires some skill. To answer the question in hand we can segment the image, autocorrelate the parts, and, if the autocorrelations are not in agreement, the presumption of uniformity can be dropped and some other hypothesis can be tried. This seems unsatisfactory and much messier than elegant theory but is quite normal when you are trying to discover something. Research procedures with data are learned by apprenticeship in each field and cannot be elaborated on here. Instead, some further examples of the variety of noise images will be presented.

THE FAMILIAR MAZE

Mazes are familiar as puzzles and date back to medieval cathedrals and the labyrinth in Crete in Mycenean times, three and a half millennia ago. From the perspective of modern image analysis, the maze belongs to a class of textural surfaces including brick walls, random masonry, parquetry and tiling, and various structural and machine finishes involving straight lines that run in two directions only. Such structures possess autocorrelation functions and power spectra, may be spatially uniform, or may exhibit spatial gradients of their properties. In addition, the patterns may possess chirality. That is, they may be distinguishable as left or right handed. A maze with this property can be generated by arranging for the weaving path to veer more often to the left than to the right. Chirality is an important textural feature; a little thought (or trial) will show that a chiral pattern and its mirror image have identical autocorrelation functions, just as a waveform in which the arrow of time is clearly apparent nevertheless has the same autocorrelation as its reverse.

Figure 17-35 shows a maze that was created by directing the pen on a random traverse. After each step a new direction is selected at random but is not drawn if the destination is previously occupied. If all three possible directions are occupied, a new traverse is started from a fresh starting point. Tee junctions are formed only where fresh starts are made from occupied points. One way of keeping track of occupied points is to build an explicit binary array as nodes are traversed; alternatively, the bitmap register that

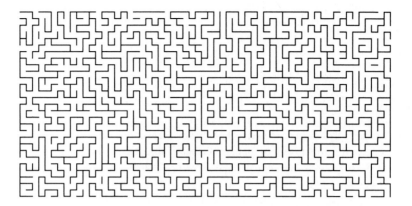

Figure 17-35 A maze constructed of a sequence of traverses that turn left, right, continue on, or fork at random but are self-avoiding.

Figure 17-36 A texture whose basis is obscure, except perhaps to a crazy chip designer.

is used for refreshing the screen, and already contains this array, can be queried with a byte-read command such as BREAD A$,1 which puts one eight-digit binary string, or one byte, into A$. A command such as WHERE x,y that keeps track of where the pen is and assigns the coordinates to x and y is also helpful. If the maze is large, or many are required, some interactive help is desirable in finding the last few starting points.

Though complicated, the maze is comprehensible to the eye. If it is shifted half a unit and added to itself, it becomes interpretable only as a texture; the brain recognizes it as an entity with uniform statistics, but the composition is inscrutable (Fig. 17-36).

Figure 17-37   Random parquetry.

Random Parquetry

A pleasant textured background, related to the maze, resembles parquetry work as shown in Fig. 17-37. The method of construction is to step along row 1 marking each fourth pixel and at the same time marking a pixel in row 2 in a position that is either ahead or behind, choosing at random. Then rows 3 and 4 are marked in the same way but staggered two pixels horizontally. The essential program segment is as follows.

RANDOM PARQUETRY

```
N=128                              length of row
FOR i=0 TO N*(N-1)-2 STEP 2
    y=2*(i DIV N)
    x=2*(i MOD N)+y MOD 4
    CALL "PSET" (x,y)
    CALL "PSET" (x+SGN(RND-0.5),y+1)
NEXT i
```

The operation x DIV y, sometimes written x\y, consists of integer division, yielding the integer part of the quotient. It is useful where explicit type declaration is not *required* (personal computing as distinct from community programming). Since the program depends on a random choice between ahead and behind the choice may be easily biased spatially. For example, if one replaces the expression RND-0.5 by RND-L, where the threshold L drops from 0.5 at the center to zero at the edges, the curious effect of Fig. 17-38 results.

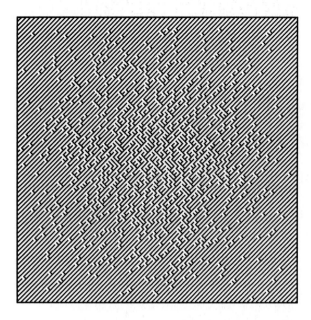

Figure 17-38 A continuous texture gradient at constant grey level controlled by a space-dependent threshold.

From the construction, the grey level is about 25 percent, and a limited number of other levels can be generated by increasing the step length and by reversing black with white. The phenomenon of space-variable disorder exhibited in Fig. 17-38 does not change the grey level and would allow the design of an alphabet or other patterns with the same grey level as their background, whose boundaries would nevertheless be apparent both to the eye and to a tailor-made-edge detection algorithm working on texture.

Random parquetry gives an impression of surface relief, which changes a little when the page is rotated 45°.

THE DRUNKARD'S WALK

Lord Rayleigh investigated the progress of a man who starts at a lamp post, takes a step, falls down, takes another step in a random direction, falls down again, and so on. It is required to find out, theoretically, the probability that after n steps he will be located at a distance between r and $r + dr$ steps from the starting point. The Rayleigh distribution $2rn^{-1}\exp(-r^2/n)$ [Rayleigh, 1905], which originated in a problem of the composition of acoustic vibrations, is connected with Brownian motion, the fading of radio waves, the probability distribution of the waveform of laser light, and many other phenomena. Figure 17-39 traces such a random walk out to a distance of 1000 steps, starting from an origin marked by a cross. As the development of this path is watched on a screen, there are intervals when the drunkard does not seem to be going anywhere and other intervals when he seems to take determined initiatives. This is not apparent from

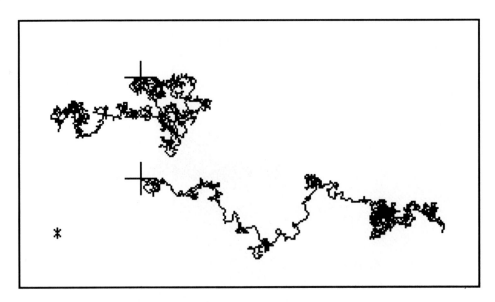

Figure 17-39   A walk of 1000 steps in random directions (above) and the same walk
superposed on a drift to the right (below).

the upper drawing, because the path continually turns back over itself; so to illustrate
the phenomenon, the lower drawing presents the case of a drunken fish in a steady current that carries it one-quarter of a step to the right at each step. The knots show the
intervals of indecision. The star marks the position that would have been reached if the
current were not carrying the fish to the right. Although the locations reached are truly
random, they have little in common with a random scatter of points; an ensemble of random walks, all starting from the same origin, would have more in common with a random
scatter.

The technique of combining a random walk with a drift was used in Fig. 17-5 and
also in Fig. 17-3, which invoked a curvilinear drift.

Figure 17-40 generates a two-dimensional distribution by superposing 20 independent walks, all taking 1000 steps from the same starting point. The result is remarkably different from the two-dimensional normal distribution, which fades away nebulously toward
its outer regions (Fig. 17-31). There are a few protruding paths along which excursions
have been made from what is otherwise a more-or-less compact body. The distance from
each endpoint to the origin follows a Rayleigh distribution with mean $\sqrt{1000} = 31.6$ and
standard deviation 16.5. The ticks on the figure, which are spaced 50 units, show that the
maximum excursion barely reaches three standard deviations beyond the mean. This type
of diagram has generated theoretical interest in connection with mapping the boundaries
of animal populations [Larralde et al., 1992].

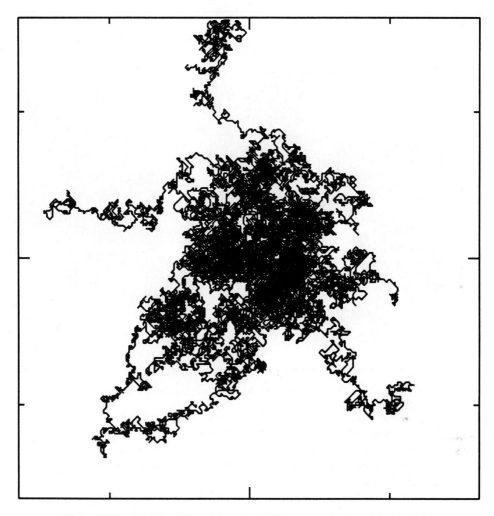

Figure 17-40   A family of 20 random walks of 1000 steps starting from the same origin.

FRACTAL POLYGONS

An important class of geometrical entities, which has received much attention through the efforts of Mandelbrot (1982), is exemplified by closed polygons, one of whose distinctions is to have lengths not proportional to the square root of the area enclosed. We begin by introducing some particular examples with information on their construction. Fig. 17-41 shows the dragon curve, a polygon that is constructed by traverse instructions calling for equal steps in the cardinal directions, starting from an origin that is marked by marginal ticks. The terminus is also indicated. The polygon is self-avoiding in the sense that previous sides are not retraced, but previous vertices may be, and mostly are, revisited. This

Figure 17-41 The dragon polygon. The corner inset is enlarged in the next figure.

self-avoiding behavior is conspicuous when one watches the polygon developing on the screen but is lost in the static outcome.

In the program segment that follows, IDRAW stands for an incremental draw command whose parameters are ± 1 or 0 to the east and ± 1 or 0 to the north.

DRAGON POLYGON

```
E(0)=1 @ N(0)=0          easting, northing
E(1)=0 @ N(1)=1
E(2)=-1 @ N(2)=1
E(3)=0 @ N(3)=-1
q=0                      direction in quadrants, CCW from east
C=0                      counts the sides
nextside:  C=C+1 @ c=C
again:     IF c MOD 2=0 THEN c=c/2 @ GOTO again
           IF (c-1)/2 MOD 2=1 THEN q=(q+1) MOD 4 ELSE q=(q-1) MOD 4
           IDRAW E(q),N(q)
GOTO nextside
```

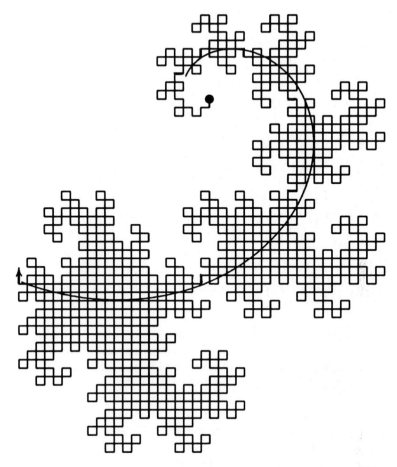

Figure 17-42   Enlargement of the first few coils of the dragon and the spiral $r = e2^{2\theta/\pi}$.

Narrow necks that articulate the main closed masses into which the dragon is coiled lie close to a logarithmic spiral $r = e^{2\theta/\pi}$. Successive coils subtend about $\pi/4$ at the origin, progressively rotating by $\pi/4$, expanding in dimensions by $\sqrt{2}$, and doubling in area. However, the fine detail on the edges remains at the same scale throughout. Thus each coil is grossly similar to its predecessor except that more and more detail appears at the edges. Any attempt to squeeze many coils onto the page encounters a resolution limit. In Fig. 17-41 the line width equals the length of a side, so the smallest bays and promontories are lost. The corner inset, which uses the finest reproducible line, does suggest the lost detail, but in order to study the fine structure one needs an enlargement that is restricted to fewer coils.

Figure 17-42 shows the first eight coils to a larger scale, shows how the polygon starts, and the relation to the spiral that expands by $\sqrt{2}$ every 45°. Because of the irregularity of the outline, and the changes from coil to coil, no smooth curve with only a few

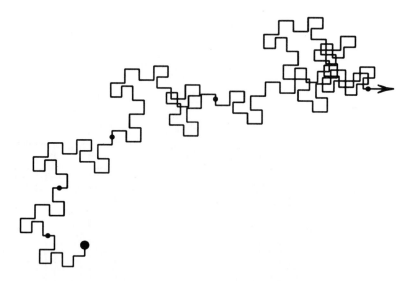

Figure 17-43   Enlarged dragon with superposed drift which unravels the time dependence.

parameters follows the necks exactly. The number of steps between adjacent necks measures the area of the coil in-between, subject to the same irregularity. The numbers of steps from the start to different necks in turn are 12, 26, 51, 101, 204, 410, 819, 1637—numbers which are useful for stopping the process at a desired point.

How to convey the kinematics of dragon development on a static diagram is a legitimate question of graphics presentation in its own right. A solution that enables the polygon to be traced by the finger is given in Fig. 17-43, which uses the principle of extending in the direction of a third axis (time) and projecting obliquely back onto the plane; or you could say that a slow drift to the north-east has been superposed. By tracing out the course from the start, you can share the sensation of a long snake slithering out of its hole at the origin and winding back and forth around itself on the flat desert, never crossing itself, and you can contemplate the algorithm that guides it.

A different sort of fractal is shown in Fig. 17-44, a leaf with 27 pinnae, each with up to a dozen or so pinnules, which themselves have up to four or five pinnatifid divisions. This is a single leaf of the Silky Oak (*Grevillea robusta*), a member of the Proteaceae. The hierarchy of scale sizes seen in the dragon is seen here, but only to three levels.

A computed leaf (Fig. 17-45) made of random dots under the control of a fractal transformation resembles the fractal structure of a natural object, but the strict discipline is noticeably more severe than in a natural product. The following program segment is for a frame 2400 pixels wide, has the origin centered at the base of the leaf petiole, and will require rescaling for frames of other dimensions. Four different affine transformations are incorporated (Stevens, 1990). The one which is performed 85 percent of the time produces a train of dots on a curve heading for the leaf apex. On random interruptions, shorter trains of dots are produced along the pinnae, pinnules, or midribs, continuing from the current location. The program begins by reading the four sets of six affine coefficients.

Figure 17-44 A fractal leaf.

FRACTAL FERN LEAF

```
FOR i=0 TO 3 @ READ a(i),b(i),c(i),d(i),e(i),f(i) @ NEXT i
x=0 @ y=0
FOR i=1 TO 30000
    R=RND
    k=(R>0.01)+(R>0.08)+(R>0.85)
    tmp=a(k)*x+b(k)*y+c(k)
    y=d(k)*x+e(k)*y+f(k)
    x=tmp
    u=1760*x
    v=44700*y DIV 28
    CALL "PSET" (u,v)
NEXT i
DATA 0,0,0,0.16,0,0
DATA 0.2,-0.26,0.23,0.22,0,0.2
DATA -0.15,0.28,0.26,0.24,0,0.2
DATA 0.85,0.04,-0.04,0.85,0,0.2
```

A well-known fractal shown in Fig. 17-46 is known as the Sierpiński gasket, published by Wacław Sierpiński (1882–1969) in 1916. It can be imagined as being dissected indefinitely into smaller and smaller components, though only a finite number of levels can be illustrated. The procedure has been explained in terms of affine transformations by

Figure 17-45	A computed fractal leaf.

Barnsley (1988), one of the discoverers of algorithms for simulating plants and landscapes; also see Peitgen et al. (1992).

A different way of looking at the diagram finds the same structure associated with the odd numbers in Pascal's triangle, as shown in Fig. 17-47 Instead of constructing the gasket by starting from a closed boundary and working inward, we can start at a vertex and expand to a triangle of desired size row by row. The binomial coefficients $C(n, k)$ or $\binom{n}{k}$ can be computed from the factorial function (NumRec 1986) but of course can also be computed element by element from the row above by convolution with {1 1}, which amounts to only one addition per new entry, an algorithm that is hard to beat. A gasket

Figure 17-46   A Sierpiński gasket computed to level 7.

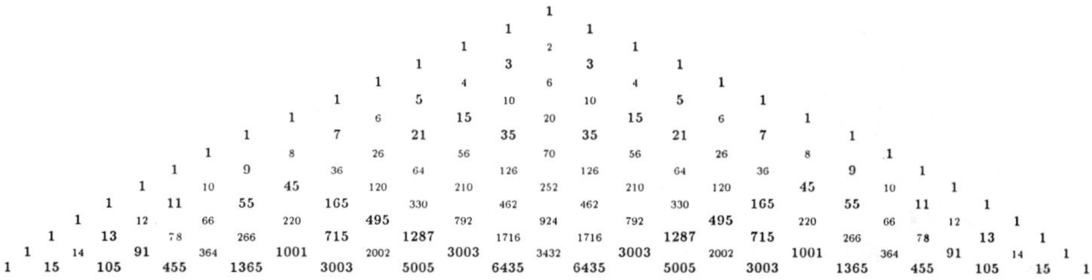

Figure 17-47   Pascal's triangle, showing odd numbers in bold type.

constructed on a screen by this method, which is shown in Fig. 17-48, has a rather crisp appearance.

SIERPINSKI GASKET FROM PASCAL TRIANGLE

```
FOR n=0 to 63
   FOR k=0 TO n
      x=2*k-n
      y=64-n
      IF C(n,k) MOD 2=1 THEN CALL "PSET" (x,y)
   NEXT k
NEXT n
```

The array of binomial coefficients $C(n, k)$ for n up to 63 needs to be filled in advance. One way is as follows.

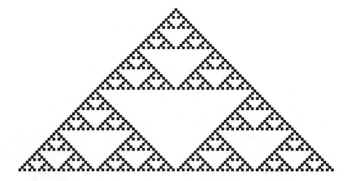

Figure 17-48 A gasket composed of square pixels at the location of the odd numbers in Pascal's triangle.

BINOMIAL COEFFICIENTS BY ADDITION

```
C(0,0)=1 @ C(1,0)=1 @ C(1,1)=1
FOR n=2 TO 63
   C(n,0)=1
   FOR k=1 TO n
      C(n,k)=C(n-1,k-1)+C(n-1,k)
   NEXT k
NEXT n
```

To extend beyond the 16 rows illustrated requires allowance for the growth of $C(n,k)$ beyond 10^{17} before row 63 is reached. This can be handled by computing the sums $C(n-1, k-1) + C(n-1, k)$ modulo 2, since use is to be made only of the evenness or oddness of the coefficient. If necessary, memory overflow at large n can be coped with by reusing the rows $C(0, k)$ and $C(1, k)$.

For a different procedure, one that is kinematically interesting as viewed on a screen, start from some inked point such as a point on the side of the bounding triangle. Plot the point halfway to one vertex; then plot the point halfway from there to a randomly chosen vertex, and continue ad infinitum. The program for developing a Sierpiński gasket in this way is, of course, rather brief. In the initial stages the pattern develops rapidly, slowing down exponentially as more and more shots hit old targets.

SIERPINSKI GASKET BY RANDOM HALVING

```
x(1)=-128 @ x(2)=0 @ x(3)=128
y(1)=0 @ y(2)=128 @ y(3)=0
x=64 @ y=64              start at middle of right side
i=1
FOR c=1 TO 36001        count
   x=(x+x(i))/2
   y=(y+y(i))/2
   CALL "PSET" (x,y)
```

```
   R=RND
   i=1+(R>1/3+(R>2/3)        select random vertex
NEXT c
```

This program does not appear to start from the largest triangle and proceed to the smaller triangles but seems to work on all sizes at the same time. To understand this, notice that the whole pattern of Fig. 17-46 is assembled from a triplicate copy of itself and that each part is subdivided in the same way. Halfway from any inked point to a vertex is also entitled to ink; thus the bisection procedure continually makes half-scale copies of larger triangles, point by point. However, about eight bisections reach the resolution limit of the medium, and round-off provides a fresh point on a larger triangle. Even the external boundary is well populated with points, despite the fact that the algorithm is nominally placing points on smaller and smaller triangles.

As a final example of a fractal curve take the Hilbert space-filling locus (Fig. 17-49). Watching the creation of this serpentine locus on the screen adds a dimension to the static presentation. For programs see Morrison (1984).

The Significance of Fractals

We have seen several examples of fractals, all of which exhibit the property of approximate self-similarity in the sense that the magnified whole contains, or at least looks like, the original. There is a conceptual difficulty, seeing that none of the illustrations really is self-similar. One stumbling block, thinking about the infinitely small, is avoided by turning to the Pascal triangle (Fig. 17-47). If the bold numbers are circled on a tracing placed over the table, then the tracing may be moved down eight rows, and it will be found that the circles fall on numbers that are also bold. What happens at infinity conveniently happens off the page. If, on the other hand, one enlarges the dragon by a factor $\sqrt{2}$ and rotates $45°$, the main coils do not agree exactly, and what to do about the origin is perplexing. If you care to try the tracing-paper experiment with the Pascal table, a glitch will be found on the row below the longest row printed here, so even the Pascal demonstration lacks perfection. Considerations of the limit of resolution in actual practice suggest that one is better off focusing on actual finite fractals than on theoretical objects. Even when only two or three levels are involved, the actual objects are interesting.

So far the fractals displayed—and one can read about many more—seem to have little practical application beyond the simulation of natural objects, such as landscapes with trees, mountains, sea and sky, where some truly stunning art work has been displayed. I am reminded of other work that has exploded rapidly during my span of attention, with little indication of ultimate application, and we may be in the middle of such a phenomenon. To participate, it will be a good idea to gain practical experience with constructing existing fractals and thinking of new ones. The programs provided are to encourage this.

It appears that fluency with the affine transform, as introduced in Chapter 2, is essential. Affine coefficients appearing in data statements make for brief code but make inscrutable recipes; thinking with magnification, rotation, shear, shifts, and vanishing points needs to be cultivated by experiments on random polygons.

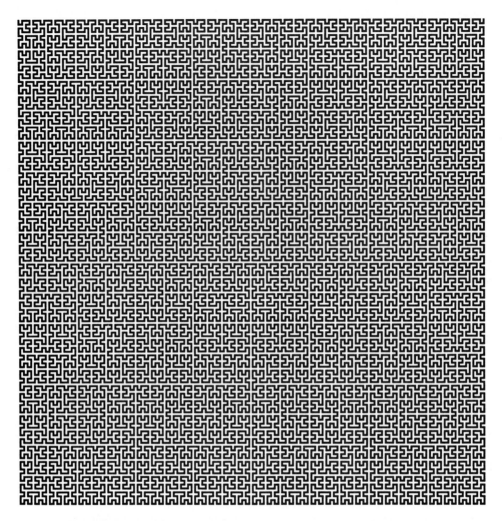

Figure 17-49   The Hilbert space-filling locus.

CONCLUSION

Clearly noise in images is a very rich and developing subject. Innumerable branches wait to be explored and to be applied in more and more basic and applied fields as computer presentation of data spreads. The examples given here are intended to emphasize the variety of subject matter and convey a sense of unfinished business. By comparison with digital processing of time signals, a mature mathematical subject built on decades of skilled attention, the study of digital images containing elements of randomness is in a state of flux.

The human visual system equips us to apprehend two-dimensional phenomena that for the time being challenge theoretical treatment. It is possible that traditional computer algorithms will prove too cumbersome for some problems and that the way to go is to incorporate the human operator into interactive, screen-based procedures. For example, the fundamental tiling arrangement that uniquely characterizes a scatter of points requires several hundred lines of code to solve something that is largely obvious to the eye. Yet the lengthy and abstract code does not allow for nodes of variable strength, size, shape, quality, and other spatially variable parameters that the eye is able to absorb quite readily in many cases, using measures and criteria that no one knows how to quantify for programming purposes. This ability can be harnessed (Samadani and Han, 1993).

The mathematical developments regarding statistical properties of images that have taken place in astronomy, biometrics, and computer graphics are impressive and can be recommended to students interested in randomness in images. At the same time one needs to cultivate awareness of opportunities for new initiatives.

LITERATURE CITED

R. J. ADLER (1981), *The Geometry of Random Fields*, John Wiley & Sons, New York.

M. F. BARNSLEY (1988), "Fractal modelling of real world images," in H-O Peitgen and D. Saupe, eds, *The Science of Fractal Images*, Springer-Verlag, New York.

M. S. BARTLETT (1975), *The Statistical Analysis of Spatial Pattern*, Chapman and Hall, London.

B. DELAUNAY (1934), "Sur la sphère vide," *Bull. Acad. Sci. USSR*, Classe des Sciences Mathématiques et Naturelles, Series 7, pp. 793–800.

G. L. DIRICHLET (1850), "Uber die Reduktion der positiven quadratischen Formen mit drei unbestimmten ganzen Zahlen," *Journal für Reine und Angewandte Mathematik*, vol. 40, pp. 209–227.

W. A. GARDNER, (1988) *Statistical Spectral Analysis, A Nonprobabilistic Approach*, Prentice Hall, Englewood Cliffs, N.J.

P. J. GREEN AND R. SIBSON (1978), "Computing Dirichlet tessellations in the plane," *The Computer Journal*, vol. 21, pp. 168-173.

P. HOLGATE (1972), "The use of distance methods for the analysis of spatial distribution of points," in P. A. W. Lewis, ed., *Stochastic Point Processes*, John Wiley & Sons, New York.

V. ICKE AND VAN DER WEYGAERT (1991), "The galaxy distribution as a Voronoi foam," *Quarterly J. Roy. Astronom. Soc.*, vol. 32, pp. 85–112.

B. JULESZ (1960), "Binocular depth perception of computer-generated patterns," *Bell Syst. Tech. J.*, vol. 39, pp. 1125-1162.

N. E. THING (1993), "Magic Eye," Viking Penguin, New York.

C. W. TYLER AND CLARKE (1990), "The autostereogram," *Proc. Int. Soc. Opt. Eng.*, vol. 1256, pp. 182-197.

H. LARRALDE et al. (1992), "Territory covered by N diffusing particles," *Nature*, vol. 355, pp. 423–426.

B. B. MANDELBROT (1982), *The Fractal Geometry of Nature*, W. H. Freeman, New York.

R. MORRISON (1984), "Low cost computer graphics for micro computers," *Software – Practice and Experience*, vol. 12, pp. 767–776.

A. OKABE, B. BOOTS, AND K. SUGIHARA (1992), *Spatial Tessellations: Concepts and Applications of Voronoi Diagrams*, John Wiley & Sons, Chichester, U.K.

A. PAPOULIS (1965), *Probability, Random Variables, and Stochastic Processes*, McGraw-Hill, New York.

H-O PEITGEN, H. JURGENS, AND D. SAUPE (1992), *Chaos and Fractals: New Frontiers of Science*, Springer-Verlag, New York.

LORD RAYLEIGH (1905), "The problem of the random walk," *Nature*, vol. 72, p. 318.

B. D. RIPLEY (1981), *Spatial Tessellations*, John Wiley & Sons, New York.

C. A. ROGERS (1964), *Packing and Covering*, Cambridge University Press, Cambridge.

R. SAMADANI AND C. HAN (1993), "Computer-assisted extraction of boundaries from images," *Conference Proceedings on Storage and Retrieval for Image and Video Databases*, 2–3 February 1993, San Jose, California, Society of Photo-Optical Engineers, vol. 1908, pp. 219–224.

J. G. SKELLAM (1952), *Biometrika*, vol. 39, pp. 346–362.

R. T. STEVENS (1990), *Fractal Programming in Turbo Pascal*, Redwood City, M&T Publishing.

D. STOYAN, W. S. KENDALL AND J. MECKE (1987), *Stochastic Geometry and Its Applications*, Akademie-Verlag, Berlin.

G. VORONOÏ (1908), "Nouvelles applications des paramètres continus à la théorie des formes quadratiques, deuxième mémoire, recherches sur les paralleloèdres primitifs," *Journal für Reine und Angewandte Mathematik*, vol. 134, pp. 198–287.

PROBLEMS

17-1. *Random scatter.* Examine Fig. 17-1(a) closely to see whether the dots are indeed scattered at random.

17-2. *Poisson distribution.* The mean number of dots per unit area in a random scatter is 10.
   (a) List the probabilities $p(n) = 10^n e^{-10}/n!$ that the exact number of dots n will be 0, 1, ... 20.
   (b) Compare these values with the normal distribution $(1/\sqrt{20\pi}) \exp\{-[(n-10)^2/20]\}$ with mean 10 and variance 10. ☞

17-3. *Binary noise.* Construct a binary object like Fig. 17-7, but with a spatially constant noise level; damage only one pixel in three or four. Experiment with erosion-dilation (opening) and the reverse, using various structure functions, including [1 **1** 1] and $\begin{bmatrix} 1 \\ \mathbf{1} \\ 1 \end{bmatrix}$. Compare the results with linear filtering as implemented by convolution with small arrays to remove high frequencies. The original object has 272×47 pixels; the letters are three pixels thick and 20 pixels high.

17-4. *The dragon's coils.* Show that the number of steps to the separations between the dragon's coils are 5, 12, 26, 51, 101, 204, 410, 819, 1637, 3276, 6554, 13107, 26213, 52428, 104858, 209715, 418429, 838860, etc.

17-5. *Natural features.*
   (a) In the illustration showing how creosote bushes in the desert tend to space themselves an effort was made to avoid the appearance of an artifact by arranging for no two bushes to have the same shape. Think about how you would do this, and inquire whether regular

pentagonal bushes all oriented the same way would create a nonnatural impression, even though the bushes are small.

(b) If the size of the field is 10×10, what is the average spacing of the bushes?

17–6. *Contiguous nodes.* Frame a 32×32 section of graph paper. Mark 25 of the small unit squares in irregular locations, spaced in the way you think trees might grow in a natural forest. Starting from what you deem to be the lower left tree (no. 1), draw a chain of four connecting links terminating at the lower right tree (no. 5). Do the same on the top, left, and right to complete an outline. As a guiding principle for the judgment required, try to keep the link length around five units. Complete a grid of interior chains, all with four links. Assign serial numbers to the trees (21 to 25 in the top chain).

(a) Explain how to form an array of numbers occupying 25 rows and 9 columns, the first entry in each row being the serial number of a tree and the remaining eight being the serial numbers of trees that are likely candidates as contiguous nodes.

(b) Test your method.

(c) Sort the eight dependent row entries by Euclidean distance from the reference tree at the start of the row and find out how successful you have been at identifying the nearest neighbor to each tree. ☞

17–7. *Judging standard deviation.* One would often like to judge the standard deviation of a data set at a glance, but the arithmetic is too complicated to do mentally even for small sets. According to F. Schlesinger (*Astronom. J.*, vol. 46, p. 161, 1936), the standard deviation σ can be estimated from the range $R = y_{max} - y_{min}$ by the formula $s = kR$, where s is the estimate of σ and, for small numbers, n, the factor k is given by

n	3	4	5	6
k	0.526	0.451	0.414	0.385

(a) Test this claim for $n = 4$ using as data the values 8, 10, 6, 17 for the number of dots in 1-cm squares of the random scatter illustrated in the text.

(b) Comment on the utility of this rough-and-ready rule of thumb. ☞

17–8. *CCD Array.* A 500×500 array of charge-coupled light detectors is composed of elements that have a linear response over a dynamic range of 1000 to one, but the individual elements vary randomly by 10 percent in gain. Devise a plan for calibrating every element by making three images of the twilight sky, or of a laboratory-generated uniformly bright object, pointing successively at $(0, 0)$, $(0, 1)$, and $(1, 0)$, where the coordinate unit is one pixel. ☞

17–9. *Nonlinear noise addition.* A digitized photograph represents the integrated density over square patches about the size of the image of a star. Stars fainter than a certain limit are not recorded, because a threshold must be exceeded in order that the incident light releases enough silver atoms for a grain in the photographic emulsion. Create an imaginary constellation having stars both brighter than and fainter than the limit L and represent the constellation by a matrix. Now add independent random numbers normally distributed about a mean $L/2$ with standard deviation $L/5$. These random numbers represent a faint uniform background of light. Replace values less than L by zero to arrive at a representation of the photographic image. A photographer says that you can photograph fainter stars when there is faint light from the sky than when the sky is dark. Does your calculation confirm this? A student of statistics says the number of false alarms makes the proposal useless in practice. Report on this disagreement.

17–10. *Three-dimensional scatter.* Points are distributed at random within a cube. Naturally, there are volumetric vacancies and concentrations on curved surfaces analogous to the vacancies and collineations in two dimensions. Consider a plan view in which each point in space projects to a point vertically below on the base of the cube. How could you tell that the plane distribution was derived from a three-dimensional one?

17–11. *Maze specification.* Show that a maze constructed on an $M \times N$ array can be specified by MN bits. Would fewer suffice?

17–12. *Autocorrelation of maze.* An $M \times N$ maze is constructed of unit-strength line impulses.

(a) Does its autocorrelation function contain line impulses or point impulses?

(b) If the autocorrelation functions of K different random mazes were computed and averaged, what could be said about the average autocorrelation as $K \to \infty$? ☞

17–13. *Octal steps.* Figure 17-40 superposes many random walks, each composed of steps ± 1 in either coordinate or both. If it were a map of the drainage system in a city, you would get the impression that north is not at the top. Why is that?

17–14. *Rayleigh distribution.* Show that the probability distribution

$$p(R) = \frac{2R}{\Sigma^2} e^{-R^2/\Sigma^2}$$

has mean $m = E\{R\} = 0.5\sqrt{\pi}\Sigma = 0.8862\Sigma$, standard deviation $\sigma = E\{R - E\{R\}\} = \sqrt{1 - \pi/4}\Sigma = 0.4633\Sigma$, and mode $= \Sigma/\sqrt{2}$. ☞

17–15. *Random flights to a fixed destination.*

(a) Explain how to construct a path from $(0, 0)$ to $(256, 0)$ for which the step Δx is always unity but Δy may be -1, 0, or 1 at random.

(b) If many such paths are superposed, is there an envelope within which the flights are contained?

17–16. *Self-avoiding spiral.* Certain burrowing wasps explore the vicinity of their burrow following a path rather like that shown in Fig. 17-50. How can you decide whether the illustration was digitized from photography in the field or was purely a computational product?

17–17. *What is a fractal?* Get some articles on fractals and copy out the sentences where the author says what a fractal is.

17–18. *Self-similar object.* Consider the two-dimensional object $\delta(r - e^\theta)$, $0 < \theta < \infty$ and magnify it by a factor $e^{2\pi}$. There is no need to rotate or translate the enlargement to test self-similarity because the enlargement falls directly on the original. But are the two entities identical? ☞

17–19. *Self-similarity of Pascal's triangle.* If Pascal's triangle is moved down eight rows and shifted eight steps to the left odd numbers fall on odd numbers as mentioned in the text and as demonstrable on the table printed. How could you convince a student that this property of odd-numbered self-superposition breaks down?

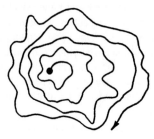

Figure 17-50   Exploratory path of a burrowing wasp.

Solutions to Problems

This appendix presents the solutions to those problems that are marked in the text with the symbol ☞.

CHAPTER 2

2-1A. The grey level varies linearly with latitude $\arcsin(z/R)$ from black at the bottom of the sphere to white at the top.

2-2A. These experiments, if done with artificial light, will not agree with what is seen under low-angle sunlight. For one thing, the brightness discrimination of the eye is lowered in artificial light, and for another thing there is no sky light.

(a) On the vertical line at $\beta = 60°$ away from the sun the angle of incidence i would be $60°$ for horizontal sunlight, but for low-angle sunlight it is a little greater. At an angle of elevation e, the angle of incidence is given by $\tan i = \sqrt{\tan^2 e + \sin^2 \beta} / \cos \beta$. If the angle of elevation was noted by measuring the length of shadow cast by the paper, this correction can be taken into account. A reference surface of half intensity can be made by standing a flat sheet of paper to make the angle of incidence $60°$, or by using grey paper made by printing with a suitable screen. The major thing one learns from this experiment is that the perceived brightness of adjacent areas is dependent on the surroundings. At a white-black boundary the white looks whiter and the black shadow

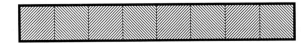

Figure A-1 Grey squares that are distinguishable by texture rather than by shade.

looks darker near the edge than it does in the interior of the shadow; the transition zone seems narrow. Turning the head on its side may make a difference.

(b) The side with the smaller angle of incidence looks distinctly brighter. Lateral brightness variation makes the vertical fold brighter than the vertical edges, and, because of brightness contrast, the bright zone down the fold may seem to be somewhat sharply bounded. This is due to reflected light, as can be verified by varying the experiment, for example by cutting the top of one side down part way. Variation from top to bottom is to be expected because of reflected light from the table.

(c) The very strong effect convinces you that reflected light is really important.

(d) Ability to see the blue depends on skill.

(e) Color photography in sunlight can record blue in the shadows – a full hemisphere of clear blue sky helps. Any psychological effect, of which brightness contrast is one, is apt to be recorded so that the photograph does not look like the scene. The shadow zone on the cylinder and the reflected light in the fold present examples. A painter can paint, within the compressed brightness range offered by pigments, so that the painting evokes the visual scene (and thus differs from the photograph of that scene).

2-3A. Not only does fresh snow look brighter than the sky behind it, but a photograph will confirm the visual impression. One factor is that an overcast sky may be about twice as bright overhead as it is at low elevatuions, because of the atmosphere. To compare snow on the skyline with the sky behind it is to ignore the effective illuminant, which is overhead. The physical brightness difference is also enhanced visually by the contrast gradient. See J. J. Koenderink and W. A. Richards, "Why is snow so bright?" *J. Optical Soc. Amer. A*, vol. 9, pp. 643–648, May 1992.

2-4A. The zone boundaries are not conspicuous, but the 15 uniform zones can be distinguished by viewing from a distance. The boundaries vanish if the dot density probability is taken to vary continuously with position instead of stepwise as illustrated.

2-5A. Innovative textures can readily be devised. One can conduct the proposed experiments with alternating squares hatched at $+45°$ and $-45°$ See Fig. A-1.

2-7A. (a) See graduated scale of flog $10x$ in Fig. A-2.

(b) The function flog x is continuous and has a continuous derivative at $x = 1$. It is adaptable; for example, if values below 0.01 require special attention, a scale of flog $100x$ can be tried.

2-8A. (a) The traditional design of Fig. A-3, as evolved for carpenters' and engineers' rules, is elegant and suitable for close use by eye, as attested by experience. This design could be altered for photographic use in the field by reducing the clutter (removing all lines that are not actually needed) and helping the nomographic function (conversion of units) by hanging the graduations from a central line. One can use the traditional rule for conversion of units, but the single-line arangement is better.

(b) The only unnecessary line is the horizontal line; you can omit it but you won't like the look. Using a thick line gives the design a lift. You can try thickening the integer scale markings, too; this is not recommended for serious mensuration and is also in

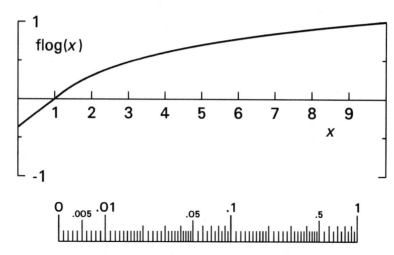

Figure A-2   An example of a scale from 0 to 1 graduated according to the funny log function. The position of the zero graduation is fixed by the property that the tick spacing does not jump at 0.01.

disrepute because some commercial graph paper that is printed this way is inaccurate as a result. In any case, you can try it to see whether you like it (I did).

(c) If the scale card is a small item in the photograph, or the photographic print is blurry, the fine graduations will lose their value. The topic of resolution of scales under adverse conditions has been addressed seriously in connection with geodetic survey. See a textbook of surveying, or look at a leveling staff, for the product of much experience. Here is an application to the present exercise. Fine graduations, at a minimum, have a width of about 5 mils, but a vertical line thickness of effectively zero is achieved at a boundary between black and white areas. The apparently coarse features resulting from use of this idea, as illustrated, are thus in fact more precise than fine ruling. The tenths of an inch can reliably be subdivided by eye into hundredths of an inch, and the half-centimeters into millimeters. Such accuracy is hard to reach with graduations in 16ths, 32nds, or millimeters, partly because of the width of the graduations. Finally, the numerals and the names of the units are omissible because the 2.54:1 ratio identifies the units.

(d) A much-reduced photocopy of the three designs will make its own comment.

2-9A. There must be another saddle point at an altitude between 10 and 20. A traveler going from the one bottom to the nearest point on the shore by the route that climbed to the least altitude would pass over this saddle. It fails to show on the map as a self-intersecting contour because its height does not correspond with a contour level; the same must be true of most saddle points on contour maps.

2-11A. If you are standing on a hilltop in daylight, you are in no doubt of your situation. On a pitch-black night though, would it suffice to verify that you are on a summit by noting that a single step in each of the four directions, north, south, east, and west leads downhill? Obviously not; you might be standing at A on a sharp ridge descending to the southwest (Fig. A-4). Likewise, the condition you were about to give for a saddle suffers from the

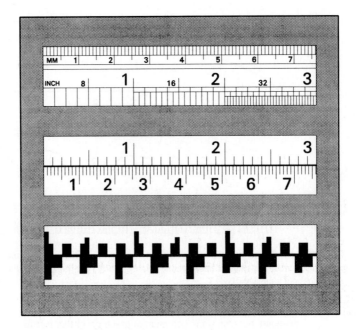

Figure A-3　Scales for field photography.

presumption that the directions of principal curvature are paralled to the rows and columns. The point B is a saddle point but does not meet the proposed condition.

2-12A. The critic's assertion is correct, as witnessed by the examples in Fig. A-5, where the same underlying function is digitized both ways. In case

 (a) the central value (7) is a row maximum and column minimum. But in the case

 (b) the same saddle in the underlying function is a minimum in both its row and its column.

2-13A. (a) See Fig. A-6.

 (b) The contour routine indicates tops of height 51 at (5, 9) and (8, 6). There is another of height 50 at (3, 3), and one of height 47 at (3, 9.5). The full list, for a total of seven, is as follows.

Height	51	51	50	49	47	47	46
Position	(5, 9)	(8, 6)	(3, 3)	(9, 9)	(3, 9.5)	(7, 5)	(7, 10)

 The additional tops can be found by careful inspection, or can be put in evidence by use of a finer contour interval.

 (c) The diagram with contour interval 2 indicates four bottoms, of height 40 at (8, 8), 34 at (6, 7), 33 at (4, 8) and 32 at (5, 6). Bottoms are indicated by downward-pointing ticks on the highest contour that drains solely into that bottom.

　　This exercise is designed to be done by hand to get the feel of contouring from data, but there is no objection to using a computer. Hand contouring generally produces smooth contours, while computing commonly produces polygons, but it is well within the range

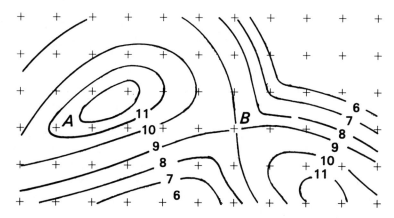

Figure A-4 The proposal that summits are higher than their four adjacent points is contradicted by *A*, which has this property but is not a summit. The point *B* contradicts the proposal for saddles.

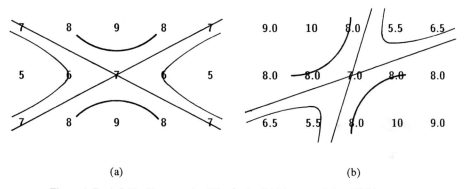

(a)   (b)

Figure A-5 A field with a central saddle of value 7 (a) is resampled at 45° (b).

of a do-it-yourself programmer to substitute a smooth cubic spline for a polygon. A topic for discussion is whether smooth contours are better than polygons. One view is that rough data should be reflected in rough contours and not artificially, and possibly misleadingly, smoothed. The sales view might be different.

2-15A. One simple rule is to search for values that exceed those to the N, S, E, and W. This gives 10 tops with the following values and locations:

52(6, 9) 52(10, 5) 51(9, 9) 49(10, 3) 48(9, 7) 47(4, 8) 46*(5, 7) 41(5, 3) 39(7, 4) 34*(3, 4).

Those marked with an asterisk would be hard to justify. There are 17 values exceeding their four diagonal neighbors; even so, two of the above are not captured, so the diagonal criterion is not a good one. Values that exceed all eight neighbors give the same result as inspection of the contours. However, all these criteria are too simple. Equal adjacent values defeat them, and the substitution of ≥ for > gives false hills. This general approach

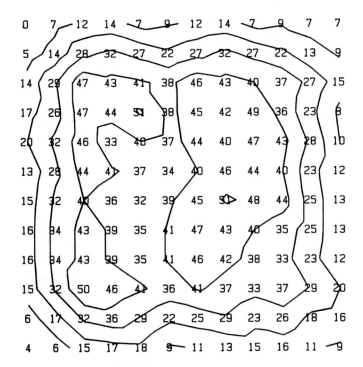

Figure A-6   Contours for 10 [10] 50.

is likely to be acceptable with some particular kind of data, but elaborating the conditions to apply to data in general is not practical (although fun to try).

2-21A. **(a)** Television signals are distinctive by virtue of their raster characteristic. The signal consists of a train of "lines" of uniform length. Successive lines are often the same or nearly the same except at the end of a frame. Then, however, the next line is often nearly the same as the first line of the previous frame because successive frames are often the same. There may also be repetitive synchronizing pulses to punctuate the lines and frames. TV signals are unmistakable. There may be other ways of building up TV pictures that are not in use on earth, for example circular pictures could be built up by spiral scanning. The nature of the repetition would again be the key to presentation.

(b) Television standards in one country may be incompatible with those in other countries (lines per frame = 525 in the United States and environs; 625 in Africa, Australia, much of Europe and the USSR, 1050, 1160, and 1250 for high-definition television), but conversion is commonly practiced. An automatic probe passing through the solar system could be programmed by its senders to determine terrestrial frequency allocations for television and the standard in use on each frequency and might surprise us by accommodating.

(c) Transmission of a constellation diagram is an excellent way to establish agreement on these conventions and on the distinction between left hand and right hand. Without knowing whether a picture was back to front or not, one would have trouble conveying

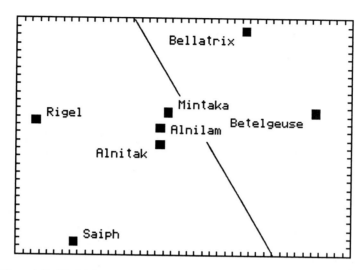

Figure A-7   The brief number string can be decoded into an unmistakable picture of the constellation Orion, to the accuracy permitted by a 41 × 39 matrix. The orientation is such that the galactic meridians are vertical. The Arabic names of the stars, in alphabetical order, are Betelgeuse, Rigel, Bellatrix, Mintaka, Alnilam, Alnitak and Saiph.

information about handedness, such as the information that earth dwellers are mostly right-handed.

(d) Taking each number as an abscissa and the serial number as an ordinate, we obtain the picture shown in Fig. A-7, which is unmistakably Orion. The celestial equator is added for reference. (Can you deduce anything from this?)

This exercise is an impressive example of pictorial communication. Working on the hint that television is involved, one notices that the message length of 1247 symbols factorizes into the primes 29 and 43. Rearranging into rows of length 43 gives the recognizable picture in Fig. A-7. The alternative presentation in rows of length 29 is not recognizable. In few more than a thousand symbols the probe has conveyed the message "I am an extraterrestrial visitor from Orion" in terms that are intelligible to an astronomer speaking any language. By repeating the message at intervals, with and without the 1 near the middle, the probe could specify ζ Orionis (Mintaka) as the home star by making it blink. The probe designers would not need to know the coordinate systems used by terrestrials or terrestrial star names. The received message is rather sparse in information; the data-compressed version 73, 493, 512, 519, 578, 664, 1169 is equivalent in a sense. The difference is that we could not interpret it in the precontact era. The signal as received tells the TV line length. Once this protocol is grasped, data compression would efficiently permit the transmission of other constellations, including faint ones, perhaps telling where more distant civilizations were to be found.

2-22A. **(a)** A topographic surface $h(x, y)$ and an image described by a brightness distribution $b(x, y)$ may be called equivalent over a stated domain if, at every point (x, y), height h and brightness b are in a constant ratio (say $b = kh$). Since brightness (i) cannot be negative and (ii) must be single-valued, the topographic surface, to be equivalent, must also have these two properties. Take the y-axis to be a boundary of the domain and suppose that the surface is tilted by an angle θ about the y-axis. The tilt angle θ must (i) lie in a range that does not produce negative heights and (ii) not be so large that $\arctan(\partial f/\partial x)_{max} + \theta$ equals or exceeds $\frac{1}{2}\pi$.

If you take your topographic surface to be equivalent to a *reflectance* distribution, you will have to meet a third condition (iii) that no reflectances greater than unity are generated.

If the tilt is small, then the values of $h(x, y)$ are increased to $h(x, y) + x \sin \theta$. The equivalent operation on the brightness distribution is to increase $b(x, y)$ to $b(x, y) + kx \sin \theta$; in other words, a linear gradient of brightness is added.

If the tilt is not small, features of the brightness image are not only brightened, but also displaced slightly in a direction perpendicular to the y-axis. It is as though a square grid of latitude and longitude lines ruled on a plane surface were compressed to a slightly nonsquare grid, each point (x, y) being transformed into the point $(x \cos \theta, y)$. Since a topographic surface is not ordinarily plane, the displacement depends a little on height, as well as on x, so that instead of $x \cos \theta$ we have $(x^2 + h^2)^{1/2} \cos[\theta + \arctan(h/x)]$. The lines of constant longitude of the compressed grid will not only have moved in slightly toward the y-axis, but will also have become wavy. This effect is important in aerial mapping, where the plan-location of property lines or other boundaries or sharp features cannot be measured from an aerial photograph by simple scaling but requires knowledge of the heights of the features. (This is interesting, because it seems to imply that to make a contour map of hilly terrain, you cannot put creeks in their correct place until you have the heights, which is what you want, not what you are given.)

(b) One good example is the total curvature at a point. Others are the radii of curvature in the coordinate directions and the visibility of one surface point from another.

2-23A. (a) In the terminology of mechanics, which is where the term centroid comes from, (i) refers to the centroid, or center of mass, of a nonuniform plane lamina whose area density in kg m^{-2} is $f(x, y)$; (ii) refers to N point masses, the ith of which, situated at (x_i, y_i), has mass m_i; (iii) refers to N unit point masses at (x_i, y_i). Case (i) is appropriate to the image formed on a plane by a lens, case (ii) suits a printed image made of black dots of varying size or, more closely, a matrix, while case (iii) corresponds to a cathode-ray tube image formed of discrete elements that are turned either on or off.

(b) Yes. Discussion topic: Is this conclusion deducible mathematically, or is it a conclusion of experimental physics that if a plane lamina balances on a needle point, then that point is unique?

(c) Define a median line as one that separates equal masses; e.g. the vertical median line $x = x_m$ is where $\int_{-\infty}^{\infty} \int_{-\infty}^{x_m} f(x, y) \, dx \, dy = \frac{1}{2} M$. The horizontal median line is at $y = y_m$. The intersection at (x_m, y_m) could be called the median point, but a simple example such as three point masses placed at random, shows that inclined median lines do not in general pass through (x_m, y_m). However, with a large number of mass points, the sharpness of location of a median point might be acceptable relative to digitizing grid error. This would need to be tested with a class of given images. In circumstances where the centroids of very large numbers of facets needed to be found, the median, which is faster to compute than the centroid, might be advantageous.

2-26A. (a) Picture the circle as centered at the origin and inscribed in a square formed by lines $x = $ const and $y = $ const. After projection, the square becomes a trapezoid whose angles are determined by the selected value of θ; the ellipse must be that one which fits into the trapezoid and is tangent to the four sides.

(b) If the eye is at infinity b/a will have to equal $\tan \frac{1}{2}\theta$.

(c) If b/a is too big the circle will appear to be lying on a plane that is tilted toward the viewer.

(d) For $\theta = 60°$ we need $b/a = \tan 30° = 0.577$, and the elevation angle α is such that $\sin \alpha = 0.577$, so $\alpha = 35°$.

(e) Yes.

(f) All horizontal circles project into ellipses with horizontal major axis regardless of location in or above the (x, y)-plane. This property is independent of closeness of the center of perspective.

2-27A. If the ground is plane, and a good camera is pointed vertically down, no correction is needed. When the camera is pointed away from the vertical, a horizontal grid on the ground is photographed as two sets of convergent lines, each set having a vanishing point. The line on the photograph joining the two vanishing points is the horizon (land-sky boundary). If the camera tilts toward a cardinal point of the compass, the vanishing points will be at infinity. Since there is a one-to-one correspondence between the plane ground and the plane photograph, there is a second perspective projection that removes the convergence. Thus the photograph can be rephotographed, or projected onto an oblique screen, to remove the distortion. For the photography, place the camera so that the film plane contains the two vanishing points; this projects the vanishing points to infinity on the film plane, thereby making the converging pencils parallel again. For projection, place the screen in the plane of the vanishing points of the transparency as mounted in the projector. Of course, the vanishing points that are in the plane of the photograph do not have to be inside the boundary of the print.

2-30A. Bonne's equal-area projection consists of concentric circles, along each of which meridian crossings occur at equal intervals proportional to the cosine of the latitude. The height and width of small cells are thus made equal by construction. All the parallels cross the central meridian at right angles. In the corners the intersections deviate from a right angle by only $1.43°$.

2-31A. **(a)** If R is the radius of the earth and $6.5R$ is the distance of a synchronous satellite from the center of the earth, then $x = RM \cos b \cos l$, $y = RM \sin b$, where the magnification factor M is $\sqrt{6.5^2 - 2 \times 6.5 \times \cos b \cos l}/(6.5 - \cos b \cos l)$, and x and y are in the plane through the center of the earth perpendicular to the earth-satellite line. If the satellite is taken to be effectively at infinity, then $x = R \cos b \sin l$ and $y = R \sin b$.

(b) To make a mosaic it is sufficient to expand each photograph in longitude by a factor $(\cos b \sin l)^{-1}$. The distortion in latitude would not affect piecing together.

2-32A. **(a)** It is not obvious to everyone why the factor sec az occurs in one of the expressions. Think of an east-west wall somewhat to the south. The top of the wall has maximum elevation angle directly to the south, and the elevation of the level top falls off both to east and west. But y must not fall off; a straight line such as the top of a wall projects into a straight line on the film, because a plane intersects a plane on a straight line. Hence the quantity tan alt, which declines to left and right, needs a factor to keep y independent of azimuth.

(b) The alt-az system enables the diagram to go from east (E) to west (W) and up to the zenith; in fact there is nothing to prevent the azimuth from reaching all the way to north on both sides. Azimuth and altitude change when the local vertical changes; for

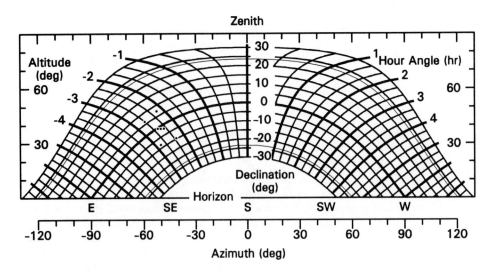

Figure A-8   How objects within the zodiacal belt move across the southern sky, in az-el coordinates, as seen from latitude 37.5° N. The thin solstitial lines at declination ±23.5° bound the zone within which the sun is found.

full precision the influence of gravity on the vertical and perturbations of the vertical as the earth rotates on its axis would need to be taken into account.

(c) At any sky location the diagram more resembles what you would record with a camera pointed to that direction rather than to the fixed south point (S). The character and amount of the distortion can be judged from the appearance of the constellation Orion, shown in the southeast as on a January evening. From this diagram one can read off the approximate duration of daylight, which is 9^h at the winter solstice and 15^h at the summer solstice (dotted paths), and the azimuth of sunrise and sunset. When the sun rises exactly in the east, its path makes an angle of 37.4° with the horizon (equal to my latitude). At a rate of 1° in 4 minutes the sun moves through its own diameter in 2 minutes. Therefore it takes $2/\sin 37.4° = 3$ minutes to rise; but in winter, when the inclination of the path to the horizon is less, it takes longer. There may be an azimuth where the declination curve is inclined at 90° to the horizon. Stars at this declination remain in about the same azimuth for an hour or so after rising and before setting and were used by Pacific Island navigators to steer by. The diagram has other uses; for example, you can measure the height of a tree or an antenna by pacing its shadow and noting the date and time of day. You can use it for marking out the figure-eight path traced in the sky by a passive geosynchronous satellite, a phenomenon that a telecommunications engineer needs to be familiar with. Just as the attitude of Orion can be displayed, so also can the attitude of an earth satellite. The diagram is absolutely basic because it can be adjusted to translate between any pair of spherical polar coordinate systems. It is neither conformal nor area preserving. It is fundamentally a nomogram for conversion, over a finite range and to a certain precision, between numerical values in different coordinate systems.

2-33A. The two poles subtend 90° at the center of the circle of inversion.

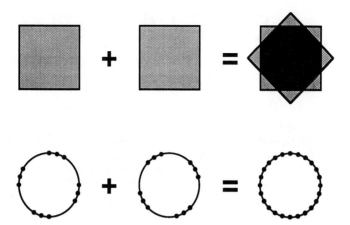

Figure A-9 The sum of two functions may have much higher symmetry than either alone.

2-39A. Starting from

$$u + iv = \frac{\alpha + \beta x + i\beta y}{\gamma + \delta x + i\delta y},$$

split the right-hand side into its real and imaginary parts. Then

$$u + iv = \frac{(\alpha + \beta x + i\beta y)(\gamma + \delta x - i\delta y)}{(\gamma + \delta x)^2 - \delta^2 y^2}$$

$$= \frac{(\alpha + \beta x)(\gamma + \delta x) - \beta\delta y^2 + i\beta y(\gamma + \delta x) - i(\alpha + \beta x)\delta y}{(\gamma + \delta x)^2 - \delta^2 y^2},$$

which reduces to the results given.

2-40A. The image has been translated by an amount (a, b), i.e., $g(x, y) = f(x - a, y - b)$. This is a result from dilatation theory, a part of affine geometry. Dilatation transforms each line of a geometrical figure into a parallel line; consequently, rotation by half a turn is a dilatation. Two half turns in succession also constitutes a dilatation, namely a translation.

2-41A. The composite function has two-fold symmetry.

2-42A. The composite function will have at least four-fold symmetry. But it may have much higher symmetry, as in Fig. A-9.

2-44A. (a) No symmetry: **B E F G J K L P Q R S**. Twofold symmetry only about axis perpendicular to paper: **N Z**. Vertical symmetry axis only: **A H M T U V W Y**. Horizontal symmetry axis only: **C D**. Both horizontal and vertical symmetry axes: **I O X**.
  (b) **MATHEMATICAL**.
  (c) **AUTOMATA**.
  (d) **DIOXIDE**.

2-45A. St. Cyril (c. A.D.826–869) provided the Slavs with their alphabets of 40 or more letters. There are quite a number of signs that could in principle be Roman capitals (Fig. A-10)

ᚱᚪᛏᛒᚻᛁᚾᚦᚩᚹᚱ

Figure A-10 Samples from the millions of 5 × 7 signs that could have been letters. It is noticeable that the principles of economy and data compression led the originators to the efficient signs – at least as regards the original alphabetic medium (possibly wood).

and you can invent letters that might have evolved from scribing on palm leaves as with those Eastern alphabets, such as Kanarese, that avoid straight lines. Other possibilities not illustrated include Russian letters, punctuation, arithmetical and other well-known symbols, patterns that are not simply connected (as is the case with = ! Ξ), including the binary complements of Latin letters, and letters with thick and thin bars (like those printed with magnetic ink on checks). This exercise begins by calling for innovation. Since only 26 letters are called for, a design aspect arises as one chooses a set that is homogeneous. In view of the diffusion of type designing among the denizens of the computing and information sciences worlds, to design an exotic alphabet that is internally harmonious and provided with lowercase forms would be a meritorious and instructive effort.

2-46A. If n is even, the same; if n is odd, then the autocorrelation function has two-fold rotational symmetry.

2-47A. The property quoted is possessed by many figures such as a centrally situated rectangle or equilateral triangle. But the letter **Z** has $n = 2$, and you cannot find two axes of "flip symmetry."

2-48A. Yes. Suppose that one can think of a one-to-one correspondence between points on the square and points on the line segment from 0 to 1. Then $f(x, y)$ over the square would be equivalent to a certain $\psi(z)$, $0 < z < 1$. A standard example of such a correspondence is this. The point $x = .314159...$, $y = .271828...$ on the square defines the point $z = .321741185298...$ on the line, and conversely the point $z = .91919190900...$ corresponds to the point $x = .99999$, $y = .111$ on the square. This idea can be adapted for computer graphics in various ways.

2-49A. The function is $f(r) = \pi^{-1}\mathrm{arcosh}(1/r)\,\mathrm{rect}(r/2)$.

2-50A. Both projections are parabolic:
(a) $\mathcal{P}_0 p(x, y) = (1 - x^2)\,\mathrm{rect}(x)$,
(b) $\mathcal{P}_{\pi/4} p(x, y) = (\sqrt{2} - x')^2 \sqrt{2}$, where x' is measured in the diagonal direction. Further question: Are the projections in all directions parabolic?

2-51A. The cross section of the cone along a line $x =$ const is the hyperbola $z = 1 - (x^2 + y^2)^{1/2}$ (Fig. A-11). To get the area under this hyperbola, which will be equal to the projection at the given value of x, we have to evaluate the integral

$$\int_{-(1-x^2)^{1/2}}^{(1-x^2)^{1/2}} [1 - (x^2 + y^2)^{1/2}]\,dy.$$

The best way to do this is to look in the shortest table of integrals or computer equivalent that you think might list the integral of $(a^2 + x^2)^{1/2}$. The answer is

$$\mathcal{P}_0 c(r) = (1 - x^2)^{1/2} - x^2 \cosh^{-1}(x^{-1}).$$

Another approach is to make the substitution of variables $y = a \sinh u$ (treating x as constant) and do the integration yourself. This will use more of your time than will be

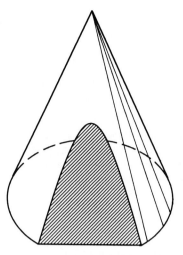

Figure A-11 Hyperbolic cross section of a cone.

used if you go looking for a table of integrals, especially if you make the checks necessary to ensure that your answer is right. A third approach would be to look up the area of a hyperbola in a geometrical context. You will soon find that the area of a parabola is two-thirds of the rectangle and that the area of an ellipse is $\pi/4$ times the rectangle; but because the area of a hyperbola does not lend itself to such simple rules you are likely to conclude that it would have been better to look for a table of integrals.

2-52A. The projection is $(2\pi)^{-1/2}\exp(-x^2/2\sigma^2)$, the correctly normalized normal distribution in one dimension.

2-53A. A cross section of the hemisphere at $x =$ const is a semicircle of radius $(a^2 - x^2)^{1/2}$ whose area $0.5\pi(a^2 - x^2)$ is the desired projection. The area of the parabola, two-thirds of the containing rectangle, is $(2\pi/3)a^2$, which is indeed equal to the volume of the hemisphere of radius a.

2-55A. **(a)** Let the ellipse have parametric equations $x = a\cos t$, $y = b\sin t$, where $a = 1.5$ and $b = 1$. Then the equations for the center of the circle of radius r are $x = a\cos t - r\sin A$, $y = b\sin t + r\cos A$, where $\tan A = -b/a\tan t$ is the the slope of the ellipse.
   (b) The curve is the locus of points lying on normals to the ellipse at distance r from the ellipse; thus from any point on the curve the distance to the nearest point on the ellipse is r. From any point on the ellipse, however, the distance to the curve is in general less than r, because the normal to the ellipse is not normal to the curve.

2-56A. Let $f(x, y) = f_{\text{polar}}(r, \theta)$. Then $g_\theta(R_1) = R_1 \int_0^{2\pi} f_{\text{polar}}(r, \theta)d\theta$. The result is simply an integral of $f(x, y)$ around a circular section of radius R_1. (b) In the expression $g_\theta(R)$ the variable R is a cartesian coordinate that assumes both positive values; to write $g(R, \theta)$ risks implying that R is a radial polar coordinate such as the r in (r, θ). Mixed coordinate systems such as (x, θ) are generally avoided, unless there is a strong reason for living with nonorthogonality and nonsinglevaluedness.

2-60A. In general there is no such point. To prove this, it is sufficient to construct a single example. In Fig. A-12, if there is a point of $f(x, y)$ that projects into the median, regardless of direction, it must lie on line L. By symmetry it must lie on lines M. But line N does not bisect the area. Therefore the supposed point does not exist in these examples. However, if

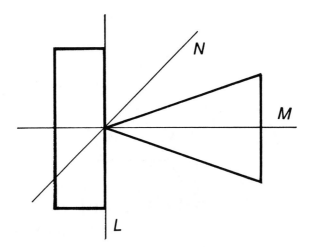

Figure A-12   A function $f(x, y)$ possessing no areal median.

$f(x, y)$ possesses twofold rotational symmetry about a point P, not necessarily the origin, then P is a point that always projects into the median. Of course, the center of symmetry is also the centroid. Can you prove that a point P can project into the median regardless of direction *only* if P is a center of twofold symmetry?

2-62A. There are two differences from the situation introduced in the text. First, the x-rays are divergent, not parallel. Therefore, at a distance s from the x-ray source, the obstructing mass is proportional to $s^2 \rho(x, y, z)$. Second, the gold farthest from the source reduces the transmitted x-ray intensity less than the same volume of gold near the source, because the x-ray intensity is already heavily reduced by passage through many centimeters of gold. Therefore a second factor $\exp[-\kappa \int_0^s \rho(x, y, z)\, ds]$ needs to be included in the integrand. With a chest x-ray, where it is desired to see features equally well regardless of depth inside the body, the emerging intensity is only fractionally less than at entry. You can verify this by looking at the impression made by the ribs, parts of which are near the entry and parts near the exit of the rays. The constant κ is the mass absorption coefficient.

2-64A. Equation (a) is true only if $f(x, y) = 0$. The reason is that a function of x and y generated by the full back-projection operator $\mathcal{B}$ is independent of y' and the back projection can therefore be equal to the original $f(x, y)$ only if $f(x, y)$ also is independent of y'. So suppose that $f(x, y) = \text{const}$. If the constant is zero, the equation is satisfied; otherwise it is not, because the projection of a nonzero constant is infinite. The second "theorem" also fails for the reason that the integral of a constant is infinite, unless the constant is zero. The equations also fail for truncated back projection but can be true for one value only of θ for conservative back projection.

2-66A. (a) From the triad of coastal points $A(x_i, y_i)$, $B(x_{i+1}, y_{i+1})$, $C(x_{i+2}, y_{i+2})$, derive three consecutive boundary points $P(u_{2i-1}, v_{2i-1})$, $Q(u_{2i}, v_{2i})$, $R(u_{2i+1}, v_{2i+1})$, as follows. Put P three miles out on the perpendicular bisector of AB, and place R similarly relative to BC. Put Q three miles from B on the bisector of the perpendicular bisectors. Move on to the next triad B, C, D and place the two further points S and T, move on to C, D, E, and so on.

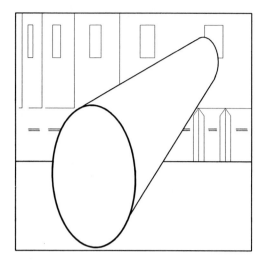

Figure A-13 Computer drawing of a construction site.

2-68A. **(a)** The wanted guide shape is obtained by dilating the cycloid with a circle of radius equal to that of the ball.

(b) The parametric equations of the track, for a rolling ball of radius b, are $x \approx \phi + \sin\phi + b\sin\phi/\sqrt{2 + 2\cos\phi}$, $y \approx 1 - \cos\phi - b(1 + \cos\phi)/(\sqrt{2 + 2\cos\phi})$.

2-69A. There is no rotational symmetry.

2-70A. A television frame contains about $440 \times 525 = 231,000$ pixels and is refreshed 30 times a second. Each pixel, if it can assume 8 different intensity levels, requires 3 bits, so the total bit rate is about 10^7 per second. Consequently, dozens of TV programs could be accommodated.

2-71A. The author of the paper has a library of subprograms such as TRANSLATE, ROTATE, FLIP, STRETCH, and so on, and is accustomed to writing lines of computer code like

```
CALL "TRANSLATE" (k,l,f(,),M,N,fshifted(,))
```

The parameters in such a line remind us of the finite width M and height N of a digital image and force attention to the margins. The subprogram listing would tell us what the user normally does about values that are shifted outside the range of the original matrix $f(i, j)$. This is the explanation. It is, of course, not graceful to mix computer code with algebra.

2-74A. If the center of the circle is put on a grid point the smallest image has a diameter of one centipoint, the next smallest consists of nine dots inside diameter $D = 1 + 2\sqrt{2} = 3.83$ centipoints, the next has $D = 1 + 4\sqrt{2} = 6.66$ centipoints, and so on. The available image diameters are $1, 3.8, 6.7, 9.5, 12.3, \ldots$ centipoints or $3.5, 13.4, 23.3, 33.2, 43.1 \ldots \mu$m.

2-76A. The speakers are all careful to make correct statements but, as the artist says, the long dimension of the cut end should look perpendicular to the pipe. The programmer was not available to explain how the mistake occurred.

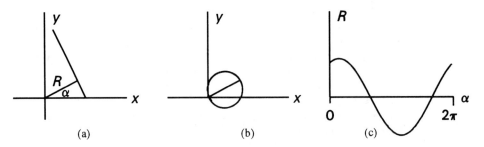

(a)   (b)   (c)

Figure A-14 (a) Representing a uniform line impulse passing at distance R from the origin and normal inclined at angle α to the x-axis. (b) The nonuniform line impulse $\delta(x_1 \cos \alpha + y_1 \sin \alpha - R)$ interpreted in polar coordinates (R, α). (c) A different interpretation taking (R, α) as cartesian coordinates.

CHAPTER 3

3-1A. The charge density is $\sigma(x, y) = Q^2 \delta(x, y-1)$ or $\sigma(x, y) = Q\delta(x)\delta(y-1)$. The units are as follows: $[\sigma] = C\,m^{-2}$, $[Q] = C$, $[x]$ and $[y] = m$, $[{}^2\delta] = m^{-2}$, $[\delta] = m$.

3-5A. The function $z_1(x, y) = y - mx - c$ cuts the (x, y)-plane along the straight line $y = mx + c$ with a slope $\partial z_1 / \partial n$ that depends on m, the coordinate n being measured in the (x, y)-plane in the direction of steepest slope, i.e., perpendicular to $y = mx + c$. On the other hand, the function $z_2(x, y) = (x^2 + y^2)^{1/2} - R$ represents a conical funnel that cuts the (x, y)-plane with a $45°$ slope at all points on the circle of intersection and regardless of the radius R of the circle. When we replace $\delta(\cdot)$ by $\tau^{-1} \text{rect}(\cdot/\tau)$ it is this slope that determines the width of the rectangular wall and therefore the linear density.

3-6A. **(a)** The expression denotes a line impulse confined to a line whose closest approach to the origin is at distance R and which has the foot of the perpendicular from the origin at $(R \cos \alpha, R \sin \alpha)$; see Fig. A-14 (a).
   (b) The strength of the line impulse is unity.
   (c) If R and α are polar coordinates, then $\delta(x \cos \alpha + y \sin \alpha - R)$ is a nonuniform ring impulse confined to the circle that passes through the origin and through the point whose rectangular coordinates are (x, y). There are other possibilities. One might plot R against α in a cartesian framework. Then $\delta(x \cos \alpha + y \sin \alpha - R)$ would be a nonuniform line impulse lying along the sinusoid $R = x \cos \alpha + y \sin \alpha$.

3-7A. Simplify the question to whether ${}^2\delta(x, y)\delta(y)$ is equivalent to ${}^2\delta(x, y)$, i.e., focus on just one of the spikes. Expanding into rectangle functions according to Rule 1, we discuss

$$\tau^{-1} \text{rect}(x/\tau)\tau^{-1} \text{rect}(y/\tau)\tau^{-1} \text{rect}(y/\tau),$$

which describes a rectangle function of base area τ^2 and height τ^{-3}. The volume is τ^{-1}, which approaches infinity as $\tau \to 0$, whereas unit volume would be required if ${}^2\delta(x, y)$ were to be simulated.

3-12A. By rule 1, convert to

$$\int_{-1}^{1} \int_{-1}^{1} \tau^{-1} \text{rect}\left(\frac{x^2 - y^2}{\tau}\right) dx\, dy.$$

The integrand is nonzero on two diagonal bands whose edges are where the argument

$(x^2 - y^2)/\tau$ equals $\pm\frac{1}{2}$. Inspection of a plot for some value of τ, say 0.05, shows that the region of integration is covered by a cross of amplitude $\tau^{-1} = 20$, and area a little bit more than the total length of the crossarms ($4\sqrt{2}$) times the width of the arms ($\sqrt{2}\tau$). Under rule 2 we evaluate the area as $8\tau + \epsilon(\tau)$, and the volume described by the integral as $[8\tau + \epsilon(\tau)] \times \tau^{-1}$. Proceeding to the limit under rule 3 gives a value 8 for the integral. The limit of $\epsilon(\tau)\tau^{-1}$ can be obtained easily by algebra, if you draw the diagram. But if you have to draw the diagram to understand how to set the limits of integration, why not draw two, for two values of τ? It will then be apparent that $\epsilon(\tau) \sim \tau^{-2}$ and that exact integration is unnecessary.

3-19A. In polar coordinates the radial and azimuthal components of the gradient of a scalar are given by $\mathrm{grad}_r\, f = \partial f/\partial r$ and $\mathrm{grad}_\theta\, f = r^{-1}\partial f/\partial\theta$. The total gradient is thus $[(\partial f/\partial r)^2 + r^{-2}(\partial f/\partial\theta)^2]^{1/2}$, which is equal to the corresponding expression $[(\partial f/\partial x)^2 + (\partial f/\partial y)^2]^{1/2}$ in cartesian coordinates Since it is the reciprocal of the gradient that measures the strength of a line impulse, the expression stated in the problem is confirmed.

3-20A. **(a)** The nonzero locus of the image is where the argument is zero, namely where $y^3 - 3x^2y = 0$. One's natural approach to solving this equation for y or x may lead to cancellation of significant factors; the best approach is to factor into $y(y^2 - 3x^2) = 0$. It follows that the locus on which the impulsive entity resides is where $y = 0$ and where $y = -\sqrt{3}x$ and where $y = \sqrt{3}x$. In short,

$$\delta(y^3 - 3x^2y) = \delta(y) + \delta(y + \sqrt{3}x) + \delta(y - \sqrt{3}x).$$

This represents three uniform spokes extending outward in six equispaced directions from the origin.

(b) Interchanging x and y provides six additional interlaced spokes, represented by

$$\delta(y^2 - 3x^2y) + \delta(x^2 - 3y^2x),$$

which can be compacted into $\delta[(y^2 - 3x^2y)(x^2 - 3y^2x)]$ or $\delta(10x^2y^2 - 3x^4 - 3y^4)$. The most compact form is

$$\delta(x^4 - 3\tfrac{1}{3}x^2y^2 + y^4).$$

Spokes can be represented by δ-functions of angle θ, but the spokes will not be uniform.

3-28A. A fan of eleven radial impulses spaced 0.1 radians of strength varying radially as $r/10$.

3-31A. The main difference compared with using the rectangle-function approach of the text is that the integrals are harder to evaluate. There is also a risk, in the case of the sinc function, that $f(x, y)$ behaves off in the distance where your attention is not focused in such a way that the integral does not exist. Any problem associated with irregular behavior of $f(x, y)$ at the origin is the same regardless of approach.

3-32A. The first expression represents a unit-strength line impulse on the square bounding $\mathrm{rect}(x, y)$. The second represents a line impulse on the circle bounding $\mathrm{rect}\, r$ of strength 2.

3-33A. If the real Fourier transform variable s is allowed to be complex (as is customary with the Laplace transform), then the Fourier transform of $\cos 2\pi bx$ can be written $\frac{1}{2}\delta(s + b) + \frac{1}{2}(s - b)$, while the Fourier transform of $\cosh 2\pi bx$, which under the ordinary definition does not exist, assumes the representation $\frac{1}{2}\delta(s + ib) + \frac{1}{2}\delta(s - ib)$. Such ideas are explored by I. M. Gel'fand and G. E. Shilov, *Generalized Functions*, Academic Press, New York, 1964.

3-34A. **(a)** One possibility is $z = \phi(x, y)$; another is $f(x, y, z) = 0$.

(b) $\delta[f(x, y, z)]$.

(c) Generalizing from two dimensions, where strength means line density, in three dimensions the strength of a surface impulse would mean surface density. For example, if $\delta[f(x, y, z)]$ were a distribution of density measured in kg m^{-3}, then the strength at a point (x, y, z) would be the surface density in kg m^{-2} at that point.

(d) The strength σ is given by

$$\sigma = |\operatorname{grad} f| = |(\partial f/\partial x)^2 + (\partial f/\partial y)^2 + (\partial f/\partial z)^2|^{-1/2}.$$

3-36A. **(a)** Instead of $\rho(x, y, z) = Q^3\delta(x, y, z)$ one could write $\rho(x, y, z) = Q\delta(\mathbf{r})$, where $\mathbf{r} = (0, 0, 0)$.

(b) The notation $\delta(0)$ would seem dangerous but could be rescued by defining a null vector $\mathbf{0} = (0, 0, 0)$.

(c)

$$\int\int\int \delta(\mathbf{r}) f(\mathbf{x}) \, dx \, dy \, dz = f(\mathbf{r}).$$

The vector argument notation does not extend to a surface charge such as $\sigma(y, z) = \sigma_0\delta(x)$ or to line charges.

CHAPTER 4

4-1A. **(a)** Let the general corrugation be $A \cos[2\pi q_1(x' - X)]$, where x' is a coordinate measured normal to the ridge lines, making an angle ϕ_1 with the x-axis. Thus $x' = x \cos\phi_1 + y \sin\phi_1$. The amplitude A is immaterial; let it be unity. However, the shift X determines whether the corrugation will be expressible by even, odd, or mixed quilt functions and will be retained. For brevity let $\omega_x = 2\pi q_1 \cos\phi_1$ and $\omega_y = 2\pi q_1 \sin\phi_1$. These are the radian spatial frequency components of the given corrugation. Then

$$\cos[2\pi q_1(x' - X)] = \cos 2\pi q_1 x' \cos 2\pi q_1 X + \sin 2\pi q_1 x' \sin 2\pi q_1 X$$

$$= \cos 2\pi q_1 X \cos[2\pi q_1(\cos\phi_1 x + \sin\phi_1 y)] + \sin 2\pi q_1 X \sin[2\pi q_1(\cos\phi_1 x + \sin\phi_1 y)]$$

$$= C \cos(\omega_x x + \omega_y y) + S \sin(\omega_x x + \omega_y y)$$

$$= C \cos \omega_x x \cos \omega_y y - C \sin \omega_x x \sin \omega_y y + S \sin \omega_x x \cos \omega_y y + S \cos \omega_x x \sin \omega_y y.$$

These are the four quilt patterns that compose the corrugation.

(b) Only two corrugations are needed to form an even quilt pattern $\cos \omega_x x \cos \omega_y y$, but four would be needed for a quilt pattern having a general displacement with respect to the origin.

4-6A. $\frac{1}{2}F(u + 1, v) + \frac{1}{2}F(u - 1, v) - \frac{1}{2}F(u, v + 1) + \frac{1}{2}F(u, v - 1)$.

4-7A. (a)

$$F_1(u, v) = \exp(-i6\pi u)\,\text{sinc}\,u\,\text{sinc}\,v + \exp(i6\pi u)\,\text{sinc}\,u\,\text{sinc}\,v = 2\cos(6\pi u)\,\text{sinc}\,u\,\text{sinc}\,v.$$

$$F_2(u, v) = 2\cos(10\pi v)\,\text{sinc}\,u\,\text{sinc}\,v.$$

$$F_3(u, v) = \exp(-i2\pi\ 0.4u)\,\text{sinc}\,u\,\text{sinc}\,v - \exp(i2\pi\ 0.4u)\,\text{sinc}\,u\,\text{sinc}\,v$$

$$= -2i\sin(2\pi\ 0.4u)\,\text{sinc}\,u\,\text{sinc}\,v.$$

$$F_4(u, v) = \exp(-i4\pi u)\exp(-i4\pi v)\,\text{sinc}\,u\,\text{sinc}\,v + \exp(i4\pi u)\exp(i4\pi v)\,\text{sinc}\,u\,\text{sinc}\,v$$

$$= 2\cos[4\pi(u + v)]\,\text{sinc}\,u\,\text{sinc}\,v.$$

(b) Isometric representation clearly representing plan position coordinates and function values.

(c) Staggered-profile and contour representations for comparison.

4-8A. (a)

$$F_1(u, v) = Hab\exp\{-\pi[(au)^2 + (bv)^2]\},$$

$$F_2(u, v) = \exp(-\pi q^2) + \exp[-\pi(Rq)^2],$$

$$F_3(u, v) = \exp(i2\pi au)\exp(-\pi q^2) + \exp(-i2\pi au)\exp[-\pi(Rq)^2].$$

(b) Representation by two contours only (at one-tenth and one-half peak).

(c) $F_1(0, 0) = H|ab|$, $F_2(0, 0) = 2$, $F_3(0, 0) = 2$. In each case the volume under the function is correctly given, as can be verified at a glance by knowing that the volume under the Gaussian hump $H\exp\{-\pi[(x/a)^2 + (y/b)^2]\}$ is $H|ab|$.

4-10A. (a) $6^2\text{sinc}(3u, 2v)$,

(b) $\exp(-i6\pi u)^2\text{sinc}(u, v)$,

(c) $\exp[i2\pi(-4u - 5v)]^2\text{sinc}(u, v)$,

(d) $0.5\exp[i2\pi(-5u, -3.5v)]\,^2\text{rect}(u, v/2)]$,

(e) $\exp(-i16\pi u)^2\text{sinc}(u, v/3)$,

(f) $3\,^2\text{rect}(u - 8, 3v)$.

4-15A. The function can be expressed as $^2\text{rect}(x, y) - \,^2\text{rect}(2x, 2y)$. The Fourier transform is $^2\text{sinc}(u, v) - \frac{1}{4}\,^2\text{sinc}(x/2, y/2)$.

4-16A. (a) $\text{rect}\,u\,\text{sinc}\,v$. The transform is the same as the function but rotated through a right angle.

4-21A. If the program segment is intended to convert (x, y) to (x', y') as prescribed by $x' = ax + by + c$, $y' = dx + ey + f$, and to rename (x', y') to (x, y), then it is correct. Changing xprime to xtemp would make the code more readable. The programmer seems to have been an expert on the wrong topic.

4-22A. Transforming $f(x, y) = g(x, y) + G(x, y)$ gives $F(u, v) = G(u, v) + g(u, v)$, showing that $f(\ ,\ )$ is the same as $F(\ ,\ )$. (b) In general, when the Fourier transformation is applied twice in succession to $f(x, y)$ the outcome is $f(-x, -y)$; in other words, the new function is the reverse of $f(\ ,\ )$. But if the function is even [$f(x, y) = f(-x, -y)$], whether complex or not, transforming twice returns the original function. Now $g(x, y) + g(-x, -y)$ is always even; therefore $g(x, y) + g(-x, -y) + \,^2\mathcal{F}[g(x, y) + g(-x, -y)]$ is its own transform. See FTA (1978), p. 405, and A.W. Lohmann and D. Mendlovic, "Image Formation of Self-Fourier Objects," *Angewandte Optik*, Annual Report of the Physikalisches Institut der Universität Erlangen-Nürnberg, p. 24, 1991.

4-25A. **(a)** The meaning of $\mathcal{F}^2 \operatorname{sinc} x$ is $\mathcal{F}\mathcal{F} \operatorname{sinc} x$, as a matter of convention in operator notation. The expression $\mathcal{F}^2 \operatorname{sinc} x$ does not specify whether $\mathcal{F}$ means $^1\mathcal{F}$ or $^2\mathcal{F}$. If the first, then $\mathcal{F}^2 \operatorname{sinc} x = \operatorname{sinc} x$. If the second, then $^2\mathcal{F} \operatorname{sinc} x = \operatorname{rect} u\delta(v)$ and $\mathcal{F}^2 \operatorname{sinc} x = {}^2\mathcal{F}[\operatorname{rect} u\delta(v)] = \operatorname{sinc} x$ also. Thus $\mathcal{F}^2 \operatorname{sinc} x = \operatorname{sinc} x$ and $^2\mathcal{F} \operatorname{sinc} x = \operatorname{rect} u\delta(v)$.

(b) Superscripts following the operator are intended to follow the rules of algebra; thus $\mathcal{F}^2\mathcal{F}^{-1} = \mathcal{F}$ and $\mathcal{F}\mathcal{F}^{-1} = \mathcal{I}$, the identity operator that leaves its operand unchanged. So $^2\mathcal{F}^{-1}$ is the inverse two-dimensional Fourier transform operator and $^2\mathcal{F}^{-1} \operatorname{sinc} x = \operatorname{rect} u\delta(v)$.

4-33A. This problem has the earthy flavor of what can happen when you go to work and the problems are not guaranteed to have clean mathematical solutions. Refer to "Textile Industry" in the *Encyclopædia Britannica* for an illustration of weaves. Plain weave reveals a square array of holes under a lens, while twill has diagonal rows of holes, the rows being more than two hole spacings apart. Satin has an interesting oblique pattern which may not show many holes. For plain weave having square holes of diameter $1/n$ times the pitch the Fraunhofer pattern is $[^2\mathrm{III}(u, v)^2 \operatorname{sinc}(nu, nv)]^2$ when scaled so that the bright spots are at unit spacing. Since the holes are not actually square, or any other definite shape, this expression is proposed only inside the first minimum of the sinc function. See C. Shao and E. Kuze, "Detection of Fabric Defects by Analyzing the Spatial Spectra of the Diffraction Patterns," *J. Textile Machinery Society of Japan*, vol. 42, no. 11, pp. 53–62, 1989.

CHAPTER 5

5-1A. The result is $\operatorname{rect} x \ \operatorname{rect} y$.

5-3A. The answer is another two-dimensional Gaussian function, oriented with its axes along the coordinate axes, $H \exp\{-\pi[(x/X)^2 + (y/Y)^2]\}$. The equivalent widths are given by $X^2 = a^2 + c^2$ and $Y^2 = b^2 + d^2$. The volume HXY is the product of the two volumes ab and cd. Therefore the height $H = abcd/XY$.

5-4A.

```
AUTOCORRELATION DONE SERIALLY

FOR j=1 TO N
    FOR i=1 TO N
        k=X(i)-X(j)+rangeX+1
        l=Y(i)-Y(j)+rangeY+1
        A(k,l)=A(k,l)+1
    NEXT i
NEXT j
```

The variables `rangeX` and `rangeY` stand for $X_{\max} - X_{\min}$ and $Y_{\max} - Y_{\min}$ respectively. The program may be understood by drawing a cursive character on the screen and replicating it N times, each time using $-X(i)$, $-Y(i)$ as the next origin. For the autocorrelation one accumulates deposits in the bins corresponding to pixels which successive drawings revisit. The final picture seen on the screen is the clipped autocorrelation function. To make the clipped autocorrelation matrix, change the line `A(k,l)=A(k,l)+1` to read `A(k,l)=1`.

5-6A. Coastline is not normally digitized at regular intervals; the information is compressed

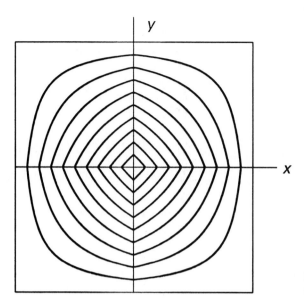

Figure A-15   Contours of lazy pyramid at 0.1 interval.

by the use of less frequent sampling where the coastline is simpler. Therefore, a matrix formulation where 1s represent land and 0s represent water is not appropriate (but not absolutely excluded). A simple scheme is to plot the whole list of points with respect to an origin that is located at point number one; then plot the whole set of dots again, referred to each of the other coastal points in turn. The composite plot should be suitable for most purposes as it stands. If the user says a continuous line has to be drawn around the assemblage of dots, do it by hand. (The computer program to do it is a lot of fun, but make sure the time saved exceeds the time taken to write it.)

5-7A. The lower boundary is defined by $y = x^2 + 2x$ $(-2 < x < -1)$, $y = -1$ $(-1 \leq x \leq 1)$, and $y = x^2 - 2x$ $(1 < x < 2)$, and the upper half is symmetrical.

5-8A. The procedure is to determine the slope of the closed curve as a function of some suitable parameter such as θ (azimuth from some origin) or s (arc distance from an origin). For each point P on the curve find another point Q that has the same slope. Then the vector $OP - OQ$ gives a point on the desired boundary.

5-10A. If both f and g have twofold rotational symmetry, or if both have antisymmetry, then $f \star\!\star g = g \star\!\star f$. If f is neither symmetrical nor antisymmetrical, the two cross-correlations will be the same if g is identical to f (or is proportional to f). If one of the functions is neither symmetrical nor antisymmetrical but the other function is a null function, the cross-correlations will be equal (identically zero).

5-12A. The volume under the squared function is 2π; therefore, $C(0, 0) = 2\pi$. The boundary is like a racetrack of length 8 and width 4, with circular ends and two notches on the sides. The volume under the autocorrelation is $4\pi^2$.

5-20A. All the contours have corners, but only the zero contour has corners on the diagonals (Fig. A-15). The other contours, being composed of four hyperbolic segments, have corners on the axes. The lower contours have very mild corners; they are rounded at the diagonals, where they change direction by rather less than 90°, and could aptly be described as squarish. The higher contours are by no means circular, being almost square, but at 45° to the lower "squares." The inner contours approximate to a square pyramid.

5-22A. **(a)** The desired array, referred to elsewhere as the 25-element binomial array, is

$$\begin{bmatrix} 1 & 4 & 6 & 4 & 1 \\ 4 & 16 & 24 & 16 & 4 \\ 6 & 24 & \mathbf{36} & 24 & 6 \\ 4 & 16 & 24 & 16 & 4 \\ 1 & 4 & 6 & 4 & 1 \end{bmatrix}$$

(b) Horizontal and diagonal cross sections agree closely and thus do not support the suggestion of noncircularity.

(c) $\sigma^2 = \tau^2 = 1$ and $\nu = 0$. To verify the value of σ^2 note that $\sigma^2 = 4\sigma_0^2$, where $\sigma_0^2 = 1/4$ is the variance of $[\begin{smallmatrix} 1 & 1 \\ 1 & 1 \end{smallmatrix}]$.

5-25A. The cross-correlation of $S(\ ,\ )$ on $D(\ ,\ )$ is obtainable with the following subprogram.

```
S( , ) SMOOTHES D( , )

SUB "CCF" (S( , ),Lx,Ly,D( , ),LX,LY,C( , ))
FOR I=1 TO Lx+LX-1
   FOR J=1 TO Ly+LY-1
      C=0
      FOR X=1 TO Lx
         FOR Y=1 to Ly
            X9=Lx-X
            Y9=Ly-Y
            K=I-X9
            L=J-X9
            inside=K<=LX AND L<=LY AND K>0 AND L>0
            IF inside THEN C=C+S(1+X9,1+Y9)* D(K,L)
         NEXT Y
      NEXT X
      C(I,J)=C
   NEXT J
NEXT I
SUBEND
```

5-27A. Once you have the program running, you can soon verify that $X > 2$ is too coarse but that any value of X less than or equal to 2 leads to the same sum (provided you take comparable numbers of terms) and that the sum correctly represents the exact integral. The theoretical explanation lies with the sampling theorem, but it is interesting to see that fearless experiment can lead to surprising economy when it comes to coarseness of tabulation. In two dimensions the computer savings are even greater, amounting to a factor of 4 compared with sampling at the Nyquist rate.

5-28A. A convolution formula exists that puts vertices at the sample points and one additional vertex at each cell center. The formula is $\mathrm{pyr}(x, y) ** [f(x, y)^2 \mathrm{III}(x, y)]$, where $\mathrm{pyr}(x, y)$ is the square pyramid function of unit height with volume = 4/3 and base area = 4, oriented with its base sides parallel to the x- and y-directions. The facets are triangular, but not equilateral, with area $\frac{1}{4}$.

CHAPTER 6

6-1A. **(a)** The system might have an entrance pupil of radius R such that when $a^2 + b^2 > R$ there is no output. Less likely possibilities also exist as a result of the fact that the system was not tested for *all* $f(x, y)$. For example, noncircularly symmetrical inputs could undergo shift-dependent rotation that would not be revealed by the circular input used.

(b) If you interpret the steady output as response to the variable input, you would say the system fails to be space invariant, but you would have trouble explaining why the output is present when the input is zero. If you consider the response to the input to be zero, then the system is space-invariant.

6-2A. **(a)** If the system were linear in the sense of the strict definition, which is mathematical in nature, then the obstruction would not affect the linearity. But if we are dealing with a camera, especially one presenting its output on film, then it would be considered linear only if the illumination was not too low or too high. Halving the illumination with a checkerboard mask might impair the linearity. If you took a picture in faint light without the mask, you might find that, when the mask is installed, doubling the exposure time is not enough to compensate the factor of one-half.

(b) A distant mask in front of the scene changes the input object rather than the system, thus leaving the original judgment about the system essentially unchanged.

6-3A. Express the functions as convolutions with impulse functions and apply the convolution theorem to obtain the following results.

$$^2\mathrm{rect}(x, y) ** [^2\delta(x - A, y) + {}^2\delta(x + A, y)] \supset 2\,{}^2\mathrm{sinc}(u, v)\cos 2\pi Au,$$

$$^2\mathrm{rect}(x, y/W) ** [^2\delta(x - A) + {}^2\delta(x + A)] \supset 2W\,{}^2\mathrm{sinc}(u, Wv)\cos 2\pi Au,$$

$$^2\mathrm{rect}(x, y) ** [^2\delta(x - A, y) + {}^2\delta(x, y) + {}^2\delta(x + A, y)] \supset 4\,{}^2\mathrm{sinc}(u, v)\cos^2 \pi Au.$$

6-5A. Start from the basic property, "If $f \overset{2}{\supset} F$ then $F \overset{2}{\supset} f(-, -)$ and conversely." Thus $f \overset{2}{\supset} F$, $F \overset{2}{\supset} f(-, -)$, $G \overset{2}{\supset} g(-, -)$. The convolution theorem says, "If $f \overset{2}{\supset} F$ and $g \overset{2}{\supset} G$ then $f ** g \overset{2}{\supset} FG$." We wish to show that $fg \overset{2}{\supset} F ** G$, which is the same as saying that $F ** G$ has inverse transform fg, or $F ** G \overset{2}{\supset} fg$. By the convolution theorem $F ** G \overset{2}{\supset} f(-, -)g(-, -)$, since $F \overset{2}{\supset} f(-, -)$ and $G \overset{2}{\supset} g(-, -)$. Therefore by the converse of the basic property, $fg \overset{2}{\supset} F ** G$.

6-6A. Since $e(x) \supset 1 + i2\pi s$ we can say that in two dimensions $e(x) \overset{2}{\supset} (1 + i2\pi u)\delta(v)$. Then $e(x) ** e(y) \overset{2}{\supset} (1 + i2\pi u)\delta(v)(1 + i2\pi v)\delta(u) = \delta(u)\delta(v) = {}^2\delta(u, v)$. Thus the desired convolution is independent of x and y with value unity.

6-7A. **(a)** The half-cycle cosine pulse $h(x) \supset 0.5\,\mathrm{sinc}(s + 0.5) + 0.5\,\mathrm{sinc}(s - 0.5)$; thus $f(x, y) = h(x)h(y) \overset{2}{\supset} 0.25[\mathrm{sinc}(u + 0.5) + \mathrm{sinc}(u - 0.5)][\mathrm{sinc}(v + 0.5) + \mathrm{sinc}(v - 0.5)] = F(u, v)$.

(b) $V_f = \langle x^2 \rangle = (\pi/2)\int_{-0.5}^{0.5} x^2 \cos \pi x\, dx = (\pi^2 - 8)/4\pi^2 = 0.047$.

(c) $\Phi(u, v) = [F(u, v)]^2 = 0.0625[\mathrm{sinc}(u + 0.5) + \mathrm{sinc}(u - 0.5)]^2[\mathrm{sinc}(v + 0.5) + \mathrm{sinc}(v - 0.5)]^2$.

(d) $C(u') = 0.625[\mathrm{sinc}(0.707u' + 0.5) + \mathrm{sinc}(0.707u' - 0.5)]^4$.

(e) A Gaussian function with variance V has a transform with variance $1/4\pi^2 V$, and, since variance is additive under convolution, $\phi(x, y)$ has x-variance $2V_f$ and $\Phi(u, v)$ has u-variance $1/8\pi^2 V_f$. We expect circular symmetry in $\Phi(u, v)$ near the origin so the u'-variance should be about the same as the u-variance, i.e. $\langle u'^2 \rangle \approx 1/8\pi^2 V_f = 0.267$.

(f) Computing $C(u')/C(0)$, we find it falls slightly below $\exp(-u'^2/0.533)$ by as much as 2 percent near $u' = 1$.

6-8A. The scanning aperture may be represented by $^2\Pi(x, y)$. Thus the light transmitted by a transparency $t(x, y)$ is $^2\Pi(x, y) ** t(x, y)$, whose FT, by the convolution theorem, is $^2\text{sinc}(u, v)T(u, v)$.

(a) The test grating has a spatial period of 20 mm; thus its fundamental spatial frequency is 0.05 cycle/mm. Since $\text{sinc}(0.05) = 0.9959$, the response ratio for a sinusoidal test pattern would be $Y_{\max}/Y_{\min} = 1.9959/0.0041 = 487$, but for the bar pattern $Y_{\max}/Y_{\min}$ is infinite.

(b) Since $\text{sinc}(0.75) = 0.3$, we might expect 0.75 bar/mm to be a rough estimate of the resolution (bar width = 0.67 mm). In fact it turns out to be exactly right, because $Y_{\max}$ is two-thirds of the full response (1-mm hole centered on 0.67 mm transparent strip) and $Y_{\min}$ is one-third (hole centered on 0.67 mm opaque bar).

(c) A steady output equal to one-half full response.

(d) The contrast is again 3 db, because $Y_{\max}$ is 0.67 (hole centered on 0.33 mm bar) and $Y_{\min} = 0.33$ (hole centered on 0.33 mm transparent strip).

6-9A. (a) When the period is equal exactly to the width of the hole, the excess light passing through any element where the transparency exceeds 0.5 is compensated by the deficit at a corresponding element one-half period away, where the transparency is equally below 0.5.

(b) The transparency distribution $t(x, y)$ has a transform $T(u, v)$ that has a spike at the origin but is otherwise nonzero only on the square of side 2 centered on the origin of the (u, v)-plane. But $^2\text{sinc}(u, v)$ is zero on this square. Therefore there is no response to the structure. *Further problem:* The actual transparency is really $t(x, y)$ $^2\Pi(x/128, y/128)$, whose transform is slightly spread to each side of the square. But only the infinitesimally thin lines forming the square are suppressed. How then can the response be exactly zero? It clearly is zero because the 1-mm hole is incapable of knowing whether $t(x, y)$ extends to infinity or whether it is limited to a 128-mm square.

6-10A. The project engineer knows there is not much you can do and wisely engages in more productive work. The consultant earns his money by devising an aperture whose transform is free from the null lines that characterize $^2\text{sinc}(u, v)$. It might not make much difference in practice whether a particular spatial frequency is reduced to a very low value or whether it is actually suppressed.

6-13A. The diagonal comma-pair, represented by $k(x, y)$, cannot be expressed as a convolution between $c(x, y)$ and some other function. But it does have twofold rotational symmetry and so, therefore, will $K(u, v)$. Proceed by transforming the third-quadrant comma $c(x + 1, y + 1)$ to get $K_1(u, v) = \exp[i2\pi(u + v)]$. Rotate $c(x + 1, y + 1)$ through $180°$ to get the first-quadrant inverted comma; its transform will be $K_1(u, v)$ rotated through $180°$. To perform the rotation we simply change the sign of both u and v to get $K_2(u, v) = \exp[-i2\pi(u + v)]C(-u, -v)$. Hence

$$K(u, v)$$
$$= K_1(u, v) + K_2(u, v) = \exp[i2\pi(u + v)]C(u, v) + \exp[-i2\pi(u + v)]C(-u, -v).$$

6-15A. Since in one dimension $\operatorname{sgn} x \supset -1/\pi s$, it follows by the separable product theorem that $\operatorname{sgn} u \operatorname{sgn} v \,^2\!\supset (1/\pi u)(1/\pi v)$.

6-16A. (a) The dark bands are straight and perpendicular to the rulings and expand away from the center of rotation, remaining parallel, as $\theta \to 0$. Under uniform rotation $\theta = \omega t$, the velocity of the first moving band goes as t^{-2}, which causes an explosive appearance as $\theta \to 0$.

(b) Superimposing the transparency alters the object by multiplication; the transparent zones multiply the brightness by unity and the black bands multiply by zero. Therefore the spectrum is altered by convolution. The part of the image spectrum on the v-axis shows two spots A representing the low-frequency fundamental of the moiré bands; the remaining spots determine the (triangular) brightness distribution over the bands.

(c) An excellent idea. Furthermore, by counting fringes as they go by you can have a digital angle readout for a wide range of angle. The rulings would need to be made and installed with care, but irregularities could be reduced by sensing the bands through slits running the full length.

6-18A. Since $f_1(kx, y) \supset k^{-1}F_1(u/k, v)$ and $f_2(kx, y) \supset k^{-1}F_2(u/k, v)$, it follows that $F_4(u, v) = k^{-2}F_1(u/k, v)F_2(u/k, v)$. Now $F_3(u, v) = F_1(u, v)F_2(u, v)$, so $F_4(u, v) = k^{-2}F_3(u/k, v)$ and $f_4(x, y) = k^{-1}f_3(kx, y)$. In other words, convolving the compressed functions produces a result that is similarly compressed and also reduced in strength by a factor k^{-1}.

6-19A. The answer is a north-south ridge with the same axis-ratio:

$$f_2(x, y) = \exp[-\pi(x^2 + y^2/W^2)].$$

Since $F_1(u, v) = W \exp[-\pi(W^2 u^2 + v^2)]$ and $F_2(u, v) = W \exp[-\pi(u^2 + W^2 v^2)]$, it follows that $F_1(u, v)F_2(u, v) = W^2 \exp\{-\pi[(W^2 + 1)u^2 + (W^2 + 1)v^2]\}$. Hence $f_1 ** f_2 = [W^2/(W^2 + 1)] \exp\{-\pi[(x^2 + y^2)/(W^2 + 1)]\}$, which has circular symmetry.

6-20A. (a) $F(u, v) = \sum a^2 \exp[-i2\pi(x_i u + y_i v)]\,^2\!\operatorname{sinc}(au, av)$.

(b) The factor $^2\!\operatorname{sinc}(au, av)$ represents a reduction in higher spatial frequencies that makes positioning of the photometer less critical.

(c) The magnitude so derived is the photographic magnitude. The density integrals are estimated by passing a square beam of light through the grey squares on the photographic transparency.

6-21A. (a) The other two point-impulse pairs are $^2\delta(x - c, 0) + {}^2\delta(x + c, 0)$ and $^2\delta(x, y - c) + {}^2\delta(x, y + c)$, where $2c$ is the side of the square. The two further factors of $F(u, v)$ are $2\cos 2\pi cu$ and $2\cos 2\pi cv$.

(b) So far we know that

$$F(u, v) = 8 \cos 2\pi au \, \cos 2\pi bv \, \cos 2\pi cu \, \cos 2\pi cv \, \Phi(u, v),$$

where $\Phi(u, v)$ is an unknown remaining factor. It is conceivable that $\Phi(u, v) = 1$, a possibility that is not excluded when we check to see whether $\int\int f(x, y)\,dx\,dy$ equals $F(0, 0)$ (both are equal to 8). In fact we can see that the known factors suffice by observing that

$$\begin{bmatrix} \bullet \\ \bullet \end{bmatrix} ** \begin{bmatrix} \bullet & \bullet \end{bmatrix} ** \begin{bmatrix} & \bullet \\ \bullet & \end{bmatrix}$$

fully accounts for $f(x, y)$.

6-23A.

$$h(r) ** H(x) = \int_{-\infty}^{\infty} \int_{-\infty}^{x} \exp[-\pi(x'^2 + y'^2)/W^2]\,dx'\,dy'$$

$$= \int_{-\infty}^{x} \exp(-\pi x'^2/W^2)\,dx' \int_{-\infty}^{\infty} \exp(-\pi y'^2/W^2)\,dy'$$

$$= W^2 \int_{-\infty}^{x/W} \exp -\pi x''^2\,dx''$$

$$= W^2 I(x/W).$$

So the cross section of the fuzzy edge that is blurred two-dimensionally is the same Gaussian integral that arises in one dimension. The scale of x is stretched by a factor W as appropriate. The amplification factor W^2 arises from the fact that the $h(r)$ introduced had volume W^2 rather than unit volume. The 10 and 90 percent points are at $x = \pm 0.51W$.

6-24A. **(a)** Center $h(r)$ at a distance s from the corner measured along the diagonal inside the corner. Then the response is

$$\int_{-\infty}^{s/\sqrt{2}} \int_{-\infty}^{s/\sqrt{2}} W^{-2} \exp[-\pi(x^2 + y^2)/W^2]\,dx\,dy$$

$$= W^{-2} \int_{-\infty}^{s/\sqrt{2}} \exp(-\pi x^2/W^2)dx \int_{-\infty}^{s/\sqrt{2}} \exp(-\pi y^2/W^2)\,dy$$

$$= [I(s/\sqrt{2}W)]^2.$$

Hence the 50 percent contour occurs where $I(s/\sqrt{2}W) = 0.707$ or at $s/\sqrt{2}W = 0.22$ or at $(x, y) = (0.22W, 0.22W)$. Likewise the 10 and 90 percent contours occur where I is $\sqrt{0.1}$ and $\sqrt{0.9}$, respectively, or at $(x, y) = (-0.19W, -0.19W)$ and $(0.65W, 0.65W)$. The offsets from P, measured diagonally, are $-0.58W$ and $0.61W$. (See Fig. A-16.)

CHAPTER 7

7-1A. Let $\phi(x', y')$ be equal to $f(x, y)$ at corresponding points in the two coordinate systems. Then $x = x'/\sqrt{2} - y'/\sqrt{2}$, $y = y'/\sqrt{2} - x'/\sqrt{2}$, and $y - x = \sqrt{2}y'$. Thus

$$\int_{-\infty}^{\infty} \int_{-\infty}^{\infty} f(x, y)\delta(y - x)\,dx\,dy = \int_{-\infty}^{\infty} \int_{-\infty}^{\infty} \phi(x', y')\delta(\sqrt{2}y')\,dx'\,dy'$$

$$= \int_{-\infty}^{\infty} \left[\int \phi(x', y')\,dx'\right]\delta(\sqrt{2}y')\,dy'$$

$$= \frac{1}{\sqrt{2}} \int_{-\infty}^{\infty} \phi(x', 0)\,dx' = \frac{1}{\sqrt{2}} \int_{-\infty}^{\infty} f(\frac{x'}{\sqrt{2}}, \frac{y'}{\sqrt{2}})\,dx'$$

$$= \sqrt{2} \int_{-\infty}^{\infty} f(x, x)\,dx,$$

since $dx' = \sqrt{2}\,dx$.

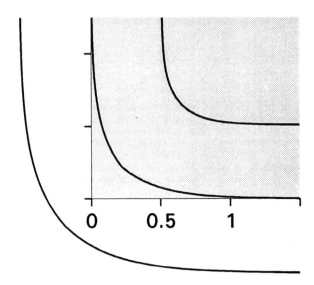

Figure A-16   The 10, 50 and 90 percent contours of a fuzzy corner (shaded) for $W = 1$.

7-6A. (a) The result follows from $^2\delta(x, y) = \delta(x)\delta(y)$. Think of the patch where $\tau^{-1}\,\text{rect}(x/\tau)$ $\times \tau^{-1}\,\text{rect}(y/\tau)$ is nonzero (taking $\tau > 1$). As $\tau \to 0$, each square intersection in the replicated pattern is the base for a prismatic tower of unit volume (width = breadth = τ, height $= \tau^{-2}$), a suitable sequence for associating with a unit impulse situated at the center of the intersection.

   (b) The grille $\text{III}(y - x)$ is made up of line impulses spaced $1/\sqrt{2}$. The oblique intersection of the pattern $\tau^{-1}\,\text{rect}(x/\tau) \times \tau^{-1}\,\text{rect}[(y - x)/\tau]$ has area $\sqrt{2}\tau^2$. The reduced strength is compensated exactly by the enhanced area, so the volume of the lozenge-shaped tower is unity. Hence the proposition is correct.

7-7A. Take the critical interval to be unity. Sampling with $\text{III}(x/2)\text{III}(2y)$ means convolution, in the (u, v)-domain, with $\text{III}(2u)\text{III}(v/2)$. The transform $F(u, v)$ of the band-limited function $f(x, y)$ is nonzero only within the unit square where $|u| < 0.5$ and $|v| < 0.5$. Since

$$f(x, y)\text{III}(x/2)\text{III}(2y) \ ^2\!\supset\ F(u, v) * *[\text{III}(2a)\text{III}(v/2)],$$

we would be able to recover the original function $f(x, y)$ if $F(u, v)$ were recoverable from the RHS. But there is inextricable overlapping of the islands constituting the RHS. Only in special cases, for example if $f(x, y)$ were symmetrical about the y-axis, could $F(u, v)$ be recovered.

7-8A. (a,b) Cross-sections of $F(u, v)$ along the u- and v-axes are zero because the projections of $f(x, y)$ onto the x- and y-axes are zero.

   (c) $F(u, u) = 4(\cos 2\pi u - 1)^2$

   (d) $f(x, y) ** [\alpha x + \beta y + g(x, y)] = f(x, y) ** g(x, y)$.

$$\begin{bmatrix} 1 & -2 & 1 \\ -2 & 4 & -2 \\ 1 & -2 & 1 \end{bmatrix} = [1 \quad -2 \quad 1] ** \begin{bmatrix} 1 \\ -2 \\ 1 \end{bmatrix} = \begin{bmatrix} 1 & -1 \\ -1 & 1 \end{bmatrix} ** \begin{bmatrix} 1 & -1 \\ -1 & 1 \end{bmatrix}.$$

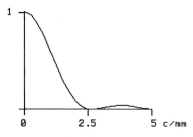

Figure A-17   A bar code $f(x, y)$ and its power spectrum $\mathbf{F}(u, 0)$.

(e) Four two-element factors can also be found.

7-12A. One of the common bar code patterns is 40mm long and consists of black lines 0.4, 0.8 and 1.2mm wide separated by white lines of the same widths. The power spectrum shown is calculated for sharp-edged bars and therefore would not cut off; but since the thinnest bar, of the form $\text{rect}(x/0.4)$ has power spectrum $0.6 \, \text{sinc}^2 \, 0.4u$ with a first null at 2.5 cycles per mm and little content beyond the second null at 5 cycles per mm, adopt 5 as the cutoff frequency (Fig. A-17). The corresponding sampling interval is half a period, or 0.1 mm. The requirements on a physical device that could reliably make four or five samples within the thickness of a thin bar as printed on a carton are much more stringent than for existing bar code readers and would not make commercial sense. The false antithesis made by the TA fails to recognize that the image is already a discrete binary code with the built-in resistance to degradation of binary coding. Conversely, the digital ASCII code belongs just as much to the analogue world as the bar code; sensors, encoders, and electromagnetic transmission lines are where digital signals come from. The bar code spectrum uncannily resembles $\text{sinc}^2 \, 0.4u$. Is this also true of your example?

7-13A. **(a)** The modulation frequency is really present and its amplitude is straightforward to calculate. There is also a hierarchy of higher frequencies.

(b) It follows that filtering can remove the fundamental modulation frequency; the eye may be happy with this in some applications where the bar spacing is very close. However, the width of the bars will remain modulated; grey will appear between the bars, and rectification effects will appear when the filtered pattern is printed because it will go negative in places.

7-18A. **(a)** Consider the sample at (i, j) and the radial line such that $\tan \theta = j/i$. Equispaced values have been measured and can be used for sinc-function interpolation, using $2N + 1$ coefficients, to obtain the wanted value. Values at other locations (mi, mj) can be obtained at the same time.

(b) With values equispaced on a circle, circular interpolation by sinc functions of arc will give a rigorous result for each wanted $\theta = \arctan(j/i)$. Allowance has to be made for circular periodicity by using $N^{-1} \sin(N\pi\theta)/ \sin \pi\theta$ to replace the infinite sum of sinc functions spaced 2π. Near the origin, samples of a cosinusoidal corrugation taken closely spaced on a ring do not contain high frequencies that would invalidate the sampling theorem, but the spectrum does spread to include lower frequencies. The material on spin averaging will show the spectrum to be that of a zero-order Bessel function.

7-22A. **(a)** The coefficients are $h(\pm 0.5) = 0.5981$, $h(\pm 1.5) = -0.1196$, $h(\pm 2.5) = 0.0239$, $h(\pm 3.5) = -0.0024$.

(b) The transfer function is unity at the origin and goes to zero at spatial frequency 0.5, and is fairly flat, drooping by about 1 dB at $s = 0.3$ and by 3 dB at $s = 0.4$. The values are $H(0.1) = 1.000$, $H(0.2) = 0.995$, $H(0.3) = 0.926$, $H(0.4) = 0.615$.

7-23A. The proposal resembles convolving with the Gaussian and resampling at the grid points. The procedure does not interpolate, as evidenced by the fact that, when a sample point coincides with a grid point, one does not recover the measured value. Rather, the scheme returns samples of a smoothed sampling density. The moral is that it is hard for beginners to compete in a mature field polished by Newtons and Lagranges.

7-24A. **(a)** A fast method assigns to each cartesian point the nearest measured value in the polar coordinate system (in the first quadrant).

POLAR TO CARTESIAN (FAST)

```
P2=2*PI
FOR Y=1 TO W
   FOR X=1 TO W
      R=INT(0.5+SQR(X^2+Y^2))
      A=INT(0.5+ATN2(Y,X)*N/P2)
      C(X,Y)=P(R,A*P2/N)
   NEXT X
NEXT Y
```

Use this program as a standard of comparison. An improvement is to take four measured values at the vertices of the trapezoid containing the cartesian point and interpolate using the four-point ("twisted-plane) formula. As the trapezoid is not a rectangle, a small correction to the fraction q would be in order.

POLAR TO CARTESIAN (IMPROVED)

```
da=2*PI/N
FOR Y=1 TO W
   FOR X=1 TO W
      r=SQR(X^2+Y^2)
      a=ATN2(Y,X)
      R=INT(r) @ p=r-R
      A=int(a/da) @ q=a/da-A
      C=P(R,A*da)
      D=P(R+1,A*da)
      E=P(R+1,(A+1)*da)
      F=P(R,(A+1)*da)
      C(x,y)=(1-p)*(1-q)*C+p*(1-q)*D +q*(1-p)*E +p*q*F
   NEXT X
NEXT Y
```

(b) Accuracy can be tested on some extreme case such as the sinusoidal corrugation $100 \sin \pi x$, which is independent of y, and has the least period (2) to be expected in data sampled at unit interval. However, the algorithm does not need to do very well by this test to be adequate for some general purpose. A more appropriate test would be on the class of images to be operated on, which may very well be rich in low spatial frequencies where accuracy will be better than for the extreme sinusoid. This sort of test also accommodates to the prevailing error level, or signal-to-noise ratio. If a feature of an image can just be discerned under the ruling noise level, and can still be discerned after an operation on the data, then it does not matter much if the procedure exhibits, let us say, 10 percent errors under some artificial test. It would certainly be unreasonable to incur a computing time penalty for an illusory gain. Valid perfectionism in this context is to seek the coarsest and fastest algorithm that achieves a *goal stated in advance*.

7-25A. Since the beat frequency is not present in either of the gratings taken separately, interleaving them will simply add the separate spectra without generating new frequencies. Even squaring the composite pattern will not generate the beat frequency. By accident, bars may occasionally overlap (an example of this occurs in the center of the figure), introducing a trace of beat frequency. If a short length of the pattern is analyzed, a trace may also be found. However, one can arrange that even these traces are reduced to zero. We then have the situation of a clearly perceived periodicity that is not seen by Fourier analysis.

CHAPTER 8

8-1A. The second expression comes from noting that the convolution operator has a factor $\frac{1}{3}[1\ 1\ 1]$ horizontally and a corresponding factor vertically. The first matrix has transform $\frac{1}{3}(1 + 2\cos 2\pi u)$ and the second $\frac{1}{3}(1 + 2\cos 2\pi v)$. By the separable product theorem the overall transfer function is the product of these separate transforms. The first expression can be converted to the second using the identity $\sin 3\theta = 3\cos^2 \theta - \sin^3 \theta$.

8-2A. (a) $\sigma^2 = 0.8$,
(b) $\sigma^2 = 1.6$.

8-4A. The printer does not have a transfer function. However, we are often asked to answer a question which, strictly speaking, does not have an answer, and we are judged by our response. The response of the printer to an input impulse depends on where the impulse is with respect to the square lattice on which the dots are positioned – this is a failure of space invariance. Also, if the strength of the impulse is doubled, the strength of the response is not doubled without change of form; instead the dot increases in area (possibly, but not probably, doubling). This is a failure of linearity. Transfer functions pertain to the realm of linearity and space invariance. Let us guess what sort of information the questioner was after. First of all, assume that the response to an impulse is a single dot, of area proportional to the strength of the impulse, situated on the lattice point nearest to the impulse. An input sinusoid of amplitude of 0.1 on a uniform background of level 0.1 would be reproduced as a lattice of dots ranging in diameter from almost zero to 0.2 cell areas (dot diameter ~ 0.5). To interpret this as a sinusoid we need to blur, let us say with a sinc function. The result will be sinusoidal with the correct amplitude. The main contribution to transfer loss may then be that due to the phase jitter of the samples. A curve showing the fall-off of amplitude of response in the close vicinity of the cut-off frequency 150 cycles per inch could be established empirically. Before doing this work, we ought to enquire whether this is really the intent of the question and to warn that frequencies just above that would not be ignored but aliased to lower frequencies.

8-5A. To say that the phase of $H(u, v)$ is zero is to say that $\Im[H(u, v)]$ is zero, since $\text{pha}[H(u, v)] = \arctan\{[H(u, v)]/\Re[H(u, v)]\}$. To put it briefly, a zero-phase filter is one whose transfer function is real, with no imaginary part. Consequently, $h(x, y)$ is real and even.

8-6A. Represent the convolving array

$$\frac{1}{9}\begin{bmatrix} 1 & 1 & 1 \\ 1 & 1 & 1 \\ 1 & 1 & 1 \end{bmatrix}$$

by $\frac{1}{9}\Pi(x/3)\Pi(y/3)$. The FT is sinc $3u$ sinc $3v$. The first nulls are at $u, v = \pm\frac{1}{3}$, so $\frac{1}{3}$ cycle/map spacing unit is suppressed horizontally and vertically. At 45°, the frequency $\sqrt{2}/3 = 0.47$ cy/unit is suppressed.

8-7A. Ordinary aliasing effects are seen. Any structure with period less than 2 is converted to a period greater than 2. Even if the random sequence (a) seems to be reasonably represented, close examination confirms that patches of fine structure are falsified. Trying the set

$$\{0 \; 1 \; -1 \; 0 \; 1 \; -1 \; \ldots\},$$

which has period 3, shows that the amplitude is reduced as compared with larger periods.

8-8A. Note that $S(u) = (3 \, \text{rect} \, 3u)^{*3}$. The threefold self-convolution of rect u, namely $(\text{rect} \, u)^{*3}$, removes the discontinuities of rect u and also the discontinuities of slope, but has support of breadth 3; so we have changed u to $3u$ to make the support of $S(u)$ equal to unity. The FT of $S(u)$ is $\text{sinc}^3(x/3)$ and of $T(u, v)$ is $\text{sinc}^3(x/3)\,\text{sinc}^3(y/3)$, which falls to zero on the square $x = \pm 3$, $y = \pm 3$ and is very small outside.
(a) The contours of $T(u, v)$ are reasonably circular, except at low levels.
(b) One choice of 9 values to simulate $a(x, y)$ is to put 1 in the center, the four values $\text{sinc}^3(1/3) = 0.57$ on the axes, and the four values $\text{sinc}^3(2/3) = 0.07$ in the corners. An approximate representation in small integers is

$$\frac{1}{50}\begin{bmatrix} 1 & 8 & 1 \\ 8 & 14 & 8 \\ 1 & 8 & 1 \end{bmatrix}.$$

The transfer function $A(u, v)$ associated with this convolving array is

$$A(u, v) = \frac{1}{50}\{14 + 16\cos(2\pi u/3) + 16\cos(2\pi v/3) + 2\cos(2\pi(u + v)/3]$$
$$+ 2\cos(2\pi(u - v)/3]\}.$$

To test the suitability of the compact, small-integer array that we have designed, compare $A(u, 0)$ with $T(u, 0)$ (the cross sections of the two transfer functions in the cardinal directions), and compare $A(u, u)$ with $T(u, u)$ (the diagonal directions).

8-9A. A function which has the property that $\phi(x) \sim x$ for small x, and is thus linear at small amplitudes, is

$$\phi(x) = \begin{cases} \ln(1 + x), & x \geq 0 \\ -\ln(1 - x), & x < 0 \end{cases}$$

8-10A. (a) The horizontal and vertical variances of the four-point pattern are $\sigma_x^2 = \sigma_y^2 = \frac{1}{2}$. The variance σ^2 with respect to an axis perpendicular to the paper is $\sigma^2 = \sigma_x^2 + \sigma_y^2 = 1$.

With respect to coordinates (x', y') making an angle with (x, y) the result is the same, so the smoothing is effectively isotropic.

(b) In this case $\sigma^2 = 2$, so the smoothing is stronger.

8-14A. The given matrix is binomial north-south and binomial-difference east-west. The processed image would reflect the east-west difference of a well-smoothed version involving $6 \times 7 = 42$ values or, roughly speaking, the general easterly upward slope. The corresponding operator rotated $90°$ allows the northerly slope of the smoothed image to be obtained. The maps of slope magnitude could be used (cautiously) to locate "edges." The matrix can be expressed as the convolution of $[1 \ -1]$ with the two-dimensional smoothing array

$$\begin{bmatrix} 0 & 1 & 2 & 2 & 1 & 0 \\ 1 & 6 & 12 & 12 & 6 & 1 \\ 3 & 15 & 30 & 30 & 15 & 3 \\ 4 & 20 & 40 & 40 & 20 & 4 \\ 3 & 15 & 30 & 30 & 15 & 3 \\ 1 & 6 & 12 & 12 & 6 & 1 \\ 0 & 1 & 2 & 2 & 1 & 0 \end{bmatrix}$$

This 6×7 array was generated by convolving the six-element row matrix of binomial coefficients with the seven-element column, dividing by 5, and rounding to the nearest integer:

$$\approx [1 \quad 5 \quad 10 \quad 10 \quad 5 \quad 1] ** \begin{bmatrix} 1 \\ 6 \\ 15 \\ 20 \\ 15 \\ 6 \\ 1 \end{bmatrix} \div 5$$

8-15A. By summing diagonally we see that the operation is like convolving one-dimensionally in the $45°$ direction with $[-1 \ -2 \ 0 \ 2 \ 1]$, which does not look much like the $[1 \ -1]$ that elicits the ordinary first difference. The result produced at $(1, 1)$ by the two-dimensional operation, namely -12, would need to be multiplied by $-\sqrt{2}/6$ for agreement with the algebraic value $2\sqrt{2}$. There will still be a discrepancy in other locations due to the noticeable smoothing introduced in the northwest direction. For example, at $(2,2)$ the northeast gradient doubles to become $4\sqrt{2}$, while the result from the "mask" jumps from -12 to -32.

8-17A. Positive values of $\partial^2 f(x, y)/\partial x \, \partial y$ result at the origin when the function is up in the first and third quadrants and down in the second and fourth, as is the case with xy. The function $f(x, y) = x^2 - y^2$, is level along the diagonals and therefore has $\partial^2 f/\partial x \, \partial y = 0$. The "strength" of the saddle could be found by differentiating with respect to rotated axes (x', y').

8-18A. (a) The implicit smoothing matrices are

$$\begin{bmatrix} 1 & 1 \\ 1 & 1 \\ 1 & 1 \end{bmatrix}, \quad \begin{bmatrix} 1 & 1 \\ 2 & 2 \\ 1 & 1 \end{bmatrix}, \quad \begin{bmatrix} 1 & 1 \\ \sqrt{2} & \sqrt{2} \\ 1 & 1 \end{bmatrix}.$$

(b) There is a computing advantage with matrix operators that have columns of zeros; a column of ones is good too. A matrix with an odd number of elements along each side also offers the convenience that the output falls on a grid point.

(c) The matrices as given are intended for convolution, which requires rotation through half a turn. If you were doing the convolution manually, you could copy the rotated matrix on a transparency, lay it over the matrix to be operated on, and sum the products of corresponding elements. It is this transparency, mask, or template that is sometimes published; it is convenient for visualization. Care is needed when factoring or other convolution manipulations are to be carried out.

(d) All three smoothing operators can be further factored; the factor [1 1], which they all possess, represents a two-element running sum; the smoothing in the y-direction is in all cases stronger, but in varying degrees. This anisotropic feature is not immediately apparent, seeing that the composite matrices are square. The corresponding operators for the y-gradient involve stronger smoothing in the x-direction.

8-19A. The smoothing that you incorporate changes the direction of the gradient, since we are talking about somewhat different maps (yours is smoothed). Our gradients will be roughly in the same direction despite the different way we have disposed the minus signs, because I am convolving and you are cross-correlating. And yet, the direction of slope can range from 0 to 2π, while arctangent is either ambiguous or restricted to $-\frac{1}{2}\pi$ to $\frac{1}{2}\pi$, so the formula given for θ is bad. The correct way to compute θ is as `ATN2(fy(i,j),fx(i,j))`, where the two-variable `ATN2( , )` function ranges from 0 to 2π. Since you are using cross-correlation, you will need to remember to put minus signs before both `fx( , )` and `fy( , )`.

8-20A. **(a)** The vertical separation of the fringes is 60 pixels and the dark fringes have a width of 50 pixels. The amplification asked for is 10.

(b) The vertical profile of a dark fringe is a stepped pyramid represented by a 50-element column matrix

$$[11111111112222222222233333333333222222222221111111111].$$

A suitable matrix to convolve with, in the presence of noise, would have 128 such columns.

8-22A. **(a)** Let $\theta = \omega t$. Then $d(\tan\theta)/dt = \omega d(\tan\theta)/d\theta = \omega \sec^2\theta$. Since $\sec 0 = 1$, the desired ratio is $\sec^2\theta_m$.

(b) Cause the printer to make a uniform grille of north-south lines spaced two or three mm, copy to a transparency, and superpose with a small inclination after shifting plus or minus half a page width.

8-25A. If your software was designed by a textbook writer, it will exchange endpoints after you have specified a reversal of direction and will erase the original lines. But when a curve is being traced as a sequence of straight segments, it is time consuming to continually switch ends just so that all segments are painted left to right; consequently, you may find that $y = x^2$ plotted left to right in increments of a few pixels in x fails to be fully erased when plotted right to left.

8-26A. The low frequencies associated with the step lengths are reminiscent of aliasing, which produces low frequencies (by conversion from preexisting high-frequency components) in the presence of sampling. Aliasing is also linear, preserving amplitude. Apart from the outcome (appearance of low frequencies) the staircase mechanism has little in common with aliasing in its established usage. Sensitivity of the step length to orientation of the line reminds one of the moiré effect, another mechanism that produces low frequencies. Parallel straight lines of gentle slope superposed on a grille of horizontal parallel lines will reproduce the staircase frequency, with the same angle dependence. Discretizing such a drawing onto a fine grid of pixels introduces some irregularity without affecting the staircase frequency. The staircase effect is thus a limiting case of moiré effect.

8-28A. The 34 vertices are, starting from (0,0) in the lower left corner: 0,0 8,0 8,1 0,1 0,2 8,2 8,3 0,3 0,4 8,4 8,5 0,5 0,6 8,6 8,7 0,7 0,8 8,8 8,0 7,0 7,8 6,8 6,0 5,0 5,8 4,8 4,0 3,0 3,8 2,8 2,0 1,0 1,8 0,8 and return to (0,0). With this scheme the lower left corner will be black as required.

8-32A. For this definition of the direct sum, or Minkowski sum, see L. A. Zadeh and C. A. Desoer, *Linear System Theory: The State Space Approach*, McGraw-Hill, New York, 1963.

8-33A. **(a)** E(i,j)=B(i,j) OR B(i−1,j+1) AND NOT B(i,j).
(b) The editor's suggestion does give a lift to many simple drawings.
(c) The suggested formula works on simple perforations.

8-34A.

$$D = (A \oplus B_1) \oplus B_2 = A \oplus B_3$$

where $B_3 = B_1 \oplus B_2$.

$$E = (A \ominus B_1) \ominus B_2 = A \ominus B_3$$

where $B_3 = B_1 \oplus B_2$.

8-36A. Some of the features found would be crosses. Define an upright tee junction by $B =$ [0 0 0 / 1 **1** 1 / 0 1 0 / 0 1 0]. Erode with this structure function to get E. Cross-correlate with [1 1 1/1 **1** 1/1 1 1/1 1 1] to get C. The locations where $E = 1$ and $C = 5$ (the sum of the elements of B) is where the tee junctions are. You may redefine B to have a longer crossbar or stem if you wish.

CHAPTER 9

9-1A. The first expression approximates the spin average integral for $3J_0(r)$, which is known to be zero for $r = 2.4048255577$, by averaging only three values of the integrand, at $15°, 45°$, and $75°$. The result is -3.4×10^{-8}, a creditable result for such a coarse spacing as $30°$. The second expression uses the same $30°$ spacing but the angles are $0°, 30°, 60°$, and $90°$. One-half of the first and last integrands are used. The result is the same.

If we use two values only, say at $22.5°$ and $77.5°$, which may seem incredibly coarse, we get -0.0002.

To explain this very favorable numerical situation, note that the period of cos(r cosα) is $180°$ in α. An interval of $30°$ is therefore adequate to sample the third cosinusoidal harmonic. For larger values of r additional maxima appear in cos($r$cosα), higher harmonics would need to be sampled, and therefore more terms would be needed.

9-3A. The student has done very well in taking a creative step of imagination. It would remain for him to verify his conjecture. His friend did very well to be skeptical, but he also left the business unfinished. It remains for us to support or reject the proposed theorem. It is often easier to disprove than to prove, for the simple reason that one counterexample suffices to disprove whereas all cases must be verified by a proof. Therefore we turn immediately to a table of Hankel transforms and see at once that the easy pairs to check are in conformity with the theorem.

$$
\begin{array}{ccccc}
f(r): & e^{-\pi r^2} & {}^2\delta(x, y) & e^{-r} & \mathrm{rect}(r) \\
F(q): & e^{-\pi q^2} & 1 & 2\pi(r\pi^2 q^2 + 1)^{-3/2} & \mathrm{jinc}\, q
\end{array}
$$

That means we have to attempt a proof.

Let $F(q)$ be the Hankel transform of $f(r)$ and let $F(u, v)$ be the two-dimensional Fourier transform. Since $F(q)$ is a cross section of $F(u, v)$ taken through the origin, it follows that, in particular

$$
\begin{aligned}
F(u) &= F(u, 0) \\
&= \int_{-\infty}^{\infty}\int_{-\infty}^{\infty} f(r)e^{-i2\pi ux}\, dx\, dy \\
&= \int_{-\infty}^{\infty}\left\{\int_{-\infty}^{\infty} f(r)\, dy\right\} e^{-i2\pi ux}\, dx
\end{aligned}
$$

Taking one-dimensional transforms of both sides,

$$
\int_{-\infty}^{\infty} F(u)e^{i2\pi ux}\, du = \int_{-\infty}^{\infty} f(r)\, dy.
$$

Putting $x = 0$ we find $\int_{-\infty}^{\infty} F(u)\, du = \int_{-\infty}^{\infty} f(y)\, dy$.

9-7A. **(a)** Let the ring impulse be $\delta(r - a)$. Then $2\pi a J_0(2\pi aq)$ is the Fourier transform and the asymptotic form for large q is $(4a/q)^{1/2}\cos(2\pi aq - \pi/4)$. The narrow annulus of width ϵ may be represented as

$$
\delta(r - a) ** [(4\epsilon^2 - r^2)^{1/2}\,\mathrm{rect}(r/\epsilon)],
$$

the Fourier transform of which is

$$
2\pi a J_0(2\pi aq)\,\mathrm{sinc}\,\epsilon q.
$$

Since the factor $\mathrm{sinc}\,\epsilon q$ is everywhere less than one, except at $q = 0$, all fringes will be reduced in amplitude relative to the central value.

(b) The asymptotic expression will now contain a further factor q^{-1}, and so the fringe amplitude for the narrow annulus will decay as $q^{-3/2}$, regardless of the width ϵ. To understand the apparent paradox we need only note that when a very small value of ϵ is chosen, we have to go to a very large value of $q \gg \epsilon^{-1}$ before the sharper decay as $q^{-3/2}$ sets in. Figure A-18 shows that the upper envelope tends to follow slope $-\frac{1}{2}$ out to a certain spatial frequency, but turns down to follow the $-\frac{3}{2}$ envelope eventually. The dips in the envelope are associated with the slit width.

9-8A. By definition the Hankel transform $F(q)$ is given by

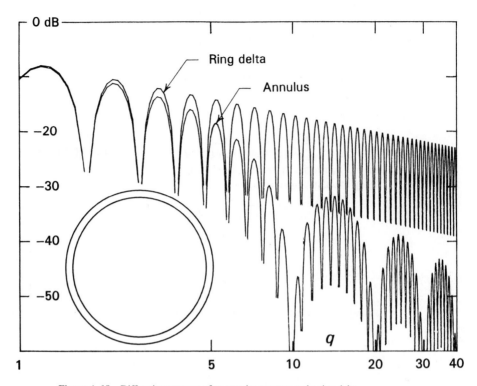

Figure A-18 Diffraction patterns of an annular aperture and a ring delta.

$$\mathbf{F}(q) = \int_0^\infty \delta'(r-a)J_0(2\pi qr)r\,dr$$

$$= 2\pi \int_0^\infty \delta'(r-a)A(r)\,dr$$

$$= -2\pi A'(a)$$

$$= -2\pi \frac{d}{dr}[rJ_0(2\pi qr)]_{r=a}$$

$$= -2\pi[J_0(2\pi aq) - 2\pi aqJ_1(2\pi aq)]$$

since $(xJ_0)' = x'J_0 + xJ_0' = J_0 - xJ_1$.

9-10A. Since $(d/dx)\,\text{jinc}\,x = (\pi/2x)J_0(\pi x) - x^{-2}J_1(\pi x)$,

$$(d/dx)(x^2\,\text{jinc}\,x) = x^2\,\text{jinc}\,x + 2x\,\text{jinc}\,x$$

$$= \tfrac{1}{2}\pi x J_0(\pi x) - J_1(\pi x) + J_1(\pi x)$$

$$= \tfrac{1}{2}\pi x J_0(\pi x).$$

In another way of stating the result,

$$\frac{d}{dx}xJ_1(\pi x) = \pi xJ_0(\pi x).$$

9-13A. Express the pattern as

$$\text{rect}(r/2\sqrt{9}) - \text{rect}(r/2\sqrt{8}) + \text{rect}(r/2\sqrt{7}) - \text{rect}(r/2\sqrt{6}) + \cdots - \text{rect}(r/2\sqrt{2}) + \text{rect}(r/2)$$

$$= \sum_{j=1}^{4}\left[\text{rect}(r/2\sqrt{2j+1}) - \text{rect}(r/2\sqrt{2j})\right] + \text{rect}(r/2).$$

Then the power spectrum is

$$\left[\sum_{j=1}^{4}(2j+1)\,\text{jinc}(2\sqrt{2j+1}q) - 8j\,\text{jinc}(2\sqrt{2j}q)\right]^2 + 4\,\text{jinc}\,4q,$$

which is readily computable.

9-18A. The convolution integral

$$\int_{-\infty}^{\infty} K(x^2 - \rho)\hat{f}(\rho)d\rho$$

$$= \int_{x^2}^{\infty} \frac{\hat{f}(\rho)d\rho}{\sqrt{\rho - x^2}} = \int_{x}^{\infty} \frac{f(r)2r\,dr}{\sqrt{r^2 - x^2}} = f_A(x).$$

9-21A. The value of $f_A(0)$ should be $\int_{-1}^{1} f(r)\,dr = 1$, while $f_A(1)$ should be zero. The results are

x	0	0.2	0.4	0.6	0.8	1
$f_A(x)$	1	1.07	1.17	1.19	1.04	0

Remember that without a plot, or further computed values, you do not know whether the results given reasonably represent the run of the transform, in this example especially at the steep end.

9-24A. Note that $\sin r$ is a circularly symmetric function with a downward pointing conical vertex at the origin. From the definition,

$$f_S(r) = \int_0^{\pi/2} |r\cos\alpha| J_0(r\cos\alpha)\,d\alpha$$

$$= r\int_0^{\pi/2} \cos\alpha\, J_0(r\cos\alpha)\,d\alpha.$$

Change variables by putting $u = \cos\alpha$; then

$$f_S(r) = r\int_0^1 \frac{u J_0(ru)}{\sqrt{1 - u^2}}du$$

$$= \sin r.$$

The integral is no. 6.5542 in GR (1965) or may be recognized in terms of the Hankel transform of $(1 - u^2)^{-1/2}\,\text{rect}(u/2)$, which is a sinc function of r. Then multiplication of the integral by r will give the result.

9-25A. (a) Since integration is carried out from 0 to $\frac{1}{2}\pi$, the presence of the term $\delta(x + a)$ in $f_1(x)$ has no effect.

(b) Averaging over the full circle would leave the spin average of $f_1(x)$ unchanged, but the spin average of $f_2(x)$ would be halved.

(c) The spin average of $\delta(x)$ would not be changed, because $\delta(x)$ is even. However, this raises a question about $(2/\pi) \int_0^{\pi/2} \delta(r \cos \alpha) \, d\alpha$, where it appears that the line $\alpha = \frac{1}{2}\pi$ on the (x, y)-plane appears to be split like a hair when $\delta(x)$ is displayed on the (x, y)-plane. This matter is resolved in the usual way by replacing $\delta(x)$ by $\tau^{-1} \text{rect}(x/\tau)$.

9-26A. Expand both $f_s(r)$ and $f(x)$ in Taylor series and equate coefficients. Note that $Sx^n = k_n r^n$ where the coefficients k_n are known: $k_0 = 1$, $k_1 = 2\pi$, $k_2 = 1/2$, etc. (see Table 9-2).

$$f_s(r) = f_s(0) + r f_s'(0) + (r^2/2) f_s''(0) + r^3/6) f_s'''(0) \cdots$$
$$= f(0) + k_1 r f'(0) + k_2 r^2 f''(0) + k_3 r^3 f'''(0) + \cdots .$$

Thus $f_s^{(n)}(0) = k_n f^{(n)}(0)$.

9-27A. The algebra depends on the trigonometric relations

$$\cos 2\theta = 2\cos^2\theta - 1$$
$$\cos 3\theta = 4\cos^3\theta - 3\cos\theta$$
$$\cos 4\theta = 8\cos^4\theta - 8\cos^2\theta + 1 \cdots$$

which are deducible from $\Re(\cos\theta + i\sin\theta)^n = (C + iS)^n$. For example,

$$\Re(C + iS)^4 = \Re[C^4 + 4C^3 iS + \frac{4.3}{1.2}C^2(iS)^2 + \frac{4.3.2}{1.2.3}C(iS)^3 + (iS)^4]$$
$$= C^4 - 6C^2 S^2 + S^4 = C^4 - 6C^2(1 - C^2) + (1 - C^2)^2$$
$$= C^4 - 6C^2 + 6C^4 + 1 - 2C^2 + C^4 = 8C^4 - 8C^2 + 1.$$

This rather tedious old-fashioned development is obsoleted by the less transparent but eminently computer-oriented expression $\cos(n \arccos x)$ which lets you program Chebyshev polynomials in one line (Fig. A-19).

9-29A. **(a)** The standard transform pair

$$\delta(r - a)e^{in\theta} \overset{2}{\supset} (-1)^n 2\pi a e^{in\phi} J_n(2\pi aq)$$

is derived from the polar form of the transform definition (FTA p. 247)

$$F(q, \phi) = \int_0^\infty \int_0^{2\pi} f(r, \theta) e^{-i2qr\cos(\theta - \phi)} r \, dr \, d\theta$$

by substituting $f(r, \theta) = \delta(r - a) \exp in\theta$ and using the identity (A&S 1964, p. 360)

$$(-1)^n 2\pi J_n(2\pi qr) = \int_0^{2\pi} e^{-i2\pi qr\cos\theta} \cos n\theta \, d\theta.$$

(b,c) Since

$$\delta(r - a)\Theta(\theta) = \delta(r - a)(\cdots c_1 e^{i\theta} + c_2 e^{i2\theta} + \cdots)$$

it follows that

$$\delta(r - a)\Theta(\theta) \overset{2}{\supset} [\cdots - 2\pi ac_1 J_1(2\pi aq) + 4\pi ac_2 J_2(4\pi aq) + \cdots$$
$$+ 2\pi anc_n J_n(2\pi anq) + \cdots]e^{in\phi}.$$

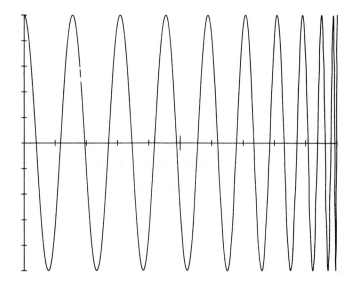

Figure A-19   The Chebyshev polynomial $T_{40}(|x|)$, an even function, for $0 \leq x \leq 1$. The function can be viewed as the projection on the x-axis of the azimuthally modulated ring impulse $\cos 40\theta \, \delta(r - 1)$.

Put $a = 1$ for part (b). For part (c) integrate over values of a from 0 to ∞.

(d) The complicated integral of a sum simplifies drastically when $f(r, \theta)$ is confined to one or a few concentric rings, or when the azimuthal dependence does not involve r, especially when the azimuthal dependence is as $\cos n\theta$.

9-30A. The result follows from the general formulas

$$\delta(r - a) \cos n\theta \overset{2}{\supset} (-1)^n 2\pi a \, J_n(2\pi a q) \cos n\phi$$

$$\delta(r - a) e^{in\theta} \overset{2}{\supset} (-1)^n 2\pi a \, J_n(2\pi a q) e^{in\phi},$$

which are deducible as follows. Remember that

$$\frac{1}{i^n \pi} \int_0^\pi e^{iz \cos \theta} \cos n\theta \, d\theta = J_n(z).$$

Now, for $n = 1$,

$$F(q, \phi) = \delta(r - a) \cos \theta \overset{2}{\supset} \int_0^\infty \int_0^{2\pi} \delta(r - a) e^{-i2\pi q r \cos(\theta - \phi)} \cos \theta r \, dr \, d\theta$$

$$= \int_0^{2\pi} a e^{-i2\pi a q \cos(\theta - \phi)} \cos \theta \, d\theta$$

using the sifting theorem. Putting $\beta = \theta - \phi$,

$$F(q, \phi) = 2a \int_0^\pi e^{-i2\pi aq \cos\beta} \cos(\beta + \phi) \, d\beta$$

$$= 2a \int_0^\pi e^{-i2\pi aq \cos\beta} \cos\beta \cos\phi \, d\beta$$

$$= 2a \cos\phi \int_0^\pi e^{-i2\pi aq \cos\beta} \cos\beta \, d\beta$$

$$= 2a \cos\phi (\pi/i) J_1(2\pi aq)$$

$$= -i2\pi a J_1(2\pi aq) \cos\phi.$$

9-31A. The result follows from the projection-slice theorem. In particular

$$J_0(2\pi x) \supset \pi^{-1}(1 - s^2)^{-1/2} \operatorname{rect}(s/2)$$

and

$$J_2(2\pi x) \supset -\pi^{-1}(2s^2 - 1)(1 - s^2)^{-1/2} \operatorname{rect}(s/2).$$

In general,

$$J_{2n}(ax) \supset 2(-1)^n (a^2 - 4\pi^2 s^2)^{-1/2} T_{2n}(2\pi s/a) \operatorname{rect}(s/2).$$

CHAPTER 10

10-1A. (a) The two-dimensional autocorrelation of $\cos\pi x \, \Pi(x)$ is $C(x) = [(2\pi)^{-1} \sin\pi \mid x \mid$ $+0.5(1- \mid x \mid) \cos\pi x] \Pi(x/2)$. The required two-dimensional autocorrelation is $C(x, y) = C(x) \Lambda(y)$. Note that $C(0) = 0.5$.

(b) The aperture distribution over a horn of width W and height H is

$$\cos(\pi\lambda x/W) \Pi(\lambda x/W) \Pi(\lambda y/H).$$

The autocorrelation of this would be $(WH/\lambda^2) C(\lambda x/W, \lambda y/H)$. Normalizing the unity at the origin, so that $T(u, v) = 1$, and changing the names of the variables to u and v, we have

$$T(u, v) = 2C(\lambda u/W, \lambda v/H).$$

Putting $W/\lambda = 10$ and $H/\lambda = 8$, $T(u, v) = 2C(u/10, v/8)$.

(c) $T(3, 2.4) = 2C(0.3, 0.3) = 2[\sin 54°/2\pi - 0.5 \times (1 - 0.3)] \Lambda(0.3) = -0.154 \times 0.7$ $= -0.11$.

10-2A. (a) Let Y be the maximum and y the minimum response. Then $C = Y/y$ and $m = (Y - y)/(Y + y) = (C - 1)/(C + 1)$, as shown in Fig. A-20 on the left. Consequently, $C = (1 + m)/(1 - m)$.

(b) Assume, as a standard condition, a nonnegative sinusoidal stimulus $1 + \sin 2\pi sx$, whose amplitude remains the same as frequency s varies. The periodic response of a linear, x-invariant system is $1 + T(s) \sin(2\pi sx + \alpha)$. Then the maximum response is $Y = 1 + |T(s)|$ and the minimum is $y = 1 - |T(s)|$. Thus the contrast ratio $C = [1 + |T(s)|]/[1 - |T(s)|] = (1 + \operatorname{sinc}^2 s)/(1 - \operatorname{sinc}^2 s)$ and $m = |T(s)|$. The absolute-value sign is needed in case $T(s)$ goes negative. Then C and m are as shown on the right.

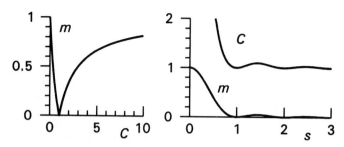

Figure A-20 Relation between modulation coefficient m and contrast ratio C (left) and the dependence of C and m on frequency s for a transfer function $T(s) = \mathrm{sinc}^2 s$ in the presence of background level of unit strength (right).

Contrast ratio max/min is in a sense simpler than modulation coefficient m, which however, is widely used, for example in the optical term *modulation transfer function* (MTF), which refers to m rather than to C, and is equal to $|T(s)|$. The MTF does not take into account the phase shift α. The spectral sensitivity function $T(s) \exp i\alpha$, which originated in radio astronomical interferometry, is the more general concept (also known as optical transfer function).

10-3A. The result can be expressed in terms of trigonometric functions but is rather complicated. The principal cross sections are $T(u, 0) = \frac{1}{2}$ chat u and $T(0, v) = \frac{1}{2}$ chat $2v$. The island boundary is formed of segments of the straight lines $v = \pm\frac{1}{2}$ and the two circles $(u \pm \frac{1}{2})^2 + v^2 = \frac{1}{4}$.

10-4A. (a)

$$T(u, v) = \begin{cases} 2 \times \text{semicircle OTF} & |u| < \frac{1}{2} \\ \text{self-convolution of semicircle} & \frac{1}{2} < |u| < 1 \\ \text{chat } \sqrt{(u - \frac{1}{2})^2 + v^2} & |u| > 1 \end{cases}$$

(b) In the first quadrant the boundary falls on the circle $(x - \frac{1}{2})^2 + y^2 = 1$ and on the circle $x^2 + (y - \frac{1}{2})^2 = \frac{1}{4}$) and the line segment from $(\frac{1}{2}, \frac{1}{2})$ to $(\frac{1}{2}, 1)$.

CHAPTER 11

11-1A. The Hankel transforms were listed in Chapter 9 as $2^{n-1}n! J_{n+1}(\pi q)/\pi^n q^{n+1}$. When $n = 0$, we have jinc q. The remaining functions of the sequence have progressively smaller sidelobes as both the aperture distribution and the transform approach the Gaussian form. The results are

n	0	1	2	3
dB	−18	−25	−32	−36

When n is half an odd integer the Bessel function reduces to a finite number of terms containing sines and cosines—for example, $J_{3/2}(x) = \sqrt{2/\pi x}(x^{-1} \sin x - \cos x)$

and $J_{5/2}(x) = \sqrt{2/\pi x}[(3x^{-2} - 1)\sin x - 3x^{-1}\cos x]$. These expressions may be more convenient to compute with in an exploratory exercise of the present kind.

11-2A. Three planes define a point (unless the planes have a common line).

11-3A. Consider the whole of space filled by just two traveling plane monochromatic waves inclined to each other. On some selected transverse plane the cosinusoidal distribution $\cos 2\pi a x$ can be found, while the wave interpretation in terms of exactly two directions $^2\delta(u - u_0, v) + {}^2\delta(u + u_0, v)$ can also be made. The two interpretations exist together throughout the whole space. Divide the space into two halves by a screen and remove the fields on one side; what remains is either an energy flow away from the screen (launching of two waves by a cosinusoidal distribution) or, if the direction of time is reversed, the production of a cosinusoidal interference pattern by the arrival of a pair of inclined waves. On this description it is clear how one and the same Fourier pair describes what is essentially just one physical situation. However, this description contains no pinholes. So it is indeed true that the convenient laboratory arrangements, pinholes in the one case, and a periodic diffraction grating in the other, are distinct.

11-7A. For a focal length of 10 inches, and 1:1 copying, the window-to-lens distance u and the lens-to-image distance v should both be 20 inches. Raising the original by 0.2 inches increases u by 1 percent and decreases v by 1 percent. Thus the magnification v/u is reduced 2 per cent. Consequently the image will now be 4 percent smaller than the original, and slightly out of focus. It is not certain that this reasoning applies to every photocopier. The best way to get a reliable answer is by experiment.

11-9A. Refraction by a prism breaks a beam of light into its constituent colors; diffraction breaks a collimated beam of nonmonochromatic light up into different directions. Water droplets are not small enough for diffraction to much affect the appearance of the rainbow.

11-10A. **(a)** Calculate the diameter d_0 from $d_0 = 2q_0 f = 2.44\lambda f/D$. The f/D ratio is 3.33. This makes the spot diameter 3.2 μm at the blue end of the spectrum ($\lambda = 400$ nm) and 6.4 μm at the red end ($\lambda = 800$ nm). The telescope is not used under conditions where such a degree of chromatic spread would occur, because time-varying atmospheric refraction spreads the image to a diameter of about 25 μm. This is the image size expected for a modest aperture of about 10 cm.

(b) The theoretical depth of focus given by $2(f/D)^2\lambda$ is 9 μm ($\lambda = 400$ nm) or 17 μm ($\lambda = 800$ nm).

11-14A. The aperture distribution $E(\widehat{x})$ produced on the emergent side of the screen is $e^{-i\pi/2}H(\widehat{x})$ $+e^{-i3\pi/2}H(-\widehat{x})$ or, if the origin of time is advanced half a period,

$$E(\widehat{x}) = e^{i\pi/2}H(\widehat{x}) + e^{-i\pi/2}H(-\widehat{x}) = i[H(\widehat{x}) - H(-\widehat{x})] = i\,\mathrm{sgn}\,\widehat{x}.$$

The Fourier transform is $-i/\pi u$. It follows that the transparent straightedge acts as a line source of radiation strongly concentrated in the forward direction, but in antiphase across the median plane. The singularity at $u = 0$ disappears (and is replaced by a null) when the screen is given a large but finite width, so that the aperture distribution becomes $i\,\mathrm{sgn}\,\widehat{x}\,\mathrm{rect}(\widehat{x}/\widehat{W})$. The transform is then $(-i/\pi u) ** \widehat{W}\,\mathrm{sinc}\,\widehat{W}u$.

11-15A. From the Cornu spiral we see that a point P at $(-0.46, -0.06)$ has the property that $J'P = J'J/\sqrt{10}$. The parameter $\widehat{x}$ read from the spiral is 0.47, which corresponds to a distance $x = \sqrt{(D\lambda/2)}\widehat{x}$ from the geometrical shadow edge, or $\sqrt{250 \times 0.3/2} \times 0.47 = 4$ m. Allowing for the height of the window above the ground, say 3 m, we need a total of

7 m immersion into the shadow. As the wall is only one-sixth of the distance from tire to window, a wall 7/6 = 1.2 m high is indicated.

11-16A. Presumably what you see is unexpected, so you are not in a good position to deduce the appearance theoretically. If you are the sort of theoretician who claims not to be an experimenter you may learn something of interest to you by giving this exercise a try; after you have tried, write down what it was, precisely, that prevented you from predicting the appearance. The explanation can be arrived at by further experiment. Some of the actions you might take are (i) rotating the can about its long axis, with and without corresponding rotation of the head, (ii) piercing a hole of somewhat different diameter, (iii) finding a can of different length, (iv) blocking direct sunlight from falling on the pinhole, (v) repeating with a dark-adapted eye, (vi) narrowing the eye, (vii) moistening the cornea by blinking, (viii) focusing the eye at infinity, (ix) averting the vision, (x) rubbing the eyelid, (xi) partially closing the hole with a razor blade, (xii) catching a solar image on paper at the mouth of the can. Notice that these actions represent creative intellectual activity rather than manual dexterity, as does the reasoning based on the outcome of each experiment. A theoretically inclined person would not wish to opt out of this type of higher intellectual activity.

CHAPTER 12

12-1A. The concept pertains to a radiation source, the choice of frequency and bandwidth made by the observer, the choice of observing points, the choice of polarization, and to the medium intervening between the source and the observing points.

12-2A. Suppose the device is placed on the skin and we consider propagation in muscle with a velocity 1570 m s^{-1}. Then $\lambda = 1570/3.5 \times 10^6 = 0.45$ mm. A 2.5 mm aperture is then $2.5/0.45 = 5.6$ wavelengths across. To get into the far field we need to go about $5.6 \times 2.5 = 14$ mm, where the beamwidth of a single element will be about $10°$, but there is little prospect of focusing all the beams at one point, because different velocities in the medium make it difficult to get phase agreement. Round-trip attenuation to a depth of 60 mm would be about 50 db plus losses due to reflection at discontinuities. The explorable volume would be a few centimeters deep times whatever skin area gives a clear look between ribs, lungs, etc. The mode of operation is to use the device like a comb with 10 teeth which independently reach to a depth of a few inches. Range is measurable by time delay to a scattering element and back. Transverse resolution is at best 2.5 mm, deteriorating as the beam from a single element diverges. Depth resolution of 4.5 mm would be achievable with a pulse 10 wavelengths long (duration $10/(3.5 \times 10^6) = 3 \ \mu s$).

12-6A.

    (a) Since we know that the information in the radiation field resides in the correlation between spaced points, once receiving points have been occupied at the ends of a certain baseline one could not extract information that belongs to a different spacing. It is true that the lobes are twice as narrow if frequency doublers are inserted, but there is a one-to-one equivalence between the two interferograms illustrated. If there were a pair of point sources so spaced in direction that the ordinary interferometer just could not resolve them, then the record made with frequency doublers would also not quite permit the sources to be resolved. Observations at double the baseline would permit resolution.

(b) Wideband operation does indeed spread the accessible spacings (measured in wavelengths) and will therefore permit increased resolution. The response to a point source will show a maximum in the direction where the path difference is zero. But the first-order maximum (p.d. = λ) will be reduced in magnitude and higher-order maxima will be smeared even more, more so as the band is wider. If a source looks different at different frequencies, then a map made with a wideband two-element interferometer will not look the same as any monochromatic map.

CHAPTER 13

13-2A. The correction is

$$g(x, y) - 1.9 \frac{g(x, y + 1) + g(x - 1, y) + g(x + 1, y) + g(x, y - 1)}{4}.$$

The factor 1.9 comes from comparing $a(x, y)/41$ with $a(x, y) ** a(x, y)/41^2$.

13-3A. If we note that the proposed solution may be written in the form

$$g(x) * [\delta'(x - \tfrac{1}{2}) + \delta'(x - 1\tfrac{1}{2}) + \delta'(x - 2\tfrac{1}{2}) + \cdots]$$

then convolution with the expression in square brackets should annul convolution with $\Pi(x)$. In other words (using $\Pi(\,)$ instead of rect) if

$$g(x) = \Pi(x) * f(x)$$

has a solution

$$f(x) = Q(x) * g(x)$$

then

$$Q(x) * \Pi(x) = \delta(x).$$

So we examine

$$[\delta'(x - \tfrac{1}{2}) + \delta'(x - 1\tfrac{1}{2}) + \cdots + \delta'(x - n + \tfrac{1}{2})] * \Pi(x),$$

which we find is equal to

$$\delta(x) + \delta(x - n).$$

Hence the proposed operation does not yield the solution, in general, but it will if $g(x)$ covers a finite range of x, say W. Then if n is chosen larger than W (remembering that the problem refers to a rectangle function of unit width), the terms $g'(x - \tfrac{1}{2})$ and $g'(x - \tfrac{1}{2} - n)$ associated, respectively, with the terms $\delta(x)$ and $\delta(x - n)$ will not overlap, and $g'(x)$ will be determined.

    Mr. Wu's logic was impeccable. Even so, something remained to be said.

    Mr. Cho no doubt had practical problems with derivatives. However, if one evaluated

$$g(x) * [\delta(x - \tfrac{1}{2}) + \delta(x - 1\tfrac{1}{2}) + ...]$$
$$= g(x - 1/2) + g(x - 1\tfrac{1}{2}) + g(x - 2\tfrac{1}{2}) + ...$$

which is a very simple thing to do, one obtains the integral of the desired $f(x)$. Then, of course, to differentiate it numerically there is no choice but to use finite differences.

An alternate way of deriving the solution is this. Given $g(x) = \Pi(x) * f(x)$ then $G(s) = \text{sinc } s F(s)$, so $F(s) = G(s)/\text{sinc } s$, $\text{sinc } s \pm 0$. Writing the sinc function as follows, $\text{sinc } s = (1/2\pi i s) \exp i\pi s [1 - \exp(-2\pi i s)]$, gives

$$F(s) = 2\pi i s \exp(-\pi i s) G(s)[1 - \exp(-2\pi i s)]^{-1},$$
$$F(s) = 2\pi i s G(s)[\exp(-\pi i s) + \exp(-3\pi i s) + \exp(-5\pi i s) + \cdots]$$

and

$$f(x) = g'(x - \tfrac{1}{2}) + g'(x - 1\tfrac{1}{2}) + g'(x - 2\tfrac{1}{2}) + \cdots.$$

13-4A. **(a)** The convergence criterion is $| 1 - A(s) | < 1$ for all s. In this case $a(x) = \Pi(x)$ and $A(s) = \text{sinc } s$. We see that $| 1 - A(s) | < 1$ in some bands of s ($0 < s < 1, 2 < s < 3$, etc.) but not others ($1 < s < 2, 3 < s < 4$, etc.) Therefore the convergence criterion for the method of successive substitutions is not satisfied.

(b) A numerical trial is likely to show that the first stage of restoration produces improvement.

(c) *Further investigation:* The most obvious further investigation is to calculate further stages of restoration. Even these will often show improvement, but in the long run the improvement will stop and the agreement will deteriorate. *Wise comment:* Since the inequality is satisfied in some bands but not others, there will be an improvement of the spectrum in some bands and deterioration in others. If the spectral content of the signal predominates in the good bands (say in the band $0 < s < 1$), then there will be a net improvement for that stage. But with further stages, continued deterioration in the bad bands will ultimately triumph. If there is no spectral content at all in the bad bands, then convergence will occur.

13-7A. The curve $f(x)$ in Fig. A-21 is the original from which the curve $g(x)$ was calculated by first convolving with the given point response $a(x)$ and then adding the noise waveform shown separately as $n(x)$. Almost any method of restoration should agree with the author's $f(x)$ quite well, even if no attention is paid to the noise. If one stage of successive substitutions is used, then the noise component will still be noticeable but perhaps not unacceptable. However, since there are many noise peaks in the width of the point response, it must be possible work against the noise while not degrading the restoration much, as explained in the text. (b) The peak signal is 47, while the r.m.s. noise is about 0.4. Therefore you could answer the inquiry by saying the signal-to-noise ratio is about 100. (c) The curve $g(x)$ was constructed as explained by adding a noise waveform $n(x)$ that was stationary and independent of signal strength. These properties could be verified from the curve $g(x)$ alone by drawing a smooth curve through $g(x)$ and subtracting. To the eye, there does not appear to be much fluctuation on the steep parts of $g(x)$, but that is a natural property of addition; the noise fluctuations reappear on the crest and plateau where the slope is less steep.

13-8A. The results are shown in Fig. A-22. In the limiting case

(a) of noiseless data the restoring function becomes infinite. With noise

(b) the restoring factor reaches a maximum and then turns down to reach zero at the cutoff frequency. With more noise

(c) the turnover frequency is not as close to the cutoff and the amplification factor is less.

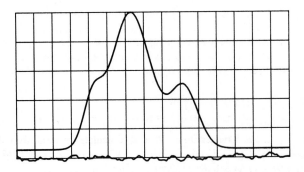

Figure A-21 Showing the true function $f(x)$ and the artificial noise $n(x)$ from which the problem was constructed. A restoration should reasonably resemble the sum $f(x) + n(x)$.

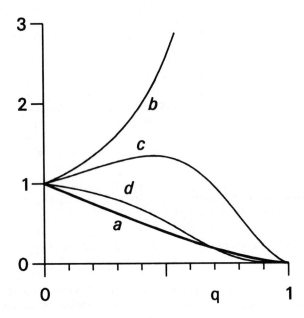

Figure A-22 (a) The transfer function $T(q) = (4/\pi)$ chat q. (b) The ideal restoring function $1/T(q)$. (c) The modified restoring function $[1/T(q)]\{T(q)/[T(q)]^2 + 0.2\}$. (d) The resultant net transfer function $T(q)/\{[T(q)]^2 + 0.2\}$.

13-9A. (a) Given a point-spread function $h(x, y) = \Pi(x/3)\Pi(y/3)$, we have a normalized transfer function $H(u, v)/H(0, 0) = $ sinc $3u$ sinc $3v$.

  (b) The successive substitutions convergence criterion $|1 - H(u, v)| < 1$ is not met in negative areas of the (u, v)-plane, e.g., near $u = 0.5$, $v = 0$.

  (c) After two convolutions with $h(x, y)$ the transfer function becomes sinc^2 $3u$ sinc^2 $3v$, which no longer has negative regions. The convergence condition is met everywhere except on the null lines, but there is no signal content there. Consequently, the suggestion is a good one. Since the deliberate blurring operation weakens the signal spectrum, it would be necessary to take care that further noise was not introduced.

CHAPTER 14

14-2A. The result follows from the fact that a single annular slit of radius R described by $\delta(r - R)$ has Hankel transform $2\pi R J_0(2\pi Rq)$. (b) and (c) See D. Tichenor and R. N. Bracewell, "Fraunhofer Diffraction of Concentric Annular Slits," *J. Opt. Soc. Amer.*, vol. 63, pp. 1620–1622, 1973.

14-5A. We need the integral of the Abel transform of $f(r) = I_0[1 - \frac{1}{4}(r/R)^2]\operatorname{rect}(r/2R)$. The Abel transform is $f_A(x) = I_0[1.5(R^2 - x^2)^{1/2} + \frac{1}{3}R^{-2}(R^2 - x^2)^{3/2}]\operatorname{rect}(x/2R)$. The desired quantity is $\int_x^R f_A(x')\,dx'$, as x ranges from $-R$ to R. Calculate the result from the standard integrals $\int(R^2 - x^2)^{1/2}\,dx = \frac{1}{2}x(R^2 - x^2)^{1/2} + \frac{1}{2}R^2\arcsin(x/R)$ and $\int(R^2 - x^2)^{3/2}dx = \frac{1}{4}x(R^2 - x^2)^{3/2} + (3/8)R^2x(R^2 - x^2)^{1/2} + (3/8)R^4\arcsin(x/R)$.

14-10A. The matrix and projections are

$$\begin{bmatrix} 12 & 13 & 19 & 13 & 12 \\ 14 & 16 & 25 & 16 & 14 \\ 24 & 31 & 99 & 31 & 24 \\ 14 & 16 & 25 & 16 & 14 \\ 12 & 13 & 19 & 13 & 12 \end{bmatrix}$$

$\theta = 0°$:	0	76	89	**187**	89	76	0
$\theta = 30°$:	12	51	92	**182**	117	51	12
$\theta = 60°$:	12	46	121	**159**	121	46	12
$\theta = 90°$:	0	69	85	**209**	85	69	0

CHAPTER 15

15-2A. **(a)** The ray length is the difference of the semiellipse $2(50^2 - y^2)^{1/2}\Pi(y/100)$ associated with a solid 100-mm shaft and the corresponding semiellipse for the hollow of variable radius $R(z)$. Thus ray length for $0 < z < 200$ is

$$L(z, y) = 100(50^2 - y^2)^{1/2}\Pi(y/100) - R(z)\{[R(z)]^2 - y^2\}^{1/2}\Pi(y/2R(z)),$$

where $R(z) = 40 + 5z/200$.

(b) If the incident intensity is I_0, the ray length in the steel is L and the absorption coefficient is α, then the intensity I is given by $I = I_0\exp(-\alpha L)$. In this question, $L = L(100, y)$.

(c) Let the x-ray source be at a distance H from the center of a solid shaft of radius R and let the ray length inside the solid shaft be g. Other variables are as shown. In terms of elevation angle θ,

$$g = 2(R^2 - p^2)^{1/2} = 2(R^2 - H^2\sin^2\theta)^{1/2}.$$

Now convert to q as a variable instead of θ; then the ordinate y on the photographic plate will be $y = q(R + H)/H$. Note that $q = H\tan\theta$. To eliminate θ use $\sin^2\theta = (1 + \cot^2\theta)^{-1}$. Then

$$g = 2\{R^2 - H^2[1 + (H/q)^2]^{-1}\}^{1/2}\Pi(q/2q_{max})$$
$$= 2\{R^2 - H^2[1 + ((H + R)/y)^2]^{-1}\}^{1/2}\Pi(y/2y_{max}).$$

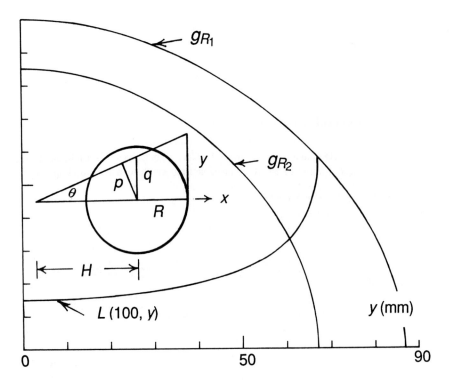

Figure A-23   Ray lengths g_{R_1}, g_{R_2} and $L(100, y)$.

With the hollow shaft, $L(z, y) = g_{R_1} - g_{R_2}$. With $R_1 = 50$ mm, $R_2 = 42.5$ mm (at $z = 100$ mm), the result is as shown in Fig. A-23.

15-5A. (a) A function which, when subjected to a stated operation (or transformation), gives nothing out. Alternatively, given an operator $\mathcal{T}$, then f is an invisible distribution if $\mathcal{T}f = 0$.

(b) No. Because if the Radon transform were zero, it would mean that all slices through the 2D FT were zero and therefore that the presumed invisible distribution was itself identically zero. Only in the trivial case of a 2D null function (massless, nonnegative, and nonzero on a region of zero measure) is there an exception. The function $\widehat{\rho}(x, y) = \delta(y + 1) - \delta(y - 1)$ (two parallel blades of strength ± 1, respectively) is invisible from almost all directions, but has a nonzero Fourier transform along $u = 0$. It has a nonzero projection on the y-axis.

15-10A. Let $f_o(x, y)$ be a solution. Then $\psi(x, y) = f_o(x, y) + f_{inv}(x, y)$ will have the same projections, and thus be a different solution, provided $f_{inv}(x, y)$ has null projections in the three given directions and provided $\psi(x, y)$ is composed of positive unit impulses. Invisible distributions other than null functions must have zero mass and thus have negatives as well as positives. As a strategy of approach let us try to find a counterexample to the student's claim. An invisible distribution for projection at $0°$, $45°$, and $90°$ is shown in Fig. A-24.

$$
\begin{array}{ccccc}
 & +1 & & -1 & \\
-1 & & & & +1 \\
 & & & & \\
+1 & & & & -1 \\
 & -1 & & +1 &
\end{array}
$$

Figure A-24 An invisible distribution.

If such a pattern is added to $f_o(x, y)$, negatives will result, unless $f_o(x, y)$ contains rings of impulses on which invisible rings can be exactly superimposed. But in that case, the resultant impulses will not be equal. We have not disproved the student's claim, but we have found that the answer to the professor is "No." Failure to disprove does not constitute proof, so the question remains open.

This conclusion is of considerable philosphical interest. We have found that a certain theorem is false but gives results with useful frequency. Furthermore, if projections were chosen at say $0°$, $25°$, and $90°$, the number of problem cases would drop even lower.

Even the professor's question, which could be logically dismissed, needs reconsideration. It would evidently be a mistake to assume that a pattern such as a constellation of unequally bright stars could not be reconstructed from three projections, because obviously every one of the 88 constellations could be so reconstructed. Only cunningly contrived artificial constellations would foil you, and even then, the contriver would need to find out in advance the orientation of each one of your projections.

Suppose an earth satellite has to control its own orientation automatically. An image of part of the sky is formed by a lens preparatory to digitization. Instead of doing a raster scan, it would be faster to scan the image with three differently oriented slits (knife edges might be even better). The digital description of the image could be easily computed, and the image could then be searched for in memory to determine the three angular coordinates of the satellite.

15-11A. Let $\rho_1(x, y)$ be a density distribution that is unity inside the octagonal boundary given and zero outside, and let $\rho_2(x, y)$ be a second such distribution with the same projections. Then $I(x, y) = \rho_2(x, y) - \rho_1(x, y)$ is an invisible distribution, one whose projections in the two directions are identically zero. Clearly $I(x, y)$ would have to have zero mean and negative as well as positive values if ρ_2 is not the same as ρ_1. The simplest invisible distribution $I_0(x, y)$ comprises four impulses at the vertices of a rectangle:

$$
I_0(x, y) = {}^2\delta(x + a, y + b) - {}^2\delta(x - a, y + b) + {}^2\delta(x - a, y - b) - {}^2\delta(x + a, y - b).
$$

The most general is a sum over a and b of a combination of such tetrads with arbitrary amplitudes and arbitrary translations; a convolution containing $I_0(x, y)$ as a factor would be such a combination. No way can be found of combining an invisible distribution with the given octagon without destroying its property of being a uniform lamina. Consider an invisible distribution comprising very small polygons within which the function values are $+1$ or -1. The polygons must be at the corners of a rectangle with sides parallel to the coordinate axes if the distribution is to be invisible. Such could be added to $\rho_1(x, y)$ to form a new function $\rho_2(x, y)$ if the positive polygons were externally and the negative ones internally tangent to the boundary of $\rho_1(x, y)$. But, in the case of $\rho_1(x, y)$, four places in the necessary rectangular relation cannot be found on the boundary. (The necessary relation is

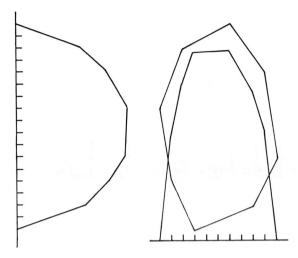

Figure A-25   The octagon with the given horizontal and vertical projections.

different but not simpler if more complicated invisible distributions are tried.) This conclusion would be true of almost all polygonal laminae. However, special octagons can be imagined that are not uniquely determined by their horizontal and vertical projections. Of course, the octagons have to be tailored and oriented with a foreknowledge of the two directions of projection. Randomly oriented, each such octagon would be determinate. (See Fig. 15-12A).

15-13A. The vertical and horizontal projections of the letters are shown in Fig. A-26. There are no two letters with identical projections. Indeed, it appears that there are no two letters whose horizontal projections are the same; there must be more information in the vertical structure than in the horizontal. In fact this is well known. If you cover the lower half of a line of print, you can still read it. If you cover the top half, you cannot read it. Scanning top to bottom while reading left to right is possible but awkward. A good suggestion is to scan left to right with an inclined slit.

15-25A. Use the nonzero numbers as column numbers and the serial number of each nonzero number as a row number and plot the seven points to get a picture of the constellation Orion. In Fig. A-7 Orion came out reversed, which tells us the raster convention of the aliens. In the present problem the picture would be reversed if serial number was used as an ordinate. The celestial equator, if marked in, proves to be inclined at $28°$ to the columns, telling us that the aliens are using the galactic equator for reference.

15-29A. **(a)** The desired curve touches the ellipse at the ends of the principal axes (Fig. A-27) and is described in general in polar coordinates by $r^2 = (a \cos \theta)^2 + (b \sin \theta)^2$.

(b) A highly eccentric ellipse results as $b \to 0$, in which case the pedal curve, or Radon boundary, becomes a pair of touching circles.

CHAPTER 16

16-1A. The phase of the echo signal advances one cycle per half-wavelength, even though the frequency of the echo is unchanged. When the echo is added to a signal of fixed frequency and phase, the resultant steps through a rise and fall, pulse by pulse, in a stepwise manner, ex-

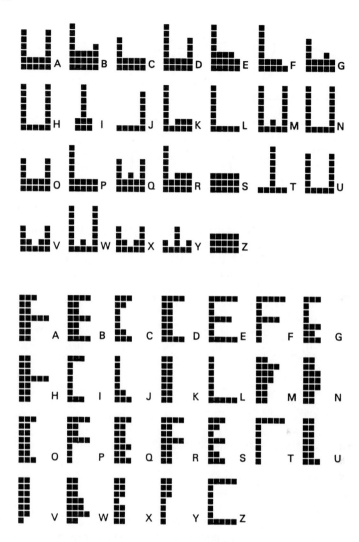

Figure A-26 The projections of the dot-matrix capitals. When this coarsest possible pixelization of the alphabet is implemented with the round dots characteristic of actual printers instead of theoretical square pixels, the legibility is really excellent. The **R** illustrated shows that the available dot size loosely controls the character size.

hibiting the same variation that would be noted if the target were in motion. Consequently, the radar would detect and correctly measure the speed of the "smart" robot.

16-2A. (a) The result follows by substituting $\dot{r} = \frac{1}{2}\dot{P}$ and $\sin\theta = y/r$ in $\dot{P} = -2v\sin\theta\cos\delta$.

(b) Use 20 km as the horizontal scale modulus and ± 300 knots as the vertical scale modulus. Then the horizontal asymptotes will be at $\dot{r} = 300$, and the curve will have a slope -1 at the origin. *Note:* The sign of the ordinate will not come out right if y is taken to be the ordinate of the aircraft. The target is at (x, y) relative to a plane that is carrying the origin of coordinates with it.

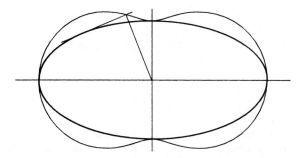

Figure A-27   The pedal curve (light line) of the ellipse (heavy line) whose origin is at the center.

CHAPTER 17

17-2A. In the following list the value of n is followed by the Poisson distribution value and the approximate normal-distribution value in parentheses.

0: 0 (0.001); 1: 0 (0.001); 2: 0.002 (0.005); 3: 0.008 (0.011); 4: 0.019 (0.021); 5: 0.038 (0.036); 6: 0.063 (0.057); 7: 0.09 (0.08); 8: 0.113 (0.103); 9: 0.125 (0.12); 10: 0.125 (0.126); 11: 0.114 (0.12); 12: 0.095 (0.103); 13:0.073 (0.08); 14: 0.052 (0.057); 15: 0.035 (0.036); 16: 0.022 (0.021); 17: 0.013 (0.011); 18: 0.007 (0.005); 19: 0.004 (0.002); 20: 0.002 (0.001).

The agreement is evidently quite good, the discrepancy amounting at the most to 0.01. However, it is also clear that for reasonable percentage accuracy, even as many as 10 dots per unit area is not enough for the normal distribution to be a good substitute for the Poisson distribution. The mean and standard deviation are not sufficient to specify the shape.

17-6A. Consider the following segment.

```
CANDIDATES FOR CONTIGUITY

FOR S=1 TO 25
    FOR J=-1 TO 1
        FOR I=-1 TO 1
            col=(S-1) MOD 5 +1+i
            row=(S-1) DIV 5 +1+j
            inside=col>0 AND col<6 AND row>0 AND row<6
            IF inside THEN PRINT col+5*(row-1); ELSE PRINT 0;
        NEXT I
    NEXT J
    PRINT
NEXT S
```

The semicolons suppress the carriage return, S stands for serial number; other names will be needed where col and row are reserved words. The printout places the serial number of the reference tree in the middle column rather than in the leading column; the middle may be a better place for it.

17-7A. **(a)** Since $y_{max} = 17$, $y_{min} = 6$ and $R = 11$, we get $s = 0.451 \times 11 = 4.96$.
   (b) Since y_{max} and y_{min} are integers R is subject to a total roundoff error of 1. The

sample standard deviation computed from all the data is $s_{all} = 4.75$, which is in fair agreement. Schlesinger reports that the estimate s is more stable than s_{all} for $n < 7$.

17-8A. From the calculable first differences a relative gain map can be produced in theory. In practice, zero-point response (to a black object) needs to be subtracted, and several images need to be recorded in various positions so that a least-mean-squares reduction can be made. See J. R.Kuhn, H. Lin, and D. Loranz, "Gain Calibrating Nonuniform Image-Array Data Using Only the Image Data," *Proc. Astronom. Soc. Pacific*, vol. 103, pp. 1097–1108, October 1991.

17-12A. There is a point impulse at the origin, line impulses in the form of an egg-crate, and a constant 1, all multiplied by a lazy pyramid $\Lambda(2x/M)\Lambda(27/N)$. In the limit, the line impulses, prior to multiplication, approach triangular.

17-14A. Let $x = R/\Sigma$ and $y = x\exp(-x^2)$. Then $dy/dx = (1 - 2x^2)\exp(-x^2)$, which equals zero at $x = 1/\sqrt{2}$. Hence the mode is at $R = \Sigma/\sqrt{2}$. For the mean calculate the first moment $\int_0^\infty x^2\exp(-x^2)dx = 0.5\sqrt{\pi}$. Hence $m = 0.5\sqrt{\pi}\,\Sigma$. For the standard deviation, first calculate the second moment $\int_0^\infty x^3\exp(-x^2)dx = 1$, subtract the square of the mean, and take the square root to get $\sqrt{1 - 0.25\pi}$. Hence $\sigma = \sqrt{1 - 0.25\pi}\,\Sigma$.

17-18A. No. The enlargement $\delta(r - e^{2\pi}e^\theta)$ does not possess the part of $\delta(r - e^\theta)$ that lies inside the circle of unit radius. If the object is redefined for θ to range from $-\infty$ to ∞, the answer is still no, as revealed by application of the rules for delta functions: the strengths of the two entities are not equal at common points.

Index

Pages with a '*p*' suffix locate entries in the *Problems* sections of the text. Pages with an '*l*' suffix locate authors cited in the text and listed in the *Literature Cited* sections.